Forms Containing $u^2 \pm a^2$

38. $\displaystyle\int \frac{du}{a^2 - u^2} = \frac{1}{2a} \ln \left| \frac{a + u}{a - u} \right| + C$

39. $\displaystyle\int \frac{du}{u^2 - a^2} = \frac{1}{2a} \ln \left| \frac{u - a}{u + a} \right| + C$

40. $\displaystyle\int \frac{du}{\sqrt{u^2 \pm a^2}} = \ln|u + \sqrt{u^2 \pm a^2}| + C$

41. $\displaystyle\int \frac{du}{u\sqrt{a^2 \pm u^2}} = -\frac{1}{a} \ln \left| \frac{a + \sqrt{a^2 \pm u^2}}{u} \right| + C$

42. $\displaystyle\int \frac{du}{au^2 + c} = \frac{1}{\sqrt{ac}} \mathrm{Tan}^{-1} \left(u\sqrt{\frac{a}{c}} \right) + C, \quad ac > 0$

43. $\displaystyle\int \frac{du}{c^2 - u^2} = \frac{1}{2c} \ln \left| \frac{c + u}{c - u} \right| + C, \quad c^2 > u^2$

44. $\displaystyle\int \sqrt{u^2 \pm a^2}\, du = \frac{1}{2}[u\sqrt{u^2 \pm a^2} \pm a^2 \ln(u + \sqrt{u^2 \pm a^2})] + C$

45. $\displaystyle\int u\sqrt{u^2 \pm a^2}\, du = \frac{1}{3}(u^2 \pm a^2)^{3/2} + C$

46. $\displaystyle\int \frac{du}{\sqrt{u^2 \pm a^2}} = \ln(u + \sqrt{u^2 \pm a^2}) + C$

47. $\displaystyle\int \frac{du}{u\sqrt{u^2 - a^2}} = \frac{1}{a} \mathrm{Sec}^{-1} \left(\frac{u}{a} \right) + C$

48. $\displaystyle\int \frac{du}{u\sqrt{u^2 + a^2}} = -\frac{1}{a} \ln \left(\frac{a + \sqrt{u^2 + a^2}}{u} \right) + C$

49. $\displaystyle\int \frac{\sqrt{u^2 + a^2}}{u}\, du = \sqrt{u^2 + a^2} - a \ln \left(\frac{a + \sqrt{u^2 + a^2}}{u} \right) + C$

50. $\displaystyle\int \frac{\sqrt{u^2 - a^2}}{u}\, du = \sqrt{u^2 - a^2} - a\, \mathrm{Sec}^{-1} \left(\frac{u}{a} \right) + C, \quad a > 0$

51. $\displaystyle\int \frac{du}{\sqrt{(u^2 \pm a^2)^3}} = \frac{}{a^2\sqrt{u^2 \pm a^2}} + C$

52. $\displaystyle\int \frac{\sqrt{u^2 \pm a^2}}{u^2}\, du = -\frac{\sqrt{u^2 \pm a^2}}{u} + \ln(u + \sqrt{u^2 \pm a^2}) + C$

53. $\displaystyle\int \frac{\sqrt{u^2 - a^2}}{u^3}\, du = -\frac{\sqrt{u^2 - a^2}}{2u^2} + \frac{1}{2a} \mathrm{Sec}^{-1} \left(\frac{u}{a} \right) + C, \quad a > 0$

54. $\displaystyle\int \sqrt{a^2 - u^2}\, du = \frac{1}{2} \left[u\sqrt{a^2 - u^2} + a^2\, \mathrm{Sin}^{-1} \left(\frac{u}{a} \right) \right] + C, \quad a > 0$

55. $\displaystyle\int \frac{du}{\sqrt{a^2 - u^2}} = \mathrm{Sin}^{-1} \left(\frac{u}{a} \right) + C$

56. $\displaystyle\int \frac{du}{u\sqrt{a^2 - u^2}} = -\frac{1}{a} \ln \left(\frac{a + \sqrt{a^2 - u^2}}{u} \right) + C$

57. $\displaystyle\int \frac{du}{\sqrt{(a^2 - u^2)^3}} = \frac{u}{a^2\sqrt{a^2 - u^2}} + C$

58. $\displaystyle\int u^2\sqrt{a^2 - u^2}\, du = -\frac{u}{4}\sqrt{(a^2 - u^2)^3}$
$\qquad + \frac{a^2}{8} \left(u\sqrt{a^2 - u^2} + a^2\, \mathrm{Sin}^{-1} \frac{u}{a} \right) + C, \quad a > 0$

59. $\displaystyle\int \frac{u^2\, du}{\sqrt{a^2 - u^2}} = -\frac{u}{2}\sqrt{a^2 - u^2} + \frac{a^2}{2}\, \mathrm{Sin}^{-1} \left(\frac{u}{a} \right) + C, \quad a > 0$

60. $\displaystyle\int \frac{du}{u^2\sqrt{a^2 - u^2}} = -\frac{\sqrt{a^2 - u^2}}{a^2 u} + C$

61. $\displaystyle\int \frac{\sqrt{a^2 - u^2}}{u^2}\, du = -\frac{\sqrt{a^2 - u^2}}{u} - \mathrm{Sin}^{-1} \left(\frac{u}{a} \right) + C, \quad a > 0$

Forms Containing $a + bu^2$

62. $\displaystyle\int \frac{du}{a + bu^2} = \frac{1}{\sqrt{ab}} \mathrm{Tan}^{-1} \frac{u\sqrt{ab}}{a} + C$

63. $\displaystyle\int \frac{du}{(u^2 + a^2)^2} = \frac{1}{2a^3} \mathrm{Tan}^{-1} \frac{u}{a} + \frac{u}{2a^2(u^2 + a^2)} + C$

64. $\displaystyle\int \frac{du}{u(a + bu^2)} = \frac{1}{2a} \ln \left| \frac{u^2}{a + bu^2} \right| + C$

Forms Involving $2au - u^2$

65. $\displaystyle\int \sqrt{2au - u^2}\, du = \frac{1}{2} \left[(u - a)\sqrt{2au - u^2} + a^2\, \mathrm{Sin}^{-1} \left(\frac{u - a}{a} \right) \right] + C$

66. $\displaystyle\int \frac{du}{\sqrt{2au - u^2}} = 2\, \mathrm{Sin}^{-1} \sqrt{\frac{u}{2a}} + C, \quad a > 0$

67. $\displaystyle\int \frac{du}{(2au - u^2)^{3/2}} = \frac{u - a}{a^2\sqrt{2au - u^2}} + C$

68. $\displaystyle\int \frac{du}{\sqrt{2au + u^2}} = \ln|u + a + \sqrt{2au + u^2}| + C$

(Continued on back endsheet)

Calculus

of Several Variables

Calculus
of Several Variables

SECOND EDITION

DENNIS D. BERKEY
BOSTON UNIVERSITY

SAUNDERS COLLEGE PUBLISHING
New York   Chicago   San Francisco
Philadelphia   Montreal   Toronto
London   Sydney   Tokyo

Text Typeface: Times Roman
Compositor: York Graphic Services, Inc.
Acquisitions Editor: Robert Stern
Developmental Editors: Sandi Kiselica, Sarah Evans
Art Director: Carol Bleistine
Art Assistant: Doris Roessner
Text Designer: Emily Harste
Cover Designer: Lawrence R. Didona
Text Artwork: J&R Technical Services, Inc.
Production Manager: Merry B. Post

Cover Credit: A fractal—a set carried to infinity. Professor Larry Guseman and Robert Sparks, Texas A&M University.

Printed in the United States of America

CALCULUS OF SEVERAL VARIABLES, 2nd edition

0-03-026338-7

Library of Congress Catalog Card Number: 88-11945

789   061   987654321

PREFACE

This text on multivariable calculus is intended for use in a one-quarter or one-semester sequel to the traditional course on the calculus of a single variable. It is intended primarily for students of mathematics, science and engineering. It reflects my own philosophy that calculus should be taught so as to produce skilled practitioners who have some real feeling for the mathematical issues underlying the techniques they have acquired. It was written in order to provide students of widely varying interests and abilities with a highly readable exposition of the principal results, including ample motivation, numerous well-articulated examples, a rich discussion of applications, and a nontrivial description and use of numerical techniques. In particular, I have tried to indicate, in an unobtrusive manner, how computers can be used both to illustrate the theory and to provide approximate solutions to problems for which more elegant techniques break down.

LEVEL AND RIGOR: Nearly all topics in the traditional multivariable calculus curriculum are treated. The discussion is informal, and geometric arguments are used wherever possible. Definitions and properties of vectors are introduced first for vectors in the plane, then generalized to vectors in space. Differentiation and integration of vector-valued functions and functions of several variables are introduced as straightforward generalizations of differentiation and integration of functions of a single variable. The theorems of Green and Stokes, as well as the Divergence Theorem, are fully discussed.

STRATEGY/SOLUTION FORMAT IN EXAMPLES: An unusually large number of examples (more than 350) has been included. In many of these the solutions have been written in a two-column format, with one of the columns labelled ''Strategy.'' In this column the student will find, in very abbreviated form, a description of the principal steps involved in the fuller solution. Here I have attempted to help students identify the more general aspects of the particular solution, and to develop problem-solving strategies of their own.

ORGANIZATION AND CHANGES FROM THE FIRST EDITION: The order of topics is consistent with those of most popular texts. Many sections have been extensively rewritten from the first edition, and the treatments of several topics have been reorganized in response to user comments.

The theory of infinite series has been reorganized from three chapters in the first edition into the two Chapters 12 and 13. Chapter 12 presents infinite sequences and series of constants while Chapter 13 discusses Taylor polynomials, Taylor series, and power series.

Vectors are introduced in Chapter 15, together with the dot and cross product and the standard equations for lines and planes in space. The calculus in higher dimensions is treated in Chapters 16 through 19. This material has been extensively edited and considerably shortened. The treatment, however, is essentially that of the first edition.

Topics on differential equations, which appeared at various points throughout the first edition, have been consolidated into the concluding Chapter 20. This chapter contains new material on nonhomogeneous differential equations and numerical approximations to solutions of differential equations.

EXERCISES: More than 2,000 exercises are included, ranging from drill to challenging in type, and including many applied exercises from a broad range of disciplines. The extensive review exercises at the end of each chapter reflect the range of topics included in that chapter.

While many theorems, particularly those with instructive proofs, can and should be presented and proved, the time available to most of us for this task is not sufficient to allow a careful treatment of the least upper bound axiom, uniform continuity, differentiability of power series, or several other topics where statements of fact must simply be made. Honesty about these omissions, together with the right picture here and a good heuristic discussion there, can result in a presentation that is both factual and effective, and one that allows us to succeed in sharing with students the excitement of the triumphs of this classic subject.

COMPUTER-GENERATED COLOR FIGURES: New to this edition is the four-color insert in the multivariable calculus section. These computer-generated, four-color figures provide students with the ability to better view detailed, mathematically correct, three-dimensional surfaces. Some of the figures are rotated to show more than one perspective. These figures also appear within the text in line art form. The corresponding computer-drawn four-color figures are referenced in the line art captions.

ACKNOWLEDGMENTS: Many individuals played instrumental roles in the development of this text. It is my pleasure to acknowledge some of these here.

Twenty-nine teachers of the calculus scrutinized one or more drafts of the first edition, both on matters of content and to identify errors. These were

Paul Baum, Brown University
David Bellamy, University of Delaware
George Blakley, Texas A&M University
Jan List Boal, Georgia State University
Alan Candiotti, Drew University
George Feissner, SUNY at Cortland
Hebert Gindler, San Diego State University
Arthur Goodman, Queen's College, CUNY
Barry Granoff, Boston University
Douglas Hall, Michigan State University
Peter Herron, Suffolk Community College
Laurence Hoffman, Claremont Men's College
Adam Hulin, University of New Orleans
Frank Kocher, Pennsylvania State University
Carlon Krantz, Kean College

Paul Kumpel, SUNY at Stony Brook
Ed Landesman, University of California, Santa Cruz
Robert Lohman, Kent State University
Eldon Miller, University of Mississippi
Daniel Moran, Michigan State University
Richard Porter, Northeastern University
Thomas Rishel, Cornell University
Rainer Sachs, University of California, Berkeley
Philip Schaefer, University of Tennessee
Mark Schilling, University of Southern California
Donald Sherbert, University of Illinois
John Thorpe, SUNY at Stony Brook
William Wheeler, Indiana University
Robert Zink, Purdue University

The following mathematicians served as readers for the second edition:

Al Boggess, Texas A&M University
Phyllis Boutillier, Michigan Technological University
Lloyd Davis, College of San Mateo
Don Edmondson, University of Texas at Austin
Garrett Etgen, University of Houston
John Higgins, Brigham Young University
Ken Kramer, Queens College, CUNY
Steven Krantz, Washington University
Nicholas Krier, Colorado State University
Robert McFadden, Northern Illinois University
Robert Moreland, Texas Tech University
James Reeder, Honolulu Community College
Nathaniel Silver, University of Hartford
Richard Thompson, University of Arizona
Donald Wilken, SUNY, Albany
Stephen Willard, University of Alberta

Galley and page proofs were scrutinized for content and accuracy by Leon Gerber of St. John's University, Charles Stone of DeKalb College, Kathleen Hollowell of Newton High School (MA), Al Boggess, Steven Krantz, and Stephen Willard. Engineering applications were reviewed by Jeff Laible of the University of Vermont, Edgar Tacher of the University of Tulsa and Arthur Tiedmann of the University of Wisconsin. Kathleen Hollowell provided special assistance in this revision through thoughtful critique of the first edition and a careful review of all exercise sets.

Each of these individuals worked meticulously to ensure the accuracy of the text. Whatever errors might remain are, of course, the sole responsibility of the author. Any comments on correcting or improving the text will be gratefully acknowledged.

At Boston University I remain indebted to Tom Orowan for typing near-perfect drafts of the manuscript, and to Lisa Doherty for managing the flow of materials between Boston and Philadelphia. The BASIC programs included in the appendix were used by students on the University's time-sharing system, as well as on the author's personal computer.

Forewarned about hazards in dealings with publishers, I was delighted to experience a warm, professional, and highly supportive relationship with each of several key individuals at Saunders College Publishing: Mathematics Editor Robert Stern, Developmental Editors Sandi Kiselica and Sarah Evans, Project Editor Sally Kusch, and Publisher Don Jackson. Their commitment to excellence was a strong guiding force throughout the development of this edition.

The historical notes, which provide an important human contrast, were written by Professor Duane Deal of Ball State University.

The computer-drawn figures for the four-color insert were kindly provided by Larry Guseman and Robert Sparks of Texas A&M University. These figures were generated using a Silicon Graphics Iris 3130 computer.

On a more general and personal level, the writing of this text was supported by three very special groups of people. First, my students at Boston University, who have encouraged this project and helped sharpen my thinking about teaching and the calculus for the past decade. Second, my colleagues in the faculty and administration of Boston University, who believe deeply in the importance of effective teaching. And, most importantly, my family, who understood my need to write this book and shared fully and willingly in the sacrifices that were required. To all I am truly grateful.

Boston, Massachusetts   **Dennis D. Berkey**

CONTENTS OVERVIEW

CONTENTS

UNIT 5

THE THEORY OF INFINITE SERIES

Brook Taylor

Colin Maclaurin

Joseph Fourier

Karl Weierstrass

It was noted in the introduction to Unit I that Newton and Leibniz both found themselves expressing functions as infinite sums. Later workers in analysis, particularly Leonhard Euler and Johann Bernoulli, also dealt with these infinite series. This early work was not on a firm mathematical footing, however, and operations were sometimes performed that just happened to work because of the circumstances in a particular problem. The development of a logical theory was yet to appear, a process that unfolded over a long period of time.

Brook Taylor (1685–1731) and Isaac Newton knew each other through their membership in the Royal Society of London. Newton was its president for many years, and Taylor served for several years as its secretary. Taylor was brought up in a well-to-do home, and music was an important part of his early life. In fact, Taylor and Newton jointly wrote a work entitled *On Musick,* although it was never completed or published. Taylor thought highly of the new Newtonian calculus and attempted to clarify the subject in his writings. His writing was rather murky, however, and his exposition not very successful.

In this unit we write functions in the form of a special type of infinite series called Taylor's Series. Taylor developed the series as a result of a chance remark made by a friend in a coffeehouse, and some have questioned whether Taylor should receive full credit. Although he published the concept in 1713, it was so badly written that it had little immediate impact: it took Euler's work 40 years later to make the series concept well known.

Also introduced in this unit is Maclaurin's series, a special case of Taylor's series which was used by Taylor (as was acknowledged by Maclaurin in 1742). Colin Maclaurin (1698–1746), a Scotsman, was the most outstanding British mathematician in the generation following Newton. He was a prodigy, matriculating at the University of Glasgow at the age of eleven. He received his M.A. degree at 15. By age 19 he was a college mathematics teacher, and at 21 he published his first mathematical work of importance.

In 1719 Maclaurin met Isaac Newton, and he quickly became a disciple of Newtonian calculus. When Bishop George Berkeley wrote a tract attacking Newton's fluxions, Maclaurin responded in 1742 with his *Treatise on Fluxions,* the first complete and systematic presentation of Newton's calculus. Although it was not a textbook and was still not the final answer to rigor in calculus, it stood as a standard for nearly a century. Maclaurin was not comfortable with the limit concept, and based calculus on geometry instead. An unfortunate consequence of this emphasis was that his writing failed to make clear the useful applications of the subject. Further, the English mathematicians continued to use Newton's inadequate symbolism instead of Leibniz's superior symbolism, and during the 18th century British mathematics lost its preeminence.

Among Maclaurin's mathematical discoveries was what is now known as Cramer's rule for evaluating determinants. There is some poetic justice to this, for Maclaurin didn't discover Maclaurin's series just as Cramer didn't discover Cramer's rule. (In fairness, it should be noted that Cramer's notation was better than Maclaurin's.)

Joseph Fourier (1768–1830) was the only mathematician ever to serve as Governor of Lower Egypt. He had supported the French Revolution, and was rewarded with an appointment to the École Polytechnique. However, he had always wanted to be an army officer, a career denied him because he was the son of a tailor. When the opportunity came to accompany Napoleon on a military campaign in Egypt, Fourier resigned his teaching position and went along, and was appointed Governor in 1798. When the British took Egypt in 1801, Fourier returned to France.

Fourier studied the flow of heat in metallic plates and rods. The theory that he developed now has applications in industry and in the study of the temperature of the earth's interior. He discovered that many functions could be expressed as infinite sums of sine and cosine terms, now called a trigonometric series, or Fourier series. A paper that he submitted to the Academy of Science in Paris in 1807 was studied by several eminent mathematicians and rejected because he failed to prove his claims. They suggested that he reconsider and refine his paper, and even made heat flow the topic for a prize to be awarded in 1812. Fourier won the prize, but the Academy still declined to publish his paper because of its lack of rigor. (When Fourier became the secretary of the Academy, in 1824, the 1812 paper was published without change.)

As Fourier grew older, he developed at least one peculiar notion. Whether influenced by his stay in the heat of Egypt or by his own studies of the flow of heat in metals, he became obsessed with the idea that extreme heat was the natural condition for the human body. He was always heavily bundled in woolen clothing, and kept his rooms at high temperatures. He died in his sixty-third year, as expressed by Howard Eves in *An Introduction to the History of Mathematics,* "thoroughly cooked."

Most creative mathematicians have shown their genius at an early age. A notable exception was Karl Weierstrass (1816–1897), who did not really become a mathematician until he was 40 years old. He had studied law at the University of Bonn at his father's insistence, but he spent much of his time fencing and drinking. He did not complete his studies, and did not get his degree even after four years. Instead he turned to mathematics, but did not complete his degree in that subject, either. Eventually Weierstrass became a *gymnasium* (high school) teacher in a variety of subjects including not only mathematics and physics, but also German, history, botany, geography, gymnastics, and calligraphy. The mathematician in him was struggling to break through, and he did some mathematical research, although he had no contact with other mathematicians. He did manage to publish a few minor papers that attracted some attention. After thirteen years of secondary level teaching, Weierstrass obtained a position at a technical school, and was then called to the University of Berlin. Here his mathematical and pedagogical abilities surfaced, and he is considered the greatest teacher of higher mathematics of the nineteenth century, as judged by the number of his students who became significant researchers. Judged on his own mathematical creativity, he is sometimes called the leading analyst of his time, and the father of modern analysis.

Weierstrass worked in several areas, but his principal contribution was the study of functions of complex numbers through power series, which will be mentioned in this unit. This was an extension of the work of such earlier mathematicians as Maclaurin, Taylor, and Euler, but with a new rigor.

With all this creativity, Weierstrass did not publish many papers, and much of what he accomplished is known only through notes taken in class by his students. Indeed, he was uninterested in publicity or fame, and seemingly did not mind that some of his students used material from his lectures as their own. Hence, the record of the mathematical achievements of Karl Weierstrass has never been completely sorted out.

(Photographs from the David Eugene Smith Collection, Rare Book and Manuscript Library, Columbia University.)

Chapter 12
The Theory of Infinite Series

We have seen several instances involving a *sequence* of values, one corresponding to each of the positive integers $n = 1, 2, 3, \ldots$. For example,

(1) The definite integral $\int_a^b f(x)\,dx$ has been interpreted as the limit of a sequence S_n of Riemann Sums

$$S_n = \sum_{j=1}^n f(t_j)\,\Delta x_j, \qquad n = 1, 2, 3, \ldots .$$

(2) The amount $P(n)$ on deposit in a savings account paying interest annually at the rate r (in decimal form) n years after an initial deposit of P_0 dollars is

$$P(n) = (1 + r)^n P_0, \qquad n = 1, 2, 3, \ldots .$$

(3) Approximations to a zero of a differentiable function f are given by the formula for Newton's Method:

$$x_{n+1} = x_n - \frac{f(x_n)}{f'(x_n)}, \qquad n = 1, 2, 3, \ldots$$

where x_1 is the initial approximation.

The purpose of this unit is to develop a formal theory for dealing with expressions such as these which are defined only for integer values of the independent variable n. Our principal interest is in knowing how to calculate limits as $n \to \infty$ of such expressions and in applying this knowledge to answer practical questions.

Chapter 12 concerns infinite *sequences* and infinite *series of constants* (i.e., numbers). Chapter 13 applies the results of Chapter 12 to the important problems of approximating differentiable functions by polynomials and representing functions by infinite series.

12.1 INFINITE SEQUENCES

Up to this point we have rather casually defined an infinite sequence to be an unending string of numbers of the form

$$a_1, a_2, a_3, \ldots, a_n, \ldots \tag{1}$$

It is important to note, however, that expression (1) indicates an *order* in which these numbers appear in the string (the subscripts) as well as the numbers themselves (the a_n's). A more precise notion of an infinite sequence is therefore the following.

DEFINITION 1	An infinite sequence is a function whose domain is the positive integers.

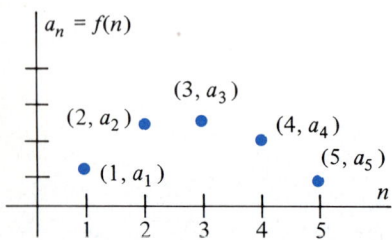

Figure 1.1 Plotting the sequence $\{a_1, a_2, a_3, \ldots\}$ where $a_n = f(n)$.

For example, the infinite sequence $\{1^2, 2^2, 3^2, 4^2, 5^2, \ldots, n^2, \ldots\}$ can be viewed as the set of values of the function $f(n) = n^2$, where $f(1) = 1^2$ is the first term, $f(2) = 2^2$ is the second term, and so on. Using the function concept, we can graph sequences on a coordinate plane. Figure 1.1 shows the graph of an arbitrary sequence $\{a_n\}$.

Usually, a sequence will be specified by a rule of the form $a_n = f(n)$ that determines the nth term of the sequence for each integer n, just as functions are usually specified by an equation of the form $y = f(n)$. For example, the rule $a_n = 2^n$ determines the sequence

$$\{2^n\} = \{2^1, 2^2, 2^3, 2^4, 2^5, \ldots\}$$
$$= \{2, 4, 8, 16, 32, \ldots\},$$

while the rule $a_n = (-1)^n$ determines the sequence

$$\{(-1)^n\} = \{-1, 1, -1, 1, -1, 1, -1, \ldots\}.$$

In such cases the term $a_n = f(n)$ is referred to as the **general term** of the sequence. Note that we use braces $\{\ \}$ to denote the entire sequence.

Example 1

Write out the first few terms of the sequences whose general terms are

(a) $a_n = 2n + 1$,

(b) $a_n = 2 + \dfrac{(-1)^n}{n}$.

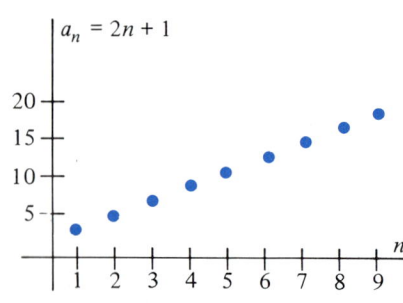

Figure 1.2 Graph of sequence $\{a_n\} = \{2n + 1\}$.

Solution: In part (a) we have

$$a_1 = 2(1) + 1 = 3, \qquad a_2 = 2(2) + 1 = 5, \qquad a_3 = 2(3) + 1 = 7, \text{ and so on,}$$

so

$$\{2n + 1\} = \{3, 5, 7, 9, 11, 13, 15, \ldots\},$$

while in (b) we have

$$a_1 = 2 + \frac{(-1)}{1} = 1, \qquad a_2 = 2 + \frac{(-1)^2}{2} = \frac{5}{2},$$

$$a_3 = 2 + \frac{(-1)^3}{3} = \frac{5}{3}, \qquad \text{and so on,}$$

so

$$\left\{2 + \frac{(-1)^n}{n}\right\} = \left\{1, \frac{5}{2}, \frac{5}{3}, \frac{9}{4}, \frac{9}{5}, \frac{13}{6}, \frac{13}{7}, \frac{17}{8}, \frac{17}{9}, \ldots\right\}.$$

Graphs of these two sequences appear in Figures 1.2 and 1.3.

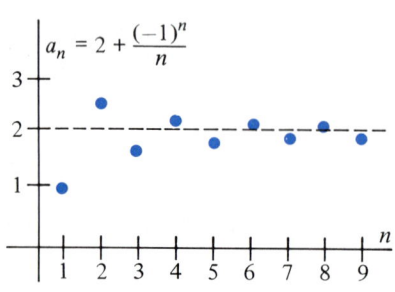

Figure 1.3 Graph of the sequence $\{a_n\} = \left\{2 + \dfrac{(-1)^n}{n}\right\}$.

There is an important difference between the sequence $\{2n + 1\}$ in Figure 1.2 and the sequence $\left\{2 + \dfrac{(-1)^n}{n}\right\}$ in Figure 1.3. In the first of these, the terms of the sequence increase uniformly, not approaching any particular number. In fact, in the language of Chapter 4, we would say that

$$\lim_{n \to \infty} \{2n + 1\} = +\infty$$

since the terms of the sequence increase without bound as $n \to \infty$.

However, the terms of the sequence $\left\{2 + \dfrac{(-1)^n}{n}\right\}$ "approach" the number $L = 2$ as $n \to \infty$, which we wish to write as

$$2 = \lim_{n \to \infty} \left\{2 + \dfrac{(-1)^n}{n}\right\}.$$

This notion of the *limit* of a sequence is just the notion of $\lim_{x \to \infty} f(x) = L$ developed for more general functions in Chapter 4. That is,

---

$L = \lim\limits_{n \to \infty} a_n$ means that the numbers a_n approach the number L as n increases without bound.

Using this working definition, together with some simple algebra, we can evaluate many types of limits by the same techniques used in Chapter 4 to evaluate $\lim_{x \to \infty} f(x)$.

Example 2

Find $\lim\limits_{n \to \infty} \dfrac{6n^3 + 5n^2 + 7}{4n^3 - 2n + 2}$.

Strategy

Divide all terms by n^3 (the highest power of n in the denominator).

Use fact that

$$\dfrac{c}{n^k} \to 0 \qquad \text{as} \qquad n \to \infty$$

if $k > 0$.

Solution

$$\lim_{n \to \infty} \dfrac{6n^3 + 5n^2 + 7}{4n^3 - 2n + 2} = \lim_{n \to \infty} \dfrac{6 + \dfrac{5}{n} + \dfrac{7}{n^3}}{4 - \dfrac{2}{n^2} + \dfrac{2}{n^3}}$$

$$= \dfrac{6 + 0 + 0}{4 - 0 + 0} = \dfrac{3}{2}.$$

$\diamond$

Example 3

Find $\lim\limits_{n \to \infty} \left[\ln(n + 4) - \dfrac{1}{2} \ln(n) \right]$.

Strategy
Use properties of $\ln x$ to reduce expression to the logarithm of a single number.

Solution

$$\lim_{n \to \infty} [\ln(n + 4) - 1/2 \ln(n)] = \lim_{n \to \infty} [\ln(n + 4) - \ln(n^{1/2})]$$

$$= \lim_{n \to \infty} \ln\left(\dfrac{n + 4}{\sqrt{n}}\right)$$

Divide both terms in numerator by $\sqrt{n}$.

$$= \lim_{n\to\infty} \ln\left(\sqrt{n} + \frac{4}{\sqrt{n}}\right)$$

Use fact that $\ln x \to \infty$ as $x \to \infty$.

$$= \infty,$$

since $\sqrt{n} \to \infty$ and $\dfrac{4}{\sqrt{n}} \to 0$ as $n \to \infty$.   ◇

Example 4

Find $\displaystyle\lim_{n\to\infty} (-1)^n \left(\frac{n+1}{n}\right).$

Strategy

Because of the factor $(-1)^n$, examine even and odd terms separately.

Solution

For even integers n, $(-1)^n = 1$, so we find that

$$\lim_{n\to\infty} (-1)^n \left(\frac{n+1}{n}\right) \qquad \text{(even integers only)}$$

Divide through by n.

$$= \lim_{n\to\infty} (1)\left(1 + \frac{1}{n}\right) = 1.$$

However, for odd integers n, $(-1)^n = -1$, so

$$\lim_{n\to\infty} (-1)^n \left(\frac{n+1}{n}\right) \qquad \text{(odd integers only)}$$

Divide through by n.

$$= \lim_{n\to\infty} (-1)\left(1 + \frac{1}{n}\right) = -1.$$

Since this shows that the terms do not approach a *single* number as $n \to \infty$, this limit does not exist (see Figure 1.4).   ◇

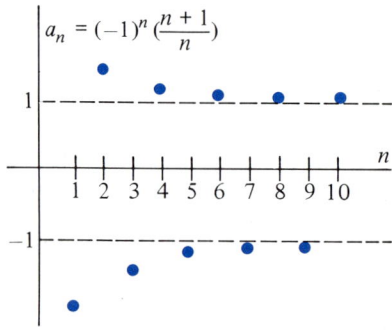

Figure 1.4 Terms of sequence $(-1)^n\left(\dfrac{n+1}{n}\right)$ approach both 1 and -1 as $n \to \infty$. The limit does not exist.

If $L = \displaystyle\lim_{n\to\infty} a_n$ exists, we say that the sequence $\{a_n\}$ **converges.** Otherwise the sequence is said to **diverge.** Note from Examples 3 and 4 that a sequence may diverge either because a_n becomes infinite as $n \to \infty$ or because a_n, remaining bounded, fails to approach a *single* number as $n \to \infty$.

We next state a formal definition of the limit of a sequence. Note the similarity with the formal definition for $L = \lim\limits_{x \to \infty} f(x)$ given in Chapter 4.

DEFINITION 2
Formal Definition of Limit

We say that $L = \lim\limits_{n \to \infty} a_n$ if and only if for each number $\epsilon > 0$ there exists an integer N so that

$$\text{if} \quad n > N, \quad \text{then} \quad |a_n - L| < \epsilon.$$

Definition 2 says this: If $L = \lim\limits_{n \to \infty} a_n$, we will find all terms of the sequence $\{a_n\}$, beyond the Nth term, lying within ϵ units of the number L. Since the integer N, in general, depends upon the number ϵ, we expect to have to look further along the sequence to observe this "closeness" as ϵ decreases in size. Figures 1.5 through 1.7 illustrate choices of N corresponding to three different values of ϵ for a typical sequence $\{a_n\}$.

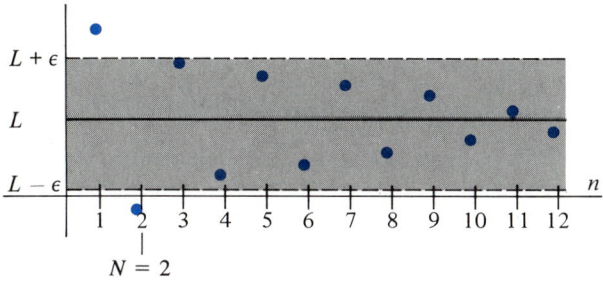

Figure 1.5 Large ϵ.

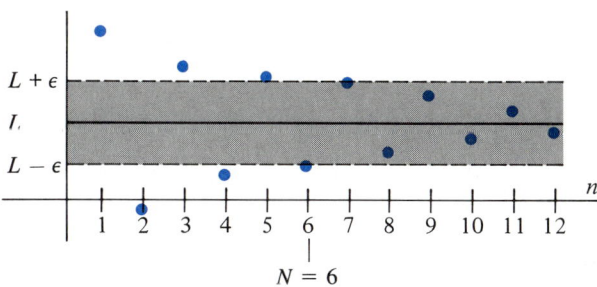

Figure 1.6 Medium ϵ.

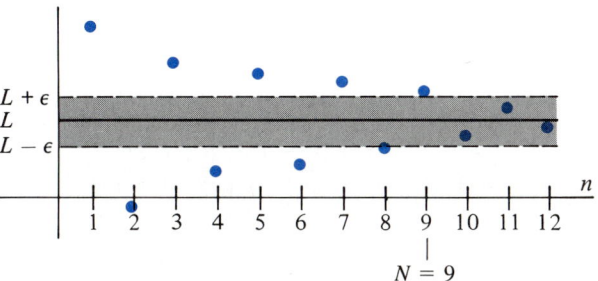

Figure 1.7 Small ϵ.

Definition 2 allows us to rigorously prove statements of the form $L = \lim\limits_{n \to \infty} a_n$ once we have found L. It does not, however, tell us how L is determined from the general term for the sequence $\{a_n\}$. For this, the intuitive notion of limit and familiarity with examples such as Examples 2–4 above are essential.

The following examples illustrate how Definition 2 is used.

Example 5

Prove that $\lim\limits_{n \to \infty} \dfrac{1}{n} = 0$.

Strategy

Set up the inequality

$$|a_n - L| < \epsilon$$

and solve for n to find a relationship between n and ϵ.

Solution

According to Definition 2, we allow $\epsilon > 0$ to be any fixed positive number. Since $a_n = 1/n$ and $L = 0$, we must determine how large to choose n to guarantee that

$$\left| \frac{1}{n} - 0 \right| = \frac{1}{n} < \epsilon. \tag{2}$$

Solving inequality (2) for n we see that it is equivalent to the inequality

$$n > \frac{1}{\epsilon}. \tag{3}$$

Choose N large enough that the inequality holds whenever $n > N$.

We therefore take N to be any integer larger* than $1/\epsilon$. Then, inequality (3) holds, whenever $n > N$, which guarantees that inequality (2) holds. In other words, if $N > 1/\epsilon$ then

$$\left| \frac{1}{n} - 0 \right| < \epsilon \qquad \text{whenever} \qquad n > N,$$

as required by Definition 2. $\diamond$

Example 6

Prove that $\displaystyle \lim_{n \to \infty} \frac{2n - 1}{n + 2} = 2$.

Strategy

Set up the inequality

$$|a_n - L| < \epsilon.$$

Solution

We assume $\epsilon > 0$ to be an arbitrary fixed number. Since $a_n = \dfrac{2n - 1}{n + 2}$ and $L = 2$, we must determine how large to choose n to guarantee that

$$\left| \frac{2n - 1}{n + 2} - 2 \right| < \epsilon. \tag{4}$$

Solve this inequality for n.

Inequality (4) is equivalent to the inequality

$$\left| \frac{2n - 1 - 2(n + 2)}{n + 2} \right| < \epsilon,$$

or

$$\frac{5}{n + 2} < \epsilon. \tag{5}$$

Solving inequality (5) for n we find that it is equivalent to the inequality

$$n > \frac{5}{\epsilon} - 2. \tag{6}$$

$$\frac{5}{n + 2} < \epsilon$$
$$\Leftrightarrow 5 < \epsilon(n + 2)$$
$$\Leftrightarrow \frac{5}{\epsilon} < n + 2$$
$$\Leftrightarrow n > \frac{5}{\epsilon} - 2.$$

Choose n sufficiently large that the desired inequality holds.

We therefore take N to be any integer larger than $\dfrac{5}{\epsilon} - 2$. Then, inequality (6)

---

*Since ϵ is a positive number, so is $1/\epsilon$. It is a property of the real numbers that, given any positive number a, an integer N can be found with $N > a$.

holds whenever $n > N$. Since inequality (6) is equivalent to inequality (4), this shows that, for $N > \dfrac{5}{\epsilon} - 2$, we may conclude that

$$\text{if} \quad n > N \quad \text{then} \quad \left| \frac{2n - 1}{n + 2} - 2 \right| < \epsilon,$$

as required by Definition 2.   ◇

Definition 2 can be used to prove several theorems giving rules by which limits of sequences may be calculated. Since the proofs of these theorems are similar to the proofs given in Chapters 2 and 4 for the corresponding theorems on limits of functions, we leave them as exercises.

THEOREM 1
Properties of Limits of Sequences

If $\lim\limits_{n \to \infty} a_n = L$, $\lim\limits_{n \to \infty} b_n = M$ and c is any real number, then

(i) $\lim\limits_{n \to \infty} (a_n + b_n) = L + M$,

(ii) $\lim\limits_{n \to \infty} (ca_n) = cL$,

(iii) $\lim\limits_{n \to \infty} (a_n b_n) = LM$,

(iv) $\lim\limits_{n \to \infty} \left(\dfrac{a_n}{b_n} \right) = \dfrac{L}{M}$, $\qquad b_n \neq 0$, $\qquad M \neq 0$.

Theorem 1 together with the proof in Example 5 makes legitimate the calculations in Example 2. The next theorem addresses a situation that occurred in Example 3.

THEOREM 2

Suppose that $\lim\limits_{n \to \infty} a_n = L$ and each number a_n lies in the domain of the function f. If f is continuous at $x = L$ then

$$\lim_{n \to \infty} f(a_n) = f(L).$$

Example 7

Since $f(x) = \tan x$ is continuous for $-\pi/2 < x < \pi/2$,

$$\lim_{n \to \infty} \tan\left(\frac{\pi n^2 + 1}{3 - 4n^2} \right) = \tan\left[\lim_{n \to \infty} \left(\frac{\pi n^2 + 1}{3 - 4n^2} \right) \right] = \tan\left(-\frac{\pi}{4} \right) = -1. \qquad ◇$$

Example 8

Since $f(x) = \sqrt{x}$ is continuous for $x \geq 0$,

$$\lim_{n \to \infty} \sqrt{\frac{4n + 1}{n}} = \left(\lim_{n \to \infty} \frac{4n + 1}{n} \right)^{1/2} = \sqrt{4} = 2. \qquad ◇$$

The following is the analogue for sequences of the Pinching Theorem for functions.

THEOREM 3
Pinching Theorem

Let $\{a_n\}$, $\{b_n\}$, and $\{c_n\}$ be sequences and let P be a positive integer. Let $a_n \leq b_n \leq c_n$ for all integers $n \geq P$. If

$$\lim_{n \to \infty} a_n = L = \lim_{n \to \infty} c_n,$$

then $\lim_{n \to \infty} b_n = L$ also.

Example 9

Show that $\displaystyle\lim_{n \to \infty} \frac{1}{n^p} = 0$ if $p \geq 1$.

Solution: If $p \geq 1$, $n^p \geq n$ for all $n = 1, 2, 3, \ldots$. Thus

$$0 \leq \frac{1}{n^p} \leq \frac{1}{n}, \qquad n \geq 1.$$

Since $\displaystyle\lim_{n \to \infty} \{0\} = 0 = \lim_{n \to \infty} \left\{\frac{1}{n}\right\}$, the conclusion follows by the Pinching Theorem.

◇

Example 10

Find $\displaystyle\lim_{n \to \infty} \frac{\sin n}{n}$.

Strategy

Find bounds on $\sin n$.

Solution

We have $|\sin n| \leq 1$ for all n, that is,

$$-1 \leq \sin n \leq 1, \qquad n \geq 1.$$

Thus

Divide by n to find bounds on $\dfrac{\sin n}{n}$.

$$-\frac{1}{n} \leq \frac{\sin n}{n} \leq \frac{1}{n}.$$

Apply Pinching Theorem.

Since $\displaystyle\lim_{n \to \infty}\left(-\frac{1}{n}\right) = 0 = \lim_{n \to \infty}\left(\frac{1}{n}\right)$,

$$\lim_{n \to \infty} \frac{\sin n}{n} = 0$$

by the Pinching Theorem.

We summarize the ideas of this section by noting that finding the limit of sequence $\{a_n\}$ is similar to finding horizontal asymptotes for the function f with $f(n) = a_n$, $n = 1, 2, \ldots$. The principal difference is that $\{a_n\}$ is a function defined only for positive integers. Thus $L = \lim_{n \to \infty} a_n$ if $L = \lim_{x \to \infty} f(x)$, but the converse need not be true (see Exercise 35).

Exercise Set 12.1

In each of Exercises 1–28, write out the first four terms of the given sequence and determine whether the sequence converges or diverges. If the sequence converges, find its limit.

1. $\left\{\dfrac{n}{2n+1}\right\}$

2. $\left\{\dfrac{2n-1}{n+3}\right\}$

3. $\left\{\dfrac{n-4}{n^2+2}\right\}$

4. $\left\{\dfrac{n^2+1}{3n(n+2)}\right\}$

5. $\left\{\dfrac{1}{1+n^2}\right\}$

6. $\left\{\dfrac{1}{e^n}\right\}$

7. $\{\sqrt{5}\}$

8. $\left\{\dfrac{(n-1)(n+1)}{2n^2+2n+2}\right\}$

9. $\left\{\dfrac{20n}{1+\sqrt{n}}\right\}$

10. $\left\{\dfrac{6-n^{3/2}}{(\sqrt{n}+1)^2}\right\}$

11. $\left\{\dfrac{3+(-1)^n\sqrt{n}}{n+2}\right\}$

12. $\{(-1)^n \sin n\}$

13. $\left\{\sqrt{1+\dfrac{1}{n}}\right\}$

14. $\left\{1+\dfrac{(-1)^n}{2^n}\right\}$

15. $\left\{\cos\left(\dfrac{n-1}{n^2}\right)\right\}$

16. $\left\{\dfrac{n+1}{n}\right\}$

17. $\left\{\dfrac{n^{3/2}+2}{2n^{3/2}}\right\}$

18. $\left\{\dfrac{e^n-e^{-n}}{e^n+e^{-n}}\right\}$

19. $\left\{\dfrac{1}{n}-\dfrac{1}{n+1}\right\}$

20. $\left\{\dfrac{2^n}{5^{n+2}}\right\}$

21. $\{\sqrt{n+1}-\sqrt{n}\}$

22. $\left\{\dfrac{\cos^2 n}{n}\right\}$

23. $\left\{\dfrac{\sqrt{2n^2+1}}{n}\right\}$

24. $\left\{\tan^{-1}\left(\dfrac{n+2}{2}\right)\right\}$

25. $\left\{n\cdot\sin\dfrac{\pi}{2n}\right\}$

26. $\left\{\ln\dfrac{n^2+1}{(n+2)(n+3)}\right\}$

27. $\left\{\left(1+\dfrac{1}{n}\right)^n\right\}$

28. $\left\{\left(1-\dfrac{1}{n}\right)^n\right\}$

In Exercises 29–33, use the definition of limit to prove that $\lim\limits_{n\to\infty} a_n = L$.

29. $a_n = \dfrac{3}{n}$; $L = 0$

30. $a_n = \dfrac{1}{2n+1}$; $L = 0$

31. $a_n = \dfrac{n}{3n+1}$; $L = \dfrac{1}{3}$

32. $a_n = \dfrac{3n-1}{n+1}$; $L = 3$

33. $a_n = \dfrac{n^2+2n+3}{1+n^2}$; $L = 1$

34. Use Theorem 2 to show that $\lim\limits_{n\to\infty} \sqrt[n]{a} = \lim\limits_{n\to\infty} e^{\ln a^{1/n}} = 1$ if $a > 0$.

35. Let $a_n = \sin \pi n$, $n = 1, 2, \ldots$, and let $f(x) = \sin \pi x$. Then $f(n) = a_n$. Show that $\lim\limits_{n\to\infty} a_n = 0$ but that $\lim\limits_{x\to\infty} f(x)$ does not exist.

36. Prove that if $\lim\limits_{n\to\infty} a_n$ exists, this limit is unique. (*Hint:* Assume that both $\lim\limits_{n\to\infty} a_n = L$ and $\lim\limits_{n\to\infty} a_n = M$. Then show $L = M$ by examining the inequality $|L - M| \le |L - a_n| + |a_n - M|$.)

37. A sequence is called **bounded** if there is a number M so that $|a_n| \le M$ for all terms a_n of the sequence. Prove that a convergent sequence must be bounded.

38. Give an example showing that a bounded sequence need *not* converge.

39. Prove that the sum of two bounded sequences is again bounded. What about the product of two bounded sequences? The quotient?

40. Prove that $\lim\limits_{n\to\infty} a_n = L$ if and only if $\lim\limits_{n\to\infty} |a_n - L| = 0$.

41. Prove that $\lim\limits_{n\to\infty} \dfrac{1}{a^n} = 0$ if $a > 1$, using the Pinching Theorem.

42. Prove part (i) of Theorem 1.

43. Prove part (ii) of Theorem 1.

44. Prove part (iii) of Theorem 1.

45. Prove part (iv) of Theorem 1.

46. Prove Theorem 2.

47. Prove Theorem 3.

12.2 MORE ON INFINITE SEQUENCES

If $\{a_n\}$ is a sequence and f is a function for which $f(n) = a_n$, $n = 1, 2, \ldots$, then we can conclude that $\lim_{n \to \infty} a_n = L$ whenever $\lim_{x \to \infty} f(x) = L$. (See Figure 2.1.)

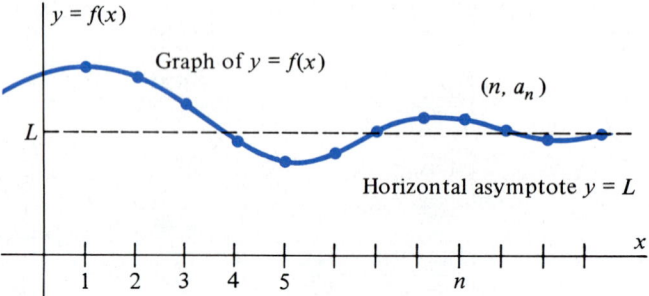

Figure 2.1 If $f(n) = a_n$ and $\lim_{x \to \infty} f(x) = L$, then $\lim_{n \to \infty} a_n = L$ also.

This observation allows us to reformulate the problem of finding $\lim_{n \to \infty} a_n$ as the problem of finding $\lim_{x \to \infty} f(x)$. In the latter problem, l'Hôpital's Rule may often be successfully applied to calculate an otherwise ambiguous limit.

Example 1

Find $\lim_{n \to \infty} \dfrac{n}{e^n}$.

Solution: The limit is not obvious because both numerator and denominator become infinite as $n \to \infty$. Using the above notion together with l'Hôpital's Rule gives

$$\lim_{n \to \infty} \frac{n}{e^n} = \lim_{x \to \infty} \frac{x}{e^x} \qquad \left(\frac{\infty}{\infty} \text{ form}\right)$$

$$= \lim_{x \to \infty} \frac{\dfrac{d}{dx}(x)}{\dfrac{d}{dx}(e^x)} \qquad \text{(l'Hôpital's Rule)}$$

$$= \lim_{x \to \infty} \frac{1}{e^x}$$

$$= 0. \qquad\qquad\qquad\qquad \diamond$$

We next establish several limits that we shall use frequently.

$$\lim_{n \to \infty} x^n = 0 \qquad \text{if} \qquad |x| < 1. \tag{1}$$

Proof: Let ϵ be an arbitrary positive number. Then, since

$$\lim_{n \to \infty} \frac{1}{n} = 0,$$

$\lim_{n \to \infty} \epsilon^{1/n} = \epsilon^0 = 1$. Thus, since $|x| < 1$, there exists an integer N, by Definition 2, such that $\epsilon^{1/n} > |x|$ whenever $n > N$. Thus

$$|x|^n < (\epsilon^{1/n})^n = \epsilon, \qquad n > N.$$

Equivalently, noting that $|x|^n = |x^n|$, we have

$$|x^n - 0| < \epsilon \qquad \text{whenever} \qquad n > N.$$

Thus,

$$\lim_{n \to \infty} x^n = 0 \qquad \text{by Definition 2.} \qquad \blacklozenge$$

$$\lim_{n \to \infty} \frac{x^n}{n!} = 0, \qquad -\infty < x < \infty. \tag{2}$$

Proof: To establish this limit we shall show that

$$\lim_{n \to \infty} \left| \frac{x^n}{n!} \right| = 0, \qquad -\infty < x < \infty. \tag{3}$$

(See Exercise 28.)

Let x be given and let N be an integer such that $N > |x|$. Write

$$J = \left| \left(\frac{x}{1} \right) \left(\frac{x}{2} \right) \left(\frac{x}{3} \right) \cdots \left(\frac{x}{N-1} \right) \right|$$

and note that J is constant since N and x are fixed. Then for $n > N$ we have

$$\left| \frac{x^n}{n!} \right| = \left| \left(\frac{x}{1} \right) \left(\frac{x}{2} \right) \left(\frac{x}{3} \right) \cdots \left(\frac{x}{N-1} \right) \left(\frac{x}{N} \right) \cdots \left(\frac{x}{n} \right) \right|$$

$$= J \left| \left(\frac{x}{N} \right) \left(\frac{x}{N+1} \right) \cdots \left(\frac{x}{n} \right) \right| \qquad (n - N + 1) \text{ factors}$$

$$\leq J \cdot \left| \frac{x}{N} \right|^{(n-N+1)}$$

Since $|x| < N$, we have $\left| \dfrac{x}{N} \right| < 1$, so $\lim\limits_{n \to \infty} \left| \dfrac{x}{N} \right|^n = 0$. Thus,

$$\lim_{n \to \infty} J \cdot \left| \frac{x}{N} \right|^{(n-N+1)} = J \cdot \left| \frac{x}{N} \right|^{(-N+1)} \cdot \lim_{n \to \infty} \left| \frac{x}{N} \right|^n = 0.$$

This establishes (3) by the preceding inequality. $\qquad \blacklozenge$

$$\lim_{n \to \infty} \frac{\ln(n)}{n} = 0. \tag{4}$$

Proof: $\lim_{n \to \infty} \dfrac{\ln(n)}{n} = \lim_{x \to \infty} \dfrac{\ln x}{x} = \lim_{x \to \infty} \dfrac{\frac{1}{x}}{1} = 0$, by l'Hôpital's Rule.   ◆

$$\lim_{n \to \infty} \sqrt[n]{n} = 1. \tag{5}$$

Proof: $\lim_{n \to \infty} \ln(n^{1/n}) = \lim_{n \to \infty} \dfrac{\ln(n)}{n} = 0$, by (4).

Thus,

$$\lim_{n \to \infty} n^{1/n} = e^0 = 1.$$ ◆

Example 2

$$\lim_{n \to \infty} \left(\frac{5}{n}\right)^{1/n} = \lim_{n \to \infty} \frac{\sqrt[n]{5}}{\sqrt[n]{n}}$$

$$= \frac{\lim_{n \to \infty} \sqrt[n]{5}}{\lim_{n \to \infty} \sqrt[n]{n}}$$

$$= \frac{1}{1} = 1$$

by statement (5).   ◇

We shall have need of one further theorem on sequences, which we develop here.

A set S of numbers is said to be *bounded* if there exists a number M so that $|x| \le M$ for every number $x \in S$. This number M is called an *upper bound* for S. A fundamental property of the real number system states that among all such upper bounds M there can always be found a smallest bound. This property is referred to as the **completeness axiom,** and it is formally stated as follows.

Completeness Axiom for Real Numbers: If S is any nonempty bounded set of real numbers there exists a least upper bound L for S. That is, there exists a number L for which

(i) $x \le L$ for every $x \in S$, and
(ii) if M is any upper bound for S, then $L \le M$.

For example, the set $S = \{x \mid -3 \le x \le 3\}$ is bounded since $|x| \le 3$ for any $x \in S$. The number 7 is an *upper bound* for S since $x < 7$ for all $x \in S$. The numbers 10, 37, and 1001 are also upper bounds for S. The number $L = 3$ is the *least* upper bound for S, because any *other* upper bound for S is greater than 3.

The completeness axiom and its role in the definition of the real number system are topics for more advanced courses. We shall use this axiom to establish a theorem about *increasing* sequences. As for functions, we define the sequence $\{a_n\}$ to be **increasing** if $a_n > a_m$ whenever $n > m$.

THEOREM 4   Every bounded increasing sequence converges.

Proof: We use the definition of the limit of a sequence and the completeness axiom. Let $\epsilon > 0$ be given, and let $\{a_n\}$ denote a bounded increasing sequence. Then the set of numbers $\{a_1, a_2, a_3, \ldots\}$ is bounded, so it has a least upper bound L according to the completeness axiom. We will complete the proof by showing that $\lim_{n\to\infty} a_n = L$.

Since $\epsilon > 0$, we must have $L - \epsilon < L$, so $L - \epsilon$ cannot be an upper bound for the sequence. Thus, there must exist an integer N such that

$$a_N > L - \epsilon. \qquad \text{(Figure 2.2.)}$$

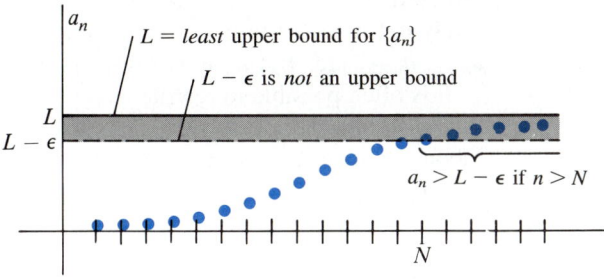

Figure 2.2   A bounded increasing sequence converges to its least upper bound.

But since $\{a_n\}$ is increasing we must have $a_n > a_N$ for all $n > N$. Thus we have the following five numbers in increasing order:

$$L - \epsilon < a_N < a_n \leq L < L + \epsilon, \qquad n > N.$$

It now follows from $L - \epsilon < a_n < L + \epsilon$ that

$$|a_n - L| < \epsilon \qquad \text{whenever} \qquad n > N.$$

Thus, $L = \lim_{n\to\infty} a_n$ according to Definition 2. ◆

There is an obvious extension of Theorem 4 to bounded *decreasing* sequences. If $\{a_n\}$ is a bounded decreasing sequence, then the sequence $\{-a_n\}$ is a bounded *increasing* sequence, which, by Theorem 4, has a limit, say L. By Theorem 1 we then have

$$\lim_{n\to\infty} a_n = -\lim_{n\to\infty} \{-a_n\} = -L.$$

We summarize these remarks as follows.

COROLLARY 1   Every bounded decreasing sequence converges.

Recursively Defined Sequences

In many applications, particularly in computer science, sequences are defined *recursively* rather than as functions of the integer n. One of the most famous recursively defined sequences is the sequence of **Fibonacci numbers**

$$1, 1, 2, 3, 5, 8, 13, 21, 34, \ldots . \qquad (6)$$

This sequence was first discovered by the Italian mathematician Leonardo of Pisa (who also went by the name Fibonacci) around the year 1200 A.D. (Fibonacci is regarded by many as the most brilliant of the pre-Renaissance mathematicians).

The Fibonacci sequence $\{F_n\}$ in (6) is determined by the rules

$$F_1 = 1, \tag{7}$$

$$F_2 = 1, \quad \text{and} \tag{8}$$

$$F_{n+2} = F_n + F_{n+1}, \quad n = 1, 2, 3, 4, \ldots . \tag{9}$$

That is, every term in the sequence beyond the second is found by adding the two preceding terms. This is what we mean by saying that the sequence $\{a_n\}$ is **recursively defined**: the term a_n is a function of one or more preceding terms, such as a_{n-1}, a_{n-2}, etc. But a_n is *not* written as an explicit function of n.

It is often possible to rewrite a recursively defined sequence as an explicit function of n, and vice versa. For example, it has been shown that the nth term of the Fibonacci sequence can be written

$$F_n = \frac{1}{\sqrt{5}} \left(\frac{1 + \sqrt{5}}{2} \right)^{n+1} - \frac{1}{\sqrt{5}} \left(\frac{1 - \sqrt{5}}{2} \right)^{n+1} .$$

(See Exercise 20 for a biological interpretation of the Fibonacci sequence.) Another such example is the factorial sequence

$$\{a_n\} = \{n!\} = \{1!, 2!, 3!, \ldots\} = \{1, 2, 6, 24, \ldots\}.$$

It can be defined recursively as the sequence $\{f_n\} = \{f_1, f_2, f_3, \ldots\}$ where

$$f_1 = 1$$
$$f_n = nf_{n-1}, \quad n = 2, 3, 4, \ldots .$$

However, not every sequence $\{a_n\}$ can be defined recursively.

Recursively defined sequences occur frequently in the analysis of computer algorithms. In such situations one is concerned only with how to determine the next term in the sequence given the present term (and, possibly, several preceding terms), not with the correspondence between integers n and the terms a_n. Exercises 20–24 in this section concern recursively defined sequences.

Exercise Set 12.2

In each of Exercises 1–18, find the indicated limit, if it exists.

1. $\lim\limits_{n \to \infty} n \sin\left(\dfrac{2}{n}\right)$

2. $\lim\limits_{n \to \infty} \dfrac{\sin^3 n}{n}$

3. $\lim\limits_{n \to \infty} \sqrt[n]{4n}$

4. $\lim\limits_{n \to \infty} \dfrac{\ln(n)}{\sqrt{n}}$

5. $\lim\limits_{n \to \infty} (n + 1)e^{-n}$

6. $\lim\limits_{n \to \infty} \dfrac{n}{e^n}$

7. $\lim\limits_{n \to \infty} \dfrac{n^n}{n!}$

8. $\lim\limits_{n \to \infty} n^{3/n}$

9. $\lim\limits_{n \to \infty} (n + \pi)^{1/n}$

10. $\lim\limits_{n \to \infty} \left(1 - \dfrac{3}{n}\right)^n$

11. $\lim\limits_{n \to \infty} \dfrac{3^n}{(n + 3)!}$

12. $\lim\limits_{n \to \infty} \left(\dfrac{e}{n} \ln \dfrac{e}{n}\right)$

13. $\lim\limits_{n \to \infty} \dfrac{n^2 \ln(n)}{2^n}$

14. $\lim\limits_{n \to \infty} n^{\sin(\pi/n)}$

15. $\lim\limits_{n \to \infty} \left(\dfrac{n + 3}{n}\right)^n$

16. $\lim\limits_{n \to \infty} \left(1 + \dfrac{1}{n^2}\right)^n$

17. $\lim\limits_{n \to \infty} \sqrt[n]{n^3}$

18. $\lim\limits_{n \to \infty} \dfrac{n - \sin n}{n + \cos n}$

19. Prove that $\lim\limits_{n\to\infty} x^n$ does not exist if $|x| > 1$.

20. The Fibonacci sequence $\{1, 1, 2, 3, 5, 8, 13, 21, 34, \ldots\}$ arises as a mathematical model for the size of a population of rabbits under the following conditions. We assume that we begin with a single pair of rabbits, that this and each other pair of rabbits become fertile one month after birth, that each pair of fertile rabbits gives birth to one new pair of rabbits each month, and that no rabbits die. If F_n represents the number of pairs of rabbits in the population after n months, show that
  a. $F_1 = 1$,
  b. $F_2 = 1$,   and
  c. $F_{n+2} = 2F_n + (F_{n+1} - F_n)$. (*Hint:* The term $2F_n$ is explained as follows. Every pair of rabbits that was present two months ago is still present along with one pair of offspring. The second term accounts for the fact that those rabbits which were fertile two months ago have produced *two* pairs of offspring since then, one of which is not counted in the first term.)
  d. Conclude from (c) that $F_{n+2} = F_n + F_{n+1}$.

21. Let $\{a_n\}$ be a sequence recursively defined by the equations

$$a_0 = 1$$
$$a_n = 2a_{n-1}, \qquad n = 1, 2, \ldots .$$

Show that the general term for this sequence is $a_n = 2^n$.

22. Find the general term for the sequence $\{a_n\}$ recursively defined by the equations

$$a_0 = 4$$
$$a_n = a_{n-1} + 1, \qquad n = 1, 2, \ldots .$$

23. Find the general term for the sequence $\{a_n\}$ recursively defined by the equations

$$a_0 = -5$$
$$a_n = a_{n-1} + 2, \qquad n = 1, 2, \ldots .$$

24. For the Fibonacci sequence (Exercise 20) show that
  a. $F_{n+3} = 2F_{n+1} + F_n$
  b. $F_{n+4} = 3F_{n+1} + 2F_n$
  c. $F_{n+p} = F_p F_{n+1} + F_{p-1} F_n, \qquad p = 3, 4, \ldots .$

25. Give an example of a bounded sequence that does not converge.

26. Give an example of an increasing sequence that does not converge.

27. Must every convergent sequence be bounded and either increasing or decreasing?

28. Prove that if $\lim\limits_{n\to\infty} |a_n| = 0$, then $\lim\limits_{n\to\infty} a_n = 0$.

12.3 INFINITE SERIES

By an *infinite series* we mean an expression of the form

$$\sum_{k=1}^{\infty} a_k = a_1 + a_2 + a_3 + \cdots. \tag{1}$$

Unlike the situation for finite sums, we cannot associate a "sum" with an infinite series simply by "adding up" the terms $a_1, a_2, \ldots$ because this would require that we perform an infinite number of additions, something not even a high-speed computer can accomplish in a finite amount of time.

The method for evaluating improper integrals of the form $\int_a^\infty f(x)\, dx$ provides the idea by which we shall determine whether an infinite series has a sum—that is, we first find the *partial sums* $S_n = \sum\limits_{k=1}^{n} a_k$ of the series in line (1) and then ask whether

the limit $\lim\limits_{n\to\infty} S_n = \lim\limits_{n\to\infty} \sum\limits_{k=1}^{n} a_k$ of these partial sums exists. If it does, this limit is

what we shall call the sum of the infinite series.

DEFINITION 3

An **infinite series** is an expression of the form

$$\sum_{k=1}^{\infty} a_k = a_1 + a_2 + a_3 + \cdots.$$

The infinite series

$$\sum_{k=1}^{\infty} a_k$$

is said to **converge** to the **sum** S if

$$S = \lim_{n \to \infty} S_n,$$

where S_n denotes the nth **partial sum**

$$S_n = a_1 + a_2 + a_3 + \cdots + a_n = \sum_{k=1}^{n} a_k.$$

If the limit S does not exist, the series $\sum_{k=1}^{\infty} a_k$ is said to **diverge.**

Example 1

The repeating decimal $0.6666\overline{6}$ may be interpreted as the infinite series

$$0.6666\overline{6} = .6 + (.06) + (.006) + (.0006) + \cdots$$

$$= \frac{6}{10} + \frac{6}{10^2} + \frac{6}{10^3} + \frac{6}{10^4} + \cdots$$

$$= \sum_{k=1}^{\infty} \frac{6}{10^k}.$$

Let us verify that this interpretation is consistent with the ordinary notion that $0.66\overline{6} = 2/3$. The nth partial sum of this series is

$$S_n = \frac{6}{10} + \frac{6}{10^2} + \frac{6}{10^3} + \cdots + \frac{6}{10^n}. \tag{2}$$

Now note that each term of the sum is 10 times the following term. Multiplying both sides of equation (2) by $\frac{1}{10}$ gives

$$\frac{1}{10} S_n = \frac{6}{10^2} + \frac{6}{10^3} + \frac{6}{10^4} + \cdots + \frac{6}{10^{n+1}}. \tag{3}$$

Subtracting corresponding sides of equation (3) from those of equation (2) gives

$$S_n - \frac{1}{10} S_n = \frac{6}{10} - \frac{6}{10^{n+1}},$$

so

$$S_n = \frac{10}{9}\left(\frac{6}{10} - \frac{6}{10^{n+1}}\right) = \frac{2}{3}\left(1 - \frac{1}{10^n}\right).$$

According to the definition of an infinite series, the sum of the series is therefore

$$S = \lim_{n\to\infty} S_n = \lim_{n\to\infty} \frac{2}{3}\left(1 - \frac{1}{10^n}\right) = 2/3.$$

This shows that

$$\frac{2}{3} = \sum_{k=1}^{\infty} \frac{6}{10^k} = .66666\overline{6}. \qquad \diamondsuit$$

Example 2

The infinite series

$$\sum_{k=1}^{\infty} (-1)^k = -1 + 1 - 1 + 1 - 1 + \cdots$$

does not converge. To see why, observe that the partial sums are

$$S_1 = -1$$
$$S_2 = -1 + 1 = 0$$
$$S_3 = -1 + 1 - 1 = -1$$
$$S_4 = -1 + 1 - 1 + 1 = 0$$
$$\cdot$$
$$\cdot$$
$$\cdot$$

$$S_{2n-1} = -1 + 1 - 1 + \cdots + 1 - 1 = -1 \qquad (2n - 1 \text{ terms})$$
$$S_{2n} = -1 + 1 - 1 + \cdots + 1 - 1 + 1 = 0 \qquad (2n \text{ terms}).$$

Thus, the terms of the sequence $\{S_n\}$ are alternately -1 or 0, so $\lim_{n\to\infty} S_n$ does not exist. $\qquad \diamondsuit$

Example 3

Determine whether the infinite series

$$\sum_{k=1}^{\infty} \frac{1}{k(k+1)}$$

converges. If it does, find its sum.

Solution: By the method of partial fractions we can show that

$$\frac{1}{k(k+1)} = \frac{1}{k} - \frac{1}{k+1}, \qquad k = 1, 2, 3, \ldots .$$

We can therefore write the partial sum S_n for this series as

$$S_n = \sum_{k=1}^{n} \frac{1}{k(k+1)} = \frac{1}{1 \cdot 2} + \frac{1}{2 \cdot 3} + \frac{1}{3 \cdot 4} + \cdots + \frac{1}{(n-1)n} + \frac{1}{n(n+1)}$$

$$= \left[\frac{1}{1} - \frac{1}{2}\right] + \left[\frac{1}{2} - \frac{1}{3}\right] + \left[\frac{1}{3} - \frac{1}{4}\right]$$

$$+ \cdots + \left[\frac{1}{n-1} - \frac{1}{n}\right] + \left[\frac{1}{n} - \frac{1}{n+1}\right]$$

$$= 1 - \frac{1}{n+1}$$

since all other terms in this "telescoping sum" cancel. This shows that

$$\sum_{k=1}^{\infty} \frac{1}{k(k+1)} = \lim_{n \to \infty} S_n = \lim_{n \to \infty} \left(1 - \frac{1}{n+1}\right) = 1.$$

Thus the series converges, and its sum is $S = 1$. ◇

REMARK: Up to this point we have been indexing all infinite series so that the first term has index $k = 1$. This is not necessary. For example, we could rewrite

$$\sum_{k=1}^{\infty} a_k \qquad \text{as} \qquad \sum_{k=2}^{\infty} b_k$$

where $b_k = a_{k-1}$, since in this case we would have

$$\sum_{k=2}^{\infty} b_k = b_2 + b_3 + b_4 + b_5 + \cdots$$

$$= a_{(2-1)} + a_{(3-1)} + a_{(4-1)} + a_{(5-1)} + \cdots$$
$$= a_1 + a_2 + a_3 + a_4 + \cdots.$$

Similarly, we can write

$$\sum_{k=1}^{\infty} a_k \qquad \text{as} \qquad \sum_{k=0}^{\infty} c_k, \qquad \text{where} \qquad c_k = a_{k+1}.$$

Note also that *the convergence of an infinite series does not depend on any finite number of terms.* For example, if $\sum_{k=1}^{\infty} a_k$ is a convergent series with sum S, then the series $\sum_{k=0}^{\infty} a_k$ and $\sum_{k=2}^{\infty} a_k$ also converge, and their sums are

$$\sum_{k=0}^{\infty} a_k = a_0 + \sum_{k=1}^{\infty} a_k = a_0 + S$$

and

$$\sum_{k=2}^{\infty} a_k = \sum_{k=1}^{\infty} a_k - a_1 = S - a_1.$$

Another way to say this is that *modifying a convergent series by adding or subtracting a finite number of terms always gives another convergent series,* but with a (possibly) different sum.

Because of these observations we shall often express an infinite series as just Σa_k without making the starting value of the index k explicit.

Geometric Series

In elementary algebra one encounters the formula for the sum of a **geometric progression:**

$$1 + x + x^2 + x^3 + \cdots + x^{n-1} = \frac{1 - x^n}{1 - x}, \qquad x \neq 1. \tag{4}$$

(Equation (4) may be verified by multiplying both sides by $1 - x$. On the left side all terms cancel except $1 - x^n$.)

In equation (4) the variable x, referred to as the **ratio term,** can represent any number except $x = 1$. This formula leads directly to the definition of the **geometric series** with ratio term x:

$$\sum_{k=0}^{\infty} x^k = 1 + x + x^2 + x^3 + \cdots + x^k + \cdots.$$

The formula for the partial sum of this series is given by equation (4):

$$S_n = \frac{1 - x^n}{1 - x}, \qquad x \neq 1.$$

To determine the numbers x for which the geometric series converges, we apply Theorem 1 to conclude that

$$\lim_{n \to \infty} S_n = \lim_{n \to \infty} \frac{1 - x^n}{1 - x} \tag{5}$$

$$= \frac{1 - \lim_{n \to \infty} x^n}{1 - x}.$$

Now by equation (1), Section 12.2, $\lim_{n \to \infty} x^n = 0$ if $|x| < 1$ and, by Exercise 19, Section 12.2, $\lim_{n \to \infty} x^n$ does not exist if $|x| > 1$. Thus (5) shows that the geometric series $\Sigma_{k=0}^{\infty} x^k$

(i) converges to $S = \lim_{n \to \infty} S_n = \dfrac{1}{1 - x}$ if $|x| < 1$, and

(ii) diverges if $|x| > 1$.

The two remaining cases are $x = \pm 1$. If $x = 1$, the geometric series becomes the constant series

$$\sum_{k=0}^{\infty} 1^k = 1 + 1 + 1 + \cdots,$$

which diverges. If $x = -1$, the series is

$$\sum_{k=0}^{\infty} (-1)^k = 1 - 1 + 1 - 1 + \cdots,$$

which diverges, as in Example 2.

This completely determines the convergence properties of the geometric series, as summarized in the following theorem.

THEOREM 5
Convergence of Geometric Series

If $|x| < 1$, the geometric series converges to the sum

$$\sum_{k=0}^{\infty} x^k = \frac{1}{1-x}.$$

If $|x| \geq 1$, the geometric series $\sum_{k=0}^{\infty} x^k$ diverges.

Example 4

The series

$$1 + \frac{2}{3} + \frac{4}{9} + \frac{8}{27} + \cdots + \frac{2^k}{3^k} + \cdots = \sum_{k=0}^{\infty} \frac{2^k}{3^k}$$

is a geometric series with $x^k = \dfrac{2^k}{3^k} = \left(\dfrac{2}{3}\right)^k$ and $x = \dfrac{2}{3}$. Since $\left|\dfrac{2}{3}\right| < 1$, the series converges and its sum, by Theorem 5, is

$$S = \sum_{k=0}^{\infty} \left(\frac{2}{3}\right)^k = \frac{1}{1 - \dfrac{2}{3}} = 3. \qquad \diamond$$

Example 5

The series

$$\frac{1}{2} + \frac{1}{4} + \frac{1}{8} + \frac{1}{16} + \cdots + \frac{1}{2^k} + \cdots$$

has the form of the geometric series with $x^k = \dfrac{1}{2^k} = \left(\dfrac{1}{2}\right)^k$ except that the term $1 = \left(\dfrac{1}{2}\right)^0$ is missing. In order to use Theorem 5, we write

$$\sum_{k=1}^{\infty} \frac{1}{2^k} = \frac{1}{2} + \frac{1}{4} + \frac{1}{8} + \frac{1}{16} + \cdots + \frac{1}{2^k} + \cdots$$

$$= \left\{ 1 + \frac{1}{2} + \frac{1}{4} + \frac{1}{8} + \frac{1}{16} + \cdots + \frac{1}{2^k} + \cdots \right\} - 1$$

$$= \left(\sum_{k=0}^{\infty} \frac{1}{2^k} \right) - 1$$

$$= \left(\frac{1}{1 - \frac{1}{2}} \right) - 1$$

$$= 1. \qquad\qquad\qquad\qquad \diamond$$

The Algebra of Convergent Infinite Series

The following theorem shows that a limited amount of algebra may be performed on convergent infinite series.

THEOREM 6

Suppose that the series Σa_k and Σb_k both converge, with sums

$$S = \sum a_k \qquad \text{and} \qquad T = \sum b_k.$$

Then

(i) the series $\Sigma(a_k + b_k)$ converges, with sum

$$\sum (a_k + b_k) = S + T, \qquad \text{and}$$

(ii) for any real number c the series $\Sigma c a_k$ converges, with sum

$$\sum c a_k = cS.$$

Proof: To prove part (i) we let S_n denote the nth partial sum for Σa_k and we let T_n denote the nth partial sum for Σb_k. Then $(S_n + T_n)$ is a partial sum for the series $\Sigma(a_k + b_k)$, since

$$\sum_{k=1}^{n} (a_k + b_k) = (a_1 + b_1) + (a_2 + b_2) + \cdots + (a_n + b_n)$$

$$= (a_1 + a_2 + \cdots + a_n) + (b_1 + b_2 + \cdots + b_n)$$
$$= S_n + T_n.$$

Applying Definition 3 and Theorem 1, part (i), we see that

$$\sum_{k=1}^{\infty} (a_k + b_k) = \lim_{n \to \infty} (S_n + T_n)$$

$$= \lim_{n \to \infty} S_n + \lim_{n \to \infty} T_n$$

$$= S + T.$$

The proof of part (ii) is similar and is left as an exercise.   ◆

REMARK: We may paraphrase Theorem 6 by writing

$$\sum(a_k + b_k) = \sum a_k + \sum b_k \tag{6}$$

and

$$\sum c a_k = c \sum a_k. \tag{7}$$

Equations (6) and (7) must be read from right to left—if the sums on the right exist then so do the sums on the left, and equality holds.

Example 6

To find the sum of the series

$$\sum_{k=0}^{\infty} \frac{3 \cdot 2^k + 3^k}{5^k} = 4 + \frac{9}{5} + \frac{21}{25} + \frac{51}{125} + \cdots$$

we use equations (6) and (7) together with Theorem 5:

$$\sum_{k=0}^{\infty} \frac{3 \cdot 2^k + 3^k}{5^k} = \sum_{k=0}^{\infty} \left(\frac{3 \cdot 2^k}{5^k} + \frac{3^k}{5^k} \right)$$

$$= 3 \sum_{k=0}^{\infty} \frac{2^k}{5^k} + \sum_{k=0}^{\infty} \frac{3^k}{5^k}$$

$$= 3 \sum_{k=0}^{\infty} \left(\frac{2}{5} \right)^k + \sum_{k=0}^{\infty} \left(\frac{3}{5} \right)^k$$

$$= 3 \left(\frac{1}{1 - \dfrac{2}{5}} \right) + \left(\frac{1}{1 - \dfrac{3}{5}} \right)$$

$$= 3 \left(\frac{5}{3} \right) + \frac{5}{2}$$

$$= \frac{15}{2}. \qquad \diamond$$

Necessary Condition for Convergence

The following theorem gives a condition which must be satisfied by *any* convergent series.

THEOREM 7
Necessary Condition for Convergence

If the infinite series $\sum a_k$ converges then $\lim_{k \to \infty} a_k = 0$.

Proof: Let S_k denote the kth partial sum for Σa_k. If Σa_k converges there exists a sum S for the series. That is, $\lim\limits_{k\to\infty} S_k = S$. Then $\lim\limits_{k\to\infty} S_{k-1} = S$ also. But $a_k = S_k - S_{k-1}$, so

$$\lim_{k\to\infty} a_k = \lim_{k\to\infty} (S_k - S_{k-1})$$
$$= \lim_{k\to\infty} S_k - \lim_{k\to\infty} S_{k-1}$$
$$= S - S$$
$$= 0.$$

$\blacklozenge$

REMARK: It is important to note that Theorem 7 is *not* useful in demonstrating that the series Σa_k converges. Although the condition $\lim\limits_{k\to\infty} a_k = 0$ is **necessary** for convergence (meaning all convergent series have this property), it is **not sufficient** to guarantee convergence (meaning some divergent series also have this property). However, Theorem 7 does establish the divergence of a series Σb_k for which $\lim\limits_{k\to\infty} b_k \neq 0$. The Venn diagram in Figure 3.1 illustrates the type of series to which Theorem 7 applies.

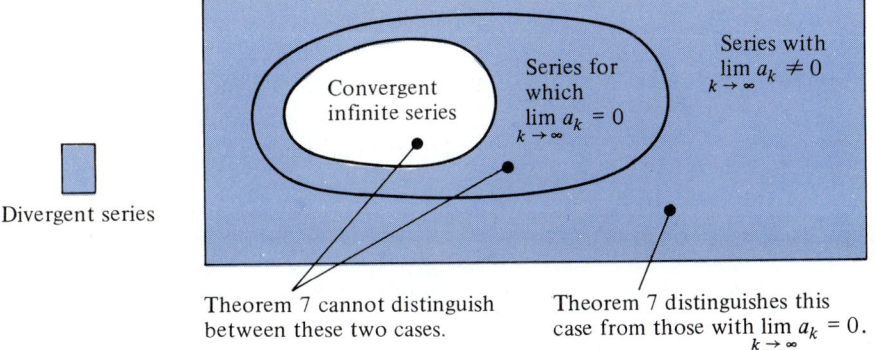

Figure 3.1 Diagram on the applicability of Theorem 7.

The following equivalent formulation of Theorem 7 summarizes these remarks and is the one you will generally find useful in your work.

COROLLARY 2

If $\lim\limits_{k\to\infty} a_k \neq 0$, the series Σa_k diverges.

Example 7

$$\sum_{k=1}^{\infty} \frac{k}{k+2}$$

diverges, by application of Corollary 2:

$$\lim_{k\to\infty} a_k = \lim_{k\to\infty} \frac{k}{k+2} = 1 \neq 0.$$

$\diamondsuit$

Example 8

$$\sum_{k=1}^{\infty} \cos(\pi k)$$

diverges, by Corollary 2, since $\lim_{k\to\infty} a_k = \lim_{k\to\infty} \cos(\pi k)$ does not exist. ◇

Example 9

Whether the series

$$\sum_{k=1}^{\infty} \frac{1}{k}$$

converges cannot be determined using Theorem 7 or Corollary 2, since $\lim_{k\to\infty} \frac{1}{k} = 0$.

However, the test discussed in the next section will show that this series diverges. ◇

Unfortunately, about the only types of infinite series yielding explicit formulas for their sums are the geometric series and those series whose partial sums telescope, as in Example 3. However, in most applications of the ideas developed in this chapter, we shall need only to determine whether a particular infinite series converges or diverges. The remaining sections of this chapter concern how this determination is made for various types of infinite series.

Exercise Set 12.3

In Exercises 1–6, write out the first four terms of the infinite series

1. $\displaystyle\sum_{k=1}^{\infty} \frac{\cos \pi k}{2^k}$

2. $\displaystyle\sum_{k=1}^{\infty} \frac{\sqrt{k} \sin \pi k}{k + 1}$

3. $\displaystyle\sum_{k=1}^{\infty} \frac{2^k + 1}{3^k + 2}$

4. $\displaystyle\sum_{k=0}^{\infty} \tan\left(\frac{\pi k}{3}\right)$

5. $\displaystyle\sum_{k=1}^{\infty} \ln\left(\frac{k}{k + 1}\right)$

6. $\displaystyle\sum_{k=1}^{\infty} k^k$

In Exercises 7–24, determine whether the given series converges or diverges. If it converges find its sum.

7. $\displaystyle\sum_{k=0}^{\infty} \frac{1}{7^k}$

8. $\displaystyle\sum_{k=1}^{\infty} \frac{1}{3^k}$

9. $\displaystyle\sum_{k=0}^{\infty} 4^k$

10. $\displaystyle\sum_{k=1}^{\infty} \frac{7^k + 3^k}{5^k}$

11. $\displaystyle\sum_{k=0}^{\infty} \frac{2^{2k}}{3^{3k}}$

12. $\displaystyle\sum_{k=2}^{\infty} \frac{-1}{3^k}$

13. $\displaystyle\sum_{k=0}^{\infty} \frac{1}{(2 + x)^k}, \quad |x| < 1$

14. $\displaystyle\sum_{k=2}^{\infty} \frac{1}{k(k + 1)}$

15. $\displaystyle\sum_{k=1}^{\infty} \left[\frac{1}{k + 2} - \frac{1}{k + 1}\right]$

16. $\displaystyle\sum_{k=2}^{\infty} \frac{2^{k+1} + 2 \cdot 7^k}{9^k}$

17. $\displaystyle\sum_{k=1}^{\infty} \cos \pi k$

18. $\displaystyle\sum_{k=2}^{\infty} \frac{1}{k^2 - 1}$

19. $\displaystyle\sum_{k=1}^{\infty} \frac{1}{k^2 + 5k + 6}$

20. $\displaystyle\sum_{k=4}^{\infty} \frac{1}{k^2 - 9}$

21. $\displaystyle\sum_{k=1}^{\infty} \ln\left(\frac{k}{k + 1}\right)$

22. $\displaystyle\sum_{k=1}^{\infty} \frac{2^{k-2} + 3^{k+1}}{5^k}$

23. $\displaystyle\sum_{k=0}^{\infty} \frac{2^{k/2}}{3^k}$

24. $\displaystyle\sum_{k=1}^{\infty} \frac{3^k}{3^{k/2}}$

In Exercises 25–30, write the given decimal fraction as **(a)** an infinite series, and **(b)** the quotient of two integers.

25. $0.33\overline{3} . . .$

26. $0.77\overline{7} . . .$

27. $0.9292\overline{92}\ldots$

28. $0.3215151\overline{5}\ldots$

29. $0.412412\overline{412}\ldots$

30. $0.0213434\overline{34}\ldots$

In Exercises 31–34, use Theorem 5 on the convergence of geometric series to establish the stated fact.

31. $\displaystyle\sum_{k=0}^{\infty}(-1)^k x^k = \frac{1}{1+x}$ if $|x| < 1$

32. $\displaystyle\sum_{k=0}^{\infty}x^{2k} = \frac{1}{1-x^2}$ if $|x| < 1$

33. $\displaystyle\sum_{k=0}^{\infty}\frac{x^k}{y^k} = \frac{y}{y-x}$ if $|x| < |y|$

34. $\displaystyle\sum_{k=1}^{\infty}x^k = \frac{x}{1-x}$ if $|x| < 1$.

35. When dropped from a height h, a ball rebounds to a height $\frac{2}{3}h$. Write an infinite series expressing the total distance travelled by the ball as it bounces an infinite number of times. What is this distance?

36. Let Σa_k be the series $\displaystyle\sum_{k=0}^{\infty}\left(1 + \frac{1}{2^k}\right)$ and let Σb_k be the series $\displaystyle\sum_{k=0}^{\infty}(-1)$. Show that the statement $\Sigma(a_k + b_k) = \Sigma a_k + \Sigma b_k$ is false, but that $\Sigma(a_k + b_k)$ converges.

37. If $\Sigma c a_k$ converges for a particular number c, must the series Σa_k converge? Why or why not?

38. Prove Theorem 6, part (ii).

39. Prove that if Σa_k diverges, so must $\Sigma c a_k$ for any $c \neq 0$.

40. Prove that if $\displaystyle\sum_{k=1}^{\infty}a_k$ converges, then $\displaystyle\sum_{k=p}^{\infty}a_k$ converges for any $p \geq 1$. $\left(\textit{Hint:}\text{ If } S_n \text{ is the } n\text{th partial sum for }\displaystyle\sum_{k=1}^{\infty}a_k,\text{ then } S_{p+n-1} - S_{p-1}\text{ is the } n\text{th partial sum for }\displaystyle\sum_{k=p}^{\infty}a_k.\right)$

41. Find a formula that defines the partial sum S_n for the series Σa_k *recursively*, in terms of the partial sum S_{n-1} and the nth term a_n.

42. *(Computer)* Program 6 in Appendix I is a BASIC program that computes partial sums for the geometric series $\displaystyle\sum_{k=p}^{\infty}ax^k$.

For example, partial sums for the series $\displaystyle\sum_{k=2}^{\infty}7\cdot\left(\frac{4}{5}\right)^k$ are obtained by specifying $p = 2$, $a = 7$, and $x = \frac{4}{5}$. Results appear in Table 3.1.

Table 3.1

n	$S_n = \displaystyle\sum_{k=2}^{n}7\left(\frac{4}{5}\right)^k$
5	13.224959
10	19.393521
25	22.294217
50	22.399598
100	22.399997
200	22.399998
500	22.399998

Show that $\displaystyle\lim_{n\to\infty}S_n = \sum_{k=2}^{\infty}7\left(\frac{4}{5}\right)^k = 22.4$.

In Exercises 43–46, use Program 6 to find the partial sums S_5, S_{10}, S_{20}, S_{50}, and S_{100}. Then find $\displaystyle\lim_{n\to\infty}S_n$.

43. $\displaystyle\sum_{k=0}^{\infty}\left(\frac{2}{3}\right)^k$

44. $\displaystyle\sum_{k=3}^{\infty}4\left(\frac{9}{10}\right)^k$

45. $\displaystyle\sum_{k=5}^{\infty}-2\left(\frac{5}{7}\right)^k$

46. $\displaystyle\sum_{k=1}^{\infty}6\left(\frac{3}{2}\right)^k$

12.4 THE INTEGRAL TEST

In this section and in Sections 12.5 and 12.6 we present five tests for convergence for series *with positive terms*. It is especially important to note that these tests apply only to series Σa_k for which $a_k > 0$ for all k.

The Integral Test

This test exploits the relationship between an infinite series of positive terms and the improper integral of a positive continuous function.

THEOREM 8
Integral Test

Let $a_k > 0$ for all $k = 1, 2, \ldots$ and let f be a continuous decreasing function defined on $[1, \infty)$ so that $f(k) = a_k$ for each $k = 1, 2, \ldots$. Then

$$\Sigma a_k \text{ converges} \quad \text{if and only if} \quad \int_1^\infty f(x)\, dx \text{ converges.}$$

(See Figures 4.1 and 4.2.) That is, the series and the improper integral either both converge or both diverge.

Since $f(x) > 0$ for all $x \in [1, \infty)$, we may interpret these two conclusions geometrically by saying that

(i) If the area of the region R_f, bounded by the graph of f and the x-axis for $1 \le x < \infty$, is finite, so is the area, R_s, of the region enclosed by the rectangles of area $a_1, a_2, a_3, \ldots$ (Figure 4.1).

(ii) If the area of the region R_f is infinite, so is the area of the region R_s (Figure 4.2).

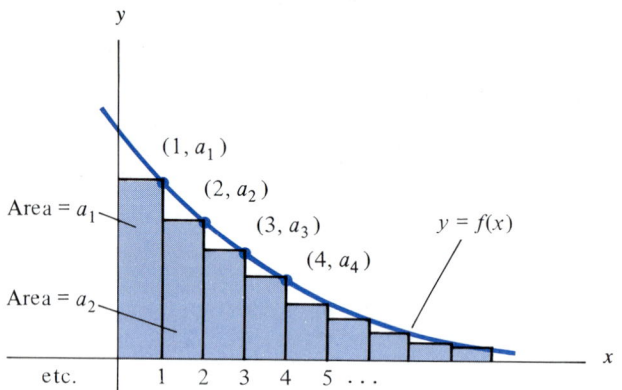

Figure 4.1 $\sum a_k$ converges if $\int_1^\infty f(x)\, dx < \infty$.

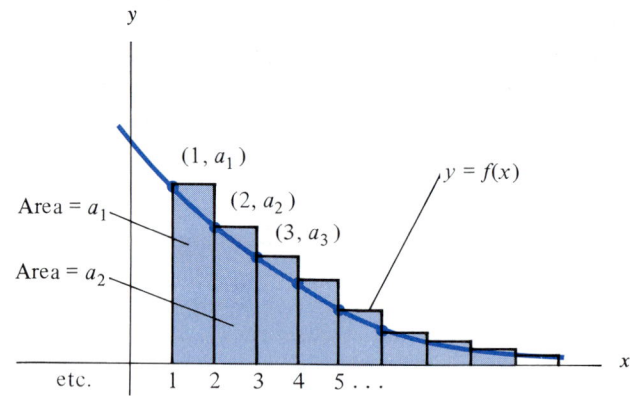

Figure 4.2 $\sum a_k$ diverges if $\int_1^\infty f(x)\, dx = \infty$.

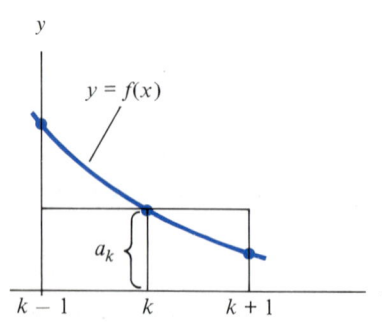

Figure 4.3 $f(x) \ge a_k$, $x \le k$; $f(x) \le a_k$, $x \ge k$.

Proof of Theorem 8: Since f is decreasing on $[1, \infty)$, for any integer $k \ge 2$,

$$f(x) \ge f(k) = a_k \quad \text{for all } x \in [k - 1, k], \quad \text{and} \tag{1}$$

$$f(x) \le f(k) = a_k \quad \text{for all } x \in [k, k + 1]. \tag{2}$$

(See Figure 4.3.)

Inequalities (1) and (2) imply that

$$\int_{k-1}^k f(x)\, dx \ge a_k, \quad \text{and} \tag{3}$$

$$\int_k^{k+1} f(x)\, dx \le a_k, \quad k = 2, 3, 4, \ldots \quad \text{(Figure 4.4).} \tag{4}$$

For any integer $n > 2$ we may sum inequality (3) from $k = 2$ to n to conclude that

$$\int_1^n f(x)\, dx = \sum_{k=2}^n \int_{k-1}^k f(x)\, dx \ge \sum_{k=2}^n a_k = S_n - a_1 \tag{5}$$

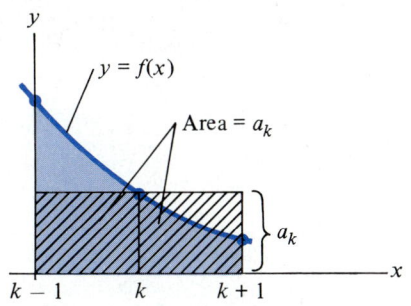

Figure 4.4

$$\int_{k-1}^{k} f(x)\,dx \geq a_k \geq \int_{k}^{k+1} f(x)\,dx.$$

The block of height a_k extending from $k-1$ to k has area a_k; so does the block from k to $k+1$.

where S_n denotes the nth partial sum $\sum_{k=1}^{n} a_k$. Thus,

$$S_n \leq a_1 + \int_1^n f(x)\,dx < a_1 + \int_1^\infty f(x)\,dx. \tag{6}$$

(The last inequality holds since $f(x) > 0$ on $[1, \infty)$.) Inequality (6) shows that if the improper integral $\int_1^\infty f(x)\,dx$ converges, the sequence $\{S_n\}$ of partial sums for the series Σa_k is a *bounded* sequence. Since $a_k > 0$ for each k, $\{S_k\}$ is also an *increasing* sequence. Since, by Theorem 4, every bounded increasing sequence converges, the sequence $\{S_n\}$ converges. That is, *if the improper integral $\int_1^\infty f(x)\,dx$ converges, so does the infinite series Σa_k.*

The converse conclusion is obtained from inequality (4). Summing both sides from $k = 1$ to $k = n$ shows that

$$\int_1^{n+1} f(x)\,dx \leq \sum_{k=1}^{n} a_k = S_n. \tag{7}$$

Now $\int_1^{n+1} f(x)\,dx$ is an increasing function of n since $f(x) > 0$ for $x \in [1, \infty)$. Thus, if the improper integral $\int_1^\infty f(x)\,dx$ diverges, we may conclude from inequality (7) that

$$\lim_{n\to\infty} S_n \geq \lim_{n\to\infty} \int_1^{n+1} f(x)\,dx = +\infty,$$

so the sequence $\{S_n\}$ diverges. This shows that *if the improper integral diverges, so does the series Σa_k.* This completes the proof. ◆

Example 1

The **harmonic series** is the series

$$\sum_{k=1}^{\infty} \frac{1}{k} = 1 + \frac{1}{2} + \frac{1}{3} + \frac{1}{4} + \cdots + \frac{1}{k} + \cdots.$$

To test this series for convergence, we use the function $f(x) = \dfrac{1}{x}$, which satisfies all hypotheses of the integral test.

Since

$$\int_1^\infty \frac{1}{x}\,dx = \lim_{t\to\infty} \int_1^t \frac{1}{x}\,dx = \lim_{t\to\infty} \ln t = +\infty$$

the harmonic series diverges. (Note that $\lim_{k\to\infty} a_k = \lim_{k\to\infty} \dfrac{1}{k} = 0$, although this series does not converge.) ◇

Example 2

To test the series

$$\sum_{k=1}^{\infty} k e^{-k^2}$$

for convergence, we use the function $f(x) = xe^{-x^2}$. Since

$$f'(x) = e^{-x^2} - 2x^2 e^{-x^2} = (1 - 2x^2)e^{-x^2} < 0 \qquad \text{if} \qquad x > \frac{\sqrt{2}}{2},$$

we are assured that f is decreasing on $[1, \infty)$. Since

$$\int_1^\infty xe^{-x^2}\, dx = \lim_{t \to \infty} \int_1^t xe^{-x^2}\, dx$$

$$= \lim_{t \to \infty} -\frac{1}{2}e^{-x^2}\Big]_1^t$$

$$= \lim_{t \to \infty} \left(\frac{1}{2e} - \frac{1}{2e^{t^2}}\right) = \frac{1}{2e}$$

the infinite series $\displaystyle\sum_{k=1}^\infty ke^{-k^2}$ converges, by the Integral Test. ◇

Example 3

A **p-series** is a series of the form

$$\sum_{k=1}^\infty \frac{1}{k^p} = 1 + \frac{1}{2^p} + \frac{1}{3^p} + \frac{1}{4^p} + \cdots, \qquad p > 0.$$

Determine for which values of p a p-series converges.

Strategy

Identify a function f with which to apply the Integral Test.

Verify that f is decreasing on $[1, \infty)$.

Set up the improper integral and evaluate, carrying along the unknown constant p.

Determine the values of p for which the improper integral converges.

Handle the remaining case ($p = 1$) directly (see Example 1).

Solution

We use the function $f(x) = \dfrac{1}{x^p} = x^{-p}$.

Since

$$f'(x) = -px^{-p-1} = \frac{-p}{x^{p+1}}$$

is negative for $x > 0$, the function f is decreasing on $[1, \infty)$.
    The improper integral for f is

$$\int_1^\infty x^{-p}\, dx = \lim_{t \to \infty} \int_1^t x^{-p}\, dx$$

$$= \lim_{t \to \infty} \frac{x^{-p+1}}{1-p}\bigg]_1^t \qquad (p \neq 1)$$

$$= \left\{\lim_{t \to \infty} \frac{t^{-p+1}}{1-p}\right\} - \frac{1}{1-p}.$$

This limit will exist only if the exponent of t is negative; that is, if

$$-p + 1 < 0, \qquad \text{or} \qquad p > 1.$$

Thus, the improper integral converges if $p > 1$, and diverges if $0 < p < 1$. In the case $p = 1$, (excluded above), the p-series becomes simply the harmonic series, which diverges.

Apply the Integral Test.

The conclusion of Example 3 is this:

> The p-series
>
> $$\sum_{k=1}^{\infty} \frac{1}{k^p} = 1 + \frac{1}{2^p} + \frac{1}{3^p} + \frac{1}{4^p} + \cdots, \qquad p > 0$$
>
> converges if $p > 1$ and diverges if $0 < p \le 1$.

◇

Example 4

The series

$$\sum_{k=1}^{\infty} \frac{1}{k^3} = 1 + \frac{1}{2^3} + \frac{1}{3^3} + \frac{1}{4^3} + \cdots$$

$$= 1 + \frac{1}{8} + \frac{1}{27} + \frac{1}{64} + \cdots$$

converges because it is a p-series with $p = 3 > 1$.

◇

Example 5

The series

$$\sum_{k=1}^{\infty} \left(\frac{1}{k}\right)^{2/3} = 1 + \left(\frac{1}{2}\right)^{2/3} + \left(\frac{1}{3}\right)^{2/3} + \cdots$$

diverges because it is a p-series $\displaystyle\sum_{k=1}^{\infty} \left(\frac{1}{k}\right)^{2/3} = \sum_{k=1}^{\infty} \frac{1}{k^{2/3}}$ with $p = \dfrac{2}{3} < 1$. ◇

Exercise Set 12.4

In Exercises 1–20, use the Integral Test to determine whether the given series converges or diverges.

1. $\displaystyle\sum \frac{1}{2k+1}$

2. $\displaystyle\sum \frac{1}{(3k+1)^2}$

3. $\displaystyle\sum \frac{1}{k \cdot \ln k}$

4. $\displaystyle\sum \frac{1}{k(\ln k)^2}$

5. $\displaystyle\sum \frac{1}{1+k^2}$

6. $\displaystyle\sum k^2 e^{-k^3}$

7. $\displaystyle\sum \frac{1}{\sqrt{k+1}}$

8. $\displaystyle\sum \frac{k+1}{k}$

9. $\displaystyle\sum k^2 e^{-k^3}$

10. $\displaystyle\sum \frac{\mathrm{Tan}^{-1} k}{1+k^2}$

11. $\displaystyle\sum \frac{k}{\sqrt{1+k^2}}$

12. $\displaystyle\sum \frac{k^2-2}{k^2+2}$

13. $\displaystyle\sum \frac{1}{\sqrt{k}}$

14. $\displaystyle\sum \frac{1}{k^2}$

15. $\displaystyle\sum k(1+k^2)^{-3}$

16. $\displaystyle\sum k^2 e^{-k}$

17. $\displaystyle\sum \frac{\ln k}{k}$

18. $\displaystyle\sum \frac{k}{1+k^2}$

19. $\displaystyle\sum \frac{k^2}{k^3+1}$

20. $\displaystyle\sum \frac{2k^2}{\sqrt{k^3+5}}$

21. Let Σa_k be a convergent infinite series that satisfies the hypotheses of the integral test. Show, using inequalities (5) and (7), that

$$\int_{n+1}^{\infty} f(x)\, dx \le \sum_{k=n+1}^{\infty} a_k \le \int_{n}^{\infty} f(x)\, dx \qquad (8)$$

where $f(k) = a_k$ and f is continuous and decreasing on $[1, \infty)$. Inequality (8) gives an estimate for the **truncation error** $R_n = \sum_{n+1}^{\infty} a_k$ which occurs in using the partial sum

$S_n = \sum_{k=1}^{n} a_k$ to approximate the sum of a convergent series

$S = \sum_{k=1}^{\infty} a_k$ when $\{a_k\}$ is a decreasing sequence with

$\lim_{k \to \infty} a_k = 0$.

22. Use the result of Exercise 21 to estimate the magnitude of the error that occurs in approximating the sum of the *p*-series

$\sum_{k=1}^{\infty} \frac{1}{k^2}$ by the partial sum $\sum_{k=1}^{10} \frac{1}{k^2}$.

23. Use the result of Exercise 21 to determine how large n must be taken so that the partial sum S_n for the series $\sum_{k=2}^{\infty} \frac{1}{k \ln^2 k}$ differs from its sum S by less than .01.

24. Use the result of Exercise 21 to determine how large n must be taken so that the partial sum S_n for the series $\sum_{k=1}^{\infty} k e^{-k^2}$ differs from its sum S by less than .02.

25. Prove that the harmonic series $\sum_{k=1}^{\infty} \frac{1}{k}$ diverges without using the Integral Test, as follows.

   a. Write the harmonic series as

   $$\sum_{k=1}^{\infty} \frac{1}{k} = \{1\} + \left\{\frac{1}{2}\right\} + \underbrace{\left\{\frac{1}{3} + \frac{1}{4}\right\}}_{2^1 \text{ terms}} + \underbrace{\left\{\frac{1}{5} + \frac{1}{6} + \frac{1}{7} + \frac{1}{8}\right\}}_{2^2 \text{ terms}}$$

   $$+ \left\{\frac{1}{9} + \cdots + \frac{1}{16}\right\} \qquad (2^3 \text{ terms})$$

   $$+ \left\{\frac{1}{17} + \cdots + \frac{1}{32}\right\} \qquad (2^4 \text{ terms})$$

   $$\cdots + \left\{\frac{1}{2^n + 1} + \cdots + \frac{1}{2^{n+1}}\right\} \qquad (2^n \text{ terms}).$$

   b. Show that the sums of the terms in each of the blocks above are greater than $\frac{1}{2}$.

   c. Conclude that $S_{2^{n+1}} \ge 1 + \frac{(n+1)}{2}$.

   d. From (c), conclude that $\lim_{n \to \infty} S_n = +\infty$.

26. Use the technique of Exercise 25 to prove the "rule of thumb" for the partial sum $S_{2^n} = \sum_{k=1}^{2^n} \frac{1}{k}$ of the harmonic series: $S_{2^n} \ge 1 + \frac{n}{2}$.

12.5 THE COMPARISON TESTS

Like the Integral Test, the next test is motivated by area considerations. However, instead of comparing terms of a given series with certain integrals, we compare them with terms of another series.

THEOREM 9
Basic Comparison Test

Let Σa_k and Σb_k be infinite series with $0 < a_k \le b_k$ for each $k = 1, 2, 3, \ldots$. Then

(i) If Σb_k converges, so does Σa_k.
(ii) If Σa_k diverges, so does Σb_k.

Proof: We let

$$S_n = \sum_{k=1}^{n} a_k \qquad \text{and} \qquad T_n = \sum_{k=1}^{n} b_k$$

be the corresponding partial sums. Since $a_k \le b_k$ for all k, we have

$$S_n \le T_n \qquad \text{for all} \qquad n = 1, 2, \ldots. \qquad (1)$$

To prove (i), we note that if Σb_k converges, $\lim\limits_{n \to \infty} T_n$ exists. Thus the sequence $\{T_n\}$ is bounded and, by inequality (1), so is the sequence $\{S_n\}$. Since $a_k > 0$ for all k, this shows that $\{S_n\}$ is a bounded increasing sequence, so by Theorem 4, $\lim\limits_{n \to \infty} S_n$ exists. That is, Σa_k converges.

To prove (ii), we note that since $\{S_n\}$ is an increasing sequence, if Σa_k diverges, we must have $\lim\limits_{n \to \infty} S_n = +\infty$. Thus, by inequality (1)

$$\lim_{n \to \infty} T_n \geq \lim_{n \to \infty} S_n = +\infty,$$

so Σb_k diverges. ◆

Example 1

The series $\displaystyle\sum \frac{1}{1 + k^3}$ converges, by comparison with the convergent p-series $\displaystyle\sum \frac{1}{k^3}$, since $\dfrac{1}{1 + k^3} < \dfrac{1}{k^3}$. ◇

Example 2

To determine the convergence of the series $\displaystyle\sum \frac{\sqrt{k} - 1}{k^2 + 2}$, we make the comparison

$$\frac{\sqrt{k} - 1}{k^2 + 2} < \frac{\sqrt{k}}{k^2 + 2} < \frac{\sqrt{k}}{k^2} = \frac{1}{k^{3/2}}.$$

Since the series $\displaystyle\sum \frac{1}{k^{3/2}}$ is a p-series with $p = 3/2 > 1$, it converges. So, therefore, does the series $\displaystyle\sum \frac{\sqrt{k} - 1}{k^2 + 2}$. ◇

Example 3

The series

$$\sum_{k=1}^{\infty} \frac{1 + \ln k}{k}$$

diverges by comparison with the harmonic series $\displaystyle\sum_{k=1}^{\infty} \frac{1}{k}$, since

$$\frac{1 + \ln k}{k} > \frac{1}{k} \qquad \text{if} \qquad k > 1.$$ ◇

REMARK 1: Successful application of the Comparison Test to the series Σa_k involves finding another series Σb_k with which you can make a useful comparison. If you suspect that Σa_k converges, you should look for a *convergent* series Σb_k with $a_k \leq b_k$. But if you suspect that Σa_k diverges, look for a divergent series Σb_k with $a_k \geq b_k$.

REMARK 2: We say that the series Σa_k *is dominated by* the series Σb_k (or, that Σb_k *dominates* Σa_k) if $a_k \leq b_k$ for all k. Using this terminology we may paraphrase Theorem 9 by saying that a positive series that is dominated by a convergent series must converge, while a series that dominates a positive divergent series must diverge.

It is important to note that two cases remain (dominated by a divergent series and dominating a convergent series) in which no valid conclusions can be drawn. Table 5.1 may help you remember which comparisons yield valid conclusions.

Table 5.1 Conclusions concerning Σa_k when compared with Σb_k.

	Σb_k converges	Σb_k diverges
$0 < a_k \leq b_k$	Σa_k converges	no valid conclusion
$a_k \geq b_k > 0$	no valid conclusion	Σa_k diverges

The Basic Comparison Test is used whenever an appropriate comparison series can be found. As the list of series with which you are familiar expands, your facility with the Basic Comparison Test will grow.

The following variation on the Basic Comparison Test is often easier to apply.

THEOREM 10
Limit Comparison Test

Let Σa_k and Σb_k be series with $a_k > 0$, $b_k > 0$ for all $k = 1, 2, \ldots$.
If the limit

$$\rho = \lim_{k \to \infty} \frac{a_k}{b_k}$$

exists, and $\rho \neq 0$, then either both series converge or both series diverge.

The Limit Comparison Test states that if the ratio of the general terms of two positive series tends to a positive limit, then the two series have the same convergence property—either both converge or both diverge.

Before proving this theorem, we consider several examples of its use.

Example 4

The series

$$\sum_{k=1}^{\infty} \frac{1}{2k + 1}$$

diverges, as can be shown by Theorem 10. Comparing this series with the harmonic series

$$\Sigma b_k = \sum_{k=1}^{\infty} \frac{1}{k},$$

we find that

$$\rho = \lim_{k \to \infty} \left(\frac{\dfrac{1}{2k+1}}{\dfrac{1}{k}} \right) = \lim_{k \to \infty} \frac{k}{2k+1} = \frac{1}{2} > 0.$$

Since the limit $\rho > 0$ exists and since the harmonic series diverges, so must the given series. ◇

Example 5

To apply the Limit Comparison Test to the series

$$\sum_{k=1}^{\infty} \frac{\sqrt{k} + 2}{k^2 + k + 1},$$

we observe that the general term is a quotient containing a highest exponent of 1/2 in its numerator and a highest exponent of 2 in its denominator. This suggests a comparison with the series whose general term is $b_k = \dfrac{k^{1/2}}{k^2} = \dfrac{1}{k^{3/2}}$. We therefore take Σb_k to be the p-series

$$\sum_{k=1}^{\infty} \frac{1}{k^{3/2}}.$$

We obtain

$$\rho = \lim_{k \to \infty} \left(\frac{\dfrac{\sqrt{k}+2}{k^2+k+1}}{\dfrac{1}{k^{3/2}}} \right) = \lim_{k \to \infty} \left(\frac{k^2 + 2k^{3/2}}{k^2 + k + 1} \right) = 1 > 0.$$

We may now conclude that, since the p-series $\displaystyle\sum_{k=1}^{\infty} \frac{1}{k^{3/2}}$ converges, so does the series

$$\sum_{k=1}^{\infty} \frac{\sqrt{k}+2}{k^2+k+1}.$$ ◇

Example 6

Determine whether the series

$$\sum_{k=1}^{\infty} \sin\left(\frac{\pi}{k}\right)$$

converges.

Solution: The Limit Comparison Test may be applied by recalling that we have previously established limit

$$\rho = \lim_{k \to \infty} \frac{\sin\left(\dfrac{\pi}{k}\right)}{\left(\dfrac{\pi}{k}\right)} = \lim_{x \to 0^+} \frac{\sin x}{x} = 1.$$

Thus, by Theorem 10, the series $\sum\limits_{k=1}^{\infty} \sin\left(\dfrac{\pi}{k}\right)$ has the same convergence property as does the series $\sum\limits_{k=1}^{\infty} \dfrac{\pi}{k}$. Since $\sum\limits_{k=1}^{\infty} \dfrac{\pi}{k} = \pi \sum\limits_{k=1}^{\infty} \dfrac{1}{k}$ is a multiple of the (divergent) harmonic series, the series $\sum\limits_{k=1}^{\infty} \sin\left(\dfrac{\pi}{k}\right)$ diverges. ◇

Proof of Theorem 10: Since all terms of Σa_k and Σb_k are positive, and $\rho \neq 0$, we must have

$$\rho = \lim_{k \to \infty} \frac{a_k}{b_k} > 0. \tag{2}$$

Now let $\epsilon = \dfrac{\rho}{2} > 0$. Then by (2) and the definition of the limit of a sequence there exists an integer N such that

$$\left| \frac{a_k}{b_k} - \rho \right| < \epsilon \qquad \text{whenever} \qquad k > N. \tag{3}$$

Statement (3) is equivalent to the statement

$$\rho - \epsilon < \frac{a_k}{b_k} < \rho + \epsilon, \qquad k > N. \tag{4}$$

Substituting $\epsilon = \dfrac{1}{2}\rho$ and multiplying through by $b_k > 0$ in (4) gives

$$\frac{1}{2}\rho \cdot b_k < a_k < \frac{3}{2}\rho \cdot b_k, \qquad k > N. \tag{5}$$

Using the left half of inequality (5), we conclude that Σa_k dominates a multiple of Σb_k, since $\Sigma \dfrac{1}{2}\rho b_k = \left(\dfrac{1}{2}\rho\right)\Sigma b_k$. Thus Σb_k converges if Σa_k converges (Theorem 9). The right half of inequality (5) shows that Σa_k is dominated by $\Sigma \dfrac{3}{2}\rho b_k = \left(\dfrac{3}{2}\rho\right)\Sigma b_k$, so Σb_k diverges if Σa_k diverges. These two conclusions taken together establish Theorem 10. ◆

Before concluding this section, we need to repeat an observation which applies to each of the tests embodied by Theorems 8, 9, and 10. We have previously noted that if the series $\Sigma_{k=N}^{\infty} a_k$ converges, then so does the series $\Sigma_{k=1}^{\infty} a_k$, and vice versa.

This observation means that we need not require that the conditions of Theorems 8 through 10 hold on the full series $\sum_{k=1}^{\infty} a_k$ but rather only on any "tail" of the series of the form $\sum_{k=N}^{\infty} a_k$.

For example, the series $\sum_{k=1}^{\infty} \dfrac{7+k}{1+k^3}$ may be successfully compared to the series

$$\sum_{k=1}^{\infty} \frac{2}{k^2} = 2\sum_{k=1}^{\infty} \frac{1}{k^2} \text{ for } k > 7 \text{ since}$$

$$\frac{7+k}{1+k^3} < \frac{k+k}{1+k^3} = \frac{2k}{1+k^3} < \frac{2k}{k^3} = \frac{2}{k^2}$$

whenever $k > 7$. The Comparison Test then shows that the "tail" $\sum_{k=8}^{\infty} \dfrac{7+k}{1+k^3}$ con-

verges, since the p-series $\sum_{k=1}^{\infty} \dfrac{1}{k^2}$ converges. Thus the original series $\sum_{k=1}^{\infty} \dfrac{7+k}{1+k^3}$ also converges.

Exercise Set 12.5

In Exercises 1–6, use the Basic Comparison Test to determine whether the given series converges or diverges.

1. $\sum \dfrac{1}{1+k^2}$

2. $\sum \dfrac{1}{k^{1/2}+k^{3/2}}$

3. $\sum \dfrac{\sqrt{k}}{1+k^3}$

4. $\sum \dfrac{\sqrt{k}}{1+k}$

5. $\sum \dfrac{3}{\sqrt{k}+2}$

6. $\sum (k-1)e^{-k}$

In Exercises 7–10, use the Limit Comparison Test to determine whether the given series converges or diverges.

7. $\sum \dfrac{k+3}{2k^2+1}$

8. $\sum \dfrac{k^2-4}{k^3+k+5}$

9. $\sum \dfrac{k+\sqrt{k}}{k+k^3}$

10. $\sum \dfrac{2k+2}{\sqrt{k^3+2}}$

In Exercises 11–30, determine whether the series converges or diverges and state which test you used.

11. $\sum \dfrac{\cos^2 \pi k}{k^2}$

12. $\sum \dfrac{k(k+1)}{(k+2)(k^2+1)}$

13. $\sum \cos\left(\dfrac{\pi k}{4}\right)$

14. $\sum \dfrac{2k^2+3k-1}{k^4-6k+10}$

15. $\sum \dfrac{\sin(\pi k)}{k}$

16. $\sum \dfrac{2^k}{3^k+1}$

17. $\sum \dfrac{1}{\sqrt{4k(k+1)}}$

18. $\sum \dfrac{2k+2}{\sqrt{k^3+2}}$

19. $\sum \dfrac{1}{\sqrt[3]{k^2+2k}}$

20. $\sum \dfrac{k+1}{2\ln k}$

21. $\sum \dfrac{\text{Tan}^{-1} k}{k^2}$

22. $\sum \dfrac{k+1}{\sqrt{k^{3/2}+1}}$

23. $\sum \dfrac{\ln k}{1+\ln k}$

24. $\sum \dfrac{\ln(k+1)}{k+2}$

25. $\sum \dfrac{n^2+3}{2n^4+n-6}$

26. $\sum \dfrac{\ln n}{n^3}$

27. $\sum \dfrac{3^k}{k+7}$

28. $\sum \dfrac{k+3}{(k+2)2^k}$

29. $\sum \dfrac{\sqrt{k}}{\cos(2k-6)+k^2}$

30. $\sum \dfrac{\text{Tan}^{-1} \sqrt{k}}{\pi+6k^2}$

31. Prove that if Σa_k and Σb_k are positive term series and $\lim\limits_{k\to\infty} \dfrac{a_k}{b_k} = 0$, then Σa_k converges if Σb_k converges.

32. Prove that if Σa_k and Σb_k are positive term series and $\lim\limits_{k\to\infty} \dfrac{a_k}{b_k} = +\infty$, then Σa_k diverges if Σb_k diverges.

33. Prove that if Σa_k converges, then $\lim\limits_{N\to\infty} \sum_{k=N}^{\infty} a_k = 0$. That is, if Σa_k converges, its "tails" tend toward zero.

12.6 THE RATIO AND ROOT TESTS

In this section we discuss two additional tests for series with positive terms. The first involves the limit of the ratio of successive terms of the series.

THEOREM 11
Ratio Test

Let $a_k > 0$ for all $k = 1, 2, 3, \ldots$ and let

$$\rho = \lim_{k \to \infty} \frac{a_{k+1}}{a_k}.$$ (1)

Then, provided this limit exists,

(i) If $\rho < 1$, the series $\sum\limits_{k=1}^{\infty} a_k$ converges.

(ii) If $\rho > 1$, the series $\sum\limits_{k=1}^{\infty} a_k$ diverges.

(iii) If $\rho = 1$, no conclusion may be drawn.

Before proving Theorem 11, we discuss several examples in which the Ratio Test applies.

Example 1

For the series $\sum \dfrac{k^2}{3^k}$, the kth term is $a_k = \dfrac{k^2}{3^k}$, so $a_{k+1} = \dfrac{(k+1)^2}{3^{k+1}}$. The limit in (1) is therefore

$$\rho = \lim_{k \to \infty} \frac{\dfrac{(k+1)^2}{3^{k+1}}}{\dfrac{k^2}{3^k}} = \lim_{k \to \infty} \frac{3^k(k+1)^2}{3^{k+1}k^2} = \lim_{k \to \infty} \frac{1}{3}\left(\frac{k+1}{k}\right)^2 = \frac{1}{3}.$$

Since $\rho = \dfrac{1}{3} < 1$, the series $\sum \dfrac{k^2}{3^k}$ converges by the Ratio Test. $\diamond$

REMARK: The Ratio Test is often useful in dealing with series involving factorial expressions. Recall that

$$k! = k(k-1)(k-2) \cdots \cdots 3 \cdot 2 \cdot 1.$$

Thus, ratios of factorials can be simplified as follows:

$$\frac{(k+1)!}{k!} = \frac{(k+1)k!}{k!} = k + 1$$

$$\frac{(k+2)!}{k!} = \frac{(k+2)(k+1)k!}{k!} = (k+2)(k+1)$$

$$\frac{k!}{(k+p)!} = \frac{k!}{(k+p)(k+p-1) \cdot \ldots \cdot (k+1) \cdot k!}$$

$$= \frac{1}{(k+p)(k+p-1) \cdot \ldots \cdot (k+1)}.$$

Example 2

For the series $\sum \dfrac{k^2}{(k+1)!}$ the limit (1) is

$$\rho = \lim_{k\to\infty} \frac{\dfrac{(k+1)^2}{[(k+1)+1]!}}{\dfrac{k^2}{(k+1)!}} = \lim_{k\to\infty} \frac{(k+1)!(k+1)^2}{(k+2)!(k^2)}$$

$$= \lim_{k\to\infty} \left(\frac{1}{k+2}\right)\left(\frac{k+1}{k}\right)^2$$

$$= 0 \cdot 1^2 = 0.$$

Thus, the series $\sum \dfrac{k^2}{(k+1)!}$ converges, by the Ratio Test. ◇

Example 3

For the series $\sum \dfrac{k!}{k^k}$ the ratio limit is

$$\rho = \lim_{k\to\infty} \frac{\dfrac{(k+1)!}{(k+1)^{k+1}}}{\dfrac{k!}{k^k}} = \lim_{k\to\infty} \frac{(k+1)!k^k}{k!(k+1)^{k+1}}$$

$$= \lim_{k\to\infty} (k+1)\left[\frac{k^k}{(k+1)^{k+1}}\right]$$

$$= \lim_{k\to\infty} \left(\frac{k+1}{k+1}\right)\left[\frac{k^k}{(k+1)^k}\right]$$

$$= \lim_{k\to\infty} \left(\frac{k}{k+1}\right)^k$$

$$= \lim_{k\to\infty} \frac{1}{\left(1+\dfrac{1}{k}\right)^k} = \frac{1}{e}.$$

The series $\sum \dfrac{k!}{k^k}$ therefore converges, by the Ratio Test. ◇

Example 4

If we attempt to apply the Ratio Test to the series $\sum \dfrac{k}{k^2+1}$ we find that

$$\rho = \lim_{k\to\infty} \frac{\left(\dfrac{k+1}{(k+1)^2+1}\right)}{\left(\dfrac{k}{k^2+1}\right)}$$

$$= \lim_{k\to\infty} \frac{(k+1)(k^2+1)}{k[(k+1)^2+1]} = \lim_{k\to\infty} \frac{k^3+k^2+k+1}{k^3+2k^2+2k} = 1,$$

so the Ratio Test is inconclusive. However, we can handle this series by a limit comparison with the harmonic series $\Sigma b_k = \Sigma\left(\dfrac{1}{k}\right)$ (Theorem 10). We find

$$\lim_{k\to\infty}\left(\frac{a_k}{b_k}\right) = \lim_{k\to\infty}\frac{\left(\dfrac{k}{k^2+1}\right)}{\left(\dfrac{1}{k}\right)} = \lim_{k\to\infty}\frac{k^2}{k^2+1} = 1.$$

Thus, since the harmonic series diverges, so does the series $\displaystyle\sum\frac{k}{k^2+1}$. ◇

In Chapter 13 we shall concern ourselves with series whose terms contain variables. For such series the Ratio Test is often helpful in determining those values of the variable for which the series converges. The following is one such example.

Example 5

For which numbers $x > 0$ does the infinite series $\Sigma x^k/k$ converge?

Solution: Applying the Ratio Test we find that

$$\rho = \lim_{k\to\infty}\left(\frac{\dfrac{x^{k+1}}{k+1}}{\dfrac{x^k}{k}}\right) = \lim_{k\to\infty}\frac{x^{k+1}\cdot k}{x^k(k+1)}$$

$$= \lim_{k\to\infty} x\left(\frac{k}{k+1}\right)$$

$$= x.$$

Since $\rho = x$, the series will converge for $\rho = x < 1$, by the Ratio Test, and diverge for $x > 1$. The case $\rho = x = 1$ is inconclusive by the Ratio Test, but if $x = 1$ the series becomes simply the harmonic series, which diverges. Thus, the series $\Sigma x^k/k$, $x > 0$, converges only if $0 < x < 1$. ◇

Proof of Theorem 11: To prove statement (i), ($\rho < 1$ implies convergence), we will compare the series $\Sigma_{k=1}^{\infty} a_k$ with a geometric series. Since $\rho < 1$, we can find a number γ so that $\rho < \gamma < 1$.
Since

$$\lim_{k\to\infty}\frac{a_{k+1}}{a_k} = \rho,$$

there exists an integer N so that

$$\frac{a_{k+1}}{a_k} \le \gamma \qquad \text{whenever} \qquad k > N. \tag{2}$$

(To see this take $\epsilon = \gamma - \rho > 0$ in Definition 2.)
Inequality (2) shows that

$$a_{N+1} \le \gamma a_N$$
$$a_{N+2} \le \gamma a_{N+1} \le \gamma^2 a_N$$

and, in general, that

$$a_{N+k} \le \gamma^k a_N, \qquad k = 1, 2, 3, \ldots . \tag{3}$$

Now let $b_k = \gamma^k a_N$ for $k = 1, 2, \ldots$. Then $\Sigma b_k = \Sigma \gamma^k a_N = a_N \Sigma \gamma^k$ is a convergent series since $\gamma < 1$ and the series $\Sigma \gamma^k$ is geometric. Thus, inequality (3) shows that the series $\Sigma_{k=1}^{\infty} a_{N+k}$ is dominated by the convergent series $\Sigma_{k=1}^{\infty} b_k$. The series $\Sigma_{k=1}^{\infty} a_{N+k} = \Sigma_{k=N+1}^{\infty} a_k$ therefore converges, by the Comparison Test. This shows that the series Σa_k converges.

The proof of statement (ii) is similar (except that $\rho > 1$, so the comparison is with a divergent geometric series) and is left as an exercise.

To prove statement (iii) note that

(a) the harmonic series $\sum \dfrac{1}{k}$ diverges, although

$$\rho = \lim_{k \to \infty} \frac{\dfrac{1}{k+1}}{\dfrac{1}{k}} = \lim_{k \to \infty} \frac{k}{k+1} = 1, \qquad \text{and}$$

(b) the p-series $\sum \dfrac{1}{k^2}$ converges, although

$$\rho = \lim_{k \to \infty} \frac{\dfrac{1}{(k+1)^2}}{\dfrac{1}{k^2}} = \lim_{k \to \infty} \frac{k^2}{(k+1)^2} = 1. \qquad \blacklozenge$$

The Root Test

The following test for convergence uses information about the kth root of the general term for Σa_k.

| THEOREM 12 |
| Root Test |

Let $a_k > 0$ for all $k = 1, 2, 3, \ldots$ and let $\rho = \lim_{k \to \infty} \sqrt[k]{a_k}$. Then

(i) If $\rho < 1$, the series $\displaystyle\sum_{k=1}^{\infty} a_k$ converges.

(ii) If $\rho > 1$, the series $\displaystyle\sum_{k=1}^{\infty} a_k$ diverges.

(iii) If $\rho = 1$, no conclusions may be drawn.

The Root Test is primarily used in series involving kth powers.

Example 6

The series $\Sigma e^k / k^k$ converges. This can be shown using the Root Test since

$$\rho = \lim_{k \to \infty} \left(\frac{e^k}{k^k} \right)^{1/k} = \lim_{k \to \infty} \frac{e}{k} = 0 < 1. \qquad \diamond$$

Example 7

For the series $\Sigma k^2/2^k$ we find that

$$\rho = \lim_{k\to\infty} \left(\frac{k^2}{2^k}\right)^{1/k} = \lim_{k\to\infty} \frac{(k^{1/k})^2}{2} = \frac{1}{2} < 1$$

so the series converges.   ◇

Proof of Theorem 12:   We proceed as in the proof of the Ratio Test. If $\rho < 1$, there exists a constant γ so that $\rho < \gamma < 1$. Then, since $\lim_{k\to\infty} \sqrt[k]{a_k} = \lim_{k\to\infty} a_k^{1/k} = \rho$, there exists a constant N so that

$$a_k^{1/k} < \gamma \qquad \text{whenever} \qquad k > N. \tag{4}$$

(To see this take $\epsilon = \gamma - \rho$ in Definition 2.) Inequality (4) may be written

$$a_k < \gamma^k, \qquad k \geq N,$$

which shows that the series $\Sigma_{k=N}^{\infty} a_k$ is dominated by the geometric series $\Sigma_{k=N}^{\infty} \gamma^k$. Since $\gamma < 1$, this geometric series converges. Thus the series $\Sigma_{k=N}^{\infty} a_k$ converges by comparison with a geometric series. This proves statement (i).

The proof of statement (ii) is similar to that of (i) and is left as an exercise. The proof of statement (iii) proceeds as in the proof of Theorem 11. We note that

(a) $\rho = \lim_{k\to\infty} \left(\frac{1}{k}\right)^{1/k} = \lim_{k\to\infty} \frac{1}{\sqrt[k]{k}} = 1$, so the series $\displaystyle\sum \frac{1}{k}$ diverges.

(b) $\rho = \lim_{k\to\infty} \left(\frac{1}{k^2}\right)^{1/k} = \lim_{k\to\infty} \left(\frac{1}{\sqrt[k]{k}}\right)^2 = 1$, so the series $\displaystyle\sum \frac{1}{k^2}$ converges.   ◆

Exercise Set 12.6

In each of Exercises 1–20, determine whether the given series converges or diverges and state the test used to verify your conclusion.

1. $\displaystyle\sum \frac{2^k}{k+2}$

2. $\displaystyle\sum \frac{k3^k}{(k+1)!}$

3. $\displaystyle\sum k^{10}e^{-k}$

4. $\displaystyle\sum \frac{k!}{2^{k+2}}$

5. $\displaystyle\sum \frac{\ln k}{e^k}$

6. $\displaystyle\sum \frac{(3k)!}{(k!)^3}$

7. $\displaystyle\sum \frac{k+2}{1+k^3}$

8. $\displaystyle\sum \left(\frac{k}{2k+1}\right)^k$

9. $\displaystyle\sum \frac{1}{(\ln k)^k}$

10. $\displaystyle\sum \frac{k!}{k^k}$

11. $\displaystyle\sum \left(1+\frac{2}{k}\right)^k$

12. $\displaystyle\sum \left(\frac{k}{k+1}\right)^k$

13. $\displaystyle\sum \frac{3^k}{k^3 2^{k+2}}$

14. $\displaystyle\sum \frac{k^3 2^{k+3}}{2^{2k}}$

15. $\displaystyle\sum \frac{(k!)^2}{(3k)!}$

16. $\displaystyle\sum \left(\frac{k}{1+k^3}\right)^k$

17. $\displaystyle\sum \frac{(k+2)!}{4!k!2^k}$

18. $\displaystyle\sum \frac{1}{(k+1)!}$

19. $\displaystyle\sum \frac{k!}{e^{3k}}$

20. $\displaystyle\sum \frac{1}{k\sqrt{\ln k}}$

In Exercises 21–24, find those numbers $x > 0$ for which the given series converges.

21. $1 + x + \dfrac{x^2}{2!} + \dfrac{x^3}{3!} + \cdots = \displaystyle\sum_{k=0}^{\infty} \frac{x^k}{k!} \qquad (0! = 1)$

22. $1 + x^2 + x^4 + \cdots = \displaystyle\sum_{k=0}^{\infty} x^{2k}$

23. $1 + \dfrac{x^2}{2} + \dfrac{x^4}{4} + \cdots = 1 + \displaystyle\sum_{k=1}^{\infty} \dfrac{x^{2k}}{2k}$

24. $1 + \dfrac{x^2}{2} + \dfrac{x^4}{3} + \cdots = \displaystyle\sum_{k=0}^{\infty} \dfrac{x^{2k}}{k+1}$

25. Prove Theorem 11, part (ii). (*Hint:* If $\rho > 1$, let γ be a constant so that $1 < \gamma < \rho$. Then, if $\dfrac{a_{k+1}}{a_k} > \gamma$, it follows that $a_k > \gamma^k$. Proceed as in the proof of part (i).)

26. Prove Theorem 12, part (ii). (*Hint:* If $\rho > 1$, let γ be a constant so that $1 < \gamma < \rho$. Then if $\sqrt[k]{a_k} > \gamma$, it follows that $a_k > \gamma^k$. Proceed as in the proof of part (i).)

27. For the *sequence* $\{a_k\}$ with $a_k > 0$, prove that if $\displaystyle\lim_{k\to\infty} \dfrac{a_{k+1}}{a_k} < 1$, then $\displaystyle\lim_{k\to\infty} a_k = 0$.

28. For the *sequence* $\{a_k\}$ with $a_k > 0$, if $\displaystyle\lim_{k\to\infty} \dfrac{a_{k+1}}{a_k} > 1$, what can you say about $\displaystyle\lim_{k\to\infty} a_k$?

12.7 ABSOLUTE AND CONDITIONAL CONVERGENCE

Each of the convergence tests discussed thus far applies only to series with strictly positive terms. In this section we discuss two criteria by which series containing both positive and negative terms may be tested for convergence.

Absolute Convergence

One straightforward procedure for testing the series Σa_k for convergence is simply to replace each term by its absolute value and then test the resulting series, $\Sigma |a_k|$, for convergence. This is the idea of *absolute convergence*.

DEFINITION 4

The series Σa_k is said to **converge absolutely** if the series of absolute values, $\Sigma |a_k|$, converges.

Example 1

The series

$$\sum_{k=1}^{\infty} \frac{(-1)^{n+1}}{k^2} = 1 - \frac{1}{4} + \frac{1}{9} - \frac{1}{16} + \frac{1}{25} - \cdots$$

converges absolutely since

$$\sum_{k=1}^{\infty} \left| \frac{(-1)^{n+1}}{k^2} \right| = \sum_{k=1}^{\infty} \frac{1}{k^2}$$

is a convergent *p*-series.

Example 2

The series

$$\sum_{k=1}^{\infty} (-1)^{k+1} = 1 - 1 + 1 - 1 + \cdots$$

does not converge absolutely since $\displaystyle\sum_{k=1}^{\infty} |(-1)^{n+1}| = 1 + 1 + 1 + \cdots$ diverges.

The following theorem shows that the terminology of Definition 4 is well chosen—absolutely convergent series indeed converge.

THEOREM 13

If the series $\Sigma|a_k|$ converges then so does the series Σa_k. That is, every absolutely convergent series converges.

Proof: Assume that the series $\Sigma|a_k|$ converges. Since the inequality

$$-|a_k| \le a_k \le |a_k|$$

holds for all terms a_k, we may add $|a_k|$ to all terms in this inequality, to get

$$0 \le a_k + |a_k| \le 2|a_k|, \qquad k = 1, 2, 3, \ldots \tag{1}$$

Since the series $\Sigma|a_k|$ converges, so does the series $2\Sigma|a_k| = \Sigma 2|a_k|$. Thus inequality (1) shows that the series $\Sigma(a_k + |a_k|)$ is a series with positive terms* that is dominated by a convergent series. The Comparison Test therefore guarantees that the series $\Sigma(a_k + |a_k|)$ converges. We may now apply Theorem 6 (Section 12.3) to conclude that the series

$$\Sigma a_k = \Sigma(a_k + |a_k|) - \Sigma|a_k|$$

converges. This completes the proof. ◆

Theorem 13 is simple to apply. If the series Σa_k contains negative terms, we test the positive series $\Sigma|a_k|$ for convergence. If $\Sigma|a_k|$ converges, so does Σa_k. However, if $\Sigma|a_k|$ diverges, we can draw no conclusions about the original series Σa_k.

Example 3

To test the series

$$\sum_{k=0}^{\infty} \frac{(-1)^k k^2}{(k + 1)!}$$

for convergence, we apply the Ratio Test to the series

$$\sum_{k=0}^{\infty} \left| \frac{(-1)^k k^2}{(k + 1)!} \right| = \sum_{k=0}^{\infty} \frac{k^2}{(k + 1)!}.$$

We find that

$$\rho = \lim_{k \to \infty} \frac{a_{k+1}}{a_k} = \lim_{k \to \infty} \frac{\dfrac{(k + 1)^2}{(k + 2)!}}{\dfrac{k^2}{(k + 1)!}} = \lim_{k \to \infty} \frac{(k + 1)^2}{k^2(k + 2)} = 0.$$

The series $\displaystyle\sum_{k=0}^{\infty} \frac{(-1)^k k^2}{(k + 1)!}$ therefore converges absolutely. ◇

---

*If $a_k < 0$, then $a_k + |a_k| = 0$, so this term drops out of the series. This is why $\Sigma(a_k + |a_k|)$ may be regarded as a series with positive terms.

Example 4

The series

$$\sum_{k=0}^{\infty} \frac{\cos \pi k}{2^k} = 1 - \frac{1}{2} + \frac{1}{4} - \frac{1}{8} + \cdots$$

contains both positive and negative terms. To test for absolute convergence we consider the series (which turns out to be a geometric series)

$$\sum_{k=0}^{\infty} \left| \frac{\cos \pi k}{2^k} \right| = \sum_{k=0}^{\infty} \frac{|\cos \pi k|}{2^k} = \sum_{k=0}^{\infty} \frac{1}{2^k}$$

$$= \sum_{k=0}^{\infty} \left(\frac{1}{2}\right)^k = \frac{1}{1 - 1/2} = 2 \qquad \text{(geometric)}.$$

The series $\sum_{k=0}^{\infty} \dfrac{\cos \pi k}{2^k}$ therefore converges absolutely. $\diamondsuit$

Example 5

For the series

$$\sum_{k=1}^{\infty} \frac{(-1)^{k+1}}{k} = 1 - \frac{1}{2} + \frac{1}{3} - \frac{1}{4} + \cdots$$

the series of absolute values is the harmonic series $\sum_{k=1}^{\infty} 1/k$, which diverges. Theorem 13 therefore gives no conclusion concerning the convergence of the series $\sum_{k=1}^{\infty} \dfrac{(-1)^{k+1}}{k}$. This question will be resolved shortly. $\diamondsuit$

Example 6

If the variable x is unrestricted, the series

$$\sum_{k=1}^{\infty} \frac{x^k}{k} = x + \frac{x^2}{2} + \frac{x^3}{3} + \cdots$$

contains negative terms for $x < 0$. To determine the numbers x for which this series converges we consider the series of absolute values

$$\sum_{k=1}^{\infty} \left| \frac{x^k}{k} \right| = \sum_{k=1}^{\infty} \frac{|x|^k}{k}.$$

Applying the Ratio Test to this series, we find that, for $x \neq 0$,

$$\rho = \lim_{k \to \infty} \frac{|a_{k+1}|}{|a_k|} = \lim_{k \to \infty} \frac{\dfrac{|x|^{k+1}}{k+1}}{\dfrac{|x|^k}{k}} = \lim_{k \to \infty} |x| \left(\frac{k}{k+1}\right) = |x|.$$

Thus, $\rho < 1$ only if $|x| < 1$. This shows that $\displaystyle\sum_{k=1}^{\infty} x^k/k$ converges if $|x| < 1$. However, we obtain no conclusions for $|x| \geq 1$ by this procedure.   ◇

Alternating Series

Several types of infinite series containing negative terms can be shown to converge even though they do not converge absolutely. Among these are certain of the *alternating series*.

DEFINITION 5

An **alternating series** is a series of the form

$$\sum_{k=0}^{\infty} (-1)^k a_k = a_0 - a_1 + a_2 - a_3 + a_4 - \cdots$$

where $a_k > 0$ for each $k = 0, 1, 2, 3, \ldots$.

The following theorem indicates the conditions under which an alternating series converges. We will demonstrate its use before we give a proof.

THEOREM 14
Alternating Series Test

Let $a_k > 0$ for each $k = 0, 1, 2, \ldots$. The alternating series

$$\sum_{k=0}^{\infty} (-1)^k a_k$$

converges if both the following hold:

 (i) The terms of the sequence $\{a_k\}_{k=0}^{\infty}$ are decreasing, and
 (ii) $\displaystyle\lim_{k \to \infty} a_k = 0$.

Note that requirement (ii) in Theorem 14 is just a restatement of the necessary condition for convergence given by Theorem 7.

Example 7

For the series

$$\sum_{k=1}^{\infty} \frac{(-1)^{k+1}}{k} = 1 - \frac{1}{2} + \frac{1}{3} - \frac{1}{4} + \cdots$$

the general term is $(-1)^{k+1} a_k$, where $a_k = \dfrac{1}{k} > 0$. Since the sequence $\left\{\dfrac{1}{k}\right\}$ is decreasing and $\displaystyle\lim_{k \to \infty} \frac{1}{k} = 0$, the series converges by Theorem 14.   ◇

Combining the results of Examples 5 and 7, we see that the **alternating harmonic series** $\displaystyle\sum_{k=1}^{\infty} \frac{(-1)^{k+1}}{k} = 1 - \frac{1}{2} + \frac{1}{3} - \frac{1}{4} + \cdots$ converges, but does not converge absolutely. (This shows that the converse of Theorem 13 is not true.) Such series are said to *converge conditionally*.

DEFINITION 6	The series Σa_k is said to **converge conditionally** if Σa_k converges and $\Sigma|a_k|$ diverges.

Example 8

Determine whether the series

$$\sum_{k=1}^{\infty} \frac{(-1)^{k+1}k}{1 + k^2}$$

converges absolutely, converges conditionally, or diverges.

Strategy

First, test for absolute convergence.

Use the Integral Test on the series of absolute values.

Since the series is alternating but does not converge absolutely, apply the alternating series test.

Solution

The series of absolute values is

$$\sum_{k=1}^{\infty} \frac{k}{1 + k^2}.$$

We may test this series by the Integral Test. Since

$$\int_{1}^{\infty} \frac{x}{1 + x^2} \, dx = \lim_{t \to \infty} \int_{1}^{t} \frac{x}{1 + x^2} \, dx$$

$$= \lim_{t \to \infty} \frac{1}{2} \ln(1 + t^2) - \frac{1}{2} \ln 2 = +\infty,$$

the series of absolute values diverges. Thus, the original series does not converge absolutely.

We next test for conditional convergence using Theorem 14. Since

$$\frac{d}{dx}\left(\frac{x}{1 + x^2}\right) = \frac{1 + x^2 - x(2x)}{(1 + x^2)^2} = \frac{1 - x^2}{(1 + x^2)^2}$$

is negative for $x > 1$, the sequence $\left\{\dfrac{k}{1 + k^2}\right\}$ is decreasing. Also, $\lim_{k \to \infty} \dfrac{k}{1 + k^2} = 0$. Thus, both conditions of Theorem 14 are met, so the series converges conditionally. ◇

Example 9

For which numbers x does the series

$$\sum_{k=1}^{\infty} \frac{x^k}{k}$$

converge?

Solution: We have already determined that the series converges absolutely for $|x| < 1$ (Example 6). If $|x| > 1$, we may apply l'Hôpital's Rule (differentiating with respect to k) to conclude that

$$\lim_{k \to \infty} \frac{|x|^k}{k} \qquad \left(\frac{\infty}{\infty} \text{ form} \right)$$

$$= \lim_{k \to \infty} \frac{|x|^k \ln|x|}{1} = +\infty.$$

Thus, $\displaystyle\lim_{k \to \infty} \frac{x^k}{k}$ does not exist if $|x| > 1$, so $\displaystyle\sum_{k=1}^{\infty} \frac{x^k}{k}$ diverges for these values of x, by Theorem 7. The remaining cases are $x = 1$ and $x = -1$.

If $x = 1$, the series is the harmonic series $\sum_{k=1}^{\infty} 1/k$, which diverges. If $x = -1$, the series is

$$\sum_{k=1}^{\infty} \frac{(-1)^k}{k} = -1 + \frac{1}{2} - \frac{1}{3} + \frac{1}{4} - \cdots .$$

This is the negative of the alternating harmonic series, which converges conditionally (Example 7).

We have therefore shown that the series $\sum_{k=1}^{\infty} x^k/k$ converges for $x \in [-1, 1)$ and diverges for all other numbers x. In particular, the series converges absolutely for $x \in (-1, 1)$ and converges conditionally for $x = -1$.   ◇

Proof of Theorem 14:   For the alternating series

$$\sum_{k=0}^{\infty} (-1)^k a_k,$$

the even partial sums have the form

$$S_{2n} = (a_0 - a_1) + (a_2 - a_3) + \cdots + (a_{2n-2} - a_{2n-1}) + a_{2n}.$$

Since the sequence $\{a_k\}$ is decreasing, each of the terms in parentheses is positive. This shows that all even partial sums are positive. Also, since $a_{2n+1} > a_{2n+2}$

$$\begin{aligned} S_{2n+2} &= S_{2n} - a_{2n+1} + a_{2n+2} \\ &= S_{2n} - (a_{2n+1} - a_{2n+2}) \\ &< S_{2n}. \end{aligned} \qquad (2)$$

Thus, the sequence of even partial sums $\{S_{2n}\}$ is decreasing. In particular, this shows that the sequence of positive terms $\{S_{2n}\}$ is bounded above by $S_0 = a_0$ and below by zero. Thus $\{S_{2n}\}$ is a bounded decreasing sequence which, by Corollary 1, must converge. We denote the limit of this sequence by $S = \lim_{n \to \infty} S_{2n}$.

Next, since $\lim_{n \to \infty} a_{2n+1} = 0$, we see that

$$\lim_{n \to \infty} S_{2n+1} = \lim_{n \to \infty} (S_{2n} - a_{2n+1})$$

$$= S - 0$$
$$= S,$$

so both even and odd partial sums converge to S. Thus $S = \lim_{n \to \infty} S_n$ and the series $\sum_{k=0}^{\infty} (-1)^k a_k$ converges.   ◆

Estimating Remainders for Alternating Series

For convergent alternating series satisfying the hypotheses of Theorem 14, there is a simple way to obtain both an estimate of the sum and a bound on the error associated with that estimate.

As in line (2) in the proof of Theorem 14, we may show that the odd partial sums of an alternating series $\sum_{k=0}^{\infty} (-1)^k a_k$ form an *increasing* sequence:

$$S_{2n+1} = S_{2n-1} + (a_{2n} - a_{2n+1}) \qquad (3)$$
$$> S_{2n-1}, \qquad n = 1, 2, \ldots$$

since $a_{2n} > a_{2n+1}$.

If we write

$$S = \sum_{k=0}^{\infty} (-1)^k a_k,$$

inequalities (2) and (3) together imply that

$$S_{2p-1} < S < S_{2m} \qquad (4)$$

for any positive integers p and m. That is, *any* even partial sum is larger than *any* odd partial sum, and the sum of the series lies between them (see Figure 7.1).

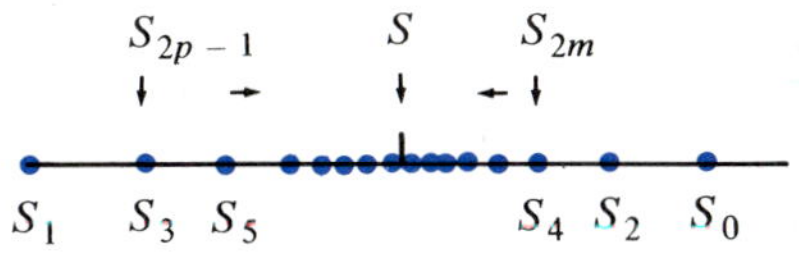

Now let

$$S = S_n + R_n \qquad (5)$$

where

$$R_n = \sum_{k=n+1}^{\infty} (-1)^k a_k.$$

S_{2p-1} S S_{2m}

$S_1 \quad S_3 \ S_5 \qquad S_4 \ S_2 \ S_0$

Partial sums (S_{2p-1}) increase to S.

Partial sums (S_{2m}) decrease to S.

Figure 7.1

Then R_n is the **truncation error,** or **remainder,** associated with the approximation of S by the partial sum S_n. From (4) and (5) it follows, by setting $m = n$ and $p = n + 1$, that

$$|R_{2n}| = |S - S_{2n}| < |S_{2n+1} - S_{2n}| = a_{2n+1}. \qquad (6)$$

Similarly, by setting $m = p = n$ we find

$$|R_{2n-1}| = |S - S_{2n-1}| < |S_{2n} - S_{2n-1}| = a_{2n}. \qquad (7)$$

Thus, regardless of whether we truncate an alternating series after an even or an odd number of terms, *the absolute value of the remainder (or tail) is less than the absolute value of the next term in the series*. This result is sufficiently important that we carefully summarize what we have proved in the form of a theorem.

THEOREM 15
Truncation Error

The absolute value of the error associated with approximating the sum of the convergent alternating series satisfying the hypotheses of Theorem 14

$$\sum_{k=0}^{\infty} (-1)^k a_k$$

by the partial sum

$$S_n = \sum_{k=0}^{n} (-1)^k a_k$$

is less than the first truncated term, a_{n+1}.

Example 10

Find an approximation to the sum of the alternating harmonic series

$$\sum_{k=0}^{\infty} (-1)^k \left(\frac{1}{k+1} \right) = 1 - \frac{1}{2} + \frac{1}{3} - \frac{1}{4} + \cdots$$

accurate to within .05.

Solution: If we approximate this sum by the partial sum

$$S_n = \sum_{k=0}^{n} (-1)^k \left(\frac{1}{k+1} \right),$$

the error in this approximation is less than $a_{n+1} = \dfrac{1}{n+2}$. We therefore choose n sufficiently large to satisfy the inequality

$$\frac{1}{n+2} \le .05 = \frac{1}{20}.$$

This gives

$$n + 2 \ge 20, \qquad \text{or} \qquad n \ge 18.$$

The partial sum S_{18} therefore has the desired accuracy. It is

$$S_{18} = 1 - \frac{1}{2} + \frac{1}{3} - \frac{1}{4} + \cdots + \frac{1}{19} \approx 0.72. \qquad \diamond$$

Example 11

How many terms must be included in the partial sum for the alternating series

$$\sum_{k=0}^{\infty} \frac{(-1)^k}{1+k^2} = 1 - \frac{1}{2} + \frac{1}{5} - \frac{1}{10} + \cdots$$

to ensure that the partial sum approximates the sum of the series accurately to within 0.01?

Solution: It is easier to work with S_{n-1}, so that the absolute value of the first truncated term is $a_n = \dfrac{1}{1+n^2}$. We therefore need

$$\frac{1}{1+n^2} \le \frac{1}{100}, \qquad \text{or} \qquad n^2 \ge 99.$$

The choice $n = 10$ therefore will suffice, and the partial sum S_9 will have the desired accuracy. $\qquad \diamond$

Exercise Set 12.7

In each of Exercises 1–30, determine whether the series converges absolutely, converges conditionally, or diverges.

1. $\displaystyle\sum_{k=1}^{\infty} \frac{(-1)^k}{k^3}$

2. $\displaystyle\sum_{k=1}^{\infty} \frac{(-1)^k}{2k+1}$

3. $\displaystyle\sum_{k=1}^{\infty} \frac{(-1)^{k+1}k^2}{(k+2)!}$

4. $\displaystyle\sum_{k=2}^{\infty} (-1)^{k+1} \frac{k}{\ln k}$

5. $\displaystyle\sum_{k=1}^{\infty} (-1)^k \frac{k!}{(2k+1)!}$

6. $\displaystyle\sum_{k=1}^{\infty} \frac{(-k)^3}{3^k}$

7. $\displaystyle\sum_{k=1}^{\infty} \frac{(-1)^k}{1+\sqrt{k}}$

8. $\displaystyle\sum_{k=1}^{\infty} \frac{\cos \pi k}{k}$

9. $\displaystyle\sum_{k=1}^{\infty} \frac{\sin\left(\frac{\pi k}{2}\right)}{k}$

10. $\displaystyle\sum_{k=1}^{\infty} \frac{(-1)^k}{\sqrt{k(k+2)}}$

11. $\displaystyle\sum_{k=1}^{\infty} \frac{(-1)^k \cdot k^3}{2^{k+2}}$

12. $\displaystyle\sum_{k=2}^{\infty} \frac{k^2(-1)^{k+1}}{\ln k}$

13. $\displaystyle\sum_{k=2}^{\infty} \frac{k^2(-1)^{k+1}}{\ln k}$

14. $\displaystyle\sum_{k=1}^{\infty} \frac{(2k+1)(-1)^k}{6k+2}$

15. $\displaystyle\sum_{k=1}^{\infty} \frac{(-1)^k \sqrt{k}}{k+1}$

16. $\displaystyle\sum_{k=1}^{\infty} \frac{(-1)^{k+1}2^k}{k^3 \cdot 3^{k+2}}$

17. $\displaystyle\sum_{k=2}^{\infty} \frac{(-1)^k}{k \ln^2 k}$

18. $\displaystyle\sum_{k=1}^{\infty} \frac{(-1)^{k+1}k!}{6^k}$

19. $\displaystyle\sum_{k=1}^{\infty} \frac{(-1)^k}{1+k^2}$

20. $\displaystyle\sum_{k=1}^{\infty} \sin\left(\frac{k\pi}{4}\right)$

21. $\displaystyle\sum_{k=2}^{\infty} \frac{(-1)^k k}{\ln \sqrt{k}}$

22. $\displaystyle\sum_{k=1}^{\infty} \frac{(-1)^k k^2}{(2k+1)(k+3)}$

23. $\displaystyle\sum_{k=1}^{\infty} \frac{(-1)^k \sinh k}{3e^{2k}}$

24. $\displaystyle\sum_{k=1}^{\infty} \frac{(-1)^k \operatorname{Tan}^{-1} k}{\sqrt{k}}$

25. $\displaystyle\sum_{k=1}^{\infty} \frac{\cos \pi k}{k+2}$

26. $\displaystyle\sum_{k=2}^{\infty} \frac{(-1)^k}{\sqrt[k]{\ln k}}$

27. $\displaystyle\sum_{k=2}^{\infty} \frac{k \sin\left[\frac{(2k+1)\pi}{2}\right]}{\sqrt{1+e^{\sqrt{k}}}}$

28. $\displaystyle\sum_{k=2}^{\infty} \frac{(-1)^k(k^{3/2}+3k)}{7-k^2+2k^{5/2}}$

29. $\displaystyle\sum_{k=1}^{\infty} \frac{\sinh k - \cosh k}{k}$

30. $\displaystyle\sum_{k=2}^{\infty} \frac{(-1)^k \cosh 2k}{ke^k}$

In Exercises 31–34, determine the numbers x for which the given series converges absolutely.

31. $\displaystyle\sum_{k=0}^{\infty} \frac{x^k}{k!} = 1 + x + \frac{x^2}{2!} + \frac{x^3}{3!} + \cdots$

32. $\displaystyle\sum_{k=0}^{\infty} \frac{x^k}{2k+1} = 1 + \frac{x}{3} + \frac{x^2}{5} + \cdots$

33. $\displaystyle\sum_{k=0}^{\infty} (-1)^k \frac{x^{2k+1}}{(2k+1)!} = x - \frac{x^3}{3!} + \frac{x^5}{5!} - \frac{x^7}{7!} + \cdots$

34. $\displaystyle\sum_{k=0}^{\infty} (-1)^k \frac{x^{2k}}{(2k)!} = 1 - \frac{x^2}{2!} + \frac{x^4}{4!} - \frac{x^6}{6!} + \cdots$

35. How large must n be taken so that the partial sum $S_n = \displaystyle\sum_{k=0}^{n} (-1)^k\left(\frac{1}{1+k}\right)$ approximates the sum of the alternating harmonic series accurate to within .01?

36. With what accuracy does the sum

$$\sum_{k=0}^{10} \frac{(-1)^k}{(k+1)^3}$$

approximate the sum of the series

$$\sum_{k=0}^{\infty} \frac{(-1)^k}{(k+1)^3}?$$

37. Prove that if Σa_k and Σb_k converge absolutely, then so does $\Sigma(a_k + b_k)$.

38. Prove that if Σa_k converges absolutely and c is any real number, the series $\Sigma c a_k$ converges absolutely.

39. Prove for the (not necessarily positive) series Σa_k that if

$$\rho = \lim_{k \to \infty} \left| \frac{a_{k+1}}{a_k} \right|,$$

   (i) Σa_k converges absolutely if $0 \le \rho < 1$;
   (ii) Σa_k diverges if $\rho > 1$;
   (iii) no conclusions may be drawn if $\rho = 1$.

40. Prove that Σa_k^2 converges if Σa_k converges absolutely. Is the converse true?

SUMMARY OUTLINE OF CHAPTER 12

◆ An **infinite sequence** $\{a_n\}$ is a function whose domain is the set of positive integers. (page 510)

◆ $L = \lim\limits_{n\to\infty} a_n$ if, given $\epsilon > 0$, there exists an integer N so that $|a_n - L| < \epsilon$ whenever $n > N$. (page 513)

◆ The sequence $\{a_n\}$ is said to **converge** if $L = \lim\limits_{n\to\infty} a_n$ exists. (page 512)

◆ The nth **partial sum** of the infinite series $\sum\limits_{k=1}^{\infty} a_k$ is the sum of the first n terms: $S_n = \sum\limits_{k=1}^{n} a_k$. (page 524)

◆ S is called the **sum** of the infinite series if $S = \lim\limits_{n\to\infty} S_n = \lim\limits_{n\to\infty} \sum\limits_{k=1}^{n} a_k$. (page 524)

◆ The series Σa_k is said to **converge** if the sum S exists. (page 524)

◆ The **geometric series** $\sum\limits_{k=0}^{\infty} x^k = 1 + x + x^2 + x^3 + \cdots$ converges, if $|x| < 1$, to the sum $S = \dfrac{1}{1-x}$, and diverges otherwise. (page 527)

◆ The **harmonic series** $\sum\limits_{k=1}^{\infty} \dfrac{1}{k} = 1 + \dfrac{1}{2} + \dfrac{1}{3} + \dfrac{1}{4} + \cdots$ diverges. (page 535)

◆ The **p-series** $\sum\limits_{k=1}^{\infty} \dfrac{1}{k^p} = 1 + \dfrac{1}{2^p} + \dfrac{1}{3^p} + \dfrac{1}{4^p} + \cdots$ converges if $p > 1$ and diverges otherwise. (page 536)

◆ A **necessary condition** for Σa_k to converge: $\lim\limits_{k\to\infty} a_k = 0$ (however, this condition does not guarantee convergence). (page 530)

◆ *Tests for convergence of Σa_k when $a_k > 0$ for all k:*
  (1) (Comparison Test)   Let $0 < a_k \le b_k$ for all k. (page 538)
      a. If Σb_k converges, Σa_k converges.
      b. If Σa_k diverges, Σb_k diverges.

  (2) (Integral Test)   If $f(k) = a_k$ for all k and f is decreasing on $[1, \infty)$ then Σa_k converges if and only if $\displaystyle\int_1^{\infty} f(x)\, dx$ converges. (page 534)

  (3) (Ratio Test)   If $\rho = \lim\limits_{k\to\infty} \dfrac{a_{k+1}}{a_k}$ exists, then (page 544)
      a. Σa_k converges if $\rho < 1$.
      b. Σa_k diverges if $\rho > 1$.
      c. No conclusion may be drawn if $\rho = 1$.

  (4) (Root Test)   If $\rho = \lim\limits_{k\to\infty} \sqrt[k]{a_k}$ exists, same conclusions as for the Ratio Test. (page 547)

  (5) (Limit Comparison Test)   If $b_k > 0$ for all k and $\rho = \lim\limits_{k\to\infty} \dfrac{a_k}{b_k} > 0$ exists, then Σa_k and Σb_k either both converge or both diverge. (page 540)

◆ The series Σa_k **converges absolutely** if $\Sigma |a_k|$ converges. (page 549)

◆ *Theorem:*   If $\Sigma |a_k|$ converges, then Σa_k converges. (page 550)

◆ An **alternating series** is a series of the form $\sum\limits_{k=0}^{\infty} (-1)^k a_k$ with all $a_k > 0$. (page 552)

◆ **Theorem:** An alternating series converges if   (page 552)
  a. $a_k > a_{k+1}$ for all k,   and
  b. $\lim\limits_{k \to \infty} a_k = 0$.

◆ The absolute value of the **error** associated with the approximation of the sum of the alternating series $\sum\limits_{k=0}^{\infty} (-1)^k a_k$ by   (page 555)

the partial sum $S_n = \sum\limits_{k=0}^{n} (-1)^k a_k$ is less than the first discarded term, a_{n+1}.

REVIEW EXERCISES—CHAPTER 12

In Exercises 1–10, determine whether the given sequence converges. If it does, find its limit.

1. $\left\{ \sin \dfrac{n\pi}{2} \right\}$

2. $\left\{ \dfrac{1}{\sqrt{n+1}} \right\}$

3. $\left\{ \dfrac{\sqrt{n}}{\ln(\sqrt{n}+1)} \right\}$

4. $\left\{ \left(1 - \dfrac{2}{n}\right)^{2n} \right\}$

5. $\left\{ \dfrac{3^n}{n!} \right\}$

6. $\left\{ \dfrac{(-1)^{n+1}}{\sqrt{n^2+1}} \right\}$

7. $\left\{ \dfrac{(-1)^{n+1} \cdot n^3}{n(1 - n + n^2)} \right\}$

8. $\left\{ \dfrac{n^2 + 2n + 2}{n^{3/2}} \right\}$

9. $\{ e^{2 \ln(n)} \}$

10. $\{ \ln(n+1) - \ln(n-1) \}$

In Exercises 11–14, find the sum of the given geometric series.

11. $1 + \dfrac{1}{6} + \dfrac{1}{36} + \dfrac{1}{216} + \cdots$

12. $10 + \dfrac{10}{2} + \dfrac{10}{4} + \dfrac{10}{8} + \cdots$

13. $\dfrac{2}{3} + \dfrac{4}{9} + \dfrac{8}{27} + \cdots$

14. $1 - \dfrac{1}{5} + \dfrac{1}{25} - \dfrac{1}{125} + \cdots$

In Exercises 15 and 16, express the repeating decimal as an infinite series and find its rational form.

15. $.37\overline{37} \ldots$

16. $.0263\overline{263} \ldots$

In Exercises 17–20, find the sum of the series.

17. $\sum\limits_{k=3}^{\infty} \left(\dfrac{1}{4} \right)^k$

18. $\sum\limits_{k=1}^{\infty} \dfrac{3^k + 2^k}{5^k}$

19. $\sum\limits_{k=2}^{\infty} \dfrac{20}{3^{k+1}}$

20. $\sum\limits_{k=1}^{\infty} \left[\dfrac{3}{2^k} + \dfrac{1}{(k+1)(k+2)} \right]$

In Exercises 21–58, determine whether the given series converges or diverges and state which test you used.

21. $\sum\limits_{k=1}^{\infty} \dfrac{1}{k(k+1)}$

22. $\sum\limits_{k=1}^{\infty} \dfrac{3^k}{k^3}$

23. $\sum\limits_{k=3}^{\infty} \dfrac{2^k}{k^5}$

24. $\sum\limits_{k=1}^{\infty} \dfrac{(-1)^k}{2k+1}$

25. $\sum\limits_{k=1}^{\infty} \dfrac{k^3}{k!}$

26. $\sum\limits_{k=0}^{\infty} \dfrac{1}{10k+2}$

27. $\sum\limits_{k=2}^{\infty} \dfrac{\cos \pi k}{\sqrt{k}}$

28. $\sum\limits_{k=2}^{\infty} \dfrac{k^k}{k!}$

29. $\sum\limits_{k=3}^{\infty} \dfrac{\ln k}{k^2}$

30. $\sum\limits_{k=0}^{\infty} \dfrac{2^k + k}{(k+1)!}$

31. $\sum\limits_{k=1}^{\infty} \dfrac{1}{9 + k^2}$

32. $\sum\limits_{k=2}^{\infty} \dfrac{(-1)^k k}{\ln k}$

33. $\sum\limits_{k=2}^{\infty} \dfrac{2}{k \ln(k+1)}$

34. $\sum\limits_{k=1}^{\infty} \dfrac{\sqrt{k}}{2^k}$

35. $\sum\limits_{k=0}^{\infty} \dfrac{k^4 \cdot 2^k}{(k+2)!}$

36. $\sum\limits_{k=1}^{\infty} (-1)^k \dfrac{\ln k}{k}$

37. $\sum\limits_{k=0}^{\infty} \dfrac{1}{\sqrt{4 + k^2}}$

38. $\sum\limits_{k=1}^{\infty} \left(\dfrac{k-1}{k+1} \right)^k$

39. $\sum\limits_{k=1}^{\infty} (-1)^{k+1} \dfrac{\sqrt{k}}{2k+1}$

40. $\sum\limits_{k=1}^{\infty} k^2 e^{-k^3}$

41. $\displaystyle\sum_{k=1}^{\infty} \left(\frac{k!}{k^k}\right)^k$

42. $\displaystyle\sum_{k=2}^{\infty} (-1)^k \frac{k^2 + 2}{1 + k + k^2}$

43. $\displaystyle\sum_{k=1}^{\infty} 2ke^{-k}$

44. $\displaystyle\sum_{k=0}^{\infty} \left(\frac{k}{1+k}\right)^k$

45. $\displaystyle\sum_{k=0}^{\infty} (-1)^k \sin \pi k$

46. $\displaystyle\sum_{k=2}^{\infty} \frac{(2k+1)!}{k^2(k+1)!}$

47. $\displaystyle\sum_{k=1}^{\infty} \frac{k!}{e^{2k}}$

48. $\displaystyle\sum_{k=2}^{\infty} \frac{1}{(k+6)^{3/2}}$

49. $\displaystyle\sum_{k=1}^{\infty} (-1)^k \frac{3}{\sqrt{k}}$

50. $\displaystyle\sum_{k=2}^{\infty} \frac{1}{(\ln k)^k}$

51. $\displaystyle\sum_{k=1}^{\infty} \frac{2^{2k-1}}{(2k-1)!}$

52. $\displaystyle\sum_{k=2}^{\infty} \frac{1}{\sqrt{k}(\ln k)^3}$

53. $\displaystyle\sum_{k=0}^{\infty} \frac{(-1)^k}{1+k^2}$

54. $\displaystyle\sum_{k=1}^{\infty} \text{Tan}^{-1} k$

55. $\displaystyle\sum_{k=1}^{\infty} k\left(\frac{\pi}{k}\right)^k$

56. $\displaystyle\sum_{k=0}^{\infty} \frac{(-1)^k k(k+2)}{(k+1)!}$

57. $\displaystyle\sum_{k=1}^{\infty} \left(\frac{k}{2k+1}\right)^k$

58. $\displaystyle\sum_{k=0}^{\infty} \frac{(-1)^k \text{Sin}^{-1}\left(\frac{1}{k+1}\right)}{k+1}$

In Exercises 59–68, determine whether the given series converges absolutely, converges conditionally, or diverges.

59. $\displaystyle\sum_{k=0}^{\infty} \frac{(-1)^{k+1}}{3k+2}$

60. $\displaystyle\sum_{k=2}^{\infty} \frac{(-1)^k}{k \ln k}$

61. $\displaystyle\sum_{k=1}^{\infty} (-1)^k \frac{k^k}{k!}$

62. $\displaystyle\sum_{k=0}^{\infty} \frac{\cos \pi k}{1+k^2}$

63. $\displaystyle\sum_{k=1}^{\infty} \frac{(-1)^{k+1}}{k(k+1)}$

64. $\displaystyle\sum_{k=0}^{\infty} (-1)^k \left(\frac{2}{3}\right)^k$

65. $\displaystyle\sum_{k=1}^{\infty} \frac{k!}{(-2)^k}$

66. $\displaystyle\sum_{k=1}^{\infty} \frac{\cos \dfrac{\pi k}{3}}{1+k^2}$

67. $\displaystyle\sum_{k=1}^{\infty} \frac{(-1)^k k^2}{(k+1)!}$

68. $\displaystyle\sum_{k=1}^{\infty} \frac{(-1)^{k+1}}{\sqrt{k}(k+1)}$

In Exercises 69–72, find a bound on the error associated with the approximation of the sum of the given series by the partial sum S_4 consisting of the sum of the first four terms.

69. $\displaystyle\sum_{k=1}^{\infty} \frac{2(-1)^k}{1+k^2} = -1 + \frac{2}{5} - \frac{2}{10} + \cdots$

70. $\displaystyle\sum_{k=1}^{\infty} \frac{(-1)^k}{k^4} = -1 + \frac{1}{16} - \frac{1}{81} + \cdots$

71. $\displaystyle\sum_{k=1}^{\infty} \frac{(-1)^{k+1}}{2k+1} = \frac{1}{3} - \frac{1}{5} + \frac{1}{7} - \frac{1}{9} + \cdots$

72. $\displaystyle\sum_{k=1}^{\infty} \frac{(-1)^{k+1} \sin^2\left(\dfrac{\pi}{2k}\right)}{k+1} = \frac{1}{2} - \frac{1}{6} + \frac{1}{16} - \cdots$

73. Define the sequence $\{a_k\}$ by $a_1 = c$, $a_{k+1} = a_1 \cdot a_k$, $k \geq 1$. For what values of c does $\{a_k\}$ converge?

74. Prove that if the sequence $\{a_n\}$ is not bounded it cannot converge.

75. If Σa_k converges but $\Sigma(a_k + b_k)$ diverges, what can you conclude about Σb_k?

76. Explain how one could use an infinite series to determine at what time between 1 P.M. and 2 P.M. is the minute hand on a clock directly over the hour hand. (*Hint:* The minute hand moves 12 times as fast as the hour hand.)

77. A ball is dropped from a height of 10 meters. Each time it strikes the ground, it rebounds to 5/6 of the height from which it fell. Find the total distance travelled by the ball
   a. when it has struck the ground 10 times.
   b. when it has come to rest.

78. Prove that if Σa_k diverges then $\Sigma c a_k$ diverges for any number $c \neq 0$.

79. Let Σa_k and Σb_k be series with positive terms. If $\Sigma(a_k + b_k)^2$ converges, what can you conclude about
   a. Σa_k^2?
   b. Σb_k^2?
   c. $\Sigma a_k b_k$?

80. Let $p(k)$ and $q(k)$ be polynomials in k with positive coefficients. Let n be the degree of $p(k)$ and let m be the degree of $q(k)$. Show that
   a. $\displaystyle\sum \frac{p(k)}{q(k)}$ converges if $m > n + 1$
   b. $\displaystyle\sum \frac{p(k)}{q(k)}$ diverges if $m \leq n + 1$.

Chapter 13
Taylor Polynomials and Power Series

This chapter begins by posing the *approximation problem:* given a differentiable function *f*, how can we approximate values of *f* by values of certain polynomials (the latter being easier to calculate)?

We shall answer this question by developing the notion of *Taylor Polynomials*. In doing so we shall see that Taylor Polynomials may be regarded as partial sums for *power series,* which are infinite series involving powers of an independent variable. Applying the theory and techniques developed in Chapter 12 for infinite series of constants, we shall show that a convergent power series determines a differentiable function, and that many familiar functions, including the trigonometric, exponential, and logarithmic functions, may be represented as power series.

13.1 THE APPROXIMATION PROBLEM AND TAYLOR POLYNOMIALS

The **approximation problem** is this: *Given a function f and a number x = a, how can we approximate values f(x) for x near a using polynomials, and what is the accuracy of these approximations?*

Of course, if *f* is a polynomial the answer is easy. We simply use *f* to approximate itself, with complete accuracy. Our real interest is in knowing how to approximate transcendental functions, like sin *x* or e^x, or algebraic functions like $\sqrt[5]{x}$.

The reason for our interest in approximating such functions is that although we have developed a rich theory of differentiation and integration for these functions, we do not yet have available a means for calculating their values, except in special cases. For example, how does one calculate sin 53°, $e^{0.05}$, or $\sqrt[5]{216}$? The theory of differentials (Chapter 3) is a first step in this direction, although it is quite limited. Even if you feel secure in the knowledge that your calculator is equipped with a special key for each of these functions or that a five-place table for such functions will always be available somewhere, the question remains as to how that calculator was programmed or how the entries in such tables are determined.

As stated above, our objective is to approximate arbitrary functions using *polynomials.* The reason for using polynomials is that values of polynomial functions can be easily computed with complete accuracy by simple multiplications and additions. In other words, we want to know how to approximate troublesome functions with convenient ones.

Taylor Polynomials

Given the function *f*, suppose we ask for the polynomial of lowest degree that best approximates *f(x)* near *x = a*. Since the polynomials of lowest degree are constants,

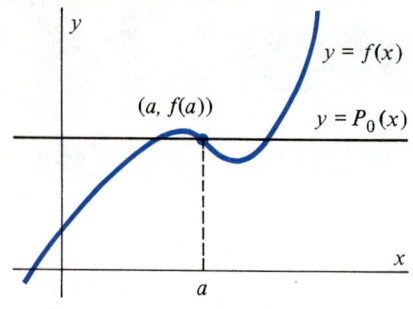

Figure 1.1 Constant polynomial $P_0 \equiv f(a)$ approximates f near a.

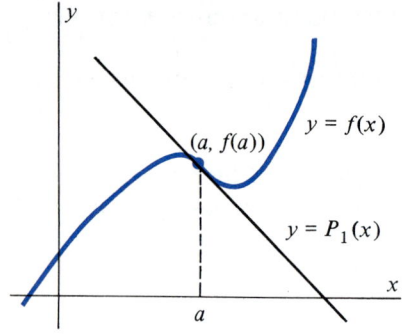

Figure 1.2 First degree polynomial $P_1(x) = f'(a)(x - a) + f(a)$.

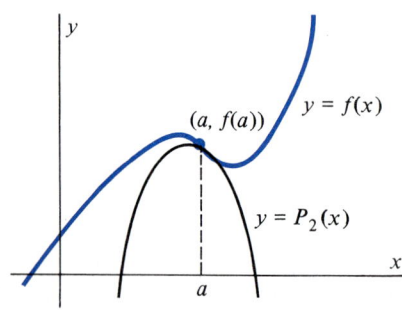

Figure 1.3 Second degree approximating polynomial P_2.

the obvious answer to this question is the constant polynomial

$$P_0(x) \equiv f(a) \qquad \text{(Figure 1.1).} \tag{1}$$

Next, let us ask for the polynomial of degree one (that is, a linear function) that best approximates f near $x = a$. We shall write this polynomial in the form

$$P_1(x) = b + c(x - a) \tag{2}$$

and seek to determine the constants b and c. Since we again want $P_1(a) = f(a)$, we obtain $b = f(a)$. Now by the theory of the derivative, the line through $(a, f(a))$ that best approximates f has slope $f'(a)$, if $f'(a)$ exists. Thus, we also want $f'(a) = P_1'(a) = c$. P_1 is therefore

$$P_1(x) = f(a) + f'(a)(x - a). \qquad \text{(See Figure 1.2.)} \tag{3}$$

Continuing in this way, we write the second degree polynomial approximating $f(x)$ as

$$P_2(x) = b + c(x - a) + d(x - a)^2.$$

The condition that $P_2(a) = f(a)$ gives $b = f(a)$, and the condition that the polynomial and the function f have the same slope at $x = a$ gives $P_2'(a) = c = f'(a)$. To determine the remaining constant d we require that P_2 and f have the same measure of concavity at $x = a$; that is, we set $P_2''(a) = f''(a)$.

This gives the equation

$$P_2''(a) = 2d = f''(a).$$

Thus, $d = \dfrac{f''(a)}{2}$, and P_2 is therefore (see Figure 1.3)

$$P_2(x) = f(a) + f'(a)(x - a) + \frac{f''(a)}{2}(x - a)^2. \tag{4}$$

We must carry this analysis one step further for the general form of these polynomials to become clear. Although we do not have available a geometric interpretation for derivatives of order higher than two, the idea is that for an approximating polynomial P_n of degree n, each derivative of order 1 through n must equal the corresponding derivative of f at $x = a$.

Write the general third-degree polynomial $P_3(x)$ in the form

$$P_3(x) = b + c(x - a) + d(x - a)^2 + e(x - a)^3.$$

Then, requiring that P_3 and f, together with their first three derivatives, agree at $x = a$ gives

$$
\begin{array}{ll}
P_3(a) = b = f(a); & b = f(a) \\
P_3'(a) = c = f'(a); & c = f'(a) \\
P_3''(a) = 2d = f''(a); & d = \dfrac{f''(a)}{2} \\
P_3'''(a) = 3 \cdot 2 \cdot e = f'''(a); & e = \dfrac{f'''(a)}{3!}.
\end{array}
$$

The polynomial P_3 is therefore (see Figure 1.4)

$$P_3(x) = f(a) + f'(a)(x - a) + \frac{f''(a)}{2}(x - a)^2 + \frac{f'''(a)}{3!}(x - a)^3. \tag{5}$$

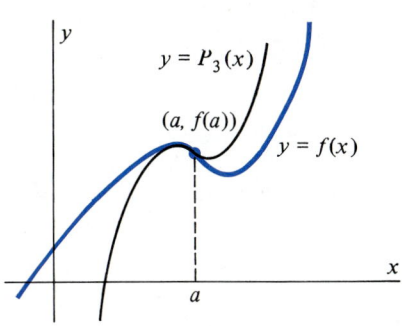

Figure 1.4 Third degree approximating polynomial P_3.

(The notation $n!$ in equation (5) is the **factorial symbol,** defined by

$$n! = n(n-1)(n-2) \cdots 3 \cdot 2 \cdot 1.$$

We also define $0! = 1$ and $1! = 1$.)

We can generalize equations (1)–(5) by stating that the approximating polynomial of degree n for the function f near $x = a$ is, for $n \geq 2$,

$$P_n(x) = f(a) + f'(a)(x-a) + \frac{f''(a)}{2}(x-a)^2 \tag{6}$$
$$+ \frac{f'''(a)}{3!}(x-a)^3 + \cdots + \frac{f^{(n)}(a)}{n!}(x-a)^n.$$

The polynomial P_n in (6) is referred to as the nth **Taylor polynomial*** for f, expanded about $x = a$. The ideas we discuss here were developed by the English mathematician Brook Taylor (1686–1731) early in the eighteenth century.

Figures 1.1 through 1.4 suggest that the polynomials P_n more closely approximate the function f near $x = a$ as the degree of P_n increases. We will verify this observation for particular choices of the function f. In Section 13.2 we will take up the question of how well the polynomial P_n approximates f. The remaining part of this section concerns finding P_n for various choices of f and a.

Example 1

Find $P_0, P_1, P_2,$ and P_3 for the function $f(x) = e^x$ at $a = 0$ and use these polynomials to approximate $e = f(1)$.

Solution: The values of $f(x) = e^x$ and the first three derivatives at $a = 0$ are as follows:

$$f(x) = e^x; \qquad f(0) = e^0 = 1$$
$$f'(x) = e^x; \qquad f'(0) = 1$$
$$f''(x) = e^x; \qquad f''(0) = 1$$
$$f'''(x) = e^x; \qquad f'''(0) = 1.$$

From this information and equation (6) we obtain

$$P_0(x) = 1$$
$$P_1(x) = 1 + x$$
$$P_2(x) = 1 + x + \frac{x^2}{2}$$
$$P_3(x) = 1 + x + \frac{x^2}{2} + \frac{x^3}{3!}.$$

Using these polynomials we may approximate the value $f(1) = e^1 = e$ by the numbers

$$P_0(1) = 1$$
$$P_1(1) = 1 + 1 = 2$$

---

*There is a slight abuse of terminology here. Since it may be the case that $f^{(n)}(a) = 0$, an nth Taylor polynomial as defined above may actually be a polynomial of lower degree (see Example 2).

$$P_2(1) = 1 + 1 + \frac{1^2}{2} = \frac{5}{2} = 2.5$$

$$P_3(1) = 1 + 1 + \frac{1^2}{2} + \frac{1^2}{6} = \frac{8}{3} = 2.666\overline{6} \ldots$$

Figure 1.5 shows graphs of these Taylor polynomials and the approximations $P_n(1)$ to the value $f(1) = e^1 = e$. Table 1.1 compares, to four decimal place accu-

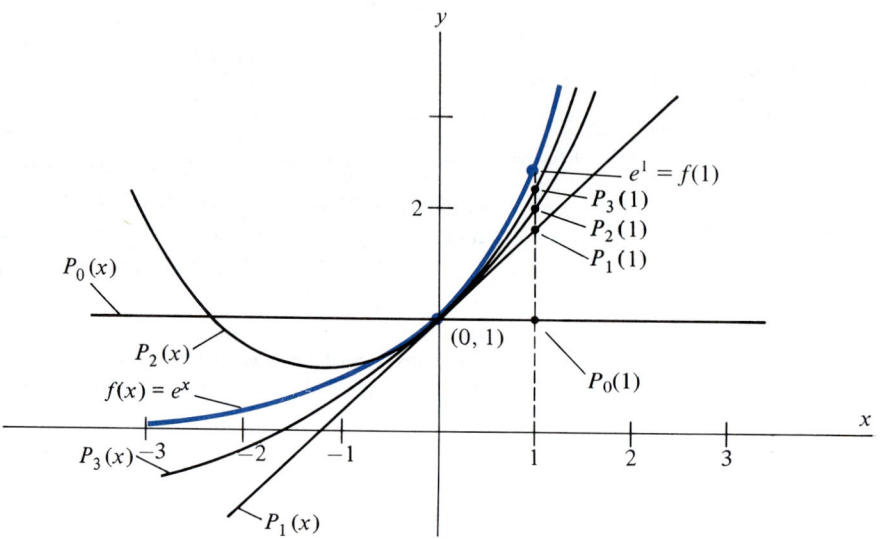

Figure 1.5 Taylor polynomials for $f(x) = e^x$ expanded about $a = 0$ and approximations $P_n(1)$ to $e^1 = e$.

racy, values of these Taylor polynomials and the function $f(x) = e^x$ for several different numbers x.

Table 1.1 Values of Taylor approximations to e^x near $a = 0$

x	$P_0(x)$	$P_1(x)$	$P_2(x)$	$P_3(x)$	e^x
-1.5	1.0	-0.5	0.6250	0.0625	0.2231
-1.0	1.0	0	0.5	0.3333	0.3679
0	1.0	1.0	1.0	1.0	1.0
0.5	1.0	1.5	1.625	1.6458	1.6487
1.0	1.0	2.0	2.5	2.6667	2.7183
1.5	1.0	2.5	3.6250	4.1875	4.4817

Example 2

Find the Taylor polynomials $P_0, P_1, P_2, \ldots, P_5$ for $f(x) = \sin x$ expanded about $a = 0$.

Solution: Here

$$f(x) = \sin x; \qquad f(0) = 0$$
$$f'(x) = \cos x; \qquad f'(0) = 1$$
$$f''(x) = -\sin x; \qquad f''(0) = 0$$
$$f'''(x) = -\cos x; \qquad f'''(0) = -1$$
$$f^{(4)}(x) = \sin x; \qquad f^{(4)}(0) = 0$$
$$f^{(5)}(x) = \cos x; \qquad f^{(5)}(0) = 1.$$

Thus, by (6),

$$P_0(x) = 0$$
$$P_1(x) = 0 + 1 \cdot x = x$$

$$P_2(x) = 0 + 1 \cdot x + \frac{0}{2} \cdot x^2 = x$$

$$P_3(x) = 0 + 1 \cdot x + \frac{0}{2}x^2 + \frac{-1}{3!}x^3 = x - \frac{x^3}{3!}$$

$$P_4(x) = 0 + 1 \cdot x + \frac{0}{2}x^2 + \frac{-1}{3!}x^3 + \frac{0}{4!}x^4 = x - \frac{x^3}{3!}$$

$$P_5(x) = 0 + 1 \cdot x + \frac{0}{2}x^2 + \frac{-1}{3!}x^3 + \frac{0}{4!}x^4 + \frac{1}{5!}x^5 = x - \frac{x^3}{3!} + \frac{x^5}{5!}.$$

(See Figure 1.6.)   ◇

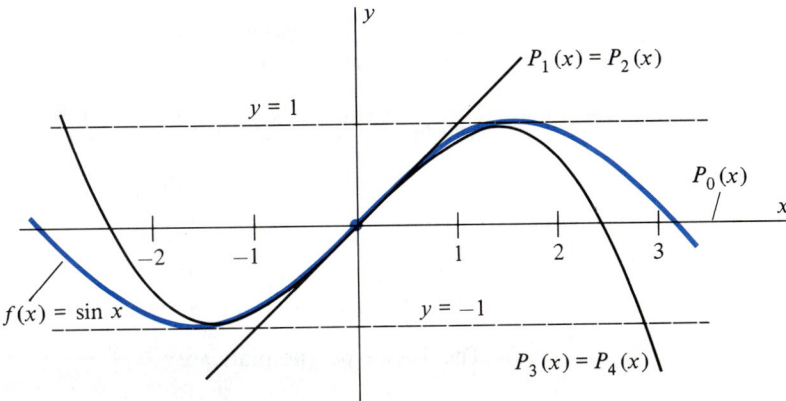

Figure 1.6 Five Taylor polynomials for $f(x) = \sin x$ at $a = 0$. Note that $P_1(x)$ and $P_2(x)$ coincide, as do $P_3(x)$ and $P_4(x)$.

Example 3

In Example 2 the even powers of x in the Taylor polynomials for $f(x) = \sin x$ at $a = 0$ dropped out, since even derivatives of $\sin x$ are zero at $a = 0$. For comparison, we next determine the fourth Taylor polynomial for $f(x) = \sin x$ expanded about $a = \pi/6$.

$$f(x) = \sin x; \qquad f(\pi/6) = \frac{1}{2}$$

$$f'(x) = \cos x; \qquad f'(\pi/6) = \frac{\sqrt{3}}{2}$$

$$f''(x) = -\sin x; \qquad f''(\pi/6) = -\frac{1}{2}$$

$$f'''(x) = -\cos x; \qquad f'''(\pi/6) = -\frac{\sqrt{3}}{2}$$

$$f^{(4)}(x) = \sin x; \qquad f^{(4)}(\pi/6) = \frac{1}{2}.$$

By (6),

$$
P_4(x) = \frac{1}{2} + \frac{\sqrt{3}}{2}\left(x - \frac{\pi}{6}\right) + \left(\frac{1}{2}\right)\left(-\frac{1}{2}\right)\left(x - \frac{\pi}{6}\right)^2
$$

$$
+ \left(\frac{1}{3!}\right)\left(-\frac{\sqrt{3}}{2}\right)\left(x - \frac{\pi}{6}\right)^3 + \left(\frac{1}{4!}\right)\left(\frac{1}{2}\right)\left(x - \frac{\pi}{6}\right)^4
$$

$$
= \frac{1}{2} + \frac{\sqrt{3}}{2}\left(x - \frac{\pi}{6}\right) - \frac{1}{4}\left(x - \frac{\pi}{6}\right)^2 - \frac{\sqrt{3}}{12}\left(x - \frac{\pi}{6}\right)^3
$$

$$
+ \frac{1}{48}\left(x - \frac{\pi}{6}\right)^4.
$$

$\Diamond$

Example 4

Find the Taylor polynomials $P_0, P_1, \ldots, P_4$ for the function $f(x) = \cos x$ expanded about $a = 0$.

Solution: The required derivatives are as follows:

$$
\begin{aligned}
f(x) &= \cos x; & f(0) &= 1 \\
f'(x) &= -\sin x; & f'(0) &= 0 \\
f''(x) &= -\cos x; & f''(0) &= -1 \\
f'''(x) &= \sin x; & f'''(0) &= 0 \\
f^{(4)}(x) &= \cos x; & f^{(4)}(0) &= 1.
\end{aligned}
$$

The Taylor polynomials are

$$P_0(x) = 1$$
$$P_1(x) = 1 + 0 \cdot x = 1$$

$$P_2(x) = 1 + 0 \cdot x + \frac{-1}{2}x^2 = 1 - \frac{x^2}{2}$$

$$P_3(x) = 1 + 0 \cdot x + \frac{-1}{2}x^2 + \frac{0}{3!}x^3 = 1 - \frac{x^2}{2}$$

$$P_4(x) = 1 + 0 \cdot x + \frac{-1}{2}x^2 + \frac{0}{3!}x^3 + \frac{1}{4!}x^4 = 1 - \frac{x^2}{2} + \frac{x^4}{4!}.$$

$\Diamond$

Example 5

Find the Taylor polynomials $P_0, P_1, \ldots, P_5$ expanded about $a = 1$ for the function $f(x) = \ln x$ and use these polynomials to approximate $\ln 2$.

Solution: We have

$$f(x) = \ln x; \qquad f(1) = \ln 1 = 0$$

$$f'(x) = \frac{1}{x} = x^{-1}; \qquad f'(1) = 1^{-1} = 1$$

$$f''(x) = -x^{-2}; \qquad f''(1) = -1^{-2} = -1$$
$$f'''(x) = 2x^{-3}; \qquad f'''(1) = 2 \cdot 1^{-3} = 2$$
$$f^{(4)}(x) = -6x^{-4}; \qquad f^{(4)}(1) = -6 \cdot 1^{-4} = -6$$
$$f^{(5)}(x) = 24x^{-5}; \qquad f^{(5)}(1) = 24 \cdot 1^{-5} = 24.$$

Thus,

$$P_0(x) = 0$$
$$P_1(x) = 0 + 1 \cdot (x - 1) = x - 1$$

$$P_2(x) = 0 + 1 \cdot (x - 1) + \frac{-1}{2}(x - 1)^2 = (x - 1) - \frac{1}{2}(x - 1)^2$$

$$P_3(x) = 0 + 1 \cdot (x - 1) + \frac{-1}{2}(x - 1)^2 + \frac{2}{3!}(x - 1)^3$$

$$= (x - 1) - \frac{1}{2}(x - 1)^2 + \frac{1}{3}(x - 1)^3$$

$$P_4(x) = 0 + 1 \cdot (x - 1) + \frac{-1}{2}(x - 1)^2 + \frac{2}{3!}(x - 1)^3 + \frac{-6}{4!}(x - 1)^4$$

$$= (x - 1) - \frac{1}{2}(x - 1)^2 + \frac{1}{3}(x - 1)^3 - \frac{1}{4}(x - 1)^4$$

and

$$P_5(x) = 0 + 1 \cdot (x - 1) + \frac{-1}{2}(x - 1)^2 + \frac{2}{3!}(x - 1)^3 + \frac{-6}{4!}(x - 1)^4$$

$$+ \frac{24}{5!}(x - 1)^5$$

$$= (x - 1) - \frac{1}{2}(x - 1)^2 + \frac{1}{3}(x - 1)^3 - \frac{1}{4}(x - 1)^4 + \frac{1}{5}(x - 1)^5.$$

Since $\ln 2 = f(2)$, we obtain approximations to $\ln 2$ from these polynomials by setting $x = 2$:

$$P_0(2) = 0$$
$$P_1(2) = (2 - 1) = 1$$

$$P_2(2) = (2 - 1) - \frac{1}{2}(2 - 1)^2 = 1 - \frac{1}{2} = \frac{1}{2} = 0.5$$

$$P_3(2) = (2 - 1) - \frac{1}{2}(2 - 1)^2 + \frac{1}{3}(2 - 1)^3 = 1 - \frac{1}{2} + \frac{1}{3} = \frac{5}{6} \approx 0.8333$$

$$P_4(2) = (2 - 1) - \frac{1}{2}(2 - 1)^2 + \frac{1}{3}(2 - 1)^3 - \frac{1}{4}(2 - 1)^4$$

$$= 1 - \frac{1}{2} + \frac{1}{3} - \frac{1}{4} = \frac{7}{12} \approx 0.5833$$

$$P_5(2) = (2-1) - \frac{1}{2}(2-1)^2 + \frac{1}{3}(2-1)^3 - \frac{1}{4}(2-1)^4 + \frac{1}{5}(2-1)^5$$

$$= 1 - \frac{1}{2} + \frac{1}{3} - \frac{1}{4} + \frac{1}{5}$$

$$= \frac{47}{60} \approx 0.7833.$$

By comparison, the exact value of ln 2, to four decimal places, is 0.6931. ◇

Exercise Set 13.1

In Exercises 1–16, find the nth Taylor polynomial for the function f expanded about $x = a$.

1. $f(x) = e^{-x}$, $a = 0$, $n = 4$

2. $f(x) = e^x$, $a = 1$, $n = 4$

3. $f(x) = \cos x$, $a = \pi/4$, $n = 6$

4. $f(x) = \sin x$, $a = \pi/3$, $n = 5$

5. $f(x) = \ln(1 + x)$, $a = 0$, $n = 4$

6. $f(x) = \dfrac{1}{1 + x}$, $a = 0$, $n = 4$

7. $f(x) = \tan x$, $a = 0$, $n = 3$

8. $f(x) = \dfrac{1}{1 + x^2}$, $a = 0$, $n = 2$

9. $f(x) = e^{x^2}$, $a = 0$, $n = 3$

10. $f(x) = \sqrt{x}$, $a = 2$, $n = 2$

11. $f(x) = \sqrt{1 - x}$, $a = 0$, $n = 3$

12. $f(x) = x \sin x$, $a = 0$, $n = 7$

13. $f(x) = \sec x$, $a = \pi/4$, $n = 3$

14. $f(x) = \operatorname{Tan}^{-1} x$, $a = 0$, $n = 3$

15. $f(x) = \dfrac{1}{1 + e^x}$, $a = 0$, $n = 3$

16. $f(x) = x^4$, $a = 2$, $n = 4$

17. Compare the results of Exercises 5 and 6 and those of Exercises 8 and 14. Formulate a conjecture on the relationship between the Taylor polynomial for f and that for f' and test your conjecture on a few examples.

18. What is the relationship between the Taylor polynomial for f at $a = 0$ and that for $h(x) = xf(x)$? Can you prove this?

19. True or false? If f is an even function, the Taylor polynomial for f expanded about $a = 0$ will contain only even powers of x. Explain.

20. The Taylor polynomial P_n is defined for f if each of the first n derivatives of f exists at $x = a$. Does this imply that $f(x)$ is defined for all x for which $P_n(x)$ is defined? (*Hint:* Consider this question for the function in Example 5.)

13.2 TAYLOR'S THEOREM

If P_n denotes the nth Taylor polynomial for the function f at $x = a$, our claim is that $P_n(x)$ may be used to approximate $f(x)$ for x near a. In order to determine the accuracy of this approximation we write

$$f(x) = P_n(x) + R_n(x). \tag{1}$$

Since $R_n(x) = f(x) - P_n(x)$, the term $R_n(x)$ represents the *error* made in approximating $f(x)$ by $P_n(x)$ (see Figure 2.1). It is referred to as the **remainder** in a Taylor approximation.

Taylor's Theorem provides a way to estimate the size of $R_n(x)$ when making a Taylor approximation.

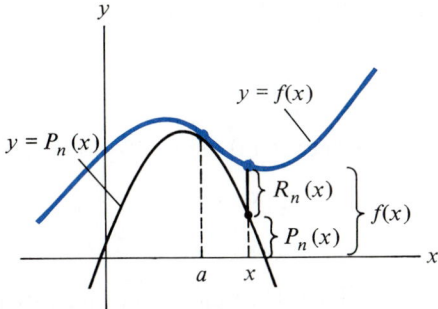

Figure 2.1 $R_n(x) = f(x) - P_n(x)$.

THEOREM 1
Taylor's Theorem

Let n be an integer and let f be a function for which $f^{(n+1)}(x)$ exists for each x in the closed interval $[\alpha, \beta]$. Let a be any number in the open interval (α, β). Then for each $x \in [\alpha, \beta]$ there exists a number c between a and x so that

$$f(x) = f(a) + f'(a)(x - a) + \frac{f''(a)}{2!}(x - a)^2 + \cdots$$

$$+ \frac{f^{(n)}(a)}{n!}(x - a)^n + \frac{f^{(n+1)}(c)}{(n+1)!}(x - a)^{n+1}.$$

(This is called Taylor's formula with remainder, or just Taylor's formula.) In other words,

$$f(x) = P_n(x) + R_n(x)$$

where $P_n(x)$ is the Taylor polynomial of degree n and

$$R_n(x) = \frac{f^{(n+1)}(c)}{(n+1)!}(x - a)^{n+1}. \tag{2}$$

We may paraphrase Taylor's Theorem by saying that the error involved in approximating a value of an $(n + 1)$ times differentiable function by the value of its Taylor polynomial $P_n(x)$ is just the term we would add to $P_n(x)$ to obtain $P_{n+1}(x)$, except that the coefficient $\dfrac{f^{(n+1)}(c)}{(n+1)!}$ is evaluated at a specific number *between* a and x rather than at $x = a$.

We will see in the next section how to determine a bound for $R_n(x)$.

Example 1

Write Taylor's formula using the third Taylor polynomial for $f(x) = e^x$ expanded about $a = 0$.

Solution: From Example 1 of Section 13.1 we have

$$P_3(x) = 1 + x + \frac{x^2}{2!} + \frac{x^3}{3!}.$$

For $f(x) = e^x$, $f^{(4)}(x) = e^x$, so $f^{(4)}(c) = e^c$. Thus, by Taylor's Theorem,

$$R_3(x) = \frac{e^c}{4!}(x - 0)^4$$

where c lies between 0 and x. Taylor's formula is therefore

$$e^x = 1 + x + \frac{x^2}{2} + \frac{x^3}{3!} + \frac{e^c}{4!}x^4.$$

◇

Since Taylor's Theorem is both important and useful, it is unfortunate that no easily motivated proof has been found. We present one of the standard proofs in the remainder of this section.

Proof of Theorem 1: In addition to the number a, choose a number $b \in [\alpha, \beta]$. Then we can define a number C (which depends on a, b, and n) by the equation

$$f(b) = f(a) + f'(a)(b - a) + \cdots + \frac{f^{(n)}(a)}{n!}(b - a)^n + C. \tag{3}$$

That is, C is the difference $C = f(b) - P_n(b)$, where $P_n(b)$ is the Taylor polynomial of degree n expanded about a.

First note that if $b = a$, then (3) collapses to $f(a) = f(a)$, and $C = 0$. Thus, we consider $b \neq a$ (it makes no difference whether $a < b$ or $b < a$) and attempt to prove that C has the form of remainder $R_n(x)$ given in the theorem.

In order to do this, we construct a rather odd-looking function of x, whose purpose is to allow us to apply Rolle's Theorem on the interval between a and b:

$$F(x) = f(b) - \left\{ f(x) + f'(x)(b - x) + \cdots + \frac{f^{(n)}(x)}{n!}(b - x)^n \right.$$
$$\left. + C\frac{(b - x)^{n+1}}{(b - a)^{n+1}} \right\}. \tag{4}$$

Now we show the following facts about F:

(i) If $x = a$, then

$$F(a) = f(b) - \left\{ f(a) + f'(a)(b - a) + \cdots \right.$$
$$\left. + \frac{f^{(n)}(a)}{n!}(b - a)^n + C\frac{(b - a)^{n+1}}{(b - a)^{n+1}} \right\}$$
$$= 0 \quad \text{(by (3))}.$$

(ii) If $x = b$, then

$$F(b) = f(b) - \left\{ f(b) + f'(b)(b - b) + \cdots \right.$$
$$\left. + \frac{f^{(n)}(b)}{n!}(b - b)^n + C\frac{(b - b)^{n+1}}{(b - a)^{n+1}} \right\}$$
$$= f(b) - f(b) = 0.$$

(iii) Because $f^{(n+1)}(x)$ exists for each $x \in [\alpha, \beta]$, F is differentiable on the entire interval, and

$$F'(x) = -\frac{f^{(n+1)}(x)}{n!}(b - x)^n + (n + 1)C\frac{(b - x)^n}{(b - a)^{n+1}}.$$

To see that this is true, we simply differentiate F in equation (4), using the Product Rule, and note that many terms cancel:

$$F'(x) = 0 - \left\{ f'(x) + f''(x)(b - x) - f'(x) \right.$$
$$+ \frac{f'''(x)}{2}(b - x)^2 - f''(x)(b - x)$$
$$+ \frac{f^{(4)}(x)}{3!}(b - x)^3 - \frac{f'''(x)}{2}(b - x)^2$$
$$+ \cdots + \frac{f^{(n)}(x)}{(n - 1)!}(b - x)^{n-1} - \frac{f^{(n-1)}(x)}{(n - 2)!}(b - x)^{n-2}$$
$$+ \frac{f^{(n+1)}(x)}{n!}(b - x)^n - \frac{f^{(n)}(x)}{(n - 1)!}(b - x)^{n-1} - (n + 1)C\frac{(b - x)^n}{(b - a)^{n+1}} \left. \right\}$$
$$= -\frac{f^{(n+1)}(x)}{n!}(b - x)^n + (n + 1)C\frac{(b - x)^n}{(b - a)^{n+1}}.$$

From statements (i) and (ii) and Rolle's Theorem, it follows that there is a number c between a and b such that $F'(c) = 0$. Using statement (iii), we may write this condition as

$$\frac{f^{(n+1)}(c)}{n!}(b - c)^n = (n + 1)C\frac{(b - c)^n}{(b - a)^{n+1}}.$$

Solving for C now gives

$$C = \frac{f^{(n+1)}(c)}{(n + 1)!}(b - a)^{n+1}. \tag{5}$$

Since we have placed no restriction on b other than $b \neq a$ (and we already know that $C = 0$ if $b = a$), we can substitute x for b in (5), which yields exactly $C = R_n(x)$. This completes the proof. ◆

Exercise Set 13.2

In Exercises 1–14, write Taylor's formula (Theorem 1) for f using the nth Taylor polynomial expanded about $x = a$.

1. $f(x) = e^{-x}$, $\quad n = 3$, $\quad a = 0$

2. $f(x) = \sin x$, $\quad n = 3$, $\quad a = 0$

3. $f(x) = \cos x$, $\quad n = 4$, $\quad a = 0$

4. $f(x) = \sin x$, $\quad n = 3$, $\quad a = \dfrac{\pi}{4}$

5. $f(x) = \text{Tan}^{-1} x$, $\quad n = 3$, $\quad a = 0$

6. $f(x) = \sqrt{x}$, $\quad n = 3$, $\quad a = 4$

7. $f(x) = \dfrac{1}{1 + x^2}$, $\quad n = 2$, $\quad a = 1$

8. $f(x) = 3x^4 + 2x + 2$, $\quad n = 3$, $\quad a = 2$

9. $f(x) = \sec x$, $\quad n = 2$, $\quad a = \dfrac{\pi}{4}$

10. $f(x) = \ln(1 + x^2)$, $\quad n = 3$, $\quad a = 0$

11. $f(x) = \sinh x$, $\quad n = 4$, $\quad a = 0$

12. $f(x) = x \sinh x$, $\quad n = 3$, $\quad a = 0$

13. $f(x) = \cosh x$, $\quad n = 3$, $\quad a = \ln 2$

14. $f(x) = (1 + x)^{3/2}$, $\quad n = 3$, $\quad a = 0$

15. Let f be a polynomial of degree n. Use Theorem 1 to prove that $f = P_n$.

16. Refer to the precise hypotheses of Rolle's Theorem in Chapter 4. Then show that the hypothesis of Theorem 1 that $f^{(n+1)}(x)$ exists for each x in $[\alpha, \beta]$ can be relaxed to the condition that $f^{(n)}$ is continuous on $[\alpha, \beta]$ and $f^{n+1}(x)$ exists for each x in (α, β).

17. Use Taylor's Theorem to justify Newton's Method for approximating the root of a function.

18. Let P_n be the nth Taylor polynomial for $f(x) = \sin x$ expanded about $a = 0$. Show that $\lim_{n \to \infty} R_n(x) = 0$ if $|x| < 1$.

19. By multiplying both sides of the equation by $1 - x$, show that

$$\frac{1}{1-x} = 1 + x + x^2 + x^3 + \cdots$$

$$+ x^n + \frac{x^{n+1}}{1-x}, \qquad x \neq 1.$$

Conclude that $R_n(x) = \dfrac{x^{n+1}}{1-x}$. Does this contradict Taylor's Theorem? Use Taylor's Theorem to verify that, for $f(x) = \dfrac{1}{1-x}$ and $a = 0$, the nth Taylor polynomial for $f(x) = \dfrac{1}{1-x}$ is

$$P_n(x) = 1 + x + x^2 + \cdots + x^n.$$

20. By substituting $-x$ for x in the equation in Exercise 19 conclude that

$$\frac{1}{1+x} = 1 - x + x^2 - x^3 + \cdots$$

$$+ (-1)^n x^n + \frac{(-1)^{n+1} x^{n+1}}{1+x}, \qquad x \neq -1.$$

Using Taylor's Theorem with $a = 0$, verify that the nth Taylor polynomial for $f(x) = \dfrac{1}{1+x}$ is

$$P_n(x) = 1 - x + x^2 - x^3 + \cdots + (-1)^n x^n.$$

21. Replace x by x^2 in Exercise 20 to conclude that

$$\frac{1}{1+x^2} = 1 - x^2 + x^4 - \cdots$$

$$+ (-1)^n x^{2n} + \frac{(-1)^{n+1} x^{2n+2}}{1+x^2}.$$

What is the nth Taylor Polynomial P_n, expanded about $a = 0$, for $f(x) = \dfrac{1}{1+x^2}$?

22. Replace x by t in Exercise 20 and integrate between 0 and x to conclude that

$$\int_0^x \frac{1}{1+t} \, dt = \int_0^x \left(1 - t + t^2 - t^3 + \cdots \right.$$

$$\left. + (-1)^{n-1} t^{n-1} + \frac{(-1)^n t^n}{1+t} \right) dt$$

$$= x - \frac{x^2}{2} + \frac{x^3}{3} - \frac{x^4}{4} + \cdots + \frac{(-1)^{n-1} x^n}{n}$$

$$+ (-1)^n \int_0^x \frac{t^n}{1+t} \, dt.$$

What is the nth Taylor polynomial, expanded about $a = 0$, for $f(x) = \ln(1 + x)$?

23. Let P_n be as in Exercise 18. Show that $\lim_{n \to \infty} R_n(x) = 0$ for *all* x. $\left(Hint: \text{ You must first show that } \lim_{n \to \infty} \dfrac{x^n}{n!} = 0. \right)$

24. Use Taylor's Theorem to prove the Binomial Theorem:

$$(x + a)^n = a^n + na^{n-1}x + \frac{n(n-1)}{2}a^{n-2}x^2 + \cdots$$

$$+ \binom{n}{r} a^{n-r} x^r + \cdots + nax^{n-1} + x^n$$

where n is a positive integer and $\binom{n}{r}$ is the **binomial coefficient**

$$\binom{n}{r} = \frac{n!}{r!(n-r)!}.$$

13.3 APPLICATIONS OF TAYLOR'S THEOREM

According to Taylor's Theorem, if we approximate $f(x)$ by the value of the nth Taylor polynomial $P_n(x)$ expanded about $x = a$, the error (remainder) in the calculation is

$$R_n(x) = \frac{f^{(n+1)}(c)}{(n+1)!}(x - a)^{n+1} \qquad (1)$$

where c lies between a and x. Since x is given, the choices that influence the size of the error involve the numbers a and n. Obviously, we will want to choose a close to

x so that the factors $(x - a)$ in $R_n(x)$ are small. Also, we will want to choose a so that the function value $f(a)$ and the various derivatives $f^{(k)}(a)$ are easy to compute.

Once a convenient value for a is chosen, the magnitude of the error depends only on the integer n. Since we seldom know the number c, we usually cannot determine $R_n(x)$ precisely. However, in most applications we are concerned only with knowing the approximate size of $R_n(x)$. For example, if we are to approximate $\ln(1.2)$ to within 0.01, we need only to be able to demonstrate that $|R_n(1.2)| < 0.01$.

In general, if we wish to achieve a Taylor approximation for $f(x)$ to within ϵ, the objective will be to find n sufficiently large to guarantee that the inequality

$$\left| \frac{f^{(n+1)}(c)}{(n + 1)!}(x - a)^{n+1} \right| < \epsilon$$

holds. By (1) this will assure the desired accuracy.

Example 1

Use the Taylor polynomial of degree 3 for $f(x) = \ln x$ expanded about $a = 1$ to approximate $\ln(1.5)$ and find the accuracy of this approximation.

Strategy

Find P_3 as in Section 13.2.

Solution

We have

$$f(x) = \ln x; \qquad f(1) = 0$$

$$f'(x) = \frac{1}{x}; \qquad f'(1) = 1$$

$$f''(x) = -\frac{1}{x^2}; \qquad f''(1) = -1$$

$$f'''(x) = \frac{2}{x^3}; \qquad f'''(1) = 2.$$

Thus

$$P_3(x) = 1(x - 1) - \frac{1}{2}(x - 1)^2 + \frac{2}{3!}(x - 1)^3.$$

Evaluate P_3 at $x = 1.5$ to obtain approximation $P_3(1.5)$.

The approximation is therefore

$$\ln(1.5) \approx P_3(1.5) = (.5) - \frac{1}{2}(.5)^2 + \frac{1}{3}(.5)^3$$

$$= .416\overline{6}.$$

Since $f^{(4)}(x) = -6x^{-4}$ for $f(x) = \ln x$, by (1) the error in the approximation is

Write expression for $|R_3(1.5)|$ using (1).

$$|R_3(1.5)| = \left| \frac{-6c^{-4}}{4!}(1.5 - 1)^4 \right| = \left| \frac{-6c^{-4}}{24}(.5)^4 \right|$$

$$= |c^{-4}| \cdot \frac{(.5)^4}{4}.$$

Since c is unknown, we must determine the largest possible value for $|f^{(4)}(c)|$ to obtain an upper bound on the error. We know that c is between $a = 1$ and $x = 1.5$.

Since $|c^{-4}| < 1$ for $1 < c < 1.5$, we have the inequality

$$|R_3(1.5)| < \frac{(.5)^4}{4} = .015625.$$

We may conclude only that the approximation is accurate to one decimal place. ◇

REMARK: When we say that a number is accurate to k decimal places, we mean that the error is less than $5 \times 10^{-(k+1)}$. That is, accuracy to one decimal place means an error less than $5 \times 10^{-2} = .05$, accuracy to two decimal places means an error less than .005, and so forth. Thus we could claim that the approximation in Example 1 was accurate to one decimal place, and the approximation in Example 2 (below) is accurate to three decimal places.

Example 2

Suppose we wish to use the approximation

$$\sin x \approx x$$

for numbers satisfying the inequality $|x| < \pi/45$. What is the maximum possible error associated with this approximation?

Strategy

Determine the Taylor polynomial $P_n(x)$ associated with the approximation.

Once $f(x)$, n, and a are known, use equation (1).

Estimate the maximum possible size of $|f''(c)|$. Use to obtain bound on $|R_2(x)|$.

Solution

For $f(x) = \sin x$, both $P_1(x) = x$ and $P_2(x) = x$ when $a = 0$. We may therefore take $n = 2$ as corresponding to the above approximation. Since $f'''(c) = -\cos(c)$, the error is

$$|R_2(x)| = \left| \frac{-\cos(c)}{3!}(x - 0)^3 \right|$$

$$= \frac{|\cos(c)|}{3!}|x|^3.$$

Since $|\cos(c)| \leq 1$ for all numbers c and since $|x| < \pi/45$, we have

$$|R_2(x)| \leq \frac{1}{3!}|x|^3 < \frac{1}{3!}\left(\frac{\pi}{45}\right)^3 < .00006.$$

The maximum error in the approximation is $.00006 = 6 \times 10^{-5}$. Note that if we had used $n = 1$ in these calculations, we would have had $|f''(c)| = |-\sin(c)| \leq 1$ and we could have claimed only that

$$|R_1(x)| \leq \frac{1}{2!}|x|^2 < \frac{1}{2}\left(\frac{\pi}{45}\right)^2 < .0025.$$
◇

Example 3

How large must n be chosen so that $\cos 48°$ is approximated with four decimal place accuracy using a Taylor polynomial for $f(x) = \cos x$ expanded about $a = 45° = \pi/4$?

Strategy

Set up the expression for $|R_n(x)|$ using

$$x = 48° = \frac{4\pi}{15}.$$

Find an upper bound for $|R_n(x)|$.

Solution

For $f(x) = \cos x$, $f^{(n+1)}(c)$ is either $\pm \sin c$ or $\pm \cos c$. In either case, $|f^{(n+1)}(c)| \leq 1$. Using this inequality we obtain

$$\left| R_n\left(\frac{4\pi}{15}\right) \right| = \left| \frac{f^{(n+1)}(c)}{(n+1)!}\left(\frac{4\pi}{15} - \frac{\pi}{4}\right)^{n+1} \right| \tag{2}$$

$$< \frac{1}{(n+1)!} \cdot \left(\frac{\pi}{60}\right)^{n+1}$$

Require that the upper bound be less than the desired degree of accuracy.

To ensure that $\left| R_n\left(\dfrac{4\pi}{15}\right) \right| < 5 \times 10^{-5}$, we need to find n large enough so that

$$\frac{1}{(n+1)!} \cdot \left(\frac{\pi}{60}\right)^{n+1} < 5 \times 10^{-5}. \tag{3}$$

We now proceed simply by trial and error to find n sufficiently large that (3) holds. Values of the left side of inequality (3) for various integers n are as follows

n	$\dfrac{1}{(n+1)!}\left(\dfrac{\pi}{60}\right)^{n+1}$
1	.0013708
2	.0000240
3	.0000003

Thus $n = 2$ is sufficient to give the desired accuracy. ◇

Example 4

Find a bound on the magnitude of $|x|$ so that the approximation

$$e^x \approx P_3(x) = 1 + x + \frac{x^2}{2!} + \frac{x^3}{3!}$$

is accurate to within .001.

Strategy

Set up the expression for $|R_3(x)|$ using (1).

Find an expression for the maximum size of the factor e^c.

Since e^x is what is being approximated, the factor $e^{|x|}$ must be replaced by a "safe" upper bound.

Solve the resulting inequality for x^n.

By trial and error (or by extracting the fourth root), find a value of x satisfying the inequality.

Solution

First note that the form of the given approximation is a Taylor polynomial expanded about $a = 0$, and that $f^{(n)}(a) = e^0 = 1$ for all n.

Since $f^{(n)}(c) = e^c$ for all n, we have

$$|R_3(x)| = \left| \frac{e^c}{4!} x^4 \right| = \frac{e^c}{24} x^4.$$

Now since e^c is an increasing function and c lies between 0 and x, the maximum value of e^c for $-|x| < c < |x|$ is $e^{|x|}$. Thus

$$|R_3(x)| \le \frac{e^{|x|}}{24} x^4. \tag{4}$$

For $|R_3(x)|$ in (4) to be less than .001, we will clearly need to take $|x| < 1$. Thus we may safely bound the term $e^{|x|}$ by, say, 4, since $e^1 \approx 2.718 < 4.$* We therefore need to find the maximum value of x for which

$$\frac{4}{24} x^4 < .001, \qquad \text{or} \qquad x^4 < .006.$$

Since $(.25)^4 < .0039 < .006$, the bound $|x| < .25$ will assure the desired accuracy. ◇

The exercise set of this section deals with calculations similar to those of Examples 1–4. Before leaving this topic, however, we want to remark on two generalizations suggested by Example 3.

Since $\dfrac{\pi}{60} < 1$, we may write inequality (2) of Example 3 as

$$\left| R_n\left(\frac{4\pi}{15}\right) \right| < \frac{1}{(n+1)!}\left(\frac{\pi}{60}\right)^{n+1} < \frac{1}{(n+1)!}.$$

This shows that $R_n\left(\dfrac{4\pi}{15}\right) \to 0$ as $n \to \infty$. In other words, cos 48° can be approximated to any desired degree of accuracy by simply taking n sufficiently large. Since increasing n corresponds to taking polynomials of higher degree, this means that the *sequence* of estimates.

$$P_0\left(\frac{4\pi}{15}\right) = \frac{\sqrt{2}}{2}$$

$$P_1\left(\frac{4\pi}{15}\right) = \frac{\sqrt{2}}{2} - \frac{\sqrt{2}}{2}\left(\frac{\pi}{60}\right)$$

$$P_2\left(\frac{4\pi}{15}\right) = \frac{\sqrt{2}}{2} - \frac{\sqrt{2}}{2}\left(\frac{\pi}{60}\right) - \frac{\sqrt{2}}{4}\left(\frac{\pi}{60}\right)^2$$

$$P_3\left(\frac{4\pi}{15}\right) = \frac{\sqrt{2}}{2} - \frac{\sqrt{2}}{2}\left(\frac{\pi}{60}\right) - \frac{\sqrt{2}}{4}\left(\frac{\pi}{60}\right)^2 + \frac{\sqrt{2}}{12}\left(\frac{\pi}{60}\right)^3$$

$$\vdots$$

approaches the number $\cos\left(\dfrac{4\pi}{15}\right)$. In the language of Chapter 12 we will say that

(i) the **sequence** of approximations $\left\{P_0\left(\dfrac{4\pi}{15}\right),\ P_1\left(\dfrac{4\pi}{15}\right),\ P_2\left(\dfrac{4\pi}{15}\right), \ldots\right\}$ approaches $\cos\left(\dfrac{4\pi}{15}\right)$ **in the limit,** that is,

$$\lim_{n\to\infty} P_n\left(\frac{4\pi}{15}\right) = \cos\left(\frac{4\pi}{15}\right);$$

(ii) the **series** of terms

$$\sum_{j=0}^{n} \frac{f^{(j)}(\pi/4)}{j!}\left(\frac{\pi}{60}\right)^j$$

$$= \frac{\sqrt{2}}{2} - \frac{\sqrt{2}}{2}\left(\frac{\pi}{60}\right) - \frac{\sqrt{2}}{4}\left(\frac{\pi}{60}\right)^2 + \cdots + \frac{f^{(n)}\left(\frac{\pi}{4}\right)}{n!}\left(\frac{\pi}{60}\right)^n$$

approaches $\cos\left(\dfrac{4\pi}{15}\right)$ **in the limit** as $n \to \infty$, that is,

$$\sum_{j=0}^{\infty} \frac{f^{(j)}\left(\frac{4\pi}{15}\right)}{j!}\left(\frac{\pi}{60}\right)^j = \cos\left(\frac{4\pi}{15}\right).$$

These observations give a glimpse of what is ahead: the limit

$$\lim_{n \to \infty} P_n(x) = \sum_{k=0}^{\infty} \frac{f^{(k)}(a)}{k!}(x - a)^k$$

will be viewed as an infinite series containing powers of the variable $x - a$ which, when it converges, will "represent" the function f. Before developing this notion of *Taylor series* we discuss the more general notion of *power series* in Section 13.4.

Exercise Set 13.3

In Exercises 1–8, use Taylor's Theorem to make the indicated approximation and estimate the accuracy using equation (1).

1. $\ln(1.5)$ $f(x) = \ln(x + 1)$, $a = 0$, $n = 3$

2. $\cos 36°$ $f(x) = \cos x$, $a = \pi/4$, $n = 2$

3. $\sqrt{3.91}$ $f(x) = \sqrt{x}$, $a = 4$, $n = 2$

4. $e^{0.2}$ $f(x) = e^x$, $a = 0$, $n = 3$

5. $\cos 1$ $f(x) = \cos x$, $a = \pi/3$, $n = 2$

6. $\text{Sin}^{-1}(0.2)$ $f(x) = \text{Sin}^{-1} x$, $a = 0$, $n = 1$

7. $\sqrt[3]{10}$ $f(x) = \sqrt[3]{x}$, $a = 8$, $n = 2$

8. $\text{Tan}^{-1}\left(\dfrac{1}{2}\right)$ $f(x) = \text{Tan}^{-1} x$, $a = 0$, $n = 2$

In Exercises 9–15, determine a bound on the accuracy of the given approximation for the indicated range of x.

9. $\sin x \approx x$, $|x| < .05$

10. $\sin x \approx x - \dfrac{x^3}{3!}$, $|x| < .15$

11. $\cos x \approx \dfrac{1}{2} - \dfrac{\sqrt{3}}{2}\left(x - \dfrac{\pi}{3}\right)$, $\left|x - \dfrac{\pi}{3}\right| < .05$

12. $\tan x \approx 1 + 2\left(x - \dfrac{\pi}{4}\right)$, $\left|x - \dfrac{\pi}{4}\right| < \dfrac{\pi}{36}$

13. $\sqrt[3]{1 + x} \approx 1 + \dfrac{x}{3}$, $|x| < .025$

14. $\ln x \approx (x - 1) - \dfrac{1}{2}(x - 1)^2 + \dfrac{1}{3}(x - 1)^3$, $|x - 1| < 0.1$

15. $\sqrt{1 + x} \approx 1 + \dfrac{x}{2}$, $0 < x < .02$

In Exercises 16–20, determine how large n must be taken to ensure accuracy to four decimal places in approximating

16. $\sqrt{38}$ using $f(x) = \sqrt{x}$, $a = 36$

17. $\ln 1.3$ using $f(x) = \ln(x + 1)$, $a = 0$

18. $\sin 9°$ using $f(x) = \sin x$, $a = 0$

19. $\cos 42°$ using $f(x) = \cos x$, $a = \pi/4$

20. $e^{0.3}$ using $f(x) = e^x$, $a = 0$.

21. Approximate e correct to four decimal places.

22. (Another way to approximate $\ln x$.)
   a. Find the third Taylor polynomial with $a = 0$ for $f(x) = \ln\left(\dfrac{1 + x}{1 - x}\right)$ including the remainder term.

   b. Find the number x for which $\dfrac{1 + x}{1 - x} = 1.5$.

   c. Find the accuracy in using the polynomial in part (a) to approximate $\ln(1.5)$.

   d. Compare this accuracy with that obtained in Example 1 using the third Taylor polynomial for $f(x) = \ln(1 + x)$.

23. A scientist needing to calculate $\cos x$ for small angles, say $|x| < 6°$, wonders how much accuracy is lost in simply using the approximation $\cos x \approx 1$. What is the answer?

24. Suppose that you needed to make many hand calculations of the function $f(x) = x^5 + 3x^3 + 2x + 6$ for numbers x between 0.8 and 1.0. Explain how you might obtain the approximation

$$f(x) \approx 12 + 16(x - 1).$$

What is the accuracy to be expected from such approximations? What if, instead, you use the approximation

$$f(x) \approx 12 + 16(x - 1) + 19(x - 1)^2?$$

25. Use Taylor's Theorem to prove the second derivative test for relative extrema.

26. Show that $\sin x = x - \dfrac{x^3}{3!} + R_4(x)$ where $|R_4(x)| \le \dfrac{|x|^5}{5!}$. Use this to show, assuming that $R_4(x)$ is continuous, that

$$\int_0^1 \sin x \, dx = \int_0^1 \left(x - \dfrac{x^3}{3!} + R_4(x)\right) dx$$

$$= \dfrac{11}{24} + \int_0^1 R_4(x) \, dx.$$

Conclude that the number $\dfrac{11}{24}$ provides an approximation to the integral

$$\int_0^1 \sin x \, dx$$

with an error E no greater than

$$E = \left| \int_0^1 \sin x \, dx - \frac{11}{24} \right| = \left| \int_0^1 R_4(x) \, dx \right|$$

$$\leq \int_0^1 |R_4(x)| \, dx$$

$$\leq \int_0^1 \frac{x^5}{5!} \, dx$$

$$\leq \frac{1}{6!} \approx 0.0014.$$

In Exercises 27–30, use the method of Exercise 26 to obtain an approximation accurate to two decimal places for the given integral.

27. $\displaystyle\int_0^1 \sin x^2 \, dx$

28. $\displaystyle\int_0^1 e^{x^2} \, dx$

29. $\displaystyle\int_0^1 \cos x^2 \, dx$

30. $\displaystyle\int_0^1 \frac{\sin x}{x} \, dx$

13.4 POWER SERIES

Because the geometric series $\sum_{k=0}^{\infty} x^k$ converges for $|x| < 1$, it actually represents a *function* with domain $(-1, 1)$. From the formula for its sum, we know that this function has equation $f(x) = \dfrac{1}{1 - x}$. That is,

$$f(x) = \frac{1}{1 - x} = \sum_{k=0}^{\infty} x^k, \qquad -1 < x < 1.$$

In the remaining sections of this chapter we study certain generalizations of the geometric series, called *power series,* to determine when they converge and the properties of the functions they represent. Among these, the Taylor Series will have the special form corresponding to the Taylor Polynomials discussed in Sections 13.1–13.3.

DEFINITION 1

A power series in powers of $x - a$ is an expression of the form

$$\sum_{k=0}^{\infty} a_k(x - a)^k = a_0 + a_1(x - a) + a_2(x - a)^2 + \cdots \qquad (1)$$

$$+ a_k(x - a)^k + \cdots$$

where the coefficients $a_0, a_1, a_2, \ldots$ are constants and x is regarded as an independent variable.[*]

---

[*]Also it should be noted at the outset that any infinite series $\sum a_k(bx - c)^k$ involving powers of a *linear* function $f(x) = bx - c$ is a power series, since the general term may be written

$$a_k(bx - c)^k = a_k \left[b\left(x - \frac{c}{b} \right) \right]^k = b^k a_k \left(x - \frac{c}{b} \right)^k = A_k(x - C)^k$$

with $A_k = b^k a_k$ and $C = c/b$.

Since expression (1) may be simplified by the change of variable $u = x - a$, we will work almost exclusively with power series of the form

$$\sum_{k=0}^{\infty} a_k x^k = a_0 + a_1 x + a_2 x^2 + \cdots + a_k x^k + \cdots. \tag{2}$$

Here are several examples of power series:

$$\sum_{k=0}^{\infty} x^k = 1 + x + x^2 + x^3 + \cdots + x^k \cdots.$$

$$\sum_{k=0}^{\infty} \frac{x^k}{k!} = 1 + x + \frac{x^2}{2!} + \frac{x^3}{3!} + \cdots + \frac{x^k}{k!} + \cdots \qquad (0! = 1). \tag{3}$$

$$\sum_{k=0}^{\infty} \frac{(-1)^k}{1+k} x^k = 1 - \frac{x}{2} + \frac{x^2}{3} + \cdots + \frac{(-1)^k}{1+k} x^k + \cdots. \tag{4}$$

The techniques of Chapter 12 are available to us for determining the numbers x for which such series converge. For example, if we test the series in equation (3) for absolute convergence using the Ratio Test, we find that

$$\rho = \lim_{k \to \infty} \frac{\left| \frac{x^{k+1}}{(k+1)!} \right|}{\left| \frac{x^k}{k!} \right|} = \lim_{k \to \infty} \frac{|x|}{k+1} = 0, \qquad x \neq 0,$$

for all $x \neq 0$, so the series in equation (3) converges for all x.

To determine the numbers x for which the series in equation (4) converges, we again apply the Ratio Test to test for absolute convergence: Since

$$\rho = \lim_{k \to \infty} \frac{\left| \frac{(-1)^{k+1} x^{k+1}}{k+2} \right|}{\left| \frac{(-1)^k x^k}{k+1} \right|} = \lim_{k \to \infty} \left(\frac{k+1}{k+2} \right) |x| = |x|, \qquad x \neq 0, \tag{5}$$

this power series converges absolutely if $|x| < 1$. Also, we can see that if $x = 1$ we obtain the alternating series $\sum_{k=0}^{\infty} \frac{(-1)^k}{1+k}$, which converges, and when $x = -1$ we obtain the harmonic series $\sum_{k=0}^{\infty} \frac{1}{1+k}$, which diverges. However, this leaves the convergence for $|x| > 1$ yet to be determined. Rather than pursue specific examples in this way, we now establish two theorems that tell us the convergence properties of power series in general.

THEOREM 2

(i) If the power series $\sum a_k x^k$ converges for $x = c \neq 0$, then it converges absolutely for all x with $|x| < |c|$.

(ii) If the power series $\sum a_k x^k$ diverges for $x = d$, then it diverges for all x with $|x| > |d|$.

Proof: To prove part (i) we assume that $\Sigma a_k c^k$ converges. It is a necessary condition (Theorem 7, Chapter 12) that $\lim_{k \to \infty} a_k c^k = 0$. Thus, there exists an integer N so that $|a_k c^k| < 1$ whenever $k > N$. Now let x be any number such that $|x| < |c|$, and let $\gamma = \dfrac{|x|}{|c|} < 1$. Then whenever $k > N$ we have

$$|a_k x^k| = \frac{|a_k c^k x^k|}{|c^k|} = |a_k c^k|\left(\frac{|x|}{|c|}\right)^k < \gamma^k.$$

This shows that the series $\displaystyle\sum_{k=N}^{\infty} |a_k x^k|$ is dominated by the convergent geometric series $\displaystyle\sum_{k=N}^{\infty} \gamma^k$. Thus, by the Comparison Test, the series $\displaystyle\sum_{k=N}^{\infty} |a_k x^k|$ converges. The series $\Sigma a_k x^k$ therefore converges absolutely if $|x| < |c|$.

To prove part (ii), we assume that the series $\Sigma a_k d^k$ diverges and we let x be any number so that $|x| > |d|$. Then $\Sigma a_k x^k$ cannot converge, since the convergence of $\Sigma a_k x^k$ would imply the convergence of $\Sigma a_k d^k$, by part (i). Thus $\Sigma a_k x^k$ diverges whenever $|x| > |d|$. This completes the proof. ◆

Theorem 2 gives insight into the geometry of the set of numbers x for which a given power series converges. Given the power series $\Sigma a_k x^k$, let S be this set.

Now suppose that there exists a number $d \notin S$. Then $\Sigma a_k d^k$ diverges, so by Theorem 2, $|x| \leq |d|$ for every $x \in S$. This shows that the set S is bounded if it is not the entire real line, $(-\infty, \infty)$. By the Completeness Axiom there then exists a least upper bound r for the set S. Let us look at two cases:

(i) If $|x| < r$, then $|x|$ is not an upper bound for S, so there exists an element $c \in S$ with $|x| < c$. Since $c \in S$, $\Sigma a_k c^k$ converges. Thus $\Sigma a_k x^k$ converges absolutely, by Theorem 2.
(ii) If $|x| > r$, then $x \notin S$ so $\Sigma a_k x^k$ diverges.

Case (i) shows that S contains the interval $(-r, r)$, and that $\Sigma a_k x^k$ converges absolutely for every $x \in (-r, r)$. Case (ii) shows that the only other possible elements of S are the endpoints r and $-r$. For this reason the number r is called the **radius of convergence** of the power series. The interval $(-r, r)$, $[-r, r)$, $(-r, r]$, or $[-r, r]$ on which $\Sigma a_k x^k$ converges is called the **interval of convergence.** We summarize these findings as follows:

THEOREM 3

Given the power series $\Sigma a_k x^k$, precisely one of the following holds:

(a) The power series converges only for $x = 0$.
(b) There exists a positive number r so that the power series converges absolutely for $|x| < r$ and diverges for $|x| > r$. (The series may or may not converge for $x = \pm r$.)
(c) The power series converges for all x.

(See Figure 4.1).

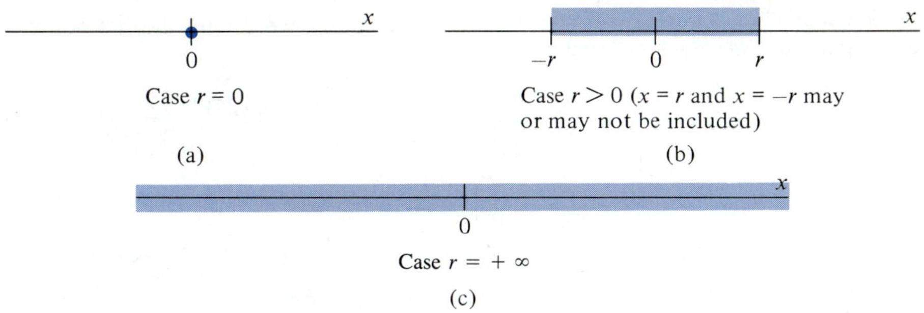

Figure 4.1 The geometry of the interval of convergence for the power series $\Sigma a_k x^k$.

Example 1

(a) The radius of convergence for the geometric power series Σx^k is $r = 1$, and the interval of convergence is $(-1, 1)$.

(b) The radius of convergence for the power series $\Sigma \dfrac{x^k}{k!}$ in equation (3) is $r = \infty$, and the interval of convergence is $(-\infty, \infty)$.

(c) To determine the radius of convergence of the power series $\Sigma \dfrac{(-1)^k}{1 + k} x^k$ in equation (4) we test for absolute convergence using the Ratio Test. We found in equation (5) that $\rho = |x|$. Thus, $\rho < 1$ if and only if $|x| < 1$, so $r = 1$ is the radius of convergence, and the power series converges absolutely for $|x| < 1$. We also checked the two values $x = r = 1$ and $x = -r = -1$, finding that the power series converges for $x = 1$ and diverges for $x = -1$. The interval of convergence for the power series $\displaystyle\sum_{k=0}^{\infty} \dfrac{(-1)^k}{1 + k} x^k$ is therefore $(-1, 1]$. $\diamondsuit$

Example 2

Find the radius and interval of convergence for the power series

$$\sum_{k=1}^{\infty} \frac{1}{k2^k} x^k.$$

Strategy

First, find the radius of convergence, using the Ratio Test. (The Root Test will also work well in this example.)

Solution

We test for absolute convergence, using the Ratio Test:

$$\rho = \lim_{k \to \infty} \frac{\left| \dfrac{1}{(k + 1)2^{k+1}} x^{k+1} \right|}{\left| \dfrac{1}{k2^k} \cdot x^k \right|}$$

$$= \lim_{k \to \infty} \left(\frac{k}{k + 1} \right) \cdot \left(\frac{1}{2} \right) \cdot |x| = \frac{1}{2} |x|.$$

Thus $\rho < 1$ if $|x| < 2$. The radius of convergence is therefore $r = 2$.

To determine the interval of convergence check the endpoints $x = r = 2$ and $x = -r = -2$.

If $x = 2$ we obtain the harmonic series

$$\sum_{k=1}^{\infty} \frac{1}{k \cdot 2^k}(2^k) = \sum_{k=1}^{\infty} \frac{1}{k} = 1 + \frac{1}{2} + \frac{1}{3} + \cdots,$$

which diverges. If $x = -2$ we obtain the *alternating* harmonic series

$$\sum_{k=1}^{\infty} \frac{1}{k \cdot 2^k}(-2)^k = \sum_{k=1}^{\infty} \frac{(-1)^k}{k} = -1 + \frac{1}{2} - \frac{1}{3} + \cdots,$$

which converges. The interval of convergence is therefore $[-2, 2)$. ◇

Example 3

Find the interval of convergence for the power series

$$\sum_{k=0}^{\infty} \frac{x^k}{3^k}.$$

Solution: Testing for absolute convergence using the Root Test we find that

$$\rho = \lim_{k \to \infty} \left\{ \left| \frac{x^k}{3^k} \right| \right\}^{1/k} = \lim_{k \to \infty} \frac{|x|}{3} = \frac{|x|}{3}.$$

Thus, $\rho < 1$ if $|x| < 3$. The radius of convergence is $r = 3$. (The Ratio Test will also work well in obtaining this conclusion.)

If $x = 3$, the series becomes

$$\sum_{k=0}^{\infty} \frac{3^k}{3^k} = \sum_{k=0}^{\infty} 1 = 1 + 1 + 1 + \cdots,$$

which diverges. When $x = -3$, we obtain the divergent series

$$\sum_{k=0}^{\infty} \frac{(-3)^k}{3^k} = \sum_{k=0}^{\infty} (-1)^k = 1 - 1 + 1 - 1 + \cdots.$$

The interval of convergence is therefore $(-3, 3)$. ◇

Example 4

The radius of convergence for the power series

$$\sum_{k=0}^{\infty} k!x^k$$

is $r = 0$ since, by the Ratio Test,

$$\rho = \lim_{k \to \infty} \frac{|(k + 1)!x^{k+1}|}{|k!x^k|} = \lim_{k \to \infty} (k + 1)|x| = \begin{cases} 0, & x = 0 \\ \infty, & x \neq 0 \end{cases}.$$

The "interval" of convergence is simply $\{0\}$. ◇

The next example involves a power series of the form $\Sigma a_k(x - a)^k$. As we stated at the outset, we will use the change of variable $u = x - a$ to bring the power series into the form $\Sigma a_k u^k$. Note, however, that the interval of convergence will be centered at $x = a$ rather than at $x = 0$.

Example 5

Find the interval of convergence for the power series

$$\sum_{k=1}^{\infty} \frac{3^k}{k}(2x - 1)^k.$$

Strategy

Use the substitution $u = 2x - 1$ to bring power series to the form of equation (2). (The interval of convergence will therefore be centered about $x = 1/2$.) Apply Ratio Test to find radius of convergence in the variable u.

Solution

Letting $u = 2x - 1$ we obtain the series

$$\sum_{k=1}^{\infty} \frac{3^k}{k}u^k.$$

Testing this series for absolute convergence using the Ratio Test we find that

$$\rho = \lim_{k \to \infty} \frac{\left| \dfrac{3^{k+1} \cdot u^{k+1}}{(k + 1)} \right|}{\left| \dfrac{3^k \cdot u^k}{k} \right|}$$

$$= \lim_{k \to \infty} 3\left(\frac{k}{k + 1} \right) |u| = 3|u|.$$

Rewrite the inequality for $|u|$ in terms of the original variable x.

The series converges absolutely for $|u| < 1/3$ or for $|2x - 1| < 1/3$. This last inequality is equivalent to

$$-\frac{1}{3} < 2x - 1 < \frac{1}{3},$$

or

$$\frac{1}{3} < x < \frac{2}{3}.$$

Test the endpoints individually.

The radius of convergence (about $x = 1/2$) is therefore $r = 1/6$. At the endpoint $x = 2/3$ we obtain the series

$$\sum_{k=1}^{\infty} \frac{3^k}{k}\left(\frac{1}{3} \right)^k = \sum_{k=1}^{\infty} \frac{1}{k},$$

which diverges. At the endpoint $x = 1/3$ we obtain

$$\sum_{k=1}^{\infty} \frac{3^k}{k}\left(-\frac{1}{3} \right)^k = \sum_{k=1}^{\infty} \frac{(-1)^k}{k},$$

which converges. The interval of convergence is therefore $[1/3, 2/3)$. $\diamond$

Exercise Set 13.4

In Exercises 1–35, find the interval of convergence of the given power series.

1. $\displaystyle\sum_{k=0}^{\infty} \frac{x^k}{k+2}$

2. $\displaystyle\sum_{k=1}^{\infty} \frac{x^k}{2k}$

3. $\displaystyle\sum_{k=0}^{\infty} \frac{(-1)^{k+1}}{k!}x^k$

4. $\displaystyle\sum_{k=0}^{\infty} \frac{2^k x^k}{(k+1)!}$

5. $\displaystyle\sum_{k=1}^{\infty} \frac{(k^2+1)}{k!}x^k$

6. $\displaystyle\sum_{k=0}^{\infty} \frac{k \cdot x^k}{2^k}$

7. $\displaystyle\sum_{k=2}^{\infty} \frac{x^k}{\ln k}$

8. $\displaystyle\sum_{k=1}^{\infty} \frac{(-1)^k e^k}{k^2}x^k$

9. $\displaystyle\sum_{k=1}^{\infty} \frac{\cos \pi k}{1+k}x^k$

10. $\displaystyle\sum_{k=1}^{\infty} \frac{(2k+1)!}{2k!}x^{2k}$

11. $\displaystyle\sum_{k=1}^{\infty} \frac{(-1)^k}{k(k+1)}x^k$

12. $\displaystyle\sum_{k=1}^{\infty} k^2 c^k x^k$

13. $\displaystyle\sum_{k=1}^{\infty} \frac{(-1)^k}{k!}(x-3)^k$

14. $\displaystyle\sum_{k=1}^{\infty} \frac{k}{3^k}(x-\pi)^k$

15. $\displaystyle\sum_{k=0}^{\infty} k!(x-1)^k$

16. $\displaystyle\sum_{k=1}^{\infty} \frac{3^k}{k^2}(2x-1)^k$

17. $\displaystyle\sum_{k=2}^{\infty} \frac{(-1)^k}{\ln k}(3x-2)^k$

18. $\displaystyle\sum_{k=2}^{\infty} \frac{1}{(\ln k)^k}(x-1)^k$

19. $\displaystyle\sum_{k=1}^{\infty} \frac{1}{k+2}x^{2k+1}$

20. $\displaystyle\sum_{k=1}^{\infty} \frac{(x+2)^k}{(k+1)3^k}$

21. $\displaystyle\sum_{k=1}^{\infty} \frac{(x-3)^k}{k(k+1)}$

22. $\displaystyle\sum_{k=0}^{\infty} \frac{x^{2k+1}}{\pi^k}$

23. $\displaystyle\sum_{k=1}^{\infty} \frac{k(x-2)^k}{e^k}$

24. $\displaystyle\sum_{k=0}^{\infty} \frac{(2x+5)^k}{\sqrt{2k+8}}$

25. $\displaystyle\sum_{k=2}^{\infty} \frac{k(7x+1)^k}{2^k}$

26. $\displaystyle\sum_{k=1}^{\infty} \frac{(x-2)^k}{3^k \cdot k^2}$

27. $\displaystyle\sum_{k=3}^{\infty} kx^k$

28. $\displaystyle\sum_{k=1}^{\infty} \frac{x^k}{\ln(k+1)}$

29. $\displaystyle\sum_{k=1}^{\infty} \frac{(x-1)^k}{3^k\sqrt{k+1}}$

30. $\displaystyle\sum_{k=0}^{\infty} \frac{(2x-1)^k}{5^k}$

31. $\displaystyle\sum_{k=1}^{\infty} \frac{(k+4)x^k}{(k+1)(k+2)e^k}$

32. $\displaystyle\sum_{k=0}^{\infty} \frac{k^3(x-2)^k}{3^k}$

33. $\displaystyle\sum_{k=0}^{\infty} \frac{(-1)^k x^{2k+1}}{2(k+1)!}$

34. $\displaystyle\sum_{k=0}^{\infty} \frac{(-1)^k x^{2k}}{(2k)!}$

35. $\displaystyle\sum_{k=1}^{\infty} \frac{x^k}{k(k+1)}$

36. Show that the radius of convergence of the power series

$$\sum_{k=1}^{\infty} \frac{1}{k}(2x-3)^k \text{ is } r = 1/2.$$

37. Prove that if the interval of convergence of the power series $\Sigma a_k x^k$ is $[-r, r)$, then the power series is conditionally convergent, but not absolutely convergent, for $x = -r$.

38. Prove that if the power series $\Sigma a_k x^k$ has radius of convergence r, then the power series $\Sigma a_k x^{ck}$ has radius of convergence $r^{1/c}$, $c > 0$.

39. Prove that if the power series $\Sigma a_k x^k$ has radius of convergence r_a and if the power series $\Sigma b_k x^k$ has radius of convergence r_b, then the series $\Sigma(a_k + b_k)x^k$ converges absolutely for $|x| < c$ where c is the smaller of r_a and r_b.

40. Show that the power series $\Sigma a_k(x-a)^k$ always converges at $x = a$.

13.5 DIFFERENTIATION AND INTEGRATION OF POWER SERIES

Within its interval of convergence, a power series represents a perfectly legitimate function of x. Thus, if the power series $\Sigma a_k x^k$ converges on the interval I_a, we may write

$$f(x) = \sum a_k x^k, \qquad x \in I_a.$$

If $\Sigma b_k x^k$ is a second power series with interval of convergence I_b, then $g(x) = \Sigma b_k x^k$ is again a function of x, but g is defined on a (possibly) different interval than is f. However, if the intervals I_a and I_b overlap (that is, if $I = I_a \cap I_b$ is not empty), we may form sums and differences of f and g as follows:

$$(f + g)(x) = f(x) + g(x)$$
$$= \sum a_k x^k + \sum b_k x^k = \sum (a_k + b_k)x^k, \qquad x \in I_a \cap I_b$$
$$(f - g)(x) = f(x) - g(x)$$
$$= \sum a_k x^k - \sum b_k x^k = \sum (a_k - b_k)x^k, \qquad x \in I_a \cap I_b.$$

(The concepts of multiplication and division for power series are more complicated and will not be pursued here.)

Example 1

The formula for the sum of a geometric series shows that the function $f(x) = \dfrac{1}{1 - x}$ may be represented as a power series for $|x| < 1$:

$$\frac{1}{1 - x} = \sum_{k=0}^{\infty} x^k = 1 + x + x^2 + x^3 + \cdots . \tag{1}$$

Multiplying both sides of equation (1) by x shows that

$$\frac{x}{1 - x} = \sum_{k=0}^{\infty} x^{k+1} = x + x^2 + x^3 + x^4 + \cdots , \qquad |x| < 1. \tag{2}$$

Replacing x by $-x$ in (1) gives

$$\frac{1}{1 + x} = \sum_{k=0}^{\infty} (-x)^k = 1 - x + x^2 - x^3 + \cdots , \qquad |x| < 1. \tag{3}$$

Adding equations (1) and (2) (which is allowed because their intervals of convergence are identical) shows that

$$\frac{2}{(1 - x)(1 + x)} = \frac{1}{1 - x} + \frac{1}{1 + x}$$

$$= \sum_{k=0}^{\infty} x^k + \sum_{k=0}^{\infty} (-x)^k = 2 + 2x^2 + 2x^4 + 2x^6 + \cdots$$

$$= 2 \sum_{k=0}^{\infty} x^{2k}, \qquad |x| < 1. \qquad \diamond$$

If you are beginning to suspect that within its interval of convergence the power series

$$f(x) = a_0 + a_1 x + a_2 x^2 + \cdots \tag{4}$$

behaves just like an "infinitely long polynomial," that is precisely the point we are trying to make. In fact, the following theorem shows that we may even differentiate expression (4) term by term, just as for polynomials, obtaining

$$f'(x) = a_1 + 2a_2 x + 3a_3 x^2 + \cdots ,$$

which is the power series representation for the derivative of the function in equation (4). Its proof is given in more advanced courses.

THEOREM 4
Differentiation of Power Series

Suppose that the power series $\Sigma a_k x^k$ has a radius of convergence $r \neq 0$ and that the function f is defined to be its sum:

$$f(x) = \sum_{k=0}^{\infty} a_k x^k = a_0 + a_1 x + a_2 x^2 + a_3 x^3 + \cdots , \qquad |x| < r.$$

Then

(i) the function f is differentiable for $x \in (-r, r)$,

(ii) the power series $\displaystyle\sum_{k=0}^{\infty} k a_k x^{k-1}$ converges absolutely for each $x \in (-r, r)$, and

(iii) $f'(x) = \displaystyle\sum_{k=0}^{\infty} k a_k x^{k-1} = a_1 + 2a_2 x + 3a_3 x^2 + 4a_4 x^3 + \cdots , \qquad |x| < r.$

Example 2

Applying Theorem 4 to the geometric series

$$\frac{1}{1-x} = \sum_{k=0}^{\infty} x^k = 1 + x + x^2 + x^3 + \cdots , \qquad |x| < 1,$$

we conclude that

$$\frac{d}{dx}\left\{ \frac{1}{1-x} \right\} = \frac{1}{(1-x)^2} = \sum_{k=0}^{\infty} k x^{k-1}$$

$$= 1 + 2x + 3x^2 + 4x^3 + \cdots , \qquad |x| < 1. \qquad \diamond$$

Example 3

Find a power series representation for the function

$$f(x) = \frac{x}{(1-x^2)^2} .$$

Solution: Since f does not resemble any of the functions for which we already know the power series representations, we look for another way to attack the problem. Here f is easily integrated, leading to the result $f(x) = \dfrac{1}{2} \dfrac{d}{dx}\left(\dfrac{1}{1-x^2} \right)$. Now we can recognize the expression in parentheses as the sum of the geometric series in x^2, which converges for $|x^2| < 1$. Then we have

$$\frac{1}{1-x^2} = \sum_{k=0}^{\infty} (x^2)^k = \sum_{k=0}^{\infty} x^{2k} = 1 + x^2 + x^4 + x^6 + \cdots , \qquad |x| < 1.$$

We conclude from Theorem 4 that

$$\frac{x}{(1-x^2)^2} = \frac{1}{2}\frac{d}{dx}\left\{\sum_{k=0}^{\infty} x^{2k}\right\} = \frac{1}{2}\sum_{k=0}^{\infty} 2kx^{2k-1} = x + 2x^3 + 3x^5 + \cdots.$$

This series converges absolutely for $|x| < 1$.   ◇

Example 4

The power series

$$\sum_{k=0}^{\infty} \frac{x^k}{k!} = 1 + x + \frac{x^2}{2!} + \frac{x^3}{3!} + \cdots$$

converges for all x, as noted in Example 1, Section 13.4. On differentiating this series we find that

$$\frac{d}{dx}\left\{\sum_{k=0}^{\infty} \frac{x^k}{k!}\right\} = \frac{d}{dx}\left\{1 + x + \frac{x^2}{2!} + \frac{x^3}{3!} + \frac{x^4}{4!} + \cdots\right\}$$

$$= \left\{1 + \frac{2x}{2!} + \frac{3x^2}{3!} + \frac{4x^3}{4!} + \cdots\right\}$$

$$= \left\{1 + x + \frac{x^2}{2!} + \frac{x^3}{3!} + \cdots\right\}$$

$$= \sum_{k=0}^{\infty} \frac{x^k}{k!}.$$

That is, if $f(x) = \sum_{k=0}^{\infty} \dfrac{x^k}{k!}$, then $f'(x) = f(x)$ for all x. Recall that we have previously shown that the only function other than $f(x) \equiv 0$ that satisfies this differential equation is the function $f(x) = Ce^x$. This shows that, for some constant C,

$$Ce^x = \sum_{k=0}^{\infty} \frac{x^k}{k!} = 1 + x + \frac{x^2}{2!} + \frac{x^3}{3!} + \cdots.$$

Setting $x = 0$ shows that $C = 1$, so we conclude that the power series representation for the function e^x is

$$e^x = \sum_{k=0}^{\infty} \frac{x^k}{k!} = 1 + x + \frac{x^2}{2!} + \frac{x^3}{3!} + \cdots$$

and that this representation is valid for all x.   ◇

The statement of Theorem 4 raises an obvious question about integrals of functions represented by power series. The answer is just what you might suspect.

THEOREM 5
Integration of Power Series

Suppose that the power series $\Sigma a_k x^k$ has radius of convergence $r \neq 0$ and that the function f is defined as its sum:

$$f(x) = \sum_{k=0}^{\infty} a_k x^k = a_0 + a_1 x + a_2 x^2 + a_3 x^3 + \cdots, \qquad |x| < r.$$

Then

(i) the power series $\displaystyle\sum_{k=0}^{\infty} \left(\frac{a_k}{k+1}\right) x^{k+1}$ converges absolutely for each $x \in (-r, r)$,

and

(ii) $\displaystyle\int f(x)\, dx = \sum_{k=0}^{\infty} \left(\frac{a_k}{k+1}\right) x^{k+1} + C$

$$= \left\{ a_0 x + \frac{a_1}{2} x^2 + \frac{a_2}{3} x^3 + \frac{a_3}{4} x^4 + \cdots \right\} + C, \qquad |x| < r.$$

In other words, a power series may be integrated term by term within its radius of convergence. The proof of Theorem 5 is left for more advanced courses. Note that, rather than writing a constant of integration for each term, we collect all constants into a single number C.

Example 5

Since $\ln(1 + x) = \displaystyle\int \frac{1}{1 + x}\, dx$, we integrate equation (3) according to Theorem 5.

$$\ln(1 + x) = \sum_{k=0}^{\infty} \left\{ \int (-x)^k\, dx \right\}$$

$$= \sum_{k=0}^{\infty} -\frac{(-x)^{k+1}}{k+1} + C$$

$$= \sum_{k=0}^{\infty} \frac{(-1)^k x^{k+1}}{k+1} + C$$

$$= \left\{ x - \frac{x^2}{2} + \frac{x^3}{3} - \frac{x^4}{4} + \cdots \right\} + C, \qquad |x| < 1.$$

Setting $x = 0$ gives $\ln 1 = 0 = C$, so

$$\ln(1 + x) = \sum_{k=0}^{\infty} -\frac{(-x)^{k+1}}{k+1}$$

$$= x - \frac{x^2}{2} + \frac{x^3}{3} - \frac{x^4}{4} + \cdots, \qquad |x| < 1. \tag{5}$$

◇

REMARK: Equation (5) provides a practical means for calculating values of $\ln a$ for $0 < a < 2$. For example, to approximate $\ln(1.2)$ we set $x = 0.2$ and apply (5) to obtain

$$\ln(1.2) \approx .2 - \frac{(.2)^2}{2} + \frac{(.2)^3}{3} - \frac{(.2)^4}{4} = .182266$$

with an error of less than $\dfrac{.2^5}{5} = .000064$ (Theorem 15, Section 12.7).

Example 6

Replacing x by x^2 in equation (3) shows that

$$\frac{1}{1 + x^2} = \sum_{k=0}^{\infty} (-x^2)^k = 1 - x^2 + x^4 - x^6 + \cdots, \qquad |x| < 1.$$

Since $\displaystyle\int \frac{1}{1 + x^2}\,dx = \text{Tan}^{-1} x + C$, Theorem 5 gives

$$\text{Tan}^{-1} x = \sum_{k=0}^{\infty} \left\{ \int (-x^2)^k\,dx \right\} + C$$

$$= \sum_{k=0}^{\infty} \frac{(-1)^k}{2k + 1} \cdot x^{2k+1} + C$$

$$= \left\{ x - \frac{x^3}{3} + \frac{x^5}{5} - \cdots \right\} + C, \qquad |x| < 1.$$

Setting $x = 0$ gives $\text{Tan}^{-1}(0) = 0 = C$. Thus

$$\text{Tan}^{-1} x = \sum_{k=0}^{\infty} \frac{(-1)^k}{2k + 1} \cdot x^{2k+1} = x - \frac{x^3}{3} + \frac{x^5}{5} - \cdots, \qquad |x| < 1. \qquad \diamondsuit$$

Exercise Set 13.5

In Exercises 1–10, find a power series representation for the given function using equation (1). State the radius of convergence for the power series obtained.

1. $\dfrac{1}{1 - 2x}$

2. $\dfrac{x^2}{1 - x}$

3. $\dfrac{1}{1 + 4x^2}$

4. $\dfrac{1}{1 - 9x^2}$

5. $\dfrac{x}{1 + x^2}$

6. $\dfrac{x}{1 - x^2}$

7. $\dfrac{x - 1}{x + 1}$

8. $\dfrac{x - 1}{1 + x^2}$

9. $\dfrac{1}{1 - x^4}$

10. $\dfrac{x}{4 + x^2}$

In Exercises 11–19, find a power series representation for the given function using Theorem 4. State the radius of convergence.

11. $f(x) = \dfrac{2}{(1 + x)^2}$ $\left(Hint: f(x) = -2\,\dfrac{d}{dx}\left(\dfrac{1}{1 + x}\right). \right)$

12. $f(x) = \dfrac{2}{(1 - x)^3}$ $\left(Hint: f(x) = \dfrac{d^2}{dx^2}\left(\dfrac{1}{1 - x}\right). \right)$

13. $f(x) = \dfrac{x}{(1 + x^2)^2}$

14. $f(x) = \dfrac{1 - x^2}{(1 + x^2)^2}$ $\left(Hint: f(x) = \dfrac{d}{dx}\left(\dfrac{x}{1 + x^2}\right). \right)$

15. $f(x) = \dfrac{1 + x^2}{(1 - x^2)^2}$

16. $f(x) = \dfrac{1}{(1 + 4x)^2}$

17. $f(x) = \dfrac{8x}{(1 + 4x^2)^2}$ (*Hint:* Use Exercise 3.)

18. $f(x) = \dfrac{1 + 2x - x^2}{(1 + x^2)^2}$ (*Hint:* Use Exercise 8.)

19. $f(x) = \dfrac{2}{(x + 1)^2}$ (*Hint:* Use Exercise 7.)

In Exercises 20–27, find a power series representation for the given function using Theorem 5. State the radius of convergence.

20. $f(x) = \ln(1 + x)$

21. $f(x) = \ln(1 - x)$

22. $f(x) = x \ln(1 + x)$

23. $f(x) = \text{Tan}^{-1}(2x)$

24. $f(x) = x \, \text{Tan}^{-1} x$

25. $f(x) = \ln(4 + x)$

26. $\displaystyle\int \dfrac{dx}{1 + x^4}$

27. $\ln(1 + x^2)$

28. Use the results of Exercises 20 and 21 to show that

$$\ln\left(\frac{1 + x}{1 - x}\right) = 2\left(x + \frac{x^3}{3} + \frac{x^5}{5} + \cdots\right), \qquad |x| < 1.$$

29. For the power series

$$\sum_{k=0}^{\infty} (-1)^k \frac{x^{2k}}{(2k)!} = 1 - \frac{x^2}{2!} + \frac{x^4}{4!} - \frac{x^6}{6!} + \cdots :$$

a. Use the Ratio Test to show that the series converges absolutely for all values of x.

b. Using Theorem 4, show that the function

$$f(x) = \sum_{k=0}^{\infty} (-1)^k \frac{x^{2k}}{(2k)!}$$

satisfies the differential equation $f''(x) = -f(x)$.

c. Show that $f(0) = 1$.

d. What function have you seen previously which satisfies both properties (b) and (c)?

30. For the power series

$$\sum_{k=0}^{\infty} (-1)^k \frac{x^{2k+1}}{(2k + 1)!} = x - \frac{x^3}{3!} + \frac{x^5}{5!} - \frac{x^7}{7!} + \cdots :$$

a. Use the Ratio Test to show that the series converges absolutely for all values of x.

b. Using Theorem 4, show that the function $g(x) = \displaystyle\sum_{k=0}^{\infty} (-1)^k \frac{x^{2k+1}}{(2k + 1)!}$ satisfies the differential equation $g''(x) = -g(x)$.

c. Show that $g(0) = 0$.

d. Show that $g'(x) = f(x)$ and $f'(x) = -g(x)$ where f is the function in Exercise 29. For what common function is $g(x)$ a power series representation?

31. Use the Chain Rule to show that if the power series

$$f(x) = \sum_{k=0}^{\infty} a_k(bx + c)^k$$

converges for $|bx + c| < r$, then

$$f'(x) = \sum_{k=0}^{\infty} kba_k(bx + c)^{k-1}, \qquad |bx + c| < r.$$

13.6 TAYLOR AND MACLAURIN SERIES

We are about to uncover a remarkable relationship between Taylor polynomials and power series, one that will provide a precise solution to the approximation problem posed in Section 13.1.

Recall that if the function f has $n + 1$ derivatives in an interval containing $x = a$, then the nth Taylor polynomial for f is

$$P_n(x) = f(a) + f'(a)(x - a)$$
$$+ \frac{f''(a)}{2!}(x - a)^2 + \cdots + \frac{f^{(n)}(a)}{n!}(x - a)^n \qquad (1)$$

$$= \sum_{k=0}^{n} \frac{f^{(k)}(a)}{k!}(x - a)^k.$$

We refer to the coefficient $\dfrac{f^{(k)}(a)}{k!}$ of $(x - a)^k$ as the **kth Taylor coefficient.** Also, if we write

$$f(x) = P_n(x) + R_n(x), \tag{2}$$

the remainder term $R_n(x)$ is bounded as follows (Taylor's Theorem):

$$|R_n(x)| \le \left| \frac{f^{(n+1)}(d)}{(n+1)!}(x-a)^{n+1} \right| \tag{3}$$

where d lies between a and x.

Now note that the right-hand side of equation (1) is precisely the nth partial sum for the power series

$$\sum_{k=0}^{\infty} \frac{f^{(k)}(a)}{k!}(x-a)^k = f(a) + f'(a)(x-a) +$$

$$\frac{f''(a)}{2!}(x-a)^2 + \frac{f'''(a)}{3!}(x-a)^3 + \cdots. \tag{4}$$

Series (4) is referred to as the **Taylor series** for f, expanded about $x = a$. It is simply the result of allowing the Taylor polynomial $P_n(x)$ to become "infinitely long." In other words,

$$\sum_{k=0}^{\infty} \frac{f^{(k)}(a)}{k!}(x-a)^k = \lim_{n \to \infty} \sum_{k=0}^{n} \frac{f^{(k)}(a)}{k!}(x-a)^k = \lim_{n \to \infty} P_n(x). \tag{5}$$

Of course, the expression in equation (4) makes sense only if the function f is infinitely differentiable. That is, the derivative $f^{(n)}(a)$ must exist for all integers $n = 1, 2, \ldots$ in order that all Taylor coefficients be defined.

The question of determining the numbers x for which a Taylor series converges to $f(x)$ may be handled by use of the remainder term, $R_n(x)$. Applying $\lim_{n \to \infty}$ to both sides of equation (2) and using (5) we see that

$$f(x) = \lim_{n \to \infty} P_n(x) + \lim_{n \to \infty} R_n(x)$$

$$= \sum_{k=0}^{\infty} \frac{f^{(k)}(a)}{k!}(x-a)^k + \lim_{n \to \infty} R_n(x).$$

This shows that *the Taylor series (4) converges to $f(x)$ if and only if $\lim_{n \to \infty} R_n(x) = 0$.*

Example 1

Find the Taylor series for the function $f(x) = \sin x$ expanded about $a = \pi/4$ and determine the numbers x for which the series converges to $f(x)$.

Strategy

Find the Taylor coefficients

$$\frac{f^{(k)}(a)}{k!}$$

with $a = \dfrac{\pi}{4}$.

Solution

We have

$$f(x) = \sin x; \qquad f\left(\frac{\pi}{4}\right) = \frac{\sqrt{2}}{2}$$

$$f'(x) = \cos x; \qquad f'\left(\frac{\pi}{4}\right) = \frac{\sqrt{2}}{2}$$

$$f''(x) = -\sin x; \qquad f''\left(\frac{\pi}{4}\right) = -\frac{\sqrt{2}}{2}$$

$$f'''(x) = -\cos x; \qquad f'''\left(\frac{\pi}{4}\right) = -\frac{\sqrt{2}}{2}$$

$$f^{(4)}(x) = \sin x; \qquad f^4\left(\frac{\pi}{4}\right) = \frac{\sqrt{2}}{2}$$

$$\vdots$$

$$f^{(2k)}(x) = (-1)^k \sin x; \qquad f^{(2k)}\left(\frac{\pi}{4}\right) = (-1)^k\frac{\sqrt{2}}{2}$$

$$f^{(2k+1)}(x) = (-1)^k \cos x; \qquad f^{(2k+1)}\left(\frac{\pi}{4}\right) = (-1)^k\frac{\sqrt{2}}{2}$$

$$\vdots$$

Insert the Taylor coefficients and $a = \pi/4$ in (4) to obtain the Taylor series.

The Taylor series is therefore

$$\sin x = \frac{\sqrt{2}}{2} + \frac{\sqrt{2}}{2}\left(x - \frac{\pi}{4}\right) - \frac{\sqrt{2}}{2 \cdot 2!}\left(x - \frac{\pi}{4}\right)^2$$
$$- \frac{\sqrt{2}}{2 \cdot 3!}\left(x - \frac{\pi}{4}\right)^3 + \frac{\sqrt{2}}{2 \cdot 4!}\left(x - \frac{\pi}{4}\right)^4 + \cdots.$$

Determine where $\lim\limits_{n\to\infty} |R_n(x)| = 0$ to find the values of x for which the series converges to $f(x)$.

Since $f^{(n+1)}(c)$ is either $\pm \sin c$ or $\pm \cos c$, we know that $\left|f^{(n+1)}(c)\right| \le 1$ for all n. Thus

$$\lim_{n\to\infty} |R_n(x)| = \lim_{n\to\infty} \left|\frac{f^{(n+1)}(c)}{(n+1)!}\left(x - \frac{\pi}{4}\right)^{n+1}\right|$$
$$\le \lim_{n\to\infty} \frac{\left|x - \dfrac{\pi}{4}\right|^{n+1}}{(n+1)!}$$
$$= 0$$

Use fact that $\lim\limits_{n\to\infty} \dfrac{x^n}{n!} = 0$ (limit (2), Section 12.2).

for all x. Thus the series converges to $\sin x$ for all x. ◇

The special case of a Taylor series expanded about $a = 0$ is referred to as a **Maclaurin series:**

$$\sum_{k=0}^{\infty} \frac{f^{(k)}(0)}{k!}x^k = f(0) + f'(0)x + \frac{f''(0)}{2!}x^2 + \cdots + \frac{f^{(k)}(0)}{k!}x^k + \cdots. \tag{6}$$

Example 2

Find a Maclaurin series for $f(x) = e^x$ and determine the numbers x for which it converges to $f(x)$.

Solution: Every derivative of $f(x) = e^x$ is the same:

$$f(x) = e^x; \qquad f(0) = 1$$

$$f'(x) = e^x; \qquad f'(0) = 1$$

$$\vdots$$

$$f^{(n)}(x) = e^x; \qquad f^{(n)}(0) = 1$$

$$\vdots$$

Thus

$$e^x = 1 + x + \frac{x^2}{2!} + \frac{x^3}{3!} + \cdots + \frac{x^k}{k!} + \cdots$$

$$= \sum_{k=0}^{\infty} \frac{x^k}{k!}.$$

Here, for all n,

$$\left| f^{(n+1)}(c) \right| = \left| e^c \right| = e^c \leq e^{|c|} \leq e^{|x|}$$

since $|c| \leq |x|$, so

$$\lim_{n \to \infty} |R_n(x)| = \lim_{n \to \infty} \left| \frac{e^c}{(n+1)!} x^{n+1} \right|$$

$$\leq \lim_{n \to \infty} e^{|x|} \frac{|x|^{n+1}}{(n+1)!}$$

$$= 0$$

for all x. The Maclaurin series for e^x therefore converges to e^x for all x. (Note that this result was proved in Example 4, Section 13.5, by a different method.)   ◇

Example 3

Find the Maclaurin series for $f(x) = \ln(1 + x)$.

Solution: Here

$$f(x) = \ln(1 + x); \qquad f(0) = 0$$
$$f'(x) = (1 + x)^{-1}; \qquad f'(0) = 1$$
$$f''(x) = -(1 + x)^{-2}; \qquad f''(0) = -1$$
$$f'''(x) = 2(1 + x)^{-3}; \qquad f'''(0) = 2$$

$$\vdots$$

$$f^{(k)}(x) = (-1)^{k+1}(k - 1)!(1 + x)^{-k}; \qquad f^{(k)}(0) = (-1)^{k+1}(k - 1)!$$

$$\vdots$$

Using (6) we obtain

$$\ln(1 + x) = 0 + 1 \cdot x - \frac{1}{2!}x^2 + \frac{2}{3!}x^3 - \cdots$$

$$+ \frac{(-1)^{k+1}(k - 1)!}{k!}x^k + \cdots$$

$$= x - \frac{x^2}{2} + \frac{x^3}{3} - \frac{x^4}{4} + \cdots + \frac{(-1)^{k+1}}{k}x^k + \cdots$$

$$= \sum_{k=1}^{\infty} \frac{(-1)^{k+1}}{k}x^k.$$

This is the power series that we showed to converge to $\ln(1 + x)$ for $|x| < 1$ in Example 5 of Section 13.5. ◇

Obviously, every Taylor or Maclaurin series is also a power series. We will now show that the converse is also true if f is represented as a power series with radius of convergence $r > 0$:

$$f(x) = a_0 + a_1x + a_2x^2 + \cdots + a_kx^k + \cdots, \qquad |x| < r \neq 0. \tag{7}$$

Using Theorem 4, we may repeatedly differentiate both sides of (7), obtaining

$$f'(x) = a_1 + 2a_2x + 3a_3x^2 + \cdots + ka_kx^{k-1} + \cdots$$
$$f''(x) = 2a_2 + 2 \cdot 3a_3x + \cdots + (k - 1)ka_kx^{k-2} + \cdots$$

$$\vdots$$

$$f^{(k)}(x) = k!a_k + (k + 1)!a_{k+1}x + \frac{(k + 2)!}{2!}a_{k+2}x^2 + \cdots. \tag{8}$$

Setting $x = 0$ in each equation causes all terms containing powers of x to vanish, so

$$f(0) = a_0, \qquad a_0 = \frac{f(0)}{0!}$$

$$f'(0) = a_1, \qquad a_1 = \frac{f'(0)}{1!}$$

$$f''(0) = 2a_2, \qquad a_2 = \frac{f''(0)}{2!}$$

$$\vdots \qquad\qquad \vdots$$

$$f^{(k)}(0) = k!a_k, \qquad a_k = \frac{f^{(k)}(0)}{k!}.$$

In other words, if f is represented by a convergent power series, as in equation (7), then the coefficients a_k must be precisely the Maclaurin coefficients

$$a_k = \frac{f^{(k)}(0)}{k!}.$$

If we replace x by $x - a$ in equation (7) and evaluate the derivatives at $x = a$, we prove the following theorem for Taylor series.

THEOREM 6

If the function f can be represented by a power series of the form

$$f(x) = a_0 + a_1(x - a) + a_2(x - a)^2 + \cdots + a_k(x - a)^k + \cdots$$

$$= \sum_{k=0}^{\infty} a_k(x - a)^k, \qquad |x| < r$$

with a radius of convergence $r > 0$, then the coefficients a_k in the power series expansion must be the Taylor coefficients

$$a_k = \frac{f^{(k)}(a)}{k!}, \qquad k = 0, 1, 2, \ldots .$$

REMARK: Be careful about what is being said. If we *know* that $f(x) = \Sigma a_k (x - a)^k$, then this series is the Taylor series for f. There are examples, however, for which the Taylor series $\Sigma a_k (x - a)^k$ for f converges, but $f(x) \neq \Sigma a_k (x - a)^k$. That is, *a convergent Taylor series for f need not converge to the value $f(x)$.* (See Exercise 37.)

Example 4

In Exercise 2 you are asked to obtain the Maclaurin series expansion

$$\sin x = x - \frac{x^3}{3!} + \frac{x^5}{5!} - \frac{x^7}{7!} + \cdots + \frac{(-1)^k x^{2k+1}}{(2k+1)!} + \cdots . \tag{9}$$

To show that this series converges to $\sin x$ for all x, we must show that $\lim_{n \to \infty} |R_n(x)| = 0$ for all x.

We do so, using Taylor's theorem, by noting that the derivative $f^{(n+1)}$ for $f(x) = \sin x$ is one of $\pm \sin x$ or $\pm \cos x$. Thus, for any x, we know that

$$|f^{(n+1)}(c)| \leq 1$$

for all c between 0 and x. Thus, by Taylor's Theorem,

$$|R_n(x)| \leq \frac{1}{(n+1)!} |x|^{n+1}, \qquad -\infty < x < \infty,$$

so

$$\lim_{n \to \infty} |R_n(x)| = \lim_{n \to \infty} \frac{|x|^{n+1}}{(n+1)!} = 0, \qquad -\infty < x < \infty$$

according to equation (2), Section 12.2.

Since the Maclaurin series (9) is a power series that converges for all x, we may differentiate this series term by term, to obtain a power series for $\cos x = \frac{d}{dx} \sin x$:

$$\cos x = 1 - \frac{3x^2}{3!} + \frac{5x^4}{5!} - \frac{7x^6}{7!} + \cdots + \frac{(-1)^k (2k+1) x^{2k}}{(2k+1)!} + \cdots$$

$$= 1 - \frac{x^2}{2!} + \frac{x^4}{4!} - \frac{x^6}{6!} + \cdots + \frac{(-1)^k x^{2k}}{(2k)!} + \cdots ,$$

which converges for all x. Theorem 6 then assures that this series is, in fact, the Maclaurin series for $f(x) = \cos x$. $\diamondsuit$

Example 5

Find a Maclaurin series for $x^3 e^{x^2}$.

Strategy

Do *not* calculate Taylor coefficients for $f(x) = x^3 e^{x^2}$. Work directly with the Maclaurin series for e^x and substitute x^2 for x.

Multiply by x^3.

Solution

Replacing x by x^2 in the series for e^x in Example 2 shows that

$$e^{x^2} = 1 + x^2 + \frac{x^4}{2!} + \frac{x^6}{3!} + \cdots = \sum_{k=0}^{\infty} \frac{x^{2k}}{k!},$$

which converges for all x. Thus

$$x^3 e^{x^2} = x^3 \left\{ 1 + x^2 + \frac{x^4}{2!} + \frac{x^6}{3!} + \cdots \right\}$$

$$= x^3 + x^5 + \frac{x^7}{2!} + \frac{x^9}{3!} + \cdots$$

$$= \sum_{k=0}^{\infty} \frac{x^{2k+3}}{k!}$$

converges for all x. By Theorem 6 this must be the Maclaurin series for $x^3 e^{x^2}$. ◇

The following examples show how Taylor and Maclaurin series may be used to calculate certain constants and integrals.

Example 6

Since the Maclaurin series

$$e^x = \sum_{k=0}^{\infty} \frac{x^k}{k!} = 1 + x + \frac{x^2}{2!} + \frac{x^3}{3!} + \cdots$$

converges to e^x for all x, we may use it to approximate e^x for any number x. Setting $x = 1$ gives

$$e = 1 + 1 + \frac{1}{2!} + \frac{1}{3!} + \cdots + \frac{1}{k!} + \cdots$$

$$= \sum_{k=0}^{\infty} \frac{1}{k!},$$

and setting $x = -1$, we obtain

$$\frac{1}{e} = 1 - 1 + \frac{1}{2!} - \frac{1}{3!} + \frac{1}{4!} - \cdots + \frac{(-1)^k}{k!} + \cdots$$

$$= \sum_{k=0}^{\infty} \frac{(-1)^k}{k!}.$$

Since this last series is an alternating series, we may approximate $\frac{1}{e}$ by, say,

$$\frac{1}{e} \approx \frac{1}{2!} - \frac{1}{3!} + \frac{1}{4!} - \frac{1}{5!} = .3667$$

with accuracy no worse than $\frac{1}{6!} = .0014$. ◇

Example 7

We may approximate the integral $\int_0^1 e^{-x^2}\,dx$ using Maclaurin series as follows. Since

$$e^x = 1 + x + \frac{x^2}{2!} + \frac{x^3}{3!} + \cdots + \frac{x^k}{k!} + \cdots$$

converges for all x, the series

$$e^{-x^2} = 1 - x^2 + \frac{x^4}{2!} - \frac{x^6}{3!} + \cdots + \frac{(-x^2)^k}{k!} + \cdots$$

also converges for all x. By Theorem 5 an antiderivative for e^{-x^2} is represented as

$$\int e^{-x^2}\,dx = \int 1\,dx - \int x^2\,dx + \int \frac{x^4}{2!}\,dx - \int \frac{x^6}{3!}\,dx + \cdots + C$$

$$= x - \frac{x^3}{3} + \frac{x^5}{5\cdot 2!} - \frac{x^7}{7\cdot 3!} + \cdots + \frac{(-1)^k x^{2k+1}}{(2k+1)k!} + \cdots + C.$$

Thus,

$$\int_0^1 e^{-x^2}\,dx = 1 - \frac{1}{3} + \frac{1}{5\cdot 2!} - \frac{1}{7\cdot 3!} + \cdots + \frac{(-1)^k}{(2k+1)k!} + \cdots.$$

Since this is an alternating series, we may approximate this series to any desired degree of accuracy by taking sufficiently many terms. For example, to obtain accuracy of .01, we find by trial and error that for $k = 4$,

$$\frac{1}{(2\cdot 4 + 1)4!} = \frac{1}{9\cdot 24} = \frac{1}{216} < \frac{1}{100}.$$

Thus, we use terms up through $k = 3$ to obtain

$$\int_0^1 e^{-x^2}\,dx \approx 1 - \frac{1}{3} + \frac{1}{5\cdot 2!} - \frac{1}{7\cdot 3!} \approx .74,$$

accurate to within .01. ◇

The Binomial Series

According to the Binomial Theorem, we know that

$$(1 + x)^n = 1 + nx + \frac{n(n-1)}{2}x^2 + \cdots$$

$$+ \binom{n}{k}x^k + \cdots + nx^{n-1} + x^n \qquad (10)$$

when n is a positive integer. In writing (10) we have used the notation $\binom{n}{k}$ for the

binomial coefficient

$$\binom{n}{k} = \frac{n!}{k!(n-k)!} = \frac{n(n-1)(n-2)\cdots(n-k+1)}{k!}.$$

We can extend equation (10) to the case where n is not a positive integer. However, the result involves an infinite series, rather than a polynomial in x. The result is called the **Binomial Series:**

$$(1 + x)^r = 1 + rx + \frac{r(r - 1)}{2}x^2 + \frac{r(r - 1)(r - 2)}{3!}x^3 + \cdots \tag{11}$$

$$= 1 + \sum_{k=1}^{\infty} \frac{r(r - 1)(r - 2) \cdots \cdot (r - k + 1)}{k!}x^k, \qquad |x| < 1.$$

(Note that the infinite series in (11) reduces to the polynomial in (10) when $r = n$ is a positive integer, since the term $(r - k + 1)$ becomes zero when k reaches $r + 1$.)

Proving the validity of (11) involves showing that the right-hand side is precisely the Maclaurin series for $f(x) = (1 + x)^r$, and that the radius of convergence for this series is one. To do so we note that

$$f(x) = (1 + x)^r \qquad \text{gives} \qquad f(0) = 1$$
$$f'(x) = r(1 + x)^{r-1} \qquad \text{gives} \qquad f'(0) = r$$
$$f''(x) = r(r - 1)(1 + x)^{r-2} \qquad \text{gives} \qquad f''(0) = r(r - 1)$$
$$\vdots$$
$$f^{(k)}(x) = r(r - 1) \cdots \cdot (r - k + 1)(1 + x)^{r-k}$$
$$\text{gives} \qquad f^{(k)}(0) = r(r - 1) \cdots \cdot (r - k + 1).$$

Thus, the kth Maclaurin coefficient for $f(x) = (1 + x)^r$ is indeed

$$\frac{f^{(k)}(0)}{k!} = \frac{r(r - 1)(r - 2) \cdots \cdot (r - k + 1)}{k!}$$

as in the Binomial Series (11).

To verify that the series converges for $|x| < 1$, we let a_k denote the kth term

$$a_k = \frac{r(r - 1) \cdots \cdot (r - k + 1)}{k!}x^k.$$

Then

$$\rho = \lim_{k \to \infty} \left| \frac{a_{k+1}}{a_k} \right| = \lim_{k \to \infty} \left| \frac{r(r - 1) \cdots \cdot (r - k)}{r(r - 1) \cdots \cdot (r - k + 1)} \cdot \frac{k!}{(k + 1)!} \cdot \frac{x^{k+1}}{x^k} \right|$$

$$= \lim_{k \to \infty} \left| \frac{(r - k)}{1} \cdot \frac{1}{(k + 1)} \cdot x \right|$$

$$= \lim_{k \to \infty} \left| \frac{r - k}{k + 1} \right| \cdot |x|$$

$$= |x|.$$

Thus, if $|x| < 1$ we have $\rho < 1$, and the series converges absolutely by the Ratio Test. We assert, but do not prove, that this series converges *to* $(1 + x)^r$ for $|x| < 1$.

Example 8

The Binomial Series, with $r = \dfrac{1}{2}$, gives

$$\sqrt{1 + x} = (1 + x)^{1/2} = 1 + \frac{1}{2}x + \frac{\dfrac{1}{2}\left(\dfrac{1}{2} - 1\right)}{2}x^2 + \frac{\dfrac{1}{2}\left(\dfrac{1}{2} - 1\right)\left(\dfrac{1}{2} - 2\right)}{3!}x^3$$

$$+ \frac{\frac{1}{2}\left(\frac{1}{2} - 1\right)\left(\frac{1}{2} - 2\right)\left(\frac{1}{2} - 3\right)}{4!} x^4 + \cdots$$

$$= 1 + \frac{1}{2}x - \frac{1}{8}x^2 + \frac{1}{16}x^3 - \frac{5}{128}x^4 + \cdots, \qquad |x| < 1. \quad \diamond$$

Example 9

Replacing x by x^3 in Example 8 gives

$$\sqrt{1 + x^3} = (1 + x^3)^{1/2}$$

$$= 1 + \frac{1}{2}x^3 - \frac{1}{8}x^6 + \frac{1}{16}x^9 - \frac{5}{128}x^{12} + \cdots, \qquad |x| < 1. \quad \diamond$$

Example 10

With $r = -\dfrac{1}{3}$ and $-x$ in place of x, (11) gives

$$\frac{1}{\sqrt[3]{1 - x}} = [1 + (-x)]^{-1/3}$$

$$= 1 + \left(-\frac{1}{3}\right)(-x) + \frac{\left(-\frac{1}{3}\right)\left(-\frac{1}{3} - 1\right)}{2}(-x)^2$$

$$+ \frac{\left(-\frac{1}{3}\right)\left(-\frac{1}{3} - 1\right)\left(-\frac{1}{3} - 2\right)}{3!}(-x)^3$$

$$+ \frac{\left(-\frac{1}{3}\right)\left(-\frac{1}{3} - 1\right)\left(-\frac{1}{3} - 2\right)\left(-\frac{1}{3} - 3\right)}{4!}(-x)^4 + \cdots$$

$$= 1 + \frac{1}{3}x + \frac{2}{9}x^2 + \frac{14}{81}x^3 + \frac{35}{243}x^4 + \cdots, \qquad |x| < 1. \quad \diamond$$

Exercise Set 13.6

In Exercises 1–18, find the Taylor or Maclaurin series for the given function expanded about the given point and determine the numbers x for which the series converges.

1. $f(x) = e^{2x}, \qquad a = 0$

2. $f(x) = \sin x, \qquad a = 0$

3. $f(x) = \cos x, \qquad a = \pi/4$

4. $f(x) = \sin x, \qquad a = \pi/6$

5. $f(x) = 1 + x^2, \qquad a = 2$

6. $f(x) = \dfrac{1}{1 + x}, \qquad a = 0$

7. $f(x) = \ln(3 + x), \qquad a = 0$

8. $f(x) = x \sin 2x, \qquad a = 0$

9. $f(x) = 2^x, \qquad a = 0$

10. $f(x) = (1 + x)^n, \qquad a = 0$

11. $f(x) = (1 + x)^{3/2}, \qquad a = 0$

12. $f(x) = \sqrt{x}, \qquad a = 4$

13. $f(x) = \sqrt{x + 1}, \qquad a = 0$

14. $f(x) = \dfrac{1}{x}, \qquad a = 2$

15. $f(x) = \dfrac{\sin x}{x}, \qquad a = 0$

16. $f(x) = x^2 e^{-x}, \qquad a = 0$

17. $f(x) = x \sin x, \qquad a = \pi/4$

18. $f(x) = x^2 \ln(1 + x), \qquad a = 0$

19. Find a Maclaurin series for $f(x) = \cos^2 x$ by use of the identity $\cos^2 x = \dfrac{1}{2}[1 + \cos 2x]$.

20. Find the first four terms of the Maclaurin series for $f(x) = \tan x$.

21. Find the first three terms of the Maclaurin series for $f(x) = \sec^2 x$, using the result of Exercise 20.

22. Find the Maclaurin series for $x \sin x^2$ using the Maclaurin series for $\sin x$.

23. Find a Maclaurin series for $\cos x^2$ using the technique of Exercise 22.

24. Find the Maclaurin series for $f(x) = \sinh x$ from the Maclaurin series for e^x and e^{-x}.

25. Find the Maclaurin series for $x \cosh x$.

26. Find the Maclaurin series for $f(x) = \sin^2 x$ (see Exercise 19).

In Exercises 27–32, use Taylor series to approximate the given quantity accurately to three decimal places.

27. $\sin 2°$

28. e^2

29. $\ln(1.1)$

30. $\displaystyle\int_0^{\pi/4} \sin x^2 \, dx$

31. $\displaystyle\int_0^{1/2} \dfrac{1}{1 + x^3} \, dx$

32. $\displaystyle\int_0^1 e^{-x^2} \, dx$

In Exercises 33–36, use the Binomial Series to find a power series for the given function and the radius of convergence.

33. $f(x) = \sqrt{1 + 2x}$

34. $f(x) = \sqrt[3]{27 + x}$

35. $f(x) = (9 + 3x)^{3/2}$

36. $f(x) = \dfrac{1}{\sqrt[3]{8 - x^2}}$

37. Define the function $f(x)$ by

$$f(x) = \begin{cases} e^{-(1/x^2)}, & x \neq 0 \\ 0, & x = 0. \end{cases}$$

a. Show that $f^{(n)}(0)$ exists and equals zero for every integer $n \geq 1$.

b. Show that the Maclaurin series for f is identically equal to zero for all values of x. Thus the function f does not equal its Maclaurin series in any interval containing zero. (*Remark:* This example shows that the existence of all derivatives is not a sufficient condition for a function to equal its Taylor series. The condition that $\lim\limits_{n \to \infty} R_n(x) = 0$ is essential.)

38. Prove that power series representations are unique. That is, prove that if

$$f(x) = \sum_{k=0}^{\infty} a_k x^k \qquad \text{and} \qquad f(x) = \sum_{k=0}^{\infty} b_k x^k,$$

then $\qquad a_0 = b_0, \, a_1 = b_1, \, \ldots, \, a_k = b_k, \, \ldots$.

(*Hint:* Set $x = 0$ to obtain $a_0 = b_0$. Then differentiate and set $x = 0$ again, and so on.)

SUMMARY OUTLINE OF CHAPTER 13

◆ The nth Taylor polynomial for f, expanded about $x = a$, is (page 563)

$$P_n(x) = f(a) + f'(a)(x - a) + \frac{f''(a)}{2!}(x - a)^2 + \cdots + \frac{f^{(n)}(a)}{n!}(x - a)^n = \sum_{j=0}^{n} \frac{f^{(j)}(a)}{j!}(x - a)^j.$$

◆ **Theorem:** If f is $(n + 1)$ times differentiable and $f(x) = P_n(x) + R_n(x)$, then $R_n(x) = \dfrac{f^{(n+1)}(c)}{(n + 1)!}(x - a)^{n+1}$ where c (page 569) lies between a and x.

◆ A **power series** has the form $\displaystyle\sum_{k=0}^{\infty} a_k(x - a)^k$. (page 578)

◆ **Theorem:** If the power series $\Sigma a_k x^k$ converges for $x = c$, it does so for all x with $|x| < |c|$. (page 579)

◆ **Theorem:** The set of all x for which a power series converges is either (i) $\{a\}$, (ii) an interval with midpoint $x = a$ (page 580) and radius $r \neq 0$, or (iii) $(-\infty, \infty)$. (The radius r is called the **radius of convergence.**)

◆ **Theorem:** If $f(x) = \Sigma a_k x^k$ with radius of convergence r, then $f'(x)$ exists, and $f'(x) = \Sigma k a_k x^{k-1}$ with radius of (page 586) convergence r.

◆ **Theorem:**   If $f(x) = \Sigma a_k x^k$ with radius of convergence r, then $\int f(x)\,dx = \Sigma\left(\dfrac{a_k}{k+1}\right)x^{k+1} + C$ with radius of   (page 588)

convergence r.

◆ A **Taylor series** is a power series of the form   (page 591)

$$\sum_{k=0}^{\infty}\frac{f^{(k)}(a)}{k!}(x-a)^k = f(a) + f'(a)(x-a) + \frac{f''(a)}{2!}(x-a)^2 + \cdots + \frac{f^{(k)}(a)}{k!}(x-a)^k + \cdots.$$

If $a = 0$, this series is called a **Maclaurin series.**

◆ The **binomial series** is   (page 598)

$$(1+x)^r = 1 + \sum_{k=1}^{\infty}\frac{r(r-1)(r-2)\cdot\,\cdots\,\cdot(r-k+1)}{k!}x^k,\ |x| < 1.$$

REVIEW EXERCISES—CHAPTER 13

In Exercises 1–10, find the Taylor polynomial of degree n for $f(x)$ expanded about $x = a$.

1. $f(x) = \sin 2x$, $a = 0$, $n = 5$

2. $f(x) = \ln(1 + x^2)$, $a = 0$, $n = 2$

3. $f(x) = x\cos x$, $a = \pi/4$, $n = 3$

4. $f(x) = \sqrt{1 + x^2}$, $a = 0$, $n = 3$

5. $f(x) = e^{x^2}$, $a = 0$, $n = 3$

6. $f(x) = \sqrt{2x + 3}$, $a = 11$, $n = 3$

7. $f(x) = \operatorname{Tan}^{-1}x$, $a = 1$, $n = 3$

8. $f(x) = x\ln(1 + x)$, $a = 0$, $n = 4$

9. $f(x) = 2^x$, $a = 0$, $n = 3$

10. $f(x) = \dfrac{1}{1 + x^3}$, $a = 0$, $n = 2$

In Exercises 11–15, find the accuracy of the approximation of the given quantity by the nth Taylor polynomial for the function $f(x)$ expanded about $x = a$.

11. $\sin 34°$, $f(x) = \sin x$, $a = \pi/6$, $n = 3$

12. $\tan(\pi/12)$, $f(x) = \tan x$, $a = 0$, $n = 2$

13. $\sqrt[5]{35}$, $f(x) = \sqrt[5]{x}$, $a = 32$, $n = 2$

14. $\sec\left(\dfrac{3\pi}{16}\right)$, $f(x) = \sec x$, $a = \dfrac{\pi}{4}$, $n = 2$

15. $e^{.25}$, $f(x) = e^x$, $a = 0$, $n = 3$

16. Show that the fourth Taylor polynomial for $f(x) = \cosh x^2$ expanded about $a = 0$ is $P_4(x) = 1 + \dfrac{x^4}{2}$.

17. By integrating the $(n-1)$st Taylor polynomial for $f(x) = \dfrac{1}{1-x}$, show that

$$\ln\frac{1}{1-x} = x + \frac{x^2}{2} + \frac{x^3}{3} + \cdots + \frac{x^n}{n} + \int_0^x \frac{t^n\,dt}{1-t}.$$

18. Show that, if $P_n(x)$ is the nth Taylor polynomial for $f(x) = \sinh x$ expanded about $a = 0$, and if $Q_{n+1}(x)$ is the Taylor polynomial of degree $n+1$ for $\cosh x$ expanded about $a = 0$, then

$$\frac{d}{dx}\,Q_{n+1}(x) = P_n(x).$$

What if $a \neq 0$?

19. Let $P_n(x)$ be the nth Taylor polynomial, expanded about $a = 0$, for $f(x) = x^{7/2}$. Show that $P_0(0) = P_1(0) = P_2(0) = P_3(0) = 0$, but that $P_n(0)$ is undefined for $n \geq 4$.

20. Let $P_n(x)$ be the nth Taylor polynomial for f expanded about $x = a$. If $P_n(x) = 0$ for all n for which $P_n(x)$ is defined, must f be the zero function? (*Hint:* See Exercise 19.)

In Exercises 21–26, find the interval of convergence for the given series.

21. $\sum k(x - 3)^k$

22. $\sum(-1)^k k^2(x - 1)^k$

23. $\sum\dfrac{\sqrt{k}(x - 2)^k}{k + 3}$

24. $\sum\dfrac{(x + 2)^k}{2^k}$

25. $\sum\dfrac{(-1)^k x^k}{k\ln k}$

26. $\sum\dfrac{\ln k(x - 1)^k}{e^k}$

27. Show that $\displaystyle\sum_{k=0}^{\infty} a_k x^{k+1} = \sum_{k=1}^{\infty} a_{k-1} x^k$.

28. Show that $\displaystyle\sum_{k=1}^{\infty} k(k+1)x^k = \sum_{k=2}^{\infty} (k-1)kx^{k-1}$.

29. Find numbers $a_1, a_2, \ldots$ so that

$$\sum na_n x^{n-1} + c\sum a_n x^n = 0, \qquad c \neq 0.$$

What is the function f whose power series is $\sum a_k x^k$?

30. Find the numbers $a_2, a_3, a_4, \ldots$ so that

$$\sum_{n=2}^{\infty} n(n-1)a_n x^{n-2} + 4\sum_{n=0}^{\infty} a_n x^n = 0.$$

31. Find a power series representation for the function $f(x) = \dfrac{1}{1+x^4}$.

32. Find a Maclaurin series for the function $f(x) = \sin x \cos x$.

33. Find a Maclaurin series for $f(x) = \sqrt[3]{1+x^2}$.

34. Find a power series for the function $f(x) = \dfrac{1}{(1-x)(2-x)}$ using the geometric series. What is the radius of convergence?

35. Find the first 3 terms of the Taylor series for $f(x) = \sin \sqrt{x}$ expanded about $x = \pi^2/4$.

36. What is the radius of convergence of a power series for $e^{\sqrt{x}}$ expanded about $a = 0$?

37. Find a Maclaurin series for $f(x) = x^2 e^{x^2}$ using the series for e^x.

38. Find the Taylor series for $f(x) = \sqrt{x+4}$ expanded about $a = 2$. For what values of x does it converge?

39. Use a Taylor series to approximate $\sqrt{e}$ accurate to three decimal places.

UNIT 6

GEOMETRY IN THE PLANE AND IN SPACE

René Descartes

Pierre de Fermat

René Descartes (1596–1650) and Pierre de Fermat (1601–1665) had at least two things in common: both were French, and neither was a professional mathematician. They shared something else of far greater importance—each independently discovered analytic geometry.

Descartes left school at the age of 16 and went to Paris, where he occasionally studied mathematics. At 21 he became a soldier-for-pay, hiring out to the armies of various nations and minor political subdivisions. He alternated military service with study and travel throughout Europe, and met a number of mathematicians. His soldiering was apparently successful; he was offered a commission as lieutenant general, but declined it since he preferred having less responsibility and more time to think. After nine years of this life, his growing interest in matters of philosophy led him to the relative peace of the Netherlands, where he studied and wrote for 20 years.

On the night of November 10, 1619, Descartes had three dreams that profoundly affected his life and the development of mathematics. He claimed that in one of these dreams he was given the key to understanding nature. Although he never revealed precisely what this was, it is widely believed that he was referring to the relation between algebra and geometry. This may therefore be said to be the founding date of analytic geometry, although publication was not to occur for 18 years.

In 1637 Descartes published *Discours de la méthode . . .* , or *Discourse on the Method of Rightly Conducting the Reason and Seeking Truth in the Sciences*. This was a program for conducting philosophical research—an attempt to reach valid conclusions in many fields by systematic reasoning. The *Discourse* included three appendices. Each was a brilliant illustration of the application of the method to a different field. The first, *La dioptrique*, included the first published statement of the law of refraction (although the Dutch mathematician Willebrord Snell (1591–1626) had discovered it earlier). The second appendix, on meteorology, included an explanation of the colors of the rainbow. The third, and most famous, was *La géométrie*. Starting with an ancient problem from the Greek mathematician Pappus (third century A.D.), Descartes applied algebra to geometry and vice versa. This was not quite today's analytic geometry, however. Coordinate axes were implied but not used explicitly, and were thought of as oblique rather than the perpendicular axes that we now label Cartesian. The notions of distance, angle between lines, and slope were not included. In no case was a curve plotted from an equation. Descartes did not consider negative numbers or negative coordinates. Only one equation was considered in detail, the general second degree equation, and that was not discussed until Chapter 15. Descartes gave the conditions on the coefficients of this equation for the curve to be an ellipse, parabola, or hyperbola. The details were omitted, and indeed the entire work was difficult to read. Readers often found it difficult to see how the appendices were related to the philosophy set forth in the main text. In spite of these inadequacies, the concepts of analytic geometry were introduced in this little work, the only book about mathematics that Descartes ever produced. It was a remarkable and striking achievement.

Descartes was lured away from the Netherlands to a position with Queen Christina of Sweden, who was interested in philosophy and wanted him to establish an academy of sciences. Unfortunately, she wanted her tutoring in philosophy at 5:00 A.M. which seriously conflicted with Descartes' lifelong practice of late rising. He was unable to maintain his health in the cold, damp Swedish climate, and died of pneumonia at 54 years of age.

Pierre de Fermat has already been mentioned in the historical note preceding Unit 2, in which his early work on differentiation (before Newton and Leibniz) is discussed. As a lawyer, politician, and judge, he apparently had sufficient time to devote to the study of his real love, mathematics. His work on analytic geometry was actually done several years before that of Descartes, but he refused to publish his mathematical discoveries. What few items were printed during his lifetime were published by friends but without Fermat's name attached. Most of his work is known either through his notes, often written in the margins of books, or through the many letters he wrote to other mathematicians. The letters frequently contained original mathematics of such merit that copies were made and then recopied for circulation to other mathematicians. But Fermat was so casual about his work that he usually failed even to keep copies of his letters and essays for himself.

Fermat's work in analytic geometry was actually written several years before Descartes' discovery. It was not published, though, until 1679, fourteen years after Fermat's death and a half century after its completion. Descartes wrote in more modern notation than did Fermat, but the latter's ideas much more closely resemble modern mathematics. The 1679 volume, titled (in translation) *Introduction to Plane and Solid Loci,* presents in archaic terminology an astonishing amount of material which is familiar to us today. His linear equation written "*D in A aequitur B in E*" we would write as $Dx = By$, and Fermat sketched it as a line—or rather, as a half-line, since he did not accept negative coordinates. He dealt with circles, parabolas, ellipses, and hyperbolas. He even extended his analytic geometry to three-dimensional space, a generalization that had not occurred to Descartes.

Fermat considered the family of equations of the form $y = x^n$ where n is a positive or negative integer. In an attempt to find the maxima and minima of the associated curves, he found the derivatives of these special functions. He also found the tangents to these curves by using what was essentially the differentiation process, though without the idea of the limit. Later Fermat found the area between such a curve and a line, which corresponds to the integral of x^n. It is curious that he did not seem to have recognized the inverse nature of these two processes. Although Fermat is sometimes called a discoverer of the calculus, he worked with only a few polynomial functions, he lacked the concept of the limit, and he failed to note the Fundamental Theorem of Calculus. Thus, he cannot be said to have reached the intellectual summit that might have allowed him to be ranked with Newton and Leibniz.

Fermat's chief mathematical interest was neither the calculus nor analytic geometry. It involved such topics as prime numbers, perfect numbers, and magic squares, which fall under the heading of the theory of numbers. He further was one of the earliest to develop the theory of probability. He was a classical scholar and could read and write fluently in several languages: French, Spanish, Italian, Greek, and Latin. He even composed poetry in Latin. But his refusal to publish his work resulted in his brilliance not being recognized until long after his death.

(Photographs from the David Eugene Smith Papers, Rare Book and Manuscript Library, Columbia University.)

Chapter 14
Polar Coordinates and Parametric Equations

Up to this point we have studied curves in the plane by regarding them as graphs of functions plotted in Cartesian (rectangular) coordinates. This chapter presents two additional ways in which we can represent curves in the plane: as graphs of functions of the form $r = f(\theta)$ plotted in *polar coordinates,* and as graphs determined by *parametric equations* for the x and y coordinates of points in Cartesian coordinates.

14.1 THE POLAR COORDINATE SYSTEM

Cartesian coordinates provide a convenient scheme to use when graphing functions whose variables represent naturally perpendicular quantities (length versus height, for example) or when attempting to approximate areas by rectangles. However, many plane curves can be more easily described in a coordinate system in which the coordinates of a point $P(r, \theta)$ measure its (radial) distance, r, along a ray from the origin that makes an angle θ with some fixed reference ray (usually the positive x-axis; see Figure 1.1).

These coordinates are especially convenient for describing curves that are symmetric with respect to the origin. An example of this situation occurs in the theory of electric potential fields. For a point charge, the lines representing equal electric potentials (called equipotentials) in any plane containing the point charge are circles with the point as center (Figure 1.2). These equipotential circles can most easily be described simply by stating their radius. Similarly, the lines of force emanating from such a point charge are rays, which are described completely by the angle they form with the positive x-axis.

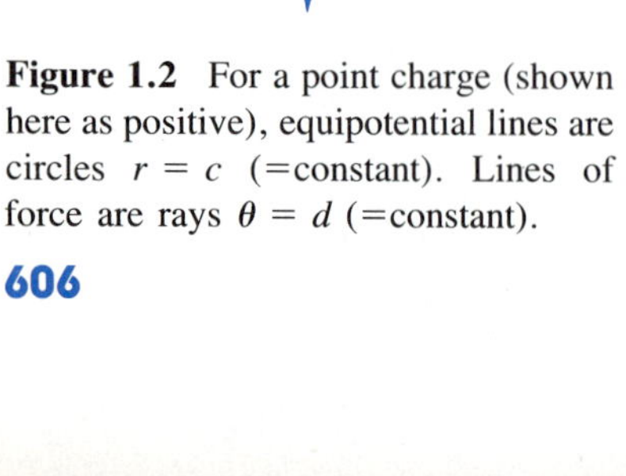

Let O be any point in the plane, called the **pole,** and let ℓ be any ray emanating from O. The ray ℓ is called the **polar axis** for the plane. The choice of a pole and a polar axis determines a **polar coordinate system** for the plane in which any point P may be assigned **polar coordinates** $P = (r, \theta)$, where

(i) r, called the **radial variable,** is the distance from O to P, and
(ii) θ, called the **angular variable,** is the angle formed between the polar axis and the ray $\overline{OP}$, measured in the counterclockwise direction (Figure 1.1).

Figure 1.3 illustrates various points plotted in polar coordinates. As in most applications of trigonometry in the calculus, we shall work in radians.

In the xy-coordinate plane, the graphs of equations of the form $x = $ constant or $y = $ constant are lines. In the polar coordinate plane, graphs of equations of the form $r = $ constant are circles, while graphs of equations of the form $\theta = $ constant are rays (see Figure 1.4).

Figure 1.1 Polar coordinates for point P.

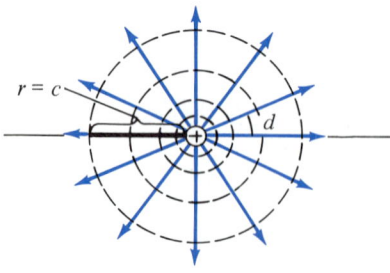

Figure 1.2 For a point charge (shown here as positive), equipotential lines are circles $r = c$ (=constant). Lines of force are rays $\theta = d$ (=constant).

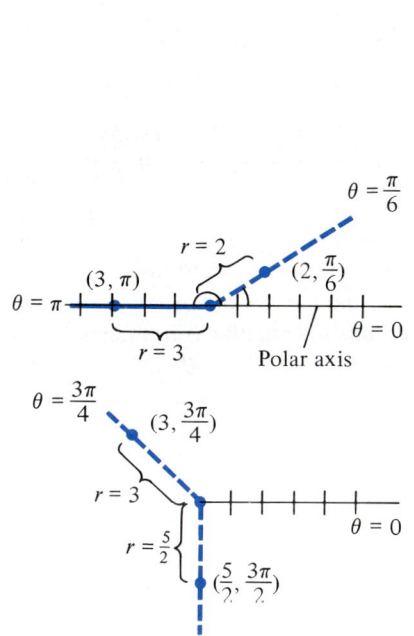

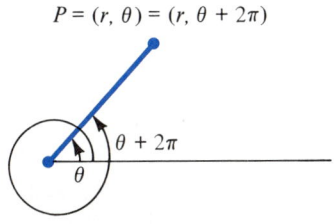

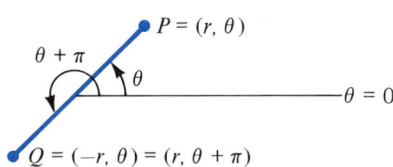

Figure 1.3 Polar coordinates for several points.

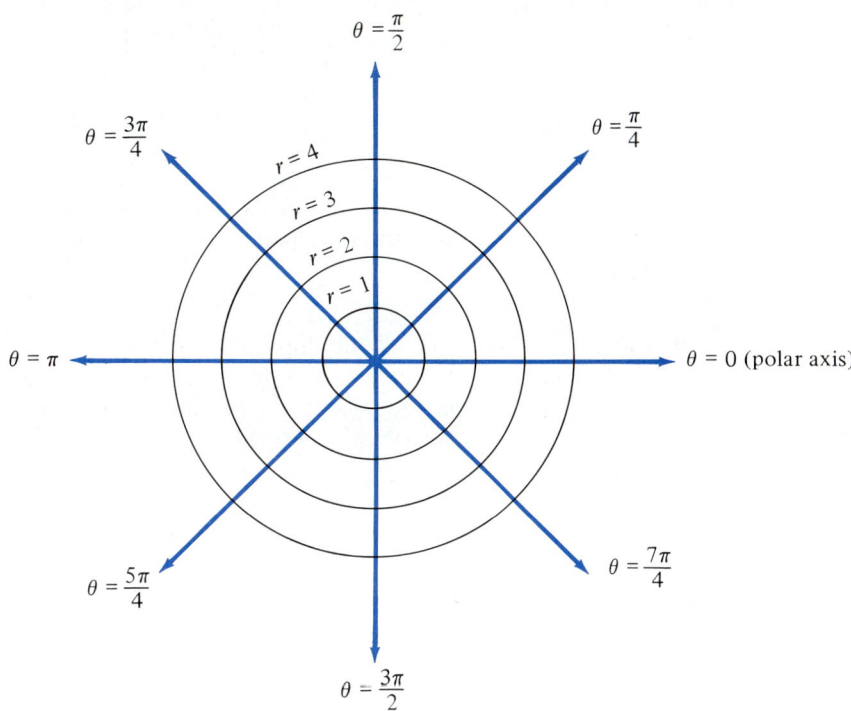

Figure 1.4 The polar coordinate plane.

You may have already noticed that polar coordinates are not unique. Indeed, since the numbers $\theta = \theta_0 + 2n\pi$, $n = 0, \pm1, \pm2, \ldots$ all determine the same ray, we have

$$(2, \pi) = (2, 3\pi) = (2, 5\pi) = (2, -\pi) = \cdots,$$

$$\left(1, \frac{\pi}{4}\right) = \left(1, \frac{9\pi}{4}\right) = \left(1, \frac{17\pi}{4}\right) = \left(1, -\frac{7\pi}{4}\right) = \cdots,$$

and, in general,

$$(r, \theta) = (r, \theta + 2n\pi), \qquad n = \pm1, \pm2, \pm3, \ldots. \tag{1}$$

(See Figure 1.5.) Moreover, it is customary to extend the range of values for the radial variable r to include negative numbers via the identity

$$(-r, \theta) = (r, \theta + \pi). \tag{2}$$

(See Figure 1.6.) In other words, the point with polar coordinates $(-r, \theta)$ lies r units along the ray pointing in the direction opposite that specified by the angular variable θ.

Example 1

By identities (1) and (2) we have

(a) $\left(3, \frac{\pi}{6}\right) = \left(3, \frac{13\pi}{6}\right) = \left(3, -\frac{11\pi}{6}\right),$

Figure 1.5 $(r, \theta) = (r, \theta + 2n\pi)$.

Figure 1.6 $(-r, \theta) = (r, \theta + \pi)$.

(b) $\left(-4, \dfrac{\pi}{4}\right) = \left(4, \dfrac{5\pi}{4}\right) = \left(4, -\dfrac{3\pi}{4}\right),$

(c) $\left(7, -\dfrac{\pi}{2}\right) = \left(-7, \dfrac{\pi}{2}\right) = \left(-7, \dfrac{5\pi}{2}\right) = \left(-7, -\dfrac{3\pi}{2}\right).$ ◇

The *pole O* in the polar coordinate plane has coordinates $(0, \theta)$ for every angle θ. It is the only point in the plane not associated with a unique angle θ in $[0, 2\pi)$.

Relationships Between Polar and Rectangular Coordinates

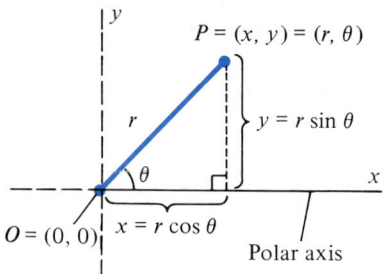

Figure 1.7 Relationships between rectangular and polar coordinates.

The relationships between polar and xy-coordinates are easily determined when the origin in xy-coordinates is identified with the pole, and the positive x-axis is taken to be the polar axis (see Figure 1.7). If the point P has rectangular coordinates (x, y) and polar coordinates (r, θ), and if $r > 0$, the definitions of $\sin \theta$ and $\cos \theta$ give

$$x = r \cos \theta \quad \text{and} \quad y = r \sin \theta. \tag{3}$$

In Exercise 54, you are asked to show that equations (3) hold for $r < 0$ as well. It follows from the equations in (3) (or from Figure 1.7) that

$$r = \sqrt{x^2 + y^2}; \quad \tan \theta = \frac{y}{x}, \quad x \neq 0. \tag{4}$$

The equations in line (3) are used in changing from polar to rectangular coordinates while the equations in line (4) allow us to change from rectangular to polar coordinates.

Example 2

To find the rectangular coordinates corresponding to the polar coordinates $(2, 2\pi/3)$, we use equations (3) with $r = 2$, $\theta = 2\pi/3$:

$$x = 2 \cos \frac{2\pi}{3} = 2\left(-\frac{1}{2}\right) = -1,$$

$$y = 2 \sin \frac{2\pi}{3} = 2\left(\frac{\sqrt{3}}{2}\right) = \sqrt{3}.$$

(See Figure 1.8.) ◇

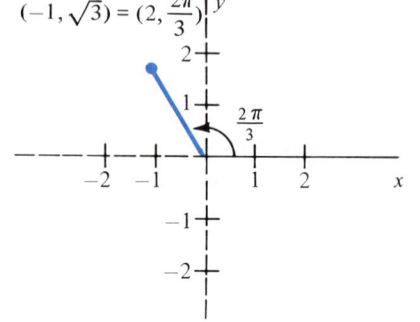

Figure 1.8

Example 3

To find the polar coordinates for the point with rectangular coordinates $(1, -1)$, we use the equations in (4):

$$r = \sqrt{x^2 + y^2} = \sqrt{1^2 + 1^2} = \sqrt{2}, \tag{5}$$

$$\tan \theta = \frac{y}{x} = \frac{-1}{1} = -1. \tag{6}$$

Now r is clearly determined by equation (5), but θ is *not* uniquely determined by equation (6), since both $\theta = 3\pi/4$ and $\theta = 7\pi/4$ satisfy $\tan \theta = -1$ (as do *all* an-

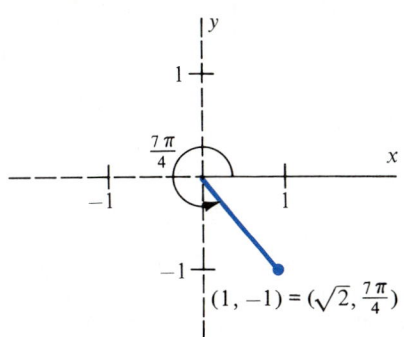

Figure 1.9

gles of the form $\theta = \dfrac{3\pi}{4} \pm n\pi$, $n = 1, 2, 3, \ldots$). The correct angle, $\theta = 7\pi/4$, is identified by noting that the point $(1, -1)$ lies in the fourth quadrant (Figure 1.9).

◇

REMARK: It is important to note, as in Example 3, that equations (4) do not uniquely determine θ for given values of x and y. This is because each value of $\tan \theta$ occurs twice for $\theta \in [0, 2\pi)$. This difficulty is overcome most simply by noting the quadrant in which the point (x, y) lies. An alternative approach is to first determine r from equation (4), and then determine θ as the (unique) simultaneous solution of equations (3): $x = r \cos \theta$; $y = r \sin \theta$.

Example 4

Graph the polar equation $r = 1 + \cos \theta$ and find the rectangular form for the equation.

Solution: One approach to graphing polar equations is to select several (convenient) values of θ (beginning with $\theta = 0$), calculate the corresponding values of r, and plot the points (r, θ) obtained. The graph is then completed by sketching a curve connecting these points. For the equation $r = 1 + \cos \theta$, we obtain the following points.

θ	0	$\dfrac{\pi}{6}$	$\dfrac{\pi}{4}$	$\dfrac{\pi}{3}$	$\dfrac{\pi}{2}$	$\dfrac{2\pi}{3}$	$\dfrac{3\pi}{4}$	$\dfrac{5\pi}{6}$	π
r	2	$\dfrac{2 + \sqrt{3}}{2}$	$\dfrac{2 + \sqrt{2}}{2}$	$\dfrac{3}{2}$	1	$\dfrac{1}{2}$	$\dfrac{2 - \sqrt{2}}{2}$	$\dfrac{2 - \sqrt{3}}{2}$	0

θ	$\dfrac{7\pi}{6}$	$\dfrac{5\pi}{4}$	$\dfrac{4\pi}{3}$	$\dfrac{3\pi}{2}$	$\dfrac{5\pi}{3}$	$\dfrac{7\pi}{4}$	$\dfrac{11\pi}{6}$	2π
r	$\dfrac{2 - \sqrt{3}}{2}$	$\dfrac{2 - \sqrt{2}}{2}$	$\dfrac{1}{2}$	1	$\dfrac{3}{2}$	$\dfrac{2 + \sqrt{2}}{2}$	$\dfrac{2 + \sqrt{3}}{2}$	2

The graph is the **cardioid** in Figure 1.10. In Section 14.2 we shall discuss a technique for graphing polar equations which addresses the problem of ensuring that points obtained as above are connected in the correct manner.

To find the rectangular form of the equation $r = 1 + \cos \theta$, we multiply both sides by r to obtain

$$r^2 = r + r \cos \theta.$$

Using the equations in (3) and (4), we obtain

$$x^2 + y^2 = \sqrt{x^2 + y^2} + x.$$

Clearly the original equation is simpler to graph than its rectangular counterpart.

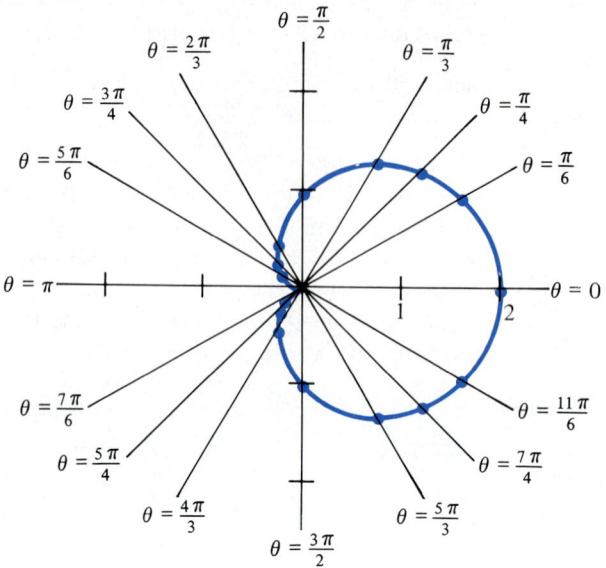

Figure 1.10 Cardioid $r = 1 + \cos\theta$.

Symmetry in Polar Coordinates

Perhaps you noticed in Example 4 that the graph of the equation $r = 1 + \cos\theta$ is symmetric about the x-axis. This is due to the identity $\cos(-\theta) = \cos\theta$, or $\cos(2\pi - \theta) = \cos\theta$, $0 \le \theta \le \pi$. More generally, graphs of polar equations $r = f(\theta)$ will be

(i) *symmetric about the x-axis* if $(r, -\theta)$ lies on the graph whenever (r, θ) does (Figure 1.11);

(ii) *symmetric about the y-axis* if $(r, \pi - \theta)$ lies on the graph whenever (r, θ) does (Figure 1.12);

(iii) *symmetric about the origin* if $(-r, \theta) = (r, \theta + \pi)$ lies on the graph whenever (r, θ) does (Figure 1.13).

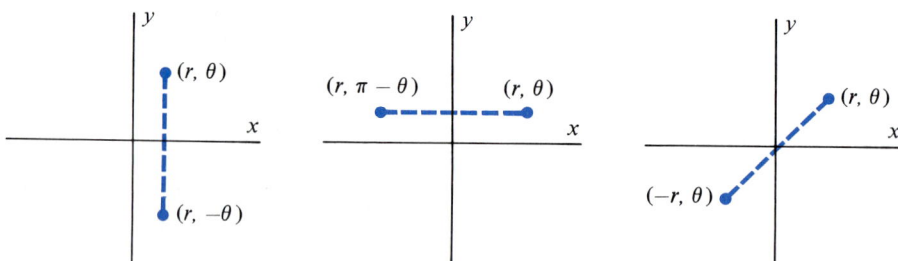

Figure 1.11 Symmetry about the x-axis.

Figure 1.12 Symmetry about the y-axis.

Figure 1.13 Symmetry about the origin.

Example 5

Graph the polar equation $r = \sin 2\theta$.

θ	0	$\dfrac{\pi}{6}$	$\dfrac{\pi}{4}$	$\dfrac{\pi}{3}$	$\dfrac{\pi}{2}$	$\dfrac{2\pi}{3}$	$\dfrac{3\pi}{4}$	$\dfrac{5\pi}{6}$	π
r	0	$\dfrac{\sqrt{3}}{2}$	1	$\dfrac{\sqrt{3}}{2}$	0	$\dfrac{-\sqrt{3}}{2}$	-1	$\dfrac{-\sqrt{3}}{2}$	0

Plotting these points gives the two leaves shown in Figure 1.14 (note that values of r are negative for $\pi/2 < \theta < \pi$).

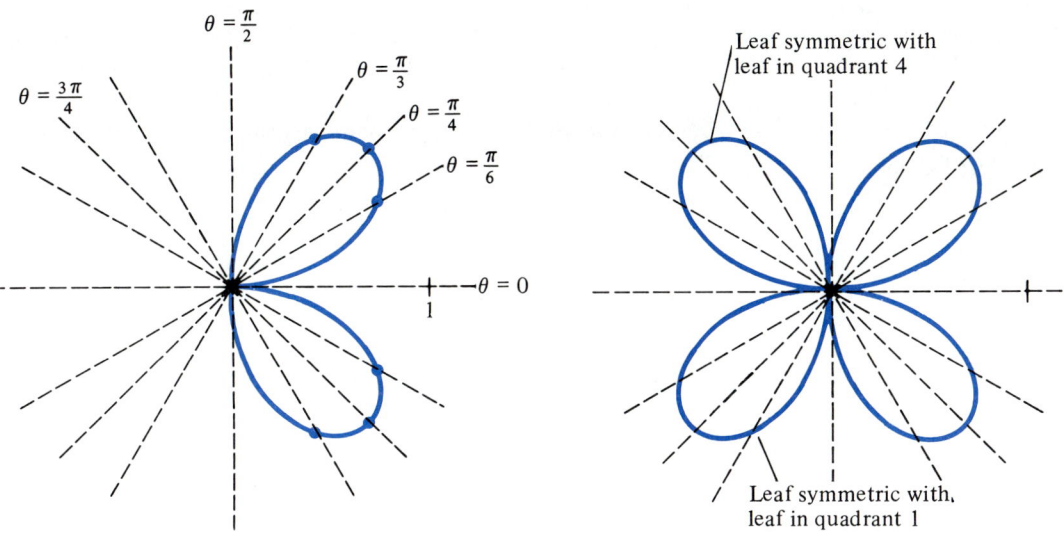

Figure 1.14 Graph of $r = \sin 2\theta$ for $0 \le \theta \le \pi$.

Figure 1.15 Graph of $r = \sin 2\theta$ for $0 \le \theta \le 2\pi$.

We can use symmetry considerations to obtain the remaining portion of the curve. To do so, we use the double angle identity to write

$$r(\theta) = \sin 2\theta = 2 \sin \theta \cos \theta.$$

Since $\sin(\theta + \pi) = -\sin \theta$ and $\cos(\theta + \pi) = -\cos \theta$, we obtain

$$r(\theta + \pi) = 2 \cdot \sin(\theta + \pi) \cos(\theta + \pi) = 2 \cdot \sin \theta \cos \theta = r(\theta).$$

This shows that $(r, \theta + \pi) = (-r, \theta)$ is on the graph whenever (r, θ) is. That is, we may sketch a leaf in the third quadrant symmetric with the leaf in the first quadrant, and a leaf in the second quadrant symmetric with the leaf in the fourth quadrant (Figure 1.15). (Actually we could obtain the entire graph by applying symmetry arguments just to the leaf obtained for $0 \le \theta \le \pi/2$. Can you see how this could be done?) The resulting figure is called a four-leaved rose. ◇

Example 6

For the polar equation $r = 2 \cos \theta$,

(a) sketch the graph;
(b) note any symmetries of the graph;
(c) find the rectangular form of the equation.

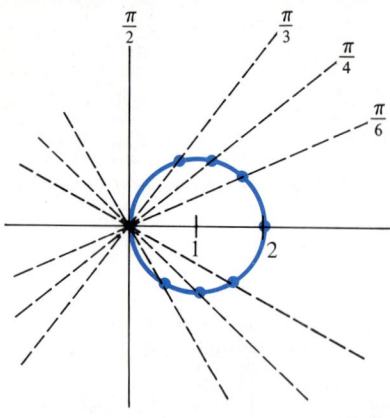

Figure 1.16 Graph of $r = 2 \cos \theta$.

Solution:

(a) For $0 \le \theta \le \pi$, we obtain the following points:

θ	0	$\dfrac{\pi}{6}$	$\dfrac{\pi}{4}$	$\dfrac{\pi}{3}$	$\dfrac{\pi}{2}$	$\dfrac{2\pi}{3}$	$\dfrac{3\pi}{4}$	$\dfrac{5\pi}{6}$	π
r	2	$\sqrt{3}$	$\sqrt{2}$	1	0	-1	$-\sqrt{2}$	$-\sqrt{3}$	-2

These points produce the curve shown in Figure 1.16.

For $\pi < \theta < 2\pi$, we note that $r(\theta_0 + \pi) = 2 \cos(\theta_0 + \pi) = -2 \cos(\theta_0) = -r(\theta_0)$ where $0 < \theta_0 < \pi$. Thus, these values simply trace over the same points a second time.

(b) Figure 1.16 suggests symmetry about the x-axis. This is indeed the case, since, for $r(\theta) = 2 \cos \theta$,

$$r(-\theta) = 2 \cos(-\theta) = 2 \cos \theta = r(\theta).$$

Thus, $(r, -\theta)$ is on the graph whenever (r, θ) is.

(c) To find the rectangular form for $r = 2 \cos \theta$, we multiply by r to obtain

$$r^2 = 2r \cos \theta.$$

Using equations (3) and (4) we obtain

$$x^2 + y^2 = 2x, \qquad \text{or} \qquad x^2 - 2x + y^2 = 0.$$

Completing the square in x gives

$$(x^2 - 2x + 1) + y^2 = 1$$

or

$$(x - 1)^2 + y^2 = 1.$$

The graph is therefore a circle of radius $r = 1$ with center $(1, 0)$. ◇

Exercise Set 14.1

In Exercises 1–8, find rectangular coordinates for the point given in polar coordinates.

1. $(1, \pi/2)$

2. $(3, \pi/6)$

3. $(0, \pi)$

4. $(-2, \pi/4)$

5. $(\sqrt{2}, -\pi/4)$

6. $\left(-1, -\dfrac{3\pi}{2}\right)$

7. $(-3, \pi)$

8. $(\pi, -\pi)$

In Exercises 9–16, find polar coordinates, subject to the stated restrictions on θ, for the point given in rectangular coordinates.

9. $(1, 1)$, $0 < \theta < \pi$

10. $(1, 1)$, $\pi < \theta < 2\pi$

11. $(-3, 0)$, $-\pi/2 < \theta < \pi/2$

12. $(1, -\sqrt{3})$, $\pi < \theta < 2\pi$

13. $(1, -\sqrt{3})$, $0 < \theta < \pi$

14. $(-2, 2)$, $0 < \theta < \pi$

15. $(-2, 2)$, $3\pi < \theta < 4\pi$

16. $(-3, 4)$, $\pi < \theta < 2\pi$

In Exercises 17–24, identify all symmetries (about the x-axis, the y-axis, or the origin) possessed by the graph of the given equation.

17. $r = 4 \cos \theta$

18. $r = 2 \sin \theta$

19. $r = 1 + \sin \theta$

20. $r^2 = \cos \theta$

21. $r^2 = \cos 2\theta$

22. $r = \sin 3\theta$

23. $r = 2$

24. $r = 4 \sin 2\theta$

In Exercises 25–32, find an equation in polar coordinates for the given equation.

25. $x^2 + y^2 = 4$

26. $y^2 = 16x$

27. $4x^2 + y^2 = 4$

28. $xy = 2$

29. $x = 6$

30. $y = 4$

31. $x^2 + y^2 + 2y = 0$

32. $y = x$

In Exercises 33–40, find an equation in rectangular coordinates for the given polar equation.

33. $r = 4 \sin \theta$

34. $r = 6$

35. $r = \tan \theta$

36. $r = \csc \theta$

37. $r = 4 \sec \theta$

38. $r = 1 + \sin \theta$

39. $r = \dfrac{1}{1 - \cos \theta}$

40. $r^2 = 4 \sec \theta$

In Exercises 41–50, sketch the graph of the given polar equation.

41. $r = \dfrac{1}{2}\theta$ (spiral of Archimedes)

42. $r = 2 \sin \theta$ (circle)

43. $r = 1 + \sin \theta$ (cardioid)

44. $r = \cos 2\theta$ (4-leaved rose)

45. $r = 1 + 2 \cos \theta$ (limaçon)

46. $r^2 = \cos 2\theta$ (lemniscate)

47. $r = e^{\theta}$, $\theta \geq 0$ (spiral)

48. $r = \dfrac{1}{e^{\theta}}$, $\theta \geq 0$

49. $r = 1 - 2 \cos \theta$ (limaçon)

50. $r = \sin 4\theta$ (8-leaved rose)

51. Prove that the graph of $r = 2 \sin \theta - 2 \cos \theta$ is a circle. Find its center and radius.

52. Prove that the graph of $r = a \sin \theta$ is a circle. Sketch the graph.

53. Prove that the graph of $r = a \cos \theta$ is a circle. Sketch the graph.

54. Show that equations (3) are valid for the case $r < 0$.

55. Show that the equation $y = ax$ of a nonvertical line through the origin is, in polar coordinates, $\tan \theta = a$.

56. The line $r = 4 \sec \theta$ is tangent to the graph of $r = a(1 + \cos \theta)$. Find a.

57. Find the area of the region enclosed by the graph of $r = 3 \cos \theta$. (*Hint:* See Exercise 53.)

58. Find the area of the region enclosed by the graph of $r^2 = 4 \cos^2 \theta$. (*Hint:* See Exercise 57.)

59. Find the vertices of the triangle determined by the lines
$$\tan \theta = 1 \quad \text{and} \quad r = 4 \sec \theta.$$
$$\theta = 0$$

60. Find the area of the triangle determined by the lines
$$\tan \theta = 2 \quad \text{and} \quad \tan \theta = -2.$$
$$r = 4 \csc \theta$$

14.2 GRAPHING TECHNIQUES FOR POLAR EQUATIONS

In Section 14.1 we graphed polar equations by plotting a few points (r_j, θ_j) and sketching in a curve. A slightly different approach is often simpler to use. That is simply to put the polar equation in the form $r = f(\theta)$, if possible, and then to "think dynamically," noting the behavior of r as θ sweeps out one or more revolutions about the pole.

For example, to graph the polar equation $r = a(1 + \cos \theta)$ we first note that the maximum value of $r = a(1 + 1) = 2a$ will occur when $\cos \theta = 1$, that is, when $\theta = 0$. The curve is therefore contained within the circle $r = 2a$ (Figure 2.1). To plot the curve we begin at the point $(2a, 0)$ and note that as θ increases from 0 to $\pi/2$, r *decreases* from length $r(0) = 2a$ to length $r(\pi/2) = a\left(1 + \cos \dfrac{\pi}{2}\right) = a$.

This observation enables us to sketch in the arc labelled 1 in Figure 2.2. We next note the behavior of r as θ increases from $\pi/2$ to π: r decreases from $r(\pi/2) = a$ to $r(\pi) = a(1 + (-1)) = 0$. This produces the arc labelled 2 in Figure 2.3. Similarly, arcs 3 and 4 are "swept out" as θ increases from π to 2π. Allowing θ to increase beyond 2π simply retraces the arcs already obtained, so the complete curve corresponding to $r = a(1 + \cos \theta)$ is obtained from $0 \leq \theta \leq 2\pi$ (Figure 2.3).

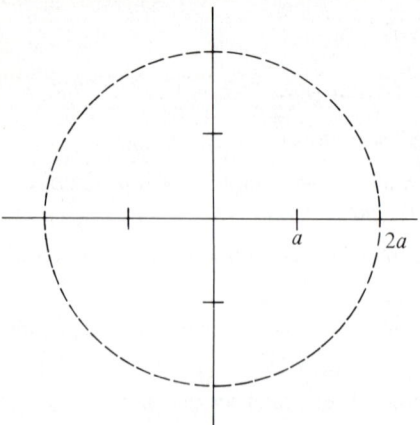

Figure 2.1 Circle $r = 2a$.

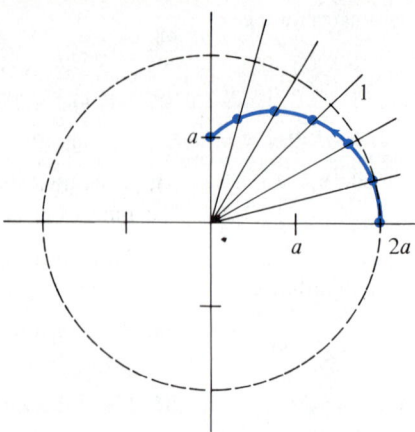

Figure 2.2 As θ increases toward $\pi/2$, r decreases from $2a$ to a.

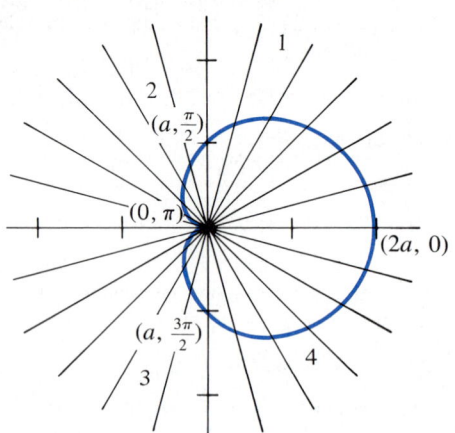

Figure 2.3 Graph of $r = a(1 + \cos\theta)$.

Example 1

Sketch the graph of the polar equation $r = 4 \sin 3\theta$.

Solution: Since $|\sin 3\theta| \leq 1$ for all θ, the curve must lie within the circle $r = 4$. Beginning at the point with $\theta = 0$, we note the following:

(a) As θ turns from 0 to $\pi/6$, 3θ turns from 0 to $\pi/2$, so r *increases* from $r(0) = 4 \sin 0 = 0$ to $r(\pi/6) = 4 \sin 3(\pi/6) = 4$. This produces arc 1 in Figure 2.4.

(b) As θ turns from $\pi/6$ to $\pi/3$, 3θ turns from $\pi/2$ to π, so r *decreases* from $r(\pi/6) = 4$ to $r(\pi/3) = 0$ (arc 2 in Figure 2.4).

(c) As θ turns from $\pi/3$ to $\pi/2$, 3θ turns from π to $3\pi/2$, so r *decreases* from $r(\pi/3) = 0$ to $r(\pi/2) = 4 \sin 3(\pi/2) = -4$, producing arc 3 in Figure 2.5.

(d) As θ increases through the next quarter turn, from $\pi/2$ to π, 3θ increases from $3\pi/2$ to 3π. Thus, r increases from -4 to 4 for $\pi/2 \leq \theta \leq 5\pi/6$, producing arcs 4 and 5, and decreases from 4 to 0 for $5\pi/6 \leq \theta \leq \pi$, producing arc 6 (Figure 2.6).

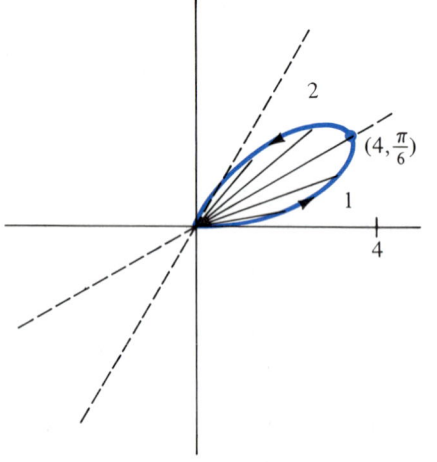

Figure 2.4 $r = 4 \sin 3\theta$, $0 \leq \theta \leq \pi/3$.

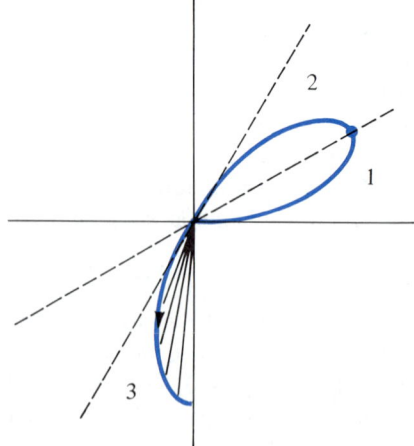

Figure 2.5 $r = 4 \sin 3\theta$, $0 \leq \theta \leq \pi/2$.

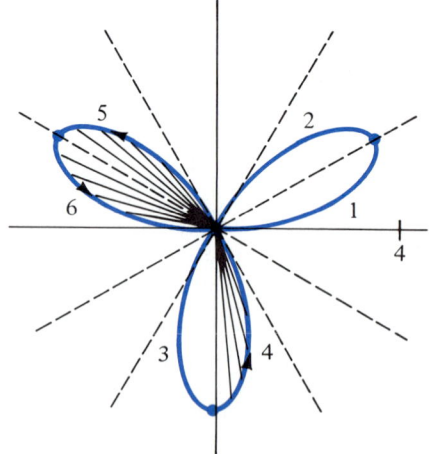

Figure 2.6 Three-leaved rose $r = 4 \sin 3\theta$, $0 \leq \theta \leq \pi$.

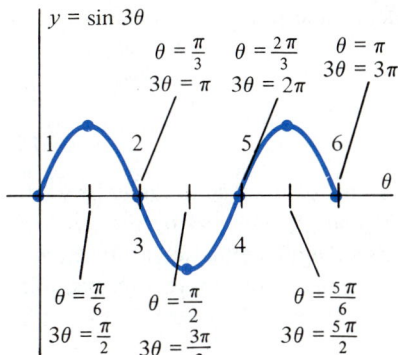

Figure 2.7 Graph of $f(\theta) = \sin 3\theta$.

By noting the behavior of r for $\theta > \pi$, we can see that the entire curve (a three-leaved rose) has been obtained with only $0 \leq \theta \leq \pi$. (Figure 2.7 recalls the graph of the function $f(\theta) = \sin 3\theta$. Note how the arcs numbered 1–6 correspond to the arcs in Figures 2.4–2.6.) ◇

Example 2

Graph the lemniscate $r^2 = a \cos 2\theta$, $a > 0$.

Solution: Taking square roots of both sides shows that the given equation corresponds to the *pair* of equations

$$r_1 = \sqrt{a \cos 2\theta} \tag{1}$$

and

$$r_2 = -\sqrt{a \cos 2\theta}. \tag{2}$$

Thus, some values of θ will correspond to two values r, while others will not correspond to any real value r.

Beginning with $\theta = 0$, we observe that

(i) As θ increases from 0 to $\pi/4$, 2θ increases from 0 to $\pi/2$. Thus,

    (a) r_1 decreases from $r_1(0) = \sqrt{a}$ to $r_1(\pi/4) = 0$ (arc 1 in Figure 2.8),
    (b) r_2 increases from $r_2(0) = -\sqrt{a}$ to $r_2(\pi/4) = 0$ (arc 2).

(ii) As θ increases from $\pi/4$ to $3\pi/4$, 2θ increases from $\pi/2$ to $3\pi/2$. Since $\cos 2\theta < 0$ for $\pi/2 < 2\theta < 3\pi/2$, equations (1) and (2) have no solutions in this interval.

(iii) As θ increases from $3\pi/4$ to π, 2θ increases from $3\pi/2$ to 2π. Thus,

    (a) r_1 increases from $r_1(3\pi/2) = 0$ to $r_1(2\pi) = \sqrt{a}$ (arc 3 in Figure 2.9),
    (b) r_2 decreases from $r_2(3\pi/2) = 0$ to $r_2(2\pi) = -\sqrt{a}$ (arc 4).

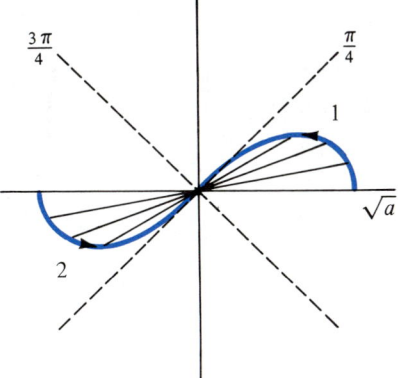

Figure 2.8 $r^2 = a \cos 2\theta$, $0 \leq \theta \leq \pi/4$.

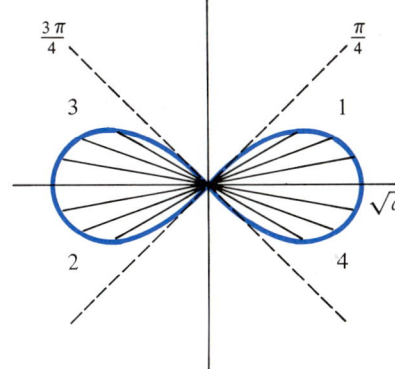

Figure 2.9 $r^2 = a \cos 2\theta$.

Checking equations (1) and (2) for values of $\theta > \pi$ shows that all points on the graph are obtained by using only $0 \leq \theta \leq \pi/4$ and $3\pi/4 \leq \theta \leq \pi$. ◇

Intersections of Graphs in Polar Coordinates

In Section 14.3, we will need to be able to determine points of intersection for graphs of pairs of equations of the form

$$r_1 = f(\theta), \tag{3}$$

$$r_2 = g(\theta). \tag{4}$$

As for equations in rectangular coordinates, points of intersection may be found by equating the right-hand sides of equations (3) and (4) and solving the resulting equation for θ. However, unlike the situation for rectangular equations, *this method will not necessarily produce all points of intersection,* as the following examples show.

Example 3

Find all points of intersection for the graphs of the equations $r_1 = 1 + \cos \theta$ and $r_2 = 3 \cos \theta$.

Solution: Setting $r_1 = r_2$ gives the equation

$$1 + \cos \theta = 3 \cos \theta,$$

so

$$\cos \theta = 1/2.$$

The solutions of this equation for $-\pi < \theta < \pi$ are $\theta_1 = \pi/3$ and $\theta_2 = -\pi/3$. The corresponding points of intersection are $(3/2, \pi/3)$ and $(3/2, -\pi/3)$. However, as Figure 2.10 illustrates, the *pole* $(0, \theta)$ is also common to both graphs. Setting

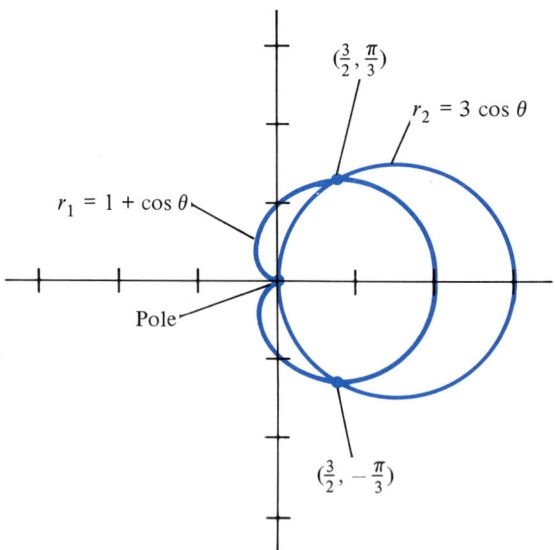

Figure 2.10 Graphs of r_1 and r_2 intersect in 3 points.

$r_1(\theta) = r_2(\theta)$ misses this point, since the pole corresponds to $\theta = \pi$ in the graph of r_1, but it corresponds to $\theta = \pi/2$ in the graph of r_2. ◇

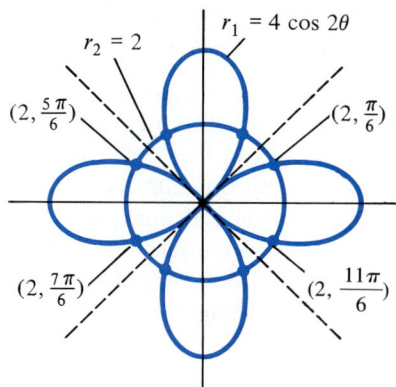

Figure 2.11 Graphs of r_1 and r_2 intersect in 8 points.

Example 4

Find all points of intersection of the four-leaved rose $r_1 = 4 \cos 2\theta$ and the circle $r_2 = 2$.

Solution: Setting $r_1 = r_2$ gives $4 \cos 2\theta = 2$, or $\cos 2\theta = 1/2$. Thus $2\theta = \pm\pi/3$ or $2\theta = \pm 7\pi/3$. The four distinct solutions of this equation in the interval $(0, 2\pi)$ are $\theta = \pi/6$, $5\pi/6$, $7\pi/6$, and $11\pi/6$. These give the four points of intersection $(2, \pi/6)$, $(2, 5\pi/6)$, $(2, 7\pi/6)$, and $(2, 11\pi/6)$. However, the method of setting $r_1 = r_2$ has failed to detect *four* additional points of intersection $(2, \pi/3)$, $(2, 2\pi/3)$, $(2, 4\pi/3)$, and $(2, 5\pi/3)$ (see Figure 2.11). The reason for this is that each of these last four points corresponds to *different* values of θ on the graph of r_1 than on the graph of r_2. ◇

Examples 3 and 4 show that simply setting $r_1 = r_2$ and solving the resulting equation $f(\theta) = g(\theta)$ for θ will not necessarily yield all points common to the graphs of equations (3) and (4). The reason for this difficulty is the nonuniqueness of polar coordinates. For example, the point $(2, \pi/3)$, common to both graphs $r_1 = 4 \cos 2\theta$ and $r_2 = 2$ in Example 4, actually occurs on the graph of r_1 in its equivalent form $(-2, 4\pi/3)$. Since these coordinates do not satisfy $r_2 = 2$, the point does not result from setting $r_1 = r_2$. Rather than trying to work out an algorithm allowing for all possible ways in which points of intersection of two polar equations can arise, *you should simply develop the habit of always sketching the two curves to ensure that all such points are found.*

Exercise Set 14.2

In Exercises 1–10, sketch the graph of the given polar equation using the method of Examples 1 and 2.

1. $r = a \cos \theta$

2. $r = a \sin \theta$

3. $r = a \sin 2\theta$

4. $r = a(1 + \cos \theta)$

5. $r = a(1 - \sin \theta)$

6. $r = 1 + 2 \sin \theta$

7. $r = 2\theta$

8. $r^2 = 4 \cos 2\theta$

9. $r = \sin \theta + \cos \theta$

10. $r = a(1 - \cos \theta)$

In Exercises 11–20, find all points of intersection of the graphs of the two given equations.

11. $r_1 = 2 \cos \theta$
$r_2 = 2 \sin \theta$

12. $r_1 = 2 \cos \theta$
$r_2 = 1$

13. $r_1 = 1 + \sin \theta$
$r_2 = 3 \sin \theta$

14. $r_1 = a \sin 3\theta$
$r_2 = a$

15. $r_1 = a(1 + \cos \theta)$
$r_2 = 3a \cos \theta$

16. $r_1 = 2a \cos 2\theta$
$r_2 = a$

17. $r_1 = a(1 + \cos \theta)$
$r_2 = a(1 - \cos \theta)$

18. $r_1 = \cos 2\theta$
$r_2 = \dfrac{\sqrt{2}}{2}$

19. $r = \dfrac{1}{\theta}$
$\theta = \pi/4$

20. $r_1 = a \sin 4\theta$
$r_2 = a$

21. Show that the graphs of the polar equations
$$r_1 = \frac{a}{1 \pm \cos \theta}; \qquad r_2 = \frac{a}{1 \pm \sin \theta}$$
are parabolas. (*Hint:* Convert to rectangular equations.)

22. Show that the graphs of the polar equations
$$r_1 = \frac{ab}{1 \pm b \cos \theta}; \qquad r_2 = \frac{ab}{1 \pm b \sin \theta}$$
are ellipses if $0 < b < 1$. (*Hint:* Convert to rectangular equations.)

In Exercises 23–28, find a rectangular form for the given polar equation, identify the graph as either a parabola or an ellipse, and sketch the graph. (See Exercises 21 and 22.)

23. $r = \dfrac{1}{1 + \cos \theta}$

24. $r = \dfrac{4}{1 - \sin \theta}$

25. $r = \dfrac{2}{1 + \dfrac{1}{2} \cos \theta}$

26. $2r = \dfrac{4}{2 - \sin \theta}$

27. $r - r \cos \theta = 2$

28. $3r + r \cos \theta - 6 = 0$

14.3 CALCULATING AREA IN POLAR COORDINATES

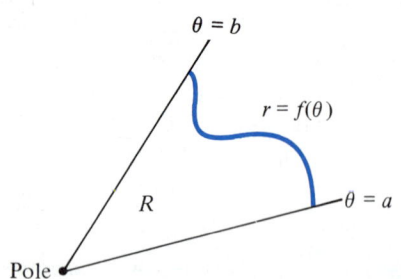

Figure 3.1 Region determined by $r = f(\theta)$, $a \leq \theta \leq b$.

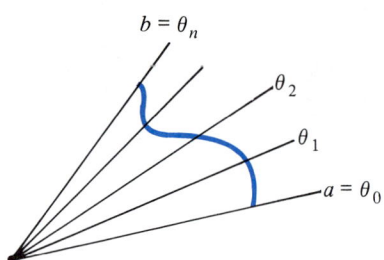

Figure 3.2 Partition $\theta_0 < \theta_1 < \cdots < \theta_n$ slices R into wedge-shaped regions.

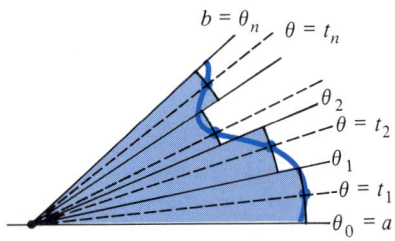

Figure 3.3 Area of R approximated by areas of sectors of circles.

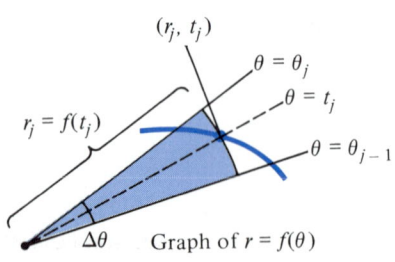

Figure 3.4 $\Delta A_j = \dfrac{1}{2} r_j^2 \, \Delta\theta$.

When a region R in the plane is determined by the graph of an equation written in the *polar* form $r = f(\theta)$, the area of R cannot be calculated by a direct application of the equations developed in Chapter 6. The difficulty lies in the fact that these formulas were obtained for functions $y = g(x)$ written in *rectangular* coordinates. To calculate areas of regions determined by the graphs of polar equations, we might try rewriting such equations in rectangular coordinates, using the substitutions $x = r \cos \theta$ and $y = r \sin \theta$, as in Section 14.1. However, this strategy often leads to one of two undesirable consequences. Either the resulting rectangular equations are much more complicated than the original polar equations, or, even worse, the resulting equations might not describe either variable as a *function* of the other.

A more satisfactory approach is to return to the basic theory of the definite integral and determine the formulas appropriate for calculating area in polar coordinates. The small amount of time spent in this effort will allow us to avoid the step of converting to rectangular coordinates for each area problem in polar coordinates.

Let us suppose that a region R is bounded by the graph of the function $r = f(\theta)$ for θ between the numbers $\theta = a$ and $\theta = b$ (Figure 3.1). Since θ is the independent variable, we partition the interval $[a, b]$ into n equal subintervals by using the increment $\Delta\theta = \dfrac{b - a}{n}$ and the endpoints

$$\theta_0 = a, \qquad \theta_1 = a + \Delta\theta, \qquad \theta_2 = a + 2\,\Delta\theta, \ldots, \qquad \theta_n = a + n\,\Delta\theta = b.$$

The rays $\theta = \theta_j$, $j = 0, 1, 2, \ldots, n$ then "slice" the region R into n wedge-shaped subregions (Figure 3.2).

We now approximate the area of each of the subregions. To do so we choose one number t_j in each interval $[\theta_{j-1}, \theta_j]$, $j = 1, 2, \ldots, n$ (Figure 3.3). We then approximate the area of the jth subregion by the area of the circular **sector** of constant radius $r_j = f(t_j)$ and angle $\Delta\theta$ (see Figure 3.4).

Since the area of the circle of radius r_j is πr_j^2, the area of the circular sector comprising the fractional part $\dfrac{\Delta\theta}{2\pi}$ of the entire circle is

$$\Delta A_j = \pi r_j^2 \left(\frac{\Delta\theta}{2\pi} \right) = \frac{1}{2} r_j^2 \, \Delta\theta = \frac{1}{2} [f(t_j)]^2 \, \Delta\theta.$$

Summing these approximations, one for each subregion, gives an approximation to the area A of the whole region R of the form

$$A \approx \sum_{j=1}^{n} \Delta A_j = \sum_{j=1}^{n} \frac{1}{2} [f(t_j)]^2 \, \Delta\theta. \tag{1}$$

If the function $r = f(\theta)$ is continuous for $a \leq \theta \leq b$, the sum on the right-hand side of approximation (1) is a *Riemann* sum for the function $\dfrac{1}{2}[f(\theta)]^2$ which "converges" to the definite integral

$$\int_a^b \frac{1}{2} \cdot [f(\theta)]^2 \, d\theta = \lim_{n \to \infty} \sum_{j=1}^{n} \frac{1}{2} [f(t_j)]^2 \, \Delta\theta \tag{2}$$

as $n \to \infty$. We therefore argue that as $n \to \infty$, the circular sectors provide an increas-

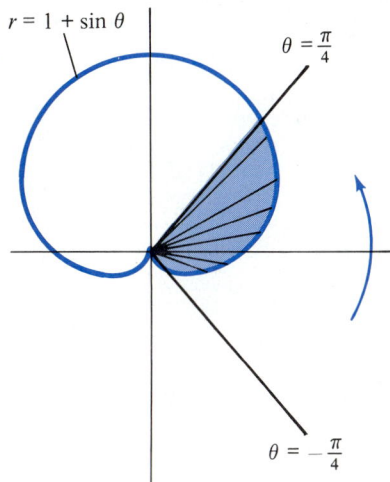

$r = 1 + \sin \theta$

$\theta = \dfrac{\pi}{4}$

$\theta = -\dfrac{\pi}{4}$

Figure 3.5 $-\pi/4 \le \theta \le \pi/4$.

ingly accurate approximation to the region R and that the area of R is correctly defined as follows.

The area A of the region bounded by the graph of the continuous function $r = f(\theta)$ between the rays $\theta = a$ and $\theta = b$, $a < b$, is given by the definite integral

$$A = \int_a^b \frac{1}{2}[f(\theta)]^2 \, d\theta.$$
(3)

The same cautions should be observed in using equation (3) as in calculating areas in rectangular coordinates. The region should first be carefully sketched, so that the proper limits of integration may be determined.

Example 1

Find the area of the region bounded by the rays $\theta = -\pi/4$ and $\theta = \pi/4$, and the graph of the equation $r = 1 + \sin \theta$.

Solution: The region is the portion of the cardioid swept out by the radius of length $r = 1 + \sin \theta$ as θ increases from $\theta = -\pi/4$ to $\theta = \pi/4$ (Figure 3.5). By equation (3) the area is

$$A = \int_{-\pi/4}^{\pi/4} \frac{1}{2}[1 + \sin \theta]^2 \, d\theta$$

$$= \int_{-\pi/4}^{\pi/4} \frac{1}{2}[1 + 2 \sin \theta + \sin^2 \theta] \, d\theta$$

$$= \int_{-\pi/4}^{\pi/4} \frac{1}{2}\left[1 + 2 \sin \theta + \left(\frac{1}{2} - \frac{1}{2} \cos 2\theta\right)\right] d\theta \quad \left(\text{using the identity}\right.$$

$$= \int_{-\pi/4}^{\pi/4} \frac{1}{2}\left[\frac{3}{2} + 2 \sin \theta - \frac{1}{2} \cos 2\theta\right] d\theta \quad \left. \sin^2 \theta = \frac{1}{2} - \frac{1}{2} \cos 2\theta\right)$$

$$= \frac{1}{2}\left[\frac{3\theta}{2} - 2 \cos \theta - \frac{1}{4} \sin 2\theta\right]_{-\pi/4}^{\pi/4}$$

$$= \frac{1}{2}\left\{\left(\frac{3\pi}{8} - \sqrt{2} - \frac{1}{4}\right) - \left(-\frac{3\pi}{8} - \sqrt{2} + \frac{1}{4}\right)\right\}$$

$$= \frac{3\pi}{8} - \frac{1}{4} \approx 0.928. \qquad \diamond$$

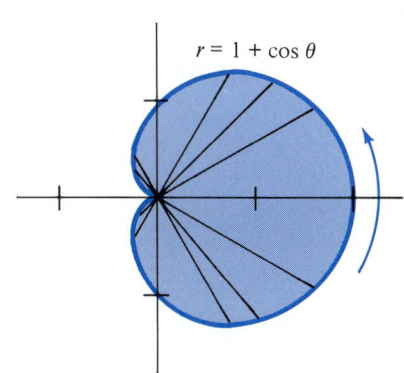

$r = 1 + \cos \theta$

Figure 3.6 $-\pi \le \theta \le \pi$.

Example 2

Find the area of the region enclosed by the graph of the cardioid $r = 1 + \cos \theta$.

Strategy

Sketch the region.

Find a pair of smallest and largest values of θ required to sweep out the region. These are the limits of integration. (Note that $0 \le \theta \le 2\pi$ works just as well as $-\pi \le \theta \le \pi$.)

Solution

The cardioid is sketched in Figure 3.6. Since the cardioid is the graph of the function $f(\theta) = 1 + \cos \theta$ as θ increases from $\theta = -\pi$ to $\theta = \pi$, the limits of integration are $a = -\pi$ and $b = \pi$. By equation (3) the area is

$$A = \int_{-\pi}^{\pi} \frac{1}{2}[1 + \cos \theta]^2 \, d\theta$$

Set up the integral given by equation (3).

$$= \frac{1}{2} \int_{-\pi}^{\pi} [1 + 2 \cos \theta + \cos^2 \theta] \, d\theta$$

Square $f(\theta)$ in the integrand and simplify, using the identity

$$\cos^2 \theta = \frac{1}{2} + \frac{1}{2} \cos 2\theta.$$

$$= \frac{1}{2} \int_{-\pi}^{\pi} \left[1 + 2 \cos \theta + \left(\frac{1}{2} + \frac{1}{2} \cos 2\theta \right) \right] d\theta$$

$$= \frac{1}{2} \int_{-\pi}^{\pi} \left[\frac{3}{2} + 2 \cos \theta + \frac{1}{2} \cos 2\theta \right] d\theta$$

Integrate.

$$= \frac{1}{2} \left[\frac{3\theta}{2} + 2 \sin \theta + \frac{1}{4} \sin 2\theta \right]_{-\pi}^{\pi}$$

$$= \frac{1}{2} \left\{ \left(\frac{3\pi}{2} + 2 \cdot 0 + \frac{1}{4} \cdot 0 \right) - \left(\frac{3}{2}(-\pi) + 2 \cdot 0 + \frac{1}{4} \cdot 0 \right) \right\}$$

$$= \frac{3\pi}{2}. \qquad \diamond$$

Often, symmetry considerations may be used to simplify the calculation of area, as the following example illustrates.

Example 3

Find the area of the region enclosed by the graph of the polar equation $r = a \sin 3\theta$.

Solution: The region is a three-leaved rose, as illustrated in Figure 3.7. Since the 3 leaves are congruent, and since the leaf determined by $0 \leq \theta \leq \pi/3$ is symmetric about the ray $\theta = \pi/6$, we may obtain the area of the entire figure by calculating the area of the half leaf determined by $0 \leq \theta \leq \pi/6$ and multiplying the result by 6. We obtain

$$A = 6 \cdot \int_0^{\pi/6} \frac{1}{2} [a \sin 3\theta]^2 \, d\theta$$

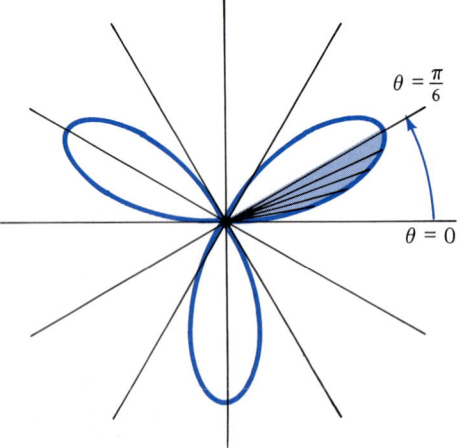

Figure 3.7 Area of three-leaved rose is 6 times area of region determined by $0 \leq \theta \leq \pi/6$.

$$= 3a^2 \int_0^{\pi/6} \sin^2 3\theta \, d\theta$$

$$= 3a^2 \int_0^{\pi/6} \left[\frac{1}{2} - \frac{1}{2} \cos 6\theta \right] d\theta$$

$$= 3a^2 \left[\frac{\theta}{2} - \frac{1}{12} \sin 6\theta \right]_0^{\pi/6}$$

$$= \frac{a^2 \pi}{4}. \qquad \diamond$$

As happens with equations in rectangular coordinates, an equation in polar coordinates may fail to determine r as a function of θ. The following example shows how symmetry may be used to overcome this difficulty.

Example 4

Find the area of the region enclosed by the lemniscate $r^2 = \cos 2\theta$.

Strategy

Sketch the figure. (Recall Example 2, Section 14.2.)

Take square roots to solve for r. Obtain a *function* $r = f(\theta)$.

Solution

The lemniscate is sketched in Figure 3.8. By extracting square roots on both sides of the equation $r^2 = \cos 2\theta$ we obtain the equation $r = \pm\sqrt{\cos 2\theta}$. By choosing the positive sign we obtain the function

$$f(\theta) = \sqrt{\cos 2\theta}.$$

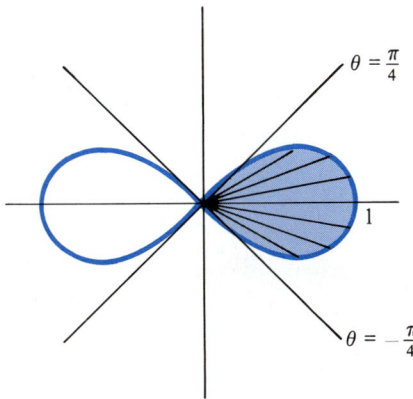

Figure 3.8 $-\pi/4 \le \theta \le \pi/4$ determines half the area of the lemniscate $r^2 = \cos 2\theta$.

Compare region determined by the function with that of the original equation.

This function is defined only for

$$-\frac{\pi}{4} \le \theta \le \frac{\pi}{4} \qquad \text{and} \qquad \frac{3\pi}{4} \le \theta \le \frac{5\pi}{4}.$$

Use symmetry to simplify calculation.

However, as θ ranges through these two intervals the two lobes of the lemniscate are traced out. We may therefore integrate over $-\pi/4 \le \theta \le \pi/4$ to obtain the area of one lobe and double the result. We obtain

Apply equation (3).

$$A = 2 \int_{-\pi/4}^{\pi/4} \frac{1}{2} [\sqrt{\cos 2\theta}]^2 \, d\theta$$

$$= \int_{-\pi/4}^{\pi/4} \cos 2\theta \, d\theta$$

$$= \frac{1}{2} \sin 2\theta \Big]_{-\pi/4}^{\pi/4}$$

$$= 1.$$

◇

Area Between Two Curves

When a region R lies between the graphs of two polar equations, as in Figure 3.9, we may calculate the area of R by subtracting the area enclosed by the inner curve from area enclosed by the outer curve. That is, if R lies inside the graph of $r = f(\theta)$ and outside the graph of $r = g(\theta)$, for $a \le \theta \le b$, the area A of R is given by the integral

$$A = \int_a^b \frac{1}{2} [f(\theta)]^2 \, d\theta - \int_a^b \frac{1}{2} [g(\theta)]^2 \, d\theta. \tag{4}$$

Example 5

Find the area of the region lying inside the circle $r = 3 \cos \theta$ and outside the cardioid $r = 1 + \cos \theta$.

Strategy

Sketch the region.
Determine the points of intersection.
This gives the limits of integration.

Solution

The graphs of the two equations are sketched in Figure 3.10. By solving the two equations simultaneously, we find the points of intersection to be

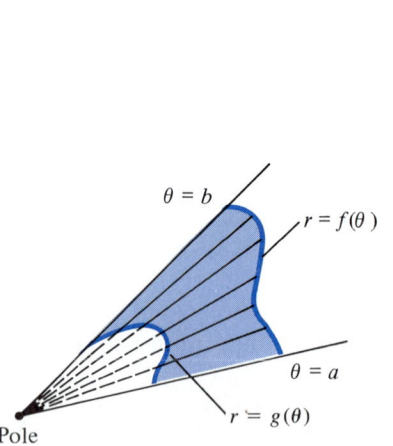

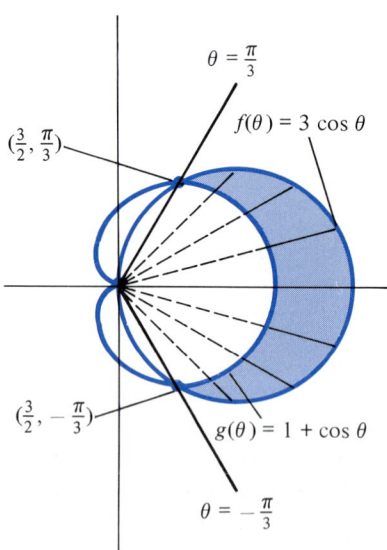

Figure 3.9 Region between graphs of $r = f(\theta)$ and $r = g(\theta)$ for $a \le \theta \le b$.

Figure 3.10

(3/2, $\pm\pi/3$) (see Example 3, Section 14.2). The limits of integration are therefore $-\pi/3 \leq \theta \leq \pi/3$.

Determine which is the outer curve.

In this interval, the outer (greater) function is $f(\theta) = 3 \cos \theta$. The inner function is $g(\theta) = 1 + \cos \theta$. By equation (4) the area is

Apply equation (4).

$$A = \int_{-\pi/3}^{\pi/3} \frac{1}{2}[3 \cos \theta]^2 \, d\theta - \int_{-\pi/3}^{\pi/3} \frac{1}{2}[1 + \cos \theta]^2 \, d\theta$$

Simplify integrand.

$$= \int_{-\pi/3}^{\pi/3} \frac{1}{2}\{9 \cos^2 \theta - (1 + 2 \cos \theta + \cos^2 \theta)\} \, d\theta$$

$$= \int_{-\pi/3}^{\pi/3} \frac{1}{2}[8 \cos^2 \theta - 1 - 2 \cos \theta] \, d\theta$$

$\cos^2 \theta = \dfrac{1}{2} + \dfrac{1}{2} \cos 2\theta.$

$$= \int_{-\pi/3}^{\pi/3} \frac{1}{2}\left[8\left(\frac{1}{2} + \frac{1}{2} \cos 2\theta\right) - 1 - 2 \cos \theta\right] d\theta$$

$$= \int_{-\pi/3}^{\pi/3} \left(\frac{3}{2} + 2 \cos 2\theta - \cos \theta\right) d\theta$$

$$= \frac{3\theta}{2} + \sin 2\theta - \sin \theta \Big]_{-\pi/3}^{\pi/3}$$

$$= \pi. \qquad \Diamond$$

Exercise Set 14.3

In Exercises 1–8, find the area of the region determined by the given equations and inequalities.

1. $r = 1 + \sin \theta,$ $\quad -\pi/2 < \theta < \pi/2$

2. $r = a \sin 2\theta,$ $\quad 0 \leq \theta \leq \pi/2$

3. $r = 2 \sin \theta,$ $\quad 0 \leq \theta \leq \pi$

4. $r = \theta,$ $\quad 0 \leq \theta \leq \pi$

5. $r = 2 + \sin \theta,$ $\quad 0 \leq \theta \leq \pi$

6. $r = a \cos 3\theta,$ $\quad \pi/2 < \theta < \dfrac{2\pi}{3}$

7. $r = 4 + \sin \theta,$ $\quad \pi/4 \leq \theta \leq \dfrac{3\pi}{4}$

8. $r = 3 \cos 2\theta,$ $\quad -\dfrac{\pi}{4} \leq \theta \leq \dfrac{\pi}{4}$

In Exercises 9–14, find the area of the region enclosed by the graph of the given equation.

9. $r = 2 \cos \theta$

10. $r = 1 + \cos \theta$

11. $r = a \sin 2\theta$

12. $r = a \sin 4\theta$

13. $r = a \cos 3\theta$

14. $r^2 = \cos \theta$

In Exercises 15–20, find the area of the region described.

15. The region inside the cardioid $r = 1 + \sin \theta$ and outside the circle $r = 2 \sin \theta$.

16. The region inside the cardioid $r = 1 + \cos \theta$ and outside the circle $r = 3 \cos \theta$. (*Hint:* You will need to treat the intervals $\left[\dfrac{\pi}{3}, \dfrac{\pi}{2}\right]$ and $\left[\dfrac{\pi}{2}, \pi\right]$ separately.)

17. The regions common to both cardioids $r = a(1 + \cos \theta)$ and $r = a(1 - \cos \theta)$.

18. The region outside the three-leaved rose $r = 4 \sin 3\theta$ and inside the circle $r = 4$.

19. The region common to the circles $r = 2 \cos \theta$ and $r = 2 \sin \theta$.

20. The region common to the circle $r = \dfrac{\sqrt{2}}{2}$ and the lemniscate $r^2 = \cos 2\theta$.

21. Find the area of the region bounded by the graphs of the spirals $r_1 = \theta$ and $r_2 = e^\theta$ for $0 \leq \theta \leq \pi$.

22. Find the area of the region inside the graph of $r = 3 + \sin \theta$ and outside the graph of $r = 4 \sin \theta$.

23. Find the area of the region outside the graph of $r = 1 + \cos \theta$ and inside the graph of $r = 2 - \cos \theta$.

24. Find the area of the region bounded by the graph of the equation $r - r \cos \theta = 2$ and the line $\theta = \dfrac{\pi}{2}$.

25. Show that the result of calculating the area of the region enclosed by the graph of $r = 2 \cos \theta$ in polar coordinates agrees with the result of calculating the area of the same region in rectangular coordinates.

26. Let $f(\theta) \geq g(\theta)$ for all θ. Sketch the region whose area is given by the integral $\int_a^b \frac{1}{2}[f(\theta) - g(\theta)]^2 \, d\theta$. Compare this region with the region whose area is given by the integral in equation (4). Show that, in general, the two integrals are not equal.

27. *(Computer)* Program 7 in Appendix I is a BASIC program that approximates $\int_a^b \frac{1}{2}[f(\theta)]^2 \, d\theta$ for the function $f(\theta) = 1 + \sqrt{\sin \theta}$.

a. Use Program 7 to approximate the integral $\int_0^{\pi/2} \frac{1}{2}[1 + \sqrt{\sin \theta}]^2 \, d\theta$ with $n = 10$, 100, and 200.

b. Use Program 7 to approximate the area of the region enclosed by the graph of $f(\theta) = 1 + \sqrt{\sin \theta}$ for $0 \leq \theta \leq \pi$.

c. Modify Program 7 to approximate the area of the region enclosed by the graph of $f(\theta) = 1 - \sqrt{\cos \theta}$, $0 \leq \theta \leq \pi/2$.

14.4 PARAMETRIC EQUATIONS

Parametric equations provide a means of describing a curve in the plane by functions, even though the curve may not represent the graph of a function of the form $y = f(x)$. Rather than describing either coordinate as a function of the other, we allow both coordinates to be functions of a third (independent) variable, called a **parameter,*** which is usually denoted by t. That is, we describe the curve as the collection of points $P(t) = (x(t), y(t))$ where the parameter t ranges through some specified domain.

For example, the graph of the equation

$$x^2 + y^2 = 1 \tag{1}$$

is the familiar unit circle in the plane. However, equation (1) does not determine y as a function of x, since two distinct values of y correspond to each x with $-1 < x < 1$. We may use our knowledge of trigonometry to obtain parametric equations, or a **parameterization,** for the unit circle as follows. Let t denote the angle formed between the positive x-axis and the radius from $(0, 0)$ to (x, y), measured in the counterclockwise direction. Then

$$x = \cos t$$

$$y = \sin t$$

are the coordinate functions of the point (x, y). By inspection we can see that as t increases from $t = 0$ to $t = 2\pi$, the point $(x, y) = (\cos t, \sin t)$ traverses the unit circle once in the counterclockwise direction (Figure 4.1). We therefore say that a parameterization of the unit circle is given by the equations

$$\left. \begin{array}{l} x(t) = \cos t \\ y(t) = \sin t \end{array} \right\} \quad 0 \leq t < 2\pi. \tag{2a}\tag{2b}$$

It is important to note that parameterizations for curves are not unique. For example, each of the following pairs of equations also represents a parameterization

---

*The term *parameter* denotes a variable that often has no geometric or physical interpretation. Here the parameter t has no geometric interpretation with respect to the curve it helps describe. In some applications the parameter t represents time, while the variables x and y represent coordinates of points in the plane.

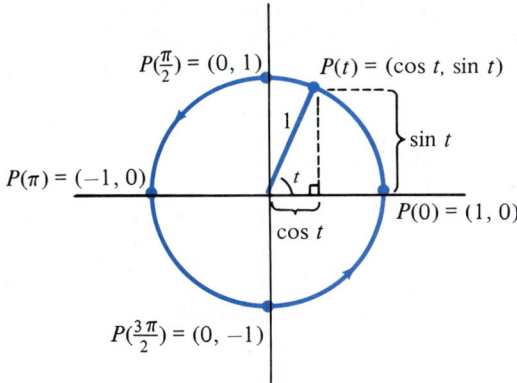

Figure 4.1 Parameterization of unit circle: $(x(t),$ $y(t)) = (\cos t, \sin t), 0 \leq t < 2\pi.$

of the unit circle

$$\left.\begin{array}{l} x(t) = \cos 2t \\ y(t) = \sin 2t \end{array}\right\} \; 0 \leq t < \pi,$$

$$\left.\begin{array}{l} x(t) = \cos t \\ y(t) = \sin(-t) \end{array}\right\} \; 0 \leq t < 2\pi.$$

Example 1

Find parametric equations for the line ℓ containing the point $P_0 = (2, 4)$ and with slope 3.

Solution: We first find the equation for ℓ in xy-coordinates, using the point-slope form. We obtain

$$y - 4 = 3(x - 2), \tag{3}$$

so

$$y = 3x - 2. \tag{4}$$

If we set $x = t$ in equation (4), we find that $y = 3t - 2$, and we obtain the parametric equations

$$\left.\begin{array}{l} x(t) = t \\ y(t) = 3t - 2 \end{array}\right\} \; -\infty < t < \infty.$$

A different parameterization for ℓ may be obtained by setting $t = x - 2$ in equation (3). Then, $y - 4 = 3t$, so $y = 3t + 4$, and the parameterization is

$$\left.\begin{array}{l} x(t) = t + 2 \\ y(t) = 3t + 4 \end{array}\right\} \; -\infty < t < \infty. \qquad \diamond$$

Up to this point, we have considered only the problem of finding parameterizations for curves described by equations in xy-coordinates. Just as important is the problem of describing a curve that is presented in parametric form. One technique that is always available is to sketch the curve simply by selecting various values for t, calculating the corresponding coordinates of the point $P(t) = (x(t), y(t))$, and sketching in a curve passing through the various points obtained (in order!). However, through a familiarity with the *xy*-coordinate forms of various curves you can often convert the parametric equations directly into their *xy* counterparts.

Example 2

A point moves along a path with coordinates

$$(x(t), y(t)) = (a \cos t,\ b \sin t)$$

at time *t*. Describe the path.

Solution: By squaring both coordinate functions we observe that

$$[x(t)]^2 = a^2 \cos^2 t; \qquad [y(t)]^2 = b^2 \sin^2 t.$$

We obtain

$$\frac{[x(t)]^2}{a^2} + \frac{[y(t)]^2}{b^2} = \cos^2 t + \sin^2 t = 1.$$

The curve is, therefore, the ellipse $\dfrac{x^2}{a^2} + \dfrac{y^2}{b^2} = 1$ seen in Figure 4.2.   ◇

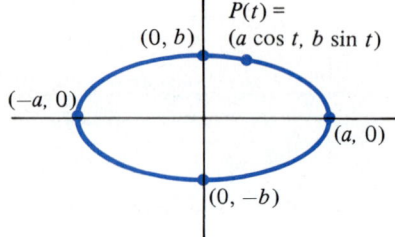

Figure 4.2 Parameterization for the ellipse $\dfrac{x^2}{a^2} + \dfrac{y^2}{b^2} = 1$.

Example 3

To convert the parametric equations

$$x(t) = t^2 - 2$$

$$y(t) = t^2 + 1$$

to an equation in *x* and *y* alone, we begin by noting that

$$y = t^2 + 1 = (t^2 - 2) + 3 = x + 3.$$

But the resulting equation is not simply $y = x + 3$. This is because the range of the function $x(t) = t^2 - 2$ is $[-2, \infty)$. (That is, we cannot have points (x, y) with $x < -2$, since $t^2 \geq 0$ for all *t*.) We must therefore restrict the domain by writing

$$y = x + 3, \qquad x \geq -2.$$

(See Figure 4.3.)   ◇

Example 4

Parametric equations are useful in describing the motion of objects in space. For example, we may use them to determine the motion of a bullet fired horizontally from an altitude of 2 meters with an initial velocity of v_0 m/s.

If we let $(x(t), y(t))$ denote the position of the bullet *t* seconds after it is fired, as in Figure 4.4, we will show in Chapter 16 that (neglecting the effect of air resistance) the coordinate *functions x* and *y* are

$$x(t) = v_0 t,$$

$$y(t) = 2 - 4.9t^2.$$

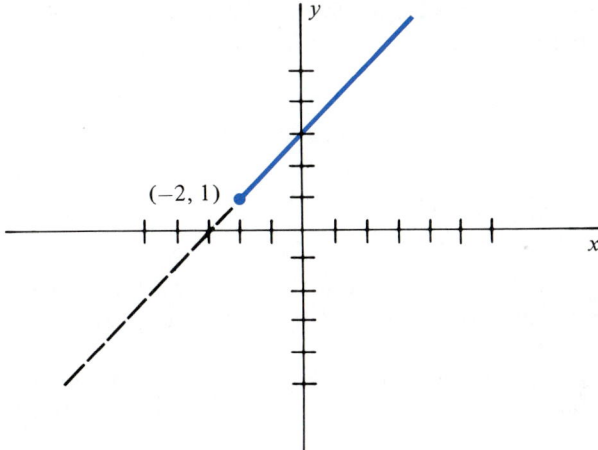

Figure 4.3 Graph of parametric equations $x(t) = t^2 - 2$, $y(t) = t^2 + 1$.

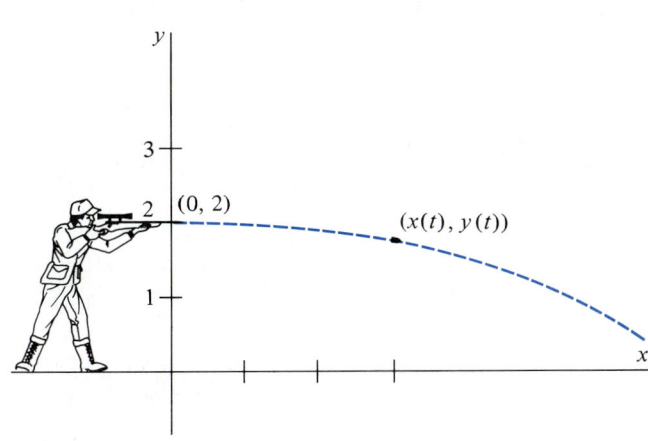

Figure 4.4 Bullet fired by a marksman actually follows a parabolic path.

To see that these equations, for $t \geq 0$, describe an arc of a parabola, we note that

$$y(t) = 2 - 4.9t^2$$

$$= 2 - \frac{4.9}{v_0^2}(v_0 t)^2$$

$$= 2 - \left(\frac{4.9}{v_0^2}\right)[x(t)]^2.$$

That is, $y = kx^2 + 2$ with $k = -\dfrac{4.9}{v_0^2}$. ◇

Having conveyed the general notion of parametric equations, we should pause at this point to sharpen our use of the terminology **curve**. In general, a curve C in the plane is the set of all points of the form

$$C = \{(x(t), y(t) \mid t \in I\} \tag{5}$$

where I is some interval, and where the coordinate functions x and y are **continuous** functions of $t \in I$.

The requirement that x and y be continuous assures that the curve will itself be continuous (unbroken) in the sense of the graph of a continuous function $y = f(x)$. The curve C in (5) is called **smooth** if the derivatives x' and y' are continuous functions of $t \in I$.

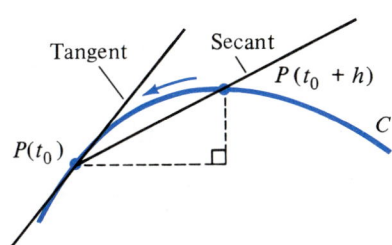

Figure 4.5 Slope of tangent is limit of slopes of secants.

Equations of Tangents

Suppose that the curve C is given in the parametric form of equation (5) and that we wish to find the slope of the line tangent to the curve at the point $P(t_0) = (x(t_0), y(t_0))$. (For example, the curve C might be the trajectory of a rocket, and we might wish to describe its angle of elevation as a function of time.) To do so, we begin by letting h be a small real number. Then $P(t_0 + h) = (x(t_0 + h), y(t_0 + h))$ is, in general, a second point on C (see Figure 4.5). If $x(t_0) \neq x(t_0 + h)$, we can write the slope of the secant through $P(t_0)$ and $P(t_0 + h)$ as

$$\text{Slope of secant} = \frac{y(t_0 + h) - y(t_0)}{x(t_0 + h) - x(t_0)}. \tag{6}$$

The limit of this slope as h tends to zero is the number m, the desired slope of the tangent at $P(t_0)$. By dividing both numerator and denominator in line (6) by $h \neq 0$, we find that

$$m = \lim_{h \to 0} \frac{\left(\dfrac{y(t_0 + h) - y(t_0)}{h} \right)}{\left(\dfrac{x(t_0 + h) - x(t_0)}{h} \right)} = \frac{y'(t_0)}{x'(t_0)}. \tag{7}$$

Obviously, m in equation (7) makes sense only if both $y'(t_0)$ and $x'(t_0)$ exist and $x'(t_0) \neq 0$. Also, recall that we needed to assume that $x(t_0 + h) \neq x(t_0)$. However, this assumption will be valid for all h near zero whenever $x'(t_0)$ exists and is not zero (see Exercise 42). Thus, we have shown that *if $x'(t_0)$ and $y'(t_0)$ exist, and if $x'(t_0) \neq 0$, the slope of the line tangent to the graph of the curve $C = \{(x(t), y(t)) \mid t \in I\}$ at $P(t_0) = (x(t_0), y(t_0))$ is*

$$m = \frac{y'(t_0)}{x'(t_0)}, \qquad x'(t_0) \neq 0. \tag{8}$$

The equation for this tangent line is therefore

$$y - y(t_0) = \frac{y'(t_0)}{x'(t_0)} (x - x(t_0))$$

or

$$x'(t_0)[y - y(t_0)] - y'(t_0)[x - x(t_0)] = 0. \tag{9}$$

If $x'(t_0) = 0$ and $y'(t_0) \neq 0$, equation (9) reduces to the equation $x = x(t_0)$ so the tangent is a vertical line through $(x(t_0), y(t_0))$. If both $x'(t_0) = 0$ and $y'(t_0) = 0$, no conclusion can be drawn about the tangent at $(x(t_0), y(t_0))$.

Example 5

Find the equation of the line tangent to the curve given by the parametric equations

$$\left. \begin{array}{l} x(t) = 2 \cos t \\ y(t) = 4 \sin t \end{array} \right\} \ 0 \leq t < 2\pi$$

at the point $(\sqrt{2}, 2\sqrt{2})$.

Strategy

Find t_0.

Solution

To find the number t_0 for which $P(t_0) = (\sqrt{2}, 2\sqrt{2})$, we set

$$x(t_0) = 2 \cos t_0 = \sqrt{2},$$

$$y(t_0) = 4 \sin t_0 = 2\sqrt{2}.$$

This gives $\cos t_0 = \sqrt{2}/2$, $\sin t_0 = \sqrt{2}/2$, so $t_0 = \pi/4$.

Find $x'(t_0)$, $y'(t_0)$.

Then

$$x'(t_0) = -2 \sin(\pi/4) = -\sqrt{2}$$

and

$$y'(t_0) = 4 \cos(\pi/4) = 2\sqrt{2}.$$

Use equation (9).

Substituting into equation (9) then gives

$$-\sqrt{2}(y - 2\sqrt{2}) - 2\sqrt{2}(x - \sqrt{2}) = 0$$

or

$$y = -2x + 4\sqrt{2}. \qquad \text{(See Figure 4.6.)}$$ ◇

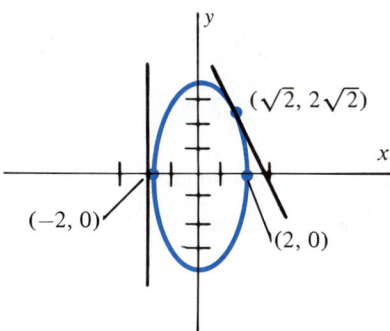

Figure 4.6   Tangents to curve in Example 5.

Example 6

Determine the points for which the curve in Example 5 has vertical tangents.

Strategy

Find t_0 so that $x'(t_0) = 0$.

Solution

Since $x'(t) = -2 \sin t$, the equation

$$x'(t) = -2 \sin t = 0$$

has solutions $t_0 = 0, \pi$ for $0 \le t < 2\pi$.

Determine if $y'(t_0) \ne 0$ for each such t_0.

Since

$$y'(0) = 4 \cos 0 = 4 \ne 0 \qquad \text{and}$$
$$y'(\pi) = 4 \cos \pi = -4 \ne 0,$$

If so, the points

$$P(t_0) = (x(t_0), y(t_0))$$

yield vertical tangents.

the points

$$P(0) = (2 \cos 0, 4 \sin 0) = (2, 0)$$

and

$$P(\pi) = (2 \cos \pi, 4 \sin \pi) = (-2, 0)$$

yield vertical tangents (see Figure 4.6).   ◇

Tangents to Polar Curves

If a curve in the plane is the graph of a polar equation of the form $r = f(\theta)$, we can use the equations

$$x = r \cos \theta \tag{10a}$$

$$y = r \sin \theta \tag{10b}$$

to obtain parametric equations for the curve. Multiplying the equation $r = f(\theta)$ on both sides by $\cos \theta$ gives the equation $r \cos \theta = f(\theta) \cos \theta$. Using (10a), we obtain

$$x(\theta) = f(\theta) \cos \theta. \tag{11a}$$

Similarly, multiplying both sides of $r = f(\theta)$ by $\sin \theta$ and using (10b), we obtain

$$y(\theta) = f(\theta) \sin \theta. \tag{11b}$$

If the derivative $f'(\theta) = \dfrac{dr}{d\theta}$ exists, we may differentiate both sides of equation (11a) and use equation (10b) to conclude that

$$x'(\theta) = f'(\theta) \cos \theta - f(\theta) \sin \theta \tag{12a}$$

$$= \frac{dr}{d\theta} \cos \theta - r \sin \theta.$$

Similarly, differentiating both sides of (11b) and using (10a), we obtain

$$y'(\theta) = f'(\theta) \sin \theta + f(\theta) \cos \theta \tag{12b}$$

$$= \frac{dr}{d\theta} \sin \theta + r \cos \theta.$$

Finally, we may apply equation (8) (with parameter θ rather than t) and equations (12a) and (12b) to conclude that *the slope of the line tangent to the graph of the polar equation $r = f(\theta)$ at the point (r, θ) is*

$$m = \frac{\dfrac{dr}{d\theta} \sin \theta + r \cos \theta}{\dfrac{dr}{d\theta} \cos \theta - r \sin \theta}. \tag{13}$$

As noted for equation (8), m in equation (13) is defined only if $x'(\theta) = \dfrac{dr}{d\theta} \cos \theta - r \sin \theta \neq 0$. If $x'(\theta) = 0$ and $y'(\theta) \neq 0$, the graph has a vertical tangent at (r, θ). If both $x'(\theta) = 0$ and $y'(\theta) = 0$, no conclusions may be drawn.

Example 7

Find the points where the four-leaved rose with equation

$$r = \sin 2\theta, \qquad 0 \leq \theta < 2\pi$$

has vertical tangents.

Strategy

Find a parameterization for the curve.

Solution

By equations (11a) and (11b), a parameterization for $f(\theta) = \sin 2\theta = 2 \sin \theta \cos \theta$ is

$$x(\theta) = 2 \sin \theta \cos^2 \theta$$

$$y(\theta) = 2 \sin^2 \theta \cos \theta.$$

Find the angles θ for which $x'(\theta) = 0$.

Thus,

$$x'(\theta) = 2 \cos^3 \theta - 4 \sin^2 \theta \cos \theta = 0$$

implies

$$\cos \theta(\cos^2 \theta - 2 \sin^2 \theta) = 0.$$

Thus, $x'(\theta) = 0$ whenever

$$\cos \theta = 0 \qquad \text{or} \qquad \cos^2 \theta = 2 \sin^2 \theta.$$

The equation $\cos \theta = 0$ has solutions $\theta = \pi/2$ and $\theta = 3\pi/2$ in the interval $[0, 2\pi)$. The equation $\cos^2 \theta = 2 \sin^2 \theta$ gives $\tan^2 \theta = \dfrac{1}{2}$ or $\tan \theta = \pm\dfrac{\sqrt{2}}{2}$, which has 4 solutions in the interval $[0, 2\pi)$, $\theta = \pm\mathrm{Tan}^{-1}\!\left(\pm\dfrac{\sqrt{2}}{2}\right)$. The 6 solutions of $x'(\theta) = 0$ are therefore $\theta = \dfrac{\pi}{2}$, $\theta = \dfrac{3\pi}{2}$, and $\theta = \pm\mathrm{Tan}^{-1}\!\left(\pm\dfrac{\sqrt{2}}{2}\right)$.

Find $y'(\theta)$.

To determine which of these yield vertical tangents we must examine

$$y'(\theta) = 4 \sin \theta \cos^2 \theta - 2 \sin^3 \theta$$
$$= 2 \sin \theta[2 \cos^2 \theta - \sin^2 \theta].$$

Determine if $y'(\theta) \neq 0$ for each θ with $x'(\theta) = 0$.

If $\theta = \pi/2$ or $3\pi/2$, then

$$y'(\theta) = -2 \sin^3 \theta \neq 0.$$

Similarly, if $\theta = \pm\mathrm{Tan}^{-1}\!\left(\pm\dfrac{\sqrt{2}}{2}\right)$, then

$$y'(\theta) = 2\left(\pm\frac{\sqrt{3}}{3}\right)\left[2 \cdot \frac{2}{3} - \frac{1}{3}\right] \neq 0.$$

If so, (r, θ) is a point that yields a vertical tangent. (See Figure 4.7.)

Thus, all 6 values of θ yield vertical tangents (note in Figure 4.7 that both $\theta = \pi/2$ and $\theta = 3\pi/2$ correspond to the origin). $\diamondsuit$

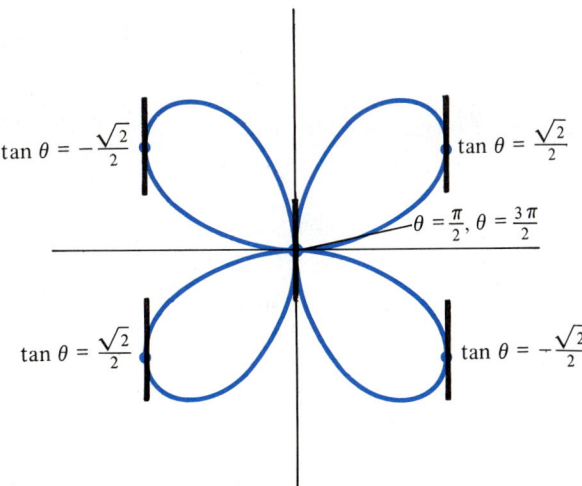

Figure 4.7 Points where graph of $r = \sin 2\theta$ has vertical tangents.

Exercise Set 14.4

In Exercises 1–12, sketch the curve described by the given parametric equations and find an equation in xy-coordinates whose graph contains the given curve.

1. $x(t) = t$
$y(t) = t + 6$

2. $x(t) = t + 2$
$y(t) = 3 - t$

3. $x(t) = 1 + 3t$
$y(t) = 2t + 2$

4. $x(t) = t$
$y(t) = t^2 + 2$

5. $x(t) = t^2 + 1$
$y(t) = t^2 - 1$

6. $x(t) = \sin t$
$y(t) = \cos^2 t$

7. $x(t) = e^t$
$y(t) = e^{3t}$

8. $x(t) = \sqrt{t}$
$y(t) = 3t - 2$

9. $x(t) = \sec t$
$y(t) = \tan t$

10. $x(t) = 3 \sin t$
$y(t) = 5 \cos t$

11. $x(t) = \sin t$
$y(t) = \sin 2t$

12. $x(t) = \cos t$
$y(t) = \sec t$

In Exercises 13–20, find the slope of the line tangent to the curve at the indicated point and the equation of the tangent.

13. $x(t) = t$
$y(t) = t^2 + 1$
$t = 2$

14. $x(t) = t^2$
$y(t) = 1 - t$
$t = 2$

15. $x(t) = \sin t$
$y(t) = \cos t$
$t = \pi/3$

16. $x(t) = \cos t$
$y(t) = \sin t$
$t = \pi/4$

17. $x(t) = \dfrac{1}{t}$
$y(t) = 3t^2 - 7$
$t = 2$

18. $x(t) = 3t^3$
$y(t) = \sin \pi t$
$t = 1$

19. $x(t) = 1 + \sqrt{t}$
$y(t) = 1 - \sqrt{t}$
$t = 4$

20. $x(t) = \sec t$
$y(t) = \tan t$
$t = \pi/4$

In Exercises 21–26, find the slope, if defined, of the line tangent to the graph of the polar equation at the point corresponding to the given value of θ.

21. $r = \cos \theta$
$\theta = \pi/4$

22. $r = 1 + \cos \theta$
$\theta = \pi/6$

23. $r = a \sin 2\theta$
$\theta = \pi/6$

24. $r = 2 \sin \theta$
$\theta = \pi/4$

25. $r = \theta$
$\theta = \pi/2$

26. $r = a \sin 3\theta$
$\theta = \pi/3$

In Exercises 27–32, find all points at which the curve described by the parametric equations has **(a)** a vertical tangent, **(b)** a horizontal tangent.

27. $x(t) = \cos t$
$y(t) = \sin t$

28. $x(t) = \sin 2t$
$y(t) = \sin t$

29. $x(t) = t^2 + 4$
$y(t) = 3t^2 - 6t + 2$

30. $x(t) = 5 + 2 \sin t$
$y(t) = 3 - \cos t$

31. $x(t) = 3t^2 + 6$
$y(t) = t - t^2$

32. $x(t) = t^{3/2}$
$y(t) = t + \cos t$

33. Find the point on the curve

$$C_1: \quad \begin{array}{l} x(t) = t \\ y(t) = 2t^2 + 3 \end{array}$$

where the tangent is parallel to the line

$$C_2: \quad \begin{array}{l} x(t) = t + 3 \\ y(t) = 4t - 10. \end{array}$$

34. Find parametric equations for the cardioid $r = 1 + \cos \theta$.

35. Find parametric equations for the three-leaved rose $r = a \sin 3\theta$.

36. Find parametric equations for the spiral $r = \theta$.

37. Find the points at which the cardioid $r = 1 + \sin \theta$ has vertical tangents.

38. Find the points at which the cardioid $r = 1 + \cos \theta$ has horizontal tangents.

39. Find the points at which the cardioid $r = 1 + \cos \theta$ has vertical tangents.

40. Show that the curve determined by the parametric equations

$$x(t) = a \cos t + h$$

$$y(t) = b \sin t + k$$

is an ellipse with center at (h, k).

41. A particle moves in the plane so that at time t it is at the point with coordinates $x(t) = t + 4$, $y(t) = 8 - t^2$. A second particle moves in the plane so that it is at the point with coordinates $x(t) = t + 4$, $y(t) = t + 6$, at time t.
  a. Find equations in xy-coordinates for each of the curves.
  b. Do the paths cross? If so, where?
  c. Do the particles collide? If so, where and at what time?

42. Prove that if $x(t)$ is differentiable at t_0 and if $x'(t_0) \neq 0$, then $x(t_0 + h) \neq x(t_0)$ for all sufficiently small h. (*Hint:* Use the differential approximation $x(t_0 + h) \approx x(t_0) + x'(t_0)h$. See Chapter 3.)

43. Show that the notion of curve is a more general concept than that of the graph of a continuous function $y = f(x)$. That is, show that every graph of a continuous function $y = f(x)$ can be expressed as a curve in parametric form, but that the converse is not true.

44. Prove that the graph of the function $y = f(x)$ is a *smooth* curve if $f'(x)$ is continuous for all x in the domain of f.

14.5 ARC LENGTH AND SURFACE AREA REVISITED

In Section 7.3 we developed formulas for arc length and surface area calculations associated with the graph of a differentiable function f. Since we have now seen that a curve is a more general notion than the graph of a function, we should ask whether the concepts of arc length and surface area can be extended to general curves. In both cases the answer is yes, and the development here parallels that of Section 7.3 very closely.

Arc Length

Suppose that C is a curve in the plane that is determined by the parametric equations

$$C: \begin{array}{l} x = x(t) \\ y = y(t) \end{array} \Big\} \ t \in I,$$

where I is some interval. Suppose further that a and b are numbers in I, with $a < b$, so that the arc of the curve from $P = (x(a), y(a))$ to $Q = (x(b), y(b))$ does not intersect itself, except possibly if $P = Q$. The problem is to define and calculate the length L of the arc of C connecting P and Q (see Figure 5.1).

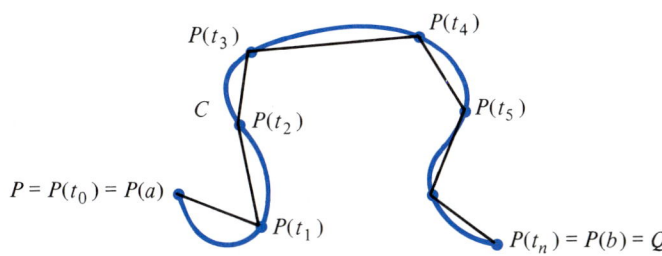

Figure 5.1 Polygonal path joining points $P(t_j) = (x(t_j), y(t_j))$ approximates C.

As in Section 2.3, we begin by partitioning the interval $[a, b]$ for the independent variable (parameter) t into n subintervals of equal length $\Delta t = \dfrac{b - a}{n}$ with endpoints

$$a = t_0 < t_1 < t_2 < \cdots < t_n = b.$$

For each integer $j = 0, 1, 2, \ldots, n$, we let $P(t_j) = (x(t_j), y(t_j))$. Then each $P(t_j)$ is a point on the arc of C joining P and Q. In each interval $[t_{j-1}, t_j]$, we use the length of the line segment from $P(t_{j-1})$ to $P(t_j)$ to approximate the length of the arc C_j connecting these two points. Using the distance formula, we can write this distance as

$$\Delta L_j = \sqrt{\Delta x_j^2 + \Delta y_j^2} \tag{1}$$
$$= \sqrt{[x(t_j) - x(t_{j-1})]^2 + [y(t_j) - y(t_{j-1})]^2}, \qquad j = 1, 2, \ldots, n.$$

(See Figure 5.2.)

If x is a differentiable function of t in each interval $[t_{j-1}, t_j]$, we may apply the Mean Value Theorem to conclude that there exists a number $c_j \in [t_{j-1}, t_j]$ so that

$$x'(c_j) = \frac{x(t_j) - x(t_{j-1})}{t_j - t_{j-1}} = \frac{x(t_j) - x(t_{j-1})}{\Delta t}.$$

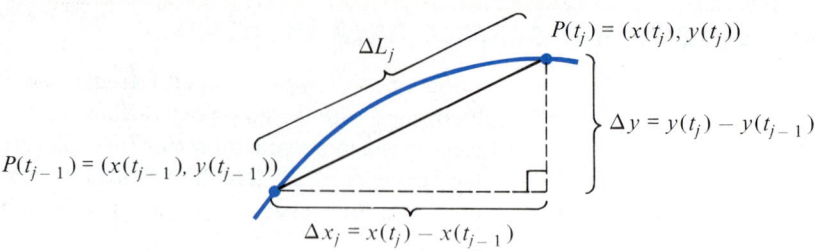

Figure 5.2 ΔL_j approximates length of arc from $P(t_{j-1})$ to $P(t_j)$.

Thus, we can write

$$x(t_j) - x(t_{j-1}) = x'(c_j)\,\Delta t, \qquad j = 1, 2, \ldots, n. \tag{2}$$

Similarly, if y is a differentiable function of t, there exist numbers $d_j \in [t_{j-1}, t_j]$ so that

$$y(t_j) - y(t_{j-1}) = y'(d_j)\,\Delta t, \qquad j = 1, 2, \ldots, n. \tag{3}$$

Combining equations (1)–(3), we obtain the length of the approximating line segment as

$$\begin{aligned}
\Delta L_j &= \sqrt{[x'(c_j)\,\Delta t]^2 + [y'(d_j)\,\Delta t]^2} \\
&= \sqrt{[x'(c_j)]^2 + [y'(d_j)]^2}\,\Delta t, \qquad j = 1, 2, \ldots, n.
\end{aligned} \tag{4}$$

Finally, we approximate the length L of the arc of C from P to Q by the length of the polygonal path connecting the points $P(t_0), P(t_1), \ldots, P(t_n)$. The latter is simply the sum of the lengths ΔL_j in line (4). We obtain the approximation

$$L \approx \sum_{j=1}^{n} \Delta L_j = \sum_{j=1}^{n} \sqrt{[x'(c_j)]^2 + [y'(d_j)]^2}\,\Delta t. \tag{5}$$

The expression on the right-hand side of equation (5) is "almost" a Riemann sum. The difficulty is that the functions x' and y' are being evaluated at two (possibly) different points in each subinterval. However, in more advanced courses, it is shown that if both x' and y' are continuous on $[a, b]$ this difficulty can be overcome, and that as $n \to \infty$ the approximating sum converges to the integral

$$L = \int_a^b \sqrt{[x'(t)]^2 + [y'(t)]^2}\,dt. \tag{6}$$

As in Section 7.3, we argue that as $n \to \infty$ the polygonal paths more closely approximate the curve C, so that the limiting value of the length of the polygonal paths provides a reasonable definition of the length of the arc. It is important to keep in mind that *both x' and y' must be continuous on $[a, b]$ for formula (6) to be valid.* That is, C must be a *smooth* curve.

Example 1

Find the length of the arc of the curve

$$C: \quad \begin{aligned} x(t) &= \cos t + t \sin t \\ y(t) &= \sin t - t \cos t \end{aligned}$$

connecting the points $P = (x(0), y(0)) = (1, 0)$ and $Q = \left(x\left(\frac{\pi}{2} \right), y\left(\frac{\pi}{2} \right) \right) = \left(\frac{\pi}{2}, 1 \right)$.

Solution: Here

$$x'(t) = -\sin t + \sin t + t \cos t = t \cos t$$

and

$$y'(t) = \cos t - \cos t + t \sin t = t \sin t.$$

By formula (6)

$$
\begin{aligned}
L &= \int_0^{\pi/2} \sqrt{(t \cos t)^2 + (t \sin t)^2} \, dt \\
&= \int_0^{\pi/2} \sqrt{t^2(\cos^2 t + \sin^2 t)} \, dt \\
&= \int_0^{\pi/2} t \, dt \\
&= \frac{t^2}{2} \Big]_0^{\pi/2} \\
&= \frac{\pi^2}{8}.
\end{aligned}
$$

◇

Arc Length in Polar Coordinates

In Section 14.4 we showed that the graph of the polar equation $r = f(\theta)$ can be parameterized, using θ as the parameter, as

$$x(\theta) = f(\theta) \cos \theta,$$

$$y(\theta) = f(\theta) \sin \theta.$$

If $f'(\theta)$ is continuous, so are the functions

$$x'(\theta) = f'(\theta) \cos \theta - f(\theta) \sin \theta$$

and

$$y'(\theta) = f'(\theta) \sin \theta + f(\theta) \cos \theta,$$

so we may apply equation (6) to determine a formula for arc length in polar coordinates. Before doing so we note that

$$[x'(\theta)]^2 = [f'(\theta)]^2 \cos^2 \theta - 2f'(\theta)f(\theta) \sin \theta \cos \theta + [f(\theta)]^2 \sin^2 \theta,$$

and

$$[y'(\theta)]^2 = [f'(\theta)]^2 \sin^2 \theta + 2f'(\theta)f(\theta) \sin \theta \cos \theta + [f(\theta)]^2 \cos^2 \theta,$$

so

$$[x'(\theta)]^2 + [y'(\theta)]^2 = [f'(\theta)]^2 + [f(\theta)]^2. \tag{7}$$

From equations (6) and (7) it follows that

$$L = \int_a^b \sqrt{[f'(\theta)]^2 + [f(\theta)]^2} \, d\theta \qquad (8)$$

gives the length of the arc of the graph of $r = f(\theta)$ from $\theta = a$ to $\theta = b$.

Example 2

Find the length of the cardioid $r = 1 + \cos \theta$.

Strategy

Determine the limits of integration. Verify that equation (8) applies.

Apply (8).

Use identity

$$\cos^2 \phi = \frac{1}{2} + \frac{1}{2} \cos 2\phi$$

in reverse to simplify integrand.

Use the symmetry of the cosine function to handle the absolute value signs. (Alternatively, we could have used the symmetry of the cardioid and integrated over $[0, \pi]$.)

Solution

Here $f(\theta) = r = 1 + \cos \theta$, and the cardioid is swept out as θ makes one complete revolution. Appropriate limits of integration are therefore $\theta = 0$ to $\theta = 2\pi$.

Since $f'(\theta) = -\sin \theta$ is continuous on $[0, 2\pi]$, we may apply formula (8):

$$L = \int_0^{2\pi} \sqrt{(1 + \cos \theta)^2 + (-\sin \theta)^2} \, d\theta$$

$$= \int_0^{2\pi} \sqrt{(1 + 2 \cos \theta + \cos^2 \theta) + \sin^2 \theta} \, d\theta$$

$$= \int_0^{2\pi} \sqrt{2 + 2 \cos \theta} \, d\theta$$

$$= \int_0^{2\pi} \sqrt{4 \left(\frac{1}{2} + \frac{1}{2} \cos \theta \right)} \, d\theta$$

$$= \int_0^{2\pi} \sqrt{4 \cos^2 \left(\frac{\theta}{2} \right)} \, d\theta$$

$$= \int_0^{2\pi} \left| 2 \cos \left(\frac{\theta}{2} \right) \right| \, d\theta$$

$$= 2 \int_0^{\pi} 2 \cos \left(\frac{\theta}{2} \right) \, d\theta$$

$$= 8 \sin \left(\frac{\theta}{2} \right) \Big]_0^{\pi} = 8. \qquad \diamond$$

Surface Area

Suppose that the region bounded by the smooth curve C with equations

$$C: \quad \begin{matrix} x = x(t) \\ y = y(t) \end{matrix} \Big\} \quad a \le t \le b$$

and the x-axis is to be revolved about the x-axis. In order to calculate the surface area of the resulting solid we shall need to require that $x'(t) \ne 0$ for all $t \in [a, b]$. (This last condition ensures that the curve will have at most one point with any given x-coordinate (see Exercise 35).)

We begin with the usual step of partitioning the interval $[a, b]$ into n equal subintervals of length $\Delta t = \dfrac{b - a}{n}$, and with endpoints

$$a = t_0 < t_1 < t_2 < \cdots < t_n = b.$$

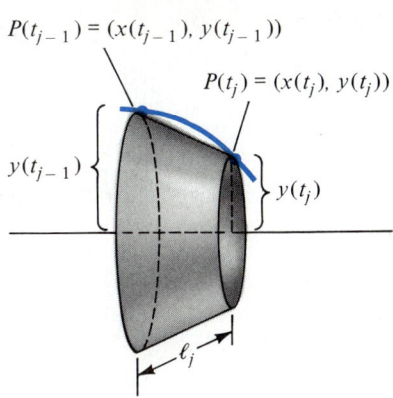

Figure 5.3 Curve to be revolved about the x-axis.

Figure 5.4 Frustum generated by $\overline{P(t_{j-1})P(t_j)}$.

Let $P(t_j) = (x(t_j), y(t_j))$. We shall approximate C by the polygonal path joining these consecutive points on C (see Figure 5.3). As each of these line segments is revolved about the x-axis, it generates the frustum of a cone, as in Figure 5.4. The radii of the jth frustum are $y(t_{j-1})$ and $y(t_j)$, so the lateral surface area of the jth frustum is

$$\Delta S_j = \pi[y(t_{j-1}) + y(t_j)]\ell_j, \qquad j = 1, 2, \ldots, n$$

where

$$\ell_j = \sqrt{(\Delta x_j)^2 + (\Delta y_j)^2}$$
$$= \sqrt{[x(t_j) - x(t_{j-1})]^2 + [y(t_j) - y(t_{j-1})]^2}. \qquad \text{(See Figure 5.4.)}$$

As in our development of the formula for arc length, we obtain the following approximation to the lateral surface area S of the volume of revolution.

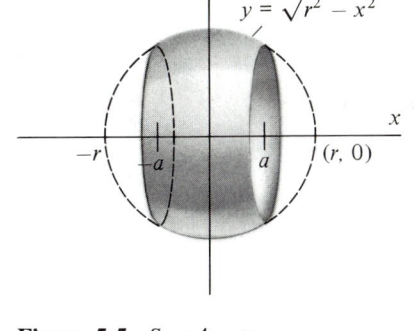

Figure 5.5 $S = 4\pi\,ar$.

$$S \approx \sum_{j=1}^{n} \Delta S_j = \sum_{j=1}^{n} \pi[y(t_{j-1}) + y(t_j)]\sqrt{[x'(c_j)]^2 + [y'(d_j)]^2}\,\Delta t$$

which converges to the definite integral

$$S = \int_a^b 2\pi y(t)\sqrt{[x'(t)]^2 + [y'(t)]^2}\,dt. \qquad (9)$$

Example 3

In Example 4, Section 7.3, it was shown that the surface area of the band of the sphere obtained by revolving the arc of the circle $x^2 + y^2 = r^2$ lying above the interval $[-a, a]$, $a < r$, about the x-axis is $S = 4\pi ar$ (Figure 5.5). Moreover, we noted in Example 5 of that section that one could not conclude from this calculation that the surface area of a sphere is $S = 4\pi r^2$.

We may now use equation (9) to demonstrate that the surface area of the sphere of radius r is indeed $S = 4\pi r^2$ (Figure 5.6). To do so we use the parameterization of the upper semicircle

$$\left.\begin{array}{l} x(t) = r\cos t \\ y(t) = r\sin t \end{array}\right\} \ 0 \le t \le \pi.$$

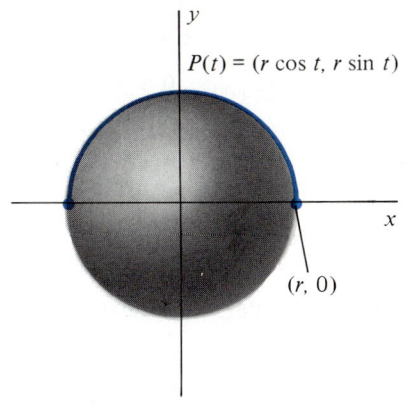

Figure 5.6 $S = 4\pi r^2$.

Then

$$x'(t) = -r \sin t; \qquad y'(t) = r \cos t.$$

Both of these derivatives are continuous for all t, in contrast to $\dfrac{dy}{dx}$ (which is discontinuous at $x = r$ and $x = -r$). Thus,

$$S = \int_0^\pi 2\pi(r \sin t)\sqrt{(-r \sin t)^2 + (r \cos t)^2}\, dt$$

$$= \int_0^\pi 2\pi r^2 \sin t\, dt$$

$$= -2\pi r^2 \cos t]_0^\pi$$

$$= -2\pi r^2(-1 - 1)$$

$$= 4\pi r^2.$$

◇

Exercise Set 14.5

In Exercises 1–11, a curve is described by a pair of parametric equations. Find the length of the given curve.

1. $\begin{aligned} x(t) &= t \\ y(t) &= t^{3/2} \end{aligned} \quad 0 \le t \le \dfrac{4}{9}$

2. $\begin{aligned} x(t) &= t + 1 \\ y(t) &= \dfrac{2}{3}t^{3/2} \end{aligned} \quad 0 \le t \le 3$

3. $\begin{aligned} x(t) &= 3t^2 \\ y(t) &= 2t^3 \end{aligned} \quad 0 \le t \le \sqrt{2}$

4. $\begin{aligned} x(t) &= \sin t \\ y(t) &= \cos t \end{aligned} \quad 0 \le t \le 2\pi$

5. $\begin{aligned} x(t) &= 2(2t + 3)^{3/2} \\ y(t) &= 3(t + 1)^2 \end{aligned} \quad 0 \le t \le 2$

6. $\begin{aligned} x(t) &= 2(1 - \cos t) \\ y(t) &= 2 \sin t \end{aligned} \quad 0 \le t \le \pi$

7. $\begin{aligned} x(t) &= t^3 - 3t^2 \\ y(t) &= 3t^2 \end{aligned} \quad 0 \le t \le 1$

8. $\begin{aligned} x(t) &= \ln \cos t \\ y(t) &= t \end{aligned} \quad 0 \le t \le \pi/3$

9. $\begin{aligned} x(t) &= \cos^3 t \\ y(t) &= \sin^3 t \end{aligned} \quad 0 \le t \le \pi$

10. $\begin{aligned} x(t) &= 3t^3 \\ y(t) &= 3t^2 \end{aligned} \quad 0 \le t \le \sqrt{5}$

11. $\begin{aligned} x(t) &= e^t \cos t \\ y(t) &= e^t \sin t \end{aligned} \quad 0 \le t \le \pi$

In Exercises 12–18, a curve is described by a pair of parametric equations. Find the surface area of the solid generated by revolving this curve about the x-axis.

12. $\begin{aligned} x(t) &= t \\ y(t) &= t^2 \end{aligned} \quad 0 \le t \le 1$

13. $\begin{aligned} x(t) &= t^2 \\ y(t) &= 4t \end{aligned} \quad 0 \le t \le 2$

14. $\begin{aligned} x(t) &= t \\ y(t) &= \dfrac{t^4}{4} + \dfrac{1}{8t^2} \end{aligned} \quad 1 \le t \le 2$

15. $\begin{aligned} x(t) &= t \\ y(t) &= \sqrt{t} \end{aligned} \quad 1 \le t \le 2$

16. $\begin{aligned} x(t) &= t \\ y(t) &= \dfrac{t^2 - 1}{2} \end{aligned} \quad 0 \le t \le 1$

17. $\begin{aligned} x(t) &= 3t^2 \\ y(t) &= 2t^3 \end{aligned} \quad 1 \le t \le 2$

18. $\begin{aligned} x(t) &= \cos^3 t \\ y(t) &= \sin^3 t \end{aligned} \quad 0 \le t \le \pi$

19. Find the length of the cardioid $r = \cos^2\left(\dfrac{\theta}{2}\right)$.

20. Find the length of the cardioid $r = 1 + \cos \theta$.

21. Find the length of the spiral $r = e^{2\theta}$, $\quad 0 \le \theta \le \pi$.

22. Find the area of the surface generated when one arch of the cycloid $x(t) = t - \sin t$, $y(t) = 1 - \cos t$ is revolved about the x-axis.

23. Find the length of the circle $x(t) = 2 \cos t$, $y(t) = 2 \sin t$, $0 \le t \le 2\pi$.

24. Find the length of the spiral $r = 2\theta$, $\quad 0 \le \theta \le \pi$.

25. Show, using equation (8), that the circumference of the circle $r = 3 \cos \theta$ is 3π.

26. Find the length of the cardioid $r = a(1 + \sin \theta)$.

27. Find the length of the graph of $r = e^\theta$ for $0 \le \theta \le 2\pi$.

28. Find the length of the graph of the equation $r = 4 \sec \theta$ for $0 \le \theta \le \dfrac{\pi}{4}$.

29. Find the surface area of the solid obtained by revolving the region bounded by the graph of $r = e^\theta$, $0 \le \theta \le \pi$, and the x-axis about the x-axis.

30. Find the area of the surface of the solid obtained by revolving the region bounded by the graph of $r = 4 \sin \theta$ about the line $\theta = 0$.

31. Determine the formula for the surface area of the volume of the solid generated when an arc of the curve determined by the parametric equations $x = x(t)$, $y = y(t)$ is revolved about the y-axis.

32. Use the result of Exercise 31 to find the surface area of the solid generated when the curve given in Exercise 13 is revolved about the y-axis.

33. Use the result of Exercise 31 to find the surface area of the solid obtained by revolving the curve in Exercise 15 about the y-axis.

34. *(Calculator/Computer)* Use Simpson's Rule to approximate the length of the ellipse
$$x(t) = 4 \cos t,$$
$$y(t) = 6 \sin t.$$

35. Show that the condition $x'(t) \ne 0$ is sufficient to guarantee that a differentiable curve C will have at most one point with any particular x-coordinate.

SUMMARY OUTLINE OF CHAPTER 14

◆ The **polar coordinates** (r, θ) for the point whose rectangular coordinates are (x, y) are determined by the equations $r = \sqrt{x^2 + y^2}$, $\tan \theta = y/x$, while $x = r \cos \theta$, $y = r \sin \theta$. (page 608)

◆ The following identities hold for polar coordinates: $(-r, \theta) = (r, \theta + \pi)$, $(r, \theta + 2n\pi) = (r, \theta)$, $n = \pm 1, \pm 2, \ldots$. (page 607)

◆ The graph of the polar equation $r = f(\theta)$ is (page 610)

1. symmetric with respect to the pole if $f(\theta + \pi) = f(\theta)$;
2. symmetric with respect to the x-axis if $f(-\theta) = f(\theta)$;
3. symmetric with respect to the y-axis if $f(\theta) = f(\pi - \theta)$.

◆ The area A of the region bounded by the graph of the function $r = f(\theta)$ and the rays $\theta = a$ and $\theta = b$ is (page 619)
$$A = \int_a^b \frac{1}{2}[f(\theta)]^2 \, d\theta.$$

◆ The slope of the line tangent to the curve determined by the **parametric equations** $x = x(t)$, $y = y(t)$, at the point $(x(t_0), y(t_0))$ is (page 628)
$$m = \frac{y'(t_0)}{x'(t_0)}$$
provided $x'(t_0)$ and $y'(t_0)$ exist and $x'(t_0) \ne 0$.

◆ The slope of the line tangent to the graph of the polar equation $r = f(\theta)$ at the point (r, θ) is (page 630)
$$m = \frac{\dfrac{dr}{d\theta} \sin \theta + r \cos \theta}{\dfrac{dr}{d\theta} \cos \theta - r \sin \theta}.$$

◆ For the arc of the smooth curve C determined by the polar equations $x = x(t)$, $y = y(t)$, between the points $(x(a), y(a))$ and $(x(b), y(b))$: (page 634)

1. The length of the arc is
$$L = \int_a^b \sqrt{[x'(t)]^2 + [y'(t)]^2} \, dt.$$

2. The surface area of the volume obtained by revolving the arc about the x-axis is
$$S = \int_a^b 2\pi y(t) \sqrt{[x'(t)]^2 + [y'(t)]^2} \, dt.$$

◆ The length of the arc of the graph of the polar equation $r = f(\theta)$, from $\theta = a$ to $\theta = b$, is   (page 636)

$$L = \int_a^b \sqrt{[f'(\theta)]^2 + [f(\theta)]^2}\, d\theta.$$

REVIEW EXERCISES—CHAPTER 14

In Exercises 1–6, sketch the curve described by the given parametric equations.

1. $\begin{aligned} x &= 5\cos\theta \\ y &= 5\sin\theta \end{aligned}$ $0 \le \theta \le 2\pi$

2. $\begin{aligned} x &= 3 + 2t \\ y &= 8 - 6t \end{aligned}$ $-\infty < t < \infty$

3. $\begin{aligned} x &= t\cos\pi t \\ y &= t\sin\pi t \end{aligned}$ $0 \le t \le 6$

4. $\begin{aligned} x &= t^2 + 1 \\ y &= t^4 - 4 \end{aligned}$ $-\infty < t < \infty$

5. $\begin{aligned} x &= \cos t \\ y &= \sin 2t \end{aligned}$ $0 \le t \le \pi$

6. $\begin{aligned} x &= 3\sqrt{t} + 1 \\ y &= 1 - \sqrt{t} \end{aligned}$ $0 \le t$

7. Eliminate the parameter in Exercise 3 to find an equation in x and y that represents the given curve.

8. Repeat Exercise 7 for the curve given in Exercise 4.

9. Repeat Exercise 7 for the curve given in Exercise 5.

In each of Exercises 10–13, find parametric equations for the given curves.

10. $x^2 + y^2 = 9$

11. $4x^2 + 9y^2 = 36$

12. $x + y = 7$

13. $x = 2y^2 + 4y + 5$

14. Show that the parametric equations $x(t) = t$, $y(t) = t^2$ describe the graph of the equation $y = x^2$ as well as the graph of the equation $y^3 = x^6$, but *not* the graph of $y^2 = x^4$.

15. Sketch the curve given parametrically in polar coordinates by the equations $r = 2t$, $\theta = \pi t^2$.

16. Sketch the curve given parametrically by the equations
$\theta = t$, $r = e^t$.

17. Show that the points on the curve determined by the parametric equations $x(t) = t^2 - 1$, $y(t) = 3t$ lie on a parabola. Find the equation for this parabola in xy-coordinates.

18. Find an equation in polar coordinates for the circle with center at the origin and radius 5.

In Exercises 19–28, sketch the graph of the given polar equation.

19. $r = \cos 2\theta$

20. $r = 3\theta$

21. $r = 5\cos 3\theta$

22. $r = |\sin 2\theta|$

23. $r = 6\sin\theta$

24. $r = a\tan\theta$

25. $r = 2 + \sin 2\theta$

26. $r = \dfrac{1}{2 - \cos\theta}$

27. $\theta = \pi/2$

28. $r = 1 - 2\sin\theta$

In Exercises 29–34, find the area of the region enclosed by the given curve.

29. $r = 2(1 + \sin 2\theta)$

30. $r = 2 - \cos\theta$

31. $r = a(1 + \sin\theta)$

32. $r = \sqrt{1 - \sin\theta}$

33. $r = 6\cos 3\theta$

34. $r = 2 + 4\sin\theta\cos\theta$

In Exercises 35–40, find the length of the curve given by the parametric equations.

35. $x = 1 + t$, $\quad y = (1 + t)^{3/2}$, $\quad 0 \le t \le 1$.

36. $x = 2t + 1$, $\quad y = t^2$, $\quad 0 \le t \le 1$

37. $x = 3t$, $\quad y = t^3$, $\quad 0 \le t \le 1$

38. $x = \sin t - t$, $\quad y = \cos t$, $\quad 0 \le t \le \pi$

39. $x = 1 - \cos t$, $\quad y = \sin t$, $\quad 0 \le t \le \pi$

40. $x = 9t^2$, $\quad y = 3t^3 - 9t^2$, $\quad 0 \le t \le 1$

In Exercises 41–44, find the length of the curve given by the polar equation.

41. $r = \cos\theta$, $\quad -\pi/4 \le \theta \le \pi/4$

42. $r = \theta^2$, $\quad 1 \le \theta \le 2$

43. $r = 1 + \sin\theta$, $\quad 0 \le \theta \le \pi/2$

44. $r = 1 - \cos\theta$, $\quad 0 \le \theta \le \pi$

45. Find the area of the region common to the circles $r = \sin\theta$ and $r = \cos\theta$.

46. Find the area of the region enclosed by the lemniscate $r^2 = 4\cos 2\theta$.

47. Find the area of the region inside the circle $r = 6\cos\theta$ and outside the cardioid $r = 2(1 + \cos\theta)$.

48. Find the area of the region inside the circle $r = 4\cos\theta$ and outside the circle $r = 2$.

49. Find the area of the region common to the circle $r = 4\sin\theta$ and the circle $r = 2$.

50. Find the slope of the line tangent to the curve given by the parametric equations $x(t) = \sin t$, $y(t) = \sin 2t$ at the point where $t = \pi/4$.

Chapter 15
Vectors in the Plane and in Space

We frequently used ordered pairs of numbers (a, b) to represent points in the plane. We shall now develop a richer interpretation for the ordered pair (a, b) that of a *vector* in the plane. We do this by defining operations of addition and multiplication by scalars (numbers) for such pairs. This will enable us to interpret the pair (a, b) as a displacement, a concept useful in physics and engineering, and to define the operations of addition and multiplication by scalars for functions that have vectors, rather than numbers, as values.

This chapter develops the concepts of vectors in the plane and in space. Chapters 16–19 use these concepts to study functions whose values are vectors, and functions of more than one variable.

15.1 VECTORS IN THE PLANE

It is useful to know about functions of the form $f(t) = (x(t), y(t))$, that is, functions whose values are points in the xy-plane rather than numbers. In order to be able to form sums and multiples of such functions, we shall need a way to add points in the plane, or multiply them by real numbers. The algebraic structure for doing this is provided by the concept of a *vector*.

DEFINITION 1

The set of all ordered pairs $\langle a, b \rangle$ of real numbers, together with the rules

(i) $\langle a_1, b_1 \rangle + \langle a_2, b_2 \rangle = \langle a_1 + a_2, b_1 + b_2 \rangle$
(ii) $c\langle a, b \rangle = \langle ca, cb \rangle$

for addition and for multiplication by real numbers c, is called the set of **vectors in the plane.** The numbers a and b are referred to as the **components** of the vector $\langle a, b \rangle$.

It is important to note that the vector $\langle a, b \rangle$ is an *ordered pair of numbers,* not a point (a, b) in the plane. We shall sometimes wish to associate the vector $\langle a, b \rangle$ with the point (a, b), but the former concept is more general.

The multiplication of the vector $\langle a, b \rangle$ by the real number c in equation (ii) is called **scalar multiplication,** and the number c is referred to as a **scalar.**

We shall observe the convention of denoting vectors by bold-faced letters. Thus if $v = \langle a_1, b_1 \rangle$ and $w = \langle a_2, b_2 \rangle$ are vectors, Definition 1 provides that

$$v + w = \langle a_1 + a_2, b_1 + b_2 \rangle,$$

etc.

Example 1

For $v = \langle 2, 5 \rangle$ and $w = \langle -3, 2 \rangle$, we have

(a) $v + w = \langle 2 + (-3), 5 + 2 \rangle = \langle -1, 7 \rangle$,
(b) $3v + w = \langle 3 \cdot 2, 3 \cdot 5 \rangle + \langle -3, 2 \rangle = \langle 3, 17 \rangle$,
(c) $4v + 2w = \langle 4 \cdot 2, 4 \cdot 5 \rangle + \langle 2(-3), 2 \cdot 2 \rangle = \langle 2, 24 \rangle$. ◇

We define the vector $-v$ as $(-1)v$. That is,

$$\text{if } v = \langle a, b \rangle, \qquad \text{then } -v = \langle -a, -b \rangle.$$

Also, we define $v - w$ as $v + (-w)$. That is, if

$$v = \langle a_1, b_1 \rangle, \qquad w = \langle a_2, b_2 \rangle,$$

then

$$v - w = v + (-w) = \langle a_1 - a_2, b_1 - b_2 \rangle.$$

The **zero vector** is defined to be the vector $\mathbf{0} = \langle 0, 0 \rangle$. Finally, we define two vectors to be **equal** if and only if they have the same components. That is,

$$\langle a_1, b_1 \rangle = \langle a_2, b_2 \rangle \qquad \text{if and only if} \qquad a_1 = a_2 \text{ and } b_1 = b_2.$$

Example 2

For $v = \langle 1, -3 \rangle$ and $w = \langle 2, 5 \rangle$,

(a) $v - w = \langle 1 - 2, -3 - 5 \rangle = \langle -1, -8 \rangle$.
(b) $2v - 3w = \langle 2, -6 \rangle - \langle 6, 15 \rangle = \langle -4, -21 \rangle$.
(c) $v - v = \langle 1 - 1, -3 + 3 \rangle = \langle 0, 0 \rangle = \mathbf{0}$. ◇

Vectors as Displacements

Figure 1.1 The displacement $\overrightarrow{PQ}$.

One interpretation of vectors is as displacements (meaning changes in location) in the plane. If P and Q are distinct points in the plane, the displacement $\overrightarrow{PQ}$ is represented by an arrow consisting of the line segment $\overline{PQ}$ to which a tip is added at Q. The tip distinguishes the *terminal point* Q from the *initial point* P. (See Figure 1.1.)

Thus, the displacement $\overrightarrow{PQ}$ has both a *length* (which we denote by $|\overrightarrow{PQ}|$ and which is the same as the length of $\overline{PQ}$) and a direction. Moreover, there is a natural way to form the "sum" of two displacements as long as the initial point of the second displacement coincides with the terminal point of the first one: $\overrightarrow{PQ} + \overrightarrow{QR} = \overrightarrow{PR}$. (See Figure 1.2.) Finally, the multiple $c\overrightarrow{PQ}$ may be defined to be the displacement that originates at P, that has either the same or the opposite direction as $\overrightarrow{PQ}$ depending on whether c is positive or negative, and that has length $|c|$ times the length of $\overrightarrow{PQ}$ (see Figure 1.3).

Here is the connection between displacements and vectors. Since a displacement indicates only a change in location (think of "walking three yards east," for exam-

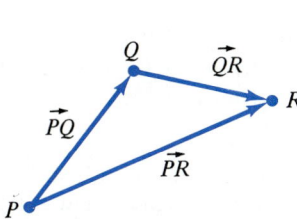

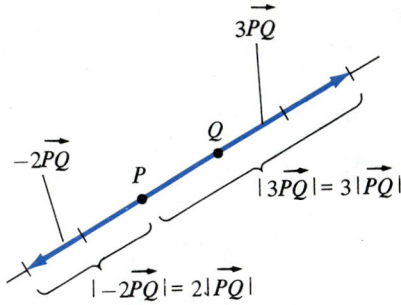

Figure 1.2 Adding two displacements: $\overrightarrow{PQ} + \overrightarrow{QR} = \overrightarrow{PR}$.

Figure 1.3 Multiples $c\overrightarrow{PQ}$ of the displacement $\overrightarrow{PQ}$.

ple), a displacement is determined only by the *changes* it causes in the x and y coordinates of the initial point, but not by the initial point itself. We therefore use vector notation to write

$$\overrightarrow{PQ} = \langle \Delta x, \Delta y \rangle, \tag{1}$$

which means that the displacement $\overrightarrow{PQ}$ consists of an increment Δx in the x-coordinate of the initial point and an increment Δy in the y-coordinate of the initial point. (See Figure 1.4.)

The point of equation (1) is that the vector $\boldsymbol{v} = \langle \Delta x, \Delta y \rangle$ represents the displacement from *any* point $P = (x_1, y_1)$ to another point $Q = (x_2, y_2)$ as long as $x_2 - x_1 = \Delta x$ and $y_2 - y_1 = \Delta y$. We therefore regard two displacements as equal if they are represented by the same vector.

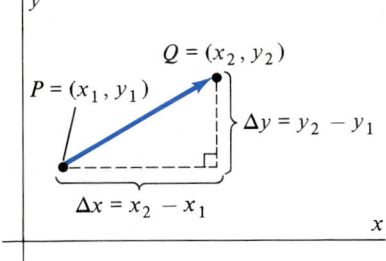

Figure 1.4 $\overrightarrow{PQ} = \langle \Delta x, \Delta y \rangle$
$= \langle x_2 - x_1, y_2 - y_1 \rangle$.

Example 3

Let $P = (0, 0)$, $Q = (2, 4)$, $R = (-3, 1)$, and $S = (-1, 5)$. Then

$$\overrightarrow{PQ} = \langle 2 - 0, 4 - 0 \rangle = \langle 2, 4 \rangle$$

and

$$\overrightarrow{RS} = \langle -1 - (-3), 5 - 1 \rangle = \langle 2, 4 \rangle,$$

so $\overrightarrow{PQ} = \overrightarrow{RS}$. (See Figure 1.5.) ◇

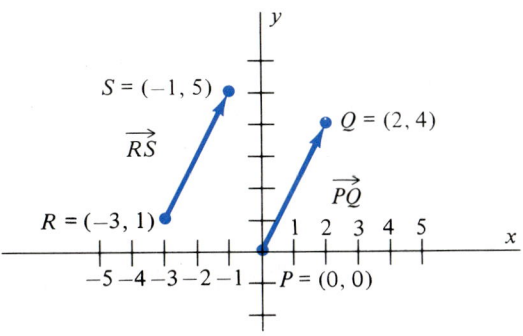

Figure 1.5 $\overrightarrow{PQ} = \overrightarrow{RS}$.

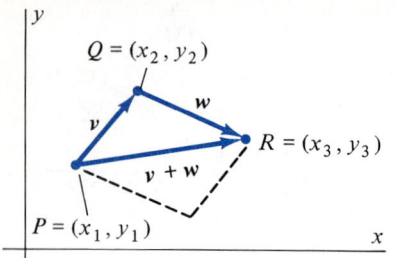

Figure 1.6 Head-to-tail addition for vectors as arrows in the plane.

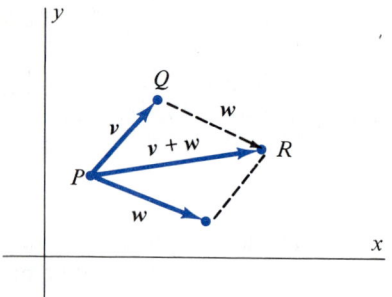

Figure 1.7 $v + w$ is the diagonal of the parallelogram determined by v and w.

Representing vectors by arrows (displacements) in the plane provides a geometric interpretation of the operations of addition and subtraction of vectors as follows.

Vector addition: If the vector $v = \langle a_1, b_1 \rangle$ is represented as an arrow originating at point $P = (x_1, y_1)$, this arrow has terminal point $Q = (x_2, y_2)$ with

$$x_2 = x_1 + a_1,$$
$$y_2 = y_1 + b_1.$$

If the vector $w = \langle a_2, b_2 \rangle$ is represented as an arrow originating at Q, this arrow has terminal point $R = (x_3, y_3)$ with

$$x_3 = x_2 + a_2 = x_1 + (a_1 + a_2),$$
$$y_3 = y_2 + b_2 = y_1 + (b_1 + b_2).$$

Since Definition 1 requires that

$$v + w = \langle a_1, b_1 \rangle + \langle a_2, b_2 \rangle = \langle a_1 + a_2, b_1 + b_2 \rangle,$$

this shows that $v + w$ is represented by the arrow originating at P and terminating at R. (See Figure 1.6.) Thus, the head-to-tail rule for adding arrows (displacements) corresponds to our formal definition of vector addition.

REMARK 1: Note in Figure 1.6 that the vector $v + w$ may also be interpreted as the diagonal of the parallelogram determined by the arrows for v and w, when positioned head-to-tail.

REMARK 2: Figure 1.7 shows that, if the arrow for w is translated (meaning moved so that its length and direction are preserved) so that it originates at P, the sum $v + w$ is still the diagonal of the parallelogram determined by v and w.

REMARK 3: More generally, the arrows for v and w may be translated to *any* initial point in the plane. As long as they originate at the same point, or are positioned head-to-tail, the sum $v + w$ is the diagonal of the parallelogram determined by v and w.

Vector Subtraction: The difference $v - w$ may be represented by an arrow which is the sum of the arrows for the vectors v and $-w$, as in Figures 1.8–1.10. That is because we have defined

$$v - w = v + (-w).$$

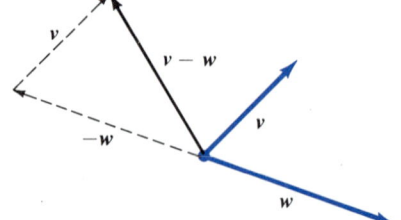

Figure 1.8 $v - w = v + (-w)$.

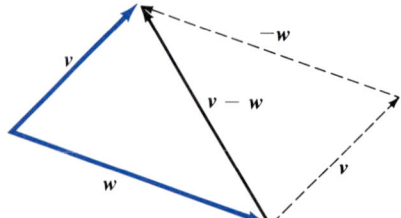

Figure 1.9 $v - w$.

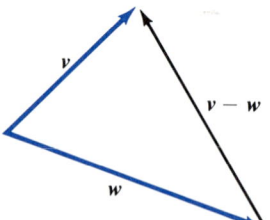

Figure 1.10 $v - w$.

Example 4

Let $P = (1, 2)$, $Q = (3, 6)$, and $R = (3, 0)$. Let $v = \overrightarrow{PQ}$ and $w = \overrightarrow{PR}$. Using the distance formula, we can see that

$$|v| = \text{length of } v = \text{length of } \overline{PQ} = \sqrt{(3-1)^2 + (6-2)^2} = 2\sqrt{5},$$
$$|w| = \text{length of } w = \text{length of } \overline{PR} = \sqrt{(3-1)^2 + (0-2)^2} = 2\sqrt{2}.$$

Also, the slope of v is $\dfrac{6-2}{3-1} = 2$, and the slope of w is $\dfrac{0-2}{3-1} = -1$. From this information it follows that

(a) v, when originating at $(3, 0)$, terminates at $(5, 4)$,
(b) w, when originating at $(3, 6)$, terminates at $(5, 4)$,
(c) $v + w$, when originating at $(1, 2)$, terminates at $(5, 4)$,
(d) $v - w$, when originating at $(3, 0)$, terminates at $(3, 6)$,
(e) $-w$, when originating at $(1, 2)$, terminates at $(-1, 4)$,
(f) $-2v$, when originating at $(1, 2)$, terminates at $(-3, -6)$,
(g) $-2v + w$, when originating at $(1, 2)$, terminates at $(-1, -8)$.

(See Figures 1.11 through 1.13.) ◇

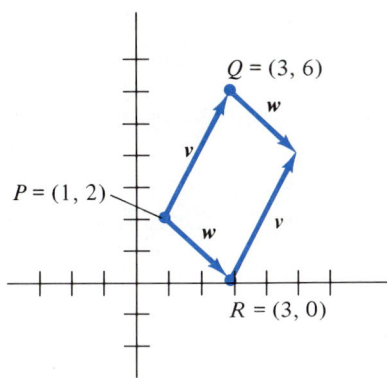

Figure 1.11

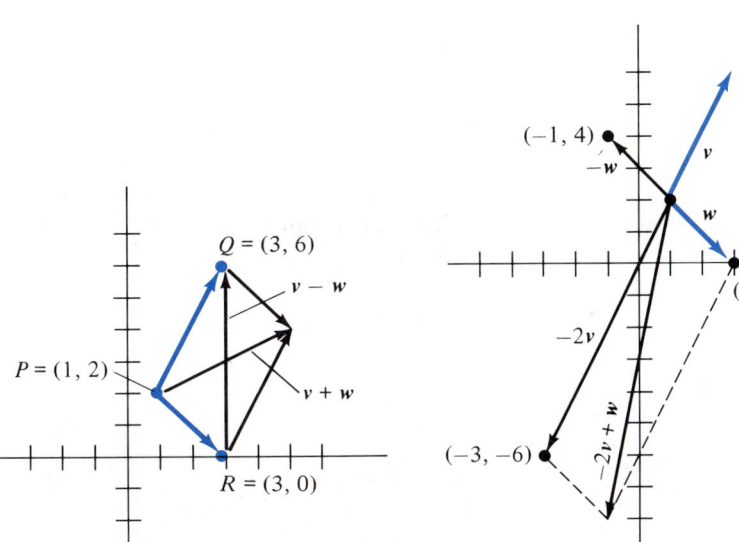

Figure 1.12   **Figure 1.13**

The next example shows how vector methods can be used in plane geometry.

Example 5

Use vectors to prove that the diagonals of a parallelogram bisect each other.

Solution: Let the vertices of the parallelogram be A, B, C, and D, as in Figure 1.14.

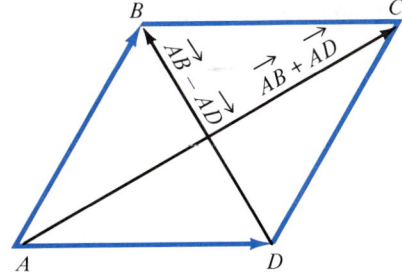

Figure 1.14 The diagonals of a parallelogram bisect each other.

Since $\overrightarrow{AC} = \overrightarrow{AB} + \overrightarrow{AD}$, the vector from A to the midpoint of diagonal $\overrightarrow{AC}$ is

$$v = \frac{1}{2}(\overrightarrow{AB} + \overrightarrow{AD}).$$

Since $\overrightarrow{DB} = \overrightarrow{AB} - \overrightarrow{AD}$, the vector from A to the midpoint of diagonal $\overrightarrow{DB}$ is

$$w = \overrightarrow{AD} + \frac{1}{2}(\overrightarrow{AB} - \overrightarrow{AD})$$

$$= \overrightarrow{AD} + \frac{1}{2}\overrightarrow{AB} - \frac{1}{2}\overrightarrow{AD}$$

$$= \frac{1}{2}(\overrightarrow{AB} + \overrightarrow{AD})$$

$$= v.$$

Since $v = w$, these midpoints are the same. ◇

The upshot of the preceding discussion is that a vector in the plane may be regarded as an ordered pair $\langle a, b \rangle$ of numbers, subject to the laws of addition and scalar multiplication given by Definition 1. Moreover, the vector $\langle a, b \rangle$ may be *represented* geometrically by an arrow originating at any point (x, y) and terminating at the point $(x + a, y + b)$ (Figure 1.15).

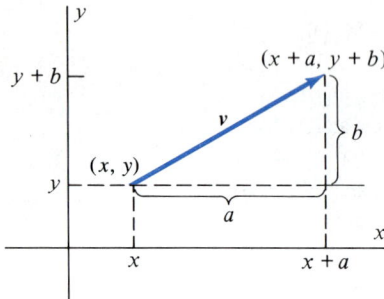

Figure 1.15 The vector $v = \langle a, b \rangle$.

Properties of Vectors

Vectors in the plane satisfy many of the same properties as do real numbers. For example, vector addition is **commutative.** This means that $v + w = w + v$ for any two vectors v and w. However, not all properties of real numbers are shared by vectors. For example, we shall not encounter a useful way to define the *product* of two vectors as another vector in the plane.

The following theorem specifies the legitimate properties of vector addition and scalar multiplication for vectors in the plane.

THEOREM 1

Let u, v, and w be vectors in the plane and let a and b be real numbers. Then

(i) $v + w = w + v$ (commutativity),
(ii) $(u + v) + w = u + (v + w)$ (associativity),
(iii) $0 + v = v$ (identity for vector addition),
(iv) $1 \cdot v = v$ (identity for scalar multiplication),
(v) $a(v + w) = av + aw$ (scalar multiplication distributes over vector addition).
(vi) $(a + b)v = av + bv$.

To prove part (i), we write the vectors v and w in component form

$$v = \langle x_1, y_1 \rangle, \qquad w = \langle x_2, y_2 \rangle$$

and apply Definition 1:

$$v + w = \langle x_1, y_1 \rangle + \langle x_2, y_2 \rangle = \langle x_1 + x_2, y_1 + y_2 \rangle, \tag{2}$$

$$w + v = \langle x_2, y_2 \rangle + \langle x_1, y_1 \rangle = \langle x_2 + x_1, y_2 + y_1 \rangle. \tag{3}$$

Since addition for real numbers is commutative, $x_1 + x_2 = x_2 + x_1$, and $y_1 + y_2 = y_2 + y_1$. Thus, the right-hand sides of equations (2) and (3) represent the same components, so $v + w = w + v$.

You are asked to prove statements (ii) through (vi) in the exercise set. Each statement is proved by writing the vector(s) in component form and using the corresponding properties of real numbers.

Length of Vectors

When represented by an arrow originating at $(0, 0)$, the vector $v = \langle a, b \rangle$ terminates at the point (a, b). The obvious way to define the length of v is to use the distance formula in the plane (see Figure 1.16).

DEFINITION 2

The **length of the vector** $v = \langle a, b \rangle$ is $|v| = \sqrt{a^2 + b^2}$.

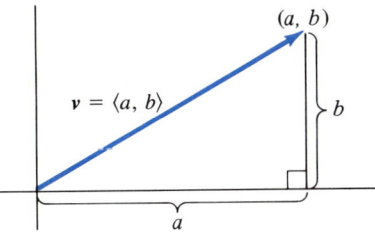

Figure 1.16 $|v| = \sqrt{a^2 + b^2}$.

Example 6

For $v = \langle 2, 1 \rangle$ and $w = \langle -1, 4 \rangle$,

(a) $|v| = \sqrt{2^2 + 1^2} = \sqrt{5}$,

(b) $|w| = \sqrt{(-1)^2 + 4^2} = \sqrt{17}$,

(c) $|2v - w| = \sqrt{(2 \cdot 2 - (-1))^2 + (2 \cdot 1 - 4)^2} = \sqrt{5^2 + 2^2} = \sqrt{29}$. ◇

The following theorem summarizes the properties of length for vectors.

THEOREM 2

Let v and w be vectors and let c be a real number. Then

(i) $|v| \geq 0$; $|v| = 0$ if and only if $v = \mathbf{0}$,

(ii) $|cv| = |c| \cdot |v|$,

(iii) $|v + w| \leq |v| + |w|$.

Properties (i) and (ii) are obvious for our geometric concept of vectors. To prove them according to Definition 1, we write v in the component form $v = \langle a, b \rangle$. Then

$$|v| = \sqrt{a^2 + b^2} \geq 0 \qquad \text{for all} \qquad a, b,$$

which proves the first part of (i). The second part of (i) is proved by noting that

$$|v| = \sqrt{a^2 + b^2} = 0 \qquad \text{if and only if} \qquad a = b = 0,$$

in which case $v = \langle a, b \rangle = \langle 0, 0 \rangle = \mathbf{0}$. Statement (ii) is just as easy:

$$|cv| = |\langle ca, cb \rangle| = \sqrt{(ca)^2 + (cb)^2} = |c|\sqrt{a^2 + b^2} = |c| \cdot |v|.$$

Statement (iii) is referred to as the **triangle inequality.** It may be interpreted geometrically as saying that the length of one leg of a triangle cannot exceed the sum of the lengths of the other two legs (Figure 1.17). A proof of statement (iii) is given in Section 16.2.

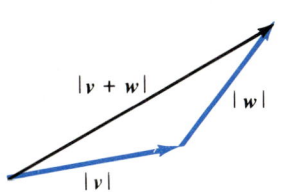

Figure 1.17 $|v + w| \leq |v| + |w|$.

Unit Vectors

A vector u with $|u| = 1$ is called a **unit vector.** In what follows we shall frequently need to find a unit vector u in the same direction as a given vector v. We may do so

using property (ii) of Theorem 2. Since

$$\left|\left(\frac{1}{|v|}\right)v\right| = \frac{1}{|v|}\cdot|v| = 1,$$

the vector

$$u = \frac{1}{|v|}v \qquad\qquad (4)$$

is a unit vector pointing in the same direction as the vector v.

Example 7

Find a unit vector in the same direction as the vector $v = \langle 6, -4\rangle$.

Solution:

$$|v| = \sqrt{6^2 + (-4)^2} = \sqrt{52} = 2\sqrt{13}.$$

Thus, by equation (4), the desired unit vector is

$$u = \frac{1}{2\sqrt{13}}\langle 6, -4\rangle = \left\langle\frac{3}{\sqrt{13}}, \frac{-2}{\sqrt{13}}\right\rangle.$$

(Use Definition 2 to verify that $|u| = 1$.) ◇

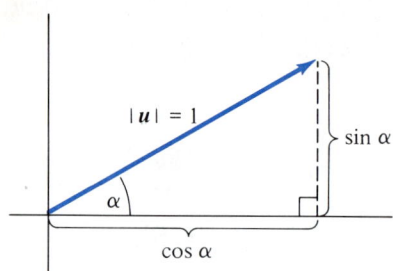

Figure 1.18 $u = \langle \cos \alpha, \sin \alpha\rangle$ is a unit vector.

A similar problem concerning unit vectors is that of finding a unit vector making an angle α with a given line or vector. For example, Figure 1.18 illustrates that *the unit vector forming an angle α with the positive x-axis is*

$$u = \langle \cos \alpha, \sin \alpha\rangle. \qquad\qquad (5)$$

Equation (5) follows from the definition of the trigonometric functions $\sin \theta$ and $\cos \theta$.

Example 8

Find

(a) a unit vector u making an angle of 60° with the positive x-axis.
(b) a vector of length 5 making an angle of 45° with the positive x-axis.

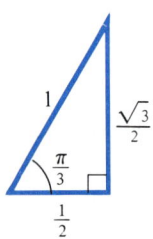

Strategy

(a) Use equation (5) and the facts

$$\cos 60° = \cos\left(\frac{\pi}{3}\right) = \frac{1}{2},$$

$$\sin 60° = \sin\left(\frac{\pi}{3}\right) = \frac{\sqrt{3}}{2}.$$

(b) First, use equation (5) to find a unit vector u in the given direction.

Solution

(a) Using equation (5) we obtain

$$u = \langle \cos 60°, \sin 60°\rangle$$
$$= \left\langle\frac{1}{2}, \frac{\sqrt{3}}{2}\right\rangle.$$

(b) By equation (5) a *unit* vector making an angle of 45° with the positive x-axis is

$$u = \langle \cos 45°, \sin 45°\rangle$$
$$= \left\langle\frac{\sqrt{2}}{2}, \frac{\sqrt{2}}{2}\right\rangle.$$

The solution is then

$$v = 5u.$$

The desired vector is therefore

$$v = 5u = \left\langle \frac{5\sqrt{2}}{2}, \frac{5\sqrt{2}}{2} \right\rangle.$$

◇

Unit Coordinate Vectors

Two special vectors enable us to develop yet another notation for representing vectors in the plane. They are the **unit coordinate vectors**

$$i = \langle 1, 0 \rangle \qquad \text{and} \qquad j = \langle 0, 1 \rangle.$$

(See Figure 1.19.) Using these two vectors, we may represent the vector $\langle a, b \rangle$ as

$$\langle a, b \rangle = a\langle 1, 0 \rangle + b\langle 0, 1 \rangle = ai + bj.$$

Geometrically, ai and bj represent the adjacent sides of the rectangle whose diagonal is $\langle a, b \rangle$ (Figure 1.20). When using this notation, we refer to the numbers a and b as the i and j components of the vector $ai + bj$, respectively.

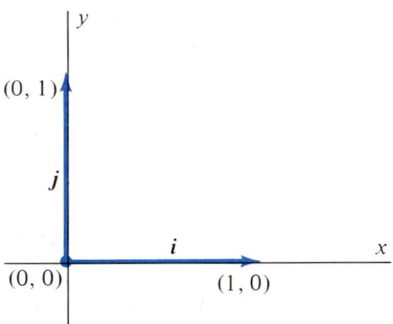

Figure 1.19 Unit coordinate vectors.

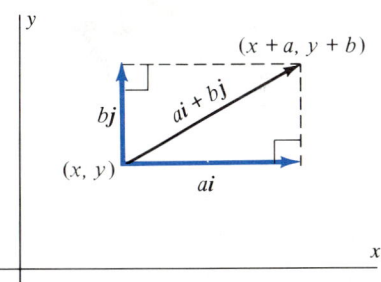

Figure 1.20 $\langle a, b \rangle = ai + bj.$

The principal advantage of the coordinate vector notation is that it helps avoid confusion between points and vectors. We shall make frequent use of this notation in what follows.

Example 9

For $v = \langle -2, 1 \rangle$ and $w = \langle 3, 4 \rangle$,

(a) $v = -2i + j; \qquad w = 3i + 4j,$
(b) $v + w = (-2i + j) + (3i + 4j) = (-2 + 3)i + (1 + 4)j = i + 5j,$
(c) $3v - 2w = (-6i + 3j) - (6i + 8j) = (-6 - 6)i + (3 - 8)j = -12i - 5j,$
(d) $|v| = |-2i + j| = \sqrt{(-2)^2 + 1^2} = \sqrt{5}.$

◇

Exercise Set 15.1

In Exercises 1–10, sketch the given vector originating at the given point P.

1. $\langle 1, 3 \rangle$ at point $P = (3, 2)$

2. $\langle -6, 2 \rangle$ at point $P = (-1, 1)$

3. $\langle -4, -2 \rangle$ at point $P = (2, 6)$

4. $\langle -2, 4 \rangle$ at point $P = (1, 5)$

5. $\langle 3, -5 \rangle$ at point $P = (1, 1)$

6. $i + j$ at point $P = (3, 5)$

7. $-i + 2j$ at point $P = (6, 0)$

8. $5i + 2j$ at point $P = (-3, 4)$

9. $-3i + 2j$ at point $P = (3, -2)$

10. $4j$ at point $P = (0, 2)$

In Exercises 11–16, let $P = (2, 5)$, $Q = (-1, 2)$, and $R = (2, -6)$. Find

11. $v = \overrightarrow{PQ}$

12. $v = \overrightarrow{PQ} + \overrightarrow{QR}$

13. $v = 2\overrightarrow{PQ} - \overrightarrow{RQ}$

14. $v = \overrightarrow{PR} + 4\overrightarrow{RQ}$

15. $v = 2\overrightarrow{RQ} - 2\overrightarrow{RP}$

16. $v = \overrightarrow{PQ} + 2\overrightarrow{QR} - 3\overrightarrow{RP}$

In Exercises 17–22, let $u = \langle 3, 1 \rangle$, $v = \langle -2, 4 \rangle$, and $w = \langle -4, -2 \rangle$. Find

17. $u + 2v$

18. $v - 2u$

19. $2u + 2v - 2w$

20. $-3u + 2v$

21. $v - u - w$

22. $7u + 3w - 6v$

In Exercises 23–30, let $u = 3i - j$, $v = 2i + 6j$, and $w = -i + j$. Find the indicated vector.

23. $u + 2v$

24. $u - 4w$

25. $u + v + w$

26. $u - v - w$

27. $3u + 4v - 2w$

28. $6u - 5w + v$

29. $3u + 3v + 3w$

30. $-u + v + 2w$

In Exercises 31–36, find the angle θ formed between the vector v and the positive x-axis.

31. j

32. $\langle \sqrt{3}, 1 \rangle$

33. $\langle -\sqrt{3}, 1 \rangle$

34. $3i - 3j$

35. $i - \sqrt{3}j$

36. $-i - j$

In Exercises 37–44, find the length of the given vector.

37. $\langle 6, -1 \rangle$

38. $\langle -3, 4 \rangle$

39. $\langle a, a^2 \rangle$

40. i

41. $i + j$

42. $3i + 4j$

43. $6i - 3j$

44. $4(i - 3j)$

45. Find a unit vector pointing in the direction opposite the positive x-axis.

46. Find a unit vector in the direction of $v = 3i + 4j$.

47. Find a unit vector making an angle of $120°$ with the positive x-axis.

48. Find a vector of length 3 in the direction opposite of $v = 2i - 3j$.

49. Find two unit vectors tangent to the graph of $y = x^3$ when positioned to originate at the point $(1, 1)$.

50. Show that the length of the vector originating at $P = (x, y)$ and terminating at $Q = (x + a, y + b)$ is $\sqrt{a^2 + b^2}$.

51. The points $A = (1, 1)$, $B = (5, 1)$, and $C = (6, 3)$ form 3 vertices of a parallelogram. Find the fourth vertex D if
a. A and C lie on a diagonal,
b. A and C lie on a common side.

52. Let $A = (0, 2)$, $B = (b, 5)$, $C = (5, 2)$, $D = (7, 5)$. Find b if $\overrightarrow{AB} + \overrightarrow{AC} = \overrightarrow{AD}$.

53. In the triangle with vertices A, B, and C, let D be the midpoint of side $\overline{AB}$ and let E be the midpoint of side $\overline{BC}$. Prove that $\overrightarrow{DE}$ is parallel to $\overrightarrow{AC}$ and half as long.

54. Prove that the midpoints of the sides of any quadrilateral form the vertices of a parallelogram.

55. Prove statements (ii) through (vi) of Theorem 1.

56. Let $v_1 = i + j$ and $v_2 = -i + j$. Show that for any vector w one can find constants c_1 and c_2 so that

$$w = c_1v_1 + c_2v_2.$$

(*Hint:* Express w in component form and obtain two linear equations for the unknowns c_1 and c_2.)

57. Generalize Exercise 56 by showing that if v_1 and v_2 are any nonzero and nonparallel vectors, then any vector w can be expressed as

$$w = c_1v_1 + c_2v_2$$

for appropriate c_1 and c_2.

58. Let P be a point in the plane and let a be the *position* vector $a = \overrightarrow{OP}$, where O is the origin. Let b be any nonzero vector in the plane. Show that the set of all terminal points of the vectors

$$r(t) = a + tb, \qquad -\infty < t < \infty$$

is a line, as follows.
a. Show that the vector $r(t)$ originates at O for each number t. For each t, let $P(t)$ be the terminal point of the vector $r(t)$.
b. For any $t_1 \neq t_2$, show that the vector $\overrightarrow{P(t_1)P(t_2)}$ is parallel to b. This shows that all points $P(t)$ lie on the same line ℓ.
c. Show that for any point Q on ℓ there is a number t_0 so that $Q = P(t_0)$. This shows that every point on the line corresponds to a vector $r(t)$.

59. In Exercise 58, let $r(t) = x(t)i + y(t)j$, $a = a_1i + a_2j$, and $b = b_1i + b_2j$. Find parametric equations for the components $x(t)$ and $y(t)$. Use these equations to provide an alternate solution to Exercise 58.

60. Use Exercise 58 to show that if P and Q are distinct points in the plane and ℓ is the line through P and Q, the point R lies

on ℓ if and only if

$$\overrightarrow{OR} = \overrightarrow{OP} + t\overrightarrow{PQ}$$

for some number t. (O denotes the origin.)

61. Under what geometric conditions does equality hold in the triangle inequality (statement (iii) of Theorem 2), that is, when does $|v + w| = |v| + |w|$?

15.2 THE DOT PRODUCT

Having defined addition and scalar multiplication for vectors in the plane, it is natural for us to ask whether one can find a useful way to define the product of two vectors. In this section we develop one such product, the *dot* product.

DEFINITION 3

The **dot product** of the vectors $v = \langle x_1, y_1 \rangle$ and $w = \langle x_2, y_2 \rangle$ is the number

$$v \cdot w = x_1 x_2 + y_1 y_2. \tag{1}$$

In other words, the dot product of two vectors is found by multiplying their corresponding components and adding the resulting products. It is important to note that the dot product $v \cdot w$ is a *number*, although each of the factors is a vector. For vectors written in unit coordinate notation, equation (1) becomes

$$(x_1 i + y_1 j) \cdot (x_2 i + y_2 j) = x_1 x_2 + y_1 y_2.$$

The dot product is also referred to as the **scalar product** or the **inner product.**

Example 1

For $v = \langle 2, 3 \rangle$ and $w = \langle -4, 5 \rangle$,

$$v \cdot w = 2(-4) + 3 \cdot 5 = 7.$$

$\diamondsuit$

Example 2

For $v = i - 3j$ and $w = 3i + 3j$,

$$v \cdot w = 1 \cdot 3 + (-3)3 = -6.$$

$\diamondsuit$

Example 3

$$i \cdot j = \langle 1, 0 \rangle \cdot \langle 0, 1 \rangle = 1 \cdot 0 + 0 \cdot 1 = 0.$$

$\diamondsuit$

Example 4

$$(i + j) \cdot (i - j) = 1 \cdot 1 + 1(-1) = 0.$$

$\diamondsuit$

We shall make frequent use of the properties of the dot product given by the following theorem.

THEOREM 3

Let u, v, and w be vectors in the plane and let c be a real number. Then

(i) $v \cdot v = |v|^2$,

(ii) $v \cdot w = w \cdot v$ ($\cdot$ is commutative),

(iii) $u \cdot (v + w) = u \cdot v + u \cdot w$ ($\cdot$ distributes over vector addition),

(iv) $(cv) \cdot w = c(v \cdot w)$ (scalars may be factored),

(v) $0 \cdot v = 0$,

(vi) $|v \cdot w| \le |v| \, |w|$ (Schwarz inequality).

Proof: The proof of the Schwarz inequality will be given later, using Theorem 4. The proofs of statements (i) through (v) follow directly from equation (1). For example, to prove statement (i) we write v in component form as $v = \langle x, y \rangle$. Then

$$v \cdot v = \langle x, y \rangle \cdot \langle x, y \rangle = x^2 + y^2 = (\sqrt{x^2 + y^2})^2 = |v|^2.$$

To prove (ii), we let $v = \langle x_1, y_1 \rangle$ and $w = \langle x_2, y_2 \rangle$. Then

$$v \cdot w = x_1 x_2 + y_1 y_2 = x_2 x_1 + y_2 y_1 = w \cdot v.$$

The proofs of statements (iii) through (v) are similar and are left as exercises. ◆

Angles Between Vectors

Examples 3 and 4 suggest that the dot product might have something to say about the angle between two vectors. In both cases the vectors are represented by perpendicular arrows, and in both cases the dot product is zero. In order to pursue this observation we need to define the concept of the angle between two vectors.

DEFINITION 4

Let the vectors $v = \langle x_1, y_1 \rangle$ and $w = \langle x_2, y_2 \rangle$ be represented by arrows in the plane, originating at the origin and terminating at the points (x_1, y_1) and (x_2, y_2), respectively. The **angle between the vectors** v and w is the smaller of the two angles formed between these arrows (Figure 2.1).

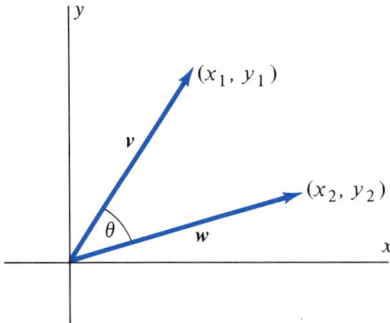

Figure 2.1 Angle between two vectors.

In other words, the angle between two vectors is simply the angle determined by their representations as arrows.

We say that two vectors are **perpendicular** if the angle between them is $\pi/2$. Perpendicular vectors are also called **orthogonal vectors.** We say that two vectors are **parallel** if the angle between them is 0 or π.

The relationship between the dot product $v \cdot w$ and the angle θ formed between v and w is given by the following theorem.

THEOREM 4

Let v and w be nonzero vectors in the plane, and let θ be the angle between v and w. Then

$$v \cdot w = |v|\,|w|\cos\theta. \tag{2}$$

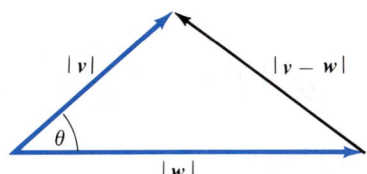

Figure 2.2 Law of Cosines: $|v - w|^2 = |v|^2 + |w|^2 - 2|v||w|\cos\theta$.

Proof: According to our geometric concept of vectors, the vectors v and w determine a triangle in the plane for which the side opposite angle θ has length $|v - w|$ (Figure 2.2). Thus, according to the law of cosines,

$$|v - w|^2 = |v|^2 + |w|^2 - 2|v|\,|w|\cos\theta. \tag{3}$$

Let's write v and w in component form as

$$v = \langle x_1, y_1 \rangle, \qquad w = \langle x_2, y_2 \rangle.$$

Then $v - w = \langle x_1 - x_2, y_1 - y_2 \rangle$, so equation (3) may be written in the form

$$(x_1 - x_2)^2 + (y_1 - y_2)^2 = (x_1^2 + y_1^2) + (x_2^2 + y_2^2) - 2|v|\,|w|\cos\theta$$

which simplifies to the equation

$$-2x_1x_2 - 2y_1y_2 = -2|v|\,|w|\cos\theta,$$

or

$$x_1x_2 + y_1y_2 = |v|\,|w|\cos\theta.$$

Since the left-hand side of this equation is $v \cdot w$, the proof is complete. ◆

When both $v \neq 0$ and $w \neq 0$, we may write equation (2) in the equivalent form

$$\cos\theta = \frac{v \cdot w}{|v|\,|w|}. \tag{4}$$

Equation (4) is useful in calculating angles between vectors, as the following example shows.

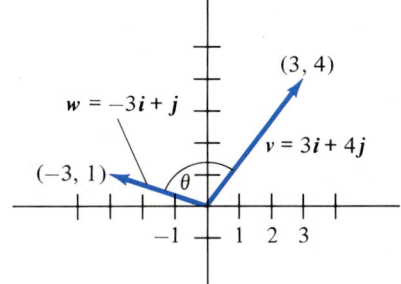

Figure 2.3

Example 5

Find the angle between the vectors $v = 3i + 4j$ and $w = -3i + j$ (Figure 2.3).

Solution: We use equation (4). Since

$$v \cdot w = 3(-3) + 4 \cdot 1 = -5,$$
$$|v| = \sqrt{3^2 + 4^2} = \sqrt{25} = 5,$$
$$|w| = \sqrt{3^2 + 1^2} = \sqrt{10},$$

we have

$$\cos\theta = \frac{-5}{5\sqrt{10}} \approx -0.3162.$$

Since we have implied in Definition 4 that $0 \le \theta \le \pi$, we have

$$\theta = \text{Cos}^{-1}\left(\frac{-5}{5\sqrt{10}}\right) \approx 0.602\pi \ (\approx 108.4°). \qquad \diamond$$

Theorem 4 provides a useful criterion for determining whether two nonzero vectors are orthogonal. If $|\boldsymbol{v}| \ne 0$ and $|\boldsymbol{w}| \ne 0$, equation (2) shows that $\boldsymbol{v} \cdot \boldsymbol{w} = 0$ if and only if $\cos \theta = 0$. Since θ is the angle between $\boldsymbol{v}$ and $\boldsymbol{w}$, $0 \le \theta \le \pi$. Thus, $\cos \theta = 0$ if and only if $\theta = \pi/2$. Combining these observations gives the following result.

COROLLARY 1

The nonzero vectors $\boldsymbol{v}$ and $\boldsymbol{w}$ are orthogonal if and only if $\boldsymbol{v} \cdot \boldsymbol{w} = 0$.

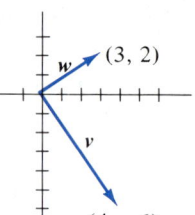

Figure 2.4

Example 6

Show that the vectors $\boldsymbol{v} = 4\boldsymbol{i} - 6\boldsymbol{j}$ and $\boldsymbol{w} = 3\boldsymbol{i} + 2\boldsymbol{j}$ are orthogonal (Figure 2.4).

Solution: We use Corollary 1. Since

$$\boldsymbol{v} \cdot \boldsymbol{w} = 4 \cdot 3 + (-6) \cdot 2 = 0,$$

the vectors are orthogonal. $\qquad \diamond$

Example 7

Show that the vector $a\boldsymbol{i} + b\boldsymbol{j}$ is orthogonal to the line ℓ with equation

$$ax + by + c = 0.$$

(See Figure 2.5.)

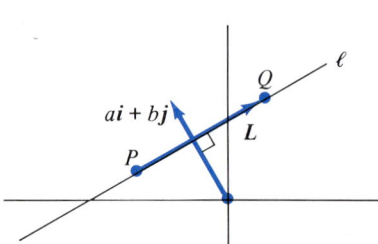

Figure 2.5

Solution: Our strategy is first to find a vector $\boldsymbol{L}$ parallel to the line ℓ and then to show that $(a\boldsymbol{i} + b\boldsymbol{j}) \cdot \boldsymbol{L} = 0$. We begin by letting $P = (x_1, y_1)$ and $Q = (x_2, y_2)$ be distinct points on ℓ. Then $\overrightarrow{PQ}$ is a segment of ℓ, so $\boldsymbol{L} = \overrightarrow{PQ}$ represents a vector parallel to ℓ, which has the component form

$$\boldsymbol{L} = (x_2 - x_1)\boldsymbol{i} + (y_2 - y_1)\boldsymbol{j}.$$

Now, since P and Q lie on ℓ, both

$$ax_1 + by_1 + c = 0 \qquad \text{and} \qquad ax_2 + by_2 + c = 0.$$

Subtracting corresponding sides of these equations gives the equation

$$a(x_2 - x_1) + b(y_2 - y_1) = 0,$$

which we can rewrite as $(a\boldsymbol{i} + b\boldsymbol{j}) \cdot \boldsymbol{L} = 0$. This shows that $a\boldsymbol{i} + b\boldsymbol{j}$ is orthogonal to $\boldsymbol{L}$, and, hence, to ℓ. $\qquad \diamond$

We next use Theorem 4 to prove the Schwarz inequality of Theorem 3. Since this inequality is so important, we restate it here as a corollary of Theorem 4.

COROLLARY 2
Schwarz Inequality

For any vectors $\boldsymbol{v}$ and $\boldsymbol{w}$,

$$|\boldsymbol{v} \cdot \boldsymbol{w}| \le |\boldsymbol{v}| \, |\boldsymbol{w}|.$$

Proof: If either $v = 0$ or $w = 0$, the inequality holds, since both sides are zero. Otherwise, an angle θ is determined between v and w. Since $|\cos \theta| \leq 1$ for all θ, we may apply Theorem 4 to conclude that

$$|v \cdot w| = |v||w| \, |\cos \theta| \leq |v| \, |w|. \qquad \blacklozenge$$

We can use the Schwarz inequality to prove the triangle inequality (Theorem 2, part (iii), Section 15.1):

$$|v + w| \leq |v| + |w|. \qquad (5)$$

The strategy will be to prove the inequality

$$|v + w|^2 \leq (|v| + |w|)^2 \qquad (6)$$

from which inequality (5) is obtained by taking square roots.

Applying Theorem 3, parts (i) through (iii), we find that

$$\begin{aligned}
|v + w|^2 &= (v + w) \cdot (v + w) \qquad (7)\\
&= v \cdot v + 2v \cdot w + w \cdot w \\
&= |v|^2 + 2v \cdot w + |w|^2.
\end{aligned}$$

Now the Schwarz inequality shows that

$$2v \cdot w \leq 2|v \cdot w| \leq 2|v| \, |w|. \qquad (8)$$

Combining (7) and (8), we obtain the inequality

$$\begin{aligned}
|v + w|^2 &\leq |v|^2 + 2|v| \, |w| + |w|^2 \\
&= (|v| + |w|)^2,
\end{aligned}$$

which proves inequality (6).

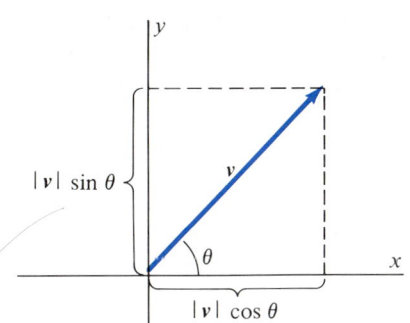

Figure 2.6 Component of v in x direction is $|v| \cos \theta = v \cdot i$.

Components

If v is a nonzero vector and θ is the angle formed between v and the unit coordinate vector i, then v can be written in the component form

$$v = |v|(\cos \theta)i + |v|(\sin \theta) \, j. \qquad (9)$$

(See Figure 2.6.) Notice that we can write the component of v in the direction of the vector i as

$$|v| \cos \theta = |v| \, |i| \cos \theta = v \cdot i \qquad \text{(Theorem 4)}. \qquad (10)$$

Now suppose we are given a second vector w, with $w \neq 0$. Let θ be the angle formed between v and w. The number $|v| \cos \theta$ is called **the component of v with respect to w**, written $\text{comp}_w \, v = |v| \cos \theta$ (Figure 2.7). The number $\text{comp}_w \, v$ may be interpreted as the change, in the direction of w, represented by the vector v. Using the dot product we may write

$$|v| \cos \theta = \frac{|v| \, |w| \cos \theta}{|w|} = \frac{v \cdot w}{|w|}$$

so

$$\boxed{\text{comp}_w \, v = \frac{v \cdot w}{|w|}.} \qquad (11)$$

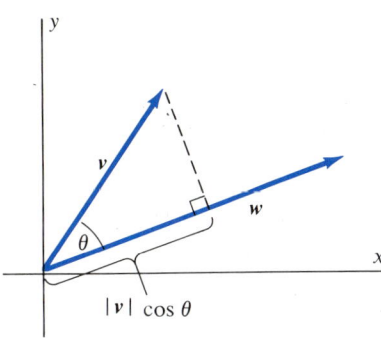

Figure 2.7 Component of v in the direction of w is $|v| \cos \theta = \dfrac{v \cdot w}{|w|}$.

Notice that $\text{comp}_i \, v = \dfrac{v \cdot i}{|i|} = v \cdot i$, so $\text{comp}_w \, v$ is a generalization of the concept of i-component in equation (10) (see Figure 2.7).

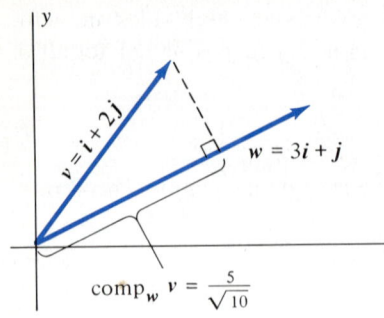

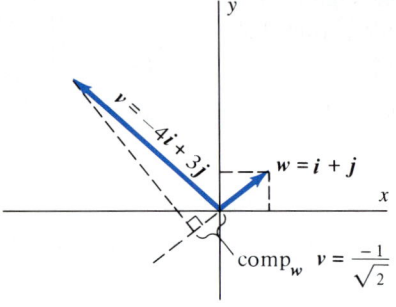

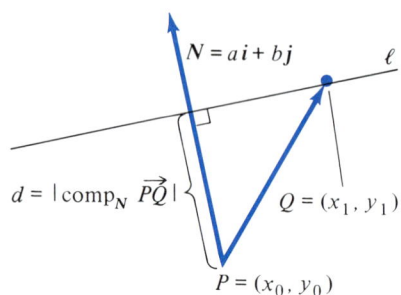

Figure 2.8

Figure 2.9

Figure 2.10 Distance from (x_0, y_0) to ℓ: $ax + by + c = 0$ is $|\text{comp}_N\ \overrightarrow{PQ}|$.

Example 8

For $v = i + 2j$ and $w = 3i + j$,

$$\text{comp}_w\ v = \frac{1 \cdot 3 + 2 \cdot 1}{\sqrt{3^2 + 1^2}} = \frac{5}{\sqrt{10}} \qquad \text{(Figure 2.8)}. \qquad \diamond$$

Example 9

For $v = -4i + 3j$ and $w = i + j$

$$\text{comp}_w\ v = \frac{(-4)(1) + 3 \cdot 1}{\sqrt{1^2 + 1^2}} = \frac{-1}{\sqrt{2}} \qquad \text{(Figure 2.9)}. \qquad \diamond$$

The notion of component can be used to develop a formula for the distance from the point $P = (x_0, y_0)$ to the line ℓ with equation ℓ: $ax + by + c = 0$. The idea is summarized by Figure 2.10: If N is a vector orthogonal to ℓ, and if $Q = (x_1, y_1)$ is any point on ℓ, then $|\text{comp}_N\ \overrightarrow{PQ}|$ is the required distance. The details go like this:
If $Q = (x_1, y_1)$ is a point on ℓ, then

$$\overrightarrow{PQ} = (x_1 - x_0)i + (y_1 - y_0)j.$$

By Example 7, we know that a vector orthogonal to ℓ is $N = ai + bj$. The required distance is therefore

$$d = |\text{comp}_N\ \overrightarrow{PQ}| = \frac{|\overrightarrow{PQ} \cdot N|}{|N|}$$

$$= \frac{|a(x_1 - x_0) + b(y_1 - y_0)|}{\sqrt{a^2 + b^2}}$$

$$= \frac{|(ax_1 + by_1) - (ax_0 + by_0)|}{\sqrt{a^2 + b^2}}.$$

Since $Q = (x_1, y_1)$ is on ℓ, we have $ax_1 + by_1 + c = 0$, so $ax_1 + by_1 = -c$. We therefore find that

$$d = \frac{|ax_0 + by_0 + c|}{\sqrt{a^2 + b^2}} \tag{12}$$

is the distance from the point $P = (x_0, y_0)$ to the line ℓ with equation $ax + by + c = 0$.

Example 10

The distance from the point $(4, -2)$ to the line with equation $3x - y + 4 = 0$ is, by equation (12),

$$d = \frac{|3(4) + (-1)(-2) + 4|}{\sqrt{3^2 + 1^2}} = \frac{18}{\sqrt{10}}. \qquad \diamond$$

Projections

Let v and w be nonzero vectors in the plane. The vector

$$\text{proj}_w\ v = \left(\frac{v \cdot w}{|w|^2}\right)w \tag{13}$$

is called the **orthogonal projection** of v onto w. Since we can write

$$\text{proj}_w \, v = \left(\frac{v \cdot w}{|w|^2} \right) w = \left(\frac{v \cdot w}{|w|} \right) \frac{w}{|w|}, \tag{14}$$

we see that $proj_w \, v$ is the product of the number $comp_w \, v$ and the unit vector $\dfrac{w}{|w|}$ in the direction of w (see Figure 2.11).

Since $\dfrac{w}{|w|}$ is a unit vector, equation (14) shows that

$$\left| \text{proj}_w \, v \right| = \left| \frac{w \cdot v}{|w|} \right| = \left| \text{comp}_w \, v \right|. \tag{15}$$

That is, the length of the *vector* $\text{proj}_w \, v$ equals the absolute value of the *number* $\text{comp}_w \, v$.

The vector

$$\text{proj}_{\perp w} \, v = v - \text{proj}_w \, v \tag{16}$$

is called the **orthogonal projection of v perpendicular to w.**

It is obvious from equation (16) that

$$v = \text{proj}_w \, v + \text{proj}_{\perp w} \, v \qquad \text{(Figure 2.12)}. \tag{17}$$

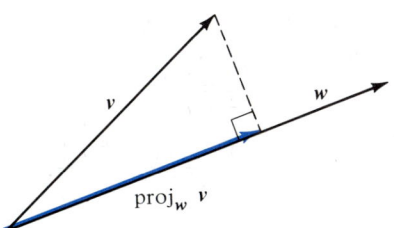

Figure 2.11 Projection of v onto w.   **Figure 2.12** $v = \text{proj}_w \, v + \text{proj}_{\perp w} \, v$.

In other words, $\text{proj}_w \, v$ and $\text{proj}_{\perp w} \, v$ may be interpreted as adjacent sides of a parallelogram for which v is the diagonal. Indeed, we can say more: the parallelogram is actually a rectangle. That is,

$$(\text{proj}_w \, v) \cdot (\text{proj}_{\perp w} \, v) = 0. \tag{18}$$

The proof of (18) is simply a calculation:

$$\begin{aligned}
(\text{proj}_w \, v) \cdot (\text{proj}_{\perp w} \, v) &= \left[\left(\frac{v \cdot w}{|w|^2} \right) w \right] \cdot \left[v - \left(\frac{v \cdot w}{|w|^2} \right) w \right] \\
&= \left(\frac{v \cdot w}{|w|^2} \right) w \cdot v - \left(\frac{v \cdot w}{|w|^2} \right)^2 w \cdot w \\
&= \left(\frac{v \cdot w}{|w|} \right)^2 - \left(\frac{v \cdot w}{|w|^2} \right)^2 |w|^2 \\
&= 0.
\end{aligned}$$

Equation (18) explains the use of the terminology "orthogonal" projection in defining $\text{proj}_w \, v$. The projections $\text{proj}_w \, v$ and $\text{proj}_{\perp w} \, v$ allow us to express a given vector v as the sum of a vector parallel to a given vector or line, and a vector orthogonal to

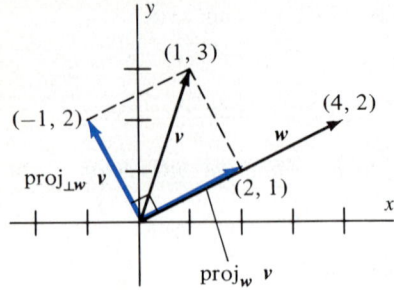

Figure 2.13 Projections of **v** onto and orthogonal to **w**.

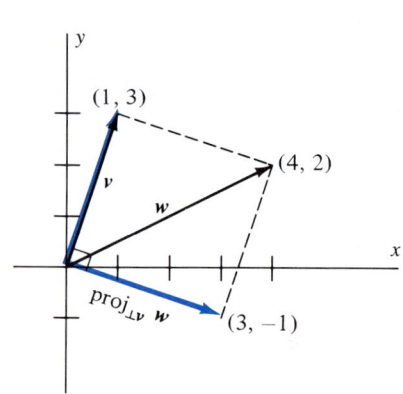

Figure 2.14 $v = \text{proj}_v \, w$.

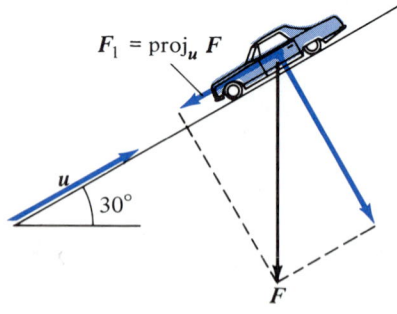

Figure 2.15 $|F_1| = |\text{proj}_u \, F|$ is the force required to prevent auto from rolling down the incline.

the given vector or line. This is a particularly useful concept in physics, where one wishes to "resolve" a force or velocity vector into **component vectors** parallel and perpendicular to a given force or velocity vector.

Example 11

For $v = i + 3j$ and $w = 4i + 2j$,

(a) $\text{proj}_w \, v = \left[\dfrac{(i + 3j) \cdot (4i + 2j)}{|4i + 2j|^2} \right] (4i + 2j)$

$\qquad = \left[\dfrac{1 \cdot 4 + 3 \cdot 2}{4^2 + 2^2} \right] (4i + 2j)$

$\qquad = 2i + j,$

and

$\text{proj}_{\perp w} \, v = (i + 3j) - (2i + j) = -i + 2j$ (Figure 2.13).

(b) $\text{proj}_v \, w = \left[\dfrac{(4i + 2j) \cdot (i + 3j)}{|i + 3j|^2} \right] (i + 3j)$

$\qquad = \left[\dfrac{4 \cdot 1 + 2 \cdot 3}{1^2 + 3^2} \right] (i + 3j)$

$\qquad = i + 3j \qquad (= v),$

and

$\text{proj}_{\perp v} \, w = (4i + 2j) - (i + 3j) = 3i - j$ (Figure 2.14). ◇

Example 12

Find the force required to hold a 5000-kilogram automobile motionless on a 30° incline, assuming that the only force that must be overcome is that due to gravity.

Solution: We may represent the force due to gravity as a vector **F** pointing vertically downward. According to Newton's second law, the magnitude of this force vector is $|F| = mg = 5000g$, where $g = 9.8 \text{ m/s}^2$ is the acceleration due to gravity. Thus

$$F = (-5000g)j.$$

By equation (5), Section 15.1, a unit vector **u** in the direction of the 30° incline is represented in Figure 2.15 and by the equation

$$u = \cos 30°i + \sin 30°j = \frac{\sqrt{3}}{2}i + \frac{1}{2}j.$$

The force due to **F** acting in the direction of the incline is therefore represented by the vector

$$F_1 = \text{proj}_u \, F = \frac{F \cdot u}{|u|^2} u = \frac{(-5000g)j \cdot \left(\frac{\sqrt{3}}{2}i + \frac{1}{2}j \right)}{1^2} \left(\frac{\sqrt{3}}{2}i + \frac{1}{2}j \right)$$

$$= -2500g \left(\frac{\sqrt{3}}{2}i + \frac{1}{2}j \right).$$

The magnitude of this force is

$$|F_1| = |\text{comp}_u\, F| = 2500g.$$

This is the force due to gravity which must be overcome to hold the automobile in place.

Example 13

When a constant force of magnitude F moves the point at which it is applied a distance d *in the direction in which the force is applied*, the work done by the force is defined to be the product

$$W = Fd. \tag{19}$$

However, if the point of application is confined to move along a line in a direction different from that of the applied force, equation (19) does not apply. If the vector F represents the magnitude and direction of the applied force, and the vector d represents the magnitude and direction of the resulting displacement, the work done is defined to be the dot product

$$W = F \cdot d. \tag{20}$$

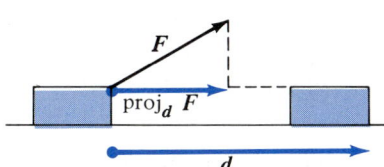

Figure 2.16 Work done by force F over the displacement d is $|\text{proj}_d\, F||d|$ $F \cdot d$.

Figure 2.16 illustrates this definition. The vector component of F acting in the direction of d is $\text{proj}_d\, F$, whose magnitude is $\text{comp}_d\, F = \dfrac{F \cdot d}{|d|}$. Since this is the force acting in the direction of motion, equation (19) gives

$$W = (\text{comp}_d\, F)(|d|) = \frac{F \cdot d}{|d|}(|d|) = F \cdot d.$$

Exercise Set 15.2

In Exercises 1–6, find the dot product $v \cdot w$.

1. $v = i + 2j, \qquad w = 3i - j$

2. $v = \langle -3, 5 \rangle, \qquad w = \langle 1, 4 \rangle$

3. $v = \langle -3, 5 \rangle, \qquad w = \langle 6, -2 \rangle$

4. $v = ai + bj, \qquad w = ci + dj$

5. $v = i + 4j, \qquad w = i - 4j$

6. $v = 2j, \qquad w = i - 3j$

7. Find the cosine of the angle between the vectors $v = i - 3j$ and $w = -4i + j$.

8. Find the cosine of the angle between the vectors $v = \langle 3, 2 \rangle$ and $w = \langle -2, 2 \rangle$.

9. For the vectors $u = -2i + 4j$, $v = 3i + 5j$, and $w = 6i - 4j$ find

  a. $u \cdot v$ **b.** $u \cdot w$

  c. $u \cdot (v + w)$ **d.** $u \cdot (v - w)$

  e. $(u + v) \cdot (v - w)$ **f.** $u \cdot (2v + 3w)$

  g. $\text{comp}_v\, u$ **h.** $\text{comp}_w\, v$

  i. $\text{proj}_u\, v$ **j.** $\text{proj}_{\perp w}\, u$

10. Let $v = ai + j$ and $w = 3i + 4j$. Find a so that

  a. $v \cdot w = 0$ **b.** $v \cdot w = 10$

  c. $\text{proj}_w\, v = v$ **d.** $\text{proj}_w\, v = 0$

11. For $v = 2i - 3j$ and $w = 5i + j$, find

  a. $\text{comp}_w\, v$ **b.** $\text{comp}_i\, v$

  c. $\text{comp}_v\, w$ **d.** $\text{proj}_w\, v$

  e. $\text{proj}_{\perp w}\, v$ **f.** $\text{proj}_{\perp v}\, w$

12. Find the interior angles of the triangle with vertices $(0, 0)$, $(3, 0)$, and $(3, 4)$.

13. Determine which of the following pairs of vectors are orthogonal.

  a. $v = 2i + 4j$ **b.** $v = 3i - 2j$
      $w = 5i - j$ $w = 10i + 15j$

  c. $v = i - 3j$ **d.** $v = i + j$
      $w = 6i + 2j$ $w = -i - j$

14. Find the number a so that the vectors $v = 3i - 5j$ and $w = ai + 3j$ are orthogonal.

15. Find two unit vectors orthogonal to $v = 2i + j$.

16. Find an example of three nonzero vectors u, v, and w for which $u \cdot v = u \cdot w$, but $v \neq w$. This shows that a "cancellation law" for dot products cannot hold.

17. Show that $v + w$ and $v - w$ are perpendicular if $|v| = |w| \neq 0$. Conclude that the diagonals of a rhombus are perpendicular. (A rhombus is a parallelogram with all sides of equal length.)

18. Express the vector $v = 3i + 4j$ as the sum of a vector parallel to $w = 3i + j$ and a vector orthogonal to w.

19. Express the unit coordinate vector i as the sum of a vector parallel to the vector $v = 2i - j$ and a vector perpendicular to v.

20. Find the distance from the point $(-4, 3)$ to the line with equation $2x - y - 3 = 0$.

21. Find the distance from the point $(3, 6)$ to the line with equation $x + y - 1 = 0$.

22. Show that the points $(1, 2)$, $(3, 4)$, and $(5, 2)$ are vertices of a right triangle. Which vertex corresponds to the right angle?

23. A child pulls a sled by exerting a 20-pound force on a rope which makes an angle of 45° with the horizontal. Find the work done by the child in pulling the sled a distance of 100 feet.

24. Prove statements (iii) through (v) of Theorem 3.

25. Assume $v \neq 0$ and $w \neq 0$. Under what conditions is $v \cdot w = |v| \cdot |w|$?

26. Find, to the nearest degree, the angle between $v = i + 4j$ and $w = 3i + 5j$.

27. Use the dot product to show that if $P = (a_1, b_1)$ and $Q = (a_2, b_2)$ are endpoints of the diameter of a circle and if $X = (x, y)$ is a point on the circle, then $\overrightarrow{XP}$ and $\overrightarrow{XQ}$ are orthogonal.

28. Use the result of Exercise 27 and the dot product to find an equation for the circle having (a_1, b_1) and (a_2, b_2) as endpoints of a diameter.

29. Prove the converse of Exercise 17: In a parallelogram, if the diagonals are perpendicular, then the parallelogram is a rhombus.

15.3 SPACE COORDINATES; VECTORS IN SPACE

Euclidean three-dimensional space provides a framework in which to discuss space curves, surfaces, and graphs of functions of two variables. By Euclidean space, we mean the coordinatizing of space by three mutually perpendicular coordinate axes, labelled as in Figure 3.1.

Scales for each of the axes are chosen so that the point of intersection of the three axes corresponds to zero on each axis. The usual convention in sketching the coordinate axes is to display the positive half of the x-axis as pointing toward the reader, the positive half of the y-axis as pointing to the right, and the positive half of the z-axis as pointing upward.

Using the scales on the three coordinate axes, we may identify each point P in space with a unique triple of numbers $P = (x_0, y_0, z_0)$ as follows: the plane through P perpendicular to the x-axis intersects the x-axis at x_0; the plane through P perpendicular to the y-axis intersects the y-axis at y_0; the plane through P perpendicular to the z-axis intersects the z-axis at z_0 (see Figure 3.2). Examples of several points in space are shown in Figure 3.3.

Figure 3.4 illustrates that the set of all points obtained by specifying two of the three coordinates (and allowing the third to range unrestricted) is a line parallel to the axis of the unrestricted variable. If one specifies only one variable, allowing all values of the other two, the figure obtained is a plane perpendicular to the axis of the restricted variable (Figure 3.5). Equations for lines and planes not parallel or perpendicular to coordinate axes are more complicated and are the subject of Section 15.5.

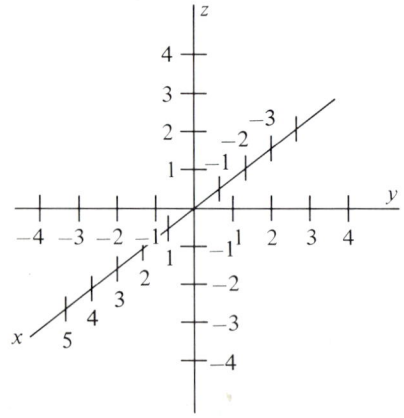

Figure 3.1 Coordinate axes for Euclidean space.

The Distance Formula

Let $P_1 = (x_1, y_1, z_1)$ and $P_2 = (x_2, y_2, z_2)$ be two points in space. To find the distance $d(P_1, P_2)$ between these points, we use the Pythagorean theorem twice.

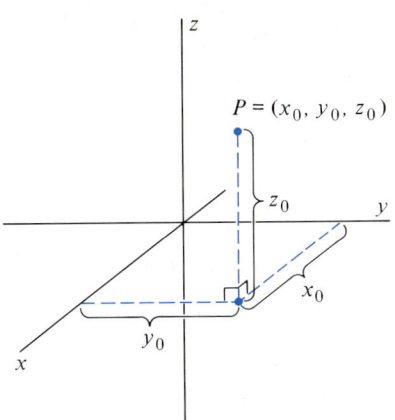

Figure 3.2 The coordinates of $P = (x_0, y_0, z_0)$.

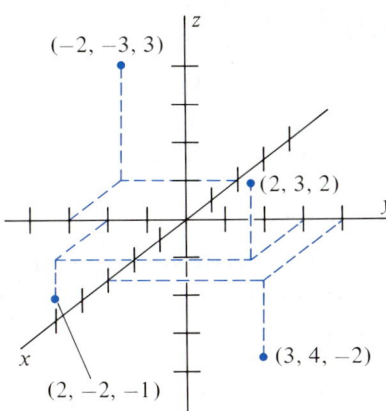

Figure 3.3 Points in space and their coordinates.

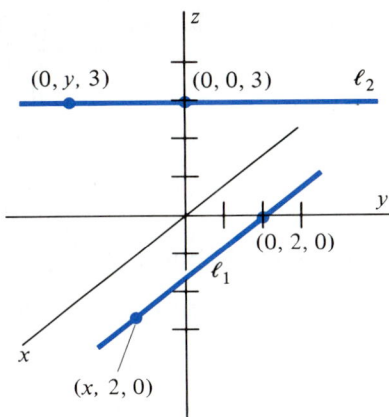

Figure 3.4 ℓ_1: The line $y = 2$, $z = 0$. ℓ_2: The line $x = 0$, $z = 3$.

Let $S = (x_1, y_2, z_1)$ and $R = (x_2, y_2, z_1)$, as in Figure 3.6. Then, since the segment $\overline{P_1 R}$ is the hypotenuse of the right triangle $\Delta P_1 RS$, the Pythagorean theorem gives

$$[d(P_1, R)]^2 = [d(S, R)]^2 + [d(P_1, S)]^2$$
$$= (x_2 - x_1)^2 + (y_2 - y_1)^2.$$

Similarly, the segment $\overline{P_1 P_2}$ is the hypotenuse of the right triangle $\Delta P_1 R P_2$. Thus, a second application of the Pythagorean theorem, together with the above equation, gives

$$[d(P_1, P_2)]^2 = [d(P_1, R)]^2 + [d(R, P_2)]^2$$
$$= (x_2 - x_1)^2 + (y_2 - y_1)^2 + (z_2 - z_1)^2.$$

Taking square roots of both sides of this equation gives the desired formula:

The distance between $P_1 = (x_1, y_1, z_1)$ and $P_2 = (x_2, y_2, z_2)$ is
$$d(P_1, P_2) = \sqrt{(x_2 - x_1)^2 + (y_2 - y_1)^2 + (z_2 - z_1)^2}.$$

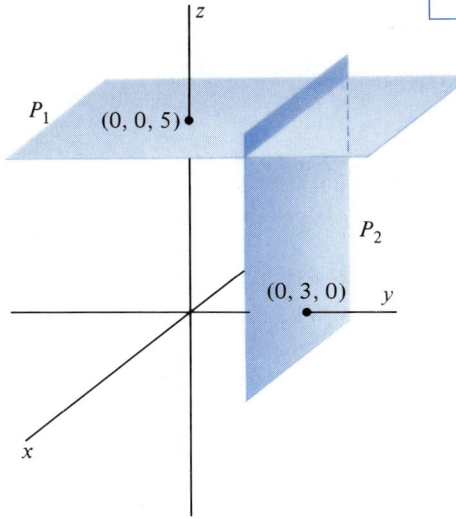

Figure 3.5 P_1: The plane $z = 5$. P_2: The plane $y = 3$.

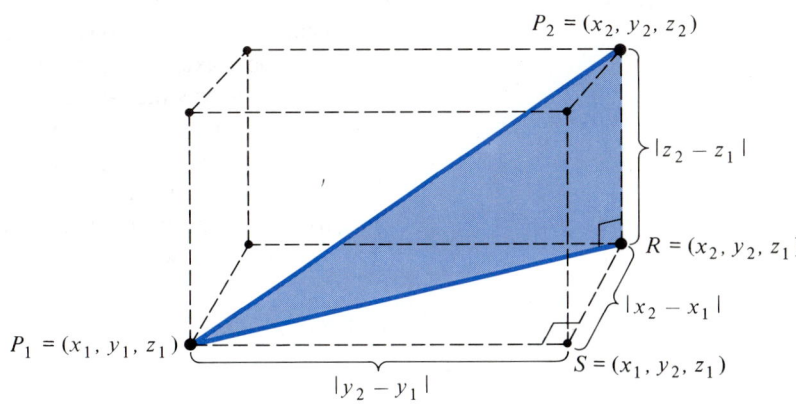

Figure 3.6 $d(P_1, P_2) = \sqrt{(x_2 - x_1)^2 + (y_2 - y_1)^2 + (z_2 - z_1)^2}$.

Example 1

The distance between $P = (-3, 0, 4)$ and $Q = (2, 5, -2)$ is

$$d = \sqrt{(2 - (-3))^2 + (5 - 0)^2 + (-2 - 4)^2}$$
$$= \sqrt{5^2 + 5^2 + (-6)^2} = \sqrt{86}.$$

◇

Equations for Spheres

The sphere with center $C = (a, b, c)$ and radius $r > 0$ is the set of points $P = (x, y, z)$ for which

$$d(P, C) = r.$$

Using the distance formula, we may rewrite this equation as

$$\sqrt{(x - a)^2 + (y - b)^2 + (z - c)^2} = r,$$

or, in standard form,

$$(x - a)^2 + (y - b)^2 + (z - c)^2 = r^2. \tag{1}$$

Example 2

The sphere with center $C = (2, -3, 1)$ and radius $r = 3$ has equation

$$(x - 2)^2 + (y + 3)^2 + (z - 1)^2 = 9 \qquad \text{(Figure 3.7)}.$$

◇

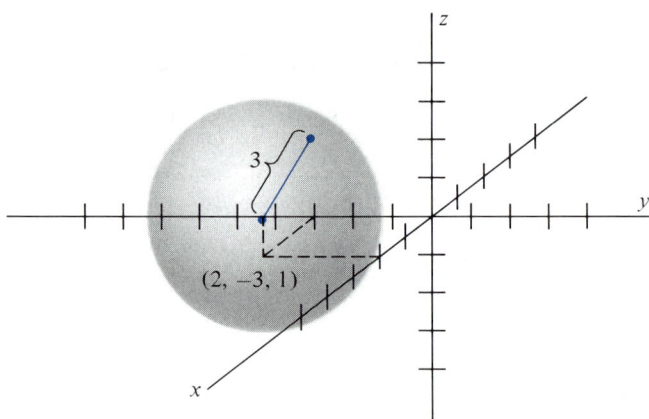

Figure 3.7 Sphere with center $(2, -3, 1)$ and radius 3.

Example 3

To determine whether the equation

$$x^2 + y^2 + z^2 - 4x + 6z - 3 = 0$$

is that of a sphere, we complete the square in each variable:

$$0 = x^2 + y^2 + z^2 - 4x + 6z - 3$$
$$= (x^2 - 4x) + y^2 + (z^2 + 6z) - 3$$
$$= (x^2 - 4x + 4) + y^2 + (z^2 + 6z + 9) - 3 - 4 - 9.$$

Thus,

$$(x - 2)^2 + y^2 + (z + 3)^2 = 16.$$

The graph is a sphere with center at $(2, 0, -3)$ and radius $r = 4$.   ◇

Vectors in Space

To define vectors in space we must work with **ordered triples,** since points in space are determined by three coordinates.

DEFINITION 5

The set of all ordered triples $\langle x, y, z \rangle$ together with the laws

(i) $\langle x_1, y_1, z_1 \rangle + \langle x_2, y_2, z_2 \rangle = \langle x_1 + x_2, y_1 + y_2, z_1 + z_2 \rangle$
(ii) $c\langle x, y, z \rangle = \langle cx, cy, cz \rangle, \qquad c \in (-\infty, \infty)$

for addition and multiplication by real numbers is the **set of vectors in space.**

For the vector $v = \langle x_1, y_1, z_1 \rangle$, the numbers $x_1, y_1,$ and z_1 are again referred to as the **components** of v. You have probably already noticed that Definition 5 prescribes that vector addition and scalar multiplication are to be carried out for vectors in space just as for vectors in the plane, except that now there is a third component to deal with. Accordingly, we define

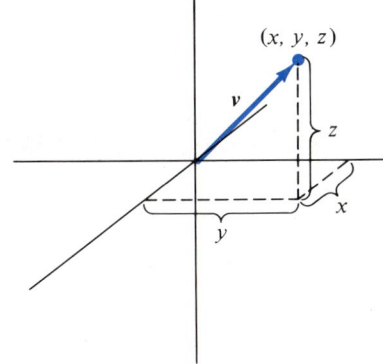

Figure 3.8 The vector $v = \langle x, y, z \rangle$.

(a) the **negative** of the vector $v = \langle x, y, z \rangle$ as

$$-v = \langle -x, -y, -z \rangle,$$

(b) the **difference** of two vectors $v = \langle x_1, y_1, z_1 \rangle$ and $w = \langle x_2, y_2, z_2 \rangle$ as

$$v - w = v + (-w) = \langle x_1 - x_2, y_1 - y_2, z_1 - z_2 \rangle,$$

(c) the **zero vector** as

$$0 = \langle 0, 0, 0 \rangle,$$

(d) **equality** of two vectors $v = \langle x_1, y_1, z_1 \rangle$ and $w = \langle x_2, y_2, z_2 \rangle$ by

$$v = w \qquad \text{if and only if} \qquad x_1 = x_2, y_1 = y_2 \text{ and } z_1 = z_2.$$

With these definitions, all properties of Theorem 1 for vectors in the plane carry over to vectors in space. The techniques of proof are the same as those used in the proof of Theorem 1, except that the third coordinate is included in all calculations (see Exercise 44).

The vector $v = \langle x_1, y_1, z_1 \rangle$ may be represented by an arrow originating at the origin and terminating at the point (x_1, y_1, z_1). More generally, if $P = (a, b, c)$ is any point in space, the vector $v = \langle x, y, z \rangle$ may be represented by an arrow originating at the point (a, b, c) and terminating at the point $(a + x, b + y, c + z)$ (see Figures 3.8 and 3.9).

As with vectors in the plane, we shall not concern ourselves with the distinction between vectors defined as ordered triples and their representations as arrows in space. For example, we say that the vector originating at the point $P = (x_1, y_1, z_1)$ and terminating at the point $Q = (x_2, y_2, z_2)$ may be written in component form as

$$\overrightarrow{PQ} = \langle x_2 - x_1, y_2 - y_1, z_2 - z_1 \rangle.$$

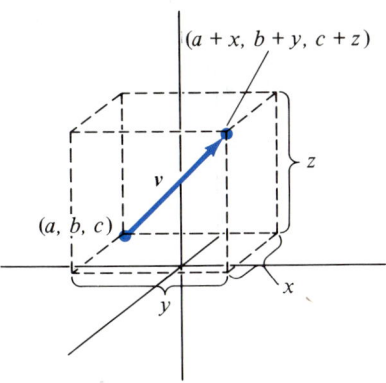

Figure 3.9 The vector $v = \langle x, y, z \rangle$.

Finally, we define the **length of the vector** $v = \langle x, y, z \rangle$ by applying the distance formula in Figure 3.6:

For $v = \langle x, y, z \rangle$,
$$|v| = \sqrt{x^2 + y^2 + z^2}.$$

(2)

With this definition of length, all properties of Theorem 2 carry over to vectors in space. For example, to prove that

$$|cv| = |c|\,|v|,$$

(3)

we express v in component form as $v = \langle x, y, z \rangle$. Then

$$
\begin{aligned}
|cv| &= \sqrt{(cx)^2 + (cy)^2 + (cz)^2} \\
&= |c|\sqrt{x^2 + y^2 + z^2} \\
&= |c|\,|v|.
\end{aligned}
$$

You are asked to verify the remaining properties in Exercise 45.

Example 4

For the vectors $v = \langle 2, -1, 4 \rangle$ and $w = \langle 0, 3, -5 \rangle$,

(a) $v + w = \langle 2 + 0, -1 + 3, 4 + (-5) \rangle = \langle 2, 2, -1 \rangle$
(b) $3v = \langle 3 \cdot 2, 3(-1), 3 \cdot 4 \rangle = \langle 6, -3, 12 \rangle$
(c) $v - w = \langle 2 - 0, -1 - 3, 4 - (-5) \rangle = \langle 2, -4, 9 \rangle$
(d) $|2v + w| = |\langle 2 \cdot 2 + 0, 2(-1) + 3, 2 \cdot 4 + (-5) \rangle|$
$= |\langle 4, 1, 3 \rangle| = \sqrt{4^2 + 1^2 + 3^2} = \sqrt{26}.$ ◇

Unit Coordinate Vectors

We define the three unit coordinate vectors in space by

$$i = \langle 1, 0, 0 \rangle, \qquad j = \langle 0, 1, 0 \rangle, \qquad k = \langle 0, 0, 1 \rangle.$$

(See Figure 3.10.) We may write the vector $v = \langle x, y, z \rangle$ in terms of $i, j,$ and k as

$$
\begin{aligned}
v = \langle x, y, z \rangle &= x\langle 1, 0, 0 \rangle + y\langle 0, 1, 0 \rangle + z\langle 0, 0, 1 \rangle \\
&= xi + yj + zk \qquad \text{(Figure 3.11).}
\end{aligned}
$$

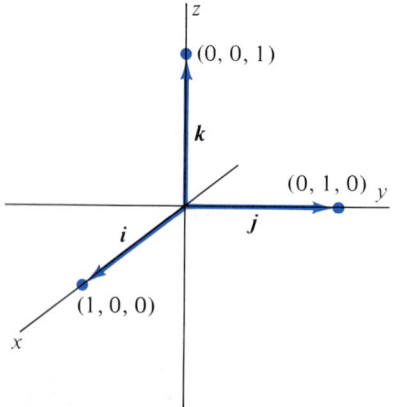

Figure 3.10 Unit coordinate vectors.

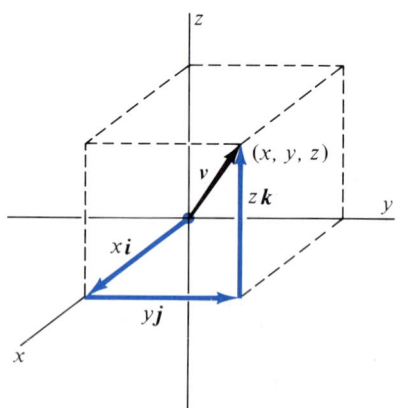

Figure 3.11 $v = xi + yj + zk$.

The Dot Product

The idea of dot product carries over to vectors in space by the obvious generalization of Definition 3:

DEFINITION 6

The **dot product** of the vectors $v = \langle x_1, y_1, z_1 \rangle$ and $w = \langle x_2, y_2, z_2 \rangle$ is the number

$$v \cdot w = x_1 x_2 + y_1 y_2 + z_1 z_2.$$

All properties of Theorems 3 and 4 concerning the dot product remain true for vectors in space, including the Schwarz inequality

$$|v \cdot w| \leq |v||w| \tag{4}$$

and the representation of the dot product in terms of the lengths of v and w and the angle θ between them:

$$v \cdot w = |v|\,|w| \cos \theta. \tag{5}$$

The proof of equation (5) is by direct analogy with that of Theorem 4, and the Schwarz inequality (4) follows from (5) as before since

$$|v \cdot w| = |v| \cdot |w||\cos \theta| \leq |v|\,|w|.$$

(See Exercises 46 and 47.)

Example 5

For the vectors $v = 3i + 2j - k$ and $w = i + 4j + k$,

(a) $v + 2w = (3i + 2j - k) + 2(i + 4j + k)$
$\qquad = (3 + 2 \cdot 1)i + (2 + 2 \cdot 4)j + (-1 + 2 \cdot 1)k$
$\qquad = 5i + 10j + k.$

(b) $v \cdot w = 3 \cdot 1 + 2 \cdot 4 + (-1)(1) = 3 + 8 - 1 = 10.$

(c) The cosine of the angle between v and w is

$$\cos \theta = \frac{v \cdot w}{|v| \cdot |w|} = \frac{10}{\sqrt{14} \cdot \sqrt{18}} = \frac{5}{3\sqrt{7}}.$$

(d) The **component** of v in the direction of w is the number

$$\text{comp}_w\, v = \frac{v \cdot w}{|w|} = \frac{10}{\sqrt{18}} = \frac{10}{3\sqrt{2}}.$$

(e) The **orthogonal projection** of v onto w is the vector

$$\text{proj}_w\, v = \frac{v \cdot w}{|w|^2} w = \frac{10}{(\sqrt{18})^2}(i + 4j + k)$$

$$= \frac{5}{9}i + \frac{20}{9}j + \frac{5}{9}k.$$

(f) A **unit vector** in the direction of w is

$$\frac{1}{|w|}w = \frac{1}{\sqrt{18}}(i + 4j + k) = \frac{1}{\sqrt{18}}i + \frac{4}{\sqrt{18}}j + \frac{1}{\sqrt{18}}k. \qquad \diamond$$

Direction Cosines

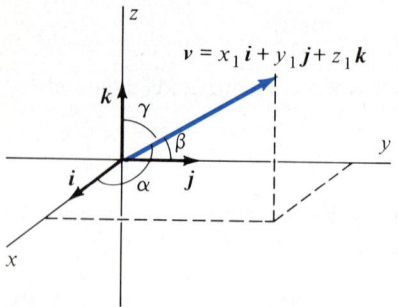

Figure 3.12 Direction angles for v.

Let $v = x_1 i + y_1 j + z_1 k$ be a nonzero vector, and let α, β, and γ denote the angles formed between v and the unit coordinate vectors i, j, and k, respectively (Figure 3.12).

We refer to α, β, and γ as the **direction angles** for v. Applying equation (5) we conclude that

$$\cos \alpha = \frac{v \cdot i}{|v| \, |i|} = \frac{v \cdot i}{|v|} = \frac{x_1}{|v|}; \qquad x_1 = |v| \cos \alpha \tag{6}$$

$$\cos \beta = \frac{v \cdot j}{|v| \, |j|} = \frac{v \cdot j}{|v|} = \frac{y_1}{|v|}; \qquad y_1 = |v| \cos \beta \tag{7}$$

and

$$\cos \gamma = \frac{v \cdot k}{|v| \, |k|} = \frac{v \cdot k}{|v|} = \frac{z_1}{|v|}; \qquad z_1 = |v| \cos \gamma. \tag{8}$$

Using (6) through (8), we may write the vector $v = x_1 i + y_1 j + z_1 k$ as

$$v = |v| \, [(\cos \alpha)i + (\cos \beta)j + (\cos \gamma)k]. \tag{9}$$

Equation (9) yields two important conclusions. First, calculating lengths on both sides of (9) shows that

$$|v| = |v| \, |\cos \alpha i + \cos \beta j + \cos \gamma k|$$
$$= |v| \sqrt{\cos^2 \alpha + \cos^2 \beta + \cos^2 \gamma}.$$

Cancelling factors of $|v| \neq 0$ and squaring both sides gives

$$\cos^2 \alpha + \cos^2 \beta + \cos^2 \gamma = 1. \tag{10}$$

Second, for a unit vector u, equation (9) becomes

$$u = (\cos \alpha)i + (\cos \beta)j + (\cos \gamma)k. \tag{11}$$

We refer to the cosines given in (6) through (8) as the **direction cosines** for the vector v. Note that these numbers are defined independent of the particular representation for v as an arrow in space. Equation (10) states that *the sum of the squares of the direction cosines is always one*. Equation (11) states that *the direction cosines of a unit vector are precisely its components*.

Example 6

Let $v = 2i + \sqrt{5}j + 4k$. The direction cosines for v are, by equations (6) through (8),

$$\cos \alpha = \frac{2}{|v|} = \frac{2}{\sqrt{4 + 5 + 16}} = \frac{2}{5};$$

$$\cos \beta = \frac{\sqrt{5}}{5}; \qquad \cos \gamma = \frac{4}{5}.$$

◇

Example 7

A vector v makes angles of $45°$ with the vector i and $60°$ with the vector j.

(a) Find the angle between v and k.
(b) Find a unit vector in the direction of v.

Solution: We are given that

$$\cos \alpha = \cos 45° = \frac{\sqrt{2}}{2}; \qquad \cos \beta = \cos 60° = \frac{1}{2}.$$

We may therefore solve for $\cos \gamma$ using equation (10):

$$\cos^2 \gamma = 1 - (\cos^2 \alpha + \cos^2 \beta)$$

$$= 1 - \left(\frac{1}{2} + \frac{1}{4}\right)$$

$$= \frac{1}{4}.$$

Thus $\cos \gamma = \pm \frac{1}{2}$, so either $\gamma = 60°$ or $\gamma = 120°$.

If $\gamma = 60°$, a unit vector u in the direction of v is, by equation (11),

$$u = \frac{\sqrt{2}}{2}i + \frac{1}{2}j + \frac{1}{2}k.$$

If $\gamma = 120°$, then $\cos \gamma = -\frac{1}{2}$, so

$$u = \frac{\sqrt{2}}{2}i + \frac{1}{2}j - \frac{1}{2}k.$$

$\diamond$

Exercise Set 15.3

1. Plot the following points: $P_1 = (-2, 1, 5)$, $P_2 = (3, 0, 2)$, $P_3 = (1, 1, 5)$, $P_4 = (2, -3, 6)$, $P_5 = (-3, -4, -5)$.

2. Sketch the following planes in coordinate space.
   a. The xy-plane
   b. The yz-plane
   c. The xz-plane
   d. The plane $x = 2$
   e. The plane $y = 4$
   f. The plane $z = -3$

3. Find the distance between the given pair of points.
   a. $P = (1, 0, 1)$, $\quad Q = (3, 2, 1)$
   b. $P = (1, 2, -3)$, $\quad Q = (-1, 4, 5)$
   c. $P = (-3, 6, 2)$, $\quad Q = (0, 3, 0)$
   d. $P = (a, b, c)$, $\quad Q = (2a, 2b, 2c)$

4. Find the midpoint of the line segment joining the points $P = (-4, 6, 1)$ and $Q = (-1, -3, -11)$.

5. Find the point one third of the distance from $P = (-4, 6, 1)$ to $Q = (-1, -3, -11)$.

6. Write an equation for the sphere with
   a. center $(0, 0, 0)$ and radius $r = 2$,
   b. center $(1, -1, 0)$ and radius $r = 1$,
   c. center $(-2, 3, -5)$ and radius $r = 3$,
   d. center $(4, 6, -2)$ and radius $r = 10$.

In Exercises 7–11, find the center and radius for the sphere with the given equation.

7. $x^2 + y^2 + z^2 - 4z = 5$

8. $x^2 + y^2 + z^2 - 2y - 4z = 4$

9. $x^2 + y^2 + z^2 - 4x + 2y - 6z = 2$

10. $x^2 + y^2 + z^2 - 4x + 4z = -4$

11. $x^2 + y^2 + z^2 - 6x + 2y + 4z = 11$

In Exercises 12–19, let $u = \langle 2, -1, 5 \rangle$, $v = \langle -3, 5, 0 \rangle$, and $w = \langle 3, 3, 1 \rangle$. Find the indicated vector or number.

12. $v + w$

13. $u + 6v$

14. $v - w$

15. $3u - 2v$

16. $\dfrac{v}{|v|}$

17. $|u + 4v|$

18. $v \cdot w$

19. $v \cdot (u + 2w)$

In Exercises 20–29, let $u = i + 2j - k$, $v = 3i - 2j + 2k$, and $w = 5i - j + 3k$. Find the indicated vector or number.

20. $u + 2v + w$

21. $v - 3w$

22. $|3v + w|$

23. $u \cdot (2v + 3w)$

24. $|v \cdot w + w \cdot u|$

25. $\dfrac{v}{|v|} + \dfrac{w}{|w|}$

26. $v \cdot w - |v|\,|w|$

27. $\text{proj}_w\, v$

28. $\text{comp}_u\, w$

29. $\text{comp}_u\, v$

30. Find a unit vector in the direction of $w = i + 4j + 3k$.

31. Find the vector $v = \overrightarrow{PQ}$ for
 a. $P = (1, 0, 1)$, $\quad Q = (3, 1, 3)$,
 b. $P = (-3, 2, 6)$, $\quad Q = (1, 1, 1)$,
 c. $P = (1, -5, 3)$, $\quad Q = (7, -2, 2)$.

32. Find the point D so that $\overrightarrow{AB} = \overrightarrow{CD}$ if $A = (3, 1, 6)$, $B = (-2, 1, 5)$, and $C = (6, -2, 2)$.

33. Find the direction cosines for the given vector.
 a. $v = 3i - j + 2k$
 b. $v = 6i - 2j + k$
 c. $\overrightarrow{PQ}$, where $P = (2, 1, 5)$ and $Q = (1, 3, 1)$

34. Which of the following sets of points are vertices of right triangles?
 a. $P = (1, 3, 2)$, $\quad Q = (4, 1, 4)$, $\quad R = (6, 5, 5)$
 b. $P = (0, 2, 5)$, $\quad Q = (1, 3, 1)$, $\quad R = (1, 4, 5)$
 c. $P = (-3, 1, 2)$, $\quad Q = (1, -3, 2)$, $\quad R = (2, -2, 2)$

35. We say that two vectors v and w in space are **parallel** if there is a constant c so that $v = cw$. The vectors are said to point in the same direction if $c > 0$, and they are said to point in opposite directions if $c < 0$.
 a. Find two vectors of length 2 parallel to the vector $v = 2i - 4j + 4k$.

 b. Find the constant a so that the vectors $v = i - 3j + 4k$ and $w = ai + 9j - 12k$ are parallel.
 c. Find a vector of length 5 in the direction opposite that of $v = i - 2j + 3k$.
 d. Find a and b so that the vectors $3i - j + 4k$ and $ai + bj - 2k$ are parallel.

36. Describe the set of all points determined by the vector $r(t) = (i + 2j + k) + t(i + j + k)$ for $-\infty < t < \infty$, where $i + 2j + k$ originates at $(0, 0, 0)$.

37. Let $\quad u = i - 2j + 3k$, $\quad v = i + 2j + k$, $\quad$ and $\quad w = i - 4j + k$. Find t so that the vector $u + tv$ is orthogonal to w.

38. Show that a nonzero vector is completely determined by its direction cosines and its length.

39. Which of the following triples can be direction angles of a single vector?
 a. $45°, 45°, 60°$
 b. $30°, 45°, 60°$
 c. $45°, 60°, 60°$

40. Use the dot product to prove the following statements.
 a. $|v + w| = |v - w|$ if and only if v and w are orthogonal.
 b. $|v + w|^2 = |v|^2 + |w|^2$ if and only if v and w are orthogonal.

41. Let v_1, v_2, and v_3 be mutually orthogonal vectors in space. Use the dot product to show that if c_1, c_2, and c_3 are scalars so that $c_1 v_1 + c_2 v_2 + c_3 v_3 = 0$, then $c_1 = c_2 = c_3 = 0$.

42. Give a geometric argument to show that three vectors v_1, v_2, and v_3 are coplanar if and only if there exist constants c_1 and c_2 so that $v_3 = c_1 v_1 + c_2 v_2$.

43. Let $v_1 = i + j$, $v_2 = j + k$, $v_3 = i + 2j + 3k$. Show that given any vector w there exist constants c_1, c_2, and c_3 so that $w = c_1 v_1 + c_2 v_2 + c_3 v_3$. (*Hint:* Express w in component form and obtain 3 linear equations for the constants c_1, c_2, and c_3.)

44. State and prove the analogue of Theorem 1 for vectors in space.

45. State and prove the analogue of Theorem 2 for vectors in space.

46. Prove the Schwarz inequality for vectors in space.

47. Prove equality (5) for vectors in space.

15.4 THE CROSS PRODUCT

In addition to the dot product, there is a second type of multiplication involving vectors. While the dot product of two vectors is a number, the *cross product* (or *vector* product) of two vectors is another vector. It is defined as follows.

DEFINITION 7

Let $v = x_1 i + y_1 j + z_1 k$ and $w = x_2 i + y_2 j + z_2 k$ be vectors in space. The **cross product**, $v \times w$, is the **vector**

$$v \times w = (y_1 z_2 - z_1 y_2) i + (z_1 x_2 - x_1 z_2) j + (x_1 y_2 - y_1 x_2) k. \qquad (1)$$

There is a geometric interpretation of expression (1) similar to that developed for the dot product. Before taking up that issue, we look to several examples for insights.

Example 1

If $v = i$ and $w = j$, then $x_1 = y_2 = 1$ and $y_1 = z_1 = x_2 = z_2 = 0$ in Definition 7. Thus, by equation (1)

$$i \times j = (0 \cdot 0 - 0 \cdot 1) i + (0 \cdot 0 - 1 \cdot 0) j + (1 \cdot 1 - 0 \cdot 0) k.$$

That is,

$$i \times j = k \qquad \text{(Figure 4.1)}. \qquad (2)$$

Similarly, you can verify that

$$j \times k = i \qquad \text{(Figure 4.2)} \qquad (3)$$

and

$$k \times i = j \qquad \text{(Figure 4.3)}. \qquad (4)$$

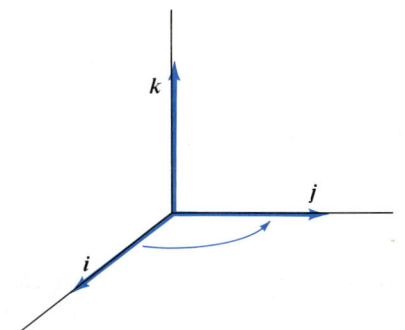

Figure 4.1 $i \times j = k$.

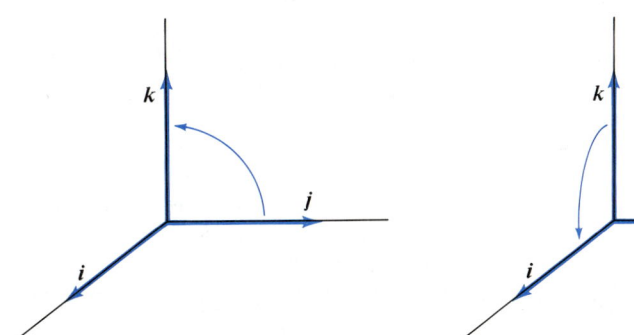

Figure 4.2 $j \times k = i$.   **Figure 4.3** $k \times i = j$.

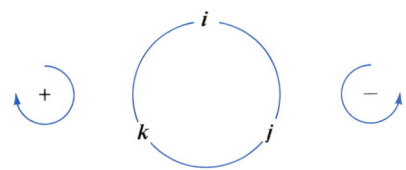

Figure 4.4 Device for remembering cross products of unit coordinate vectors.

The curved arrows in Figures 4.1 through 4.3 indicate the order of the vectors in the cross products.

However, you can also verify that the cross product is not commutative, since

$$j \times i = -k, \qquad k \times j = -i, \qquad \text{and} \qquad i \times k = -j. \qquad (5)$$

Figure 4.4 presents a device for remembering these results. By beginning at the first factor and traversing the circle in the direction leading directly to the second factor, one determines the sign of the cross product: positive if the direction is clockwise (equations (2)–(4)); negative if the direction is counterclockwise (equations (5)).   ◇

Example 2

For $v = 2i + j + 3k$ and $w = 4i - j + 2k$,

$$v \times w = [1 \cdot 2 - 3(-1)]i + [3 \cdot 4 - 2 \cdot 2]j + [2(-1) - 1 \cdot 4]k$$
$$= 5i + 8j - 6k.$$

◇

Properties of the Cross Product

It is easy to see that in each of the cross products of Example 1 the vector $v \times w$ is orthogonal to both factors, v and w. The same is true for the vectors v and w of Example 2: $v \times w = 5i + 8j - 6k$ is orthogonal to $v = 2i + j + 3k$, since $(v \times w) \cdot v = 5 \cdot 2 + 8 \cdot 1 + (-6)(3) = 0$; $v \times w$ is orthogonal to w, since $(v \times w) \cdot w = 5 \cdot 4 + 8(-1) + 2(-6) = 0$.

These observations lead to the conjecture that $v \times w$ is *always* orthogonal both to v and to w. To prove this conjecture, we write v and w in the component forms $v = x_1 i + y_1 j + z_1 k$ and $w = x_2 i + y_2 j + z_2 k$. Then, using Definition 7, we find that

$$(v \times w) \cdot v = [(y_1 z_2 - z_1 y_2)i + (z_1 x_2 - x_1 z_2)j$$
$$+ (x_1 y_2 - y_1 x_2)k] \cdot [x_1 i + y_1 j + z_1 k]$$
$$= x_1(y_1 z_2 - z_1 y_2) + y_1(z_1 x_2 - x_1 z_2) + z_1(x_1 y_2 - y_1 x_2)$$
$$= 0.$$

Similarly, you can show that $(v \times w) \cdot w = 0$. Thus, it follows from Corollary 1 (Section 16.2) that $v \times w$ is orthogonal to both v and w.

This proves one part of the following theorem. The proofs of the other statements are carried out in the same way. Simply express all vectors in component form and apply Definition 7.

THEOREM 5
Properties of the Cross Product

Let u, v, and w be vectors and let c be a scalar. Then

(i) $v \times w = -w \times v$ (anticommutative)

(ii) $u \times (v + w) = u \times v + u \times w$ (multiplication distributes over addition)

(iii) $c(v \times w) = (cv) \times w = v \times (cw)$

(iv) $(v \times w) \perp v$; $(v \times w) \perp w$

(v) $v \times v = 0$ (self-annihilating)

The Determinant Notation

There is a useful formula for remembering equation (1) that involves the concept of **determinants**. The determinant of the 2×2 matrix

$$\begin{bmatrix} a_1 & b_1 \\ a_2 & b_2 \end{bmatrix}$$

is the number

$$\det \begin{bmatrix} a_1 & b_1 \\ a_2 & b_2 \end{bmatrix} = a_1 b_2 - b_1 a_2. \tag{6}$$

For example,

$$\det \begin{bmatrix} 3 & 2 \\ 4 & 1 \end{bmatrix} = 3 \cdot 1 - 2 \cdot 4 = 3 - 8 = -5.$$

The determinant of the 3×3 matrix

$$\begin{bmatrix} a_1 & b_1 & c_1 \\ a_2 & b_2 & c_2 \\ a_3 & b_3 & c_3 \end{bmatrix}$$

is the number

$$\det \begin{bmatrix} a_1 & b_1 & c_1 \\ a_2 & b_2 & c_2 \\ a_3 & b_3 & c_3 \end{bmatrix} = a_1 \cdot \det \begin{bmatrix} b_2 & c_2 \\ b_3 & c_3 \end{bmatrix} - b_1 \cdot \det \begin{bmatrix} a_2 & c_2 \\ a_3 & c_3 \end{bmatrix}$$

$$+ c_1 \cdot \det \begin{bmatrix} a_2 & b_2 \\ a_3 & b_3 \end{bmatrix} \quad (7)$$

$$= a_1(b_2c_3 - c_2b_3) - b_1(a_2c_3 - c_2a_3) + c_1(a_2b_3 - b_2a_3).$$

For example,

$$\det \begin{bmatrix} 1 & 2 & 3 \\ 4 & 5 & 6 \\ 7 & 8 & 9 \end{bmatrix} = 1 \cdot \det \begin{bmatrix} 5 & 6 \\ 8 & 9 \end{bmatrix} - 2 \cdot \det \begin{bmatrix} 4 & 6 \\ 7 & 9 \end{bmatrix} + 3 \cdot \det \begin{bmatrix} 4 & 5 \\ 7 & 8 \end{bmatrix} \quad (8)$$

$$= (5 \cdot 9 - 6 \cdot 8) - 2(4 \cdot 9 - 6 \cdot 7) + 3(4 \cdot 8 - 5 \cdot 7)$$

$$= 0.$$

The reason for bringing up the topic of determinants is that by comparing equations (1) and (7) you can see that the memory device

$$v \times w = \det \begin{bmatrix} i & j & k \\ x_1 & y_1 & z_1 \\ x_2 & y_2 & z_2 \end{bmatrix}$$

$$= (y_1z_2 - z_1y_2)i + (z_1x_2 - x_1z_2)j + (x_1y_2 - y_1x_2)k \quad (9)$$

provides a convenient way to remember equation (1). (Of course, equation (9) is not really a proper use of the determinant defined in equation (7), since the top row of the matrix consists of vectors rather than numbers.)

Determinants play an important role in the theory of systems of linear equations. Here we are making use of only a very limited aspect of determinants.

Example 3

For $v = 3i - j - 4k$ and $w = -2i + 2j + k$,

$$v \times w = \det \begin{bmatrix} i & j & k \\ 3 & -1 & -4 \\ -2 & 2 & 1 \end{bmatrix}$$

$$= ((-1)(1) - (-4)(2))i$$

$$\quad + ((-4)(-2) - (3 \cdot 1))j + (3 \cdot 2 - (-1)(-2))k$$

$$= 7i + 5j + 4k.$$

$\diamondsuit$

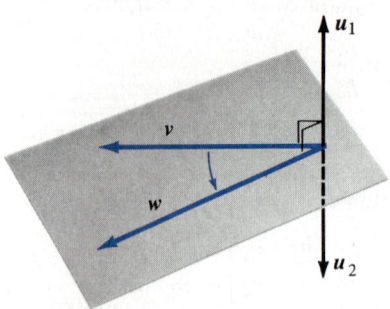

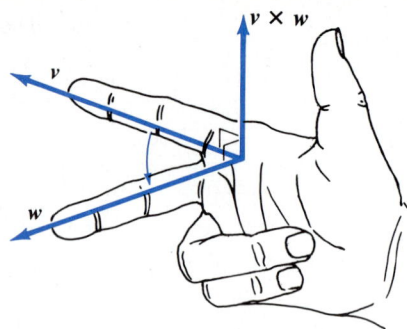

Figure 4.5 Both u_1 and u_2 are orthogonal to plane determined by v and w. Which one corresponds to $v \times w$?

Figure 4.6 Right-hand rule determines direction of $v \times w$.

The next item on our agenda is a theorem that allows us to interpret the cross product geometrically. Before stating this theorem we need to clarify an earlier point. We have already shown that $v \times w$ is perpendicular both to v and to w, so $v \times w$ is a vector perpendicular to the *plane* determined by v and w. This leaves two possible directions for $v \times w$, as Figure 4.5 illustrates.

However, from Figures 4.1 through 4.3 you can see that each of the cross products $i \times j$, $j \times k$, and $k \times i$ satisfies the **right-hand rule:** *the thumb of a right hand points in the direction of $v \times w$ when the index finger points in the direction of v and the second finger points in the direction of w* (Figure 4.6).

In checking additional examples, you will observe that this rule holds in each case.

THEOREM 6

Let v and w be nonzero vectors forming an angle θ between them. Then

$$v \times w = (|v|\,|w|\,\sin\theta)N \tag{10}$$

where N is a unit vector orthogonal to both v and w, in the direction determined by the right-hand rule.

Proof: We have already noted the direction of $v \times w$. What remains to be shown is that

$$|v \times w| = |(|v|\,|w|\,\sin\theta)N| = |v|\,|w|\,|\sin\theta|. \tag{11}$$

To prove equation (11), we write v and w in component form as

$$v = x_1 i + y_1 j + z_1 k; \qquad w = x_2 i + y_2 j + z_2 k.$$

Then, using equation (1) and the definition of the dot product we find that

$$
\begin{aligned}
|v \times w|^2 &= |(y_1 z_2 - z_1 y_2)i + (z_1 x_2 - x_1 z_2)j + (x_1 y_2 - y_1 x_2)k|^2 \\
&= (y_1 z_2 - z_1 y_2)^2 + (z_1 x_2 - x_1 z_2)^2 + (x_1 y_2 - y_1 x_2)^2 \\
&= (x_1^2 + y_1^2 + z_1^2)(x_2^2 + y_2^2 + z_2^2) - (x_1 x_2 + y_1 y_2 + z_1 z_2)^2 \\
&= |v|^2 |w|^2 - (v \cdot w)^2.
\end{aligned}
\tag{12}
$$

Now by Theorem 4,

$$(v \cdot w)^2 = |v|^2 |w|^2 \cos^2\theta. \tag{13}$$

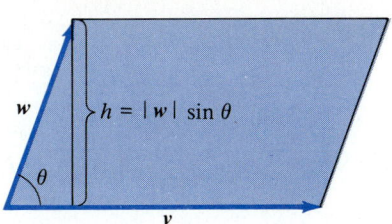

Figure 4.7 Area of parallelogram with base $|v|$ and adjacent side w is $|v \times w|$.

Combining equations (12) and (13) we have

$$
\begin{aligned}
|v \times w|^2 &= |v|^2|w|^2 - |v|^2|w|^2 \cos^2 \theta \\
&= |v|^2|w|^2(1 - \cos^2 \theta) \\
&= |v|^2|w|^2 \sin^2 \theta.
\end{aligned}
$$

Taking square roots of both sides then gives equation (11).   ◆

Figure 4.7 illustrates an immediate consequence of Theorem 6. If θ is the angle between v and w, the altitude of the parallelogram determined by the vectors v and w is $h = |w| \sin \theta$. Since the base has length $b = |v|$, the area A must be

$$A = bh = |v|\,|w| \sin \theta = |v \times w|. \qquad (14)$$

Example 4

For $v = i + 2j - k$ and $w = -3i - j + 2k$,

$$
v \times w = \det \begin{bmatrix} i & j & k \\ 1 & 2 & -1 \\ -3 & -1 & 2 \end{bmatrix} = (4 - 1)i + (3 - 2)j + (-1 + 6)k
$$

$$= 3i + j + 5k.$$

The area of the parallelogram determined by v and w is, by equation (14),

$$A = |3i + j + 5k| = \sqrt{9 + 1 + 25} = \sqrt{35} \qquad \text{(Figure 4.8)}. \qquad ◇$$

Example 5

Find the sine of the angle between the vectors v and w in Example 4.

Solution: Solving equation (14) for $\sin \theta$ gives

$$\sin \theta = \frac{|v \times w|}{|v|\,|w|}.$$

Since $|v \times w| = \sqrt{35}$, $|v| = \sqrt{1^2 + 2^2 + 1^2} = \sqrt{6}$, and $|w| = \sqrt{3^2 + 1^2 + 2^2} = \sqrt{14}$, we have

$$\sin \theta = \frac{\sqrt{35}}{\sqrt{6} \cdot \sqrt{14}} = \frac{1}{2}\sqrt{\frac{5}{3}}. \qquad ◇$$

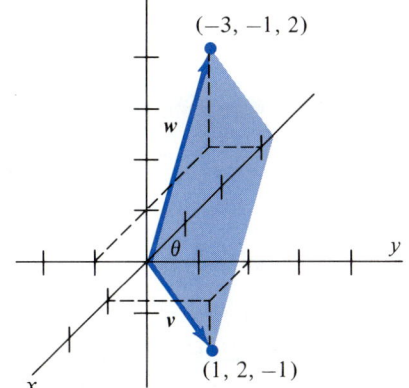

Figure 4.8

We can use equation (14) to learn something about 2×2 determinants as follows. Let $v = x_1 i + y_2 j$ and $w = x_2 i + y_2 j$. Then v and w are vectors in the plane $z = 0$, and the area of the parallelogram they determine is

$$
\begin{aligned}
A = |v \times w| &= \det \begin{bmatrix} i & j & k \\ x_1 & y_1 & 0 \\ x_2 & y_2 & 0 \end{bmatrix} \\
&= |(x_1 y_2 - y_1 x_2)k| \\
&= |x_1 y_2 - y_1 x_2| \\
&= \left| \det \begin{bmatrix} x_1 & y_1 \\ x_2 & y_2 \end{bmatrix} \right|.
\end{aligned}
$$

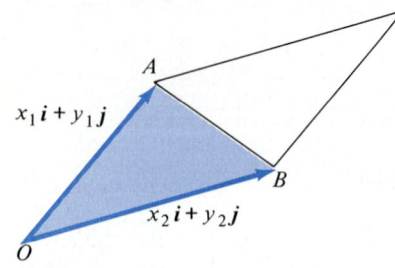

Figure 4.9 Area of ΔOAB is $\frac{1}{2}\left|\det\begin{bmatrix} x_1 & y_1 \\ x_2 & y_2 \end{bmatrix}\right|$

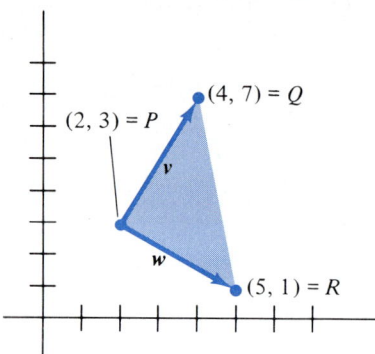

Figure 4.10

That is, the absolute value of the determinant of the matrix

$$\begin{bmatrix} x_1 & y_1 \\ x_2 & y_2 \end{bmatrix}$$

is the area of the parallelogram determined by the vectors $x_1i + y_1j$ and $x_2i + y_2j$. Since the diagonals of a parallelogram bisect the parallelogram into two congruent triangles, an even simpler interpretation of the 2×2 determinant is that *the absolute value of the determinant*

$$\det\begin{bmatrix} x_1 & y_1 \\ x_2 & y_2 \end{bmatrix}$$

is twice the area of the triangle determined by the vectors $x_1i + y_1j$ *and* $x_2i + y_2j$ (Figure 4.9).

Example 6

Find the area of the triangle with vertices $(2, 3)$, $(4, 7)$, and $(5, 1)$.

Solution: Let $P = (2, 3)$, $Q = (4, 7)$, and $R = (5, 1)$. Two of the sides may be interpreted as the vectors

$$v = \overrightarrow{PQ} = 2i + 4j$$

and

$$w = \overrightarrow{PR} = 3i - 2j \qquad \text{(Figure 4.10)}.$$

By the preceding observation, the area of the triangle is

$$A = \frac{1}{2}\left|\det\begin{bmatrix} 2 & 4 \\ 3 & -2 \end{bmatrix}\right| = \frac{1}{2} \cdot \left|2(-2) - 4(3)\right| = \frac{16}{2} = 8. \qquad \diamond$$

Clearly, the geometry associated with the dot and cross products is rich. We shall use these ideas in Section 15.5 to develop equations for lines and planes in space.

Exercise Set 15.4

In Exercises 1–6, let $u = i + j + k$, $v = 2i - j + 2k$, and $w = 3i + 4j - k$. Find the indicated vector or number.

1. $v \times w$

2. $u \times w$

3. $w \times u$

4. $u \times v$

5. $u \cdot (v \times w)$

6. $v \cdot (w \times u)$

In Exercises 7–14, let $u = 2i - j + 4k$, $v = i - 3k$, and $w = 2i + 3j + 4k$. Find the indicated vector or number.

7. $u \times v$

8. $v \times w$

9. $u \times w$

10. $3u \times 2w$

11. $u \times (v \times w)$

12. $(u \times v) \times w$

13. $u \cdot (v \times w)$

14. $w \cdot (v \times u)$

In Exercises 15–18, find the area of the parallelogram determined by the given vectors.

15. $v = 2i + j + k, \qquad w = 3i - j - 2k$

16. $v = -3i + k, \qquad w = 2i + 2j + k$

17. $v = 2i - j, \qquad w = 4i + 2j$

18. $v = -6i + 4j - 3k, \qquad w = i - j - k$

In Exercises 19–21, find the area of the triangle in space with given vertices.

19. $A = (3, 0, 1), \qquad B = (2, -1, 2), \qquad C = (1, 3, -2)$

20. $A = (5, 2, -2), \qquad B = (-1, 4, 2), \qquad C = (-4, 5, 3)$

21. $A = (0, 2, 1), \qquad B = (-4, 1, -2), \qquad C = (1, 1, -2)$

22. Find the area of the triangle with vertices $(2, 4)$, $(-3, 8)$, and $(-5, -2)$.

23. Find the area of the triangle with vertices $(1, 5)$, $(-3, -4)$, and $(4, 6)$.

24. Find a unit vector orthogonal to both $v = 2i + j - k$ and $w = i + j + 4k$.

25. Find a unit vector orthogonal to both $v = i + 3j$ and $w = 4i - j$.

26. Show that the cross product is not associative. That is, find vectors u, v, and w so that $u \times (v \times w) \neq (u \times v) \times w$.

27. Find two unit vectors orthogonal to $v = i + j$ and $w = 2i - j + 3k$.

28. Show that $|v \times w| = |v|\,|w|$ if v and w are orthogonal.

29. Show that if $u = x_1 i + y_1 j + z_1 k$, $v = x_2 i + y_2 j + z_2 k$, and $w = x_3 i + y_3 j + z_3 k$, then

$$u \cdot (v \times w) = \det \begin{bmatrix} x_1 & y_1 & z_1 \\ x_2 & y_2 & z_2 \\ x_3 & y_3 & z_3 \end{bmatrix}.$$

30. Show that the volume of the parallelepiped formed by u, v, and w is $|u \cdot (v \times w)|$. (*Hint:* $|v \times w|$ is the area of the base. Thus, $|u \cdot (v \times w)| = |v \times w| \cdot (|u|\,|\cos \theta|)$ is the area of the base times the altitude. Why?)

31. Use Exercises 29 and 30 to find the volume of the parallelepiped determined by the vectors $u = 2i + j - 4k$, $v = -i + 3j - 2k$, and $w = 4i - j + 3k$.

32. Find the volume of the parallelepiped determined by the vectors $u = -3i + j - k$, $v = 2i + 3j + 2k$, and $w = i + 4j + 2k$.

33. Prove that $u \times (v \times w) = (u \cdot w)v - (u \cdot v)w$.

34. Prove that $u \cdot (v \times w) = v \cdot (w \times u)$.

35. Prove that $u \cdot (v \times w) = w \cdot (u \times v)$.

36. Prove that $(u \times v) \times w = u \times (v \times w)$ only if $(u \times w) \times v = 0$.

37. Suppose that $u + v + w = 0$. Show that $u \times v = v \times w = w \times u$. What is the geometric interpretation of this result?

38. Why did we not try to define the cross product for vectors in the plane?

15.5 EQUATIONS FOR LINES AND PLANES

In the plane, a line ℓ is determined by its slope and one point on the line. However, a point and a slope do not provide sufficient information to determine a line in space. Indeed, even the term "slope" becomes ambiguous in space. For example, think about the problem of writing down an equation for the line determined by the beam from a searchlight as it sweeps across the sky. The beam is determined not only by a *point* (the location of the searchlight) and an *elevation* (perhaps the analogue of slope that you had in mind) but also by the angle at which the searchlight is *rotated* away from a fixed direction, say due north (Figure 5.1).

One of the simplest ways to think about how to find the general form for a line in space is to remember that two points determine a line. If P and Q are two points in space, then the vector $b = \overrightarrow{PQ}$ is parallel to the line ℓ containing P and Q. Moreover, any multiple tb is also parallel to ℓ (Figure 5.2).

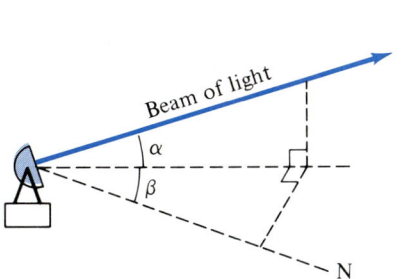

Figure 5.1 Searchlight positioned with elevation angle α and rotation angle β.

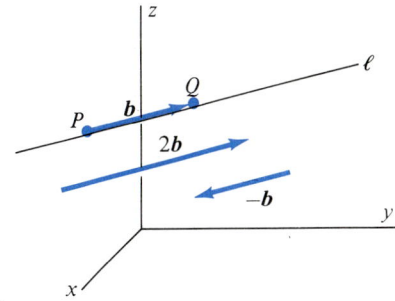

Figure 5.2 Multiples of $b = \overrightarrow{PQ}$ are parallel to ℓ.

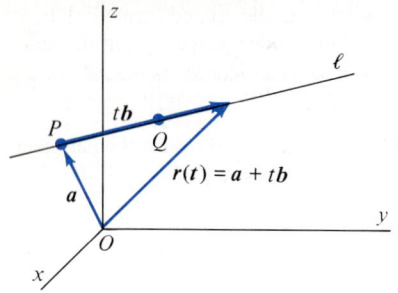

Figure 5.3 If $a = \overrightarrow{OP}$, the vector $r(t) = a + tb$ terminates on ℓ.

We can use the point P and the vector b to determine points on the line ℓ as follows. Let $a = \overrightarrow{OP}$ be the **position vector** originating at the origin O and terminating at the point P on ℓ. Then, since tb is parallel to ℓ, the vector $r(t) = a + tb$ also terminates on the line ℓ (Figure 5.3). As the values of t increase from 0 to ∞, the vector $a + tb$ terminates successively at all points on ℓ lying on one side of P, while as t decreases from 0 to $-\infty$ the vector $a + tb$ determines all points on ℓ lying on the opposite side of P (Figure 5.4). (See Exercises 58–60 of Section 15.1 for a more explicit demonstration of this concept.)

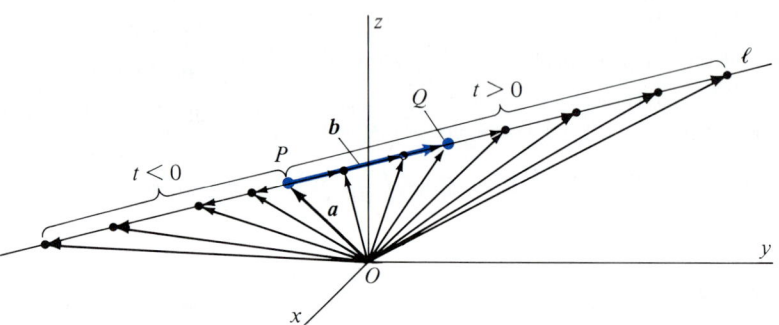

Figure 5.4 ℓ: $\{r(t) = a + tb \mid -\infty < t < \infty\}$.

We can summarize this discussion by saying that a line ℓ in space is determined by a *position vector* a, originating at the origin, and a *direction vector* b as

$$\ell: r(t) = a + tb, \qquad -\infty < t < \infty. \tag{1}$$

Equation (1) is called the **vector form** of the equation for ℓ. If P and Q are distinct points on ℓ we can always take

$$a = \overrightarrow{OP} \tag{2}$$

and

$$b = \overrightarrow{PQ} \tag{3}$$

as in the preceding discussion. In this case equation (1) becomes

$$\ell: r(t) = \overrightarrow{OP} + t\overrightarrow{PQ}, \qquad -\infty < t < \infty.$$

More generally, equation (1) allows us to interpret the line ℓ as an infinite collection of vectors, each terminating at a point on ℓ.

Example 1

Find an equation for the line containing the point $P = (2, -1, 4)$ and parallel to the vector $v = i - j + 2k$.

Solution: The position vector is

$$a = \overrightarrow{OP} = 2i - j + 4k$$

and the direction vector is given as

$$b = v = i - j + 2k.$$

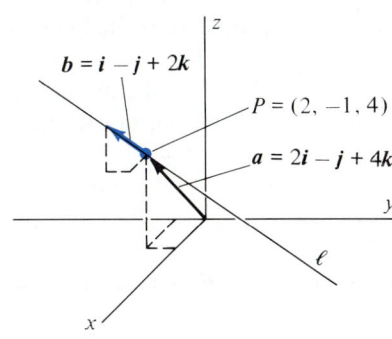

Figure 5.5

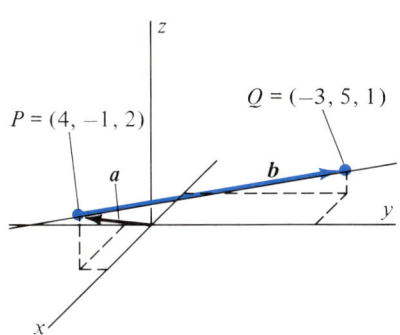

Figure 5.6

The line is, by equation (1), the set of all vectors of the form

$$r(t) = (2i - j + 4k) + t(i - j + 2k) \qquad \text{(Figure 5.5).}$$ ◇

Example 2

Find an equation for the line containing the points $P = (4, -1, 2)$ and $Q = (-3, 5, 1)$.

Solution: By equation (2) a position vector is

$$a = \overrightarrow{OP} = 4i - j + 2k$$

and, by (3), a direction vector is

$$b = \overrightarrow{PQ} = (-3 - 4)i + (5 - (-1))j + (1 - 2)k = -7i + 6j - k.$$

The line has vector equation

$$r(t) = (4i - j + 2k) + t(-7i + 6j - k) \qquad \text{(Figure 5.6).}$$ ◇

Example 3

Find the point, if it exists, where the lines

$$\ell_1: r_1(t) = (3i + 2j - k) + t(-6i + 4j + 3k)$$

$$\ell_2: r_2(t) = (5i + 4j + 7k) + t(14i - 6j + 2k)$$

intersect.

Solution: If ℓ_1 and ℓ_2 intersect at point P, it is not necessary that they do so for the same number t. The condition of intersection is therefore that

$$r_1(t_1) = r_2(t_2) \tag{4}$$

for some times t_1 and t_2. Equating i-components in equation (4) gives

$$3 - 6t_1 = 5 + 14t_2,$$

or

$$6t_1 + 14t_2 = -2. \tag{5}$$

Equating j-components in (4) gives the equation

$$2 + 4t_1 = 4 - 6t_2$$

or

$$4t_1 + 6t_2 = 2. \tag{6}$$

Solving equations (5) and (6) simultaneously (either by substitution or by elimination) gives $t_1 = 2$ and $t_2 = -1$. To ensure that equation (4) holds for these numbers t we must write out $r_1(t_1)$ and $r_2(t_2)$ and verify that the k-components are equal.
Since

$$r_1(t_1) = (3i + 2j - k) + 2(-6i + 4j + 3k) = -9i + 10j + 5k$$

and

$$r_2(t_2) = (5i + 4j + 7k) + (-1)(14i - 6j + 2k) = -9i + 10j + 5k,$$

we can see that the lines indeed intersect at the point $(-9, 10, 5)$.   ◇

The Distance from a Point to a Line in Space

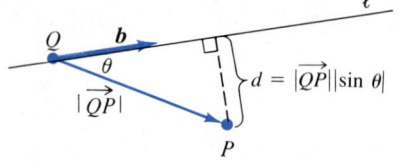

Figure 5.7 $d = |\overrightarrow{QP}||\sin \theta| = \dfrac{|\overrightarrow{PQ} \times \boldsymbol{b}|}{|\boldsymbol{b}|}$.

In space, a line does not determine a unique direction for vectors orthogonal to that line. (In fact, the family of vectors orthogonal to a given line determines an entire plane, as we shall see later in this section.) The problem of finding the distance d from a point $P = (x_0, y_0, z_0)$ to a line with equation $\boldsymbol{r}(t) = \boldsymbol{a} + t\boldsymbol{b}$, must be handled differently from the corresponding problem in the plane.

Figure 5.7 illustrates that the desired distance d is given by the expression

$$d = |\overrightarrow{QP}||\sin \theta| \qquad (7)$$

where Q is any point on ℓ and θ is the angle formed between $\overrightarrow{QP}$ and the direction vector $\boldsymbol{b}$.

The right side of equation (7) is suggestive of the cross product. In fact, we may use Theorem 6 to rewrite equation (7) as

$$d = |\overrightarrow{QP}||\sin \theta| = \frac{|\overrightarrow{QP}||\boldsymbol{b}||\sin \theta|}{|\boldsymbol{b}|} = \frac{|\overrightarrow{PQ} \times \boldsymbol{b}|}{|\boldsymbol{b}|}.$$

Since $|\boldsymbol{b}| \neq 0$ for all lines ℓ, we have the formula

$$d = \frac{|\overrightarrow{PQ} \times \boldsymbol{b}|}{|\boldsymbol{b}|} \qquad (8)$$

for the distance from a point P to a line with direction vector $\boldsymbol{b}$ and containing the point Q.

Example 4

Find the distance from the point $P = (1, -1, 2)$ to the line with vector equation

$$\ell: \boldsymbol{r}(t) = 3\boldsymbol{i} - 2\boldsymbol{j} + 4\boldsymbol{k} + t(\boldsymbol{i} - 2\boldsymbol{j} + 2\boldsymbol{k}).$$

Strategy

Find the direction vector $\boldsymbol{b}$.

Find a point Q on the line ℓ.

Form the vector $\overrightarrow{QP}$.

Apply equation (8).

Solution

Since the equation for ℓ is in the form of equation (1) the direction vector is

$$\boldsymbol{b} = \boldsymbol{i} - 2\boldsymbol{j} + 2\boldsymbol{k}.$$

Setting $t = 0$ shows that the point

$$Q = (3, -2, 4)$$

lies on ℓ. Thus,

$$\begin{aligned}\overrightarrow{QP} &= (1 - 3)\boldsymbol{i} + (-1 - (-2))\boldsymbol{j} + (2 - 4)\boldsymbol{k} \\ &= -2\boldsymbol{i} + \boldsymbol{j} - 2\boldsymbol{k}.\end{aligned}$$

The distance from P to ℓ is therefore

$$\begin{aligned}d &= \frac{1}{\sqrt{1^2 + (-2)^2 + 2^2}} \cdot \left| \det \begin{bmatrix} \boldsymbol{i} & \boldsymbol{j} & \boldsymbol{k} \\ -2 & 1 & -2 \\ 1 & -2 & 2 \end{bmatrix} \right| \\ &= \frac{1}{3}\left| -2\boldsymbol{i} + 2\boldsymbol{j} + 3\boldsymbol{k} \right| \\ &= \frac{\sqrt{17}}{3}.\end{aligned}$$

$\diamondsuit$

Parametric Equations for Lines

Suppose that the vectors a and b have components $a = a_1 i + a_2 j + a_3 k$, and $b = b_1 i + b_2 j + b_3 k$. Then if we write $r(t) = x(t) i + y(t) j + z(t) k$, the vector equation $r(t) = a + t b$ for a line has component form

$$x(t) i + y(t) j + z(t) k = (a_1 i + a_2 j + a_3 k) + t(b_1 i + b_2 j + b_3 k)$$
$$= (a_1 + t b_1) i + (a_2 + t b_2) j + (a_3 + t b_3) k.$$

Equating components in this equation gives the **parametric form** for the line ℓ with position vector $a = a_1 i + a_2 j + a_3 k$ and direction vector $b = b_1 i + b_2 j + b_3 k$:

$$\begin{cases} x(t) = a_1 + t b_1 \\ y(t) = a_2 + t b_2 \\ z(t) = a_3 + t b_3. \end{cases} \tag{9}$$

The point (x, y, z) is on ℓ if and only if the coordinates satisfy the equations in (9) for some single value of t.

Example 5

Find parametric equations for the line in Example 1.

Solution: From Example 1 we have

$$a_1 = 2, \qquad a_2 = -1, \qquad a_3 = 4, \qquad b_1 = 1, \qquad b_2 = -1, \qquad b_3 = 2.$$

Thus, by (9)

$$\begin{cases} x(t) = 2 + t \\ y(t) = -1 - t \\ z(t) = 4 + 2t \end{cases}$$

are the parametric equations for ℓ. ◇

Example 6

Find a vector parallel to the line with parametric equations

$$x = -4 + t, \qquad y = 3 - 6t, \qquad \text{and} \qquad z = 2 + 2t.$$

Solution: Comparing these equations with those in (9), we see that a direction vector for the line is

$$b = b_1 i + b_2 j + b_3 k = i - 6j + 2k.$$

This vector is parallel to the line. ◇

Symmetric Equations for Lines

Solving each of the equations in (9) for t gives the equations

$$t = \frac{x(t) - a_1}{b_1}; \qquad t = \frac{y(t) - a_2}{b_2}; \qquad t = \frac{z(t) - a_3}{b_3}.$$

Equating the right-hand sides of each of these equations gives the **symmetric** (or, **rectangular**) form of the equations for the line ℓ with position vector $a = a_1 i +$

$a_2\boldsymbol{j} + a_3\boldsymbol{k}$ and direction vector $\boldsymbol{b} = b_1\boldsymbol{i} + b_2\boldsymbol{j} + b_3\boldsymbol{k}$:

$$\frac{x - a_1}{b_1} = \frac{y - a_2}{b_2} = \frac{z - a_3}{b_3}.$$

(10)

Since the position vector $\boldsymbol{a}$ terminates at the point (a_1, a_2, a_3) on ℓ, we may also refer to equation (10) as the symmetric form for the equations of the line containing the point (a_1, a_2, a_3) and having direction vector $\boldsymbol{b} = b_1\boldsymbol{i} + b_2\boldsymbol{j} + b_3\boldsymbol{k}$.

Example 7

Find symmetric equations for the line determined by parametric equations in Example 5.

Solution: Using the values of a_j and b_j, $j = 1, 2, 3$, in Example 5, together with equations in (10), we obtain

$$\frac{x - 2}{1} = \frac{y + 1}{-1} = \frac{z - 4}{2}.$$

◇

Example 8

Find a unit vector orthogonal to the line ℓ with symmetric equations

$$\frac{x - 2}{3} = \frac{y - 4}{-2} = \frac{z + 1}{5}.$$

Strategy

Find a vector parallel to ℓ—its direction vector $\boldsymbol{b}$.

Find any vector $\boldsymbol{c}$ with $\boldsymbol{c} \cdot \boldsymbol{b} = 0$.

The desired unit vector is

$$\boldsymbol{u} = \left(\frac{1}{|\boldsymbol{c}|}\right)\boldsymbol{c}.$$

Solution

The direction vector for ℓ is, by equation (10),

$$\boldsymbol{b} = 3\boldsymbol{i} - 2\boldsymbol{j} + 5\boldsymbol{k}.$$

A vector $\boldsymbol{c} = c_1\boldsymbol{i} + c_2\boldsymbol{j} + c_3\boldsymbol{k}$ is orthogonal to $\boldsymbol{b}$ if and only if

$$\boldsymbol{b} \cdot \boldsymbol{c} = 3c_1 - 2c_2 + 5c_3 = 0.$$

One solution (among many) of this equation is

$$c_1 = 2, \qquad c_2 = 3, \qquad c_3 = 0.$$

The vector $\boldsymbol{c} = 2\boldsymbol{i} + 3\boldsymbol{j}$ is therefore orthogonal to ℓ. A *unit* vector in the same direction is

$$\boldsymbol{u} = \frac{\boldsymbol{c}}{|\boldsymbol{c}|} = \frac{\boldsymbol{c}}{\sqrt{13}} = \frac{2}{\sqrt{13}}\boldsymbol{i} + \frac{3}{\sqrt{13}}\boldsymbol{j}.$$

◇

Example 9

Find the symmetric equations for the line containing the points $P = (7, 3, -1)$ and $Q = (4, -5, -2)$.

Strategy

Find the direction vector $\boldsymbol{b} = \overrightarrow{PQ}$.

Solution

The direction vector for the line is

$$\boldsymbol{b} = \overrightarrow{PQ} = (4 - 7)\boldsymbol{i} + (-5 - 3)\boldsymbol{j} + (-2 - (-1))\boldsymbol{k}$$
$$= -3\boldsymbol{i} - 8\boldsymbol{j} - \boldsymbol{k}.$$

Use $P = (a_1, a_2, a_3)$ and substitute into equations in (10).

Using the point $P = (7, 3, -1)$ on the line and the direction vector $\boldsymbol{b}$ we obtain from the equations in (10) that

$$\frac{x - 7}{-3} = \frac{y - 3}{-8} = \frac{z + 1}{-1}.$$

◇

Equations for Planes

In geometry a plane is characterized by the following two facts:

(i) Through any point P on a line ℓ, there is precisely one plane σ perpendicular to ℓ.

(ii) The point $Q \neq P$ lies on the plane σ in (i) if and only if the line through P and Q is perpendicular to ℓ.

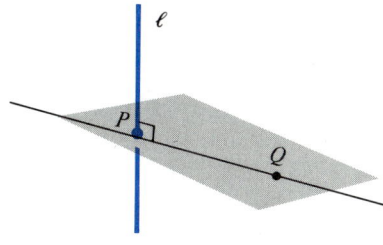

Figure 5.8 Plane σ determined by a point P and a line ℓ.

(Figure 5.8.)

We can use vectors to develop an analytic description of the plane σ as follows. If $\boldsymbol{n}$ is a direction vector for the line ℓ, the condition that the line through P and Q be perpendicular to the line ℓ is equivalent to the condition that the vectors $\overrightarrow{PQ}$ and $\boldsymbol{n}$ be orthogonal. By Corollary 1 of Section 15.2, this last condition is equivalent to the condition that $\overrightarrow{PQ} \cdot \boldsymbol{n} = 0$. We may therefore write *the plane σ determined by the point P and the vector $\boldsymbol{n}$* as

$$\sigma = \{Q \mid \overrightarrow{PQ} \cdot \boldsymbol{n} = 0\}. \tag{11}$$

The vector $\boldsymbol{n}$ in equation (11) is referred to as a **normal vector** for the plane σ (Figure 5.9).

To develop an equation in xyz-coordinates for the plane σ, we express $\boldsymbol{n}$ in component form as $\boldsymbol{n} = a\boldsymbol{i} + b\boldsymbol{j} + c\boldsymbol{k}$, and we assume the point P to have coordinates $P = (x_0, y_0, z_0)$. Let the point Q have coordinates $Q = (x, y, z)$. Then the vector $\overrightarrow{PQ}$ is

$$\overrightarrow{PQ} = (x - x_0)\boldsymbol{i} + (y - y_0)\boldsymbol{j} + (z - z_0)\boldsymbol{k}, \tag{12}$$

and the condition in line (11) is that

$$[(x - x_0)\boldsymbol{i} + (y - y_0)\boldsymbol{j} + (z - z_0)\boldsymbol{k}] \cdot (a\boldsymbol{i} + b\boldsymbol{j} + c\boldsymbol{k}) = 0. \tag{13}$$

Equation (13) simplifies to the equation

$$a(x - x_0) + b(y - y_0) + c(z - z_0) = 0,$$

or

$$ax + by + cz = ax_0 + by_0 + cz_0. \tag{14}$$

Equation (14) is *the equation for the plane with normal vector $\boldsymbol{n} = a\boldsymbol{i} + b\boldsymbol{j} + c\boldsymbol{k}$ containing the point (x_0, y_0, z_0).* Notice that the coefficients of x, y, and z are just the respective components of the normal vector $\boldsymbol{n}$, while the right-hand side is a *constant* determined from $\boldsymbol{n}$ and P.

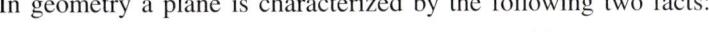

Figure 5.9 Plane σ determined by a point P and a normal vector $\boldsymbol{n}$.

Example 10

Find a vector normal to the plane with equation $3x + 5y - z = 9$.

Solution: The components of a normal vector are simply the coefficients of x, y, and z:

$$\boldsymbol{n} = 3\boldsymbol{i} + 5\boldsymbol{j} - \boldsymbol{k}.$$

◇

Example 11

Find an equation for the plane with normal vector $n = i + 4j + 2k$, and containing the point $P = (6, -3, 4)$.

Solution: Let $Q = (x, y, z)$ be a point in the desired plane. We are given

$$a = 1, \quad b = 4, \quad c = 2$$

and

$$x_0 = 6, \quad y_0 = -3, \quad z_0 = 4.$$

Substituting into equation (14) gives

$$x + 4y + 2z = 1 \cdot 6 + 4(-3) + 2 \cdot 4 = 2.$$

The desired equation is therefore

$$x + 4y + 2z = 2. \qquad \diamond$$

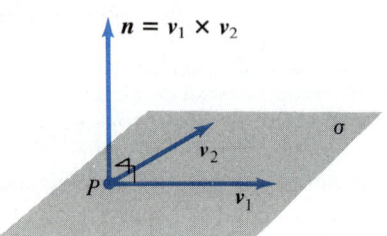

Figure 5.10 $n = v_1 \times v_2$ is orthogonal to the plane σ if v_1 and v_2 lie in σ.

If a normal vector is not specified, we can sometimes determine a normal vector for a plane from other information about the plane. In particular, if we can find two vectors, v_1 and v_2, lying in the plane, then the cross product $n = v_1 \times v_2$ is a normal to the plane, since it is orthogonal both to v_1 and to v_2 (Figure 5.10).

Example 12

Find an equation for the plane determined by the three points $P = (-4, 0, 2)$, $Q = (1, -3, 1)$, and $R = (2, -2, 6)$.

Strategy

Find two vectors.

$$v_1 = \overrightarrow{PQ}, \qquad v_2 = \overrightarrow{PR}$$

in the plane.

Solution

Two vectors in the plane are

$$v_1 = \overrightarrow{PQ} = 5i - 3j - k$$

and

$$v_2 = \overrightarrow{PR} = 6i - 2j + 4k.$$

Find a normal

$$n = v_1 \times v_2.$$

A normal vector is therefore

$$n = v_1 \times v_2 = \det \begin{bmatrix} i & j & k \\ 5 & -3 & -1 \\ 6 & -2 & 4 \end{bmatrix}$$
$$= -14i - 26j + 8k.$$

Substitute into equation (14), using n and P.

Using n and the point P, we obtain from equation (14) that

$$-14x - 26y + 8z = -14(-4) + (-26)(0) + 8 \cdot 2$$
$$= 72.$$

The desired equation is

$$-14x - 26y + 8z = 72. \qquad \diamond$$

Example 13

Find the distance from the point $P = (3, -1, 2)$ to the plane with equation $2x - y + z = 4$.

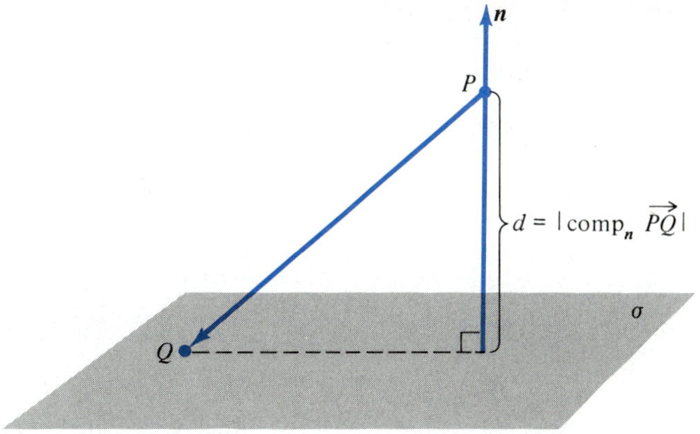

Figure 5.11 Distance from a point P to a plane σ.

Strategy

Find a normal n for the plane.

Set $x = y = 0$ and solve for z to find any convenient point Q in the plane.

Project $\overrightarrow{PQ}$ onto n. Recall that

$|\text{proj}_n \overrightarrow{PQ}| = |\text{comp}_n \overrightarrow{PQ}|$

(Section 16.2).

Solution

A normal vector for the plane is

$$n = 2i - j + k.$$

A point Q in the plane is $Q = (0, 0, 4)$. A vector parallel to n originating at P and terminating in the plane is

$$v = \text{proj}_n \overrightarrow{PQ} \qquad \text{(see Figure 5.10)}.$$

The desired distance is the length of this vector

$$|v| = |\text{comp}_n \overrightarrow{PQ}| = \frac{|\overrightarrow{PQ} \cdot n|}{|n|}$$

$$= \frac{|(-3i + j + 2k) \cdot (2i - j + k)|}{|2i - j + k|}$$

$$= \frac{5}{\sqrt{6}} \qquad \text{(Figure 5.11)}.$$

You may wish to try another convenient point Q, say $(2, 0, 0)$, and verify that the result is the same. ◇

Exercise Set 15.5

In Exercises 1–7, find a parametric form of the equations for the line described.

1. The line with direction vector $b = i + j - k$ and containing the point $P = (1, 2, 3)$.

2. The line with direction vector $b = 2i - j + 3k$ and containing the point $P = (3, -6, 2)$.

3. The line with position and direction vectors $a = b = 3i - j + 5k$.

4. The line containing the points $P = (1, 3, -1)$ and $Q = (7, -2, 5)$.

5. The line containing the points $P = (-4, 2, 1)$ and $Q = (-3, 5, 3)$.

6. The line with position vector $a = i + 2j - 6k$ and direction vector $b = 2i - j + 4k$.

7. The line through $(1, 2, 0)$ and orthogonal to *both* vectors $v = i + 3j + k$ and $w = j - 3k$.

In Exercises 8–11, find rectangular (symmetric) equations for the line determined by the given parametric equations.

8. $x(t) = 2t$
$y(t) = 3 - t$
$z(t) = t$

9. $x(t) = t + 7$
$y(t) = 4t - 6$
$z(t) = 3 - 2t$

10. $x(t) = 2t + 5$
$y(t) = t - 6$
$z(t) = 5t + 2$

11. $x(t) = t$
$y(t) = 8t - 6$
$z(t) = 4t + 4$

Find symmetric equations for

12. The line in Exercise 2.

13. The line in Exercise 3.

14. The line in Exercise 5.

In Exercises 15–18, find parametric equations for the line determined by the given symmetric equations.

15. $\dfrac{x}{3} = \dfrac{y}{2} = \dfrac{z}{5}$

16. $\dfrac{x - 2}{3} = \dfrac{y + 1}{-4} = \dfrac{z}{3}$

17. $\dfrac{x + 4}{4} = \dfrac{y - 2}{-2} = \dfrac{z + 3}{3}$

18. $x = \dfrac{y - 3}{2} = z + 1$

19. Find a vector parallel to the line in Exercise 9.

20. Find a vector parallel to the line in Exercise 16.

21. Find a direction vector for the line in Exercise 17.

22. Find a direction vector for the line in Exercise 18.

23. For the line ℓ with vector form $r(t) = a + tb$, $b \neq 0$, find the number t_0 for which $r(t_0) \perp b$. Conclude that $r(t_0) \perp \ell$.

24. Use Exercise 23 to show that the distance from the origin to the line ℓ with equation $r(t) = a + tb$ is $d = |r(t_0)|$, where t_0 is as in Exercise 23.

25. Use Exercise 24 to find the distance from the origin to the line with vector equation $r(t) = i + 2j + t(j - k)$.

26. Use Exercise 24 to find the distance from the origin to the line with symmetric equations

$$x - 1 = \dfrac{2 - y}{2} = \dfrac{z}{-2}.$$

27. Find the distance from the point $P = (2, -1, 4)$ to the line with equation $r(t) = (3i - j - k) + t(i + 2j + k)$.

28. Find the distance from the point $P = (1, 3, -2)$ to the line with equations $x(t) = 1 + t$, $y(t) = 3 - 2t$, $z(t) = 2t - 2$.

29. Find the distance from the point $P = (0, -2, 1)$ to the line with equations $\dfrac{x - 1}{4} = \dfrac{y + 3}{-2} = \dfrac{z + 1}{5}$.

30. For the line ℓ with vector form $r(t) = a + tb$, $b = b_1 i + b_2 j + b_3 k$, the numbers b_1, b_2, and b_3 are called the **direction numbers**. What condition must be satisfied by the direction numbers in order that the symmetric equations for ℓ exist?

31. Find the distance d between the lines

$$\ell_1: \dfrac{x - 2}{3} = \dfrac{y + 1}{2} = \dfrac{z - 3}{5}$$

and

$$\ell_2: \dfrac{x + 4}{1} = \dfrac{y - 3}{2} = \dfrac{z + 1}{2}$$

as follows:

a. Find direction vectors b_1 and b_2 for lines ℓ_1 and ℓ_2, respectively.

b. Observe that a vector perpendicular to both lines ℓ_1 and ℓ_2 is the cross product $n = b_1 \times b_2$.

c. Find a point P on ℓ_1 and a point Q on ℓ_2.

d. Obtain the distance d as

$$d = |\text{comp}_n \overrightarrow{PQ}|.$$

32. Use Exercise 31 to find the distance between the lines ℓ_1 and ℓ_2:

$$\ell_1: x(t) = 1 + t, \qquad y(t) = 5t, \qquad z(t) = 1 - t,$$
$$\ell_2: x(t) = 2 + t, \qquad y(t) = 2 - 3t, \qquad z(t) = 1 + 5t.$$

Two lines with vector equations $\ell_1: r_1(t) = a_1 + tb_1$ and $\ell_2: r_2(t) = a_2 + tb_2$ **intersect** if $r_1(t_1) = r_2(t_2)$ for some numbers t_1 and t_2. Find the point at which each of the following pairs of lines intersect.

33. $r_1(t) = (2i + j + 2k) + t(5i + j + 3k)$
$r_2(t) = (-4i + 7j + 10k) + t(3i - 3j - 4k)$

34. $\ell_1: x(t) = 1 + t, \qquad y(t) = 2 - 2t, \qquad z(t) = t + 5$
$\ell_2: x(t) = 2 + 2t, \qquad y(t) = 5 - 9t, \qquad z(t) = 2 + 6t$

In each of Exercises 35–39, find an equation for the plane with the given normal containing the given point.

35. $n = i + 2j - k, \qquad P = (1, 2, -3)$

36. $n = 4i - j + k, \qquad P = (-1, 3\ 5)$

37. $n = 2i - 2j + 3k, \qquad P = (0, -2, 5)$

38. $n = i + j \qquad P = (4, 2, -3)$

39. $n = i + 2j + k \qquad P = (0, -1, 3)$

In Exercises 40–44, find an equation for the plane containing the given points.

40. $P = (0, 1, -2), \qquad Q = (1, 1, 1), \qquad R = (3, 5, 1)$

41. $P = (-2, 3, -4), \qquad Q = (1, 5, 1), \qquad R = (-2, -2, -2)$

42. $P = (7, 0, -2), \qquad Q = (1, -3, 4), \qquad R = (5, 2, -3)$

43. $P = (-1, 1, 1), \qquad Q = (0, 2, -1), \qquad R = (3, 5, -2)$

44. $P = (-2, 1, 5), \qquad Q = (2, 0, -2), \qquad R = (-1, -1, 3)$

45. Find a vector normal to the plane with equation $2x - 3y + z = 5$.

46. Find a vector normal to the plane containing $(-2, 1, 3)$, $(5, 1, 5)$, and $(0, 0, 2)$.

47. Find an equation for the plane perpendicular to the line with symmetric equation

$$\frac{x-2}{3} = \frac{1-y}{6} = \frac{z+2}{2}$$

containing the point $(3, -2, 5)$.

48. Find an equation for the plane containing the point $(3, -1, -1)$ and perpendicular to the line with parametric equations $x = 3 + t$, $y = -1 + 4t$, $z = 5 + 3t$.

49. Find parametric equations for the line of intersection of the planes with equations $x + y - 2z = 6$ and $2x - 4y + z = 3$.

50. The angle between two planes is equal to the angle between vectors normal to the two planes. Find the angle between the planes with equations $x + y + z = 6$ and $3x - y + 2z = 5$.

51. Find the angle between the planes with equations $x - y + z = 2$ and $x + 3y + 3z = -4$ (see Exercise 50).

52. Show that the distance from the point $P_0 = (x_0, y_0, z_0)$ to the plane with equation $ax + by + cz = d$ is

$$d = \frac{|ax_0 + by_0 + cz_0 - d|}{\sqrt{a^2 + b^2 + c^2}}.$$

(*Hint:* See Example 13.)

53. Find the distance from the point $P = (3, -6, 5)$ to the plane with equation $6x - y + z = 2$.

54. Show that the plane with equation $ax + by + cz = 1$ intersects the coordinate axes at the respective values $x = \dfrac{1}{a}$, $y = \dfrac{1}{b}$, $z = \dfrac{1}{c}$ if $a \neq 0$, $b \neq 0$, $c \neq 0$.

55. Find parametric equations for the line containing the point $(2, 4, -3)$ and perpendicular to the plane with equation $2x + 3y - 7z = 9$.

56. Determine which of the following sets of points are coplanar.
 a. $(2, 1, 0)$, $(3, 0, 5)$, $(1, 1, 1)$, $(2, 3, -12)$
 b. $(1, -6, 2)$, $(3, -5, 11)$, $(4, 0, 4)$, $(1, 5, -2)$
 c. $(1, 7, 2)$, $(3, 4, -2)$, $(1, 5, 1)$, $(1, 9, 3)$

57. Find an equation for the plane containing the point $(3, -1, 3)$ and the line with symmetric equations

$$\frac{x-2}{3} = \frac{y+1}{5} = \frac{z}{4}.$$

15.6 CYLINDRICAL AND SPHERICAL COORDINATES

We have seen that certain equations involving two variables are more appropriately graphed using polar coordinates than rectangular (Cartesian) coordinates for the plane. Similarly, certain surfaces in space are more easily described using *cylindrical* or *spherical* coordinates than by use of rectangular coordinates. In fact, some surfaces that are easily described in one of these coordinate systems are almost impossible to determine in rectangular coordinates.

Cylindrical Coordinates

In this coordinate system, the *xy*-plane is coordinatized by polar coordinates, and the third coordinate denotes the usual rectangular *z*-coordinate. That is, if the point P has cylindrical coordinates $P = (r, \theta, z)$ and rectangular coordinates $P = (x, y, z)$, we have the equations

$$\begin{aligned} x &= r \cos \theta, & r &\geq 0 \qquad &\text{(1a)} \\ y &= r \sin \theta, & r &\geq 0 \qquad &\text{(1b)} \\ z &= z \qquad &\text{(1c)} \end{aligned}$$

giving the rectangular coordinates in terms of the cylindrical coordinates, and the equations

$$\begin{aligned} r &= \sqrt{x^2 + y^2} \qquad &\text{(2a)} \\ \tan \theta &= \frac{y}{x} \qquad &\text{(2b)} \\ z &= z \qquad &\text{(2c)} \end{aligned}$$

for the cylindrical coordinates in terms of the rectangular coordinates (Figure 6.1). (Note that we now require $r \geq 0$, unlike the case for polar coordinates in the plane.)

Figure 6.2 shows why cylindrical coordinates are named as they are. The graph of the equation $r = r_0$, r_0 constant, is a cylinder in space. The graph of $\theta = \theta_0$, θ_0 constant, is a half plane, as in Figure 6.3. The graph of $z = z_0$, z_0 constant, is a horizontal plane, as in rectangular coordinates. Several points, plotted in both rectangular and cylindrical coordinates, appear in Figure 6.4.

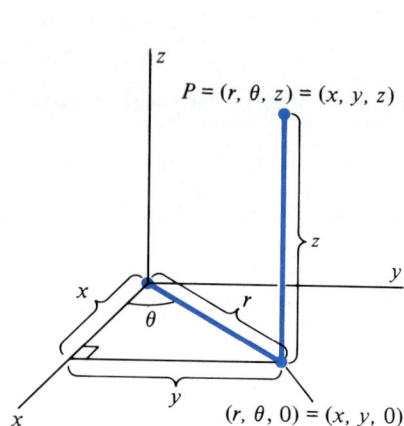

Figure 6.1 The cylindrical coordinate system.

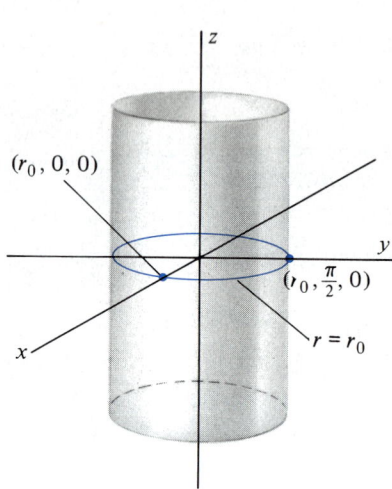

Figure 6.2 Graph of $r = r_0$ is a circular cylinder.

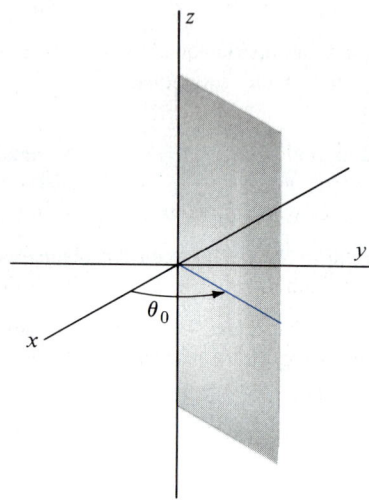

Figure 6.3 Graph of $\theta = \theta_0$ is a half plane.

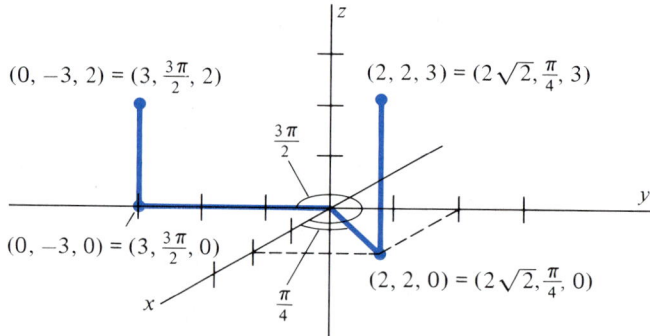

Figure 6.4 Four points in polar coordinates.

Example 1

Find cylindrical coordinates for the point with rectangular coordinates $(1, \sqrt{3}, 4)$.

Solution: Here $x = 1$, $y = \sqrt{3}$, $z = 4$. By equations (2a) through (2c),

$$r = \sqrt{1^2 + (\sqrt{3})^2} = 2,$$

$$\tan \theta = \sqrt{3} \quad \text{so} \quad \theta = \pi/3 \qquad \text{(since } x \text{ and } y \text{ place the point in the first quadrant),}$$

$$z = 4.$$

The cylindrical coordinates are $(2, \pi/3, 4)$. ◇

Example 2

Express the equation $z^2 = x^2 + y^2$ in cylindrical coordinates.

Solution: Using (1a) and (1b) the equation becomes

$$z^2 = (r \cos \theta)^2 + (r \sin \theta)^2$$
$$= r^2 \cdot (\cos^2 \theta + \sin^2 \theta) = r^2.$$

The equation therefore becomes

$$z = \pm r.$$

The graph of this equation is the cone in Figure 6.5. ◇

Example 3

Graph the equation $r^2 = a^2 \sin \theta$ in cylindrical coordinates.

Solution: Since the variable z is missing, the graph is a cylinder in the z direction whose trace in the xy-plane is the graph of the lemniscate $r^2 = a^2 \sin \theta$ (see Figure 6.6). ◇

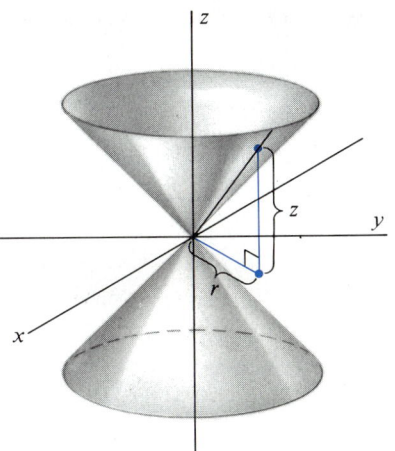

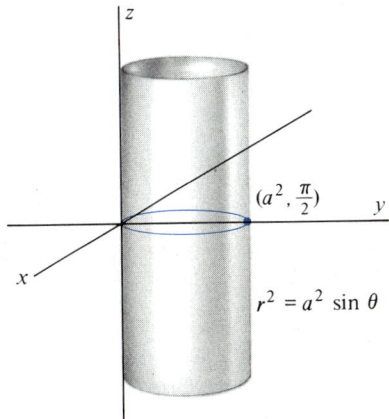

Figure 6.5 Graph of $z = \pm r$ is a cone.

Figure 6.6 Graph of $r^2 = a^2 \sin \theta$ is a cylinder in cylindrical coordinates.

Spherical Coordinates

In spherical coordinates, a point is determined by the ordered triple $P = (\rho, \theta, \phi)$ where $\rho = |\overrightarrow{OP}|$ is the distance of the point P from the origin, θ is the polar angle associated with the vertical line through P, and ϕ is the (tilt) angle between the vector $\overrightarrow{OP}$ and the positive z-axis (Figure 6.7). By convention, we require $\rho \geq 0$, $0 \leq \theta < 2\pi$, and $0 \leq \phi \leq \pi$.

Figure 6.8 illustrates the reason for the terminology ''spherical coordinates.'' The graph of the equation $\rho = \rho_0$, ρ_0 constant, is a sphere of radius ρ_0 centered at the origin. The graph of $\theta = \theta_0$, θ_0 constant, is a half plane, as in cylindrical coordinates. The graph of $\phi = \phi_0$, ϕ_0 constant, is a cone (Figure 6.9).

Figure 6.10 illustrates the relationships between the spherical coordinates (ρ, θ, ϕ) and the rectangular coordinates (x, y, z) for a point P. Since the vector $\overrightarrow{OP}$

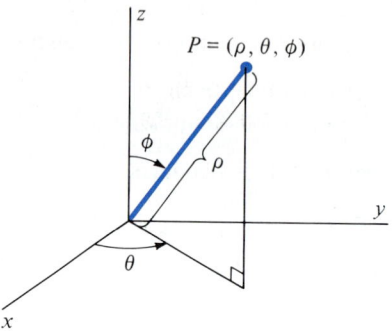

Figure 6.7 Spherical coordinates.

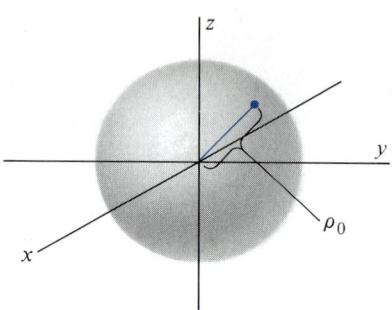

Figure 6.8 Graph of $\rho = \rho_0$ is a sphere.

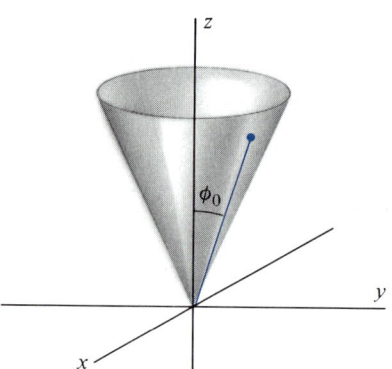

Figure 6.9 Graph of $\phi = \phi_0$ is a cone.

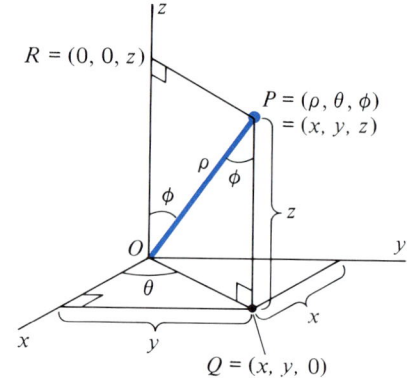

Figure 6.10 Relationships between rectangular and spherical coordinates.

is a diagonal of the rectangle $OQPR$,

$$|\overrightarrow{OQ}| = \rho \sin \phi, \quad \text{and} \quad z = \rho \cos \phi. \tag{3}$$

Also,

$$x = |\overrightarrow{OQ}| \cos \theta, \quad \text{and} \quad y = |\overrightarrow{OQ}| \sin \theta. \tag{4}$$

Combining equations (3) and (4) gives

$$x = \rho \sin \phi \cos \theta, \qquad \rho \geq 0 \tag{5a}$$
$$y = \rho \sin \phi \sin \theta, \qquad \rho \geq 0 \tag{5b}$$
$$z = \rho \cos \phi, \qquad \rho \geq 0. \tag{5c}$$

Also, from the distance formula we find

$$\rho = \sqrt{x^2 + y^2 + z^2}. \tag{6}$$

Equations for θ and ϕ may be easily derived from equations (3) and (4).

Example 4

The point P has spherical coordinates $P = (3, \pi/3, \pi/4)$. Find rectangular coordinates for P.

Solution: Here $\rho = 3$, $\theta = \pi/3$, and $\phi = \pi/4$. By equations (5a) through (5c),

$$x = 3 \sin\left(\frac{\pi}{4}\right) \cdot \cos\left(\frac{\pi}{3}\right) = 3\left(\frac{\sqrt{2}}{2}\right)\left(\frac{1}{2}\right) = \frac{3\sqrt{2}}{4},$$

$$y = 3 \sin\left(\frac{\pi}{4}\right) \cdot \sin\left(\frac{\pi}{3}\right) = 3\left(\frac{\sqrt{2}}{2}\right)\left(\frac{\sqrt{3}}{2}\right) = \frac{3\sqrt{6}}{4},$$

$$z = 3 \cos\left(\frac{\pi}{4}\right) = \frac{3\sqrt{2}}{2}.$$

The rectangular coordinates are $P = \left(\dfrac{3\sqrt{2}}{4}, \dfrac{3\sqrt{6}}{4}, \dfrac{3\sqrt{2}}{2}\right)$. ◇

Example 5

Express the equation $x^2 - y^2 + z^2 = 4$ in spherical coordinates.

Solution: Using equations (5a) through (5c) the equation becomes

$$(\rho \sin \phi \cos \theta)^2 - (\rho \sin \phi \sin \theta)^2 + (\rho \cos \phi)^2 = 4,$$

$$\rho^2 \sin^2 \phi[\cos^2 \theta - \sin^2 \theta] + \rho^2 \cos^2 \phi = 4,$$

$$\rho^2 \sin^2 \phi[1 - 2 \sin^2 \theta] + \rho^2 \cos^2 \phi = 4,$$

$$\rho^2 - 2\rho^2 \sin^2 \phi \sin^2 \theta = 4. \quad ◇$$

Exercise Set 15.6

1. The following points are given in rectangular coordinates. Find their cylindrical coordinates.
  a. $(1, 1, 0)$ **b.** $(\sqrt{3}, 1, 3)$
  c. $(-1, 1, -2)$ **d.** $(-1, \sqrt{3}, 4)$
  e. $(0, 3, -5)$ **f.** $(-\sqrt{2}, \sqrt{2}, \sqrt{2})$

2. The following points are given in cylindrical coordinates. Find their rectangular coordinates.
  a. $(2, \pi/4, -3)$ **b.** $(1, \pi/6, 4)$
  c. $(4, \pi/3, -5)$ **d.** $(2, 4\pi/3, 2)$
  e. $(1, \pi, 1)$ **f.** $(5, 5\pi/3, 5)$

3. The following points are given in rectangular coordinates. Find their spherical coordinates.
  a. $(1, 0, 0)$ **b.** $(1, 1, \sqrt{2})$
  c. $(1, -1, \sqrt{2})$ **d.** $(1, \sqrt{3}, 2)$
  e. $(-\sqrt{3}, 1, -2)$ **f.** $(-2, 2, 2\sqrt{2})$

4. The following points are given in spherical coordinates. Find their rectangular coordinates.
  a. $(2, \pi/4, \pi/3)$ **b.** $(1, \pi/2, \pi)$
  c. $(2, 3\pi/4, \pi/4)$ **d.** $(3, 3\pi/2, 2\pi/3)$
  e. $(5, \pi/6, 5\pi/6)$ **f.** $(2, 5\pi/3, 3\pi/4)$

5. The following points are given in cylindrical coordinates. Find their spherical coordinates.
  a. $(1, 0, 0)$ **b.** $(\sqrt{2}, -\pi/4, \sqrt{2})$
  c. $(2, \pi/3, 2)$ **d.** $(2, \pi/4, 2)$
  e. $(2, 5\pi/3, 0)$ **f.** $(2, \pi/6, 2)$

6. The following points are given in spherical coordinates. Find their cylindrical coordinates.
  a. $(2, \pi/4, \pi/2)$ **b.** $(2, \pi/2, \pi/4)$
  c. $(3, 2\pi/3, \pi/2)$ **d.** $(1, \pi/2, 2\pi/3)$
  e. $(4, \pi/4, 0)$ **f.** $(2, \pi/3, \pi/2)$

In Exercises 7–14, an equation in cylindrical coordinates is given. Write the equation in rectangular coordinates and sketch the graph.

7. $z = 3$

8. $r = 2$

9. $z = 2r$

10. $z = r \sin \theta$

11. $r^2 + z^2 = 4$

12. $z^2 = r^2 \cos^2 \theta - 1$

13. $\cos^2 \theta - \sin^2 \theta = \dfrac{a^2}{r^2}$

14. $z = r^2 \cos^2 \theta$

In Exercises 15–22, an equation in rectangular coordinates is given. Write the equation in cylindrical coordinates.

15. $x^2 + y^2 = 9$

16. $x^2 + y^2 + z^2 = 9$

17. $x^2 + y^2 = 9z$

18. $y^2 + z^2 = 1$

19. $x^2 + z^2 = 4$

20. $x + y + z = 4$

21. $xy - ax = 4$

22. $x^2 - y^2 = 4$

In Exercises 23–30, write the equation in the exercise referred to in spherical coordinates.

23. Exercise 15

24. Exercise 16

25. Exercise 17

26. Exercise 18

27. Exercise 19

28. Exercise 20

29. Exercise 21

30. Exercise 22

15.7 QUADRIC SURFACES AND CYLINDERS

In Chapter 14 we showed that the graph of the general second degree equation in two variables

$$Ax^2 + Bxy + Cy^2 + Dx + Ey + F = 0$$

if not degenerate, is either a line, a parabola, a circle, an ellipse, or a hyperbola in the plane. The general second degree equation in three variables has the form

$$Ax^2 + By^2 + Cz^2 + Dxy + Exz + Fyz + Gx + Hy + Iz + J = 0. \qquad (1)$$

Equation (1) is called a **quadric equation**. Graphs of quadric equations are surfaces in space, called **quadric surfaces**. In this section we shall examine six general types of quadric surfaces and a seventh special case that we refer to as the **quadric cylinder.**

The six general quadric surfaces are

1. the ellipsoid,
2. the elliptic paraboloid,
3. the elliptic cone,
4. the hyperboloid of one sheet,
5. the hyperboloid of two sheets,
6. the hyperbolic paraboloid.

We will study each of these figures by finding its *traces* in the coordinate planes and in planes parallel to the coordinate planes. To understand what this means, let us first recall the equations for the three coordinate planes:

$$z = 0 \text{ is the equation for the } xy\text{-plane}, \qquad (2a)$$

$$y = 0 \text{ is the equation for the } xz\text{-plane}, \qquad (2b)$$

$$x = 0 \text{ is the equation for the } yz\text{-plane}. \qquad (2c)$$

(See Figure 7.1).

The **trace** of a surface in any plane is simply the intersection of the surface and the plane. The equation for this trace is obtained by substituting the constant value determining the plane for the corresponding variable in the equation for the surface. For example, we have already seen (Section 15.3) that the equation for the sphere

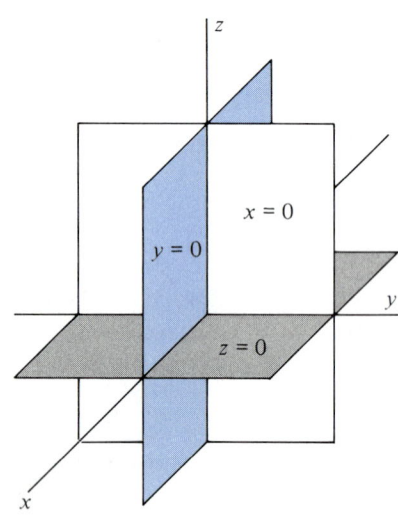

Figure 7.1   The three coordinate planes.

with center $(0, 0, 0)$ and radius r is

$$x^2 + y^2 + z^2 = r^2. \tag{3}$$

Combining equations (2a) and (3), we find that the equation for the trace of the sphere in the xy-plane is $x^2 + y^2 = r^2$. The trace is therefore the circle with center $(0, 0)$ and radius r in the xy-plane (Figure 7.2).

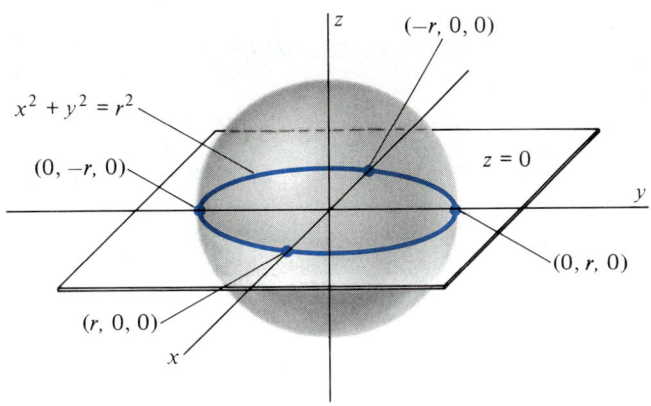

Figure 7.2 Trace of sphere $x^2 + y^2 + z^2 = r^2$ in plane $z = 0$ is circle $x^2 + y^2 = r^2$.

More generally, equations for planes *parallel* to the coordinate planes are

$$z = d, \quad \text{a plane parallel to the } xy\text{-plane}, \tag{4a}$$

$$y = d, \quad \text{a plane parallel to the } xz\text{-plane}, \tag{4b}$$

$$x = d, \quad \text{a plane parallel to the } yz\text{-plane}. \tag{4c}$$

Thus, to find the equation of the trace of the sphere in, say, the plane $z = d$, we combine equations (3) and (4a) to get $x^2 + y^2 + d^2 = r^2$, or $x^2 + y^2 = r^2 - d^2$. If $d^2 < r^2$, then the trace is again a circle (although smaller than the one in the xy-plane); if $d^2 > r^2$, there are no points of intersection (the plane is above or below the sphere).

The Ellipsoid: $\dfrac{x^2}{a^2} + \dfrac{y^2}{b^2} + \dfrac{z^2}{c^2} = 1$

Setting $z = 0$ shows that the trace in the xy-plane is the ellipse

$$\frac{x^2}{a^2} + \frac{y^2}{b^2} = 1.$$

Similarly, setting $y = 0$ and then $x = 0$ shows that the traces in the xz- and yz-plane are also ellipses. These three traces are shown in Figure 7.3.

Using Equations (4a) through (4c), we find that the traces in planes that are parallel to the coordinate planes and that intersect the ellipsoid are again ellipses. For example, setting $z = d$ gives the trace

$$\frac{x^2}{a^2} + \frac{y^2}{b^2} = 1 - \frac{d^2}{c^2}. \tag{5}$$

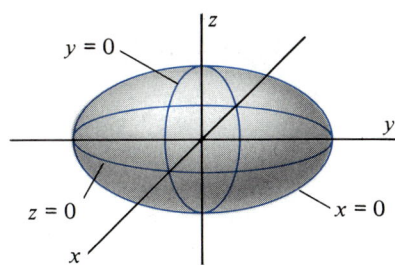

Figure 7.3 Traces of the ellipsoid $\dfrac{x^2}{a^2} + \dfrac{y^2}{b^2} + \dfrac{c^2}{d^2} = 1$ in the coordinate planes.

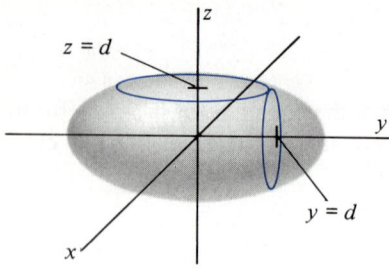

Figure 7.4 Traces of the ellipsoid in the planes $y = d$ and $z = d$.

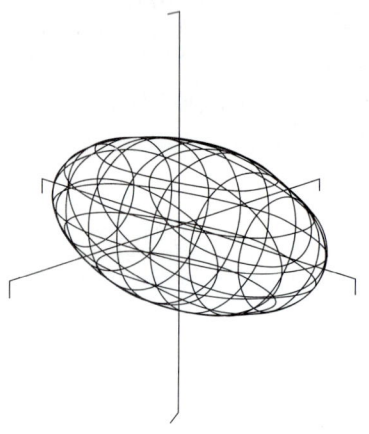

Figure 7.5 Various traces of an ellipsoid.

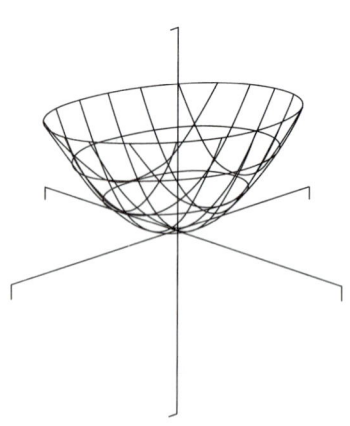

Figure 7.8 Various traces of an elliptic paraboloid.

The graph of equation (5) is an ellipse if $0 \le |d| < |c|$, it is a pair of points $(0, 0, \pm d)$ if $|d| = |c|$, and the equation has no solutions if $|d| > |c|$ (Figure 7.4). Figure 7.5 shows traces of the ellipsoid for various planes determined by equations (4a) through (4c).

The ellipsoid is symmetric with respect to each of the coordinate planes, since replacing x by $-x$, y by $-y$, or z by $-z$ does not change the equation. If $a = b = c$, the ellipsoid is, of course, a sphere.

The Elliptic Paraboloid: $z = \dfrac{x^2}{a^2} + \dfrac{y^2}{b^2}$

Setting $x = 0$ shows that the trace of this figure in the yz-plane is the parabola $z = \dfrac{y^2}{b^2}$. Similarly, the trace in the xz-plane is the parabola $z = \dfrac{x^2}{a^2}$. The trace in the xy-plane ($z = 0$) is simply the origin $(0, 0)$ (Figure 7.6).

The traces in the planes $x = d_1$ and $y = d_2$ are again parabolas:

$$z = \frac{y^2}{b^2} + \frac{d_1^2}{a^2}; \qquad z = \frac{x^2}{a^2} + \frac{d_2^2}{b^2}.$$

However, traces in the plane $z = d$ are ellipses with equations

$$\frac{x^2}{a^2 d} + \frac{y^2}{b^2 d} = 1$$

if $d > 0$. If $d < 0$ there are no traces (Figure 7.7).

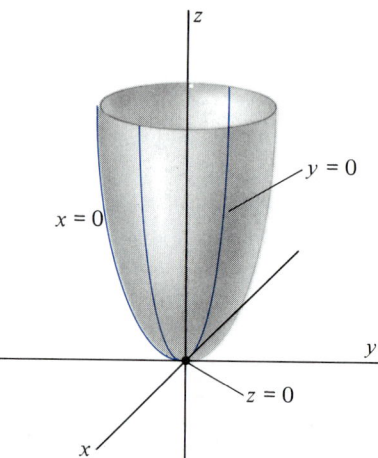

Figure 7.6 Traces of the elliptic paraboloid $z = \dfrac{x^2}{a^2} + \dfrac{y^2}{b^2}$ in the coordinate planes.

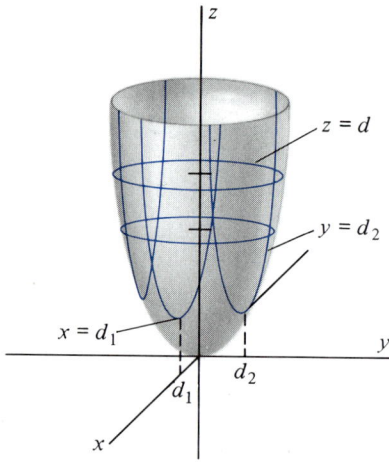

Figure 7.7 Traces of the elliptic paraboloid in planes $x = d_1$, $y = d_2$, $z = d$.

The elliptic paraboloid is symmetric with respect to the xz- and yz-planes, as you can see by replacing x by $-x$, or y by $-y$ in its equation (Figure 7.8).

The Elliptic Cone: $z^2 = \dfrac{x^2}{a^2} + \dfrac{y^2}{b^2}$.

The difference between the equation of the elliptic cone and that of the elliptic paraboloid is the presence of z^2 rather than z. As a result, the traces in the xz-plane ($y = 0$) are the lines $z = \pm\dfrac{x}{a}$, and the traces in the yz-plane ($x = 0$) are the lines $z = \pm\dfrac{y}{b}$. The trace in the xy-plane ($z = 0$) is simply the origin $(0, 0, 0)$ (Figure 7.9).

In the planes $x = d$, the traces are the hyperbolas

$$\frac{a^2z^2}{d^2} - \frac{a^2y^2}{b^2d^2} = 1.$$

Similarly, the traces in the planes $y = d$ are hyperbolas (Figure 7.10). However, in

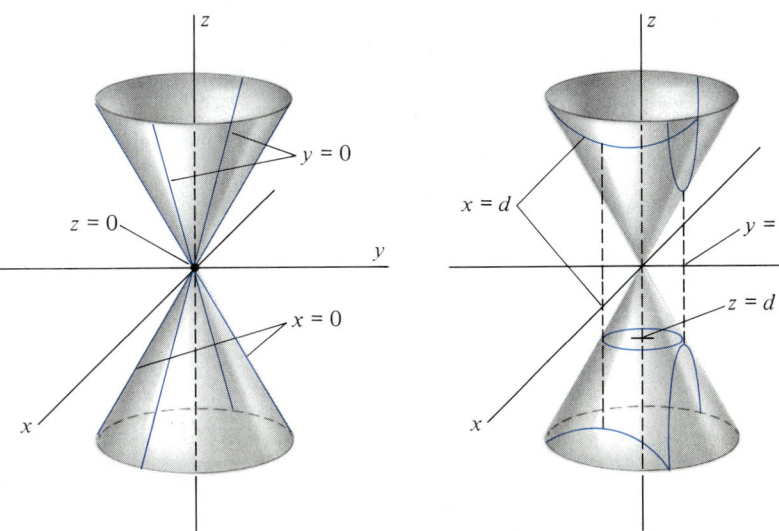

Figure 7.9 Traces of elliptic cone $z^2 = \dfrac{x^2}{a^2} + \dfrac{y^2}{b^2}$ in the coordinate planes.

Figure 7.10 Traces of elliptic cone in planes $x = d$, $y = d$, and $z = d$.

the planes $z = d$ and the traces are the ellipses with equations

$$\frac{x^2}{a^2d^2} + \frac{y^2}{b^2d^2} = 1.$$

The elliptic cone is symmetric with respect to each of the three coordinate planes. Also, it differs from the elliptic paraboloid in that it is unbounded in both the positive and negative z directions (Figure 7.11). If $a = b$, the cone is called a *right* or *circular* cone.

The Hyperboloid of One Sheet: $\quad \dfrac{x^2}{a^2} + \dfrac{y^2}{b^2} - \dfrac{z^2}{c^2} = 1$

The trace in the xy-plane ($z = 0$) is the ellipse

$$\frac{x^2}{a^2} + \frac{y^2}{b^2} = 1,$$

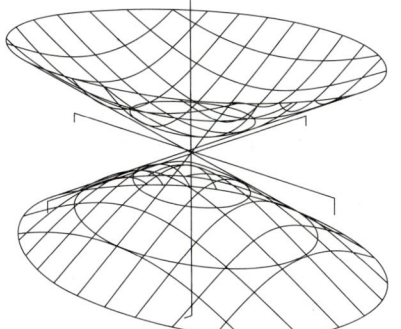

Figure 7.11 Various traces of an elliptic cone.

while traces in the other two coordinate planes are the hyperbolas

$$\frac{x^2}{a^2} - \frac{z^2}{c^2} = 1 \qquad \text{and} \qquad \frac{y^2}{b^2} - \frac{z^2}{c^2} = 1$$

(Figure 7.12). Similarly, setting $z = d$ shows that traces parallel to the xy-plane are ellipses, while the traces in planes parallel to the xz- or yz-planes are hyperbolas. The hyperboloid of one sheet is symmetric with respect to each of the coordinate axes (Figures 7.13 and 7.14).

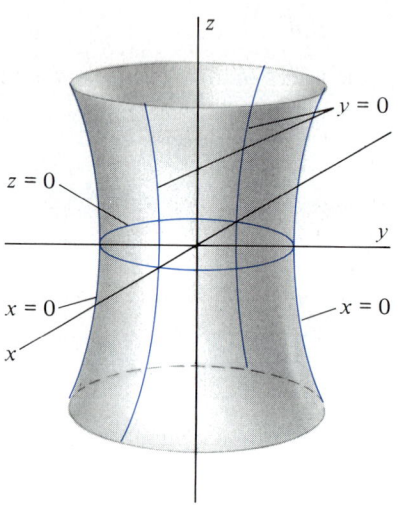

Figure 7.12 Traces of the hyperboloid of one sheet $\dfrac{x^2}{a^2} + \dfrac{y^2}{b^2} - \dfrac{z^2}{c^2} = 1$ in the coordinate planes.

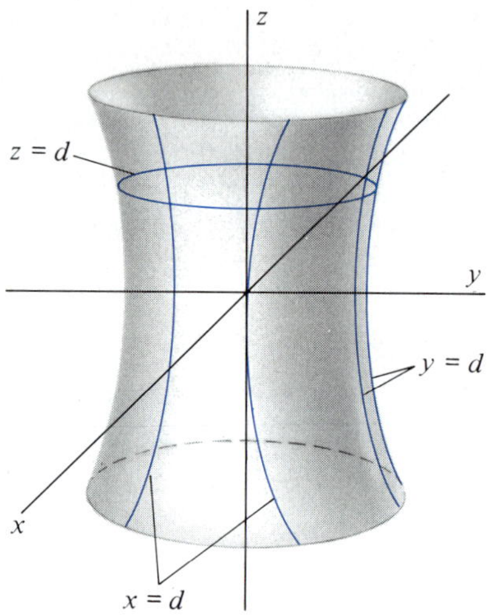

Figure 7.13 Traces of the hyperboloid of one sheet in the planes $x = d$, $y = d$, and $z = d$.

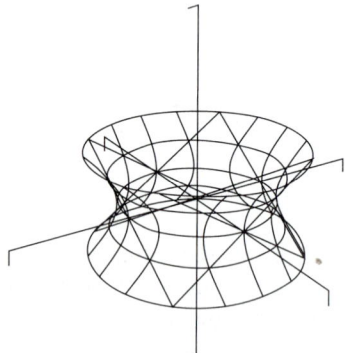

Figure 7.14 Various traces of a hyperboloid of one sheet.

The Hyperboloid of Two Sheets: $\dfrac{x^2}{a^2} + \dfrac{y^2}{b^2} - \dfrac{z^2}{c^2} = -1$

Setting $z = 0$ shows that this surface has no trace in the xy-plane. The traces in the xz- and yz-planes are the hyperbolas

$$\frac{z^2}{c^2} - \frac{x^2}{a^2} = 1 \qquad \text{and} \qquad \frac{z^2}{c^2} - \frac{y^2}{b^2} = 1,$$

respectively. The traces in planes parallel to the xz- and yz-planes are also hyperbolas. However, traces in planes parallel to the xy-plane ($z = d$) are ellipses

$$\frac{x^2}{a^2} + \frac{y^2}{b^2} = \frac{d^2}{c^2} - 1$$

provided $|d| > |c|$. The hyperboloid of two sheets is symmetric with respect to each of the coordinate planes (Figures 7.15–7.17).

The Hyperbolic Paraboloid: $z = \dfrac{y^2}{b^2} - \dfrac{x^2}{a^2}$

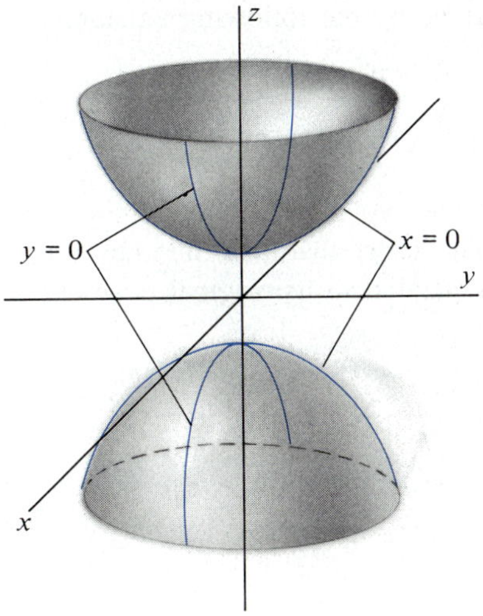

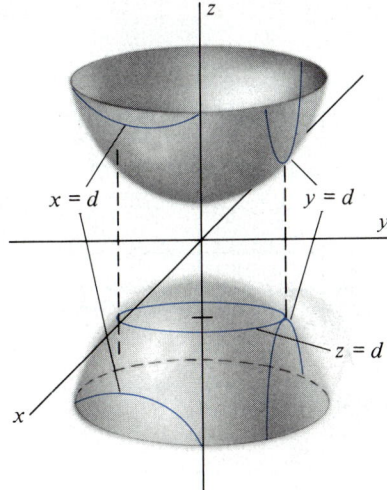

Figure 7.15 Traces of the hyperboloid of two sheets $\dfrac{x^2}{a^2} + \dfrac{y^2}{b^2} - \dfrac{z^2}{c^2} = -1$.

Figure 7.16 Traces of the hyperboloid of two sheets in planes parallel to the coordinate planes.

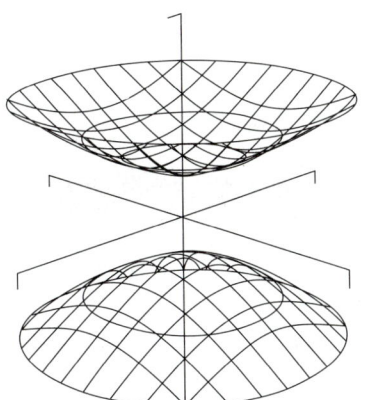

Figure 7.17 Traces of a hyperboloid of two sheets.

Setting $z = 0$ shows that the trace in the xy-plane is the pair of intersecting straight lines $y = \pm \left| \dfrac{b}{a} \right| x$. In the xz-plane ($y = 0$) the trace is the parabola $z = -\dfrac{x^2}{b^2}$ (Figure 7.18). The traces in planes parallel to the xz- and yz-planes are also parabolas. However, the traces in planes parallel to the xy-plane ($z = d$) are hyperbolas. The hyperbolic paraboloid is symmetric with respect to the xz- and yz-planes (Figure 7.19).

The six figures described here do not exhaust all possibilities for the graph of the general second degree equation (1). We shall not attempt to present an all encompassing discussion as we were able to do for the two variable case in Chapter 14. However, we do wish to point out that the technique of completing the square may often be applied to reduce a given second degree equation to one of the six types discussed above. In problems of this type it is often useful to use the **translation of**

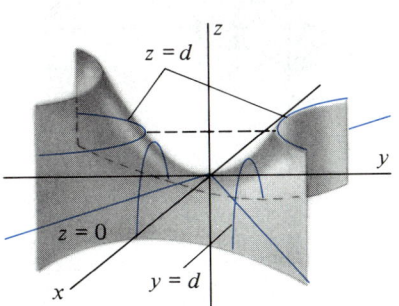

Figure 7.18 Traces of the hyperbolic paraboloid $z = \dfrac{y^2}{b^2} - \dfrac{x^2}{a^2}$.

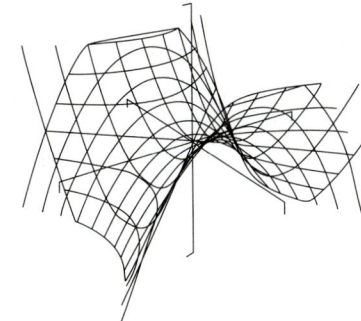

Figure 7.19 Various traces of a hyperbolic paraboloid.

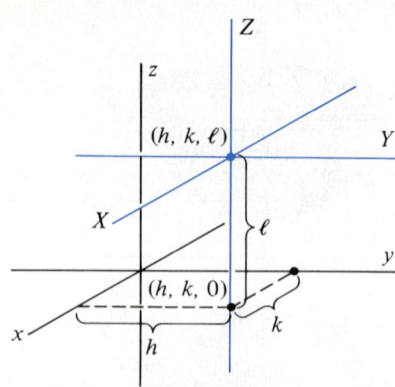

Figure 7.20 Translation of axes $X = x - h$, $Y = y - k$, $Z = z - \ell$.

axes defined by the following equations

$$X = x - h, \tag{6a}$$
$$Y = y - k, \tag{6b}$$
$$Z = z - \ell. \tag{6c}$$

As in the two variable case, these substitutions amount to a relocation of the coordinate axes so that the origin lies at (h, k, ℓ) and so that each of the coordinate axes lies parallel to its original position (see Figure 7.20).

Example 1

Describe the graph of the equation

$$x^2 + y^2 + 3z^2 - 2x + 4y - 4 = 0.$$

Strategy

Complete the square in x and in y.

Solution

$$x^2 - 2x + y^2 + 4y + 3z^2 - 4 = 0$$

$$(x^2 - 2x + 1) + (y^2 + 4y + 4) + 3z^2 - 4 - 1 - 4 = 0$$

$$(x - 1)^2 + (y + 2)^2 + 3z^2 - 9 = 0$$

Use translated variables (6a) through (6c) to simplify equation.

The given equation therefore has the form

$$X^2 + Y^2 + 3Z^2 = 9,$$

or

Identify the form of the equation obtained.

$$\frac{X^2}{3^2} + \frac{Y^2}{3^2} + \frac{Z^2}{(\sqrt{3})^2} = 1$$

with $X = x - 1$, $Y = y + 2$, and $Z = z$. The graph of this equation is an ellipsoid with center at $(1, -2, 0)$ (Figure 7.21). ◇

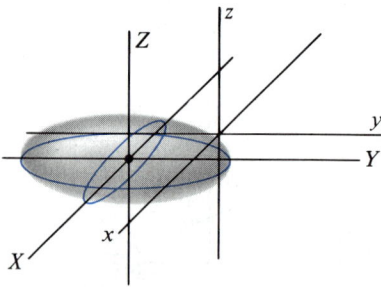

Figure 7.21 Ellipsoid $x^2 + y^2 + 3z^2 - 2x + 4y - 4 = 0$.

Cylinders

When an equation involves fewer variables than the number of axes on which it is graphed, the traces obtained by selecting various values of the missing variable must all be the same. The simplest example of this occurs when the equation $x = a$ is graphed in the xy-coordinate plane. Any trace obtained by setting $y = d$ is simply a point whose x-coordinate is a (Figure 7.22).

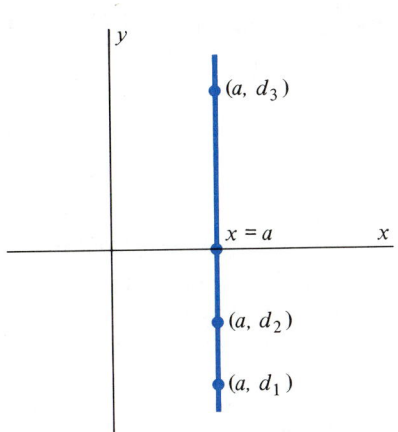

Figure 7.22 Graph of equation $x = a$ is a *cylinder* in the *xy*-plane.

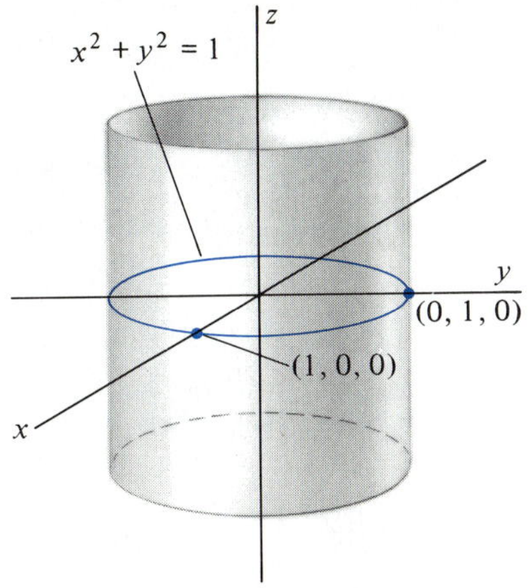

Figure 7.23 The equation $x^2 + y^2 = 1$ is a cylinder in *xyz* space.

A similar phenomenon occurs when an equation involving only two variables, say x and y, is graphed in *xyz*-space. Any trace corresponding to a fixed value of the missing variable, $z = d$, will be a "copy" of the graph of the equation in the *xy*-plane. The graph in *xyz*-space will therefore be a surface which can be thought of as an "infinite stack" of copies of the two dimensional figure. Graphs in *xyz*-space of equations with one or two variables missing are therefore called **cylinders.**

A typical cylinder is the graph of $x^2 + y^2 = 1$ in *xyz*-space (Figure 7.23). This is a special case of the **elliptic cylinder**

$$\frac{x^2}{a^2} + \frac{y^2}{b^2} = 1$$

in space. Three other examples of cylinders are

(a) a plane whose equation contains only two variables, such as $x + y = 2$ (Figure 7.24);

(b) the parabolic cylinder $y = x^2$ (Figure 7.25);

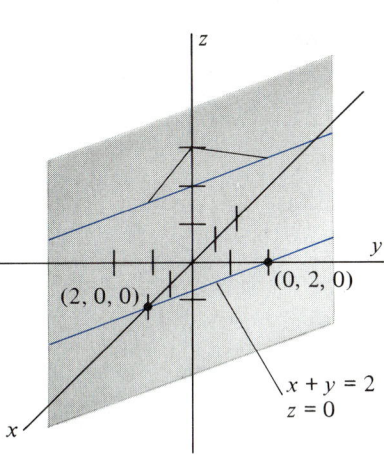

Figure 7.24 The plane $x + y = 2$ is a cylinder.

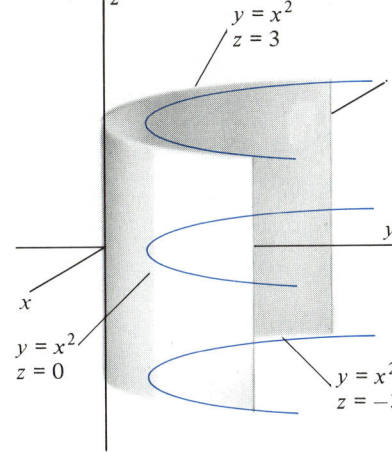

Figure 7.25 Parabolic cylinder $y = x^2$.

(c) the hyperbolic cylinder $\dfrac{y^2}{a^2} - \dfrac{x^2}{b^2} = 1$ (Figure 7.26).

Example 2

Sketch the cylinder $9x^2 + 4z^2 - 18x - 16z = 11$.

Solution: The first task is to complete the square in x and z:

$$9x^2 - 18x + 4z^2 - 16z = 11$$
$$9(x^2 - 2x) + 4(z^2 - 4z) = 11$$
$$9(x^2 - 2x + 1) + 4(z^2 - 4z + 4) = 11 + 9 \cdot 1 + 4 \cdot 4$$
$$9(x - 1)^2 + 4(z - 2)^2 = 36$$
$$\frac{(x - 1)^2}{2^2} + \frac{(z - 2)^2}{3^2} = 1.$$

This is the equation for an ellipse with center $(1, 2)$ in the xz-plane. Since the variable y is missing, the figure is an elliptic cylinder with central axis parallel to the y-axis (Figure 7.27). ◇

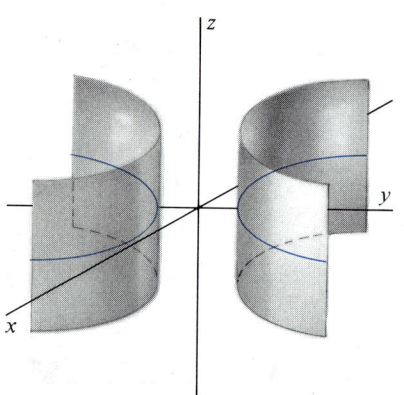

Figure 7.26 The hyperbolic cylinder $\dfrac{y^2}{a^2} - \dfrac{x^2}{b^2} = 1$.

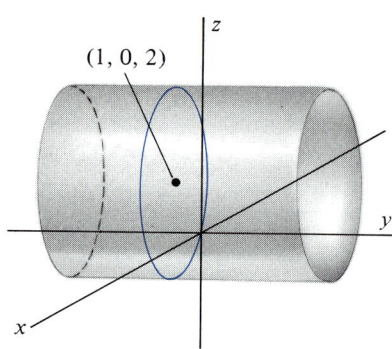

(1, 0, 2)

Figure 7.27 The elliptic cylinder $9x^2 - 18x + 4z^2 - 16z = 11$.

Exercise Set 15.7

In Exercises 1–20, describe the quadric surface that is the graph of the given equation.

1. $x^2 + y^2 + z^2 = 10$ *Ellipsoid*

2. $\dfrac{x^2}{4} + \dfrac{y^2}{9} + \dfrac{z^2}{16} = 1$

3. $\dfrac{x^2}{4} + \dfrac{y^2}{9} = z^2$ *parabloid*

4. $6x^2 + 12y^2 - 8z^2 - 24 = 0$

5. $4x^2 + 9y^2 - 36z^2 + 36 = 0$

6. $9y^2 - 6x^2 = 54$

7. $2x^2 - 3y^2 + 8x + 6y - 6z + 5 = 0$

8. $8x^2 + 4y^2 - 2z^2 + 24y - 4z + 44 = 0$

9. $x^2 + 2y^2 - 2z^2 + 6x - 4y + 7 = 0$

10. $x^2 + 4y^2 - 4z^2 + 6x - 8y - 8z + 9 = 0$

11. $6x^2 + 9y^2 + 36y - 54z + 36 = 0$

12. $4x^2 + 3y^2 + 3z^2 - 12y + 18z + 27 = 0$

13. $9z^2 + 4y^2 - 36x = 0$

14. $x^2 - 4y^2 + 4z^2 = 0$

15. $z^2 + y^2 - x^2 - 1 = 0$

16. $6x^2 - 12y^2 + 8z^2 + 24 = 0$

17. $x^2 + y^2 + z^2 - 6x + 2y + 6z + 18 = 0$

18. $9x^2 + 4y^2 + 9z^2 - 36x + 8y - 18z + 13 = 0$

19. $x^2 + y^2 + 4x - 2y - 36z + 175 = 0$

20. $y^2 - x^2 - 6x - 2y - 4z - 16 = 0$

In Exercises 21–36, describe the given cylinder in *xyz*-space.

21. $x = 3$

22. $y = 6$

23. $x + y = 6$

24. $x^2 + y^2 = 5$

25. $\dfrac{x^2}{4} - \dfrac{y^2}{9} = 1$

26. $\dfrac{x^2}{2} + \dfrac{y^2}{4} = 1$

27. $xy = 1$

28. $y = \sin x$

29. $y = 4z^2$

30. $x = \sqrt{9 - z^2}$

31. $x^2 - z^2 = 1$

32. $z^2 - y^2 = 1$

33. $z = 1 - y^2$

34. $(x - 1)^2 + (z + 3)^2 = 9$

35. $3x^2 + 12x - z + 16 = 0$

36. $9x^2 + 4z^2 + 18x - 16z - 11 = 0$

37. Write the equation for the surface obtained by revolving the graph of $y = 2x$ about the *y*-axis.

38. Describe the set of all points P for which the distance from P to the *y*-axis is twice the distance from P to the *xz*-plane.

39. Find an equation for the solid obtained by revolving the graph of $y = z^2$ about the *y*-axis.

SUMMARY OUTLINE OF CHAPTER 15

◆ A **vector** in the plane is an ordered pair of numbers $v = \langle a, b \rangle$ subject to the laws (page 641)

$\langle a_1, b_1 \rangle + \langle a_2, b_2 \rangle = \langle a_1 + a_2, b_1 + b_2 \rangle,$

$c \langle a, b \rangle = \langle ca, cb \rangle, \qquad c \in (-\infty, \infty).$

◆ A vector in space is an ordered triple $v = \langle a, b, c \rangle$ subject to the laws (page 663)

$\langle a_1, b_1, c_1 \rangle + \langle a_2, b_2, c_2 \rangle = \langle a_1 + a_2, b_1 + b_2, c_1 + c_2 \rangle,$

$d \langle a, b, c \rangle = \langle da, db, dc \rangle, \qquad d \in (-\infty, \infty).$

◆ The **unit coordinate vectors** are $i = \langle 1, 0, 0 \rangle, j = \langle 0, 1, 0 \rangle, k = \langle 0, 0, 1 \rangle.$ (page 664)

◆ *Notation:* $v = \langle a, b, c \rangle = ai + bj + ck.$ (page 664)

◆ The **length** of the vector $v = ai + bj + ck$ is $|v| = \sqrt{a^2 + b^2 + c^2}.$ (page 664)

◆ Vector length has the following properties: (page 647)

    (i) $|v| \geq 0;$ $|v| = 0$ if and only if $v = \mathbf{0},$

    (ii) $|cv| = |c|\,|v|,$

    (iii) $|v + w| \leq |v| + |w|.$

◆ A **unit vector** in the direction of v is $u = \dfrac{v}{|v|}.$ (page 648)

◆ The **dot product** of the vectors $v = x_1 i + y_1 j + z_1 k$ and $w = x_2 i + y_2 j + z_2 k$ is the number (page 665)

$v \cdot w = x_1 x_2 + y_1 y_2 + z_1 z_2.$

◆ *Theorem:* $v \cdot w = |v|\,|w| \cos \theta,$ where θ is the angle between v and $w.$ (page 653)

◆ The vectors v and w are **orthogonal** (perpendicular) if and only if $v \cdot w = 0.$ (page 654)

◆ The **component** of the vector v in the direction of the vector w is the **number** (page 655)

$\text{comp}_w\, v = \dfrac{v \cdot w}{|w|}.$

◆ The **projection** of the vector v along the vector w is the **vector** (page 657)

$\text{proj}_w\, v = \left(\dfrac{v \cdot w}{|w|^2} \right) w.$

◆ The distance from the point $P = (x_0, y_0)$ in the plane to the line with equation $ax + by + c = 0$ is $\qquad$ (page 656)

$$d = \frac{|ax_0 + by_0 + c|}{\sqrt{a^2 + b^2}}.$$

◆ The distance d between two points $P = (x_1, y_1, z_1)$ and $Q = (x_2, y_2, z_2)$ in space is $\qquad$ (page 661)

$$d = \sqrt{(x_2 - x_1)^2 + (y_2 - y_1)^2 + (z_2 - z_1)^2}.$$

◆ The **cross product** of the vectors $v = x_1 i + y_1 j + z_1 k$ and $w = x_2 i + y_2 j + z_2 k$ in space is the **vector** $\qquad$ (page 669)

$$v \times w = \det \begin{bmatrix} i & j & k \\ x_1 & y_1 & z_1 \\ x_2 & y_2 & z_2 \end{bmatrix} = (y_1 z_2 - z_1 y_2) i + (z_1 x_2 - x_1 z_2) j + (x_1 y_2 - y_1 x_2) k.$$

◆ ***Theorem:*** $v \times w = (|v|\,|w|\,\sin\theta) n$, where n is a unit vector orthogonal to both v and w in the direction determined by the right-hand rule. $\qquad$ (page 672)

◆ $|v \times w|$ is the area of the parallelogram determined by v and w. $\qquad$ (page 673)

◆ The **line** with position vector $a = a_1 i + a_2 j + a_3 k$ and direction vector $b = b_1 i + b_2 j + b_3 k$ has $\qquad$ (page 676)

(i) vector equation $r(t) = a + tb, \qquad -\infty < t < \infty.$
(ii) parametric equations $x(t) = a_1 + tb_1, \ y(t) = a_2 + tb_2, \ z(t) = a_3 + tb_3.$

(iii) symmetric equations $\dfrac{x - a_1}{b_1} = \dfrac{y - a_2}{b_2} = \dfrac{z - a_3}{b_3}.$

◆ The **distance** d from the point P to the line with direction vector b containing the point Q is $\qquad$ (page 678)

$$d = \frac{|\overrightarrow{PQ} \times b|}{|b|}.$$

◆ The **plane** with normal vector n containing the point P is the set of all points Q for which $\overrightarrow{PQ} \cdot n = 0$. $\qquad$ (page 681)

◆ If $n = ai + bj + ck$, $P = (x_0, y_0, z_0)$, and $Q = (x, y, z)$, the equation of the plane is $\qquad$ (page 681)

$ax + by + cz = ax_0 + by_0 + cz_0.$

◆ The cylindrical coordinates (r, θ, z) and the rectangular coordinates (x, y, z) for a point P are related via the equations $\qquad$ (page 685)

$x = r \cos\theta, \qquad r \geq 0, \qquad\qquad r = \sqrt{x^2 + y^2}$

$y = r \sin\theta, \qquad r \geq 0, \qquad$ and $\qquad \tan\theta = y/x,$

$z = z \qquad\qquad\qquad\qquad\qquad z = z.$

◆ The spherical coordinates (ρ, θ, ϕ) and the rectangular coordinates (x, y, z) for a point P are related via the equations $\qquad$ (page 688)

$x = \rho \sin\phi \cos\theta, \qquad \rho \geq 0 \qquad\qquad \rho = \sqrt{x^2 + y^2 + z^2}$

$y = \rho \sin\phi \sin\theta, \qquad \rho \geq 0 \qquad$ and $\qquad \tan\theta = y/x, \qquad x \neq 0$

$z = \rho \cos\phi, \qquad \rho \geq 0 \qquad\qquad \tan\phi = \dfrac{\sqrt{x^2 + y^2}}{z}, \qquad z \neq 0.$

◆ ***The Ellipsoid:*** $\dfrac{x^2}{a^2} + \dfrac{y^2}{b^2} + \dfrac{z^2}{c^2} = 1$ $\qquad$ (page 691)

◆ ***The Elliptic Paraboloid:*** $z = \dfrac{x^2}{a^2} + \dfrac{y^2}{b^2}$ $\qquad$ (page 692)

◆ ***The Elliptic Cone:*** $z^2 = \dfrac{x^2}{a^2} + \dfrac{y^2}{b^2}.$ $\qquad$ (page 692)

◆ ***The Hyperboloid of One Sheet:*** $\dfrac{x^2}{a^2} + \dfrac{y^2}{b^2} - \dfrac{z^2}{c^2} = 1$ $\qquad$ (page 693)

◆ **The Hyperboloid of Two Sheets:** $\dfrac{x^2}{a^2} + \dfrac{y^2}{b^2} - \dfrac{z^2}{c^2} = -1$ (page 694)

◆ **The Hyperbolic Paraboloid:** $z = \dfrac{y^2}{b^2} - \dfrac{x^2}{a^2}$ (page 694)

REVIEW EXERCISES—CHAPTER 15

1. Find the distance between the points $P = (1, 2, -4)$ and $Q = (2, -5, 2)$.

2. Find an equation for the set of all points $P = (x, y, z)$ equidistant from the fixed points $P_1 = (1, -2, 1)$ and $Q = (3, 4, -3)$.

3. Describe the graph of the equation $x^2 + y^2 + z^2 - 6x + 4y - 2z + 10 = 0$.

4. Let $u = i + 2j - 3k$, $v = 2i + 2j + 6k$, $w = i - 4j + 3k$. Find

  a. $u + 2v$ **b.** $u - v$

  c. $|3u + v|$ **d.** $u \cdot v$

  e. $u \cdot (v \times w)$ **f.** $|u + v - w|$

5. Find the cosine of the angle between the vectors $v = i - 3j + 2k$ and $w = 3i + 3j + 2k$.

6. Determine whether the following pairs of lines intersect and, if so, at what point.

  a. $x = t$, $y = t + 2$, $z = 2t - 4$;

      $x = 1 - t$, $y = 3 + t$, $z = 4t$

  b. $x = 1 + 2t$, $y = t - 2$, $z = 1 + t$;

      $x = 2t + 1$, $y = 4 - 2t$, $z = 5 - t$

7. Find parametric equations for the line containing the point $(2, 1, -3)$ that is perpendicular to both of these lines.

  $\ell_1: x = 2t$, $y = 3 + t$, $z = 5t$

  $\ell_2: x = t + 5$, $y = 6 - 4t$, $z = 3t + 4$

8. Find the direction cosines for the vector $v = 3i + 4j + 5k$.

9. Find an equation for the plane containing the points $(-1, 2, 1)$, $(2, 5, 3)$, and $(-4, 0, 2)$.

10. Find an equation for the plane containing the line

  $\ell_1: r(t) = i - 3j + k + t(4i + 2j - k)$

  and the point $(1, 2, 1)$.

11. Find a vector normal to the plane with equation $8x + y - 2z = 5$.

12. Find the distance from the point $P = (1, 2, -4)$ to the plane with equation $3x + 2y - 5z = 5$.

13. Find a vector equation for the line of intersection of the planes with equations $2x + 3y - z = 4$ and $x - 3y + 5z = 2$.

14. Show that if $v \times w = 0$ and $w \neq 0$ then $v = cw$ for some constant c.

15. Show that $u \times (v \times w) = (u \cdot w)v - (u \cdot v)w$.

16. Show that $(u + v) \times (u - v) = 2v \times u$.

17. Sketch the graph of the equation $y^2 = 9 + z^2$.

18. Sketch the graph of the equation $x^2 + z^2 = 1 + y$.

19. Find rectangular coordinates for the point with spherical coordinates $(2, \pi/4, \pi/3)$.

20. Find parametric equations for the line containing the points $(3, 5, -6)$ and $(-2, 3, -1)$.

21. Find an equation for the plane containing the point $(2, -5, 3)$ and perpendicular to the line with parametric equations $x = 4 + 2t$, $y = 3 - t$, $z = 5 + 6t$.

22. Find an equation for the sphere with center on the z-axis and containing the points $(5, 0, 0)$ and $(0, 0, 4)$.

23. For $v = i + 2j + 4k$ and $w = 3j + 4k$, find

  a. $\text{comp}_w\, v$ **b.** $\text{proj}_w\, v$

24. Find the distance from the point $(-4, 2, 5)$ to the line with equation $x = 3 + t$, $y = 4 - 2t$, $z = 5 + 5t$.

25. Find the area of the triangle with vertices $(1, 3, -2)$, $(1, 1, 1)$, and $(4, 0, 3)$.

26. Find a unit vector in the same direction as $v = i - 3j + 4k$.

27. Find an equation for the plane containing the points $(3, -2, 6)$ and $(4, -2, 2)$ that is perpendicular to the xz-plane.

28. Find cylindrical coordinates for the point with rectangular coordinates $(2, 2, 5)$.

29. Find symmetric equations for the line with parametric equations $x = 1 + 3t$, $y = 2 + 7t$, $z = 3 - t$.

In Exercises 30–33, find a rectangular equation for the given equation and sketch the graph.

30. $r = \cos \theta$ **31.** $\phi = \pi/4$

32. $\rho = 2 \sec \phi$ **33.** $\rho \sin \phi = 2 \cos \theta$

34. Find the vector of length 3 with direction cosines $\dfrac{\sqrt{2}}{2}$, 0, and $\dfrac{\sqrt{2}}{2}$.

35. For $v = 2i + 3j - k$ and $w = i + 3j + k$ find

  a. $\text{comp}_w\, v$ **b.** $\text{proj}_v\, w$

36. Find the distance from the point $(1, 2, 1)$ to the plane containing the points $(1, 1, 1)$, $(2, -1, 5)$, and $(3, 1, -2)$.

37. Find the equation for the plane with intercepts $x = 3$, $y = 4$, $z = 5$.

38. Find an equation for the plane perpendicular to the line with symmetric equations

$$\frac{x - 2}{3} = \frac{y + 1}{2} = \frac{z - 3}{-2}$$

and containing the point $(3, -2, 3)$.

39. Find parametric equations for the line parallel to the line of intersection of the two planes $x + y + 2z = 6$ and $x - y - 2z = 4$ and containing the point $(5, 2, -3)$.

40. Find all vectors of length 2 orthogonal to both $v = 3i + 2j + k$ and $w = i + j - 2k$.

41. Find the area of the parallelogram determined by the vectors $v = 2i + 3j + k$ and $w = i - j + 2k$.

42. Prove that the planes $a_1 x + b_1 y + c_1 z = d_1$ and $a_2 x + b_2 y + c_2 z = d_2$ are perpendicular if and only if $a_1 a_2 + b_1 b_2 + c_1 c_2 = 0$.

43. Show that the points $(1, 1, -1)$, $(0, 1, -1/2)$, $(-1, 1, 0)$, and $(0, 0, 1/4)$ all lie in the same plane.

In Exercises 44–50, describe the graph.

44. $9x^2 - 4y^2 + 36z^2 = -36$

45. $6x^2 + y^2 - 2z^2 = 6$

46. $36(x - 1)^2 + 18(y + 3)^2 + 8(z + 1)^2 = 72$

47. $-x^2 + y^2 = 9z$

48. $y^2 = 4x^2 + 2z$

49. $9x^2 + 9y^2 - 4z^2 + 18x - 16z - 43 = 0$

50. $(x + 1)(y - 3) = 1$

UNIT 7

CALCULUS IN HIGHER DIMENSIONS AND DIFFERENTIAL EQUATIONS

Josiah Willard Gibbs

George Gabriel Stokes

A host of mathematicians contributed to the several topics included in this unit: partial differentiation, multiple integration, and vector analysis. Several of these people have been discussed in other units.

Newton differentiated functions of two variables by means of formulas that we now obtain by partial differentiation. This work is recorded in his personal papers, but was not published. Leibniz also differentiated functions of two variables, but made little use of them. Jakob Bernoulli and his nephew Nicolaus used what were essentially partial derivatives around 1720, although they seemingly did not recognize that there is a difference between differentiation of a function of one variable and that for a function of two or more variables. The same symbol was used by many early writers for regular and partial derivatives, which of course led to much confusion. The "rounded d" symbol, ∂, was first used by Euler in 1776, but not in the way that we use it today. The first to use $\partial y/\partial x$ was the Frenchman Adrien-Marie Legendre in 1786, but it was more than a century before that notation came into general use.

Multiple integration was first used by Newton, but his arguments were geometrical and somewhat unclear. In the first half of the eighteenth century Euler used repeated integrations in order to integrate over a bounded domain. Joseph Louis Lagrange used a triple integral in a work on gravitation involving ellipsoids around 1775. By the nineteenth century the use of multiple integrals had become fairly common.

The concept of a vector was known in antiquity. Aristotle represented forces by vectors, and knew that two forces acting in different directions could be summed by what we call the parallelogram law. Representation of complex numbers in the plane was done by Wessel in Denmark, Argand in France, and Gauss in Germany between 1798 and 1806. It was fairly well known by 1830 that vectors could be nicely expressed as complex numbers. However, many physical situations involve more than two factors which are not all in the same plane. For many years mathematicians searched for a three-dimensional vector system which preserved all of the properties of real-number algebra, but their efforts were unsuccessful.

The problem was resolved by Ireland's greatest mathematician, William Rowan Hamilton (1805–1865). Hamilton was a largely self-taught youthful prodigy whose ability first manifested itself, as with several other mathematicians, in language. He read Greek, Latin, and Hebrew by the age of 5, and by the age of 14 he knew 14 languages including Sanskrit and Arabic. He became interested in mathematics through meeting an American calculating prodigy, and had soon read all four volumes of Newton's *Principia* and Laplace's *Celestial Mechanics,* in which he found a mathematical error. At the age of 22, still an undergraduate, he was appointed Professor of Astronomy at an Irish university and Royal Astronomer of Ireland.

While walking to Dublin, Hamilton suddenly realized the impossibility of a three-dimensional system with the desired properties, but that a *four*-dimensional system is possible. He stopped to carve the basic relations on the handrail of a bridge, and that carving can still be seen. He postulated a system with four mutually independent unit vectors, called *quaternions*. Quaternions do not obey the commutative law (that is, $A \cdot B \neq B \cdot A$); this was the first algebra in which such behavior was studied. The use of quaternions was advocated throughout the nineteenth century by Hamilton and others, and courses in the subject were taught in

British and American graduate schools well into the twentieth century. Eventually, however, it was superseded by the vector analysis pioneered by J. W. Gibbs. Hamilton did not lead a happy life; his marriage was unpleasant and his quaternions were not well-received on the Continent. He was knighted early, in 1835, but became an alcoholic in the last twenty years of his life.

Josiah Willard Gibbs (1839–1903) was an American mathematician and scientist in a day when American science was held in little esteem. Born in New Haven, Connecticut, he attended Yale College and obtained a Ph.D. in physics in 1863, one of the first doctorates to be granted in the United States. He then went to Europe for further study. Returning in 1871, he became Professor of Mathematical Physics at Yale, though he received no salary for the first nine years. He remained at Yale for the rest of his life, living in the house in which he had grown up, less than a block from the campus. In 1881 he had printed, at his own expense, a little pamphlet called *Elements of Vector Analysis* for the use of his students. In it, he created the subject much as it is known today. The pamphlet became well-known, and E. B. Wilson published a book called *Vector Analysis,* based on Gibbs' lectures, in 1901.

Oliver Heaviside (1850–1925) was another contributor to the development of vector analysis, and in fact he essentially created the subject independently of Gibbs. He had only an elementary school education in his native England, but embarked on a regimen of reading which resulted in his becoming a pragmatically successful mathematical physicist. He began his career as a telegrapher, but abandoned that work as a result of the deafness which bothered him all of his life. His creative work was disdained by many professional mathematicians particularly because of his lack of formal education and unorthodox methods. His work in electromagnetic theory was particularly brilliant when he generalized Maxwell's theory. Heaviside used ideas from Hamilton's theory of quaternions, and developed both scalar and vector multiplication. He never married, but lived with his parents until they died. He spent his last years in poverty and died in a nursing home.

Green's Theorem and Stokes' Theorem are two applications of vectors which are found in this unit. George Green (1793–1841), like Heaviside, had only an elementary school English education, and was largely self-taught. He was one of the first to treat static electricity and magnetism in mathematical terms. At the age of 36 Green decided to get a university education. He studied for four years, was admitted to Cambridge at the age of 40, and graduated four years later. His marks were not high, largely because he spent more time writing original mathematics than in studying for classes. In 1828 he had printed at his own expense his most important work, in which his mathematical treatment of electricity and magnetism appears. Here too is found what we now call Green's Theorem. It remained unknown until four years after his death when the British physicist Lord Kelvin became aware of the paper and had it reprinted in a German journal. By this time the Russian mathematician Michel Ostrogradski had also discovered the theorem; it is known by his name in the Soviet Union.

George Gabriel Stokes (1819–1903) was an Irishman who was Lucasian Professor of Mathematics at Cambridge, the same post held by Newton a century and a half earlier. Though prestigious, the position did not pay well, and Stokes had to teach at a School of Mines in order to make ends meet. Also like Newton, he served the Royal Society of London as an officer for many years—he was either president or secretary for 45 years. He studied various physical phenomena from a mathematical point of view, and through a brilliant paper established the foundations of hydrodynamics. The theorem named after him is a three-dimensional vector version of Green's Theorem.

Chapter 16
Vector-Valued Functions

In this chapter we shall discuss *vector-valued functions* of a real variable. These are functions defined on subsets of the real numbers, $(-\infty, \infty)$, but with vectors (either in the plane or in space) as values. In keeping with the vector notation of Chapter 15, we shall write the name f of a vector-valued function in boldface to remind you that its values $f(t)$ are vectors.

16.1 PROPERTIES OF VECTOR-VALUED FUNCTIONS

A **vector-valued function** is a function

$$f(t) = x(t)i + y(t)j + z(t)k \tag{1}$$

whose domain is a set of real numbers and whose values $f(t)$ are vectors.

In equation (1) we refer to the functions x, y, and z as the "component functions," to the values $x(t)$, $y(t)$, and $z(t)$ as the "scalar components," and to the vectors $x(t)i$, $y(t)j$, and $z(t)k$ as the "vector components." When we use the term "components" alone, it will always be clear from the context which type we mean.

By the **graph** of a vector function f, we shall mean the set of all points in space that are the terminal points of the arrows representing the vectors $f(t)$ *when $f(t)$ originates at the origin.* That is, we shall associate the vector $f(t) = x(t)i + y(t)j + z(t)k$ with the point $(x(t), y(t), z(t))$ in space.

We have already seen two examples of functions of this type:

(i) The **line** with vector equation

$$\begin{aligned} r(t) &= (a_1 i + a_2 j + a_3 k) + t(b_1 i + b_2 j + b_3 k) \\ &= (a_1 + tb_1)i + (a_2 + tb_2)j + (a_3 + tb_3)k \end{aligned}$$

has the form of equation (1) with $x(t) = (a_1 + tb_1)$, $y(t) = (a_2 + tb_2)$, and $z(t) = (a_3 + tb_3)$. The vector $r(t)$ is interpreted as extending from the origin to the point $P(t)$ whose coordinates are the components of the vector $r(t)$. If t denotes time, the vector function represents the motion of a particle moving in space.

(ii) The **curve** C in the plane determined by the parametric equations

$$C: \quad \begin{cases} x = x(t) \\ y = y(t) \end{cases} \qquad a \leq t \leq b$$

may be interpreted as the graph of the vector function

$$f(t) = x(t)i + y(t)j, \qquad a \leq t \leq b$$

where the vector $f(t)$ extends from the origin to the point $(x(t), y(t))$ on C.

The concept of vector function allows us to pursue generalizations of both of these examples in a unified way. By interpreting $f(t)$ as a vector from the origin to a particle in space, we may study the motion of that particle, addressing the usual issues of position, speed, velocity, acceleration, distance, and elapsed time.

In studying vector-valued functions we shall make heavy use of the fact that the vector function $f(t)$ is the sum of three vector components, each of which involves a function of a single independent variable (Figure 1.1). The laws of vector algebra

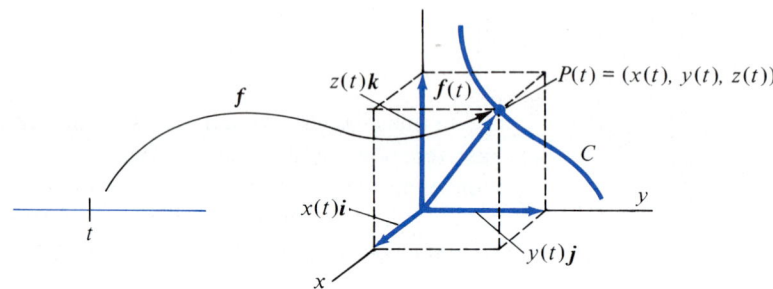

Figure 1.1 Graph of the vector-valued function $f(t) = x(t)\mathbf{i} + y(t)\mathbf{j} + z(t)\mathbf{k}$.

allow us to perform algebraic operations with the components, while the theory of the calculus (Chapters 2–10) will enable us to address many of the issues associated with derivatives and integrals of vector functions.

Example 1

Sketch the graph of the vector function

$$f(t) = a \cos t\mathbf{i} + a \sin t\mathbf{j} + bt\mathbf{k}.$$

where a and b are constants.

Solution: We note that in any plane parallel to the xy-plane the distance from the point $(a \cos t, a \sin t, bt)$ to the point $(0, 0, bt)$ on the z-axis is

$$d = \sqrt{a^2 \cos^2 t + a^2 \sin^2 t} = |a|.$$

Thus, all points lie on the graph of the circular cylinder $x^2 + y^2 = a^2$. As t increases, the z-coordinate of the point determined by $f(t)$ increases uniformly. For every point $(x(t), y(t), z(t))$ on the graph, infinitely many points $(x(t), y(t), z(t) + 2\pi nb)$, $n = \pm1, \pm2, \ldots$ also lie on the graph. The graph is the **circular helix** in Figure 1.2. ◇

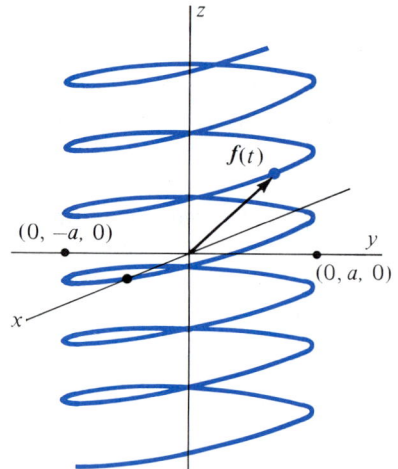

Figure 1.2 The circular helix $f(t) = a \cos t\mathbf{i} + a \sin t\mathbf{j} + bt\mathbf{k}$.

Algebra, Limits, and Continuity

Given two vector-valued functions

$$f(t) = x_1(t)\mathbf{i} + y_1(t)\mathbf{j} + z_1(t)\mathbf{k}$$

and

$$g(t) = x_2(t)\mathbf{i} + y_2(t)\mathbf{j} + z_2(t)\mathbf{k},$$

we may form the following functions, as indicated.

(i) The **sum** of f and g is the vector-valued function $f + g$ where

$$
\begin{aligned}
(f + g)(t) &= f(t) + g(t) \\
&= [x_1(t) + x_2(t)]\mathbf{i} + [y_1(t) + y_2(t)]\mathbf{j} + [z_1(t) + z_2(t)]\mathbf{k}.
\end{aligned}
$$

In other words, we add vector functions component by component.

(ii) The **scalar multiple** of f by the real number c is the vector-valued function cf where

$$(cf)(t) = cf(t) = [cx_1(t)]i + [cy_1(t)]j + [cz_1(t)]k.$$

That is, multiplication of vector functions by scalars is done component by component.

(iii) The **dot product** of f and g is the **real-valued function $f \cdot g$** where

$$(f \cdot g)(t) = f(t) \cdot g(t) = x_1(t)x_2(t) + y_1(t)y_2(t) + z_1(t)z_2(t).$$

The values of the dot product function are *numbers* rather than vectors.

(iv) The **cross product** of f and g is the vector-valued function $f \times g$ where

$$(f \times g)(t) = f(t) \times g(t)$$
$$= [y_1(t)z_2(t) - z_1(t)y_2(t)]i + [z_1(t)x_2(t) - x_1(t)z_2(t)]j$$
$$+ [x_1(t)y_2(t) - y_1(t)x_2(t)]k.$$

The values of the cross product function are *vectors*.

Example 2

From the vector-valued functions

$$f(t) = ti + t^2j + \sqrt{t}k, \qquad t \geq 0$$

and

$$g(t) = \cos ti + \sin tj + tk,$$

the following functions may be formed:

(a) $(f + g)(t) = (t + \cos t)i + (t^2 + \sin t)j + (\sqrt{t} + t)k, \qquad t \geq 0.$
(b) $3f(t) = 3ti + 3t^2j + 3\sqrt{t}\,k, \qquad t \geq 0.$
(c) $(f - 2g)(t) = (t - 2\cos t)i + (t^2 - 2\sin t)j + (\sqrt{t} - 2t)k, \qquad t \geq 0.$

(d) $(f \times g)(t) = \det \begin{bmatrix} i & j & k \\ t & t^2 & \sqrt{t} \\ \cos t & \sin t & t \end{bmatrix}$

$$= (t^3 - \sqrt{t}\sin t)i + (\sqrt{t}\cos t - t^2)j + (t\sin t - t^2\cos t)k. \qquad \diamondsuit$$

Limits of Vector-Valued Functions

The limit of a vector-valued function is defined just as the limit of a real-valued function: $\lim_{t \to a} f(t) = L$ means that $|f(t) - L|$ is small whenever $|t - a|$ is small. (See Figure 1.3.) Note, however, that since $f(t)$ and L are vectors, the expression $|f(t) - L|$ is the *length* of the vector $f(t) - L$.

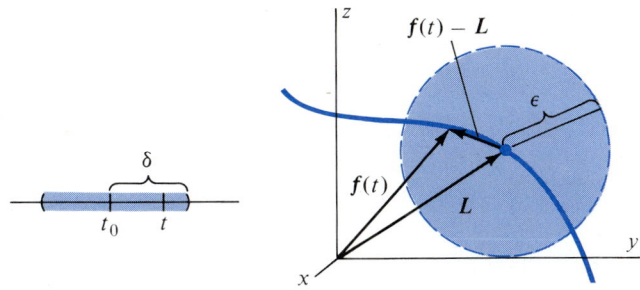

Figure 1.3 If $0 < |t - t_0| < \delta$, then $|f(t) - L| < \epsilon$.

DEFINITION 1

Let f be a vector-valued function that is defined on an open interval containing the number a, except possibly at $t = a$. We say that the vector L is the *limit* of the function f as t approaches a, written $L = \lim\limits_{t \to a} f(t)$, if, corresponding to each number $\epsilon > 0$ there is a number $\delta > 0$ so that

$$\text{if } \quad 0 < |t - a| < \delta, \qquad \text{then} \quad |f(t) - L| < \epsilon.$$

The following theorem shows that Definition 1 is equivalent to the existence of limits for each of the component functions of f. This allows us to evaluate limits of vector-valued functions by using the techniques of Chapter 2 for limits of real-valued functions. The proof is given at the end of this section.

THEOREM 1

Let f be the vector-valued function

$$f(t) = x(t)i + y(t)j + z(t)k$$

and let

$$L = L_1 i + L_2 j + L_3 k$$

be a vector. Let f be defined on an open interval containing the number a except possibly at $t = a$. Then

$$L = \lim_{t \to a} f(t)$$

if and only if

$$L_1 = \lim_{t \to a} x(t), \qquad L_2 = \lim_{t \to a} y(t), \qquad \text{and} \qquad L_3 = \lim_{t \to a} z(t).$$

Theorem 1 states that the limit of a vector-valued function is the vector whose components are the limits of the individual components. If one or more of these component limits fails to exist we say that the limit of the vector-valued function fails to exist.

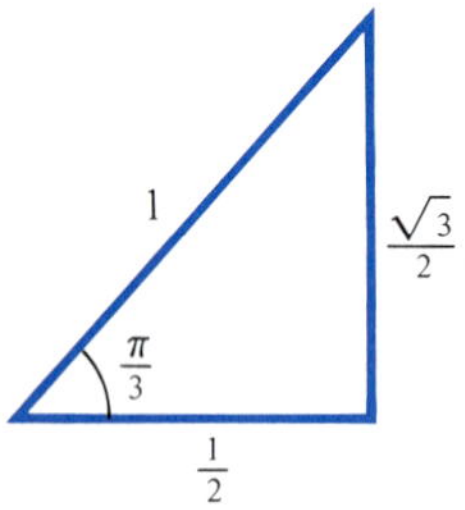

Figure 1.4

Example 3

Find $\lim\limits_{t \to \pi/3} (t^2 i + \sin tj + \cos tk)$.

Solution: The limits of the components are

$$L_1 = \lim_{t \to \pi/3} x(t) = \lim_{t \to \pi/3} t^2 = \frac{\pi^2}{9},$$

$$L_2 = \lim_{t \to \pi/3} y(t) = \lim_{t \to \pi/3} \sin t = \frac{\sqrt{3}}{2} \qquad \text{(Figure 1.4)},$$

$$L_3 = \lim_{t \to \pi/3} z(t) = \lim_{t \to \pi/3} \cos t = \frac{1}{2}.$$

Thus, by Theorem 1,

$$\lim_{t \to \pi/3} (t^2 i + \sin tj + \cos tk) = \frac{\pi^2}{9} i + \frac{\sqrt{3}}{2} j + \frac{1}{2} k.$$

Continuity

Continuity for vector-valued functions is defined in terms of limits, just as for real-valued functions. Intuitively, the statement that f is continuous at $t = a$ means that $\lim_{t \to a} f(t) = f(a)$. The formal definition is the following.

DEFINITION 2

Let f be a vector-valued function that is defined on an open interval I containing the number $t = a$. We say that f is *continuous* at a if, corresponding to each number $\epsilon > 0$ there is a number $\delta > 0$ so that

$$\text{if} \quad |t - a| < \delta, \qquad \text{then} \quad |f(t) - f(a)| < \epsilon.$$

We say that f is continuous on I if f is continuous at each number $a \in I$.

Like the definition of limit, the definition of continuity for a vector-valued function is equivalent to the requirement that each of the component functions be continuous. We state this result as a corollary of Theorem 1 and we leave its proof, which is similar to the proof of Theorem 1, as an exercise.

COROLLARY 1

The vector-valued function $f(t) = x(t)i + y(t)j + z(t)k$ is **continuous** at $t = t_0$ if each of the component functions $x(t)$, $y(t)$, and $z(t)$ is continuous at t_0.

Note that each component of f must be continuous at t_0 if f is continuous at t_0. Thus, a vector-valued function is **discontinuous** at t_0 if one or more of its component functions is discontinuous at t_0.

Example 4

Find the intervals on which the vector-valued function

$$f(t) = \frac{1}{t}i + t^2 j + \frac{2}{t^2 - 4}k$$

is continuous.

Strategy

Determine the numbers t where one or more of the component functions is discontinuous.

Solution

The i-component $x(t) = 1/t$ is discontinuous at $t = 0$.

The j-component $y(t) = t^2$ is continuous for all t.

The k-component $z(t) = \dfrac{2}{t^2 - 4} = \dfrac{2}{(t - 2)(t + 2)}$ is discontinuous for $t = -2, 2$.

The vector-valued function is continuous for all other values of t.

The vector-valued function f is therefore discontinuous at $t = -2, 0$, and 2, so f is continuous on the intervals

$$(-\infty, -2), \ (-2, 0), \ (0, 2), \ \text{and} \ (2, \infty). \qquad \diamond$$

Proof of Theorem 1: First, let us assume that

$$L_1 = \lim_{t \to t_0} x(t), \tag{2}$$

$$L_2 = \lim_{t \to t_0} y(t), \qquad \text{and} \tag{3}$$

$$L_3 = \lim_{t \to t_0} z(t). \tag{4}$$

Let $\epsilon > 0$ be given. According to the definition of the limit of a function of a single variable and equation (2), there exists a number δ_1 so that

$$\text{if} \quad 0 < |t - t_0| < \delta_1 \qquad \text{then} \qquad |x(t) - L_1| < \epsilon/3. \tag{5}$$

(You will see in a few lines why we use $\epsilon/3$ rather than just ϵ.) Similarly, it follows from equations (3) and (4) that there exist positive numbers δ_2 and δ_3 so that

$$\text{if} \quad 0 < |t - t_0| < \delta_2 \qquad \text{then} \qquad |y(t) - L_2| < \epsilon/3 \tag{6}$$

and

$$\text{if} \quad 0 < |t - t_0| < \delta_3 \qquad \text{then} \qquad |z(t) - L_3| < \epsilon/3. \tag{7}$$

If we take δ to be the smallest of the numbers δ_1, δ_2, and δ_3, then each of the inequalities on the left-hand sides of (5) through (7) is fulfilled when $0 < |t - t_0| < \delta$. In this case, using inequalities (5) through (7) and the triangle inequality, we find that

$$\begin{aligned}
|\boldsymbol{f}(t) - \boldsymbol{L}| &= |(x(t)\boldsymbol{i} + y(t)\boldsymbol{j} + z(t)\boldsymbol{k}) - (L_1\boldsymbol{i} + L_2\boldsymbol{j} + L_3\boldsymbol{k})| \\
&= |(x(t) - L_1)\boldsymbol{i} + (y(t) - L_2)\boldsymbol{j} + (z(t) - L_3)\boldsymbol{k}| \\
&\le |x(t) - L_1| + |y(t) - L_2| + |z(t) - L_3| \\
&< \frac{\epsilon}{3} + \frac{\epsilon}{3} + \frac{\epsilon}{3} \\
&= \epsilon.
\end{aligned}$$

That is,

$$\text{if} \quad 0 < |t - t_0| < \delta, \qquad \text{then} \quad |\boldsymbol{f}(t) - \boldsymbol{L}| < \epsilon. \tag{8}$$

To prove the converse, we assume that whenever $\epsilon > 0$ is given we can find a number $\delta > 0$ so that statement (8) holds, and we show that this guarantees that statements (2)–(4) hold. To do so we observe that if t is any number for which $0 < |t - t_0| < \delta$ (with δ as in equation (8)) we have

$$\begin{aligned}
|x(t) - L_1| &= \sqrt{|x(t) - L_1|^2} \tag{8} \\
&\le \sqrt{|x(t) - L_1|^2 + |y(t) - L_2|^2 + |z(t) - L_3|^2} \\
&= |\boldsymbol{f}(t) - \boldsymbol{L}| \\
&< \epsilon.
\end{aligned}$$

That is, if $0 < |t - t_0| < \delta$, then $|x(t) - L_1| < \epsilon$. This shows that $\lim_{t \to t_0} x(t) = L_1$.

Similar comparisons show that $\lim_{t \to t_0} y(t) = L_2$ and that $\lim_{t \to t_0} z(t) = L_3$. This shows that if $\boldsymbol{L} = \lim_{t \to t_0} \boldsymbol{f}(t)$, then statements (2)–(4) hold. This completes the proof. ◆

Exercise Set 16.1

In Exercises 1–5, sketch the graph of the given vector function.

1. $\boldsymbol{f}(t) = \boldsymbol{i} + \boldsymbol{j} + t\boldsymbol{k}$

2. $\boldsymbol{f}(t) = \cos t\boldsymbol{i} + \sin t\boldsymbol{j} + t\boldsymbol{k}$

3. $\boldsymbol{f}(t) = t\boldsymbol{i} + \cos t\boldsymbol{j} + \sin t\boldsymbol{k}$

4. $\boldsymbol{f}(t) = t\boldsymbol{i} + t^2\boldsymbol{j}$

5. $\boldsymbol{f}(t) = \sin t\boldsymbol{i} + \boldsymbol{j} + \boldsymbol{k}$

The **implicit domain** of a vector-valued function f is the largest set of numbers t for which each of the component functions is defined. State the implicit domain for each of the following vector-valued functions.

6. $f(t) = ti + \sqrt{t}\,j + \sin tk$

7. $f(t) = t^2i - tj + \tan tk$

8. $f(t) = \dfrac{1}{\sqrt{1-t}}i + 2tj + \sin^2 tk$

9. $f(t) = \dfrac{1}{9-t^2}i + \dfrac{2}{1+t}j + \sqrt{t}\,k$

10. $f(t) = (t^2 - 1)i + \sec tj + t^3k$

11. $f(t) = t^2i + t^{3/2}j + t^{2/3}k$

12. $f(t) = \ln ti + \cos tj + \sqrt{9 - t^2}k$

In Exercises 13–18, let $f(t) = ti + 3tj + t^2k$ and $g(t) = \sin ti + \cos tj + k$. Find the indicated functions.

13. $f + g$

14. $f - g$

15. $3f + 2g$

16. $4f - 3g$

17. $f \cdot g$

18. $f \times g$

In Exercises 19–24, let $f(t) = (1 + t)i + (1 - t^2)j + tk$, $g(t) = \sqrt{t}\,i + (1 - t)j + e^tk$, and $h(t) = \sin t$. Find the indicated functions.

19. $f + 3g$

20. hf

21. $h(f - g)$

22. $2f - 3hg$

23. $f \cdot g$

24. $3f \times 2g$

In Exercises 25–30, find the indicated limit.

25. $\lim\limits_{t\to\pi/4} (\sin ti + \tan tj + \cos tk)$

26. $\lim\limits_{t\to2}(t^2i + (3t - 2)j + 6tk)$

27. $\lim\limits_{t\to0}\left\{\left(\dfrac{\sin t}{t}\right)i + \left(\dfrac{1 - \cos t}{t}\right)j + e^{2t}k\right\}$

28. $\lim\limits_{t\to0}\left\{\dfrac{\sin 3t}{t}i + \dfrac{\tan 2t}{3t}j + \ln(1 + t)k\right\}$

29. $\lim\limits_{t\to3}\left\{\left(\dfrac{9 - t^2}{3 - t}\right)i + \left(\dfrac{t^2 + t - 12}{t - 3}\right)j + \left(\dfrac{t^3 - 13t + 12}{t - 3}\right)k\right\}$

30. $\lim\limits_{t\to2}\left\{(t + 3)i + \left(\dfrac{t^2 - 4}{t - 2}\right)j + \left(\dfrac{t^3 + t^2 - 4t - 4}{t - 2}\right)k\right\}$

In Exercises 31–37, determine the intervals on which the given vector function is continuous.

31. $f(t) = 2ti + \cos tj + \tan tk$

32. $f(t) = \sqrt{9 - t^2}i + \dfrac{1}{t}j + \sin tk$

33. $f(t) = \dfrac{1}{1 + \sqrt{t}}i + \dfrac{1}{1 - \sqrt{t}}j + t^{3/2}k$

34. $f(t) = \left(\dfrac{1}{t^2 - t + 12}\right)i + \ln(2t)j + e^{-t}k$

35. $f(t) = \ln(1 + t^2)i + \ln(1 - t^2)j + \sqrt{t}\,k$

36. $f(t) = \begin{cases} ti + (3t - 2)j + (3 - t)k, & -\infty < t < 2 \\ 2i + (2 + t)j + tk, & 2 \le t < \infty \end{cases}$

37. $f(t) = \begin{cases} t^2i + (3 - t)j + 4tk, & -\infty < t < 1 \\ ti + (4 + t)j + (1 - t^2)k, & 1 \le t < \infty \end{cases}$

38. An automobile assembly robot turns a machine screw at a constant rate of 10π radians per second. The **pitch** of the screw (the number of threads per millimeter of length) is such that the screw advances 0.5 mm for each complete revolution.
  a. Write a vector function for the motion of a paint spot on one thread of the screw.
  b. At what rate, in mm/s, must the robot arm advance?

39. The **polarization** of a light wave is determined by the motion of the tip of the associated "electric vector" $E(t)$. If the motion follows a circular helix, the light is said to be circularly polarized. If a particular light wave is circularly polarized according to $E(t) = \cos(10^3t)i + \sin(10^3t)j + (3 \times 10^8)tk$, how far does the wave advance along the z-axis during one complete revolution of the electric vector?

40. Describe the graph of the vector function $f(t) = 3\cos ti + 4\sin tj + tk$.

41. Prove that if $\lim\limits_{t\to t_0} f(t) = L$ and $\lim\limits_{t\to t_0} g(t) = M$, then $\lim\limits_{t\to t_0} (f + g)(t) = L + M$ as follows.
  a. Write $f(t)$ and $g(t)$ in component form.
  b. Obtain $(f + g)(t)$ in component form.

42. Prove that if f and g are continuous at $t = t_0$ then so is
  a. $f + g$,
  b. cf,
  c. $f \times g$.

43. Prove that if f and g are continuous vector-valued functions then so are the real-valued functions
  a. $f \cdot g$,
  b. $|f|$.

44. For each set of vector-valued functions, find the intervals on which f, g, h, and d are continuous.
  a. $f(t) = \sqrt{t}\,i - \dfrac{1}{t}j + tk$, $g(t) = -\sqrt{t}\,i + \dfrac{1}{t}j + 3tk$, $h(t) = (f + g)(t)$
  b. $f(t) = \dfrac{1}{t}i + \ln(1 + t)j + \sqrt{t}\,k$, $g(t) = ti + \sqrt{t}\,k$, $d(t) = (f \cdot g)(t)$

45. If $f + g$ is continuous, what can you say about the continuity of f and g? What conclusions can you draw from the continuity of the dot product $f \cdot g$?

46. Prove Corollary 1.

16.2 DERIVATIVES AND INTEGRALS OF VECTOR-VALUED FUNCTIONS

Let f be a vector-valued function and let t be a number in the domain of f. Also, assume that the numbers $t + h$ lie in the domain of f for h sufficiently small. The vector

$$\left(\frac{1}{h}\right)[f(t + h) - f(t)], \qquad h \neq 0 \tag{1}$$

is referred to as a **difference quotient** for the vector function $f(t)$. The vector in equation (1) may be interpreted as the average change in the vector $f(t)$ per unit change in t over an interval of length h. As for functions of a single variable, we are interested in knowing about the limiting value of this vector as $h \to 0$, which we refer to as the *derivative* of the vector-valued function f. The next three sections in part concern the geometric and physical interpretations of this derivative. Our interest here focuses on defining and calculating this particular limit.

DEFINITION 3

The vector-valued function f is said to be **differentiable** at the number t if the limit

$$f'(t) = \lim_{h \to 0} \frac{1}{h}[f(t + h) - f(t)] \tag{2}$$

exists. The vector $f'(t)$ is called the *derivative* of the vector-valued function f at t.

As for functions of a single variable, the vector-valued function f is said to be differentiable on an open interval I if $f'(t)$ in (2) exists for every $t \in I$. Thus, the derivative of a vector-valued function is again a vector-valued function, and the domain of f' is the subset of the domain of f for which the limit in line (2) exists.

The following theorem shows that the derivative f' may be calculated directly from the components of f.

THEOREM 2

Let $f(t) = x(t)i + y(t)j + z(t)k$. The vector-valued function f is differentiable if and only if each of the component functions x, y, and z is differentiable. In this case

$$f'(t) = x'(t)i + y'(t)j + z'(t)k. \tag{3}$$

Theorem 2 is proved by examining the difference quotient in Definition 3. According to Theorem 1,

$$f'(t) = \lim_{h \to 0} \frac{1}{h}[f(t + h) - f(t)]$$

$$= \lim_{h \to 0} \frac{1}{h}\{[x(t + h)i + y(t + h)j + z(t + h)k] - [x(t)i + y(t)j + z(t)k]\}$$

$$= \lim_{h \to 0} \left[\left(\frac{x(t + h) - x(t)}{h}\right)i + \left(\frac{y(t + h) - y(t)}{h}\right)j + \left(\frac{z(t + h) - z(t)}{h}\right)k\right]$$

$$= \left[\lim_{h \to 0}\left(\frac{x(t + h) - x(t)}{h}\right)\right]i + \left[\lim_{h \to 0}\left(\frac{y(t + h) - y(t)}{h}\right)\right]j$$

$$+ \left[\lim_{h \to 0}\left(\frac{z(t + h) - z(t)}{h}\right)\right]k.$$

This shows that f is differentiable if and only if each of the component functions is differentiable. Applying each of the indicated limits and using Definition 3 establishes equation (3).

Example 1

Let $f(t) = t^2 i + \cos t j + e^{3t} k$. Find $f'(t)$.

Solution: $f'(t) = \left[\dfrac{d}{dt}(t^2)\right] i + \left[\dfrac{d}{dt}(\cos t)\right] j + \left[\dfrac{d}{dt}(e^{3t})\right] k$

$\qquad\qquad = 2t i - \sin t j + 3e^{3t} k.$ ◇

If the derivative f' is itself differentiable, we define its derivative to be the *second* derivative of f. That is, $f'' = (f')'$. The second derivative of f is again a vector-valued function. Just as for real-valued functions, some or all derivatives of order higher than second may exist as well.

Example 2

Let $f(t) = \sin 2t i + \cos 2t j + \sqrt{t} k$. Find both f' and f''.

Solution: For $t > 0$ we have

$$f'(t) = 2 \cos 2t i - 2 \sin 2t j + \frac{1}{2} t^{-1/2} k,$$

and

$$f''(t) = -4 \sin 2t i - 4 \cos 2t j - \frac{1}{4} t^{-3/2} k.$$ ◇

Example 3

Show that the function $f(t) = A \sin \omega t i + B \cos \omega t j$ satisfies the **vector differential equation**

$$f''(t) + \omega^2 f(t) = 0. \tag{4}$$

Strategy
Calculate $f'(t)$ and $f''(t)$.

Solution
Here

$$f'(t) = \omega A \cos \omega t i - \omega B \sin \omega t j$$

and

$$f''(t) = -\omega^2 A \sin \omega t i - \omega^2 B \cos \omega t j.$$

Substitute f and f'' into the left-hand side of (4) and verify that **0** is obtained.

Substituting into equation (4) shows that

$$f''(t) + \omega^2 f(t) = [-\omega^2 A \sin \omega t i - \omega^2 B \cos \omega t j]$$
$$+ \omega^2 [A \sin \omega t i + B \cos \omega t j]$$
$$= 0.$$ ◇

The following theorem shows how derivatives of various other vector-valued functions may be calculated.

THEOREM 3

Let the vector-valued functions f and g and the scalar-valued function $y = h(t)$ be differentiable on appropriate intervals and let c be a number. Then the vector-valued functions $f + g$, cf, hf, $f \times g$, and $f \circ h$, and the scalar-valued function $f \cdot g$ are differentiable, and

(i) $(f + g)'(t) = f'(t) + g'(t)$,

(ii) $(cf)'(t) = cf'(t)$,

(iii) $(hf)'(t) = h(t)f'(t) + h'(t)f(t)$,

(iv) $(f \times g)'(t) = [f(t) \times g'(t)] + [f'(t) \times g(t)]$,

(v) $(f \circ h)'(t) = h'(t)f'(h(t))$ (Chain Rule),

(vi) $(f \cdot g)'(t) = f(t) \cdot g'(t) + f'(t) \cdot g(t)$.

Before commenting on the proof of Theorem 3, we consider two examples of its use.

Example 4

Let $f(t) = t^2 i + e^t k$ and $g(t) = \sin t i + \cos t j + k$. Find the derivative of the function $r = f \times g$.

Solution: The derivatives of the given functions are

$$f'(t) = 2t i + e^t k \qquad \text{and} \qquad g'(t) = \cos t i - \sin t j.$$

According to part (iv) of Theorem 3 we have

$$r'(t) = [f(t) \times g'(t)] + [f'(t) \times g(t)]$$

$$= \det \begin{bmatrix} i & j & k \\ t^2 & 0 & e^t \\ \cos t & -\sin t & 0 \end{bmatrix} + \det \begin{bmatrix} i & j & k \\ 2t & 0 & e^t \\ \sin t & \cos t & 1 \end{bmatrix}$$

$$= [e^t \sin t i + e^t \cos t j - t^2 \sin t k]$$
$$\quad + [-e^t \cos t i + (e^t \sin t - 2t)j + 2t \cos t k]$$

$$= e^t(\sin t - \cos t)i + [e^t(\sin t + \cos t) - 2t]j$$
$$\quad + (2t \cos t - t^2 \sin t)k.$$

This same result may be obtained directly by noting that

$$r(t) = f(t) \times g(t)$$

$$= \det \begin{bmatrix} i & j & k \\ t^2 & 0 & e^t \\ \sin t & \cos t & 1 \end{bmatrix}$$

$$= -e^t \cos t i + (e^t \sin t - t^2)j + t^2 \cos t k$$

and differentiating component by component, according to Theorem 2.   ◇

Example 5

Let f and g be as in Example 4. Find the derivative of the scalar-valued function $s = f \cdot g$.

Solution: By Theorem 3, part (vi),

$$s'(t) = f(t) \cdot g'(t) + f'(t) \cdot g(t)$$
$$= [(t^2 i + e^t k) \cdot (\cos ti - \sin tj)] + [(2ti + e^t k) \cdot (\sin ti + \cos tj + k)]$$
$$= t^2 \cos t + 2t \sin t + e^t.$$

To calculate this derivative directly, we first compute

$$s(t) = f(t) \cdot g(t) = t^2 \sin t + (0)(\cos t) + e^t$$
$$= t^2 \sin t + e^t.$$

Differentiating then gives the above result. $\diamondsuit$

Examples 4 and 5 suggest that Theorem 3 may be proved by writing the vector-valued functions f and g in component form and applying Theorem 2. This is indeed the case. We prove part (v) here and leave the other statements as exercises. To do so we write f in component form as

$$f(t) = x(t)i + y(t)j + z(t)k.$$

Then

$$f(h(t)) = x(h(t))i + y(h(t))j + z(h(t))k.$$

According to Theorem 2 and the Chain Rule of Chapter 3,

$$(f \circ h)'(t) = [(x \circ h)'(t)]i + [(y \circ h)'(t)]j + [(z \circ h)'(t)]k$$
$$= [x'(h(t)) \cdot h'(t)]i + [y'(h(t)) \cdot h'(t)]j + [z'(h(t)) \cdot h'(t)]k$$
$$= h'(t)[x'(h(t))i + y'(h(t))j + z'(h(t))k]$$
$$= h'(t) \cdot f'(h(t)).$$

The result of the following example will be useful in later calculations.

Example 6

Let f be a vector-valued function and let c be a constant. Prove that if f is differentiable on some interval I and if $|f(t)| = c$ for all $t \in I$, then $f(t)$ and $f'(t)$ are orthogonal for all $t \in I$.

Strategy

Express $|f(t)|^2$ as a dot product using $|v|^2 = v \cdot v.$

Solution

By Theorem 3, Chapter 15,

$$|f(t)|^2 = f(t) \cdot f(t).$$

The resulting expression equals c^2.

Thus, since $|f(t)| = c$, we have

$$f(t) \cdot f(t) = c^2, \qquad t \in I.$$

Differentiate both sides of this equation using part (vi) of Theorem 3.

Differentiating both sides of this equation with respect to t gives

$$f(t) \cdot f'(t) + f'(t) \cdot f(t) = 0$$

or,

$$2f(t) \cdot f'(t) = 0.$$

Use fact that v is orthogonal to w if and only if $v \cdot w = 0$.

Thus $f'(t) \cdot f(t) = 0$, $t \in I$, so $f(t)$ and $f'(t)$ are orthogonal for all $t \in I$. $\diamondsuit$

Integrals of Vector-Valued Functions

Just as for real-valued functions, we say that the vector-valued function F is an *antiderivative* for the vector-valued function f on the interval I if $F'(t) = f(t)$ for all t in I.

If the vector-valued function

$$F(t) = X(t)i + Y(t)j + Z(t)k$$

is an antiderivative for

$$f(t) = x(t)i + y(t)j + z(t)k$$

on the interval I, it follows that any other antiderivative G for f on I must have the form

$$\begin{aligned} G(t) &= [X(t) + c_1]i + [Y(t) + c_2]j + [Z(t) + c_3]k \\ &= X(t)i + Y(t)j + Z(t)k + (c_1 i + c_2 j + c_3 k) \\ &= F(t) + C, \qquad C = (c_1 i + c_2 j + c_3 k). \end{aligned}$$

We refer to the family of *all* antiderivatives for f on I as the *indefinite integral* for f on I, denoted by $\int f(t)\, dt$, the component form of which is given by the following theorem.

THEOREM 4

The **indefinite integral** of the vector-valued function $f(t) = x(t)i + y(t)j + z(t)k$ on the interval I is

$$\int f(t)\, dt = \left[\int x(t)\, dt\right]i + \left[\int y(t)\, dt\right]j + \left[\int z(t)\, dt\right]k \tag{5}$$

provided that the integrals of the component functions exist on I.

REMARK: It is important to note in applying equation (5) that *three* constants of integration will arise—one for each of the antiderivatives of the respective components. That is, if $X'(t) = x(t)$, $Y'(t) = y(t)$, and $Z'(t) = z(t)$, we can write equation (5) as either

$$\int f(t)\, dt = [X(t) + c_1]i + [Y(t) + c_2]j + [Z(t) + c_3]k, \tag{6}$$

or,

$$\int f(t)\, dt = X(t)i + Y(t)j + Z(t)k + C \tag{7}$$

where

$$C = c_1 i + c_2 j + c_3 k.$$

A common mistake is to write the constant C in equation (7) as a number rather than as a vector.

Example 7

Find the indefinite integral $\int f(t)\, dt$ for

$$f(t) = 2ti + e^{-3t}j + \sec^2 tk.$$

Solution: By Theorem (4)

$$\int f(t)\,dt = \left[\int 2t\,dt\right]i + \left[\int e^{-3t}\,dt\right]j + \left[\int \sec^2 t\,dt\right]k$$

$$= (t^2 + c_1)i + \left(-\frac{1}{3}e^{-3t} + c_2\right)j + (\tan t + c_3)k$$

$$= t^2 i - \frac{1}{3}e^{-3t}j + \tan t k + C, \qquad C = c_1 i + c_2 j + c_3 k. \qquad \diamondsuit$$

Example 8

Find the vector-valued function f for which

$$f'(t) = i + t^3 j + \left(\frac{1}{1 + t^2}\right)k,$$

and $f(0) = 2i - j + 3k$.

Solution: By integration we obtain

$$f(t) = \int f'(t)\,dt$$

$$= \int \left[i + t^3 j + \left(\frac{1}{1 + t^2}\right)k\right] dt$$

$$= (t + c_1)i + \left(\frac{1}{4}t^4 + c_2\right)j + (\text{Tan}^{-1} t + c_3)k.$$

Setting $t = 0$ and using the given *initial condition* $f(0) = 2i - j + 3k$ gives the equation

$$f(0) = c_1 i + c_2 j + c_3 k = 2i - j + 3k.$$

Thus, $c_1 = 2$, $c_2 = -1$, and $c_3 = 3$, so

$$f(t) = (t + 2)i + \left(\frac{1}{4}t^4 - 1\right)j + (\text{Tan}^{-1} t + 3)k$$

is the desired solution. $\diamondsuit$

Finally, the definite integral of the vector-valued function f is defined as the vector whose components are the respective definite integrals of the components of f.

DEFINITION 4

Let $f(t) = x(t)i + y(t)j + z(t)k$. If the definite integral of each of the components exists on the interval $[a, b]$, we define

$$\int_a^b f(t)\,dt = \left[\int_a^b x(t)\,dt\right]i + \left[\int_a^b y(t)\,dt\right]j + \left[\int_a^b z(t)\,dt\right]k.$$

Example 9

For $f(t) = \sin 3t\,i + \cos tj + k$, find

$$\int_0^{\pi/2} f(t)\,dt.$$

Solution: By Definition 4,

$$\int_0^{\pi/2} f(t)\,dt = \left[\int_0^{\pi/2} \sin 3t\,dt\right]i + \left[\int_0^{\pi/2} \cos t\,dt\right]j + \left[\int_0^{\pi/2} 1\,dt\right]k$$

$$= \left[-\frac{1}{3}\cos 3t \,\bigg|_0^{\pi/2}\right]i + \left[\sin t \,\bigg|_0^{\pi/2}\right]j + \left[t \,\bigg|_0^{\pi/2}\right]k$$

$$= \left[0 - \left(-\frac{1}{3}\right)\right]i + [1 - 0]j + \left[\frac{\pi}{2} - 0\right]k$$

$$= \frac{1}{3}i + j + \frac{\pi}{2}k. \qquad \diamond$$

We may summarize the results of this section by saying that for the purposes of calculating limits, derivatives, or integrals, the vector-valued function $f(t) = x(t)i + y(t)j + z(t)k$ is regarded as a sum of three component functions of a single variable. In the remaining sections of this chapter, we shall see how these results may be combined with a broader interpretation of $f(t)$ as a vector in space to develop the concepts of tangent, normal, velocity, acceleration, and curvature associated with curves in space.

Exercise Set 16.2

In Exercises 1–4, state the intervals on which the given vector-valued function is differentiable.

1. $f(t) = ti + \cos tj + \sqrt{t}\,k$

2. $f(t) = t^{-2}i + |t - 3|j + t^3 k$

3. $f(t) = \ln(3 + t)i + \left(\dfrac{1}{t+2}\right)j + e^t k$

4. $f(t) = \sqrt{1 - t^2}\,i + \ln tj + \dfrac{1}{1 - t}k$

In Exercises 5–10, find the derivative of the given vector-valued function.

5. $f(t) = ti + \sqrt{t}\,j$

6. $f(t) = i + \sin tj + \cos 2tk$

7. $f(t) = \sqrt{t}\,i + t^{-3/2}j + \ln(2t - 1)k$

8. $f(t) = \cos^2 ti + \sec tj + \text{Tan}^{-1} tk$

9. $f(t) = \text{Sin}^{-1} ti + \sqrt{1 + t^2}\,j + e^{-t^3}k$

10. $f(t) = e^{\sqrt{t}}i + 3j - \text{Cos}^{-1} 2tk$

11. Find f'' for the function f in Exercise 7.

12. Find f'' for the function f in Exercise 10.

13. Show that the function $f(t) = Ae^{\omega t}i + Be^{-\omega t}j$ satisfies the vector differential equation

$$f''(t) - \omega^2 f(t) = 0.$$

14. Show that the function

$$f(t) = C \sinh \omega ti + D \cosh \omega tj$$

satisfies the vector differential equation in Exercise 13.

In Exercises 15–20, let

$$f(t) = \sin ti + j + t^2 k,$$
$$g(t) = ti + \cos tk,$$
$$h(t) = e^{3t}.$$

Find the derivative of the given function.

15. $r(t) = (3f - 2g)(t)$

16. $r(t) = h(t)\,f(t)$

17. $r(t) = (f \times g)(t)$

18. $r(t) = (f \cdot g)(t)$

19. $r(t) = f(h(t))$

20. $r(t) = |g(t)|^2$

21. Verify that $f(t)$ and $f'(t)$ are orthogonal if f is the vector function $f(t) = \cos 2ti + \sin 2tj + 3k$.

In Exercises 22–26, find the indefinite integral for the given vector-valued function.

22. $f(t) = ti + \cos tj + \sin tk$

23. $f(t) = \sqrt{t}\, i + e^{2t}j + \dfrac{1}{t}k$

24. $f(t) = t \sin ti + \cos^2 tj + (1 - t)k$

25. $f(t) = \ln ti + \dfrac{1}{t \ln t}j$

26. $f(t) = t \sec^2(1 - t^2)i + \sqrt{1 - t}\, j + \sqrt{t}\, k$

27. Find f if $f'(t) = i + t^2j$ and $f(0) = 3i + 5j$.

28. Find f if $f'(t) = \cos 2ti + \left(\dfrac{1}{1 + t^2}\right)j + \dfrac{t}{\sqrt{1 + t^2}}k$ and $f(0) = -4i + k$.

29. Find f if $f'(t) = \cos^2 ti + \sin^2 tj + e^{-t}k$ and $f(0) = 3i - 6j + 2k$.

30. Find a solution of the vector differential equation $f'(t) = f(t)$ for which
$$f(0) = i + 2j - k.$$

31. Find a solution of the vector differential equation
$$f'(t) + 4f(t) = 0$$
satisfying the initial condition $f(0) = i + 4k$.

32. Find a family of solutions to the vector differential equation
$$f''(t) - 2f(t) = 0.$$

33. Find $\displaystyle\int_a^b (ti + t^2j - t^3k)\, dt$.

34. Find $\displaystyle\int_0^1 f(t)\, dt$ if $f(t) = t^2i - 2tj + \sqrt{t}\, k$.

35. Find $\displaystyle\int_0^{\pi/4} f(t)\, dt$ if $f(t) = \cos ti + \sin 2tj + \cos^2 tk$.

36. Find $\displaystyle\int_1^4 f(t)\, dt$ if $f(t) = e^ti + t^2j + \ln 2tk$.

37. Show that if $f'(t) = 0$ for all $t \in I$, then f is constant on the interval I.

38. Show that if $f'(t) = g'(t)$ for all $t \in I$ then $f(t) = g(t) + C$ for some vector C and all $t \in I$.

39. True or false? If f is differentiable on $[a, b]$, there exists a constant $c \in (a, b)$ so that
$$f'(c) = \frac{1}{b - a}[f(b) - f(a)].$$

40. Prove statements (i) through (iv) and (vi) of Theorem 3 by first writing the vector functions in component form.

41. Complete the following alternate proof of differentiation formula (iii) of Theorem 3:
$$(hf)'(t) = h'(t)\, f(t) + h(t)\, f'(t).$$

a. Show that we can write
$$(hf)'(t) = \lim_{h \to 0} \frac{1}{h}[h(t + h)\, f(t + h) - h(t)\, f(t)]$$
$$= \lim_{h \to 0} \frac{1}{h}[h(t + h) - h(t)]f(t + h)$$
$$+ \lim_{h \to 0} \frac{1}{h}[f(t + h) - f(t)]h(t).$$

b. Show that the limit in part (a) is
$$(hf)'(t) = h'(t)\, f(t) + h(t)\, f'(t).$$

42. Use a method similar to that of Exercise 41 to prove part (vi) of Theorem 3:
$$(f \cdot g)'(t) = f'(t) \cdot g(t) + f(t) \cdot g'(t).$$

16.3 TANGENT VECTORS AND ARC LENGTH

We have already seen that a curve C in the plane may be described by a pair of parametric equations. For example, the unit circle is parameterized by the equations
$$C: \begin{cases} x(t) = \cos t \\ y(t) = \sin t \end{cases} \quad 0 \le t < 2\pi.$$

In a similar way, we may regard the vector-valued function
$$C: \quad r(t) = x(t)i + y(t)j + z(t)k \qquad t \in I \tag{1}$$

as giving a parameterization of a curve C in space. That is, we interpret C as the set of all points $P(t) = (x(t), y(t), z(t))$ lying at the tips of the position vectors $r(t)$. (Recall that interpreting $r(t)$ as a **position vector** means that $r(t)$ originates at the origin.) As for curves in the plane, we shall say that the curve C is **differentiable** if the function r is differentiable on I, which in turn requires that each of the component functions x, y, and z is differentiable (Theorem 2). The curve C is called **smooth** if r' is continuous on I.

When the derivative $r'(t_0)$ exists, we refer to it as a *tangent vector*.

DEFINITION 5

For the differentiable curve C in equation (1), if $r'(t_0) \neq 0$, the vector

$$r'(t_0) = x'(t_0)\boldsymbol{i} + y'(t_0)\boldsymbol{j} + z'(t_0)\boldsymbol{k}, \qquad t_0 \in I$$

is called a **tangent vector** for C at the tip of the vector $r(t)$ (Figure 3.1).

To see that Definition 5 agrees with our earlier concept of tangent, observe that the vector

$$r(t_0 + h) - r(t_0) \tag{2}$$

originates at the tip of $r(t_0)$ and terminates at the tip of $r(t_0 + h)$ (see Figure 3.1). This vector therefore determines a secant, joining the points $P(t_0)$ and $P(t_0 + h)$ on C. As $h \to 0$ the secant through these same points approaches the *tangent* to C at $r(t_0)$. However, we cannot obtain a direction vector for this line by simply applying $\lim\limits_{h\to 0}$ to the vector in equation (1) since the length of this vector will approach zero as $h \to 0$. We therefore multiply the vector in (2) by the scalar $\dfrac{1}{h}$ (changing its length, but not its direction) before applying the limit. Thus, a direction vector for the tangent to C at $r(t_0)$ is

$$r'(t_0) = \lim_{h \to 0} \left\{ \frac{1}{h} [r(t_0 + h) - r(t_0)] \right\}. \tag{3}$$

(See Figure 3.2.) Since we assume $r'(t_0) \neq 0$, the tangent vector has nonzero length.

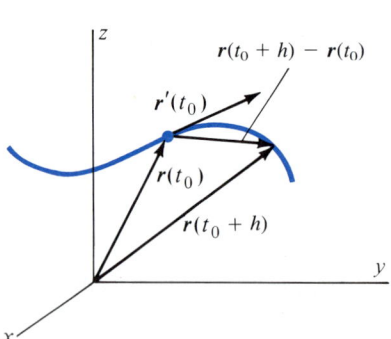

Figure 3.1 $r'(t_0)$ is tangent to C at tip of $r(t_0)$.

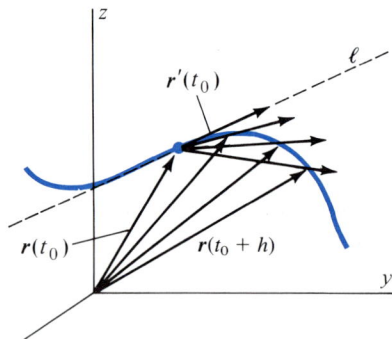

Figure 3.2 Line ℓ tangent to C at tip of $r(t_0)$ has direction vector $r'(t_0)$.

Example 1

The vector-valued function

$$r(t) = a \cos t\boldsymbol{i} + a \sin t\boldsymbol{j}, \qquad a > 0$$

has as its graph the circle in the *xy*-plane with center $(0, 0)$ and radius a. For this function the tangent vector at $\boldsymbol{r}(t)$ is

$$\boldsymbol{r}'(t) = -a \sin t\boldsymbol{i} + a \cos t\boldsymbol{j}.$$

Note that

$$\begin{aligned}
\boldsymbol{r}(t) \cdot \boldsymbol{r}'(t) &= -a^2(\cos t)(\sin t) + a^2(\sin t)(\cos t) \\
&= 0,
\end{aligned}$$

which confirms the familiar fact that the tangent to a circle at point P is perpendicular to the radius at that point. (See Figure 3.3.) ◇

The next example generalizes the observation of Example 1 to the circular helix of radius a in space.

Example 2

Show that a vector tangent to the circular helix

$$C: \quad \boldsymbol{r}(t) = a \cos t\boldsymbol{i} + a \sin t\boldsymbol{j} + b\boldsymbol{k}, \qquad 0 \le t < 2\pi$$

at $\boldsymbol{r}(t)$ is always orthogonal to the position vector $\boldsymbol{r}(t)$.

Solution: A vector tangent to C is, by Definition 5,

$$\boldsymbol{r}'(t) = -a \sin t\boldsymbol{i} + a \cos t\boldsymbol{j}.$$

Since $\boldsymbol{r}(t) \cdot \boldsymbol{r}'(t) = -a^2 \cos t \sin t + a^2 \sin t \cos t = 0$, the vectors $\boldsymbol{r}(t)$ and $\boldsymbol{r}'(t)$ are orthogonal. (Note that since $|\boldsymbol{r}(t)| = \sqrt{a^2 \cos^2 t + a^2 \sin^2 t + b^2} = \sqrt{a^2 + b^2}$ is constant, this result also follows by Example 6, Section 17.2.) (See Figure 3.4.) ◇

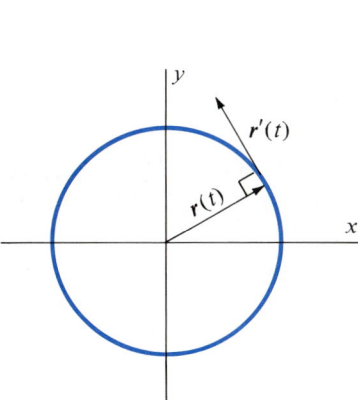

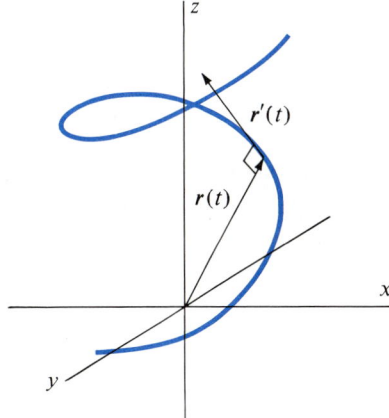

Figure 3.3 For the circle with parameterization

$$\boldsymbol{r}(t) = a \cos t\boldsymbol{i} + a \sin t\boldsymbol{j},$$

the tangent $\boldsymbol{r}'(t)$ is always orthogonal to the radius vector $\boldsymbol{r}(t)$.

Figure 3.4 For the circular helix

$$\boldsymbol{r}(t) = a \cos t\boldsymbol{i} + a \sin t\boldsymbol{j} + b\boldsymbol{k},$$

the tangent vector $\boldsymbol{r}'(t)$ is always orthogonal to the position vector $\boldsymbol{r}(t)$.

If $r'(t_0) \neq 0$, a unit vector T tangent to the curve C at $r(t_0)$ is given by

$$\boxed{T = \frac{r'(t_0)}{|r'(t_0)|} \qquad \text{(unit tangent)}.} \tag{4}$$

Example 3

Find a unit vector tangent to the circular helix $r(t) = a \cos t\,i + a \sin t\,j + bt\,k$ at the point $(0, a, b\pi/2)$.

Strategy

Find t_0 for which $r(t_0)$ terminates at the point $\left(0, a, \dfrac{b\pi}{2}\right)$.

Solution

Setting $r(t_0) = 0i + aj + \dfrac{b\pi}{2}k$ gives the equations

$$a \cos t_0 = 0, \qquad a \sin t_0 = a, \qquad bt_0 = \frac{b\pi}{2},$$

to which the solution is $t_0 = \pi/2$.

Find $r'(t_0)$.

According to Definition 5 a tangent vector at $r(t_0) = r(\pi/2)$ is

$$r'\left(\frac{\pi}{2}\right) = -a \sin\left(\frac{\pi}{2}\right)i + a \cos\left(\frac{\pi}{2}\right)j + bk$$

$$= -ai + bk.$$

Find $|r'(t_0)|$.

This vector has length

$$\left|r'\left(\frac{\pi}{2}\right)\right| = \sqrt{a^2 + b^2}.$$

Apply (4).

The desired unit tangent is

$$T = \frac{r'\left(\dfrac{\pi}{2}\right)}{\left|r'\left(\dfrac{\pi}{2}\right)\right|} = \frac{-ai + bk}{\sqrt{a^2 + b^2}}. \qquad \diamondsuit$$

To find an equation for the *line* ℓ tangent to C at $r(t_0)$, we use the fact that $r(t_0)$ is a position vector for ℓ and $r'(t_0)$ is a direction vector for ℓ. Since t is the parameter for C, we should use a different letter to parameterize ℓ, say ω. We may then use equation (1), Section 16.5, to write a parameterization for ℓ as

$$\boxed{\ell: \quad R(\omega) = r(t_0) + \omega r'(t_0).} \tag{5}$$

(Be careful when using equation (5): $R(\omega)$ is the position vector of a point on ℓ, while $r(t_0)$ and $r'(t_0)$ are fixed vectors determined by the curve C.)

Example 4

Find vector, parametric, and symmetric equations for the line ℓ tangent to the curve

$$C: \quad r(t) = t^2 i + (3 - t)j + t^3 k$$

at the point $(4, 1, 8)$.

Strategy

Find t_0 for which $r(t_0)$ terminates at the point (4, 1, 8). $r(t_0)$ is the position vector for ℓ.

Solution

Setting $r(t_0) = 4i + j + 8k$ gives the equations

$$t^2 = 4, \qquad 3 - t = 1, \qquad \text{and} \qquad t^3 = 8$$

for which the solution is $t_0 = 2$.

Since $r'(t) = 2ti - j + 3t^2k$,

Find $r'(t_0)$. This is the direction vector for ℓ.

$$r'(t_0) = r'(2) = 4i - j + 12k.$$

Use equation (5) to write the vector form of ℓ.

The vector form of ℓ is therefore

$$\ell: \quad R(\omega) = (4i + j + 8k) + \omega(4i - j + 12k).$$

The parametric equations for x, y, and z are the components of $R(\omega)$.

To find parametric equations for ℓ, we let $P(\omega) = (x(\omega), y(\omega), z(\omega))$ be a point on ℓ. Then

$$
\begin{aligned}
x(\omega) &= 4 + 4\omega & (i\text{-components}) \\
y(\omega) &= 1 - \omega & (j\text{-components}) \\
z(\omega) &= 8 + 12\omega & (k\text{-components})
\end{aligned}
$$

The symmetric equations are found from the parametric equations as in Section 15.5.

are the parametric equations for the coordinates of P. Solving each of these equations for ω and equating the results gives the symmetric equations

$$\frac{x - 4}{4} = 1 - y = \frac{z - 8}{12}$$

for ℓ. $\diamond$

Arc Length

If C is a smooth curve in space parameterized by the vector function

$$r(t) = x(t)i + y(t)j + z(t)k$$

an expression for the length L of the arc of C from $r(a)$ to $r(b)$ is obtained in a manner similar to that for plane curves (see Section 7.3). First, the interval $[a, b]$ is partitioned into n subintervals of equal length $\Delta t = \dfrac{b - a}{n}$ and with endpoints $a = t_0 < t_1 < t_2 < \cdots < t_n = b$. This determines a polygonal path connecting the tips of the vectors $r(t_0), r(t_1), \ldots, r(t_n)$ of total length

$$\sum_{j=1}^{n} \Delta L_j = \sum_{j=1}^{n} \sqrt{\Delta x_j^2 + \Delta y_j^2 + \Delta z_j^2} \tag{6}$$

where $\Delta x_j = x(t_j) - x(t_{j-1})$, $\Delta y_j = y(t_j) - y(t_{j-1})$, and $\Delta z_j = z(t_j) - z(t_{j-1})$. As before, applications of the Mean Value Theorem (once for each component) lead to an approximation of the form

$$L \approx \sum_{j=1}^{n} \sqrt{[x'(u_j)]^2 + [y'(v_j)]^2 + [z'(w_j)]^2} \, \Delta t_j. \tag{7}$$

As $n \to \infty$, $\Delta t \to 0$ and we obtain

$$L = \int_a^b \sqrt{[x'(t)]^2 + [y'(t)]^2 + [z'(t)]^2} \, dt \tag{8}$$

which we take as the definition of the arc length L. (Since this is familiar ground, we do not pursue the details of this development here.) Since the expression under the radical in equation (8) is equal to $r'(t) \cdot r'(t) = |r'(t)|^2$, the arc length may be expressed more compactly as

$$L = \int_a^b |r'(t)| \, dt.$$ (9)

Example 5

Find the length of the curve

$$C: \quad r(t) = \cos \pi t i + \sin \pi t j + t^{3/2} k$$

from $r(0) = i$ to $r(4) = i + 8k.$

Solution: Here

$$r'(t) = -\pi \sin \pi t i + \pi \cos \pi t j + \frac{3}{2}\sqrt{t}\, k$$

so

$$|r'(t)| = \sqrt{\pi^2 \sin^2 \pi t + \pi^2 \cos^2 \pi t + \frac{9}{4}t}$$

$$= \sqrt{\pi^2 + \frac{9}{4}t}.$$

By formula (9),

$$L = \int_0^4 \sqrt{\pi^2 + \frac{9}{4}t}\, dt = \frac{8}{27}\left(\pi^2 + \frac{9}{4}t\right)^{3/2}\Bigg]_0^4$$

$$= \frac{8}{27}[(\pi^2 + 9)^{3/2} - \pi^3] \approx 15.1$$

(Figure 3.5). ◇

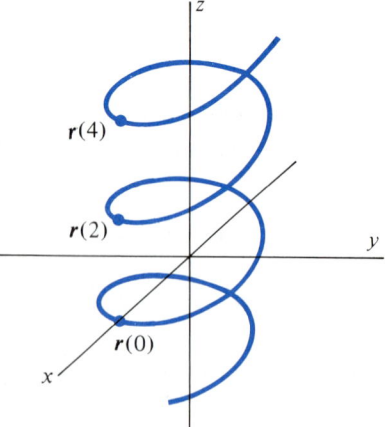

$r(4)$

$r(2)$

$r(0)$

Figure 3.5

Parameterization by Arc Length

Up to this point, we have not concerned ourselves with the *rate* at which the vector-valued function f traces out a curve in space. In order to proceed further with our analysis of curves, we will now have to do so. A simple example of this type of concern is the observation that the functions $f(t) = a + tb$ and $g(t) = a + 2tb$ both trace out the same line in space, although a particle located at the tip of $g(t)$ moves along the line at a speed twice that of a particle located at the tip of $f(t)$.

A baseline for comparing the rate at which the vector function f traces out a curve in space is provided by the notion of *parameterization by arc length*. Intuitively, this notion says that as the parameter t varies through an interval of length t_0, an arc of this same length $s = t_0$ is traced out by the vector function f (Figure 3.6). A more precise formulation of this concept is the following.

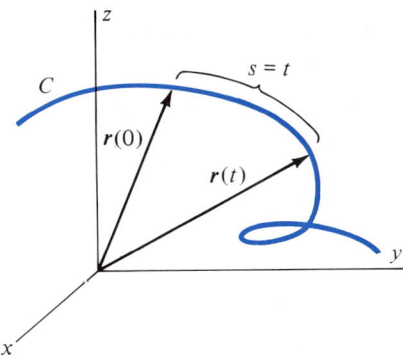

Figure 3.6 r is parameterized by arc length if $\Delta s = \Delta t$.

DEFINITION 6

Let the curve C be the graph of the differentiable function $r(t) = x(t)i + y(t)j + z(t)k$ defined on an interval I containing the number zero. The curve C (and the function r) is said to be **parameterized by arc length** if

$$\int_0^t |r'(\omega)|\, d\omega = t \qquad (10)$$

for all t in the interval I.

Since the integral in equation (10) is the length of the arc of C from $r(0)$ to $r(t)$, Definition 6 agrees with the intuitive description of parameterization by arc length given above. Differentiating both sides of equation (10) with respect to t shows that the condition

$$\boxed{|r'(t)| = 1 \qquad \text{for all} \qquad t \in I} \qquad (11)$$

is equivalent to C being parameterized by arc length. Thus, a curve is *parameterized by arc length if and only if its derivative has constant length equal to one.*

Example 6

The unit circle

$$r(t) = \cos t i + \sin t j, \qquad t \in [0, 2\pi)$$

is parameterized by arc length, since

$$|r'(t)| = |-\sin ti + \cos tj|$$
$$= \sqrt{\sin^2 t + \cos^2 t}$$
$$= 1$$

for all $t \in [0, 2\pi)$. (This should not be surprising. Both the circumference of the unit circle and the length of the interval $[0, 2\pi)$ are 2π.) ◇

Example 7

The circle of radius ρ centered at the origin is traced out by the vector function

$$r(t) = \rho \cos \alpha ti + \rho \sin \alpha tj, \qquad t \in \left[0, \frac{2\pi}{\alpha}\right), \qquad \alpha > 0.$$

Find the value of α for which this circle is parameterized by arc length.

Solution: We first find $|r'(t)|$:

$$r'(t) = -\alpha\rho \sin \alpha ti + \alpha\rho \cos \alpha tj.$$

Thus

$$|r'(t)| = \sqrt{(-\alpha\rho \sin \alpha t)^2 + (\alpha\rho \cos \alpha t)^2}$$
$$= |\alpha\rho|.$$

In order that equation (10) hold, we must have

$$|r'(t)| = |\alpha\rho| = 1.$$

We therefore take $\alpha = 1/\rho$. The circle of radius ρ is parameterized by arc length by the function

$$r(t) = \rho \cos\left(\frac{t}{\rho}\right)i + \rho \sin\left(\frac{t}{\rho}\right)j.$$ ◇

We conclude by highlighting the comment following equation (11). If the curve C determined by the vector function r is parameterized by arc length, the **unit tangent** $T(s)$ at $r(s)$ is

$$\boxed{T(s) = r'(s).}$$ (12)

Exercise Set 16.3

In Exercises 1–6, find a vector tangent to the given curve at the given point.

1. $r(t) = t^3i + \sqrt{t}\,j + 2k, \qquad t = 4$

2. $r(t) = \sin ti + \cos 2tj + e^{3t}k, \qquad t = 0$

3. $r(t) = \sec ti + \tan tj, \qquad t = \pi/4$

4. $r(t) = \sqrt{1 + t^2}\,i + \text{Tan}^{-1}\,tj + \ln(1 + t)k, \qquad t = 0$

5. $r(t) = (1 + t^2)i + (t - 4)j + 6tk$, at the point $(2, -5, -6)$

6. $r(t) = ai + btj + ct^2k$ at the point (a, b, c).

7. Find a unit vector tangent to the curve $r(t) = a \cos \omega ti + b \sin \omega tj$ at the point where $\omega t = \pi/4$.

In Exercises 8–10, let $r(t) = (t^2 + 2)i + 3tj$.

8. Find the number(s) t for which $r(t)$ and $r'(t)$ are orthogonal.

9. Find the number(s) t for which $r(t)$ and $r'(t)$ have the same direction.

10. Find the number(s) t for which $r(t)$ and $r'(t)$ have opposite direction.

11. Find an equation in vector form for the line tangent to the curve in Exercise 1 at the given point.

12. Find an equation in parametric form for the line tangent to the curve in Exercise 2 at the given point.

13. Find an equation in symmetric form for the line tangent to the curve in Exercise 5 at the given point.

14. Show that every tangent to the unit circle is orthogonal to the radius vector through the point of tangency.

15. Let $r(t) = e^t \cos t\, i + e^t \sin t\, j$. Find the angle between $r(t)$ and the tangent $r'(t)$ for
a. $t = 0$,
b. $t = \pi/4$.

In Exercises 16–20, find the length of the indicated arc.

16. $r(t) = \cos t\, i + \sin t\, j, \qquad 0 \le t \le \pi/4$

17. $r(t) = \cos t\, i + \dfrac{\sqrt{2}}{2} \sin t\, j + \dfrac{\sqrt{2}}{2} \sin t\, k, \quad 0 \le t \le \pi/2$

18. $r(t) = e^t \cos t\, i + e^t \sin t\, j, \qquad 0 \le t \le \pi/2$

19. $r(t) = t\, i + (t^2 + 1)\, j + t\, k, \qquad 0 \le t \le 1$

20. $r(t) = t^3 i + 3t^2 j + 6t\, k, \qquad 0 \le t \le 3$

21. Find a unit tangent $T(t)$ to the graph of $y = x^2 + 3$ at the point $(2, 7)$. (*Hint:* Let $x = t$. Then $y = t^2 + 3$, and a vector parameterization for the curve is $r(t) = t\, i + (t^2 + 3)\, j$.)

22. Use the method of Exercise 21 to find a unit tangent $T(t)$ to the graph of $y^3 = 1 - x^2$ at the point $(3, -2)$.

23. Determine the constant α so that the curve

$$r(t) = \cos \alpha t\, i + \sin \alpha t\, j + \alpha t\, k, \quad 0 \le t \le \frac{2\pi}{\alpha}, \quad \alpha > 0,$$

is parameterized by arc length.

24. Find a parameterization by arc length for the curve $y = 3x - 2$ so that $x = 0$ and $y = -2$ when $s = 0$.

25. Show that if $r(t) = x(t)i + y(t)j$ and $x'(t_0) \ne 0$, Definition 5 for the *tangent* at $r(t_0)$ agrees with our earlier definition of tangent for the curve C determined by the parametric equations

$$C: \begin{cases} x = x(t) \\ y = y(t). \end{cases}$$

26. Use Example 6, Section 17.2, to show that if $T(t) = \dfrac{r'(t)}{|r'(t)|}$ is differentiable on I, then $T'(t)$ is orthogonal to $T(t)$ for $t \in I$. If $T'(t) \ne 0$, we define $N(t) = \dfrac{T'(t)}{|T'(t)|}$ to be the **principal unit normal** to the curve $C = \{r(t) \mid t \in I\}$.

27. Find the principal unit normals to the curve with parameterization

$$r(t) = \sqrt{2} \cos t\, i + \sqrt{2} \sin t\, j, \qquad 0 \le t \le 2\pi.$$

(See Exercise 26.)

28. Show that the line with vector equation

$$R(\omega) = r(t_0) + \omega r''(t_0)$$

is orthogonal to the tangent vector $T(t_0)$ if $r(t)$ is a parameterization by arc length, $r(t)$ is twice differentiable at $r(t_0)$, $r'(t_0) \ne 0$, and $r''(t_0) \ne 0$. (We refer to this line as the line **normal** to the curve at $r(t_0)$.)

29. Find an equation for the line normal to the curve in Example 4 at the point $(4, 1, 8)$ (see Exercise 28).

30. Give an example to show that $r(t)$ is not always orthogonal to $r'(t)$.

16.4 VELOCITY AND ACCELERATION

If a particle moves in space so that its location at time t is the point $P(t) = (x(t), y(t), z(t))$, its motion may be described by the **position vector** function

$$r(t) = x(t)i + y(t)j + z(t)k. \tag{1}$$

That is, we think of the position of the particle as the tip of the vector $r(t)$ originating at the origin and terminating at the point $P(t)$.

In such cases the **average velocity** from time t to time $t + h$ is

$$\text{average velocity} = \frac{\text{change in position}}{\text{change in time}}$$

$$= \frac{1}{h} \Delta r$$

$$= \frac{1}{h} [r(t + h) - r(t)].$$

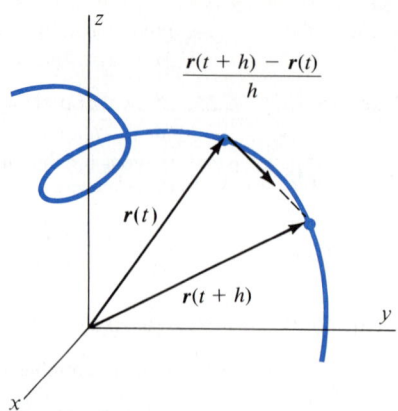

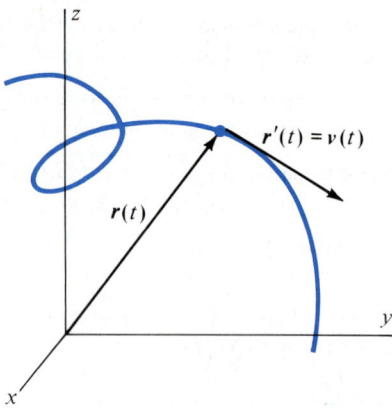

Figure 4.1 Average velocity is the vector $\dfrac{1}{h}[r(t + h) - r(t)]$.

Figure 4.2 Velocity is the vector

$$v(t) = r'(t) = \lim_{h \to 0} \frac{1}{h}[r(t + h) - r(t)],$$

and $|v(t_0)| = \text{speed}$.

(See Figure 4.1.)

If our earlier concept of (instantaneous) velocity as the limit of average velocity as $h \to 0$ is to carry over to motion in space, we should define

$$\text{velocity at time } t = \lim_{h \to 0} \frac{1}{h}[r(t + h) - r(t)]. \tag{2}$$

(See Figure 4.2.)

We have already seen that the limit on the right-hand side of equation (2), when it exists, is the derivative $r'(t)$ of the vector-valued function r. According to Theorem 2, this limit exists precisely when each of the component functions x, y, and z is differentiable. The definition of the *velocity* of a particle moving in space is therefore the following.

DEFINITION 7

Let the vector function

$$r(t) = x(t)i + y(t)j + z(t)k$$

be the position function for a particle moving in space and let the parameter t denote time. If r is differentiable at time $t = t_0$, the **velocity** of the particle at $t = t_0$ is the vector

$$v(t_0) = r'(t_0) = x'(t_0)i + y'(t_0)j + z'(t_0)k. \tag{3}$$

If r is differentiable on an interval I, equation (3) determines the **velocity function** for r on the interval I.

Since velocity is just the derivative of the position function, we have already determined that $v(t_0)$ is *tangent* to the graph of r at $r(t_0)$, and points in the direction of increasing t, provided $v(t_0) \neq 0$ (Section 17.3).

As with real-valued functions, the length of the velocity vector $|v(t_0)| = |r'(t_0)|$ is defined to be the **speed** (rate of change of distance along the path) of the particle at time t_0.

Example 1

Show that a particle moving about the unit circle in the xy-plane with position function

$$r(t) = \cos \alpha t\, i + \sin \alpha t\, j$$

moves at a constant speed.

Solution: By Definition 7, the velocity function is

$$v(t) = -\alpha \sin \alpha t\, i + \alpha \cos \alpha t\, j.$$

The speed at time t is therefore

$$|v(t)| = \sqrt{(\alpha \sin \alpha t)^2 + (\alpha \cos \alpha t)^2} = |\alpha|,$$

which is constant. For this reason, physicists call this **uniform circular motion.**
◇

As in the case of motion along a line, we define the *acceleration* of a particle moving in space to be the rate of change of velocity with respect to time.

DEFINITION 8

Let the vector function

$$r(t) = x(t)i + y(t)j + z(t)k$$

be the position function of a particle moving in space and let the parameter t denote time. If r is twice differentiable at time $t = t_0$, the **acceleration** of the particle at $t = t_0$ is the vector

$$a(t_0) = r''(t_0) = x''(t_0)i + y''(t_0)j + z''(t_0)k.$$

Thus,

$$a(t_0) = v'(t_0). \tag{4}$$

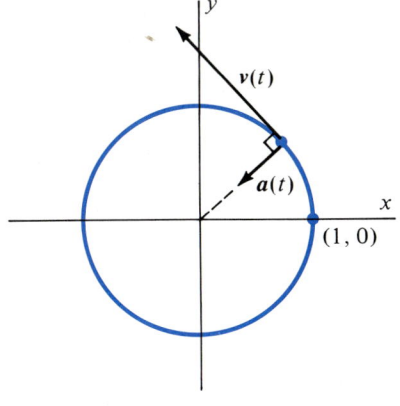

Figure 4.3 Velocity and acceleration vectors associated with uniform circular motion.

Example 2

Show that for the particle moving about the unit circle in Example 1, the acceleration vector always points toward the center of the circle and is of constant magnitude.

Solution: From the solution to Example 1 and equation (4) we have

$$\begin{aligned}
a(t) = v'(t) &= -\alpha^2 \cos \alpha t\, i - \alpha^2 \sin \alpha t\, j \\
&= -\alpha^2(\cos \alpha t\, i + \sin \alpha t\, j) \\
&= -\alpha^2 r(t).
\end{aligned}$$

Thus, $a(t)$ points in the direction opposite $r(t)$ and has length

$$|a(t)| = \sqrt{(\alpha^2 \cos \alpha t)^2 + (\alpha^2 \sin \alpha t)^2} = \alpha^2.$$

(See Figure 4.3.) Note that $v(t)$ and $a(t)$ are orthogonal, an observation that follows from Example 6, Section 16.2, since $|v(t)| = |\alpha| = $ constant.
◇

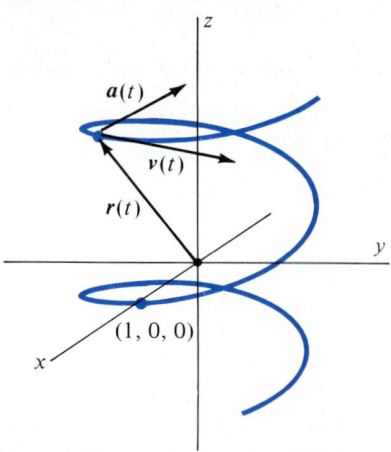

Figure 4.4 Velocity and acceleration vectors for the helix $r(t) = \cos t i + \sin t j + t^2 k$.

Example 3

Find the velocity and acceleration functions associated with the position function

$$r(t) = \cos t i + \sin t j + t^2 k$$

and show that $|a(t)|$ is constant.

Solution:

$$v(t) = r'(t) = -\sin t i + \cos t j + 2t k$$
$$a(t) = v'(t) = -\cos t i - \sin t j + 2k.$$

Since $|a(t)| = \sqrt{\cos^2 t + \sin^2 t + 4} = \sqrt{5}$, $|a(t)|$ is constant. The vectors $r(t)$, $v(t)$, and $a(t)$ are sketched in Figure 4.4. ◇

Just as velocity and acceleration can be determined by differentiating the position function r, the functions for position and velocity can be obtained by integrating the acceleration function, provided antiderivatives of each component function can be found. More precisely, equation (4) gives

$$v(t) = \int a(t)\, dt, \tag{5}$$

and equation (3) shows that

$$r(t) = \int v(t)\, dt. \tag{6}$$

In using equations (5) and (6), you must remember that one constant of integration will appear in each component in each integration. These constants may be determined when *initial conditions* are specified for the functions v and r.

Example 4

A particle starts from rest at the point $P_0 = (2, 0, -3)$ and moves with an acceleration of $a(t) = 2i + 6tj$ m/s^2. Find

(a) the velocity function for the particle,
(b) the position function for the particle,
(c) the location of the particle after 3 seconds.

Strategy
Identify the initial conditions.

Obtain v by integrating a. (Don't forget that

$$\left[\int 0\, dt\right]k = (0 + c_3)k.)$$

Apply the initial condition $v_0 = 0$ to determine c_1, c_2, and c_3.

Solution
Since the particle starts at $P_0 = (2, 0, -3)$, we have

$$r_0 = r(0) = 2i - 3k.$$

Since the particle starts from rest, we have

$$v_0 = v(0) = 0.$$

Using equation (5), we have

$$v(t) = \int (2i + 6tj)\, dt$$

$$= (2t + c_1)i + (3t^2 + c_2)j + c_3 k.$$

Setting $t = 0$ and applying the initial condition $v(0) = 0$ gives

$$0 = c_1 i + c_2 j + c_3 k,$$

so

$$c_1 = c_2 = c_3 = 0.$$

The velocity function is therefore

$$v(t) = 2t i + 3t^2 j.$$

Obtain r by integrating v.

Using equation (6), we find that

$$r(t) = \int (2t i + 3t^2 j)\, dt$$

$$= (t^2 + d_1)i + (t^3 + d_2)j + d_3 k.$$

Determine the constants d_1, d_2, and d_3 by applying the initial condition

$r(0) = 2i - 3k.$

Setting $t = 0$ and applying the initial condition $r(0) = 2i - 3k$ gives

$$2i - 3k = d_1 i + d_2 j + d_3 k$$

so

$$d_1 = 2, \qquad d_2 = 0, \qquad \text{and} \qquad d_3 = -3.$$

Thus,

$$r(t) = (t^2 + 2)i + t^3 j - 3k.$$

The location of the particle after 3 seconds is $r(3)$.

The location of the particle after 3 seconds is

$$r(3) = (3^2 + 2)i + 3^3 j - 3k$$
$$= 11i + 27j - 3k,$$

or $(11, 27, -3)$. Note that, because there is no component of a in the z direction, the entire motion takes place in the plane $z = -3$.   ◇

Projectile Motion

A frequent application of equations (5) and (6) concerns the trajectory followed by a projectile launched with a prescribed initial velocity, v_0. The term ''projectile'' refers to any object, such as a missile, a flare, or a baseball, which is not self-propelling. Thus, the trajectory followed by a projectile is completely determined by the speed and direction (i.e., the velocity) at which it is launched. In the discussion that follows, we shall assume that air resistance is negligible, so that the only force acting on the projectile after launch is the force due to gravity. Also, we shall take the y-axis as vertical and the x-axis as horizontal, with the positive x-axis pointing in the direction of the horizontal component of the initial velocity. The motion is then restricted entirely to the xy-plane.

The motion of projectiles (and, for that matter, of all moving objects) is governed by the vector form of Newton's second law of motion:

$$F = ma. \tag{7}$$

Here F is the vector sum of all forces acting on the projectile, m is its mass, and a is its acceleration.

According to our use of the term projectile, the only force acting on the object in flight is the force due to gravity,

$$F = -mg j \tag{8}$$

where $g = 9.81$ m/s^2 is the gravitational constant. (We use a negative sign since this force acts downward.)

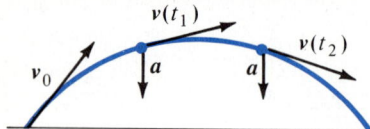

Figure 4.5 Motion of a projectile with initial velocity v_0.

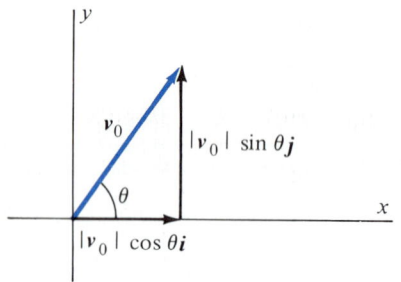

Figure 4.6 Components of initial velocity are

$$v_0 = |v_0| \cos \theta i + |v_0| \sin \theta j.$$

Combining equations (7) and (8) we find that a projectile in flight experiences the acceleration

$$a = -gj. \tag{9}$$

Figure 4.5 illustrates a typical trajectory for a projectile fired with initial velocity v_0.

Now consider the problem of determining the trajectory of a projectile fired from the origin at an angle θ with the horizontal and with an initial speed s_0. This means that the initial velocity v_0 has length $|v_0| = s_0$. We may therefore write v_0 in component form as

$$v_0 = s_0 \cos \theta i + s_0 \sin \theta j. \tag{10}$$

(See Figure 4.6.)

Applying equation (5) with $a(t)$ as in equation (9), we find

$$v(t) = \int a\, dt = \int -gj\, dt = c_1 i + (c_2 - gt)j.$$

Setting $t = 0$ and applying the initial condition given by equation (10), we obtain

$$v(0) = c_1 i + c_2 j = s_0 \cos \theta i + s_0 \sin \theta j.$$

Thus,

$$c_1 = s_0 \cos \theta; \qquad c_2 = s_0 \sin \theta,$$

and therefore

$$v(t) = s_0 \cos \theta i + (s_0 \sin \theta - gt)j.$$

According to equation (6), the position function is

$$r(t) = \int v(t)\, dt = \int [s_0(\cos \theta)i + (s_0 \sin \theta - gt)j]\, dt$$

$$= [s_0(\cos \theta)t + d_1]i + \left[s_0(\sin \theta)t - \frac{1}{2}gt^2 + d_2\right]j.$$

Since the projectile is fired from the origin, we have the initial condition $r(0) = \mathbf{0}$. Setting $t = 0$ in the above expression for $r(t)$ and applying this initial condition gives

$$r(0) = d_1 i + d_2 j = \mathbf{0},$$

so

$$d_1 = d_2 = 0.$$

The position function for the projectile motion is therefore

$$r(t) = [s_0(\cos \theta)t]i + \left[s_0(\sin \theta)t - \frac{1}{2}gt^2\right]j.$$

We summarize our findings as follows:

> The trajectory of a projectile fired from the origin with initial speed s_0 and angle of elevation θ is parameterized by the vector-valued function
>
> $$r(t) = [s_0(\cos \theta)t]i + \left[s_0(\sin \theta)t - \frac{1}{2}gt^2\right]j. \tag{11}$$

Example 5

A flare is fired from ground level at an elevation of 60° with an initial speed of 50 m/s. Find

(a) the maximum height of the trajectory,
(b) the length of time during which the flare is airborne, and
(c) the distance from the point of launch to the point of impact.

Strategy

Write the expression for $r(t) = x(t)i + y(t)j$ using (11).

Solution

With $\theta = 60° = \pi/3$ and $s_0 = 50$ m/s, $r(t)$ is

$$r(t) = 50\left(\cos \frac{\pi}{3}\right)ti + \left(50\left(\sin \frac{\pi}{3}\right)t - \frac{1}{2}gt^2\right)j$$

$$= 25ti + \left(25\sqrt{3}t - \frac{1}{2}gt^2\right)j.$$

Find the time, t_{max}, of maximum height by maximizing the j-component of $r(t)$.

The height of the flare is given by the j-component

$$y(t) = 25\sqrt{3}t - \frac{1}{2}gt^2.$$

The maximum height occurs when

$$y'(t) = 25\sqrt{3} - gt = 0,$$

or

$$t_{max} = \frac{25\sqrt{3}}{g} = \frac{25\sqrt{3}}{9.81} \approx 4.4 \text{ seconds.}$$

$y(t_{max})$ is the maximum height.

The maximum height is

$$y(t_{max}) = 25\sqrt{3}\left(\frac{25\sqrt{3}}{9.81}\right) - \frac{1}{2}(9.81)\left(\frac{25\sqrt{3}}{9.81}\right)^2$$

$$= \frac{\left(\frac{1}{2}\right)25^2 \cdot 3}{9.81} \approx 95.6 \text{ meters.}$$

Find the time, t_{end}, at which the projectile lands by setting $y(t) = 0$.

The projectile lands when the j-component reaches zero, that is, when

$$y(t) = 25\sqrt{3}t - \frac{1}{2}gt^2 = 0.$$

This equation has solutions $t = 0$ and $t = \dfrac{50\sqrt{3}}{g}$.

The time $t = 0$ corresponds to launch. The time at which the projectile returns to ground level is therefore

$$t_{end} = \frac{50\sqrt{3}}{g} = \frac{50\sqrt{3}}{9.81} \approx 8.8 \text{ seconds.}$$

Distance from launch point to impact point is $x(t_{end})$.

The distance from the point of launch to the point of impact is the i-component of $r(t_{end})$.

$$x(t_{end}) = 25(8.8) = 220 \text{ meters.} \qquad \diamondsuit$$

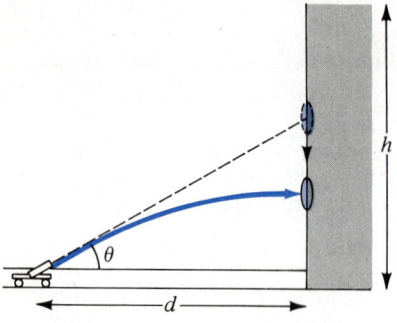

Figure 4.7 The cannon–target experiment.

Example 6

The following experiment is performed in an elementary physics class. A target is hung on a wall of a gymnasium at a height of h meters above the floor. A small cannon, which fires a lead slug, is located on the floor d meters from the wall. The muzzle velocity of the cannon is adjustable. The cannon is aimed at the bull's-eye on the target. By means of a trip mechanism, the target is released from its hanger at the precise instant at which the cannon is fired and falls toward the floor. The result of repeated trials of the experiment is that no matter what muzzle velocity is chosen, the slug always strikes the bull's-eye (assuming that d, θ, and the muzzle velocity are such that the slug does not strike the floor before reaching the wall). (See Figure 4.7.) How can this be explained?

Solution: Let θ be the angle of elevation of the barrel of the cannon when aimed at the bull's-eye, and let s_0 be the muzzle velocity of the cannon. We assume the cannon to be located at the origin of an xy-coordinate system. The trajectory of the slug is then given by $r(t)$ in equation (11).

The slug reaches the wall when the i-component of $r(t)$ satisfies the equation

$$x(t) = s_0(\cos \theta)\, t = d$$

which occurs at time $t_f = \dfrac{d}{s_0 \cos \theta}$. At this time the j-component of the slug is

$$y(t_f) = s_0 \sin \theta \left(\frac{d}{s_0 \cos \theta} \right) - \frac{1}{2} g \left(\frac{d}{s_0 \cos \theta} \right)^2$$

$$= d \tan \theta - \frac{1}{2} g t_f^2.$$

Since $\dfrac{h}{d} = \tan \theta$, $h = d \tan \theta$. Thus

$$y(t_f) = h - \frac{1}{2} g t_f^2$$

is the j-component of $r(t)$ when the slug reaches the wall.

Since the target is a freely falling body, after t_f seconds it has fallen a distance $\dfrac{1}{2} g t_f^2$, so the j-component of the bull's-eye's position vector after t_0 seconds is also $h - \dfrac{1}{2} g t_f^2$. Since the slug and the bull's-eye have the same position vector, $r(t_f) = di + (h - \dfrac{1}{2} g t_f^2)j$, at time $t = t_f$, the slug must always score a direct hit on the target. ◇

Exercise Set 16.4

In Exercises 1–7, find the velocity and acceleration functions corresponding to the given position functions.

1. $r(t) = i + 2tj$

2. $r(t) = 3ti + t^3j + (2t + 3)k$

3. $r(t) = \cos 3ti + \sin 3tj$

4. $r(t) = e^ti + e^{-t}j + \sqrt{t}k, \quad t \geq 1$

5. $r(t) = \ln t^2 i + \cos 2tj + \dfrac{1}{t}k, \quad t \geq 1$

6. $r(t) = t(t-1)i + \text{Tan}^{-1} tj + te^t k$

7. $r(t) = e^t \cos ti + e^t \sin tj + \cos tk$

8. A particle moves with position function $r(t) = \cos \pi ti + \sin \pi tj$. Find its speed at time $t = 1/4$.

9. A particle moves with position vector $r(t) = t^2 i + 2tj + t^3 k$. Find its speed at time $t = 2$.

In Exercises 10–13, find the position function r from the given velocity function and the initial condition $r(0) = r_0$.

10. $v(t) = 2ti + t^2 j, \quad r_0 = i + 4j$

11. $v(t) = \cos ti + \sin tj, \quad r_0 = 3i + 2j$

12. $v(t) = e^t i + \sqrt{t}j + 2tk, \quad r_0 = 6i + 4k$

13. $v(t) = \left(\dfrac{1}{1+t^2}\right)i + te^{t^2}j, \quad r_0 = -2i + j + 4k$

In Exercises 14–18, find the velocity function v and the position function r given the acceleration function a and the initial conditions $v_0 = v(0)$ and $r_0 = r(0)$.

14. $a = i + j, \quad v_0 = i + j, \quad r_0 = 0$

15. $a = 2i + k, \quad v_0 = 0, \quad r_0 = 3i - j + 4k$

16. $a = 6ti + e^t j, \quad v_0 = 0, \quad r_0 = 4i + j + 2k$

17. $a = \cos ti + \sin tj, \quad v_0 = k, \quad r_0 = 3i - j + 2k$

18. $a = e^t i + tj + e^{-t}k, \quad v_0 = i + k, \quad r_0 = i + 4j + k$

19. Verify that the vectors $v(t)$ and $a(t)$ of Example 2 are orthogonal for each t.

20. Prove that if a particle moves with constant speed its velocity and acceleration vectors are orthogonal.

21. A particle moves about a circle in the xy-plane with position function $r(t) = \rho \cos \alpha ti + \rho \sin \alpha tj$. Find a relationship between the radius of the circle and the speed of the particle.

22. A projectile is fired from ground level at an elevation angle of $\theta = 45°$ with an initial speed of 60 meters per second. Find
   a. the position function r of the projectile,
   b. the maximum altitude of the projectile,
   c. the flight time,
   d. the distance from the launch point to the impact point,
   e. the speed on impact.

23. Find each of the quantities in Exercise 22 for a projectile fired at an elevation angle of $\theta = 30°$ with an initial speed of 100 meters per second.

24. A bale of hay is dropped from an airplane flying at a ground speed of 200 km/h and an elevation of 1000 meters. How far from the drop point does the bale strike the ground?

25. A handgun used in a physics experiment simultaneously fires one slug through the barrel and drops a second slug mounted on the side of the barrel. Independent of the muzzle velocity, the two slugs are observed to strike the ground at the same instant when the gun is fired horizontally. Why?

26. How far along the path $r(t)$ does the particle of Example 4 travel during the first 3 seconds of its motion? What is its distance from the starting point P_0 at $t = 3$ seconds?

27. a. Write the integral that expresses the arc length of the motion of the flare in Example 5.
   b. *(Calculator/Computer)* Use Simpson's Rule with $n = 8$ to approximate the value of the integral in (a).

16.5 CURVATURE

Curvature is a measure of the rate at which a curve in space turns or twists. To make this idea more precise, we consider a curve C determined parametrically by the vector function

$$C: \quad r(s) = x(s)i + y(s)j + z(s)k, \qquad a \le s \le b. \tag{1}$$

We assume that C is parameterized by arc length, so that the tip of $r(s)$ moves along C uniformly. That is, we assume that $|r'(s)| = 1, a \le s \le b$.

Since $r'(s)$ is a unit tangent at each point $r(s)$, the second derivative r'' measures the rate at which the unit tangent changes (turns or twists) as s increases. This is analogous to the one variable case in which the second derivative f'' measures the rate at which the slope function f' is changing. However, the derivatives here are vectors rather than numbers. We therefore make the following definition.

DEFINITION 9

Let the curve C in (1) be parameterized by arc length, and let r in (1) be twice differentiable for $a \le s \le b$. The vector $r''(s)$ is called the **curvature vector** at $r(s)$. Its length, $|r''(s)|$, is called the **curvature** of C at $r(s)$.

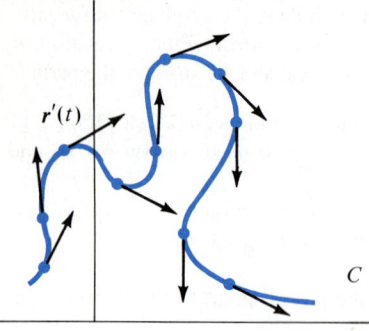

Figure 5.1 $\kappa(s)$ is large when C turns sharply.

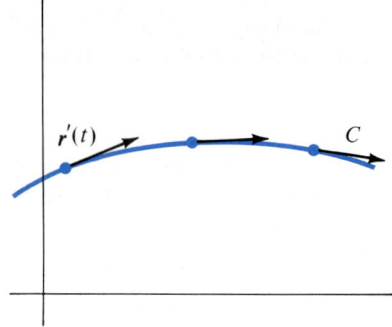

Figure 5.2 $\kappa(s)$ is small when C turns slowly.

REMARK: If C were not parameterized by arc length, then $|r''(s)|$ would depend upon both the direction and the magnitude of $r'(s)$. By insisting on parameterization by arc length, we have removed the dependence of the curvature on the "speed of the curve, $r'(s)$."

The Greek letter κ (kappa) is usually used to denote curvature. Thus, Definition 9 states that

$$\kappa(s) = |r''(s)|, \qquad a \le s \le b. \tag{2}$$

Figures 5.1 and 5.2 illustrate the essence of curvature: $\kappa(s)$ is large when C turns sharply (rapid change in $r'(s)$), and $\kappa(s)$ is small when C turns slowly (small change in $r'(s)$).

Example 1

In Example 7, Section 16.3, we determined that the circle of radius ρ centered at the origin is parameterized by arc length by the vector function

$$r(s) = \rho \cos\left(\frac{s}{\rho}\right)i + \rho \sin\left(\frac{s}{\rho}\right)j, \qquad 0 \le s < 2\rho\pi.$$

The curvature vector is therefore

$$r''(s) = -\frac{1}{\rho}\cos\left(\frac{s}{\rho}\right)i - \frac{1}{\rho}\sin\left(\frac{s}{\rho}\right)j = -\frac{1}{\rho^2}r(s)$$

and the curvature at $r(s)$ is

$$\kappa(s) = |r''(s)| = \frac{1}{\rho}.$$

When the circle is allowed to have any center, a slight generalization of this argument shows that *the curvature of a circle of radius ρ is $\kappa = 1/\rho$ at each point on the circle* (see Exercise 33). ◇

Figure 5.3 shows the curvature vectors for several circles of various radii.
For more general plane curves C parameterized by arc length, the expression $\rho(s) = \dfrac{1}{\kappa(s)}$ is called the **radius of curvature** at $r(s)$. Since $\rho(s)$ is the radius of the circle whose curvature is also $\kappa(s)$, the number $\rho(s)$ determines a circle that

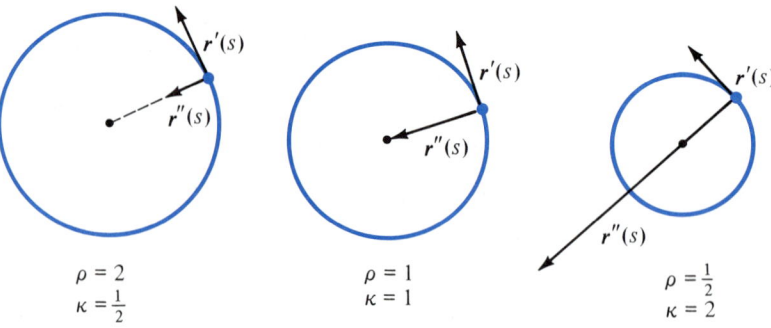

Figure 5.3 Tangent and curvature vectors for various circles.

(i) is tangent to C at $\mathbf{r}(s)$,

(ii) has radius $\rho(s) = \dfrac{1}{\kappa(s)}$,

(iii) has the same curvature vector as C at $\mathbf{r}(s)$.

This circle is called the **osculating circle,** or **circle of curvature,** at $\mathbf{r}(s)$. The center of this circle is called the **center of curvature.** Figures 5.4 and 5.5 show the osculating circles at various points on a curve C.

For plane curves that are not parameterized by arc length, the following theorem indicates how curvature may be computed.

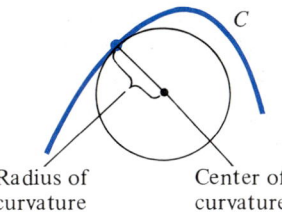

Radius of curvature   Center of curvature

Figure 5.4 Circle of curvature for a plane curve C.

Figure 5.5 Osculating circles at various points on C.

THEOREM 5

Let the curve C be determined parametrically by the twice differentiable function

$$\mathbf{r}(t) = x(t)\mathbf{i} + y(t)\mathbf{j}, \qquad a \leq t \leq b. \tag{3}$$

The curvature $\kappa(t)$ of C at the point $(x(t), y(t))$ is

$$\kappa(t) = \frac{|x'(t)y''(t) - y'(t)x''(t)|}{[(x'(t))^2 + (y'(t))^2]^{3/2}}. \tag{4}$$

Before indicating the proof of Theorem 5, we consider two examples.

Example 2

For the curve C given parametrically by the function

$$\mathbf{r}(t) = (2t^2 + 1)\mathbf{i} + (t^3 - 3)\mathbf{j}, \qquad 1 \leq t \leq 3$$

find

(a) the curvature function κ, and
(b) the radius of curvature at the point $(3, -2)$.

Solution: Here $x(t) = 2t^2 + 1$ and $y(t) = t^3 - 3$. Thus

$$x'(t) = 4t \qquad y'(t) = 3t^2,$$
$$x''(t) = 4 \qquad y''(t) = 6t.$$

According to Theorem 5, the curvature is

$$\kappa(t) = \frac{|4t \cdot 6t - 3t^2 \cdot 4|}{[(4t)^2 + (3t^2)^2]^{3/2}} = \frac{12t^2}{(16t^2 + 9t^4)^{3/2}}.$$

The point $(3, -2)$ corresponds to $t = 1$. The curvature at this point is therefore

$$\kappa(1) = \frac{12}{(16 + 9)^{3/2}} = \frac{12}{125}.$$

The radius of curvature at this point is

$$\rho(1) = \frac{1}{\kappa(1)} = \frac{125}{12}. \qquad \diamond$$

Example 3

Find the curvature function κ for the graph of the function $f(t) = \sin t$.

Solution: We can write the function $f(t) = \sin t$ in parametric (vector) form as

$$r(t) = t\mathbf{i} + \sin t\mathbf{j}.$$

Thus, $x(t) = t$ and $y(t) = \sin t$, so

$$x'(t) = 1 \qquad y'(t) = \cos t,$$
$$x''(t) = 0 \qquad y''(t) = -\sin t.$$

By Theorem 5,

$$\kappa(t) = \frac{|1 \cdot (-\sin t) - 0 \cdot \cos t|}{[1^2 + (\cos t)^2]^{3/2}} = \frac{|\sin t|}{(1 + \cos^2 t)^{3/2}}.$$

Notice that $\kappa(t)$ is a minimum (zero) at $t = 0, \pm\pi, \pm 2\pi, \ldots$ and that $\kappa(t)$ is a maximum (one) at $t = \pi/2 \pm n\pi$, $n = 1, 2, \ldots$. $\qquad \diamond$

Example 3 shows how Theorem 5 may be used to calculate the curvature for the graph of a twice differentiable function $y = f(x)$. Such functions can always be written in vector form as

$$r(x) = x\mathbf{i} + f(x)\mathbf{j}.$$

Then $x' = 1$, $x'' = 0$, $y' = f'(x)$, and $y'' = f''(x)$, so the curvature $\kappa(x)$, according to Theorem 5, is

$$\kappa(x) = \frac{|0 \cdot f'(x) - f''(x) \cdot 1|}{[1 + (f'(x))^2]^{3/2}}.$$

We state this result as a corollary of Theorem 5.

COROLLARY 2

Let f be a twice differentiable function of x. The curvature $\kappa(x)$ at the point $(x, f(x))$ on the graph of $y = f(x)$ is

$$\kappa(x) = \frac{|f''(x)|}{[1 + (f'(x))^2]^{3/2}}.$$

Example 4

Show that the curvature of a nonvertical line in the plane is zero.

Solution: The equation for a nonvertical line in the plane can be written in the form $f(x) = mx + b$ where m and b are constants. Thus $f'(x) = m$ and $f''(x) = 0$. Thus, $\kappa(x) = 0$, according to Corollary 2. ◇

Proof of Theorem 5: We will demonstrate the proof up to a point at which you can complete the argument with a calculation (Exercise 34).

The difficulty to be overcome is that the parameterization given by $r(t)$ in equation (3) is not necessarily a parameterization by arc length. We address this problem by letting ℓ denote the length of the arc of C from $r(a)$ to $r(b)$. Then the *arc length function*

$$s(t) = \int_a^t |r'(\omega)|\, d\omega$$

$$= \int_a^t \sqrt{[x'(\omega)]^2 + [y'(\omega)]^2 + [z'(\omega)]^2}\, d\omega \qquad (5)$$

gives, as for curves in the plane, the length of the arc of C from $r(a)$ to $r(t)$. (See Section 16.3.) This function has the properties that

$$s(a) = 0,$$
$$s(b) = \ell,$$

and

$$s'(t) = |r'(t)|, \qquad a \le t \le b. \qquad (6)$$

Now, suppose that the vector function

$$R(s) = X(s)i + Y(s)j, \qquad 0 \le s \le \ell$$

is an arc length parameterization for C. Then the composition of R with s in (5) is

$$R(s(t)) = X(s(t))i + Y(s(t))j, \qquad a \le t \le b.$$

It is important to note that $R(s(t)) = r(t)$ for each $t \in [a, b]$, since both vectors terminate at a point $s(t)$ units along C from $r(a)$ (see Figure 5.6).

Since $R(s)$ is a parameterization by arc length,

$$T(t) = R'(s) = \frac{d}{ds} R(s(t)) \qquad (7)$$

is a unit tangent for C at $r(t) = R(s(t))$ (equation (12), Section 17.3). Differentiating both sides of equation (7) with respect to t, gives

$$T'(t) = \frac{d}{dt}\left(\frac{d}{ds} R\right)$$

$$= \frac{d}{ds}\left(\frac{d}{ds} R\right)\frac{ds}{dt} \qquad \text{(by Chain Rule)}$$

$$= R''(s)\frac{ds}{dt}$$

$$= R''(s)\, |r'(t)| \qquad \text{(by equation (6)),}$$

so

$$|R''(s)| = \frac{|T'(t)|}{|r'(t)|}. \qquad (8)$$

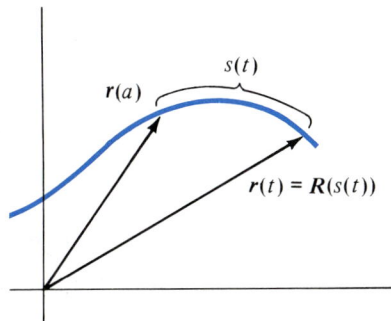

$s(t)$
$r(a)$
$r(t) = R(s(t))$

Figure 5.6 $R(s(t)) = r(t)$, $a \le t \le b$.

The expression in equation (8) is the curvature of C at $r(t) = R(s(t))$. It is left for you to show that

$$\frac{|T'(t)|}{|r'(t)|} = \frac{|x''(t)y'(t) - x'(t)y''(t)|}{[(x'(t))^2 + (y'(t))^2]^{3/2}},\qquad(9)$$

which will complete the proof. ◆

Curvature for Space Curves

If C is a space curve parameterized by arc length by the vector function

$$C:\quad r(s) = x(s)i + y(s)j + z(s)k,\qquad a \le s \le b,\qquad(10)$$

the vector $T(s) = r'(s)$ is a unit tangent to the curve C at $r(s)$. Since $|r'(s)| = 1$, the curvature vector $r''(s)$ is orthogonal to the tangent $T(s)$ (Example 6, Section 17.2). Thus, the curvature vector $r''(s)$ is **normal** to C at $r(s)$, and the **principal unit normal** to C at $r(s)$ is the vector

$$N(s) = \frac{1}{|r''(s)|}r''(s).\qquad(11)$$

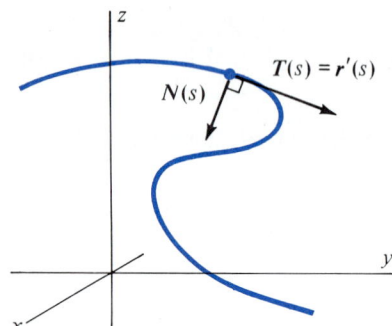

Of course, the principal unit normal in (11) is defined only when r is twice differentiable and $|r''(s)| \ne 0$ (see Figure 5.7). If C in (10) is a plane curve, the vector $N(s)$ points toward the center of curvature. Multiplying (11) through by $|r''(s)| = \kappa(s)$ and writing $r'(s) = T(s)$ gives the equation

$$T'(s) = \kappa(s)N(s),\qquad(12)$$

an equation by which we could alternatively have defined curvature for space curves.

Figure 5.7 Principal unit normal is $N(s) = \dfrac{r''(s)}{|r''(s)|}$.

Normal and Tangential Components of Acceleration

We may apply these ideas to obtain an informative expression for the velocity of a particle moving in space with position function given by

$$r(t) = x(t)i + y(t)j + z(t)k.\qquad(13)$$

We assume that r is twice differentiable (although not necessarily an arc length parameterization for the path).

Letting s denote arc length and $T = \dfrac{dr}{ds}$ the unit tangent, we apply the Chain Rule to obtain the velocity function v as

$$v(t) = \frac{dr}{dt} = \frac{dr}{ds}\frac{ds}{dt} = \frac{ds}{dt}T.$$

Similarly, the acceleration function is

$$a(t) = \frac{dv}{dt} = \frac{d}{dt}\left(\frac{dr}{ds}\frac{ds}{dt}\right)\qquad(14)$$

$$= \frac{d^2s}{dt^2}\frac{dr}{ds} + \frac{ds}{dt}\left(\frac{d^2r}{ds^2}\frac{ds}{dt}\right)$$

$$= \frac{d^2s}{dt^2}\frac{dr}{ds} + \left(\frac{ds}{dt}\right)^2\left[\frac{d}{ds}\left(\frac{dr}{ds}\right)\right]$$

$$= \frac{d^2s}{dt^2}T(s) + \left(\frac{ds}{dt}\right)^2\kappa(s)N(s)\qquad\text{(equation (12))}.$$

That is, we can express acceleration as the sum of a vector parallel to the unit tangent $T(s)$ and a vector parallel to the unit normal $N(s)$. It is customary to refer to the **tangential component of acceleration** as the coefficient $a_T = \dfrac{d^2s}{dt^2}$, and to the **normal component of acceleration** as the coefficient $a_N = \left(\dfrac{ds}{dt}\right)^2 \kappa(s)$. We may then write

$$a = a_T T + a_N N. \tag{15}$$

The components a_T and a_N have straightforward interpretations. The tangential component is

$$a_T = \frac{d^2s}{dt^2} = \frac{d}{dt}\left(\frac{ds}{dt}\right) = \frac{d}{dt}(|v(t)|),$$

so that a_T is just the rate at which speed is changing. For example, if the speed of the particle is constant, then $a_T = 0$. The normal component of acceleration can be written $a_N = \left(\dfrac{ds}{dt}\right)^2 \kappa(s) = |v|^2 \kappa$. Thus, a_N is influenced both by speed and by curvature. Since N is orthogonal to T, this explains why the force tending to pull an object (such as an automobile) from its path is influenced both by the speed of the object and by the curvature of the path.

Since $a_T = \dfrac{d}{dt}|v|$, computing a_T from r is usually simple. However, calculating a_N can be considerably more difficult. An aid to this calculation is obtained by using equation (15) and the fact that T and N are orthogonal unit vectors:

$$\begin{aligned}
|a|^2 = a \cdot a &= (a_T T + a_N N) \cdot (a_T T + a_N N) \\
&= (a_T)^2 |T|^2 + a_N |N|^2 \\
&= a_T^2 + a_N^2.
\end{aligned}$$

Thus

$$a_N = \sqrt{|a|^2 - a_T^2}. \tag{16}$$

Example 5

Find the tangential and normal components of acceleration for a particle moving along the helical path

$$r(t) = \cos t^2 i + \sin t^2 j + t k.$$

Strategy

Find $v(t) = r'(t)$.

Solution

Here

$$v(t) = -2t \sin t^2 i + 2t \cos t^2 j + k,$$

so

Find $|v(t)|$.

$$\begin{aligned}
|v(t)| &= \sqrt{4t^2 \sin^2 t^2 + 4t^2 \cos^2 t^2 + 1} \\
&= \sqrt{4t^2 + 1}.
\end{aligned}$$

Thus

$$a_T = \frac{d}{dt}(|v(t)|).$$

$$a_T = \frac{d}{dt}(\sqrt{4t^2 + 1}) = \frac{4t}{\sqrt{4t^2 + 1}}.$$

Find *a*.

To find *a* we differentiate *v*:

$$a(t) = (-2 \sin t^2 - 4t^2 \cos t^2)i + (2 \cos t^2 - 4t^2 \sin t^2)j,$$

so

Find $|a(t)|$.

$$\begin{aligned}|a(t)| &= \{4 \sin^2 t^2 + 16t^2 \sin t^2 \cos t^2 + 16t^4 \cos^2 t^2 + 4 \cos^2 t^2 - \\ &\quad 16t^2 \sin t^2 \cos t^2 + 16t^4 \sin^2 t^2\}^{1/2} \\ &= \sqrt{4 + 16t^4}.\end{aligned}$$

Apply equation (16).

Thus

$$\begin{aligned}a_N = \sqrt{|a|^2 - a_T^2} &= \sqrt{(4 + 16t^4) - \frac{16t^2}{4t^2 + 1}} \\ &= \sqrt{\frac{64t^6 + 16t^4 + 4}{4t^2 + 1}}.\end{aligned}$$

Notice that $\lim_{t \to \infty} a_T = 2$, while $\lim_{t \to \infty} a_N = +\infty$. Can you explain why this is so?

◇

Vector Form of Curvature

A drawback of our work on curvature up to this point is that $\kappa(s)$ is difficult to compute, using Definition 9, when the curve C determined by *r* is not parameterized by arc length. We shall now develop a formula for κ that is easier to apply. We imagine the parameterization *r* for C as the position function of a particle moving in space. Then, writing

$$v(t) = \frac{dr}{dt} = \frac{dr}{ds} \cdot \frac{ds}{dt} = \left(\frac{ds}{dt}\right)T$$

and using equation (14) we find that

$$\begin{aligned}v \times a &= \left(\frac{ds}{dt}\right)T \times \left[\left(\frac{d^2s}{dt^2}\right)T + \kappa\left(\frac{ds}{dt}\right)^2 N\right] \\ &= \kappa\left(\frac{ds}{dt}\right)^3 (T \times N)\end{aligned}$$

since $T \times T = 0$. Also, since T and N are orthogonal unit vectors,

$$|T \times N| = |T||N||\sin \theta| = 1.$$

Thus,

$$|v \times a| = \kappa\left(\frac{ds}{dt}\right)^3 = \kappa|v|^3,$$

so

$$\boxed{\kappa = \frac{|v \times a|}{|v|^3}.}$$

(17)

Example 6

Find the curvature of the helix

$$C: \quad r(t) = \cos t\,i + \sin t\,j + t\,k.$$

Solution: Here

$$v(t) = -\sin t i + \cos t j + k$$

and

$$a(t) = -\cos t i - \sin t j.$$

Thus

$$v \times a = \det \begin{bmatrix} i & j & k \\ -\sin t & \cos t & 1 \\ -\cos t & -\sin t & 0 \end{bmatrix}$$

$$= \sin t i - \cos t j + (\sin^2 t + \cos^2 t)k,$$

so

$$|v \times a| = \sqrt{\sin^2 t + \cos^2 t + 1} = \sqrt{2}.$$

Also,

$$|v| = \sqrt{\sin^2 t + \cos^2 t + 1} = \sqrt{2}.$$

Thus, by (17),

$$\kappa = \frac{\sqrt{2}}{(\sqrt{2})^3} - \frac{1}{2}.$$

◇

Exercise Set 16.5

In Exercises 1–4, verify that the given curve is parameterized by arc length. Then find (a) the unit tangent $T(s)$, (b) the curvature vector $r''(s)$, and (c) the curvature $\kappa(s)$.

1. $r(s) = \frac{1}{2} s i + \frac{\sqrt{3}}{2} s j$

2. $r(s) = (4 + \cos s)i + (2 + \sin s)j$

3. $r(s) = \frac{\sqrt{2}}{2} \cos s i + \frac{\sqrt{2}}{2} \sin s j + \frac{\sqrt{2}}{2} s k$

4. $r(s) = \sin\left(\frac{s}{2}\right)i + \frac{\sqrt{3}}{2} s j + \cos\left(\frac{s}{2}\right)k$

In Exercises 5–10, find the curvature function $\kappa(t)$ for the given plane curves.

5. $r(t) = 3i + t^2 j$

6. $r(t) = ti + (t^2 + 3)j$

7. $r(t) = ti + e^t j$

8. $r(t) = e^t i + e^{-t} j$

9. $r(t) = (3 + 2 \sin t)i + (5 + 2 \cos t)j$

10. $r(t) = (t - \sin t)i + (1 - \cos t)j$

In Exercises 11–16, find the curvature of the graph of the given function.

11. $f(x) = 3 + x^2$

12. $f(x) = \sqrt{x + 4}$

13. $f(x) = x^3$

14. $f(x) = \ln x$

15. $f(x) = \cos x$

16. $f(x) = e^{x^2}$

In Exercises 17–20, find the curvature $\kappa(t)$ for the space curve determined by $r(t)$.

17. $r(t) = i + tj + t^2 k$

18. $r(t) = \cos t i + \sin t j + k$

19. $r(t) = e^t \cos t i + e^t \sin t j + tk$

20. $r(t) = \ln t i + tj + tk, \qquad t > 0$

21. Find the center of curvature for the graph of $y = \sin x$ at the point $(\pi/2, 1)$.

22. Find the point on the graph of $y = \ln x$ where curvature is a maximum.

23. For the graph of $y = \sqrt{x}$, find
   a. the curvature at $(1, 1)$,
   b. the center of curvature at $(1, 1)$,
   c. $\lim_{x \to 0^+} \kappa(x)$.

24. Find the equation for the osculating circle (circle of curvature) for the graph of $y = e^{-x}$ at the point where $x = 1$.

25. Show that the curvature of a line in space is zero.

26. Show that if the curve C is the graph of the polar equation $r = f(\theta)$ and if $f''(\theta)$ exists, then the curvature $\kappa(\theta)$ is given by

$$\kappa(\theta) = \frac{|f(\theta)f''(\theta) - 2[f'(\theta)]^2 - [f(\theta)]^2|}{[(f'(\theta))^2 + (f(\theta))^2]^{3/2}}.$$

(*Hint:* C can be written as $r(\theta) = x(\theta)i + y(\theta)j$, with $x(\theta) = f(\theta) \cos\theta$, $y(\theta) = f(\theta) \sin\theta$.)

In Exercises 27–30, use the result of Exercise 26 to find the curvature function $\kappa(\theta)$.

27. $r = 1 + \sin\theta$ **28.** $r = \theta$

29. $r = 1 - \cos\theta$ **30.** $r = e^\theta$

31. For a curve C in the plane, explain why the curvature vector $r''(s)$ always points in the direction of the concave side of the curve.

32. Find the curvature of the ellipse $x^2 + 2y^2 = 4$ at the point $(2, 0)$.

33. Prove that the curvature of a circle of radius ρ is $\kappa = \dfrac{1}{\rho}$.

34. Complete the proof of Theorem 5.

35. Find the tangential and normal components of acceleration for a particle moving along the helical path $2ti + \sin t^2 j + \cos t^2 k$.

SUMMARY OUTLINE OF CHAPTER 16

◆ A **vector-valued function** f has the form   (page 705)

$$f(t) = x(t)i + y(t)j, \quad \text{or} \quad f(t) = x(t)i + y(t)j + z(t)k$$

where $x(t)$, $y(t)$, and $z(t)$ are real-valued functions.

◆ For $f(t) = x(t)i + y(t)j + z(t)k,$ (page 708)

(i) $\lim\limits_{t\to t_0} f(t) = \left[\lim\limits_{t\to t_0} x(t)\right]i + \left[\lim\limits_{t\to t_0} y(t)\right]j + \left[\lim\limits_{t\to t_0} z(t)\right]k$

(ii) $f'(t) = x'(t)i + y'(t)j + z'(t)k$

(iii) $\int f(t)\,dt = \left[\int x(t)\,dt\right]i + \left[\int y(t)\,dt\right]j + \left[\int z(t)\,dt\right]k$

(iv) $\int_a^b f(t)\,dt = \left[\int_a^b x(t)\,dt\right]i + \left[\int_a^b y(t)\,dt\right]j + \left[\int_a^b z(t)\,dt\right]k.$

(v) f is continuous at t_0 if and only if $x(t)$, $y(t)$, and $z(t)$ are all continuous at t_0.

◆ Derivatives of vector-valued functions are calculated as follows: (page 712)

(i) $(f + g)'(t) = f'(t) + g'(t)$
(ii) $(cf)'(t) = cf'(t)$
(iii) $(hf)'(t) = h(t)f'(t) + h'(t)f(t)$
(iv) $(f \times g)'(t) = [f(t) \times g'(t)] + [f'(t) \times g(t)]$
(v) $(f \circ h)'(t) = h'(t)f'(h(t))$
(vi) $(f \cdot g)'(t) = f(t) \cdot g'(t) + f'(t) \cdot g(t).$

◆ If $r'(t_0) \neq 0$, the **tangent** to the graph of r at $r(t_0)$ is $r'(t_0) = x'(t_0)i + y'(t_0)j + z'(t_0)k$.   (page 720)

◆ The **unit tangent** to the graph of r at $r(t_0)$ is $T = \dfrac{1}{|r'(t_0)|}r'(t_0)$ $r'(t_0) \neq 0.$ (page 722)

◆ The curve C is **parameterized by arc length** on the interval I if   (page 725)

$$\int_0^t |r'(\omega)|\,d\omega = t$$

for all t in I. Equivalently, $|r'(t)| = 1$, $t \in I.$

◆ For a particle moving in space with the twice differentiable **position function** r,   (page 728)

(i) the **velocity** function is $v = r'$

(ii) the **speed** is $|v(t)| = |r'(t)|$

(iii) the **acceleration** function is $a = v' = r''$.

◆ The **trajectory** of a projectile fired from the origin with initial speed s_0 and angle of elevation θ is   (page 732)

$$r(t) = s_0(\cos\theta)ti + \left(s_0 \sin\theta t - \frac{1}{2}gt^2\right)j.$$

◆ If C: $r(s)$ is parameterized by arc length, the **curvature** at $r(s)$ is $\kappa(s) = |r''(s)|$.   (page 735)

◆ The **radius of curvature** at $r(s)$ is   (page 736)

$$\rho(s) = \frac{1}{\kappa(s)}, \qquad \kappa(s) \ne 0.$$

◆ If the plane curve C is parameterized by the equations $x = x(t)$, $y = y(t)$, the **curvature** at $(x(t), y(t))$ is   (page 737)

$$\kappa(t) = \frac{|x'(t)y''(t) - y'(t)x''(t)|}{[(x'(t))^2 + (y'(t))^2]^{3/2}}.$$

◆ If the curve C is the graph of $y = f(x)$,   (page 738)

$$\kappa(x) = \frac{|f''(x)|}{[1 + (f'(x))^2]^{3/2}}.$$

◆ The **principal unit normal** to $r(s)$ is $N(s) = \dfrac{1}{|r''(s)|}r''(s)$ when $r(s)$ is parameterized by arc length.   (page 740)

◆ If r is a twice differentiable position function, its acceleration function can be written $a(t) = a_T T + a_N N$, where the **tangential component of acceleration** is   (page 741)

$$a_T = \frac{d^2 s}{dt^2}$$

and the **normal component of acceleration** is

$$a_N = \left(\frac{ds}{dt}\right)^2 \kappa(s) = |v|^2 \kappa.$$

◆ If the curve C is determined by the twice differentiable **position function r**, then   (page 742)

$$\kappa = \frac{|v \times a|}{|v|^3}$$

at each point on C.

REVIEW EXERCISES—CHAPTER 16

1. Find an equation in rectangular coordinates for the graph of the vector function $r(t) = a \cos ti + b \sin tj$.

2. Sketch the graph of the vector function $r(t) = i + \cos tj + k$.

3. Find the implicit domain of the vector function

$$r(t) = ti + \sqrt{1 - t^2}j + \frac{1}{t}k.$$

4. Show that the graph of the vector function $r(t) = \tan ti + \sec tj$, $-\pi/2 < t < \pi/2$, is one branch of the hyperbola with rectangular equation $y^2 - x^2 = 1$. What are the asymptotes for this curve?

5. Sketch the graph of the **cycloid** given by the vector function $r(t) = a(t - \sin t)i + a(1 - \cos t)j$, $t \ge 0$. Find the numbers t for which the tangent vector is zero. (The cycloid gives the path of a point on the rim of a wheel of radius a as the wheel rolls along a horizontal surface.)

In Exercises 6–8, find the indicated limit.

6. $\displaystyle\lim_{t \to 0^+}\left\{\cos 2ti + t \cos tj + \frac{t}{\sqrt{t + 2}}k\right\}$

7. $\displaystyle\lim_{t \to 0}\left\{\left(\frac{\sin 3t}{t}\right)i + \sqrt{t + 2}j + \frac{\tan 2t}{t}k\right\}$

8. $\lim_{t \to 2^+} \left\{ \sqrt{2 + t}\, i + \ln \sqrt{t}\, j + \dfrac{t^3 - 8}{t - 2} k \right\}$

In Exercises 9–11, determine the numbers t for which the function is discontinuous.

9. $r(t) = \sin t\, i + \sec t\, j, \qquad 0 \le t \le 2\pi$

10. $r(t) = \dfrac{1}{\sqrt{t}}\, i + \tan t\, j + \ln(1 + t)k, \qquad 0 \le t \le \pi$

11. $r(t) = \dfrac{1}{t^2 - 9}\, i + \dfrac{t + 2}{t^2 - 4}\, j + t k, \qquad -4 \le t \le 4$

In Exercises 12–15, find r' and r''.

12. $r(t) = e^{-2t} i + \cos\left(\dfrac{\pi t}{4}\right) j + k$

13. $r(t) = t e^t i + t e^{-t} j$

14. $r(t) = \operatorname{Tan}^{-1} 2t\, i + \sqrt{t}\, j + t k$

15. $r(t) = \ln 3t\, i + e^{\sqrt{t}}\, j$

16. Find $\displaystyle\int_0^1 r(t)\, dt$ for r in Exercise 12.

17. Find $\displaystyle\int_0^1 r(t)\, dt$ for r in Exercise 13.

18. Find a vector function r for which $r'(t) = 4r(t)$ and $r(0) = 4i - j + \pi k$.

19. Find a vector function r for which $r''(t) = 9r(t)$ and $r(0) = 4i + 3j$.

20. Find the numbers α for which the graph of the vector function

$$r(t) = \alpha t i + \cos(2\alpha t) j + \sin(2\alpha t) k, \qquad 0 \le t \le \pi$$

is parameterized by arc length.

21. Find an arc length parameterization for the circle $x^2 + y^2 = 25$.

22. Find the unit tangent $T(t)$ for the ellipse $r(t) = a \cos t\, i + b \sin t\, j$, $0 \le t < 2\pi$.

23. For the graph of the vector function $r(t) = t i + \cos t\, j + \sin t k$, find

a. the curvature, $\kappa(t)$,
b. the unit tangent $T(t)$.

24. A particle moves in space with position function $r(t) = a \cos t\, i + t b\, j + a \sin t k$. Show that
a. the velocity vector $v(t)$ and the acceleration vector $a(t)$ have constant lengths,
b. $v(t)$ and $a(t)$ are orthogonal for each t.

25. Find the length of the helix $r(t) = 3t i + \cos 2t\, j + \sin 2t k$ between the points $r(0)$ and $r(3)$.

26. Find the length of the curve determined by the position function $r(t) = \ln t\, i + t j$ between $r(1)$ and $r(2)$.

27. Find the curvature of the cubic curve $y = x^3$ at the point $(2, 8)$.

28. Find the curvature of the curve determined by the vector function $r(t) = 3i + \sqrt{t}\, j + t^2 k$ at the point $r(4)$.

29. Let C be a curve in space determined by the differentiable vector function r, $a \le t \le b$. Show that if $t_0 \ne a$, $t_0 \ne b$, and $r(t_0)$ is a point on C either nearest or farthest from the origin, then $r(t_0)$ and $r'(t_0)$ are orthogonal. (*Hint:* Consider $|r(t)|^2 = r(t) \cdot r(t)$.)

30. Use Newton's law $F = ma$ to show that if a particle moves in space subject to zero net force, then
a. the motion is along a line, and
b. both the tangential and normal components of acceleration are zero.

31. Prove that for a particle moving in space, the speed of the particle is constant if and only if its velocity and acceleration vectors are orthogonal at all points along its path.

32. Find the radius of curvature and an equation for the circle of curvature at the point $(0, 1)$ on the graph of $y = \cos x$.

33. Find the point on the graph of $y = e^x$ at which the radius of curvature is a minimum.

34. A particle moves along a curve with position function $r(t) = t \cos t\, i + t \sin t\, j$. Find its speed as a function of t.

35. A projectile is fired with an angle of elevation $\alpha = 30°$ and an initial speed $|v(0)| = 20$ m/s at a vertical wall 10 meters away. At what height does it strike the wall?

Chapter 17
Differentiation for Functions of Several Variables

The goal of this chapter is to extend the theory of differentiation to real-valued functions that involve more than one independent variable. Examples of such functions abound in nature and in the social sciences. Here are two.

(i) In chemistry, the ideal gas law relates the pressure, volume, and temperature of an ideal gas by the equation

$$PV = nRT$$

where n is the number of moles of the gas present and R is a constant. When written in the form

$$P = (nR)\left(\frac{T}{V}\right) = f(n, T, V),$$

this equation determines P as a function of n, T, and V.

(ii) In economics, one encounters the simple equation

$$R = Px$$

for the revenue obtained by selling x items at a price of P dollars per item. This equation has the form $R = f(P, x)$, meaning that one must know both the selling price and the number of sales in order to calculate revenue.

17.1 FUNCTIONS OF SEVERAL VARIABLES

By a **function of two variables,** we shall mean a rule that assigns a unique number to each *pair* (x, y) of numbers for which the rule is defined. Functions of two variables will generally be written in the form

$$z = f(x, y).$$

If the function f is defined for all ordered pairs (x, y), we may represent this situation symbolically as

$$f\colon\ \mathbb{R}^2 \to \mathbb{R}, \quad \text{or} \quad (x, y) \xrightarrow{\ f\ } z$$

where $\mathbb{R}^2$ represents the set of all pairs (x, y) with both $x \in \mathbb{R}$ and $y \in \mathbb{R}$. (You may think of $\mathbb{R}^2$ as simply the coordinate plane.) If the function f is defined only for (x, y) in a subset D of $\mathbb{R}^2$ (called the **domain** of the function), we write

$$f\colon\ D \to \mathbb{R}, \quad D \subset \mathbb{R}^2.$$

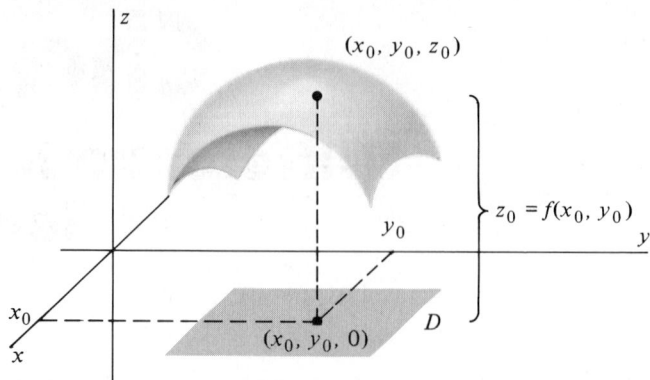

Figure 1.1 Graph of a function $z = f(x, y)$ of two variables with domain D. (See **Plates 1** and **2.**)

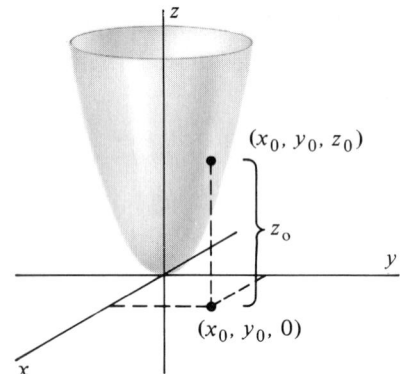

Figure 1.2 Paraboloid $z = x^2 + y^2$ is the graph of a function of two variables. (See **Plate 3.**)

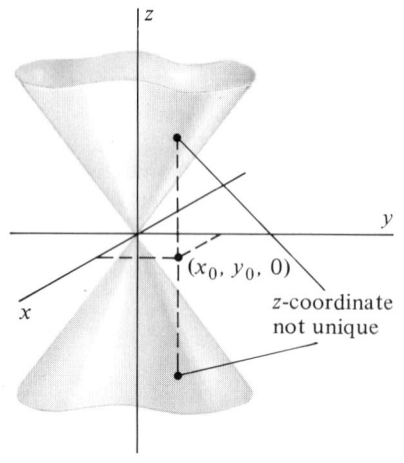

Figure 1.3 Cone $z^2 = x^2 + y^2$ is not the graph of a function of two variables. (See **Plate 4.**)

As for functions of a single variable, we will refer to the set of values $f(x, y)$, for $(x, y) \in D$, as the **range** of the function f.

Figure 1.1 shows how we may graph functions of two variables. We interpret the ordered pairs (x, y) as points in the xy-plane. We indicate the value of the function, $z = f(x, y)$, by plotting the point (x, y, z) in space. Then, the height of the point (x, y, z) above or below the point $(x, y, 0)$ represents the number (or **value**) z assigned by the function to the ordered pair (x, y). The set of all such points is called the **graph of the function** $z = f(x, y)$.

Notice that while graphs of functions of a single variable are *curves in the plane* (in general), graphs of functions of two variables are *surfaces in space*.

An example of a function of two variables is given by the equation

$$z = x^2 + y^2.$$

The graph of this function is the paraboloid in Figure 1.2. However, the equation

$$z^2 = x^2 + y^2 \tag{1}$$

does *not* describe a function. The reason is that each pair of independent variables $(x, y) \neq (0, 0)$ corresponds to *two* values of z. This can be seen by writing equation (1) in the form

$$z = \pm\sqrt{x^2 + y^2}.$$

The graph of equation (1) is the cone in Figure 1.3.

Functions of three or more independent variables are defined in the analogous way. A function w of the three independent variables x, y, and z takes the form

$$w = f(x, y, z),$$

and we write

$$f: \ \mathbb{R}^3 \to \mathbb{R}, \qquad \text{or} \qquad (x, y, z) \xrightarrow{f} w.$$

An example of a function of three variables is the temperature function for the room in which you are sitting. If you use a rectangular coordinate system to describe the space within the room, the number $w_0 = T(x_0, y_0, z_0)$ is the temperature at the point in the room with coordinates (x_0, y_0, z_0). Unfortunately, we cannot sketch a graph for a function of three variables, since all three axes are required just to plot the points in its domain.

More generally, a function of the n independent variables $x_1, x_2, \ldots, x_n$ has the form

$$w = f(x_1, x_2, \ldots, x_n).$$

Symbolically, we describe such functions by writing

$$f\colon \ \mathbb{R}^n \to \mathbb{R}, \qquad \text{or} \qquad (x_1, x_2, \ldots, x_n) \xrightarrow{f} w.$$

Graphing Techniques: Traces and Level Curves

Graphing a function of two variables is difficult, at best. Two techniques frequently give a general idea of the appearance of the graph of the function $z = f(x, y)$. The first is simply setting one of the two independent variables equal to a constant, so that we obtain a function of a single variable, whose graph can be sketched in the appropriate plane in space. Setting $x = c$ in $z = f(x, y)$ gives the equation $z = f(c, y) = g(y)$, whose graph is the intersection of the desired graph with the plane $x = c$; setting $y = c$ in $z = f(x, y)$ gives the equation $z = f(x, c) = h(x)$, whose graph is the intersection of the desired graph with the plane $y = c$.

We refer to the intersections of the graph of f with the planes $x = c$ or $y = c$ as the **traces** of f in the respective planes. By sketching traces of f in several planes, we can sometimes gain a fairly accurate picture of the graph of f.

Example 1

Sketch several traces for the graph of $z = x^2 + y^2$.

Solution: Tables 1.1 and 1.2 show the results of setting one of the independent variables equal to one of several constants.

Traces in each of these planes are indicated in Figures 1.4 and 1.5. They are combined in Figure 1.6. Figure 1.2 is a sketch of the corresponding paraboloid.

$\diamondsuit$

Example 2

For the function $f(x, y) = xy$, the equations of various traces are given in Tables 1.3 and 1.4.

Table 1.1

Plane $y = c$	Function $z = f(x, c)$
0	$z = x^2$
1	$z = x^2 + 1$
2	$z = x^2 + 4$
-1	$z = x^2 + 1$
-2	$z = x^2 + 4$

Table 1.2

Plane $x = c$	Function $z = f(c, y)$
0	$z = y^2$
1	$z = y^2 + 1$
2	$z = y^2 + 4$
-1	$z = y^2 + 1$
-2	$z = y^2 + 4$

Table 1.3

Plane $y = c$	Equation $z = f(x, c)$
0	$z = 0$
1	$z = x$
2	$z = 2x$
-1	$z = -x$
-2	$z = -2x$

Table 1.4

Plane $x = c$	Equation $z = f(c, y)$
0	$z = 0$
1	$z = y$
2	$z = 2y$
-1	$z = -y$
-2	$z = -2y$

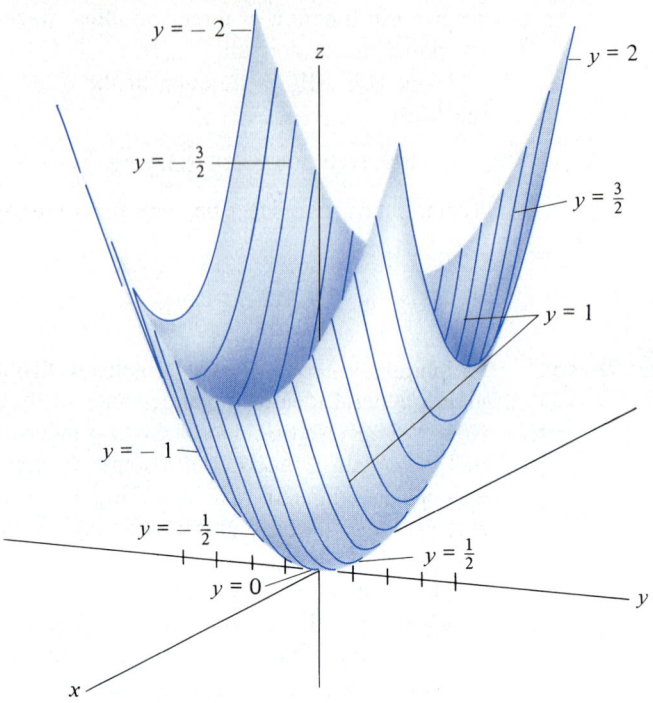

Figure 1.4 Traces of $z = x^2 + y^2$ in planes $y = c$.

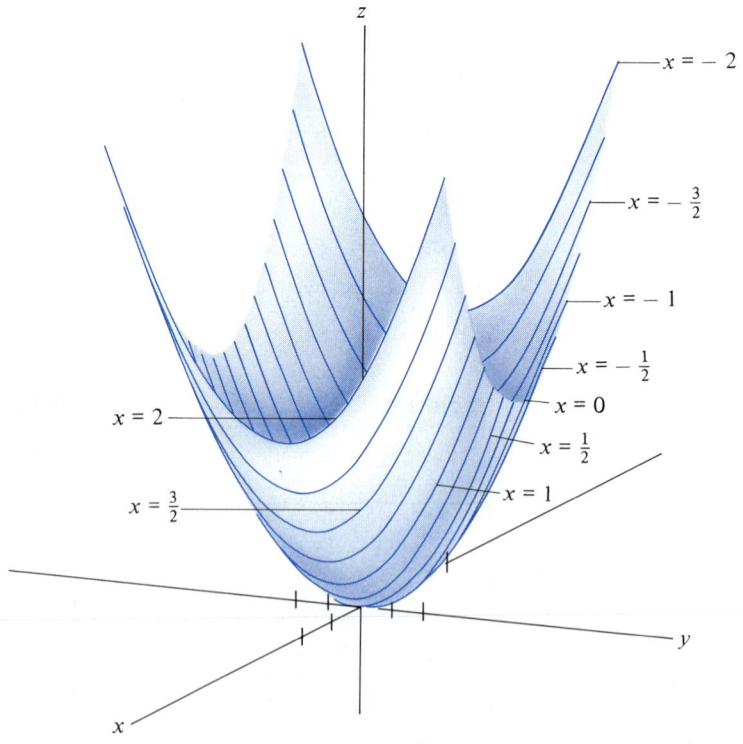

Figure 1.5 Traces of $z = x^2 + y^2$ in planes $x = c$.

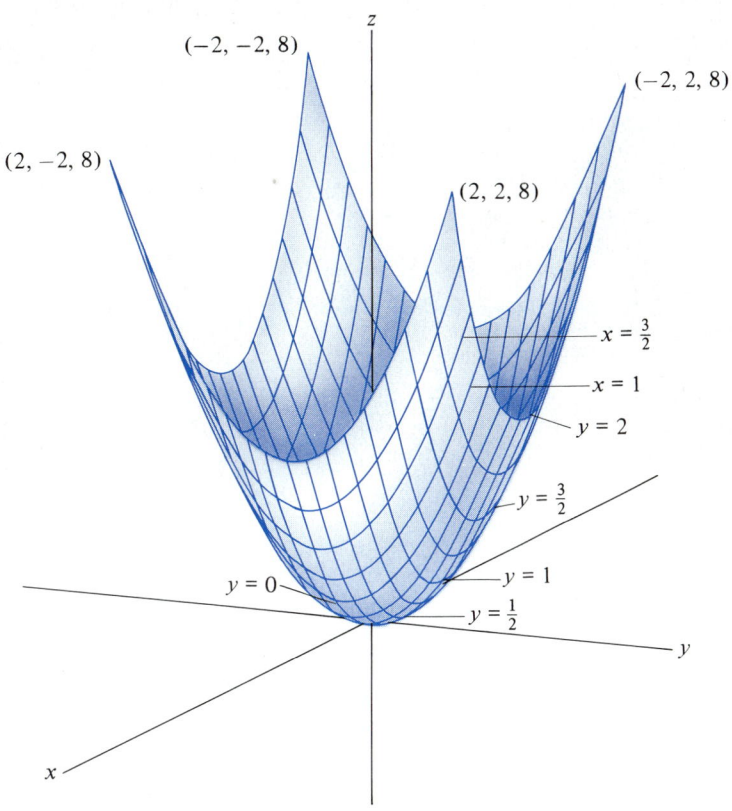

Figure 1.6 Traces of $z = x^2 + y^2$ in planes of both types.

These traces, and several others, are sketched in Figures 1.7 and 1.8. Figure 1.9 shows a portion of the graph of $f(x, y) = xy$. ◇

Example 3

For the function $f(x, y) = y^2 - x^2$, the equations of various traces are given in Tables 1.5 and 1.6. They are all parabolas.

Table 1.5

Plane $y = c$	Equation $z = f(x, c)$
0	$z = -x^2$
1	$z = 1 - x^2$
2	$z = 4 - x^2$
-1	$z = 1 - x^2$
-2	$z = 4 - x^2$

Table 1.6

Plane $x = c$	Equation $z = f(c, y)$
0	$z = y^2$
1	$z = y^2 - 1$
2	$z = y^2 - 4$
-1	$z = y^2 - 1$
-2	$z = y^2 - 4$

Figure 1.10 shows both types of traces. Figure 1.11 shows the graph of $f(x, y) = y^2 - x^2$. ◇

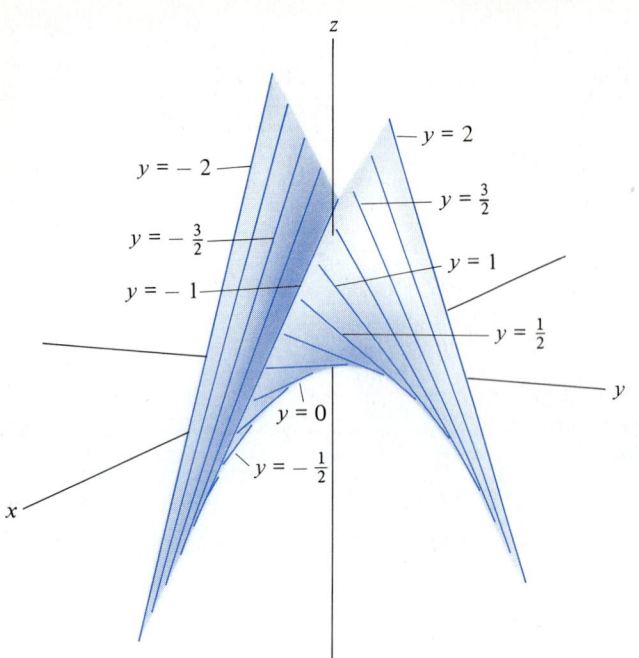

Figure 1.7 Traces of $z = xy$ in planes $y = c$.

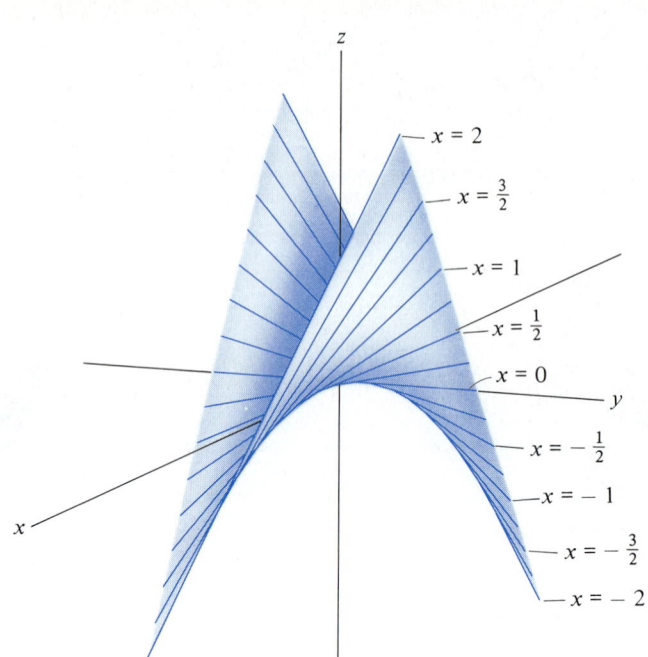

Figure 1.8 Traces of $z = xy$ in planes $x = c$.

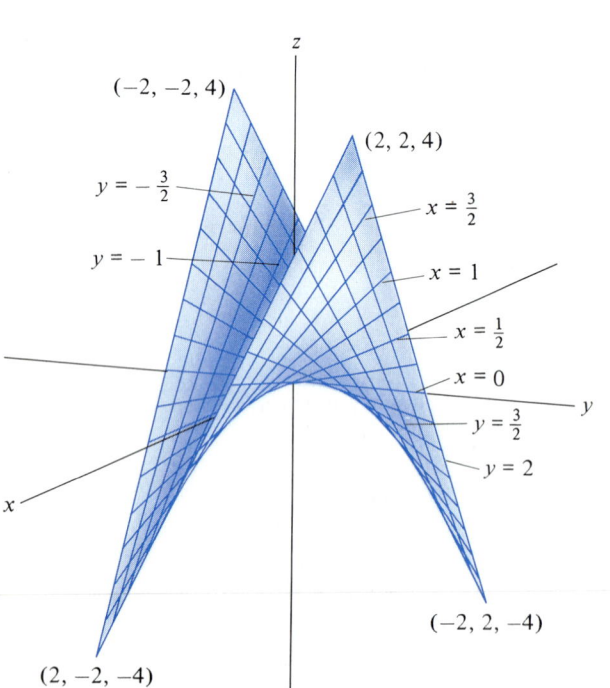

Figure 1.9 Graph of $f(x, y) = xy$.

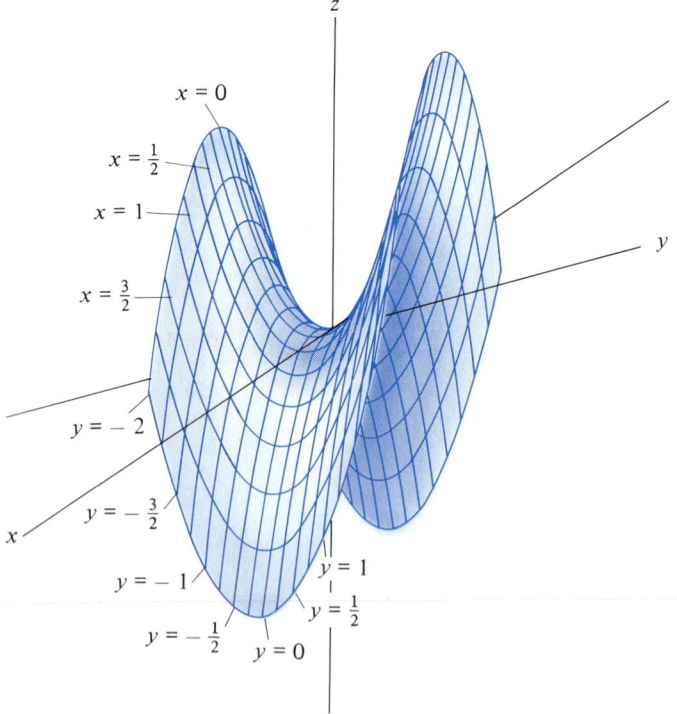

Figure 1.10 Traces of $z = y^2 - x^2$ in planes $y = c$ and $x = c$.

Level Curves and Surfaces

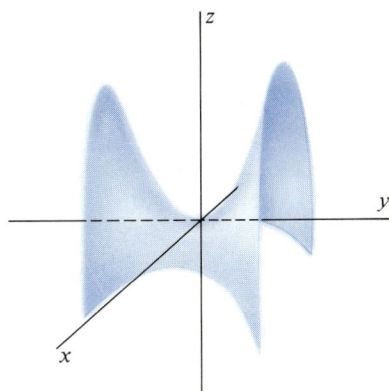

Figure 1.11 Graph of $z = y^2 - x^2$.

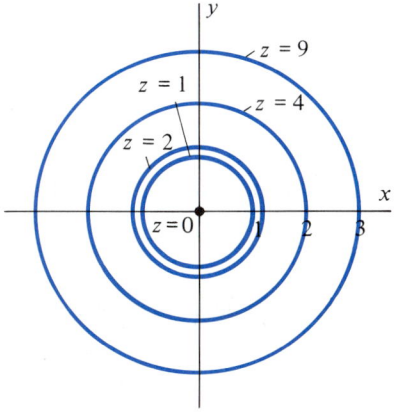

Figure 1.14 Level curves for $z = x^2 + y^2$ are circles $x^2 + y^2 = c$, $c \geq 0$.

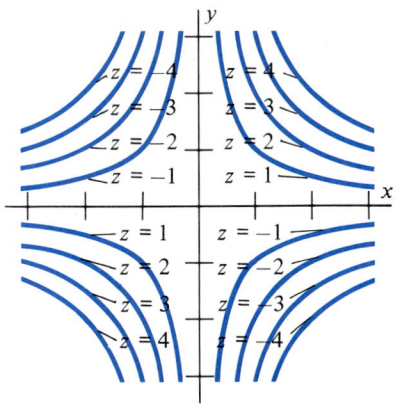

Figure 1.15 Level curves for $z = xy$ are curves $y = c/x$, $x \neq 0$. (See Figure 1.9.)

The idea behind using traces to describe the graph of $z = f(x, y)$ is that we actually try to sketch a three-dimensional graph. Another approach to representing a surface is to draw a purely two-dimensional sketch that conveys certain information about the surface. Here we want to know which points of the graph lie at various altitudes— that is, which points satisfy the equation $z = c$ for various values of c. This is the concept of **level curves.**

You have probably encountered level curves before. They are commonly used in topographical maps where one indicates the ''lay of the land'' by plotting curves of constant altitude (see Figures 1.12 and 1.13).

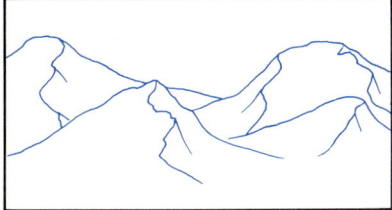

Figure 1.12 A picture of a mountainous region.

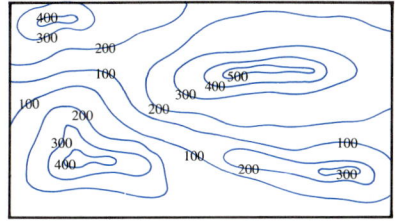

Figure 1.13 A topographical map of the region.

To use the notion of level curves to describe the graph of $z = f(x, y)$, we set $z = c$ and graph the resulting equation $f(x, y) = c$ in the xy-plane. Doing so for several values of c conveys the same type of information about the graph of f as does a topographical map. Note that although setting $z = c$ produces a trace of f, we make no attempt to produce a three-dimensional sketch by this method. Also, you should be sure to label the value of z corresponding to each level curve sketched.

Example 4

Consider the function $z = x^2 + y^2$ in Example 1. Setting $z = c$ gives the equation $x^2 + y^2 = c$, valid for $c \geq 0$. The level curves are concentric circles, as sketched in Figure 1.14. ◇

Example 5

For the function $z = xy$ in Example 2, the level curves are $xy = c$, or $y = \dfrac{c}{x}$. These are the hyperbolas in Figure 1.15. ◇

Example 6

For the function $f(x, y) = y^2 - x^2$ in Example 3, the level curves have equation $y^2 - x^2 = c$ or $y = \pm\sqrt{x^2 + c}$. Graphs of several such curves appear in Figure 1.16. ◇

Although we cannot sketch graphs of functions of three variables, we can sketch **level surfaces** for functions of the form $w = f(x, y, z)$ by setting $w = c$ and sketching the graph of the equation $f(x, y, z) = c$. For example, level surfaces for the function $w = x^2 + y^2 + z^2$ are spheres $x^2 + y^2 + z^2 = c$, $c \geq 0$. More generally,

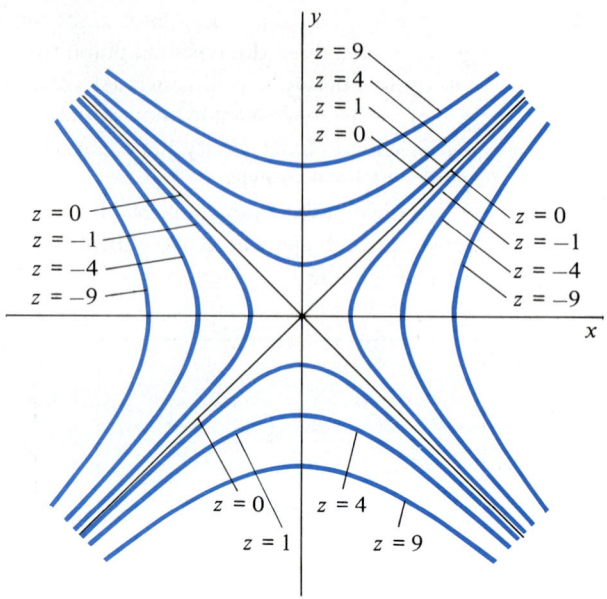

Figure 1.16 Level curves for $f(x, y) = y^2 - x^2$ are curves $y^2 - x^2 = c$. (See Figure 1.11.)

any function of two variables $z = f(x, y)$ may be regarded as a level surface for a function of three variables

$$g(x, y, z) = z - f(x, y) = 0.$$

We will exploit this observation later in finding planes tangent to graphs of functions of two variables.

A simple example of a level surface is the set of all points in your classroom at which the temperature $T(x, y, z)$ has a particular value. Such surfaces of constant temperature are called **isothermal surfaces.** Another example occurs in the theory of electricity and magnetism, where surfaces on which an electric potential is constant are called **equipotential surfaces.**

Neighborhoods

By a **neighborhood** of a point $P_0 = (x_0, y_0)$ in the plane, we shall mean an open **disc** N with center (x_0, y_0). That is,

$$N = \{(x, y) \mid \sqrt{(x - x_0)^2 + (y - y_0)^2} < r\} \tag{2}$$

where r is the radius of the disc N. (By analogy with open intervals on the real number line, the term "open" means that the boundary points are not included in the set.) Similarly, a neighborhood of a point $Q = (x_0, y_0, z_0)$ in space is an open **ball** N with center (x_0, y_0, z_0),

$$N = \{(x, y, z) \mid \sqrt{(x - x_0)^2 + (y - y_0)^2 + (z - z_0)^2} < r\}. \tag{3}$$

Using vector notation, we may generalize both (2) and (3) by saying that a neighborhood of the vector $\mathbf{x}_0$ is a set of vectors

$$N = \{\mathbf{x} \mid |\mathbf{x} - \mathbf{x}_0| < r\}. \tag{4}$$

In equation (4), r is called the **radius** of the neighborhood.

Finally, we shall define the term *deleted neighborhood* to mean all points in a neighborhood of x_0 except the vector x_0 itself. We shall use this terminology in defining $\lim\limits_{x \to x_0} f(x)$ where we shall require $f(x)$ to be defined for all x in a neighborhood of x_0 except possibly at x_0 itself.

Limits

Just as we did for functions of a single variable, we want the statement

$$\lim_{x \to x_0} f(x) = L \tag{5}$$

to mean that the values $f(x)$ of the function f "approach" the number L as the vector (point) x approaches the fixed vector x_0. However, in formulating this definition we must realize that x may approach x_0 along many different paths. This observation points out a significant difference between functions of one variable and functions of several variables. In the one-variable case, a variable can approach a constant only from one of two directions along the number line. In writing statement (5) we shall mean that $f(x) \to L$ as $x \to x_0$ *regardless of the path* along which x approaches x_0. Figure 1.17 illustrates this concept. Figure 1.18 suggests a situation in which the limit in (5) fails to exist.

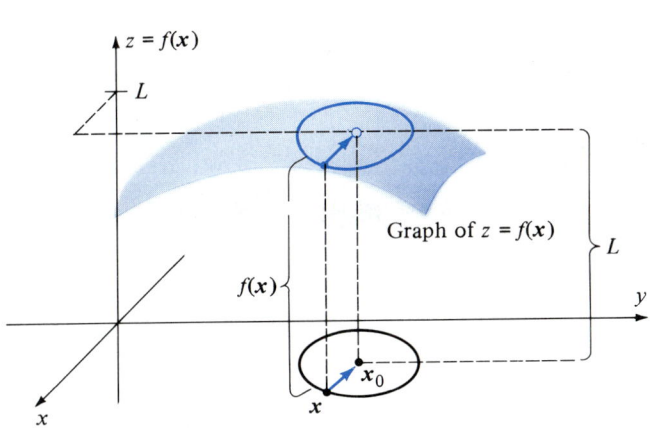

Figure 1.17 $\lim\limits_{x \to x_0} f(x) = L$ if $f(x) \to L$ as x approaches x_0 from any direction.

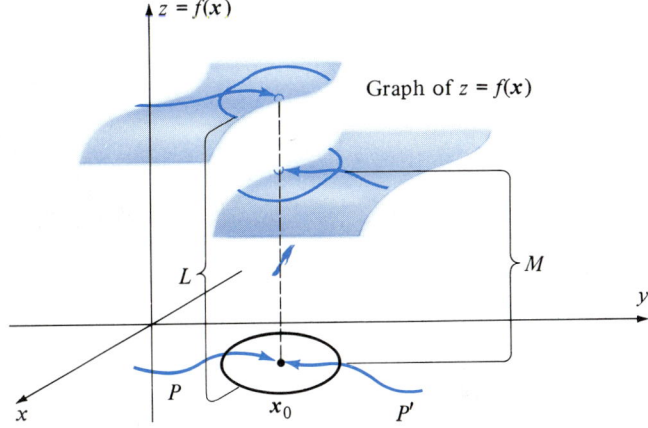

Figure 1.18 $f(x) \to L$ as $x \to x_0$ along path P, but $f(x) \to M$ as $x \to x_0$ along path P', so $\lim\limits_{x \to x_0} f(x)$ does not exist.

Before formulating a precise definition for the limit of a function of several variables, we examine several examples.

Example 7

Consider the function $f(x, y) = 4 - x - y$. We claim that

$$\lim_{(x, y) \to (1, 1)} f(x, y) = 2. \tag{6}$$

Notice first that we have not used the vector notation of statement (5) in writing statement (6), since the number and names of the independent variables are known explicitly. From the form of $f(x, y)$ it is easy to see that $f(x, y) \to (4 - 1 - 1) = 2$

as $x \to 1$ and $y \to 1$. Furthermore, we can see that this result is independent of the path along which (x, y) approaches $(1, 1)$ by setting $z = f(x, y)$ and recalling that the graph of the equation $z = 4 - x - y$ is a plane (see Figure 1.19). Regardless of how (x, y) approaches $(1, 1)$, the point (x, y, z) on the plane approaches the point $(1, 1, 2)$. ◇

Example 8

Consider the function $f(x, y) = \dfrac{xy}{x^2 + y^2}$. We claim that

$$\lim_{(x, y) \to (0, 0)} \frac{xy}{x^2 + y^2} \tag{7}$$

does not exist. On first glance this conclusion may not be obvious. For example, we may allow (x, y) to approach $(0, 0)$ along the x-axis by setting $y = 0$. We obtain

$$\lim_{(x, 0) \to (0, 0)} \frac{xy}{x^2 + y^2} = \lim_{x \to 0} \frac{x \cdot 0}{x^2 + 0^2} = 0 \qquad \text{(see Figure 1.20).} \tag{8}$$

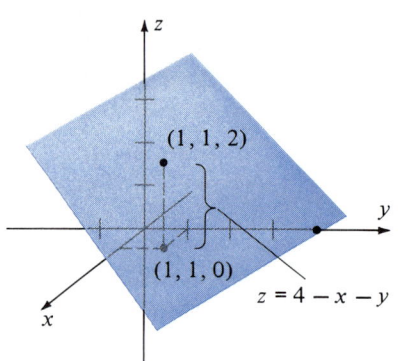

Figure 1.19 $\displaystyle\lim_{(x,y) \to (1,1)} (4 - x - y) = 2.$

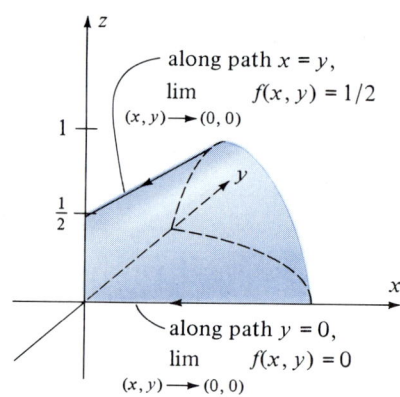

Figure 1.20 Portion of the graph of

$$f(x, y) = \frac{xy}{x^2 + y^2}.$$

$\displaystyle\lim_{(x,y) \to (0,0)} f(x, y)$ does not exist. (xy-plane has been rotated 90° for clarity.)

Similarly, allowing (x, y) to approach $(0, 0)$ along the y-axis ($x = 0$) gives

$$\lim_{(0, y) \to (0, 0)} \frac{xy}{x^2 + y^2} = \lim_{y \to 0} \frac{0 \cdot y}{0^2 + y^2} = 0. \tag{9}$$

However, if we allow (x, y) to approach $(0, 0)$ along the line $y = x$, we find that

$$\lim_{(x, x) \to (0, 0)} \frac{xy}{x^2 + y^2} = \lim_{x \to 0} \frac{x^2}{x^2 + x^2} = \frac{1}{2}. \tag{10}$$

Since the result in (10) does not agree with that in (8) or in (9), we conclude that the limit in (7) cannot exist. (See Exercise 48 for a different look at this function.) ◇

Example 9

Find $\displaystyle\lim_{(x,\,y,\,z)\to(2,\,\pi/3,\,3)} x\cos yz$.

Solution: Since $f(x, y, z) = x\cos yz$ is a function of three variables, we cannot rely on a graph to determine this limit. We must, instead, rely on our intuition about both the cosine function and the operation of multiplication. As $y\to\pi/3$ and $z\to 3$, $yz\to(\pi/3)(3) = \pi$, so $\cos yz\to\cos\pi = -1$. Thus, as $x\to 2$, $x\cos yz\to 2(-1) = -2$. That is,

$$\lim_{(x,\,y,\,z)\to(2,\,\pi/3,\,3)} x\cos yz = -2.$$ ◇

Next, we give a formal definition for the limit in line (5).

DEFINITION 1

Let x be the position vector for the point $(x_1, x_2, \ldots, x_n)$ in $\mathbb{R}^n$. Let f be a function of n variables defined for all x in a deleted neighborhood of x_0. Let L be a real number. We say that L is the **limit of the function** f as x approaches x_0, written

$$\lim_{x\to x_0} f(x) = L,$$

if and only if, for every $\epsilon > 0$, there exists a number $\delta > 0$ so that

if $0 < |x - x_0| < \delta$, then $|f(x) - L| < \epsilon$.   (11)

Note in statement (11) that the second inequality involves the *absolute value* of the *number* $f(x) - L$, while the first concerns the *length* of the *vector* $x - x_0$. When applying Definition 1 in a particular situation, you should rewrite the first inequality in (11) in the appropriate component form.

Example 10

Use Definition 1 to prove that

$$\lim_{(x,\,y)\to(0,\,0)} \sqrt{9 - x^2 - y^2} = 3.$$

Solution: Here $L = 3$, and $x_0 = \mathbf{0}$ is the position vector associated with the origin. We begin by rewriting the first inequality in (11) as

$$0 < \sqrt{(x - 0)^2 + (y - 0)^2} = \sqrt{x^2 + y^2} < \delta.$$ (12)

Also, the second inequality in (11) becomes

$$\left|\sqrt{9 - x^2 - y^2} - 3\right| < \epsilon,$$ (13)

or

$$3 - \sqrt{9 - x^2 - y^2} < \epsilon$$

since $\sqrt{9 - x^2 - y^2} \le 3$. Solving this inequality for the radical term, we obtain

$$\sqrt{9 - x^2 - y^2} > 3 - \epsilon,$$

which holds if and only if the following chain of equivalent inequalities holds:

$$9 - (x^2 + y^2) > (3 - \epsilon)^2 \qquad \text{(assuming } \epsilon < 3\text{)},$$
$$x^2 + y^2 < 6\epsilon - \epsilon^2,$$
$$\sqrt{x^2 + y^2} < \sqrt{6\epsilon - \epsilon^2}. \tag{14}$$

Now let's assume that $\epsilon > 0$ is a small ($\epsilon < 3$) positive given number. If we define the number δ to be $\delta = \sqrt{6\epsilon - \epsilon^2}$, then (12) holds whenever inequality (14) holds. Since inequality (14) is equivalent to inequality (13), this shows that

$$\text{if } 0 < \sqrt{x^2 + y^2} < \delta, \text{ then } |\sqrt{9 - x^2 - y^2} - 3| < \epsilon.$$

This proves the stated limit (see Figure 1.21).   ◇

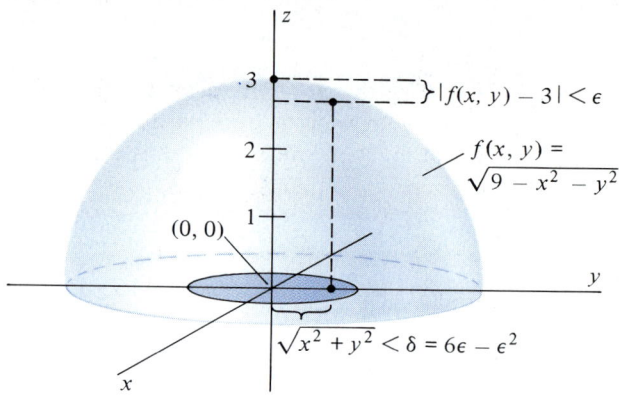

Figure 1.21 $\lim\limits_{(x,y)\to(0,0)} \sqrt{9 - x^2 - y^2} = 3.$

The algebra of limits for functions of several variables is analogous to that for functions of a single variable. The following theorem may be proved using the same ideas used to prove Theorem 1, Chapter 2 (see Exercises 50 and 51).

THEOREM 1

Let f and g be functions of two or three variables defined in a deleted neighborhood of x_0. Suppose that both $\lim\limits_{x\to x_0} f(x)$ and $\lim\limits_{x\to x_0} g(x)$ exist. Let c be any constant. Then

(i) $\lim\limits_{x\to x_0} [f(x) + g(x)] = \lim\limits_{x\to x_0} f(x) + \lim\limits_{x\to x_0} g(x),$

(ii) $\lim\limits_{x\to x_0} cf(x) = c \cdot \lim\limits_{x\to x_0} f(x),$

(iii) $\lim\limits_{x\to x_0} f(x)g(x) = \left(\lim\limits_{x\to x_0} f(x) \right)\left(\lim\limits_{x\to x_0} g(x) \right),$

(iv) $\lim\limits_{x\to x_0} \dfrac{f(x)}{g(x)} = \dfrac{\lim\limits_{x\to x_0} f(x)}{\lim\limits_{x\to x_0} g(x)}, \qquad \text{if } \lim\limits_{x\to x_0} g(x) \neq 0.$

Continuity

Continuity for functions of several variables is defined just as for functions of a single variable.

DEFINITION 2

The function f is **continuous** at $x_0 \in \mathbb{R}^n$ if

(i) $f(x_0)$ is defined,

(ii) $\lim\limits_{x \to x_0} f(x)$ exists, and

(iii) $\lim\limits_{x \to x_0} f(x) = f(x_0)$.

In other words, f is continuous at x_0 if $f(x) \to f(x_0)$ as $x \to x_0$, regardless of the direction in which x approaches x_0. Here are some examples.

1. The function $f(x, y) = 4 - x - y$ in Example 7 is continuous at $(1, 1)$ since
$$\lim_{(x, y) \to (1, 1)} (4 - x - y) = 2 = f(1, 1).$$

2. The function in Figure 1.18 is discontinuous at x_0, since $\lim\limits_{x \to x_0} f(x)$ does not exist.

3. The function $f(x, y) = \dfrac{xy}{x^2 + y^2}$ in Example 8 is discontinuous at $(0, 0)$, since $f(0, 0)$ is undefined.

4. The function $f(x, y, z) = x \cos yz$ in Example 9 is continuous at $(2, \pi/3, 3)$ since
$$\lim_{(x, y, z) \to (2, \pi/3, 3)} x \cos yz = -2 = f(2, \pi/3, 3).$$

5. The function $f(x, y) = \sqrt{9 - x^2 - y^2}$ in Example 10 is continuous at $(0, 0)$ since
$$\lim_{(x, y) \to (0, 0)} \sqrt{9 - x^2 - y^2} = 3 = f(0, 0).$$

As is the case with limits of functions of several variables, continuity is a property that is often difficult to prove; however it can usually be determined by knowing how continuous functions combine to form other continuous functions. For example, since powers, products, and sums of continuous functions of a single variable are again continuous, we would expect a polynomial function such as

$$p(x, y) = x^3 y^2 + 6xy + xy^4$$

to be a continuous function of two variables. Indeed, using Definition 2 and the ideas of Chapter 2, we can prove that sums, multiples, products, powers, and quotients (where denominators are not zero) of continuous functions of several variables are continuous (see Exercises 52 and 53). Moreover, the composition of a continuous function of a single variable with a continuous function of n variables is a continuous function of n variables (see Exercise 54).

Points of discontinuity for functions of several variables occur for the usual reasons: a denominator becomes zero, an expression underneath a radical sign of even order becomes negative, and so forth.

Exercise Set 17.1

State the implicit domain for each of the following functions.

1. $f(x, y) = \dfrac{1}{x^2 + y^2}$

2. $f(x, y) = \dfrac{x - y}{x + y}$

3. $f(x, y) = \sqrt{y - x}$

4. $f(x, y) = \sqrt{1 - xy^2}$

5. $f(x, y) = \sin xy^2$

6. $f(x, y, z) = \sqrt{4 - x^2 - y^2 - z^2}$

7. $f(x, y, z) = xyz$

8. $f(x, y) = \dfrac{1}{x - 3} + \dfrac{2}{x - y}$

9. $f(x, y, z) = \dfrac{y}{xyz}$

10. Let $f(x, y) = x^2 + y^2$, and $g(z) = 2z + 3$. Write the composite function $h(x, y) = g(f(x, y))$ as an explicit function of x and y.

11. Let $f(x, y) = x + y^2$, and $g(z) = \sqrt{z}$.
  a. Write the composite function $h(x, y) = g(f(x, y))$ as an explicit function of x and y.
  b. Find the domain of the function h.

12. Let V be the volume of a right circular cone. Write $V = V(r, h)$ as a function of the radius r and height h.

13. Let S be the total exterior surface area of a right circular cylinder with top and bottom. Express S as a function $S(r, h)$ of the radius and the height of the cylinder.

14. An object is dropped from rest h meters above the ground. Express the time required for it to fall to the ground, $T(g, h)$ as a function of h and g, where g is the acceleration due to gravity (ignore air resistance).

In Exercises 15–22, evaluate the given limit.

15. $\displaystyle\lim_{(x, y) \to (1, 3)} (x^2 - 2y)$

16. $\displaystyle\lim_{(x, y) \to (1, -1)} \sqrt{x - y}$

17. $\displaystyle\lim_{(x, y) \to (3, -1)} \dfrac{1}{\sqrt{x + y}}$

18. $\displaystyle\lim_{(x, y) \to (1, -2)} \dfrac{x + y^3}{x^2 + 2xy + y^2}$

19. $\displaystyle\lim_{(x, y) \to (\pi/2, 1)} x \cos xy$

20. $\displaystyle\lim_{(x, y) \to (1, 0)} xe^{y-x}$

21. $\displaystyle\lim_{(x, y) \to (\pi/2, 1)} \ln \sin xy$

22. $\displaystyle\lim_{(x, y) \to (2, 2)} \dfrac{\text{Tan}^{-1}(y/x)}{1 + xy}$

In Exercises 23–30, sketch several level curves for the given functions.

23. $f(x, y) = y - x^2$

24. $f(x, y) = x^2 + 4y^2$

25. $f(x, y) = 2xy$

26. $f(x, y) = x^2 - y^2$

27. $f(x, y) = xe^y$

28. $f(x, y) = x \cos y$

29. $f(x, y) = \sqrt{x + y}$

30. $f(x, y) = \sqrt{y^2 - x^2}$

In Exercises 31–38, sketch the graph of $z = f(x, y)$ by first sketching several traces of $z = f(x, y)$ in the planes $x = c$ and $y = c$.

31. $f(x, y) = x$

32. $f(x, y) = y^2$

33. $f(x, y) = x + y$

34. $f(x, y) = x \cos y$

35. $f(x, y) = x^2 + 4y^2$

36. $f(x, y) = y - x$

37. $f(x, y) = y \sin x$

38. $f(x, y) = y/x$

39. Find $\displaystyle\lim_{(x, y) \to (1, 0)} \dfrac{xy - y}{x^2 + y^2 - 2x + 1}$, if it exists.

40. Is $f(x, y) = \dfrac{4xy}{x^2 + y^2}$ continuous at $(0, 0)$? What about
$$g(x, y) = \dfrac{4xy}{\sqrt{x^2 + y^2}}?$$

41. Is the function
$$f(x, y) = \begin{cases} \dfrac{x^2 y}{x^3 + y^3}, & (x, y) \neq (0, 0) \\ 0, & (x, y) = (0, 0) \end{cases}$$
continuous at $(0, 0)$? (*Hint:* Consider the path $y = x$.)

42. Show that the function
$$f(x, y) = \begin{cases} \dfrac{x^3 y}{x^3 + y^5}, & (x, y) \neq (0, 0) \\ 0, & (x, y) = (0, 0) \end{cases}$$
is not continuous at $(0, 0)$.

43. Is the function
$$f(x, y) = \begin{cases} \dfrac{x^3}{x^2 + y^2}, & (x, y) \neq (0, 0) \\ 0, & (x, y) = (0, 0) \end{cases}$$
continuous at $(0, 0)$?

44. Let
$$f(x, y) = \begin{cases} \dfrac{\sin x \cos y}{x}, & (x, y) \neq (0, 0) \\ 0, & (x, y) = (0, 0). \end{cases}$$
Find $\displaystyle\lim_{(x, y) \to (0, 0)} f(x, y)$. (*Hint:* Write $f(x, y) = g(x, y)h(x, y)$, $(x, y) \neq (0, 0)$, with $g(x, y) = \cos y$, $h(x, y) = \dfrac{\sin x}{x}$, and apply Theorem 1.)

45. Let
$$f(x, y) = \begin{cases} x \csc x \tan y, & (x, y) \neq (0, 0) \\ 0, & (x, y) = (0, 0). \end{cases}$$
Find $\displaystyle\lim_{(x, y) \to (0, 0)} f(x, y)$ (see Exercise 44).

46. True or false? If $\lim\limits_{(x,\,y)\to(a,\,b)} f(x, y) = L$, then $\lim\limits_{x\to a} f(x, b) = L$. Explain.

47. True or false? If $\lim\limits_{x\to a} f(x, b) = L$ and $\lim\limits_{y\to b} f(a, y) = L$, then
$$\lim_{(x,\,y)\to(a,\,b)} f(x, y) = L.$$

48. Consider the function $f(x, y) = \dfrac{xy}{x^2 + y^2}$ whose graph appears in Figure 1.20.
   a. Letting $x = r \cos\theta$ and $y = r \sin\theta$, show that f may be expressed in polar coordinates as
   $$f(r, \theta) = \frac{1}{2}\sin 2\theta.$$
   b. Conclude from part (a) that the values $f(r, \theta)$ are inde-

pendent of r, depending only on θ. Explain why this shows that f is not continuous at the origin.

49. Prove that $\lim\limits_{(x,\,y)\to(0,\,0)} (x^2 + y^2) = 0$ using Definition 1.

50. Prove part (i) of Theorem 1.

51. Prove part (ii) of Theorem 1.

52. Prove that sums and constant multiples of continuous functions of several variables are continuous.

53. Prove that the product of two continuous functions of several variables is continuous.

54. Prove that if g is a continuous function of a single variable and if $y = f(x)$ is a continuous function of several variables, then the composite function $R(x) = g(f(x))$ is a continuous function of several variables.

17.2 PARTIAL DIFFERENTIATION

The *derivative* of the function f of one variable is the limit

$$f'(x) = \lim_{h\to 0} \frac{f(x + h) - f(x)}{h}, \tag{1}$$

and this limit measures the *rate of change* of $f(x)$ with respect to change in x. In the case of a function of two or more independent variables we have already seen that the question of calculating a rate of change for $z = f(x, y)$ at (x_0, y_0) by limits is a complicated issue, since the "nearby" point (x, y) may approach (x_0, y_0) along an infinite number of distinct paths.

We begin by examining rates at which $f(x, y)$ changes along paths parallel to the coordinate axes. This is the concept of **partial differentiation.**

DEFINITION 3

Let $f(x, y)$ be defined in a neighborhood of (x_0, y_0). The **partial derivative of f with respect to x** at (x_0, y_0) is the number

$$\frac{\partial f}{\partial x}(x_0, y_0) = \lim_{h\to 0} \frac{f(x_0 + h, y_0) - f(x_0, y_0)}{h}, \tag{2}$$

if this limit exists. Similarly, the **partial derivative of f with respect to y** at (x_0, y_0) is the number

$$\frac{\partial f}{\partial y}(x_0, y_0) = \lim_{h\to 0} \frac{f(x_0, y_0 + h) - f(x_0, y_0)}{h}, \tag{3}$$

provided this limit exists.

Comparing equations (1) and (2), we see that the partial derivative $\dfrac{\partial f}{\partial x}(x_0, y_0)$ is simply the result of holding the variable y constant and differentiating the function $z = f(x, y_0)$ as a function of x alone. Similarly, the partial derivative $\dfrac{\partial f}{\partial y}(x_0, y_0)$

results from treating the variable x as the constant $x = x_0$ and differentiating $z = f(x_0, y)$ as a function of y alone. Thus, partial derivatives may be calculated by the rules developed earlier for differentiating functions of a single variable.

Example 1

Calculate the partial derivatives with respect to x and y for the function $f(x, y) = x^2 y^3 + e^x + \ln y$ and evaluate each at $(1, 4)$.

Solution: In computing the partial derivative with respect to x, we consider y (and any expression containing y alone) to be a constant. Thus,

$$\frac{\partial f}{\partial x}(x, y) = \left(\frac{d}{dx}x^2\right)y^3 + \frac{d}{dx}e^x + \frac{d}{dx}\ln y$$

$$= 2xy^3 + e^x + 0.$$

Similarly, when we consider x constant,

$$\frac{\partial f}{\partial y}(x, y) = x^2\left(\frac{d}{dy}y^3\right) + \frac{d}{dy}e^x + \frac{d}{dy}\ln y$$

$$= 3x^2 y^2 + 0 + \frac{1}{y}, \qquad y \neq 0.$$

Thus,

$$\frac{\partial f}{\partial x}(1, 4) = 2 \cdot 1 \cdot 4^3 + e^1 = 128 + e$$

and

$$\frac{\partial f}{\partial y}(1, 4) = 3 \cdot 1^2 \cdot 4^2 + \frac{1}{4} = 48 + \frac{1}{4} = \frac{193}{4}. \qquad \diamond$$

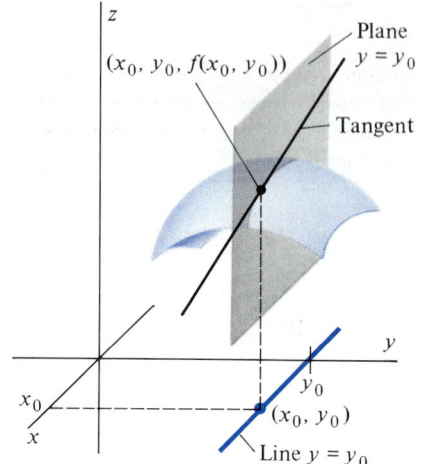

Figure 2.1 Partial derivative $\frac{\partial f}{\partial x}(x_0, y_0)$ is the slope of the line tangent to the trace of f in the plane $y = y_0$. (See **Plates 5** and **6**.)

Example 2

For $f(x, y) = \sin xy$, we apply the rule $\frac{d}{dt}\sin at = a \cos at$, with one variable playing the role of a and the other playing the role of t:

$$\frac{\partial f}{\partial x}(x, y) = y \cos xy \qquad \text{and} \qquad \frac{\partial f}{\partial y}(x, y) = x \cos xy. \qquad \diamond$$

Figures 2.1 and 2.2 illustrate the geometric interpretations of the partial derivatives in Definition 3. To interpret $\frac{\partial f}{\partial x}(x_0, y_0)$, we note that the set of points (x, y, z) in $\mathbb{R}^3$ with $y = y_0$ is a plane. The intersection of this plane with the graph of f is the *trace* of f in the plane $y = y_0$. This curve may be viewed as the graph of the function of one variable

$$h(x) = f(x, y_0), \qquad y_0 \text{ fixed.}$$

Since

$$h'(x_0) = \frac{\partial f}{\partial x}(x_0, y_0),$$

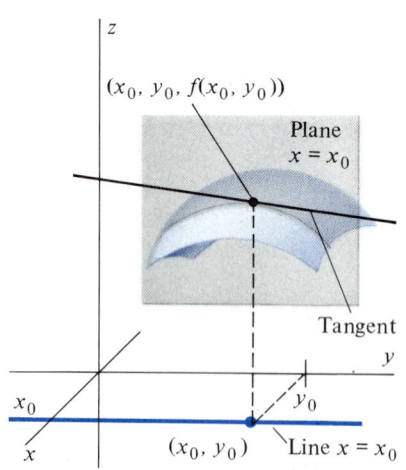

Figure 2.2 Partial derivative $\frac{\partial f}{\partial y}(x_0, y_0)$ is the slope of the line tangent to the trace of f in the plane $x = x_0$.

the partial derivative $\dfrac{\partial f}{\partial x}(x_0, y_0)$ gives the slope of the line tangent to this trace at the point $(x_0, y_0, f(x_0, y_0))$. A similar interpretation is valid for the partial derivative $\dfrac{\partial f}{\partial y}(x_0, y_0)$.

Example 3

For the function $f(x, y) = \sqrt{1 - x^2 - y^2}$ show that

(i) $\dfrac{\partial f}{\partial x}(x, y) = \dfrac{\partial f}{\partial y}(x, y)$ if and only if $y = x$, and

(ii) $\dfrac{\partial f}{\partial x}(x, y) = -\dfrac{\partial f}{\partial y}(x, y)$ if and only if $y = -x$

and interpret this result geometrically.

Solution: Differentiating with respect to x, holding y constant, gives

$$\frac{\partial f}{\partial x}(x, y) = \frac{-x}{\sqrt{1 - x^2 - y^2}}.$$

Similarly, the partial derivative with respect to y is

$$\frac{\partial f}{\partial y}(x, y) = \frac{-y}{\sqrt{1 - x^2 - y^2}}.$$

Conclusions (i) and (ii) follow immediately.

Figure 2.3 shows the geometric interpretation of this result. The graph of $z = f(x, y) = \sqrt{1 - x^2 - y^2}$ is a hemisphere. The only points on the surface of this hemisphere for which the tangent in the plane parallel to the x-axis has the same slope as the tangent in the plane parallel to the y-axis are the points lying above the line $y = x$. Similarly, the points at which these tangents have opposite slope lie above the line $y = -x$ in the xy-plane (see Figure 2.4). ◇

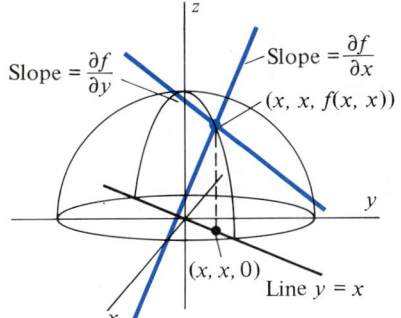

Figure 2.3 $\dfrac{\partial f}{\partial x} = \dfrac{\partial f}{\partial y}$ if $y = x$.

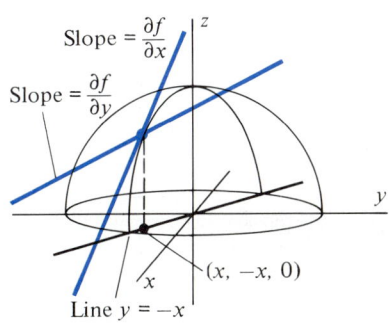

Figure 2.4 $\dfrac{\partial f}{\partial x} = -\dfrac{\partial f}{\partial y}$ if $y = -x$.

Notation for Partial Derivatives

For a function of two variables $z = f(x, y)$, the symbol $\dfrac{\partial}{\partial x}$ denotes the partial derivative with respect to x just as the Leibniz notation $\dfrac{d}{dx}$ denotes the derivative of the function of a single variable. We also use subscripts to denote partial derivatives, such as $f_x(x, y)$ or z_x for $\dfrac{\partial f}{\partial x}(x, y)$.

We may summarize the various types of notation for partial derivatives as follows: If $z = f(x, y)$, then

$$\frac{\partial f}{\partial x}(x, y) = \frac{\partial}{\partial x} f(x, y) = z_x(x, y) = z_x,$$

and

$$\frac{\partial f}{\partial y}(x, y) = \frac{\partial}{\partial y} f(x, y) = z_y(x, y) = z_y.$$

Functions of More than Two Variables

Partial derivatives are defined for functions of more than two variables by the same idea used in Definition 3: Hold all variables constant except one, and differentiate the resulting function of a single variable as before. For example, if $w = f(x, y, z)$ is a function of the three independent variables x, y, and z, the three partial derivatives are defined as follows:

$$\frac{\partial f}{\partial x}(x, y, z) = \lim_{h \to 0} \frac{f(x + h, y, z) - f(x, y, z)}{h}$$

$$\frac{\partial f}{\partial y}(x, y, z) = \lim_{h \to 0} \frac{f(x, y + h, z) - f(x, y, z)}{h}$$

$$\frac{\partial f}{\partial z}(x, y, z) = \lim_{h \to 0} \frac{f(x, y, z + h) - f(x, y, z)}{h}$$

Partial derivatives for functions of more than three independent variables are defined and calculated in an analogous way.

Example 4

Let $f(x, y, z) = \sqrt{x}\,e^{y/z}$, $\quad z \neq 0$, $\quad x \geq 0$. Then

$$\frac{\partial f}{\partial x}(x, y, z) = \left\{ \frac{d}{dx} \sqrt{x} \right\} e^{y/z} = \frac{1}{2\sqrt{x}} \cdot e^{y/z},$$

$$\frac{\partial f}{\partial y}(x, y, z) = \sqrt{x}\left\{ \frac{\partial}{\partial y}(e^{y/z}) \right\} = \sqrt{x}\,e^{y/z} \cdot \frac{1}{z} = \frac{\sqrt{x}}{z} e^{y/z},$$

$$\frac{\partial f}{\partial z}(x, y, z) = \sqrt{x}\left\{ \frac{\partial}{\partial z} e^{y/z} \right\} = \sqrt{x}\,e^{y/z}\left(\frac{-y}{z^2} \right) = \frac{-y\sqrt{x}}{z^2} e^{y/z}. \qquad \diamondsuit$$

Example 5

The radial rate of heat flow in a substance between two concentric spheres is given by the function

$$H(r, R, t, T) = \frac{(t - T)4\pi kRr}{R - r}$$

where k is a constant, and where the inner sphere has radius r and temperature t and the outer sphere has radius R and temperature T. Thus (using the Quotient Rule to differentiate with respect to r and R),

$$H_r = \frac{\partial}{\partial r} H(r, R, t, T) = \frac{(R - r)[(t - T)4\pi kR] - (-1)[(t - T)4\pi kRr]}{(R - r)^2}$$

$$= \frac{(t - T)4\pi kR^2}{(R - r)^2},$$

$$H_R = \frac{\partial}{\partial R} H(r, R, t, T) = \frac{(R - r)[(t - T)4\pi kr] - (1)[(t - T)4\pi kRr]}{(R - r)^2}$$

$$= \frac{-(t - T)4\pi kr^2}{(R - r)^2},$$

$$H_t = \frac{\partial}{\partial t} H(r, R, t, T) = \frac{\partial}{\partial t}\left[t\frac{4\pi kRr}{R - r} - T\frac{4\pi kRr}{R - r}\right] = \frac{4\pi kRr}{R - r},$$

$$H_T = \frac{\partial}{\partial T} H(r, R, t, T) = \frac{\partial}{\partial T}\left[t\frac{4\pi kRr}{R - r} - T\frac{4\pi kRr}{R - r}\right] = -\frac{4\pi kRr}{R - r}.$$ ◇

Example 6

For the vectors $\boldsymbol{x} = x_1\boldsymbol{i} + x_2\boldsymbol{j} + x_3\boldsymbol{k}$ and $\boldsymbol{y} = y_1\boldsymbol{i} + y_2\boldsymbol{j} + y_3\boldsymbol{k}$, the dot product

$$\boldsymbol{x} \cdot \boldsymbol{y} = x_1 y_1 + x_2 y_2 + x_3 y_3$$

may be viewed as a function of the six independent variables (components) x_1, x_2, x_3, y_1, y_2, y_3. Thus,

$$\frac{\partial}{\partial x_1}(\boldsymbol{x} \cdot \boldsymbol{y}) = y_1; \qquad \frac{\partial}{\partial y_1}(\boldsymbol{x} \cdot \boldsymbol{y}) = x_1,$$

$$\frac{\partial}{\partial x_2}(\boldsymbol{x} \cdot \boldsymbol{y}) = y_2; \qquad \frac{\partial}{\partial y_2}(\boldsymbol{x} \cdot \boldsymbol{y}) = x_2,$$

$$\frac{\partial}{\partial x_3}(\boldsymbol{x} \cdot \boldsymbol{y}) = y_3; \qquad \frac{\partial}{\partial y_3}(\boldsymbol{x} \cdot \boldsymbol{y}) = x_3.$$

Thus, the rate at which $\boldsymbol{x} \cdot \boldsymbol{y}$ changes with respect to change in the component x_1 is just the component y_1, and so forth. ◇

Limitations of Partial Derivatives

Before proceeding further, we should dispel a notion that students sometimes develop at this point. Although knowledge of the derivative $f'(x_0)$ completely determines the rate at which the function f changes at $x = x_0$, *knowledge of the partial derivatives $f_x(x_0, y_0)$ and $f_y(x_0, y_0)$ is not always sufficient to determine the rate at which the function of two variables $z = f(x, y)$ is changing at (x_0, y_0)*. The analogous statement for functions of more than two variables is also true.

To see this, let us consider the function

$$f(x, y) = \begin{cases} \dfrac{xy}{x^2 + y^2}, & (x, y) \neq (0, 0) \\ 0, & (x, y) = (0, 0) \end{cases}$$

Using Definition 3, we find that

$$\frac{\partial f}{\partial x}(0, 0) = \lim_{h \to 0} \frac{f(0 + h, 0) - f(0, 0)}{h} = \lim_{h \to 0}\left(\frac{0}{h}\right) = 0$$

and

$$\frac{\partial f}{\partial y}(0, 0) = \lim_{h \to 0} \frac{f(0, 0 + h) - f(0, 0)}{h} = \lim_{h \to 0}\left(\frac{0}{h}\right) = 0.$$

Thus, both partial derivatives at $(0, 0)$ are zero. But it is false to conclude that the rate of change of $f(x, y)$ at $(0, 0)$ in *every* direction is zero, as can be seen from Figure 2.5. For example, a small change from the point $(0, 0)$ to the point (h, h) causes an abrupt change in the value $f(x, y)$ from $f(0, 0) = 0$ to $f(h, h) = \dfrac{1}{2}$, no matter how small h may be.

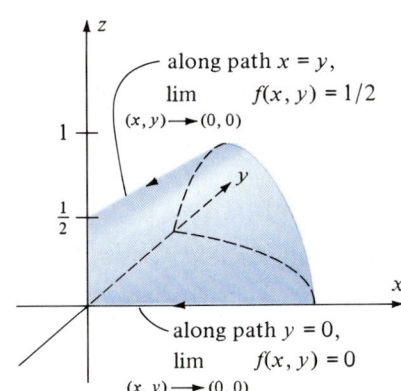

Figure 2.5 Portion of the graph of

$$f(x, y) = \frac{xy}{x^2 + y^2}.$$

$\displaystyle\lim_{(x,y)\to(0,0)} f(x, y)$ does not exist. (*xy*-plane has been rotated 90° for clarity.)

It is important to understand that partial derivatives give information about functions *only* in the directions of the coordinate axes. We must develop a richer theory of differentiation in order to obtain general conclusions about the rate of change of a function of several variables near a particular point.

Higher Order Partial Derivatives

Repeated applications of partial differentiation lead to **higher order partial derivatives.** There is nothing terribly complicated about this concept, except that we must be very careful about notation since we encounter **mixed partial derivatives,** in which one differentiation is performed with respect to a particular variable, followed by another differentiation with respect to a different variable.

We will use the following notation:

$$\frac{\partial^2 f}{\partial x^2}(x, y) = \frac{\partial^2}{\partial x^2}f(x, y) \qquad \text{means} \qquad \frac{\partial}{\partial x}\left(\frac{\partial f}{\partial x}(x, y)\right).$$

$$\frac{\partial^2 f}{\partial y \partial x}(x, y) = \frac{\partial^2}{\partial y \partial x}f(x, y) \qquad \text{means} \qquad \frac{\partial}{\partial y}\left(\frac{\partial f}{\partial x}(x, y)\right).$$

$$\frac{\partial^2 f}{\partial x \partial y}(x, y) = \frac{\partial^2}{\partial x \partial y}f(x, y) \qquad \text{means} \qquad \frac{\partial}{\partial x}\left(\frac{\partial f}{\partial y}(x, y)\right).$$

$$\frac{\partial^2 f}{\partial y^2}(x, y) = \frac{\partial^2}{\partial y^2}f(x, y) \qquad \text{means} \qquad \frac{\partial}{\partial y}\left(\frac{\partial f}{\partial y}(x, y)\right).$$

Note that the order in which the differentiations are performed is indicated by reading the "denominator" of the derivative notation from *right* to *left*. Similar definitions hold for third and higher order partial derivatives.

Example 7

For the function $f(x, y) = x^2y^3 + \cos x \sin y,$

$$\frac{\partial^2 f}{\partial x^2}(x, y) = \frac{\partial}{\partial x}\left(\frac{\partial f}{\partial x}(x, y)\right) = \frac{\partial}{\partial x}(2xy^3 - \sin x \sin y) = 2y^3 - \cos x \sin y,$$

$$\frac{\partial^2 f}{\partial y \partial x}(x, y) = \frac{\partial}{\partial y}\left(\frac{\partial f}{\partial x}(x, y)\right) = \frac{\partial}{\partial y}(2xy^3 - \sin x \sin y) = 6xy^2 - \sin x \cos y,$$

$$\frac{\partial^2 f}{\partial x \partial y}(x, y) = \frac{\partial}{\partial x}\left(\frac{\partial f}{\partial y}(x, y)\right) = \frac{\partial}{\partial x}(3x^2y^2 + \cos x \cos y) = 6xy^2 - \sin x \cos y$$

$$\frac{\partial^2 f}{\partial y^2}(x, y) = \frac{\partial}{\partial y}\left(\frac{\partial f}{\partial y}(x, y)\right) = \frac{\partial}{\partial y}(3x^2y^2 + \cos x \cos y) = 6x^2y - \cos x \sin y,$$

$$\frac{\partial^3 f}{\partial y^3}(x, y) = \frac{\partial}{\partial y}\left(\frac{\partial^2 f}{\partial y^2}(x, y)\right) = \frac{\partial}{\partial y}\left(\frac{\partial^2 f}{\partial y^2}\right) = \frac{\partial}{\partial y}(6x^2y - \cos x \sin y)$$

$$= 6x^2 - \cos x \cos y. \qquad \diamond$$

We may also use subscript notation to indicate higher order partial derivatives as follows: If $z = f(x, y),$

$$f_{xx}(x, y) \text{ or } z_{xx} \qquad \text{means} \qquad \frac{\partial^2 f}{\partial x^2}(x, y),$$

$$f_{xy}(x, y) \text{ or } z_{xy} \qquad \text{means} \qquad \frac{\partial^2 f}{\partial y \partial x}(x, y),$$

$$f_{yx}(x, y) \text{ or } z_{yx} \qquad \text{means} \qquad \frac{\partial^2 f}{\partial x \partial y}(x, y),$$

$$f_{yy}(x, y) \text{ or } z_{yy} \qquad \text{means} \qquad \frac{\partial^2 f}{\partial y^2}(x, y),$$

with similar statements holding for functions of more than two variables and for higher order derivatives.

It is important to note that, when subscripts are used, the differentiations are performed *in the order indicated by the subscripts, read from left to right.* This is just the opposite of the order in which the differentiations are indicated in Leibniz notation.

Example 8

For the function $f(x, y, z) = x^2 y^3 z^4$,

$$f_x = 2xy^3z^4; \qquad f_y = 3x^2y^2z^4; \qquad f_z = 4x^2y^3z^3$$

$$f_{xy} = \frac{\partial}{\partial y}(2xy^3z^4) = 6xy^2z^4; \qquad f_{yx} = \frac{\partial}{\partial x}(3x^2y^2z^4) = 6xy^2z^4$$

$$f_{yz} = \frac{\partial}{\partial z}(3x^2y^2z^4) = 12x^2y^2z^3; \qquad f_{zy} = \frac{\partial}{\partial y}(4x^2y^3z^3) = 12x^2y^2z^3$$

$$f_{xz} = \frac{\partial}{\partial z}(2xy^3z^4) = 8xy^3z^3; \qquad f_{zx} = \frac{\partial}{\partial x}(4x^2y^3z^3) = 8xy^3z^3$$

$$f_{xx} = 2y^3z^4; \qquad f_{yy} = 6x^2yz^4; \qquad f_{zz} = 12x^2y^3z^2$$

$$f_{xyz} = \frac{\partial}{\partial z}(6xy^2z^4) = 24xy^2z^3; \qquad f_{yzx} = \frac{\partial}{\partial x}(12x^2y^2z^3) = 24xy^2z^3$$

$$f_{xxx} = 0; \qquad f_{yyy} = 6x^2z^4; \qquad f_{zzz} = 24x^2y^3z. \qquad \diamond$$

Equality of Mixed Partials

You have no doubt observed, in Example 7, that $f_{yx} = f_{xy}$ and, in Example 8, that $f_{xy} = f_{yx}, f_{yz} = f_{zy}$, and $f_{xz} = f_{zx}$. This is not true for all functions. However, when the function f and various of its partial derivatives are continuous, these mixed partials will be equal. The following theorem makes this precise. It is typically proved in courses on advanced calculus.

THEOREM 2
Equality of Mixed Partials

If the function $z = f(x, y)$ and the partial derivatives

$$\frac{\partial f}{\partial x}, \qquad \frac{\partial f}{\partial y}, \qquad \frac{\partial^2 f}{\partial x \partial y}, \qquad \text{and} \qquad \frac{\partial^2 f}{\partial y \partial x}$$

are all continuous in a neighborhood of the point (x_0, y_0), then

$$\frac{\partial^2 f}{\partial x \partial y}(x_0, y_0) = \frac{\partial^2 f}{\partial y \partial x}(x_0, y_0). \tag{4}$$

Equation (4) plays an important role in the theory of differentiation that follows.

Exercise Set 17.2

In Exercises 1–20, find all first order partial derivatives.

1. $f(x, y) = xy$

2. $z = \sqrt{x + y^2}$

3. $z = x \tan y^2$

4. $f(x, y) = xy^3 + \sqrt{y}$

5. $f(x, y) = e^{x^2 + y^2}$

6. $z = \text{Tan}^{-1}(y/x)$

7. $f(r, \theta) = r \sin(\pi/2 - \theta)$

8. $f(s, t) = \dfrac{s - t}{s + t}$

9. $z = \ln(xy^2 + x - y)$

10. $h(u, v) = e^{u-v} + e^{v-u}$

11. $f(x, y) = x^y$

12. $z = 2^x y^2$

13. $f(r, \theta) = r^2 \cos \theta$

14. $f(x, y, z) = x^3 e^y \ln z$

15. $f(x, y, z) = xy^3 - yz^2$

16. $w = \ln(x^2 + y^2 + z^2)$

17. $f(x, y, z) = \left(\dfrac{x - y}{x + y}\right)^z$

18. $f(u, v, w) = \dfrac{\sin u}{v^3 \, \text{Tan}^{-1} w}$

19. $f(r, s, t) = \dfrac{\sqrt{r}\, s \ln t}{\sqrt{s^2 - 2r + t}}$

20. $f(u, v, w) = \dfrac{ue^{vw}}{\sin^2 u + \tan^2 w}$

21. Find $\dfrac{\partial f}{\partial x}(2, 5)$ for $f(x, y) = xy^3 - y$.

22. Find $\dfrac{\partial f}{\partial y}(1, 0)$ for $f(x, y) = e^{x-y^2}$.

23. Find $z_x(2, 1)$ for $z = \sqrt{x + y^2}$.

24. Find f_{xx}, f_{xy}, f_{yx}, and f_{yy} for $f(x, y)$ in Exercise 4.

25. Find $\dfrac{\partial^2 f}{\partial r^2}, \dfrac{\partial^2 f}{\partial r \partial \theta}$, and $\dfrac{\partial^2 f}{\partial \theta^2}$ for $f(r, \theta)$ in Exercise 13.

26. Find $f_{xx}, f_{xy}, f_{xz}, f_{yz}, f_{yy}$, and f_{zz} for $f(x, y, z)$ in Exercise 15.

27. Find $w_{xx} + w_{yy} + w_{zz}$ for $w(x, y, z)$ in Exercise 16.

28. For a particle traveling in a circular orbit of radius r, the relation between angular speed ω and velocity v is $v = \omega r$.
Find $\dfrac{\partial v}{\partial r}$ and $\dfrac{\partial v}{\partial \omega}$.

29. For the constant volume flow of an incompressible fluid through a tube of varying cross-sectional area, the equation $A_1 v_1 = A_2 v_2$ expresses the relationship between the respective cross-sectional areas and velocities at two points in the tube.
a. Express v_2 as a function of A_2, A_1, and v_1.
b. Find $\dfrac{\partial v_2}{\partial A_2}$, the rate at which v_2 changes with respect to change in A_2 alone.
c. Suppose that $A_1 = 5$ cm^2, $A_2 = 3$ cm^2, and $v_1 = 20$ cm/s. Find the rate of change of v_2 with respect to A_2 if A_1 and v_1 are held constant.

d. With A_1, A_2, and v_1 as in (c), find the rate of change of A_2 with respect to change in v_1 if A_1 and v_2 are held constant.

In Exercises 30–36, use the equations

$$x = \rho \sin \phi \cos \theta$$
$$y = \rho \sin \phi \sin \theta$$
$$z = \rho \cos \phi$$

for changing from rectangular to spherical coordinates to find the indicated partial derivative.

30. $\dfrac{\partial x}{\partial \phi}$

31. $\dfrac{\partial y}{\partial \theta}$

32. $\dfrac{\partial z}{\partial \rho}$

33. $\dfrac{\partial x}{\partial \rho}$

34. $\dfrac{\partial y}{\partial \phi}$

35. $\dfrac{\partial z}{\partial \phi}$

36. $\dfrac{\partial \phi}{\partial z}$

37. For $f(x, y) = \displaystyle\int_{x}^{x+y} \cos t^2 \, dt$ find

a. $\dfrac{\partial f}{\partial x}$ **b.** $\dfrac{\partial f}{\partial y}$

38. For $f(x, y, z) = \displaystyle\int_{z}^{x} \sqrt{t^3 + 1} \, dt - \int_{y}^{z} \sqrt{t^3 + 1} \, dt$ find

a. $\dfrac{\partial f}{\partial x}$ **b.** $\dfrac{\partial f}{\partial y}$ **c.** $\dfrac{\partial f}{\partial z}$

w respect to x then do y (show their equal)

39. Show that the function $f(x, y) = \text{Sin}^{-1}\left(\dfrac{x - y}{x + y}\right)$ is a solution of the differential equation

$$x \dfrac{\partial z}{\partial x} + y \dfrac{\partial z}{\partial y} = 0.$$

40. Show that the function $w = \ln(e^x + e^y + e^z)$ satisfies the partial differential equation

$$\dfrac{\partial w}{\partial x} + \dfrac{\partial w}{\partial y} + \dfrac{\partial w}{\partial z} = 1.$$

41. The **electric potential** at an axial point for a charged disc is given by

$$V = \dfrac{\sigma}{2\epsilon_0}(\sqrt{a^2 + r^2} - r)$$

where σ and ϵ_0 are constants, a is the radius of the disc, and r is the distance from the point to the disc.
a. Find the rate of change of V with respect to a if r is held constant.
b. Find the rate of change of V with respect to r if a is held constant.

42. The national unemployment rate u may be viewed as the dot product $u = \mathbf{v} \cdot \mathbf{w}$ where the components of the vector $\mathbf{v} = \langle v_1, v_2, \ldots, v_n \rangle$ are the percentages of employable citizens in each of the n job categories and the components of the vector $\mathbf{w} = \langle w_1, w_2, \ldots, w_n \rangle$ are the unemployment rates within each category.

a. Find $\dfrac{\partial u}{\partial v_j}$ and interpret this rate.

b. Find $\dfrac{\partial u}{\partial w_j}$ and interpret this rate.

43. Let $f(x, y, z) = e^{x+y+z}$. Show that all partial derivatives of all orders equal $f(x, y, z)$.

44. Show that a function of the form $f(x, y) = e^{kx}g(y)$ satisfies the differential equation $\dfrac{\partial f}{\partial x}(x, y) = kf(x, y)$.

45. Show that the function $f(x, y) = \sin xy$ satisfies the differential equation

$$x \frac{\partial f}{\partial x}(x, y) - y \frac{\partial f}{\partial y}(x, y) = 0.$$

46. Show that the function $f(x, y) = \sin xy$ satisfies the differential equation

$$x^2 \frac{\partial^2 f}{\partial x^2}(x, y) - y^2 \frac{\partial^2 f}{\partial y^2}(x, y) = 0.$$

47. Find a solution of the differential equation

$$x^n \frac{\partial^n f}{\partial x^n} = y^n \frac{\partial^n f}{\partial y^n}, \qquad n = 1, 2, 3, \ldots .$$

(See Exercises 45 and 46.)

48. Show that the function $y(x, t) = g(t - cx)$ satisfies the **one-dimensional wave equation**

$$\frac{\partial^2 y}{\partial t^2} = \frac{1}{c^2} \frac{\partial^2 y}{\partial x^2}.$$

⇒ same as 39)

49. **Laplace's equation** for the function $f(x, y)$ is

$$\frac{\partial^2 f}{\partial x^2} + \frac{\partial^2 f}{\partial y^2} = 0.$$

Show that the following functions satisfy Laplace's equation.

a. $f(x, y) = e^x \sin y$
b. $f(x, y) = e^{-x} \cos y$
c. $f(x, y) = \ln \sqrt{x^2 + y^2}$

50. Show that a polynomial $P(x, y)$ in x and y (like $P(x, y) = 3x^2y^3 + xy^6 - 5x^3y$) has the property that $\dfrac{\partial^2 P}{\partial x \partial y} = \dfrac{\partial^2 P}{\partial y \partial x}$.

17.3 TANGENT PLANES

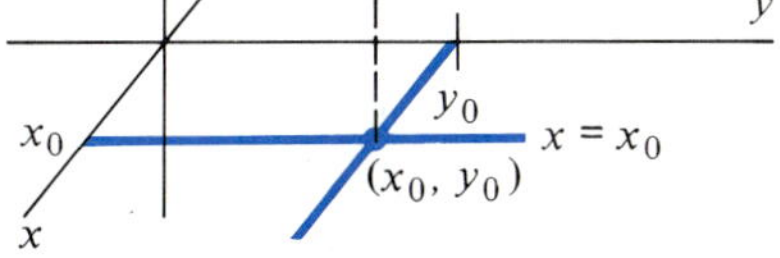

Figure 3.1 Partial derivatives determine vectors $\mathbf{u}_x = \mathbf{i} + f_x(x_0, y_0)\mathbf{k}$ and $\mathbf{u}_y = \mathbf{j} + f_y(x_0, y_0)\mathbf{k}$ tangent to graph of $z = f(x, y)$ at point P.

For the function f of a single variable, knowledge of $f(a)$ and the derivative $f'(a)$ enables us to write an equation for the line tangent to the graph of f at the point $(a, f(a))$. The purpose of this brief section is to show how we can obtain an equation for a plane tangent to the graph of a function of two variables $z = f(x, y)$ from knowledge of its partial derivatives. A complete discussion of the conditions under which a tangent plane exists is beyond the scope of this text. The issue here is simply how one goes about using partial derivatives to find such a plane, assuming that it exists.

Suppose that f is a function of two variables that is defined in a neighborhood of the point (x_0, y_0). Assume also that the partial derivatives $\dfrac{\partial f}{\partial x}(x_0, y_0)$ and $\dfrac{\partial f}{\partial y}(x_0, y_0)$ both exist. From the discussion of the geometric interpretation of partial derivatives in Section 17.2, we may conclude the following (see Figure 3.1).

(i) The partial derivative $\dfrac{\partial f}{\partial x}(x_0, y_0)$ gives the slope of the line tangent to the trace of f in the plane $y = y_0$. Since the plane $y = y_0$ is parallel to the xz-coordinate plane, a vector parallel to this tangent line is

$$\mathbf{u}_x = \mathbf{i} + \frac{\partial f}{\partial x}(x_0, y_0)\mathbf{k}.$$

(To check this statement, note that the slope of the vector, thought of as a line segment in the plane $y = y_0$, is $\dfrac{\partial f}{\partial x}(x_0, y_0)$, as required.)

(ii) The partial derivative $\dfrac{\partial f}{\partial y}(x_0, y_0)$ gives the slope of the line tangent to the trace of f in the plane $x = x_0$. Since the plane $x = x_0$ is parallel to the yz-coordinate plane, a vector parallel to this tangent line is

$$u_y = j + \frac{\partial f}{\partial y}(x_0, y_0)k.$$

Since both vectors u_x and u_y must lie in the tangent plane, they together determine a normal to the plane as

$$N = u_y \times u_x = \det \begin{bmatrix} i & j & k \\ 0 & 1 & \dfrac{\partial f}{\partial y}(x_0, y_0) \\ 1 & 0 & \dfrac{\partial f}{\partial x}(x_0, y_0) \end{bmatrix} \tag{1}$$

$$= \frac{\partial f}{\partial x}(x_0, y_0)i + \frac{\partial f}{\partial y}(x_0, y_0)j - k.$$

(See Figure 3.2.)

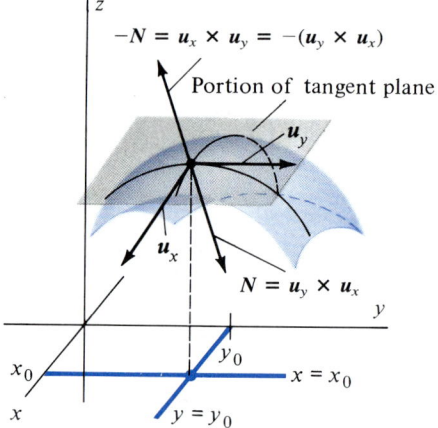

Figure 3.2 Vectors u_x and u_y determine normal $N = u_y \times u_x$ for plane tangent to graph of $z = f(x, y)$ at point P.

According to equation (11), Section 15.5, an equation for the plane with normal vector N, as above, and containing the point $(x_0, y_0, z_0) = (x_0, y_0, f(x_0, y_0))$ is

$$\frac{\partial f}{\partial x}(x_0, y_0)(x - x_0) + \frac{\partial f}{\partial y}(x_0, y_0)(y - y_0) - (z - z_0) = 0 \tag{2}$$

or

$$z = \frac{\partial f}{\partial x}(x_0, y_0)(x - x_0) + \frac{\partial f}{\partial y}(x_0, y_0)(y - y_0) + z_0. \tag{3}$$

Either equation (2) or equation (3) may be used to write the equation of the plane tangent to the graph of f at the point (x_0, y_0, z_0), where $z_0 = f(x_0, y_0)$. More generally, you may simply remember the idea used to develop these equations: the partial derivatives $\frac{\partial f}{\partial x}(x_0, y_0)$ and $\frac{\partial f}{\partial y}(x_0, y_0)$ determine **direction vectors, u_x and u_y,** whose cross product $N = u_y \times u_x$ is a normal to the desired plane.

Example 1

Find an equation for the plane tangent to the graph of $f(x, y) = x^2 + 4y^2$ at the point $(2, 1, 8)$.

Solution: The required partial derivatives are

$$\frac{\partial f}{\partial x}(2, 1) = 2x \Big|_{\substack{x=2 \\ y=1}} = 4,$$

$$\frac{\partial f}{\partial y}(2, 1) = 8y \Big|_{\substack{x=2 \\ y=1}} = 8,$$

and $x_0 = 2$, $y_0 = 1$, $z_0 = 8$. Thus, by equation (3), the equation is

$$z = 4(x - 2) + 8(y - 1) + 8$$

or

$$z = 4x + 8y - 8.$$

$\diamond$

Example 2

Find an equation for the line of intersection of the plane tangent to the graph of $z = 9 - x^2 - y^2$ at $(1, 2, 4)$ and the xy-coordinate plane.

Solution: Since

$$z_x(1, 2) = -2x \Big|_{\substack{x=1 \\ y=2}} = -2$$

and

$$z_y(1, 2) = -2y \Big|_{\substack{x=1 \\ y=2}} = -4,$$

the equation of the tangent plane, according to (3), is

$$z = -2(x - 1) - 4(y - 2) + 4$$

or

$$z = -2x - 4y + 14.$$

Setting $z = 0$ gives the line of intersection of this plane with the xy-plane as $2x + 4y - 14 = 0$ (see Figure 3.3).

$\diamond$

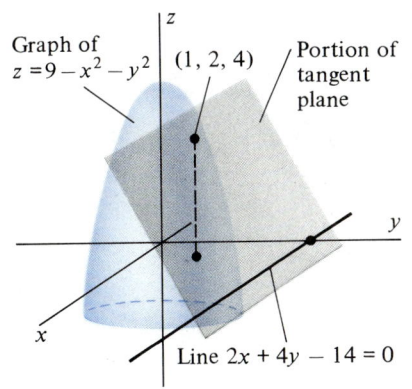

Figure 3.3 Plane tangent to graph of $y = 9 - x^2 - y^2$ at $(2, 1, 4)$ intersects xy-plane along line $2x + 4y - 14 = 0$. (See **Plate 7.**)

Normal Lines

If a surface has a tangent plane at point P, we say that a line through P is **normal** to the surface if it is normal to the tangent plane at P. Using N in (1) as a direction vector normal to the tangent plane for f at (x_0, y_0, z_0), we may write the normal line in vector form as

$$\ell: r(t) = x_0 + tN \tag{4}$$

where x_0 is the position vector $x_0 = x_0 i + y_0 j + z_0 k$. Writing $r(t)$ in component form as $r(t) = x(t)i + y(t)j + z(t)k$ and using (1), we may write (4) in the component form

$$x(t)i + y(t)j + z(t)k$$

$$= (x_0 i + y_0 j + z_0 k) + t\left(\frac{\partial f}{\partial x}(x_0, y_0)i + \frac{\partial f}{\partial y}(x_0, y_0)j - k\right). \tag{5}$$

Example 3

Find equations for the line normal to the graph of $z = e^{y-x^2}$ at the point $(1, \ln 2, 2/e)$.

Solution: For the function $f(x, y) = e^{y-x^2}$,

$$\frac{\partial f}{\partial x}(1, \ln 2) = -2xe^{y-x^2}\bigg|_{\substack{x=1 \\ y=\ln 2}} = -2e^{\ln 2-1} = \frac{-4}{e}$$

and

$$\frac{\partial f}{\partial y}(1, \ln 2) = e^{y-x^2}\bigg|_{\substack{x=1 \\ y=\ln 2}} = e^{\ln 2-1} = \frac{2}{e}.$$

Also, $x_0 = 1$, $y_0 = \ln 2$, and $z_0 = \dfrac{2}{e}$. The vector form for the normal line is therefore

$$\ell: r(t) = i + \ln 2j + \frac{2}{e}k + t\left(-\frac{4}{e}i + \frac{2}{e}j - k\right). \qquad \diamond$$

Exercise Set 17.3

In each of Exercises 1–14, find an equation for the plane tangent to the graph of the given function at point P, assuming it exists.

1. $f(x, y) = x^2 + y^2$; $P = (1, 3, 10)$

2. $f(x, y) = x^2 + 2xy + y^2$; $P = (3, -1, 4)$

3. $z = x^2 + y^2 - xy - 4x - 2y$; $P = (1, -1, 1)$

4. $f(x, y) = 2x^2 - y^2$; $P = (2, \sqrt{5}, 3)$

5. $f(x, y) = \dfrac{x-2}{y+2}$; $P = (4, -1, 2)$

6. $f(x, y) = \sqrt{9 - x^2 - y^2}$; $P = (1, -2, 2)$

7. $f(x, y) = \ln xy$; $P = (1, 1, 0)$

8. $f(x, y) = e^{-x} \sin y$; $P = (0, \pi/6, 1/2)$

9. $z = \dfrac{x}{x^2 + y^2}$; $P = (1, 1, 1/2)$

10. $f(x, y) = \text{Tan}^{-1}(y/x)$; $P = (2, 2, \pi/4)$

11. $z = \ln y^x$, $P = (1, 1, 0)$

12. $z = \dfrac{y-6}{3x^2-1}$, $P = (1, 4, -1)$

13. $f(x, y) = \ln\left(\dfrac{y-x}{y+x}\right)$, $P = (0, e, 0)$

14. $f(s, t) = \dfrac{1 - \sqrt{s}}{t(s - \sqrt{t})}$, $P = \left(4, 1, -\dfrac{1}{3}\right)$

In Exercises 15–20, find a vector equation for the line normal to the graph of the given function at the point P, as described in the stated exercise.

15. Exercise 2.

16. Exercise 3.

17. Exercise 6.

18. Exercise 8.

19. Exercise 11.

20. Exercise 14.

21. Find the point on the graph of $z = -x^2 + xy + 2y^2$ where the tangent plane is parallel to the plane with equation $x - 14y + z = 4$.

22. Find a vector equation for the line of intersection of the plane tangent to the graph of $z = x^2 + 2y^2 - 4y + 2$ at $(2, 1, 4)$ and the xy-coordinate plane.

23. Show that all lines normal to the graph of $f(x, y) = x \sin y$ at points $(x, \pi/2)$ are parallel.

24. Show that at all points $(x, \pi/2)$, the planes tangent to the graph of $f(x, y) = x \sin y$ are the same. Find an equation for this plane.

25. Find an equation for the plane tangent to the paraboloid $z = 9x^2 + 4y^2$ at the point $(1, 2, 25)$.

26. Show that the volume of the tetrahedron formed by the planes $x = 0$, $y = 0$, and $z = 0$, and any tangent to the graph of the function $f(x, y) = \dfrac{c}{xy}$, $c > 0$ is $V = \dfrac{9c}{2}$.

27. Find an equation for the plane tangent to the sphere $x^2 + y^2 + z^2 = r^2$ at the point (x_0, y_0, z_0), $z_0 \neq 0$.

28. Find an equation for the plane tangent to the graph of the function $y = f(x, y)$ at the point $(x_0, y_0, z_0) = (x_0, f(x_0, z_0), z_0)$.

29. Use the result of Exercise 28 to find the equation of the plane tangent to the graph of $y = \text{Tan}^{-1}(z/x)$ at the point where $x = z = 1$.

30. Use the result of Exercise 28 to find a vector equation for the line of intersection of the plane tangent to the graph of $y = x^2 + 2xz - z^2$ at the point $(2, -8, -2)$ and the xz-coordinate plane.

31. Prove that every normal to a sphere passes through the center.

17.4 RELATIVE AND ABSOLUTE EXTREMA

One of the principal applications of the derivative for functions of a single variable is in finding relative and absolute extrema. The purpose of this section is to show how partial derivatives may be used to find relative extrema for functions of several variables. Although the ideas discussed here are not confined to functions of just two independent variables, we will deal mainly with this case because of the opportunity to interpret the results geometrically.

We begin by defining what we mean by relative extrema.

DEFINITION 4

The number $z_0 = f(x_0, y_0)$ is called a **relative maximum** for the function f if there exists a neighborhood N of (x_0, y_0) so that $f(x, y)$ is defined for all $(x, y) \in N$, and so that

$$f(x_0, y_0) \geq f(x, y)$$

for all $(x, y) \in N$.

The number $z_0 = f(x_0, y_0)$ is called a **relative minimum** for f if there exists a neighborhood N of (x_0, y_0) so that $f(x, y)$ is defined for all $(x, y) \in N$ and so that

$$f(x_0, y_0) \leq f(x, y)$$

for all $(x, y) \in N$.

The number $z_0 = f(x_0, y_0)$ is a **relative extremum** for the function $z = f(x, y)$ if it is either a relative maximum or a relative minimum.

Thus, a relative maximum for f is the precise analogue of a relative maximum for the function of a single variable: $z_0 = f(x_0, y_0)$ is the largest value $z = f(x, y)$ for all (x, y) "near" (x_0, y_0). A similar interpretation holds for relative minima.

Example 1

The function $f(x, y) = \sqrt{x^2 + y^2}$ has a relative minimum of $z_0 = 0$ at the point $(x_0, y_0) = (0, 0)$. This is easy to see, since

$$f(x, y) = \sqrt{x^2 + y^2} \geq 0 = f(0, 0)$$

for all points (x, y) (see Figure 4.1). ◇

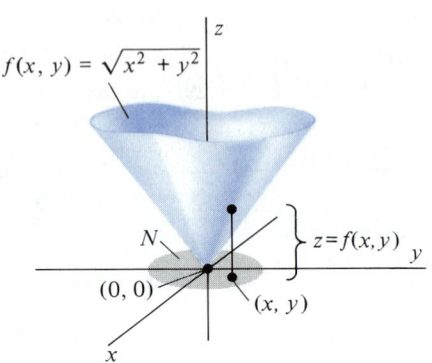

$f(x, y) = \sqrt{x^2 + y^2}$

N

$(0, 0)$

$z = f(x, y)$

(x, y)

Figure 4.1 $f(0, 0) \leq f(x, y)$ for all (x, y) near $(0, 0)$; $f(0, 0)$ is a relative minimum. (See **Plate 8.**)

One way to verify that the number $z_0 = f(x_0, y_0)$ is a relative extremum is to simply compare the number $f(x_0, y_0)$ with values of the function f for points (x, y) near (x_0, y_0). While such comparisons are sometimes difficult, if not impossible, to make, this method does handle a variety of polynomial functions in two variables. The idea is to write $x = x_0 + h$ and $y = y_0 + k$ and then to examine the sign of the difference:

$$f(x_0, y_0) - f(x_0 + h, y_0 + k). \tag{1}$$

If this difference is nonnegative for all small values of h and k, we conclude that $f(x_0, y_0)$ is a relative maximum. Similarly, if the difference in line (1) is nonpositive for all small values of h and k, we conclude that $f(x_0, y_0)$ is a relative minimum.

Example 2

Verify that $f(1, 2) = 4$ is a relative maximum for the function $f(x, y) = 2x + 4y - x^2 - y^2 - 1$.

Solution: Here $x_0 = 1$ and $y_0 = 2$, so the nearby point (x, y) is written $(x, y) = (x_0 + h, y_0 + k) = (1 + h, 2 + k)$. The difference in (1) is

$$
\begin{aligned}
f(1, 2) - f(1 + h, 2 + k) \qquad\qquad\qquad &\tag{2}\\
= 4 - \{2(1 + h) + 4(2 + k) - (1 + h)^2 - (2 + k)^2 - 1\}&\\
= 4 - 2 - 2h - 8 - 4k + 1 + 2h + h^2 + 4 + 4k + k^2 + 1&\\
= h^2 + k^2.&
\end{aligned}
$$

Since $h^2 + k^2 \geq 0$ for all h and k, the above calculation shows that the difference in (2) is nonnegative for all small h and k. Thus, $f(1, 2) = 4$ is a relative maximum for the function f (see Figure 4.2). ◇

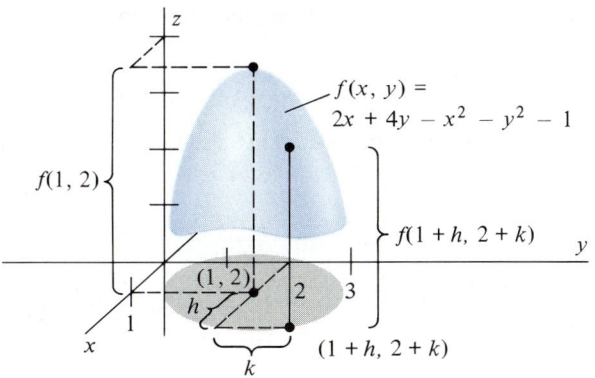

Figure 4.2 $f(1, 2) \geq f(1 + h, 2 + k)$ for small values of h and k; $f(1, 2)$ is a relative maximum. (See **Plate 9**.)

Finding Relative Extrema

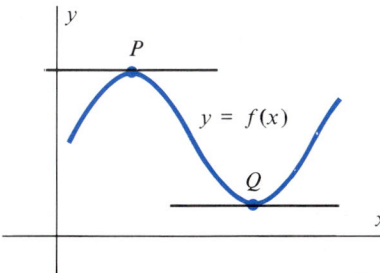

Figure 4.3 At relative extremum P or Q, $f'(x_0) = 0$ and tangent *line* is horizontal.

The preceding discussion addressed the issue of verifying that $f(x_0, y_0)$ is a relative extremum. But how do we find (x_0, y_0) to begin with? Recall the one-variable case again. If the function f has a relative extremum at $x = x_0$, then either $f'(x_0) = 0$ or else $f'(x_0)$ fails to exist. In the former case, the line tangent to the graph of f at $(x_0, f(x_0))$ is horizontal (see Figure 4.3). In the latter case, no tangent exists at $(x_0, f(x_0))$.

From a strictly geometric viewpoint, it seems that the same relationship should hold between the points (x_0, y_0), at which the function f has a relative extremum, and the *plane* tangent to the graph of f at $(x_0, y_0, f(x_0, y_0))$: At such points the tangent plane should either be horizontal or fail to exist (see Figure 4.4).

Since a tangent plane is determined at the point $(x_0, y_0, f(x_0, y_0))$ by the partial derivatives $\dfrac{\partial f}{\partial x}(x_0, y_0)$ and $\dfrac{\partial f}{\partial y}(x_0, y_0)$, these geometric observations lead to the following theorem.

THEOREM 3

If the number $z_0 = f(x_0, y_0)$ is a relative extremum for the function f at the point (x_0, y_0), one of the following two conditions must hold:

(i) $\dfrac{\partial f}{\partial x}(x_0, y_0) = \dfrac{\partial f}{\partial y}(x_0, y_0) = 0$, or

(ii) one or both of $\dfrac{\partial f}{\partial x}(x_0, y_0)$ and $\dfrac{\partial f}{\partial y}(x_0, y_0)$ fails to exist.

Proof: First consider the case where $f(x_0, y_0)$ is a relative maximum. We assume that both $\dfrac{\partial f}{\partial x}(x_0, y_0)$ and $\dfrac{\partial f}{\partial y}(x_0, y_0)$ exist, and we attempt to show that

$$\frac{\partial f}{\partial x}(x_0, y_0) = \frac{\partial f}{\partial y}(x_0, y_0) = 0.$$

Holding $y = y_0$ fixed, we define the function (of a single variable) g by

$$g(x) = f(x, y_0).$$

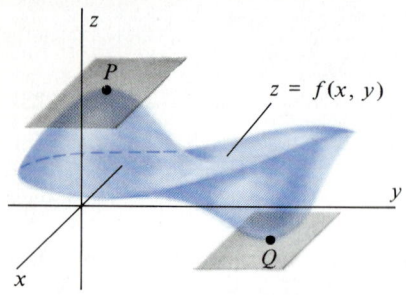

Figure 4.4 At relative extremum P or Q, $\dfrac{\partial f}{\partial x}(x_0, y_0) = \dfrac{\partial f}{\partial y}(x_0, y_0) = 0$ and tangent *plane* is horizontal.

Since $f(x_0, y_0)$ is a relative maximum, $f(x_0, y_0) \geq f(x, y)$ for all (x, y) near (x_0, y_0). Thus,

$$g(x_0) = f(x_0, y_0) \geq f(x, y_0) = g(x)$$

for all x near x_0. This shows that $g(x_0)$ is a relative maximum for the function g. Since the derivative of g is

$$g'(x) = \lim_{h \to 0} \frac{g(x + h) - g(x)}{h}$$

$$= \lim_{h \to 0} \frac{f(x + h, y_0) - f(x, y_0)}{h}$$

$$= \frac{\partial f}{\partial x}(x, y_0),$$

we know that $g'(x_0) = \dfrac{\partial f}{\partial x}(x_0, y_0)$ exists (Figure 4.5). But, if $g(x_0)$ is a relative maximum and $g'(x_0)$ exists, then $g'(x_0) = 0$. This shows that

$$\frac{\partial f}{\partial x}(x_0, y_0) = g'(x_0) = 0.$$

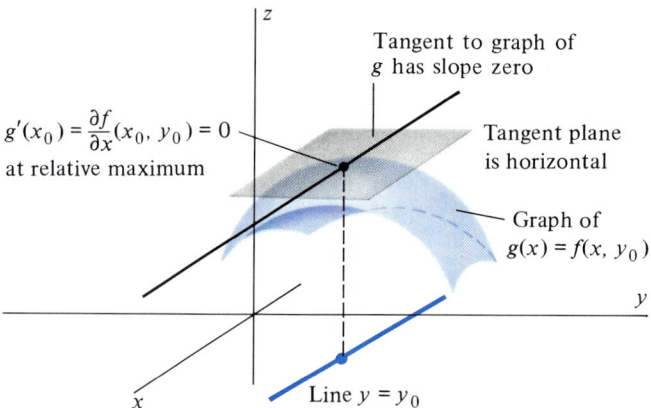

Figure 4.5 A relative maximum: If $\dfrac{\partial f}{\partial x}(x_0, y_0)$ exists, then $\dfrac{\partial f}{\partial x}(x_0, y_0) = 0.$

By repeating this argument with $x = x_0$ fixed, we can show that $\dfrac{\partial f}{\partial y}(x_0, y_0) = 0$ as well. Finally, we note that we have made no special use of the assumption that the extremum $f(x_0, y_0)$ was a relative maximum—the argument applies for relative minima as well. ◆

Theorem 3 determines a procedure for finding relative extrema for f: Find all points (x_0, y_0) where $\dfrac{\partial f}{\partial x}(x_0, y_0) = \dfrac{\partial f}{\partial y}(x_0, y_0) = 0$ or where either $\dfrac{\partial f}{\partial x}(x_0, y_0)$ or

$\dfrac{\partial f}{\partial y}(x_0, y_0)$ fails to exist. (We call these points **critical points.**) Then test each critical point to determine whether it yields a relative extremum.

Example 3

For the function $f(x, y) = 2x + 4y - x^2 - y^2 - 1$, setting both partial derivatives equal to zero gives the equations

$$\frac{\partial f}{\partial x}(x, y) = 2 - 2x = 0, \qquad \frac{\partial f}{\partial y}(x, y) = 4 - 2y = 0.$$

The (simultaneous) solution of this pair of equations is $x = 1$, $y = 2$. Thus, condition (i) in Theorem 3 yields the single critical point $(1, 2)$. We have verified that $f(1, 2)$ is a relative maximum in Example 2. Since the partial derivatives are defined for all (x, y), there are no points satisfying condition (ii) of Theorem 3. Thus, the only relative extremum for this function is the relative maximum at $(1, 2)$ (see Figure 4.2). ◇

Example 4

Find all relative extrema for the function

$$f(x, y) = \begin{cases} \sqrt{x^2 + y^2}, & (x, y) \neq (0, 0) \\ 0, & (x, y) = (0, 0). \end{cases}$$

Solution: The partial derivatives are

$$\frac{\partial f}{\partial x}(x, y) = \frac{x}{\sqrt{x^2 + y^2}}; \qquad \frac{\partial f}{\partial y}(x, y) = \frac{y}{\sqrt{x^2 + y^2}}.$$

Both partial derivatives are undefined at $(x, y) = (0, 0)$. For all other points at least one of the partial derivatives is nonzero. Since it is easy to see that $f(0, 0) = 0 < f(x, y)$ for all $(x, y) \neq (0, 0)$, $f(0, 0) = 0$ is a relative minimum. (Figure 4.1.) ◇

It is important to understand that the conditions of Theorem 3 do not *guarantee* that $f(x_0, y_0)$ is a relative extremum. Theorem 3 merely provides *necessary* conditions for an extremum. Without satisfying condition (i) or (ii) of Theorem 3, $f(x_0, y_0)$ cannot be a relative extremum. However, some critical points do not yield extrema. The following is a typical such example.

Example 5

For the function $f(x, y) = y^2 - x^2$, the partial derivatives are

$$\frac{\partial f}{\partial x} = -2x; \qquad \frac{\partial f}{\partial y} = 2y.$$

Thus, $\dfrac{\partial f}{\partial x}(0, 0) = \dfrac{\partial f}{\partial y}(0, 0) = 0$. However, the number $f(0, 0) = 0$ is neither a

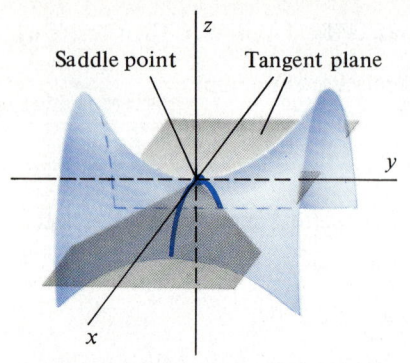

Saddle point   Tangent plane

Figure 4.6 Graph of $z = y^2 - x^2$ has a *saddle* point, but no extremum, at $(0, 0)$. (See **Plate 10.**)

relative maximum nor a relative minimum. To see this, we compare $f(0, 0)$ with $f(h, k)$ where (h, k) is a point near $(0, 0)$. Since

$$f(0, 0) - f(h, k) = 0 - (k^2 - h^2) = h^2 - k^2,$$

we see that the sign of this difference depends only on the relative sizes of $|h|$ and $|k|$. Since this difference is not of constant sign, $f(0, 0)$ is neither a maximum nor a minimum.

Figure 4.6 illustrates why $f(x, y) = y^2 - x^2$ does not have an extremum at $(0, 0)$. Although both partial derivatives are zero, the function $g(x) = f(x, 0) = -x^2$ reaches a relative *maximum* while the function $h(y) = f(0, y) = y^2$ reaches a relative *minimum* at $(0, 0)$. ◇

In Example 5, the point $(0, 0)$ is called a *saddle point* for rather obvious reasons. The surface bows upward along one axis and downward along another. More generally, a point (x_0, y_0) in the domain of a function of two variables is called a **saddle point** if (x_0, y_0) is a critical point and if $f(x_0, y_0)$ is neither a relative maximum nor a relative minimum. Figure 4.7 shows the graph of another function that has a saddle point at $(0, 0)$ (see Exercise 20).

Second Derivative Test

There is a theorem that helps sort out actual extrema from saddle points. It may be regarded as the analogue of the Second Derivative Test for functions of a single variable. A proof of this theorem may be found in texts on advanced calculus.

THEOREM 4
Second Derivative Test

Let f be a function of two variables. Suppose that all second order partial derivatives of f are continuous in a neighborhood of (x_0, y_0) and that $\dfrac{\partial f}{\partial x}(x_0, y_0) = \dfrac{\partial f}{\partial y}(x_0, y_0) = 0$. Let

$$A = \frac{\partial^2 f}{\partial x^2}(x_0, y_0), \qquad B = \frac{\partial^2 f}{\partial y \partial x}(x_0, y_0), \qquad C = \frac{\partial^2 f}{\partial y^2}(x_0, y_0)$$

and

$$D = B^2 - AC.$$

Then

(i) If $D < 0$ and $A < 0$, $f(x_0, y_0)$ is a relative maximum.
(ii) If $D < 0$ and $A > 0$, $f(x_0, y_0)$ is a relative minimum.
(iii) If $D > 0$, (x_0, y_0) is a saddle point.
(iv) If $D = 0$, no conclusions may be drawn.

Theorem 4 verifies that the critical point $(0, 0)$ in Example 5 is a saddle point, since

$$A = \frac{\partial^2 f}{\partial x^2}(0, 0) = -2, \qquad B = \frac{\partial^2 f}{\partial y \partial x}(0, 0) = 0, \qquad C = \frac{\partial^2 f}{\partial y^2}(0, 0) = 2$$

and

$$D = B^2 - AC = 4 > 0.$$

Graphic Representations of
Mathematical Surfaces in Three Dimensions

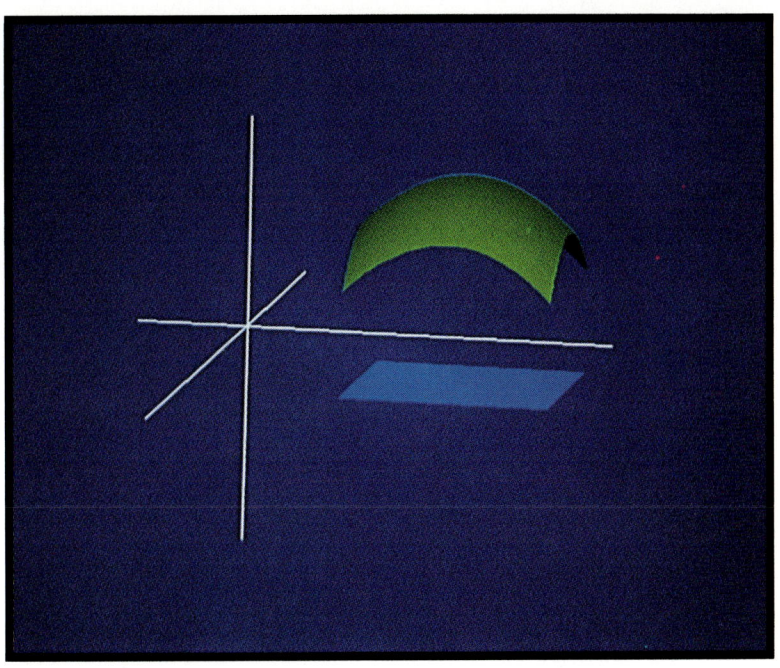

PLATE 1   Graph of a function $z = f(x, y)$ of two variables with domain D.

PLATE 2   Figure 18.1.1 (Plate 1) rotated.

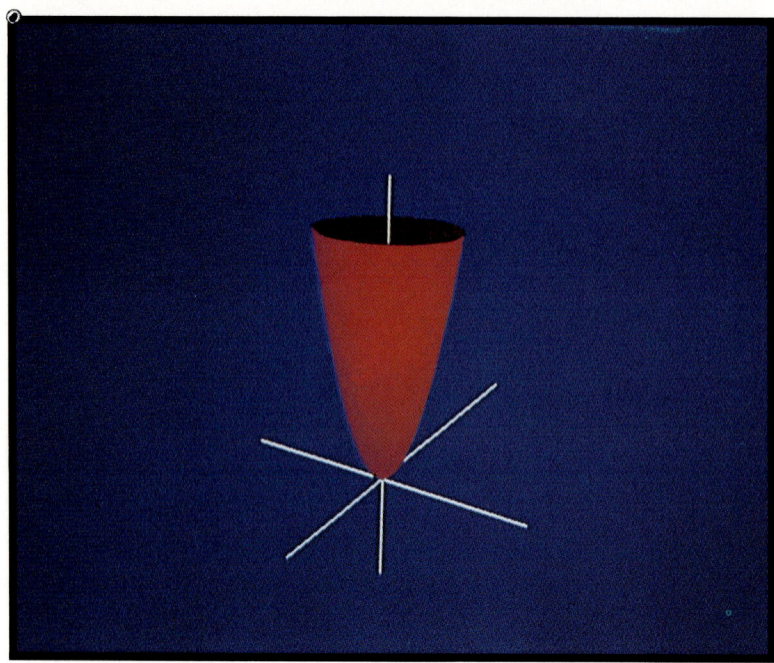

PLATE 3 Paraboloid $z = x^2 + y^2$ is the graph of a function of two variables.

PLATE 4 A cone is the graph of the equation $z^2 = x^2 + y^2$.

PLATE 5   Partial derivative $\dfrac{\partial f}{\partial x}(x_0, y_0)$ is the slope of the line tangent to the trace of $z = f(x, y)$ in the plane $y = y_0$.

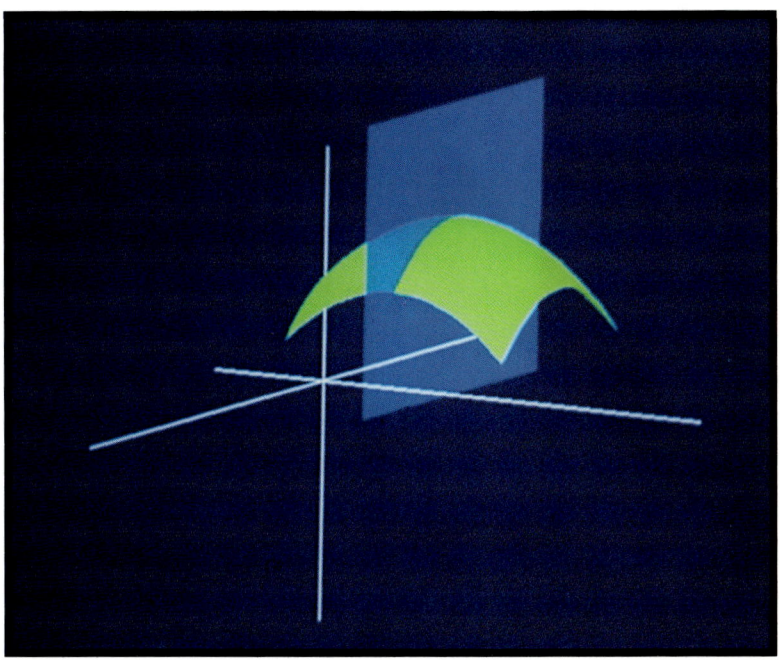

PLATE 6   Figure 18.2.1 (Plate 5) rotated.

PLATE 7 Plane tangent to the graph of $y = 9 - x^2 - y^2$.

PLATE 8 Relative minimum for $z = \sqrt{x^2 + y^2}$ is at the vertex of a cone.

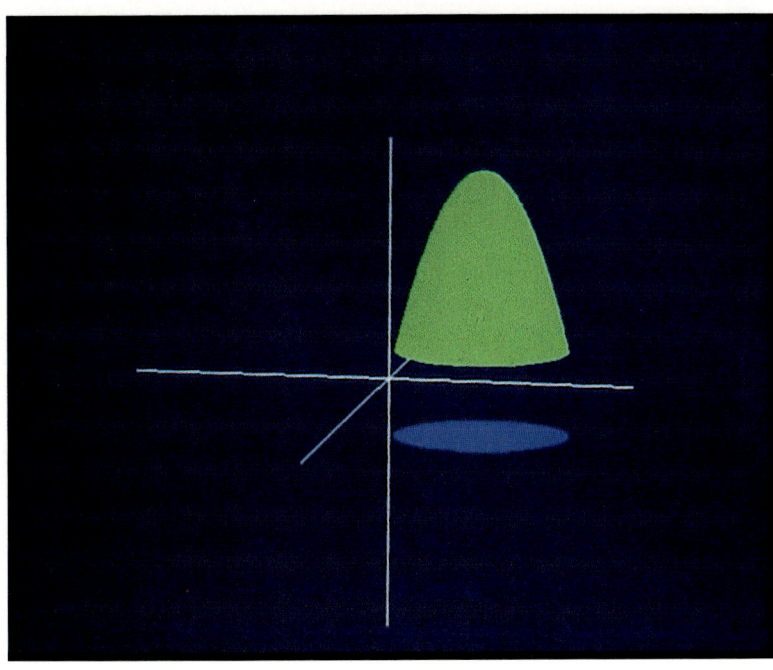

PLATE 9 Graph of a paraboloid $z = 2x + 4y - x^2 - y^2 - 1$ over a circle.

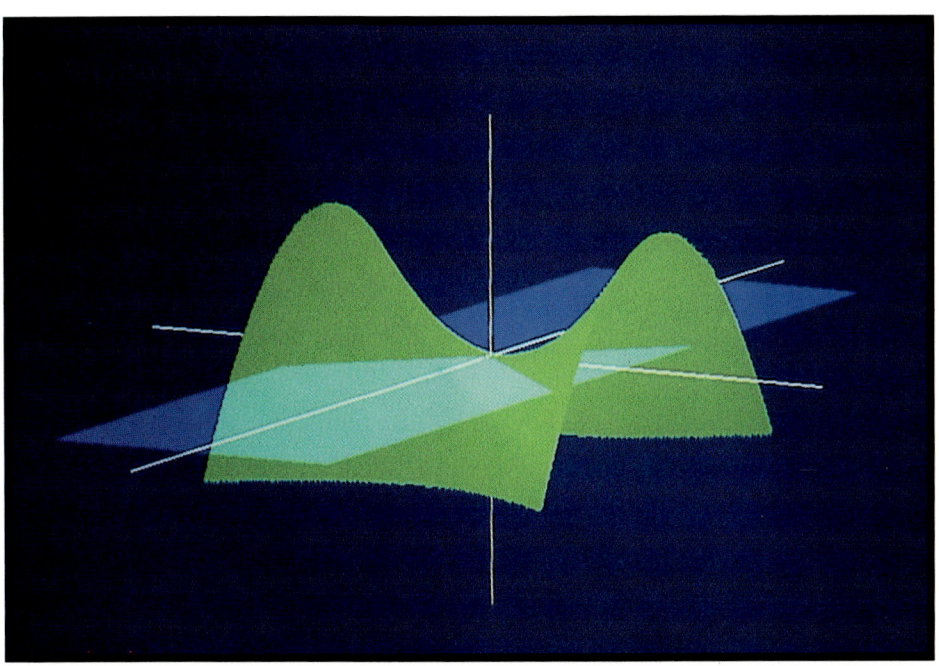

PLATE 10 Plane tangent to the graph of $z = y^2 - x^2$ at the saddle point.

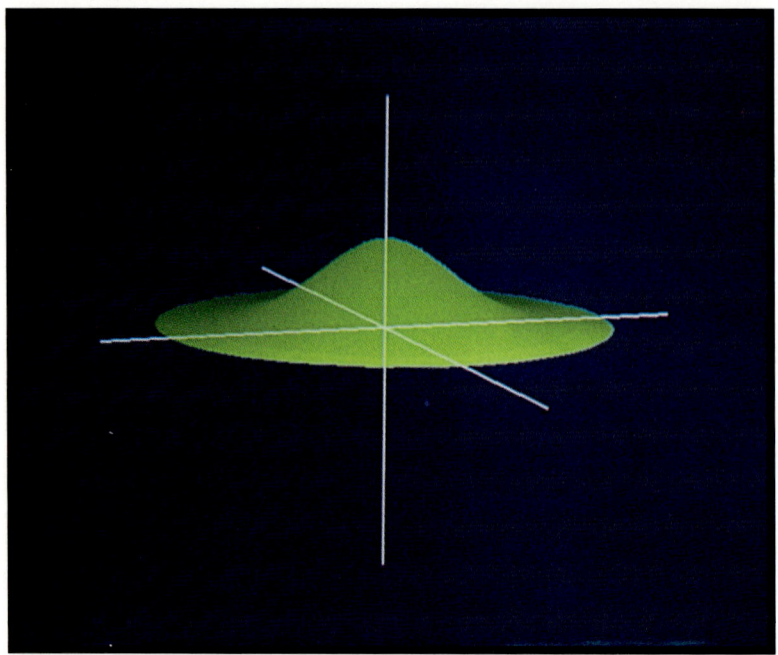

PLATE 11 **Graph of** $f(x, y) = e^{-(x^4+y^4)}$.

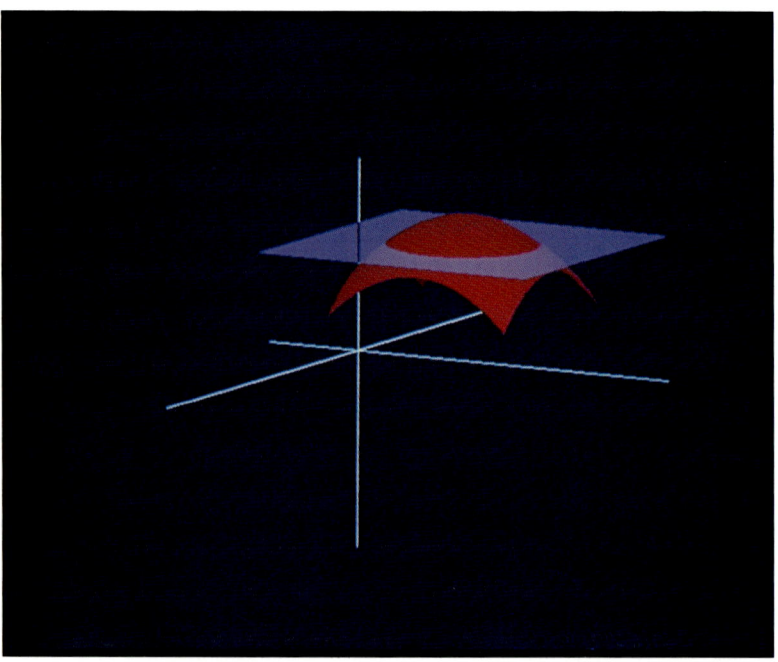

PLATE 12 **Horizontal plane intersects the graph in the** *level curve*.

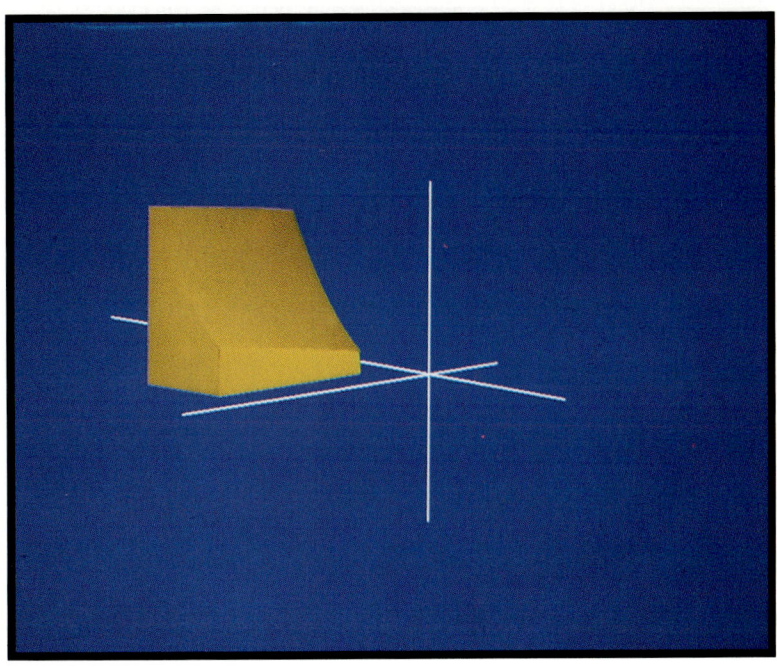

PLATE 13 Solid bounded by the graph of $f(x, y) = x^2 + 2y$ and xy plane.

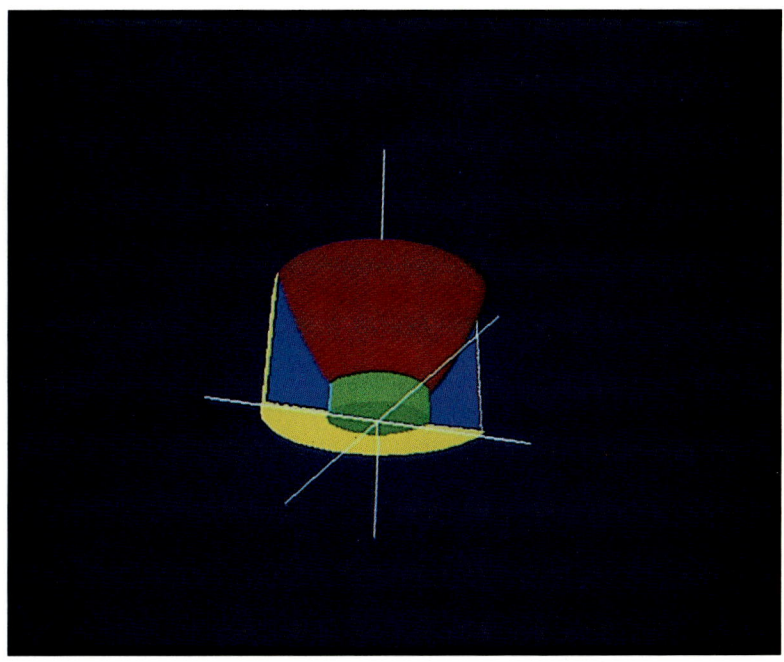

PLATE 14 One half of the bearing sleeve of Example 3, page 873.

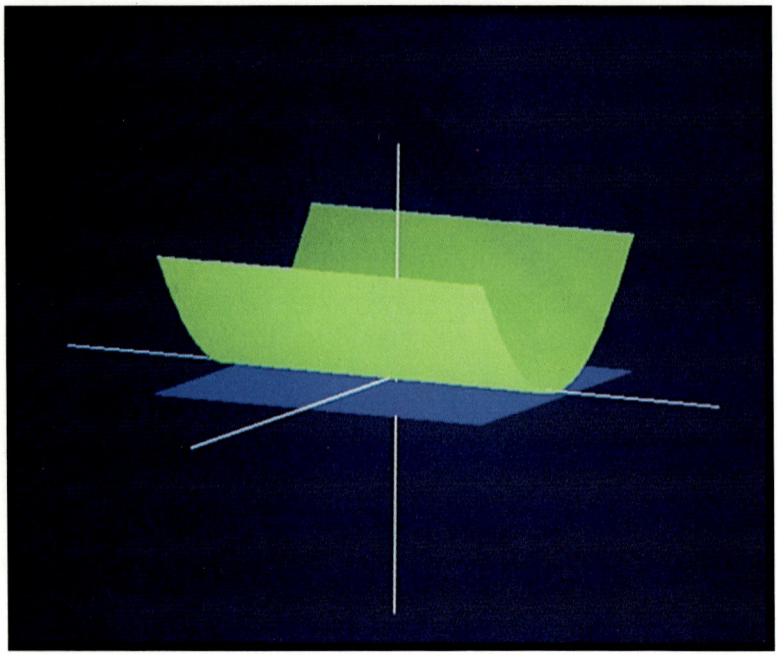

PLATE 15 **Graph of $f(x, y) = x^2$ is a *cylinder*.**

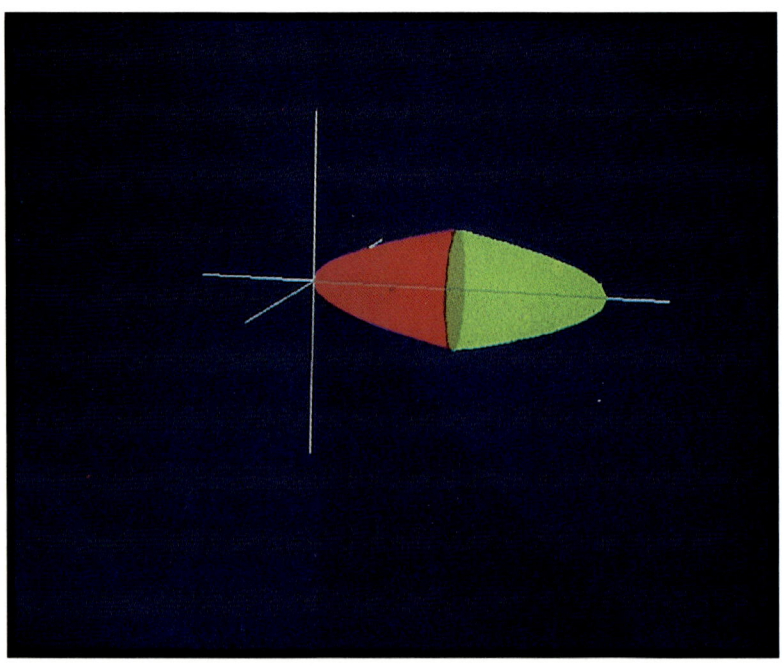

PLATE 16 **Paraboloids intersecting in a circle.**

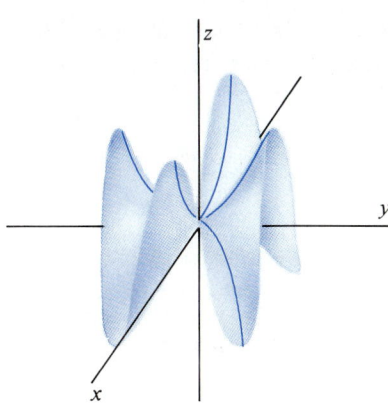

Figure 4.7 Graph of

$f(x, y) = x^4 - 3x^2y^2 + y^4$

has a saddle point at $(0, 0)$.

Similarly, for the function f in Example 3 and the critical point $(1, 2)$, we have

$$A = \frac{\partial^2 f}{\partial x^2}(1, 2) = -2, \qquad B = \frac{\partial^2 f}{\partial y \partial x}(1, 2) = 0, \qquad C = \frac{\partial^2 f}{\partial y^2}(1, 2) = -2$$

and

$$D = B^2 - AC = -4.$$

Thus, since $D < 0$ and $A < 0$, the Second Derivative Test agrees with the conclusion of Example 3, that $f(1, 2)$ is a relative maximum.

Example 6

Find and classify all relative extrema for the function $f(x, y) = x^4 + y^4 - 4xy$.

Solution: The partial derivatives are

$$\frac{\partial f}{\partial x} = 4x^3 - 4y; \qquad \frac{\partial f}{\partial y} = 4y^3 - 4x.$$

Since both partial derivatives are defined for all (x, y), the extrema can occur only at points where

$$\frac{\partial f}{\partial x} = 4x^3 - 4y = 0, \qquad \text{or} \qquad y = x^3 \tag{3}$$

and

$$\frac{\partial f}{\partial y} = 4y^3 - 4x = 0, \qquad \text{or} \qquad x = y^3. \tag{4}$$

Substituting for x in (3) using (4) gives the equation $y = y^9$. Thus, either $y = 0$, or else $y^8 = 1$, which gives $y = \pm 1$. Using (4) we find that $x = 0$ if $y = 0$, $x = 1$ if $y = 1$, and $x = -1$ if $y = -1$. The three critical points are therefore $(0, 0)$, $(1, 1)$, and $(-1, -1)$.

Next, we calculate the second order partials:

$$\frac{\partial^2 f}{\partial x^2} = 12x^2; \qquad \frac{\partial^2 f}{\partial y \partial x} = -4; \qquad \frac{\partial^2 f}{\partial y^2} = 12y^2.$$

At the critical point $(1, 1)$ we have

$$A = \frac{\partial^2 f}{\partial x^2}(1, 1) = 12, \qquad B = \frac{\partial^2 f}{\partial y \partial x}(1, 1) = -4, \qquad C = \frac{\partial^2 f}{\partial y^2}(1, 1) = 12,$$

and

$$D = B^2 - AC = 16 - 12 \cdot 12 = -128 < 0.$$

Thus, since $D < 0$ and $A > 0$, $f(1, 1) = -2$ is a relative minimum, according to Theorem 4.

At the critical point $(-1, -1)$ the values of A, B, C, and D are the same as for $(1, 1)$ so $f(-1, -1) = -2$ is also a relative minimum.

At the critical point $(0, 0)$, $A = C = 0$ and $B = -4$, so $D = B^2 - AC = 16 > 0$. Thus, $(0, 0)$ is a saddle point. ◇

In each of Examples 7, 8, and 9, we obtain $D = 0$ at the critical point. These three examples show that any of the three possible outcomes (relative maximum, relative minimum, saddle point) may result when $D = 0$.

Example 7

For the function $f(x, y) = e^{-(x^4 + y^4)}$, we have

$$\frac{\partial f}{\partial x} = -4x^3 e^{-(x^4 + y^4)} \qquad \frac{\partial f}{\partial y} = -4y^3 e^{-(x^4 + y^4)}$$

$$\frac{\partial^2 f}{\partial x^2} = (16x^6 - 12x^2)e^{-(x^4 + y^4)} \qquad \frac{\partial^2 f}{\partial y^2} = (16y^6 - 12y^2)e^{-(x^4 + y^4)}$$

$$\frac{\partial^2 f}{\partial y \partial x} = 16y^3 x^3 e^{-(x^4 + y^4)}.$$

The point $(0, 0)$ is a critical point since $\dfrac{\partial f}{\partial x}(0, 0) = \dfrac{\partial f}{\partial y}(0, 0) = 0$. At this critical point

$$A = \frac{\partial^2 f}{\partial x^2}(0, 0) = 0; \qquad B = \frac{\partial^2 f}{\partial y \partial x}(0, 0) = 0; \qquad C = \frac{\partial^2 f}{\partial y^2}(0, 0) = 0.$$

Thus, the Second Derivative Test yields no conclusion about the critical point $(0, 0)$. However, it is easy to see that the expression $x^4 + y^4$ has a minimum at $(0, 0)$, so

$$f(x, y) = e^{-(x^4 + y^4)} = \frac{1}{e^{x^4 + y^4}}$$

has a relative maximum at $(0, 0)$ (see Figure 4.8).   ◇

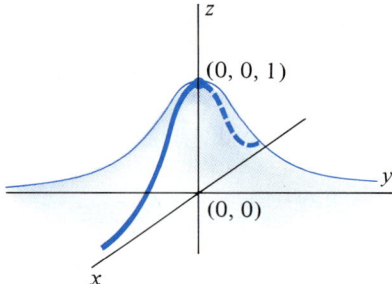

Figure 4.8 $f(x, y) = e^{-(x^4 + y^4)}$ has a relative maximum at $(0, 0)$. (See **Plate 11**.)

Example 8

The function $f(x, y) = x^4 + y^4$ obviously has a relative minimum of $f(0, 0) = 0$, since $f(x, y) > 0$ at all other points. However, we find

$$A = \frac{\partial^2 f}{\partial x^2}(0, 0) = 12x^2 \bigg|_{x=0} = 0, \qquad B = \frac{\partial^2 f}{\partial y \partial x}(0, 0) = 0,$$

$$C = \frac{\partial^2 f}{\partial y^2}(0, 0) = 12y^2 \bigg|_{y=0} = 0, \qquad D = B^2 - AC = 0.$$

Thus, the Second Derivative Test fails to classify this critical point.   ◇

Example 9

For the function $f(x, y) = x^3 - y^3$, the only simultaneous solution of the two equations

$$\frac{\partial f}{\partial x}(x, y) = 3x^2 = 0 \qquad \text{and} \qquad \frac{\partial f}{\partial y}(x, y) = -3y^2 = 0$$

is $x = y = 0$, so $(0, 0)$ is the only critical point. Since

$$\frac{\partial^2 f}{\partial x^2} = 6x, \qquad \frac{\partial^2 f}{\partial y \partial x} = 0, \qquad \frac{\partial^2 f}{\partial y^2} = -6y,$$

we have $A = B = C = D = 0$ at the critical point $(0, 0)$.

Thus, the test gives no information as to the nature of this critical point. However, since $f(x, y) = x^3 - y^3$ takes on both positive and negative values in every neighborhood of $(0, 0)$, the critical point $(0, 0)$ is a saddle point.   ◇

Functions of More Than Two Variables

The discussion of this section has focused on the question of finding relative extrema for functions of two variables. For functions of more than two variables, the statement of Theorem 3 generalizes directly:

(a) If the function f of three variables has a relative extremum at (x_0, y_0, z_0), then either

$$\frac{\partial f}{\partial x} = \frac{\partial f}{\partial y} = \frac{\partial f}{\partial z} = 0$$

at this point, or else one or more of these partial derivatives fails to exist at (x_0, y_0, z_0).

(b) More generally, the function f of n variables can have a relative extremum at the point $(a_1, a_2, \ldots, a_n)$ only if one or more of the partial derivatives fails to exist, or if

$$\frac{\partial f}{\partial x_1} = \frac{\partial f}{\partial x_2} = \cdots = \frac{\partial f}{\partial x_n} = 0$$

at this point.

However, the Second Derivative Test does not generalize quite so easily. For functions of more than two variables, the classification of critical points is considerably more difficult and will not be pursued here.

Extrema on Closed Sets

Finally, we note that we have not addressed the question of *absolute* extrema, as we did for functions of a single variable in Chapter 4. In a manner analogous to the behavior of a function of one variable on a closed interval, a continuous function f of two variables will assume both an absolute maximum and an absolute minimum on any **closed bounded set** S in the plane. By a closed set in the plane we mean (roughly speaking) a set that includes its boundary, such as the disc $D = \{(x, y) \mid x^2 + y^2 \leq 1\}$ or the square $S = \{(x, y) \mid 0 \leq x \leq 1, 0 \leq y \leq 1\}$. As you might suspect, such absolute extrema will occur either at critical points or at points on the boundary of S. Checking for extrema among boundary points can be a quite complicated task, so we have chosen to defer this topic to more advanced courses where appropriate theory and techniques are developed. (However, this issue is addressed in simple settings in Exercises 29–32.)

Exercise Set 17.4

In Exercises 1–16, find all critical points. Classify each as a relative maximum, relative minimum, or saddle point.

1. $f(x, y) = x^2 + y^2 + 4y + 4$

2. $f(x, y) = x^2 + y^2 + 4x - 2y + 11$

3. $f(x, y) = x^2 - y^2 + 6x + 4y + 5$

4. $f(x, y) = 7 - 2x + 2y - x^2 - y^2$

5. $f(x, y) = xy + 9$

6. $f(x, y) = x^2 + y^4 - 2x - 4y^2 + 5$

7. $f(x, y) = 5x^2 + y^2 - 10x - 6y + 15$

8. $f(x, y) = x^2 + y^3 - 3y$

9. $f(x, y) = x^3 - y^3$

10. $f(x, y) = e^{x^2 - 2x + y^2 + 4}$

11. $f(x, y) = x^2 - xy$

12. $f(x, y) = e^x \cos y$

13. $f(x, y) = x^3 + y^3 + 4xy$

14. $f(x, y) = x^4 + y^4 - 4xy$

15. $f(x, y) = x \cos y$

16. $f(x, y) = \dfrac{y}{x} - \dfrac{x}{y}$

17. Show that the function $z = 4 - \sqrt{x^2 + y^2}$ has a relative maximum at $(0, 0)$. (Note that none of the partial derivatives exists at this point.)

18. Show that the function $z = x^2 + y^2 - 4x - 2y + 9$ has a relative minimum at $(2, 1)$ using the method of Example 2.

19. Find the highest point on the graph of $z = 2y^3 - 3x^2 - 3xy + 9x$ if the positive z-axis is upward.

20. Show that the function $f(x, y) = x^4 - 3x^2y^2 + y^4$ has a saddle point at $(0, 0)$ (see Figure 4.7).

21. Show that the function $f(x, y) = x^2 - y^2 + 2x + 4y - 3$ has a saddle point at $(-1, 2)$ by the method of Example 2.

22. Consider the function $f(x, y) = \dfrac{x^2 + y^2}{(x + y)^2}$.

   a. Show that $\dfrac{\partial f}{\partial x}(x, y) = \dfrac{\partial f}{\partial y}(x, y) = 0$ if and only if $y = x$ and $x \neq 0$.

   b. Show that f has a relative minimum at each point (x, x) with $x \neq 0$.

   c. Show that the function $g(\lambda) = f(x, \lambda x) = \dfrac{1 + \lambda^2}{(1 + \lambda)^2}$, $\lambda \neq -1$, has a relative minimum at $\lambda = 1$.

   d. Explain the geometric and algebraic relationships between the functions g and f.

23. A rectangular box, with a top, is to hold 16 cubic meters. Find the dimensions that produce the least expensive box if the material for the side walls is half as expensive as the material for the top and the bottom.

24. Find the point on the plane $2x + 3y + z - 14 = 0$ nearest the origin.

25. Find the point on the plane with equation $ax + by + cz = d$ nearest the origin.

26. Find the dimensions of the closed rectangular box of volume $V = 8000$ cm^3 and of minimum surface area.

27. Find the dimensions of the rectangular package of largest volume that can be mailed under the restrictions that length plus girth cannot exceed 84 inches. (Girth is the perimeter of the cross section taken perpendicular to the length.)

28. A manufacturer of tape recorders intends to market x recorders through retail outlets and y recorders through wholesale outlets. The manufacturer estimates that the revenue per recorder sold retail will be $r(x, y) = 250 - \dfrac{x}{20} - \dfrac{y}{5}$, while the revenue per recorder on the wholesale market is estimated to be $w(x, y) = 200 - \dfrac{x}{50} - \dfrac{y}{20}$. The cost of producing the recorders is known to be \$100 each. Using the model Profit = Revenue − Cost, find the number of recorders that should be marketed retail and the number that should be marketed wholesale so as to maximize profit.

29. Find the *absolute* maximum and minimum values of the function $f(x, y) = 2x + 2y + 4$ on the *closed* disc $D = \{(x, y) \mid x^2 + y^2 \leq 1\}$. (*Hint:* To check for absolute extrema along the boundary $C = \{(x, y) \mid x^2 + y^2 = 1\}$ use the parameterization $x = \cos t$, $y = \sin t$, and optimize $f(x(t), y(t))$ as a function of t.)

30. Find the absolute maximum and minimum values of the function $f(x, y) = 3x - 2y^3 + 2$ on the (closed) square $S = \{(x, y) \mid 0 \leq x \leq 1, 0 \leq y \leq 1\}$ (see Exercise 29).

31. Find the absolute maximum and minimum values of the function $z = x + 2y + 6$ on the closed set consisting of the ellipse $4x^2 + 2y^2 = 4$ and its interior (see Exercise 29).

32. Find the absolute maximum and minimum values for the function $f(x, y) = x^2 + 2y^2 - 2xy - 6x + 4y$ on the rectangle $0 \le x \le 8$, $-2 \le y \le 2$.

33. Given a set of points $(x_1, y_1), (x_2, y_2), \ldots , (x_n, y_n)$, one way to define a line $y = mx + b$ that "best fits" these points is given by the Method of Least Squares. As illustrated by Figure 4.9, the distance from the data point (x_j, y_j) to the point on the line $y = mx + b$ with x-coordinate x_j is

$$|y_j - (mx_j + b)| = |y_j - mx_j - b|.$$

The Method of Least Squares defines the best fitting line $y = mx + b$ (called the **regression line**) to be the line that minimizes the sum of the squares of these individual distances:

$$S(m, b) = \sum_{j=1}^{n} (y_j - mx_j - b)^2.$$

Viewing $x_1, x_2, \ldots , x_n$ and $y_1, y_2, \ldots , y_n$ as fixed constants and m and b as independent variables, show that the values of m and b for which $S(m, b)$ is a minimum are given by the formulas

$$m = \frac{n \sum_{j=1}^{n} x_j y_j - \left(\sum_{j=1}^{n} x_j\right)\left(\sum_{j=1}^{n} y_j\right)}{n \sum_{j=1}^{n} x_j^2 - \left(\sum_{j=1}^{n} x_j\right)^2},$$

$$b = \frac{\left(\sum_{j=1}^{n} x_j^2\right)\left(\sum_{j=1}^{n} y_j\right) - \left(\sum_{j=1}^{n} x_j\right)\left(\sum_{j=1}^{n} x_j y_j\right)}{n \sum_{j=1}^{n} x_j^2 - \left(\sum_{j=1}^{n} x_j\right)^2}.$$

(*Hint:* Set $\dfrac{\partial S}{\partial m}(m, b) = 0$ and $\dfrac{\partial S}{\partial b}(m, b) = 0$ and solve the resulting system of equations.)

34. For a group of $n = 6$ calculus students, achievement test scores x_j and final course averages in calculus y_j were as follows:

j	1	2	3	4	5	6
x_j	52	46	69	54	61	48
y_j	74	66	94	91	84	80

The regression line for these data appears in Figure 4.10. Use the result of Exercise 33 to show that this regression line has equation $y = 0.95x + 29$.

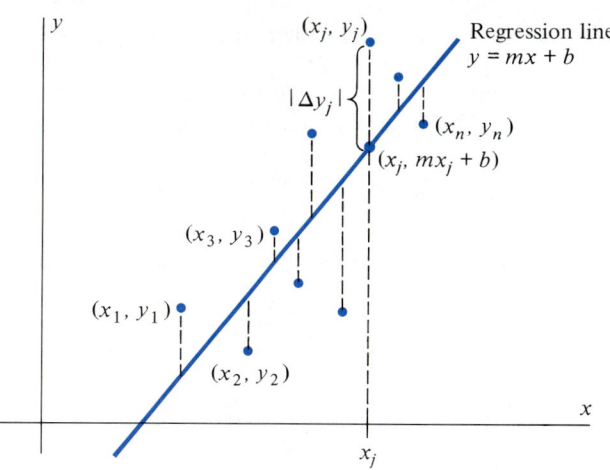

Figure 4.9 The Method of Least Squares determines the line that "best fits" the data points (x_1, y_1), $(x_2, y_2), \ldots , (x_n, y_n)$.

35. A regression line corresponding to a set of data points $(x_1, y_1), (x_2, y_2), \ldots , (x_n, y_n)$ provides a model for *predicting* a value of y corresponding to any particular value of x. Use the regression line obtained in Exercise 34 to predict the course average for a student whose achievement score is $x = 60$.

36. Given the five points whose coordinates are:

x	0	1	1	2	3
y	2	4	3	6	6

a. Find the regression line for the data using the Method of Least Squares.
b. Plot the data points and the regression line.
c. Find the predicted value for $x = 4$.

37. Repeat Exercise 36 for the data below.

x	6	8	9	10	12
y	2	5	5	7	9

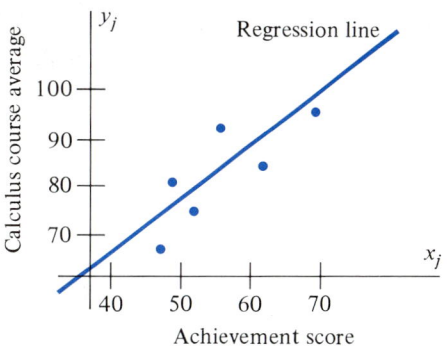

Figure 4.10 Achievement scores versus calculus course averages for six students.

38. For data consisting of ordered triples $(x_1, y_1, z_1), \ldots,$ (x_n, y_n, z_n), the analog of the regression line is the (regression) **plane** $z = ax + by + c$. This plane is defined as the plane that minimizes the sum of squares

$$S = \sum_{j=1}^{n} [z_j - (ax_j + by_j + c)]^2.$$

Show that the coefficients a, b, and c in this equation satisfy the equations

(i) $a \sum_{j=1}^{n} x_j + b \sum_{j=1}^{n} y_j + nc = \sum_{j=1}^{n} z_j$,

(ii) $a \sum_{j=1}^{n} x_j^2 + b \sum_{j=1}^{n} x_j y_j + c \sum_{j=1}^{n} x_j = \sum_{j=1}^{n} x_j y_j$,

(iii) $a \sum_{j=1}^{n} x_j y_j + b \sum_{j=1}^{n} y_j^2 + c \sum_{j=1}^{n} y_j = \sum_{j=1}^{n} y_j z_j$.

39. Recall, from Section 7.7, that the *centroid* of the n data points $(x_1, y_1), (x_2, y_2), \ldots, (x_n, y_n)$ is the point $(\bar{x}, \bar{y})$ where

$$\bar{x} = \frac{1}{n} \sum_{j=1}^{n} x_j, \qquad \bar{y} = \frac{1}{n} \sum_{j=1}^{n} y_j.$$

Show that the centroid lies on the regression line determined by the Method of Least Squares.

40. Suppose that you suspected that the n data points (x_1, y_1), $(x_2, y_2), \ldots, (x_n, y_n)$ could be "fitted" well by a curve of the form $y = m \ln x + b$ (a common situation in biology and chemistry). Explain how the Method of Least Squares could be used to find the constants m and b.

41. Prove Theorem 4 in the case of the function $f(x, y) = ax^2 + byx + cy^2$.

17.5 APPROXIMATIONS AND DIFFERENTIALS

For the function of a single variable $y = f(x)$, the existence of the derivative $f'(x)$ leads to the approximation formula

$$f(x + \Delta x) \approx f(x) + f'(x)\, \Delta x \tag{1}$$

and to the definition of the differential

$$df = f'(x)\, dx. \tag{2}$$

In this section, we take up the analogous issues for functions of several variables. The key to the entire discussion (and to the discussions of the sections that follow) is the Approximation Theorem.

THEOREM 5
Approximation Theorem

Let f be a function of two variables. Let f and its first partial derivatives $\frac{\partial f}{\partial x}(x, y)$ and $\frac{\partial f}{\partial y}(x, y)$ be continuous in an open rectangle $R = \{(x, y) \mid a_1 < x < a_2, b_1 < y < b_2\}$ in the xy-plane. Let the point (x_0, y_0) and the point $(x_0 + \Delta x, y_0 + \Delta y)$ both lie in R. Then

$$f(x_0 + \Delta x, y_0 + \Delta y) = f(x_0, y_0) + \frac{\partial f}{\partial x}(x_0, y_0)\, \Delta x + \frac{\partial f}{\partial y}(x_0, y_0)\, \Delta y \tag{3}$$

$$+ \epsilon_1\, \Delta x + \epsilon_2\, \Delta y$$

where $\lim_{\substack{\Delta x \to 0 \\ \Delta y \to 0}} \epsilon_1 = 0$ and $\lim_{\substack{\Delta x \to 0 \\ \Delta y \to 0}} \epsilon_2 = 0$.

Theorem 5 provides the relationship between the value of the function f at (x_0, y_0) and the value of this function at the "nearby" point $(x_0 + \Delta x, y_0 + \Delta y)$.

Ignoring the error term ($\epsilon_1 \, \Delta x + \epsilon_2 \, \Delta y$) in equation (3) leads to the approximation

$$f(x_0 + \Delta x, y_0 + \Delta y) \approx f(x_0, y_0) + \frac{\partial f}{\partial x}(x_0, y_0) \, \Delta x + \frac{\partial f}{\partial y}(x_0, y_0) \, \Delta y \qquad (4)$$

for functions of two variables.

Figure 5.1 provides a geometric interpretation of approximation (4). Since both $\dfrac{\partial f}{\partial x}(x_0, y_0)$ and $\dfrac{\partial f}{\partial y}(x_0, y_0)$ exist, they determine a ''tangent'' plane with equation

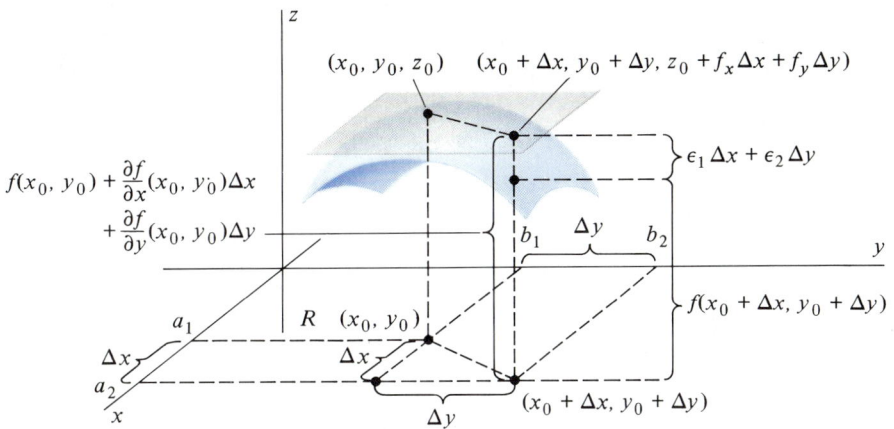

Figure 5.1 Point on tangent plane approximates point on graph of $z = f(x, y)$ above point $(x_0 + \Delta x, y_0 + \Delta y)$.

$$z = \frac{\partial f}{\partial x}(x_0, y_0)(x - x_0) + \frac{\partial f}{\partial y}(x_0, y_0)(y - y_0) + f(x_0, y_0) \qquad (5)$$

(equation (3), Section 18.3). To find the point on this plane above $(x, y) = (x_0 + \Delta x, y_0 + \Delta y)$, we set $x = x_0 + \Delta x$ and $y = y_0 + \Delta y$. Then $x - x_0 = \Delta x$ and $y - y_0 = \Delta y$, so (5) gives the z-coordinate as

$$z_T = \frac{\partial f}{\partial x}(x_0, y_0) \, \Delta x + \frac{\partial f}{\partial y}(x_0, y_0) \, \Delta y + f(x_0, y_0),$$

which is precisely the right-hand side of approximation (4). Thus, the left-hand side of approximation (4) is the z-coordinate of the point above $(x_0 + \Delta x, y_0 + \Delta y)$ *on the graph of f*, the right-hand side of (4) is the z-coordinate of the point above $(x_0 + \Delta x, y_0 + \Delta y)$ *on the tangent plane,* and the difference between these two numbers is the ''error term,'' $\epsilon_1 \, \Delta x + \epsilon_2 \, \Delta y$.

The proof of Theorem 5 is given at the end of this section.

Example 1

To see how this all works out in a particular example let us consider the specific function $f(x, y) = 2x^2 + 4y^2$ and the problem of calculating $f(1 + \Delta x, 2 + \Delta y)$. Here

$$(x_0, y_0) = (1, 2)$$
$$f(x_0, y_0) = f(1, 2) = 2 \cdot 1^2 + 4 \cdot 2^2 = 18$$

$$\frac{\partial f}{\partial x}(x_0, y_0) = 4x \Big|_{x=1} = 4$$

$$\frac{\partial f}{\partial y}(x_0, y_0) = 8y \Big|_{y=2} = 16$$

and

$$f(1 + \Delta x, 2 + \Delta y) = 2(1 + \Delta x)^2 + 4(2 + \Delta y)^2 \tag{6}$$
$$= 2(1 + 2\,\Delta x + \Delta x^2) + 4(4 + 4\,\Delta y + \Delta y^2)$$
$$= 18 + 4\,\Delta x + 16\,\Delta y + 2\,\Delta x^2 + 4\,\Delta y^2$$
$$= 18 + 4\,\Delta x + 16\,\Delta y + (2\,\Delta x)\Delta x + (4\,\Delta y)\Delta y.$$

$$\underbrace{\quad}_{f(1,\,2)} \quad \underbrace{\quad}_{\frac{\partial f}{\partial x}(1,\,2)} \quad \underbrace{\quad}_{\frac{\partial f}{\partial y}(1,\,2)} \quad \overbrace{\epsilon_1} \quad \overbrace{\epsilon_2}$$

Thus, the approximation (4) is

$$f(1 + \Delta x, 2 + \Delta y) \approx 18 + 4\,\Delta x + 16\,\Delta y. \tag{7}$$

By comparing (6) and (7), we can see that the error in this approximation is

$$\epsilon_1\,\Delta x + \epsilon_2\,\Delta y = 2(\Delta x)^2 + 4(\Delta y)^2.$$

Thus, $\epsilon_1 = 2\,\Delta x$ and $\epsilon_2 = 4\,\Delta y$ in this example. ◇

Example 2

Use approximation (4) to estimate the value of the expression $\sqrt{(3.04)^2 + (3.95)^2}$.

Solution: If we let $f(x, y) = \sqrt{x^2 + y^2}$, the problem is then to approximate $f(3.04, 3.95)$. To do so we note that for $x_0 = 3$ and $y_0 = 4$ it is easy to compute

$$f(x_0, y_0) = \sqrt{3^2 + 4^2} = \sqrt{25} = 5.$$

We therefore write

$$3.04 = x_0 + \Delta x \qquad \text{and} \qquad 3.95 = y_0 + \Delta y,$$

where

$$x_0 = 3, \qquad \Delta x = 0.04, \qquad y_0 = 4, \qquad \text{and} \qquad \Delta y = -0.05.$$

Also

$$\frac{\partial f}{\partial x}(x_0, y_0) = \frac{\partial}{\partial x}(\sqrt{x^2 + y^2}) \Big|_{(3,\,4)} = \frac{3}{\sqrt{3^2 + 4^2}} = 0.6$$

and

$$\frac{\partial f}{\partial y}(x_0, y_0) = \frac{\partial}{\partial y}(\sqrt{x^2 + y^2}) \Big|_{(3,\,4)} = \frac{4}{\sqrt{3^2 + 4^2}} = 0.8.$$

Thus, by (4),

$$\sqrt{(3.04)^2 + (3.95)^2} \approx 5 + (0.6)(0.04) + (0.8)(-0.05)$$
$$= 4.984.$$

The actual value of this expression, to four decimal places, is 4.9844. The *relative error* in the approximation is therefore

$$\frac{4.9844 - 4.984}{4.9844} \approx .00008,$$

a *percentage error* of less than 0.01%. ◇

Approximation (4) is written in a form that is useful in approximating a particular value $f(x_0 + \Delta x, y_0 + \Delta y)$. By using the notation

$$\Delta f = f(x_0 + \Delta x, y_0 + \Delta y) - f(x_0, y_0),$$

we may rewrite approximation (4) as

$$\Delta f \approx \frac{\partial f}{\partial x}(x_0, y_0)\, \Delta x + \frac{\partial f}{\partial y}(x_0, y_0)\, \Delta y. \tag{8}$$

Approximation (8) is useful in approximating the *change* in the value of the function f due to changes Δx and Δy in the independent variables.

Example 3

The ideal gas law is $PV = nRT$, where P is pressure, V is volume, n is the number of moles of gas present, R is constant, and T is temperature. Approximate, according to the ideal gas law, the change in the volume of 1000 cc of gas at temperature 300 K and pressure 780 mm of mercury if the gas is heated by 10 K and the pressure is increased by 5 mm of mercury.

Solution: We first simplify the gas law by solving for the constant nR and using the given data. We find that

$$nR = \frac{PV}{T} = \frac{780 \cdot 1000}{300} = 2600.$$

Using this constant, we may solve the gas law for V as a function of T and P:

$$V(T, P) = \frac{nRT}{P} = \frac{2600T}{P}.$$

According to approximation (8),

$$\Delta V \approx \frac{\partial}{\partial T}\left(\frac{2600T}{P}\right) \Delta T + \frac{\partial}{\partial P}\left(\frac{2600T}{P}\right) \Delta P$$

$$= \frac{2600\, \Delta T}{P} - \frac{2600T\, \Delta P}{P^2}.$$

Using the data $T = 300$, $P = 780$, $\Delta T = 10$, and $\Delta P = 5$, we obtain

$$\Delta V \approx \frac{2600(10)}{780} - \frac{2600(300)(5)}{780^2} \tag{9}$$

$$= 26.92 \text{ cc.}$$

The actual value of V corresponding to the temperature $T = 310$ and pressure $P = 785$ mm of mercury is

$$V = \frac{2600(310)}{785} = 1026.75,$$

so the actual value of ΔV is 26.75. The approximation in (9) is therefore in error by $26.92 - 26.75 = 0.17$ cc, a percentage error of only $\left(\dfrac{0.17}{26.75}\right) \times 100\% = 0.64\%$.

◇

Differentials

The preceding ideas may be written a bit more compactly using the terminology of **differentials.** Recall that in the one-variable case the differential df for the differentiable function f is defined to be

$$df = f'(x)\ dx. \tag{10}$$

The differential in (10) provides an approximation to the change Δf in $f(x)$, corresponding to the change $\Delta x = dx$ in x, obtained by following the tangent at $(x, f(x))$ rather than the graph of f.

For precisely the same reason, we define the differential of the function f of two variables to be

$$\boxed{df = \frac{\partial f}{\partial x}(x, y)\ dx + \frac{\partial f}{\partial y}(x, y)\ dy.} \tag{11}$$

As Figure 5.2 illustrates, the differential df provides an approximation to the change Δf corresponding to the changes $dx = \Delta x$ in x and $dy = \Delta y$ in y, obtained by following the tangent plane at $(x, y, f(x, y))$ rather than the graph of f. The differential in equation (11) is sometimes referred to as the **total differential** for the function $z = f(x, y)$.

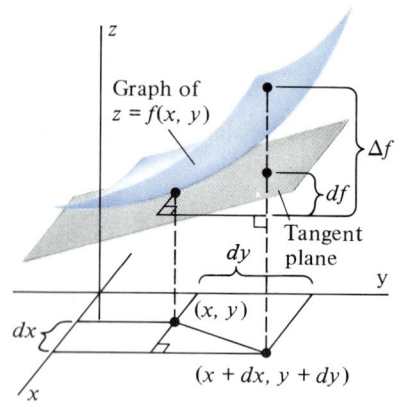

Figure 5.2 $df = f_x\ dx + f_y\ dy$ approximates change Δf by use of tangent plane.

Example 4

In economic theory, the Cobb-Douglas production function relating output production y to the input of labor L and capital K has the form

$$y = \lambda L^{\alpha} K^{1-\alpha}$$

where λ and α are positive constants. The change Δy in production resulting from a change dL in labor input and a change dK in capital input is therefore approximated by the total differential

$$dy = \frac{\partial}{\partial L}(\lambda L^{\alpha} K^{1-\alpha})\ dL + \frac{\partial}{\partial K}(\lambda L^{\alpha} K^{1-\alpha})\ dK$$

$$= \alpha \lambda L^{\alpha-1} K^{1-\alpha}\ dL + (1 - \alpha)\lambda L^{\alpha} K^{-\alpha}\ dK.$$

◇

Functions of Three Variables

The ideas of this section generalize directly to functions of three variables. The corresponding version of the Approximation Theorem (Theorem 5) is the following.

THEOREM 5′

Let f and its partial derivatives be continuous in the open rectangular box $B = \{(x, y, z) \mid a_1 < x < a_2, b_1 < y < b_2, c_1 < z < c_2\}$ in $\mathbb{R}^3$. If (x_0, y_0, z_0) and $(x_0 + \Delta x, y_0 + \Delta y, z_0 + \Delta z)$ both lie in B, then

$$f(x_0 + \Delta x, y_0 + \Delta y, z_0 + \Delta z) = f(x_0, y_0, z_0) + \frac{\partial f}{\partial x}(x_0, y_0, z_0)\Delta x$$

$$+ \frac{\partial f}{\partial y}(x_0, y_0, z_0)\Delta y + \frac{\partial f}{\partial z}(x_0, y_0, z_0)\Delta z$$

$$+ \epsilon_1 \, \Delta x + \epsilon_2 \, \Delta y + \epsilon_3 \, \Delta z$$

where $\epsilon_1 \to 0$, $\epsilon_2 \to 0$, and $\epsilon_3 \to 0$ as $\Delta x \to 0$, $\Delta y \to 0$, and $\Delta z \to 0$.

The proof of Theorem 5′ is similar to that of Theorem 5 (see Exercise 31). Finally, the definition of the differential, suggested by Theorem 5′, is

$$df = \frac{\partial f}{\partial x}(x, y, z) \, dx + \frac{\partial f}{\partial y}(x, y, z) \, dy + \frac{\partial f}{\partial z}(x, y, z) \, dz$$

or

$$dw = f_x \, dx + f_y \, dy + f_z \, dz$$

where $w = f(x, y, z)$.

Example 5

According to Newton's Law of Universal Gravitation, the force of attraction between two bodies of mass m_1 and m_2 is

$$F(m_1, m_2, r) = \frac{Gm_1m_2}{r^2},$$

where r is the distance between the bodies and G is the universal gravitation constant. The change ΔF in this force caused by changes dm_1 and dm_2 in the masses and dr in the distance is approximated by the differential

$$dF = \frac{\partial}{\partial m_1}\left(\frac{Gm_1m_2}{r^2}\right) dm_1 + \frac{\partial}{\partial m_2}\left(\frac{Gm_1m_2}{r^2}\right) dm_2 + \frac{\partial}{\partial r}\left(\frac{Gm_1m_2}{r^2}\right) dr$$

$$= \frac{Gm_2}{r^2} \, dm_1 + \frac{Gm_1}{r^2} \, dm_2 - \frac{2Gm_1m_2}{r^3} \, dr. \qquad \diamondsuit$$

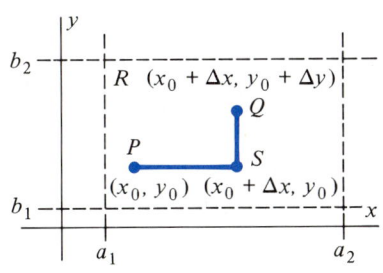

Figure 5.3 Diagram for proof of Theorem 5.

Proof of Theorem 5: Let $P = (x_0, y_0)$ and $Q = (x_0 + \Delta x, y_0 + \Delta y)$. Since R is a rectangle, the point $S = (x_0 + \Delta x, y_0)$ also lies in R, as do the line segments PS and SQ (see Figure 5.3).

Along the line segment PS the variable $y = y_0$ is fixed, so $g(x) = f(x, y_0)$ is a function of x alone, and $g'(x) = \frac{\partial f}{\partial x}(x, y_0)$ exists for all $x \in (x_0, x_0 + \Delta x)$. Thus, by the Mean Value Theorem, there exists a number $c \in (x_0, x_0 + \Delta x)$ so that

$$\frac{f(x_0 + \Delta x, y_0) - f(x_0, y_0)}{\Delta x} = \frac{\partial f}{\partial x}(c, y_0). \qquad (12)$$

Similarly, along the line segment SQ, the function $h(y) = f(x_0 + \Delta x, y)$ is a function of y alone, and $h'(y) = \frac{\partial f}{\partial y}(x_0 + \Delta x, y)$ for each $y \in (y_0, y_0 + \Delta y)$. Again by the Mean Value Theorem, there exists a number $d \in (y_0, y_0 + \Delta y)$ so that

$$\frac{f(x_0 + \Delta x, y_0 + \Delta y) - f(x_0 + \Delta x, y_0)}{\Delta y} = \frac{\partial f}{\partial y}(x_0 + \Delta x, d). \tag{13}$$

Next we multiply both sides of equation (12) by Δx and both sides of equation (13) by Δy. Adding the resulting equations gives

$$f(x_0 + \Delta x, y_0 + \Delta y) - f(x_0, y_0) = \frac{\partial f}{\partial x}(c, y_0)\, \Delta x \tag{14}$$

$$+ \frac{\partial f}{\partial y}(x_0 + \Delta x, d)\, \Delta y.$$

We now invoke the continuity of the partial derivatives. Since c lies between x_0 and $x_0 + \Delta x$, we have $c \to x_0$ as $\Delta x \to 0$. Thus, since $\frac{\partial f}{\partial x}(x, y)$ is continuous,

$$\lim_{\Delta x \to 0} \frac{\partial f}{\partial x}(c, y_0) = \frac{\partial f}{\partial x}(x_0, y_0). \tag{15}$$

Another way to write equation (15) is simply

$$\frac{\partial f}{\partial x}(c, y_0) = \frac{\partial f}{\partial x}(x_0, y_0) + \epsilon_1 \tag{16}$$

where $\epsilon_1 \to 0$ as $\Delta x \to 0$. Similarly, since d lies between y_0 and $y_0 + \Delta y$, $d \to y_0$ as $\Delta y \to 0$. Thus, since $\frac{\partial f}{\partial y}(x, y)$ is continuous, we conclude that

$$\lim_{\substack{\Delta x \to 0 \\ \Delta y \to 0}} \frac{\partial f}{\partial y}(x_0 + \Delta x, d) = \frac{\partial f}{\partial y}(x_0, y_0),$$

or, that

$$\frac{\partial f}{\partial y}(x_0 + \Delta x, d) = \frac{\partial f}{\partial y}(x_0, y_0) + \epsilon_2 \tag{17}$$

where $\epsilon_2 \to 0$ as $\Delta x \to 0$ and $\Delta y \to 0$.

Combining statements (14), (16), and (17) now gives

$$f(x_0 + \Delta x, y_0 + \Delta y) = f(x_0, y_0) + \frac{\partial f}{\partial x}(x_0, y_0)\, \Delta x + \frac{\partial f}{\partial y}(x_0, y_0)\, \Delta y$$

$$+ \epsilon_1\, \Delta x + \epsilon_2\, \Delta y$$

where $\epsilon_1 \to 0$ and $\epsilon_2 \to 0$ as $\Delta x \to 0$ and $\Delta y \to 0$ as desired. $\blacklozenge$

Exercise Set 17.5

In Exercises 1–12, find the total differential df.

1. $f(x, y) = x^2 y^4$

2. $f(x, y) = x^{2/3} y^{5/2}$

3. $f(x, y) = \sqrt{x^2 + y^4}$

4. $f(x, y) = x \sin y^2$

5. $f(x, y) = e^{\sqrt{x}} \cos y$

6. $f(x, y) = \mathrm{Tan}^{-1}(y/x)$

7. $f(x, y, z) = x^2 y z^3$

8. $f(x, y, z) = \ln(x^2 + 3y + z^3)$

9. $f(x, y, z) = \dfrac{x}{y + 3^z}$

10. $f(x, y, z) = \sqrt{\dfrac{x}{y^2 + z^2}}$

11. $f(x, y, z) = \dfrac{x - y}{x^2 + y^2 + z^2}$

12. $f(x, y, z) = ze^{x^2 - y^3}$

13. Let $f(x, y) = 2xy^2 + x^2y$, $x_0 = 2$, $y_0 = 3$, $\Delta x = 0.1$, and $\Delta y = 0.2$.
    a. Calculate $f(x_0, y_0)$ and $f(x_0 + \Delta x, y_0 + \Delta y)$.
    b. Calculate $\Delta f = f(x_0 + \Delta x, y_0 + \Delta y) - f(x_0, y_0)$.

14. *(Calculator)* Let $f(x, y) = \sin x \cos y$, $x_0 = \pi/2$, $y_0 = \pi/4$, $\Delta x = \pi/16$, and $\Delta y = -\pi/16$. Answer the questions in Exercise 13.

In Exercises 15–22, use the total differential to approximate the given number.

15. $\sqrt{(3.02)^2 + (4.08)^2}$

16. $\sqrt{9.4} + \sqrt{15.6}$

17. $\sin 88° \cos 42°$

18. $(2.02)^3 + (3.96)^3$

19. $(5.03)^2(1.02)^3$

20. $(5.94) \ln(1.15)$

21. $(6.04)(3.1)(2.96)$

22. $\sqrt{(12.1)^2 + (4.05)^2 + (2.96)^2}$

23. A cylindrical bearing of radius 2 cm and length 6 cm is dipped in a molten brass solution to produce a brass coating. If the thickness of the coating varies from 0.02 cm to 0.06 cm, find an approximation for the volume of the coated bearing.

24. The money supply in the economy is defined by the equation $M = C + D$ where C is the currency outstanding in banks and D is the amount of money in demand and term deposits. Use differentials to approximate the percentage change in the money supply caused by a 10% increase in the outstanding currency and a 5% decrease in total deposits.

25. For the Cobb-Douglas production function $P = \lambda L^3 K^{3/2}$, use differentials to approximate the percentage change in P

caused by a 4% increase in labor L and a 10% decrease in capital K.

26. For the Cobb-Douglas production function $P = \lambda L^\alpha K^\beta$, what can you say about α and β if a percentage decrease in labor of $\gamma\%$, combined with a percentage increase in capital of $\gamma\%$, produces no change in P?

27. For small oscillations, the period of a simple pendulum is given by the equation

$$T = 2\pi\sqrt{\dfrac{s}{g}}$$

where s is the length of the pendulum, g is the acceleration due to gravity, and T is time. (If the units for s and g involve meters and seconds, the units for T are seconds.) What is the approximate error in the calculation of T resulting from its calculation using $s = 10$ m and $g = 9.8$ m/s^2 if the actual values of these quantities are $s = 10.2$ and $g = 9.75$?

28. Approximate $f(x_0, y_0, z_0)$ for $f(x, y, z) = \dfrac{\sqrt{x^2 - y^2}}{\tan z}$ and $(x_0, y_0, z_0) = (5.04, 3.97, 0.78)$. (*Hint:* $\dfrac{\pi}{4} \approx 0.7853982$.)

29. When three resistors with resistances R_1, R_2, and R_3 are connected in parallel, the resistance of the resulting circuit is

$$\dfrac{1}{R} = \dfrac{1}{R_1} + \dfrac{1}{R_2} + \dfrac{1}{R_3}.$$

What is the maximum percentage change in R resulting from a change of no more than 10% in each of R_1, R_2, and R_3?

30. The work involved in an adiabatic (no heat lost to the surroundings) expansion of an ideal gas is given by the equation

$$w = nC_v(T_2 - T_1)$$

where C_v is the average heat capacity of the gas, and T_1 and T_2 are the initial and final temperatures. Show that if both T_1 and T_2 are increased by λ percent, the amount of work involved is also increased by λ percent.

31. Prove Theorem 5′, using the Mean Value Theorem three times.

17.6 CHAIN RULES

In this section, we use the Approximation Theorem to establish a generalization of the Chain Rule for functions of a single variable:

$$(f \circ g)'(x) = f'(g(x))g'(x).$$

Our generalization concerns the *composite* function

$$w(t) = f(x(t), y(t), z(t)), \qquad t \in (a, b).$$

It is important to note that this composite function is just a function of a single variable. Our interest is in knowing when the derivative $w'(t)$ exists and, if it does, how it may be calculated.

Before stating the theorem that answers this question, we describe a simple setting in which a composite function of this type occurs. Think of a region of airspace, such as that over the state of California, through which an airplane is about to fly. The region of airspace may be coordinatized with xyz-coordinates (which you may think of as representing latitude, longitude, and altitude, if you prefer), and the path of the airplane may be thought of as a curve parameterized by the vector function $r(t) = x(t)i + y(t)j + z(t)k$, $a \leq t \leq b$ giving the location of the airplane at time t. (See Figure 6.1.) Assuming a steady state temperature distribution through-

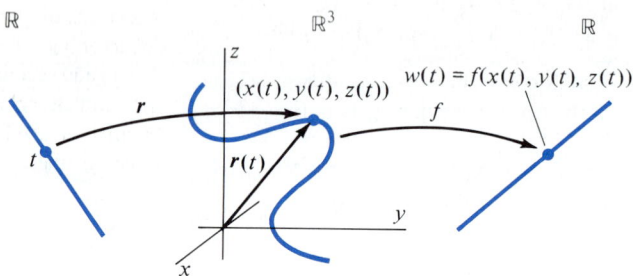

Figure 6.1 If $r: \mathbb{R} \to \mathbb{R}^3$ and $f: \mathbb{R}^3 \to \mathbb{R}$, the composite function $w(t) = f(r(t)) = f(x(t), y(t), z(t))$ maps $\mathbb{R}$ into $\mathbb{R}$.

out the region, we let $T(x, y, z)$ be the temperature at location (x, y, z) in the airspace. *The composite function*

$$w(t) = T(x(t), y(t), z(t))$$

gives the temperature outside the airplane at time t.

What factors determine the rate, $\dfrac{dw}{dt}$, at which the temperature outside the airplane changes? Obviously $\dfrac{dx}{dt}$, $\dfrac{dy}{dt}$, and $\dfrac{dz}{dt}$ are involved, since they are the components of the velocity of the airplane. But $\dfrac{dw}{dt}$ will also depend upon the partial derivatives $\dfrac{\partial T}{\partial x}$, $\dfrac{\partial T}{\partial y}$, and $\dfrac{\partial T}{\partial z}$, since these determine the rate of change of temperature with respect to change in position. The precise relationship between these rates is given by the following theorem.

THEOREM 6
Chain Rule

Let f be a continuous function with continuous partial derivatives for all (x, y, z) in an open "box"

$$Q = \{(x, y, z) \mid a_1 < x < b_1, a_2 < y < b_2, a_3 < z < b_3\}$$

in space. Assume that x, y, and z are functions of t so that $x'(t)$, $y'(t)$, and $z'(t)$ exist for each $t \in (a, b)$, and so that the point $(x(t), y(t), z(t))$ lies in Q for each $t \in (a, b)$. Then the composite function

$$w(t) = f(x(t), y(t), z(t))$$

is a differentiable function of $t \in (a, b)$, and

$$\frac{dw}{dt} = \frac{\partial f}{\partial x} \cdot \frac{dx}{dt} + \frac{\partial f}{\partial y} \cdot \frac{dy}{dt} + \frac{\partial f}{\partial z} \cdot \frac{dz}{dt}. \tag{1}$$

The proof of Theorem 6 is given at the end of this section.

Example 1

Let $f(x, y, z) = \sqrt{x}\, y^2 e^{2z}$, $x(t) = 3t^2 + 2$, $y(t) = 6t$, $z(t) = 1 - t^3$, and $w(t) = f(x(t), y(t), z(t))$. Find $w'(t)$.

Solution: We have

$$\frac{\partial f}{\partial x} = \frac{\partial}{\partial x}(\sqrt{x}\, y^2 e^{2z}) = \frac{y^2 e^{2z}}{2\sqrt{x}}$$

$$\frac{\partial f}{\partial y} = \frac{\partial}{\partial y}(\sqrt{x}\, y^2 e^{2z}) = 2\sqrt{x}\, y e^{2z}$$

and

$$\frac{\partial f}{\partial z} = \frac{\partial}{\partial z}(\sqrt{x}\, y^2 e^{2z}) = 2\sqrt{x}\, y^2 e^{2z}.$$

Also,

$$x'(t) = 6t, \qquad y'(t) = 6, \qquad \text{and} \qquad z'(t) = -3t^2.$$

According to Theorem 6 we must have

$$w'(t) = \frac{y^2 e^{2z}}{2\sqrt{x}}(6t) + 2\sqrt{x}\, y e^{2z}(6) + 2\sqrt{x}\, y^2 e^{2z}(-3t^2)$$

$$= \frac{108 t^3 e^{2(1-t^3)}}{\sqrt{3t^2 + 2}} + 72t\sqrt{3t^2 + 2}\, e^{2(1-t^3)} - 216 t^4 \sqrt{3t^2 + 2}\, e^{2(1-t^3)}$$

$$= \sqrt{3t^2 + 2}\, e^{2(1-t^3)} \left[\frac{108 t^3}{3t^2 + 2} + 72t - 216 t^4 \right]. \qquad \diamond$$

Example 2

A function giving the temperature $T(x, y, z)$ at a point (x, y, z) in space is

$$T(x, y, z) = \lambda\sqrt{x^2 + y^2 + z^2}$$

where λ is constant.

Find the rate of change of temperature with respect to t along the elliptical helix

$$\mathbf{r}(t) = a \cos t\mathbf{i} + b \sin t\mathbf{j} + ct\mathbf{k}.$$

Solution: The temperature function along the helix is the composite function $T(x(t), y(t), z(t))$ where

$$x(t) = a \cos t, \qquad y(t) = b \sin t, \qquad z(t) = ct.$$

Using equation (1), we obtain the desired rate as

$$\frac{dT}{dt} = \frac{\partial}{\partial x}(\lambda\sqrt{x^2 + y^2 + z^2}) \cdot \frac{d}{dt}(a\cos t)$$

$$+ \frac{\partial}{\partial y}(\lambda\sqrt{x^2 + y^2 + z^2}) \cdot \frac{d}{dt}(b\sin t) + \frac{\partial}{\partial z}(\lambda\sqrt{x^2 + y^2 + z^2}) \cdot \frac{d}{dt}(ct)$$

$$= \frac{\lambda x(-a\sin t)}{\sqrt{x^2 + y^2 + z^2}} + \frac{\lambda yb\cos t}{\sqrt{x^2 + y^2 + z^2}} + \frac{\lambda cz}{\sqrt{x^2 + y^2 + z^2}}$$

$$= \frac{\lambda(-a^2 + b^2)\sin t\cos t + \lambda c^2 t}{\sqrt{a^2\cos^2 t + b^2\sin^2 t + c^2 t^2}}.$$

$\diamond$

Example 3

The **electric potential** at an axial point for a charged disc is given by the equation

$$V = \frac{\sigma}{2\epsilon_0}(\sqrt{a^2 + r^2} - r)$$

where a is the radius of the disc, r is the distance from the point to the disc, and σ and ϵ_0 are constants. If the radius of the disc is increasing at a rate of 2 cm/s and the point is moving away from the disc at a rate of 5 cm/s, find the rate at which the electric potential is changing when $a = 3$ cm and $r = 4$ cm.

Solution: Since V is a function of only two variables, a and r, we simply consider the third variable in the Chain Rule (1) to be $z = 0$, so $\frac{dz}{dt} = 0$. Applying the Chain Rule we find

$$\frac{dV}{dt} = \frac{\partial}{\partial a}\left[\frac{\sigma}{2\epsilon_0}(\sqrt{a^2 + r^2} - r)\right]\frac{da}{dt} \qquad (2)$$

$$+ \frac{\partial}{\partial r}\left[\frac{\sigma}{2\epsilon_0}(\sqrt{a^2 + r^2} - r)\right]\frac{dr}{dt}$$

$$= \frac{\sigma}{2\epsilon_0}\left(\frac{a}{\sqrt{a^2 + r^2}}\right)\frac{da}{dt} + \frac{\sigma}{2\epsilon_0}\left(\frac{r}{\sqrt{a^2 + r^2}} - 1\right)\frac{dr}{dt}.$$

We are given

$$\frac{da}{dt} = 2 \text{ cm/s}, \qquad \frac{dr}{dt} = 5 \text{ cm/s}, \qquad a = 3 \text{ cm, and } r = 4 \text{ cm}.$$

Substituting these values into (2) gives

$$\frac{dV}{dt} = \frac{\sigma}{2\epsilon_0}\left(\frac{3}{\sqrt{3^2 + 4^2}}\right)(2) + \frac{\sigma}{2\epsilon_0}\left(\frac{4}{\sqrt{3^2 + 4^2}} - 1\right)(5)$$

$$= \frac{\sigma}{10\epsilon_0}.$$

$\diamond$

Differentiating Other Types of Composite Functions

Various other sorts of composite functions can be formed using functions of several variables, each of which may be differentiated using the Chain Rule. For example, if

$$x = x(s, t), \qquad y = y(s, t) \qquad \text{and} \qquad z = z(s, t)$$

are functions of two variables, and if

$$w = f(x, y, z)$$

is a function of three variables, the composite function

$$w(s, t) = f(x(s, t), y(s, t), z(s, t))$$

is again a function of two variables. To obtain $\dfrac{\partial w}{\partial s}$, we hold t constant and differen-

tiate the resulting function of a single variable $g(s) = w(s, t)$ according to Theorem

6. Similarly, the partial derivative $\dfrac{\partial w}{\partial t}$ is obtained by holding s constant and differ-

entiating with respect to t.

Example 4

Let $f(x, y) = x^2 y^3$ where the variables x and y are functions of the polar variables r and θ:

$$x(r, \theta) = r^2 \cos \theta, \qquad y(r, \theta) = r(1 - \sin \theta).$$

To find the partial derivative $\dfrac{\partial f}{\partial r}$, we treat θ as a constant, leaving x and y as

functions of the single independent variable r, and apply Theorem 6:

$$\frac{\partial f}{\partial r} = \frac{\partial f}{\partial x} \cdot \frac{\partial x}{\partial r} + \frac{\partial f}{\partial y} \cdot \frac{\partial y}{\partial r}$$

$$= \left[\frac{\partial}{\partial x} (x^2 y^3) \right] \left[\frac{\partial}{\partial r} (r^2 \cos \theta) \right] + \left[\frac{\partial}{\partial y} (x^2 y^3) \right] \left[\frac{\partial}{\partial r} (r(1 - \sin \theta)) \right]$$

$$= 2xy^3 \cdot 2r \cos \theta + 3x^2 y^2 (1 - \sin \theta)$$

$$= 7r^6 \cos^2 \theta (1 - \sin \theta)^3.$$

Similarly,

$$\frac{\partial f}{\partial \theta} = \frac{\partial f}{\partial x} \cdot \frac{\partial x}{\partial \theta} + \frac{\partial f}{\partial y} \cdot \frac{\partial y}{\partial \theta}$$

$$= 2xy^3 (-r^2 \sin \theta) + (3x^2 y^2)(-r \cos \theta)$$

$$= r^7 \cos \theta (1 - \sin \theta)^2 [2 \sin^2 \theta - 2 \sin \theta - 3 \cos^2 \theta]. \qquad \diamond$$

Higher Order Derivatives

Using Theorem 6 we can work out formulas for higher order ordinary and partial derivatives of composite functions. It is usually simpler, however, to substitute directly for the independent variables after the first order derivatives are found, and then proceed to the second order derivatives by a second application of Theorem 6, as the following example illustrates.

Example 5

Let $f(x, y) = x^2 + 2xy$, $x = r \cos \theta$, and $y = r \sin \theta$. Find (a) $\dfrac{\partial^2 f}{\partial r^2}$ and (b) $\dfrac{\partial^2 f}{\partial \theta^2}$.

Solution: (a) Since $\dfrac{\partial^2 f}{\partial r^2} = \dfrac{\partial}{\partial r}\left(\dfrac{\partial f}{\partial r}\right)$, we begin by finding

$$\frac{\partial f}{\partial r} = \frac{\partial f}{\partial x} \cdot \frac{\partial x}{\partial r} + \frac{\partial f}{\partial y} \cdot \frac{\partial y}{\partial r}$$

$$= \frac{\partial}{\partial x}(x^2 + 2xy) \cdot \frac{\partial}{\partial r}(r \cos \theta) + \frac{\partial}{\partial y}(x^2 + 2xy) \cdot \frac{\partial}{\partial r}(r \sin \theta)$$

$$= (2x + 2y)(\cos \theta) + (2x)(\sin \theta)$$
$$= (2r \cos \theta + 2r \sin \theta)(\cos \theta) + (2r \cos \theta)(\sin \theta)$$
$$= 2r \cos^2 \theta + 4r \sin \theta \cos \theta.$$

Thus

$$\frac{\partial^2 f}{\partial r^2} = \frac{\partial}{\partial r}(2r \cos^2 \theta + 4r \cos \theta \sin \theta)$$

$$= 2 \cos^2 \theta + 4 \sin \theta \cos \theta.$$

(b) Similarly, you can verify that

$$\frac{\partial^2 f}{\partial \theta^2} = 2r^2[\sin^2 \theta - 4 \sin \theta \cos \theta - \cos^2 \theta].$$ ◇

Proof of Theorem 6:   Let $t_0 \in (a, b)$ and let $\Delta t \neq 0$ be sufficiently small that $(t_0 + \Delta t)$ also lies in (a, b). Let

$$x_0 = x(t_0) \qquad\qquad \Delta x = x(t_0 + \Delta t) - x(t_0)$$
$$y_0 = y(t_0) \qquad\qquad \Delta y = y(t_0 + \Delta t) - y(t_0)$$
$$z_0 = z(t_0), \qquad \text{and} \qquad \Delta z = z(t_0 + \Delta t) - z(t_0).$$

Then, according to the Approximation Theorem (Theorem 5'),

$$w(t_0 + \Delta t) = w(t_0) + \frac{\partial f}{\partial x}(x_0, y_0, z_0)\, \Delta x + \frac{\partial f}{\partial y}(x_0, y_0, z_0)\, \Delta y \tag{3}$$

$$+ \frac{\partial f}{\partial z}(x_0, y_0, z_0)\, \Delta z + \epsilon_1 \Delta x + \epsilon_2 \Delta y + \epsilon_3 \Delta z$$

where $\epsilon_1 \to 0$, $\epsilon_2 \to 0$, and $\epsilon_3 \to 0$ as $\Delta x \to 0$, $\Delta y \to 0$, and $\Delta z \to 0$.

Subtracting $w(t_0)$ from both sides of equation (3) and dividing by Δt gives the equation

$$\frac{w(t_0 + \Delta t) - w(t_0)}{\Delta t} = \frac{\partial f}{\partial x}(x_0, y_0, z_0)\frac{\Delta x}{\Delta t} + \frac{\partial f}{\partial y}(x_0, y_0, z_0)\frac{\Delta y}{\Delta t} \tag{4}$$

$$+ \frac{\partial f}{\partial z}(x_0, y_0, z_0)\frac{\Delta z}{\Delta t} + \epsilon_1 \frac{\Delta x}{\Delta t} + \epsilon_2 \frac{\Delta y}{\Delta t} + \epsilon_3 \frac{\Delta z}{\Delta t}.$$

Now according to our definitions, we have

$$\lim_{\Delta t \to 0} \frac{\Delta x}{\Delta t} = \lim_{\Delta t \to 0} \frac{x(t_0 + \Delta t) - x(t_0)}{\Delta t} = x'(t_0).$$

Thus,

$$\lim_{\Delta t \to 0}\left[\frac{\partial f}{\partial x}(x_0, y_0, z_0)\frac{\Delta x}{\Delta t}\right] \tag{5}$$

$$= \frac{\partial f}{\partial x}(x_0, y_0, z_0) \cdot \lim_{\Delta t \to 0}\frac{\Delta x}{\Delta t} = \frac{\partial f}{\partial x}(x_0, y_0, z_0)x'(t_0)$$

and

$$\lim_{\Delta t \to 0} \epsilon_1 \cdot \frac{\Delta x}{\Delta t} = \left(\lim_{\Delta t \to 0} \epsilon_1 \right) \left(\lim_{\Delta t \to 0} \frac{\Delta x}{\Delta t} \right) = 0 \cdot x'(t_0) = 0. \tag{6}$$

Equations analogous to (5) and (6) hold for the variables y and z.

Finally, from equation (4) and the equations of the form (5) and (6) for each of the variables x, y, and z, we conclude that

$$w'(t_0) = \lim_{\Delta t \to 0} \frac{w(t_0 + \Delta t) - w(t_0)}{\Delta t}$$

$$= \frac{\partial f}{\partial x}(x_0, y_0, z_0)x'(t_0) + \frac{\partial f}{\partial y}(x_0, y_0, z_0)y'(t_0)$$

$$+ \frac{\partial f}{\partial z}(x_0, y_0, z_0)z'(t_0),$$

as required. ◆

Exercise Set 17.6

In Exercises 1–10, find the rate of change of f with respect to t, $\dfrac{df}{dt}$, along the given curves.

1. $f(x, y) = x^2 + y^2, \quad x(t) = 2t, \quad y(t) = 6 - t^2$

2. $f(x, y) = x - y^2, \quad x(t) = 3t^2 + 2, \quad y(t) = 4 + t$

3. $f(x, y) = xy^2, \quad r(t) = \cos t\, i + \sin t\, j$

4. $f(x, y) = x^2 - y^2, \quad r(t) = e^t i + e^{-t} j$

5. $f(x, y) = \sqrt{x + y}, \quad x(t) = \sqrt{t}, \quad y(t) = e^{t^2}$

6. $f(x, y) = y^2 - x^2, \quad x(t) = a \cos t, \quad y(t) = a \sin t$

7. $f(x, y, z) = xy - x + z^2, \quad r(t) = t^2 i - 2t j + \sin t\, k$

8. $f(x, y, z) = x^2 + y^2 + z^2,$
$\quad r(t) = a \cos \pi t\, i + b \sin \pi t\, j - t^2 k$

9. $f(x, y, z) = z(y^2 - x^2), \quad x(t) = a \cosh t,$
$\quad y(t) = b \sinh t, \quad z(t) = e^{-2t}$

10. $f(x, y, z) = e^{xyz}, \quad x(t) = \ln \sqrt{t}, \quad y(t) = t \sin t,$
$\quad z(t) = 2^t.$

In Exercises 11–22, find the indicated derivative(s).

11. $f(x, y) = x^2 y^3, \quad x(t) = \cos t, \quad y(t) = t \sin t.$
Find $\dfrac{df}{dt}$.

12. $f(x, y) = x^2 y^3, \quad x(s, t) = st, \quad y(s, t) = s^2 - t^2.$
Find
a. $\dfrac{\partial f}{\partial s}$ **b.** $\dfrac{\partial f}{\partial t}$

13. $f(x, y, z) = x^2 + y^2 - z^2, \quad x(s, t) = e^{st},$
$y(s, t) = st, \quad z(s, t) = s - t.$ Find
a. $\dfrac{\partial f}{\partial s}$ **b.** $\dfrac{\partial f}{\partial t}$

14. $f(x, y) = xy - y^2, \quad x(r, s, t) = rst, \quad y(r, s, t) = e^{rst}.$
Find
a. $\dfrac{\partial f}{\partial r}$ **b.** $\dfrac{\partial f}{\partial s}$ **c.** $\dfrac{\partial f}{\partial t}$

15. $f(x, y) = \sin(xy^2) - x^2 y, \quad x(s, t) = s^2 - st,$
$y(s, t) = t^2 s^2.$ Find
a. $\dfrac{\partial f}{\partial s}$ **b.** $\dfrac{\partial f}{\partial t}$

16. $f(x, y) = x^2 e^{y-x}, \quad x(s, t) = s - t, \quad y(s, t) = \sqrt{s + t}.$
Find
a. $\dfrac{\partial f}{\partial s}$ **b.** $\dfrac{\partial f}{\partial t}$

17. $f(x, y, z) = xy^3 z^2, \quad x(s, t) = s \sin t,$
$y(s, t) = t \cos(s), \quad z(s, t) = t^2 - s^2.$ Find
a. $\dfrac{\partial f}{\partial s}$ **b.** $\dfrac{\partial f}{\partial t}$

18. $f(x, y) = e^{xy}, \quad x(r, s) = \sqrt{s^2 - r^2},$
$y(r, s) = \text{Tan}^{-1}\left(\dfrac{r}{s}\right).$ Find
a. $\dfrac{\partial f}{\partial r}$ **b.** $\dfrac{\partial f}{\partial s}$

19. $f(u, v) = e^{u^2} - e^{2v}, \quad u(x, y) = xy^2, \quad v(x, y) = x^2 y.$
Find
a. $\dfrac{\partial f}{\partial x}$ **b.** $\dfrac{\partial f}{\partial y}$

20. $f(x, y, z) = x^2 - xy + yz, \quad x(r, \theta) = r \cos \theta,$
$y(r, \theta) = 2r, \quad z(r, \theta) = r \sin \theta.$ Find
a. $\dfrac{\partial f}{\partial r}$ **b.** $\dfrac{\partial f}{\partial \theta}$

21. $f(r, \theta) = r^2(1 - \cos \theta)$, $\qquad r(t) = 1 + t^3$,

$\theta(t) = \sqrt{1 + t^2}$. Find $\dfrac{df}{dt}$.

22. $f(x, y, z) = \ln(x^2 + y^2 + z^2)$, $\qquad x(u, v, w) = u \cos v$,

$y(u, v, w) = v \sin u$, $\qquad z(u, v, w) = uvw$. Find

a. $\dfrac{\partial f}{\partial u}$ $\qquad$ **b.** $\dfrac{\partial f}{\partial v}$ $\qquad$ **c.** $\dfrac{\partial f}{\partial w}$

23. The radius of the base of a cone is 6 cm and it is increasing at a rate of 2 cm/s. The height of the cone is 10 cm and it is increasing at a rate of 3 cm/s. At what rate is the volume increasing?

24. The radius of a cylinder is 8 cm and it is increasing at a rate of 2 cm/s. The height of the cylinder is 20 cm and it is increasing at a rate of 4 cm/s.
a. Find the rate at which the volume is increasing.
b. Find the rate at which the lateral surface area is increasing.

25. A particle moves along a helix with position function $\mathbf{r}(t) =$ cos $t\mathbf{i}$ + sin $t\mathbf{j}$ + $t\mathbf{k}$.
a. Find the function $D(x, y, z)$ giving the distance from the particle to the origin.
b. Find $\dfrac{dD}{dt}$, the rate at which the distance from the particle to the origin changes as a function of time.
c. Show that $\lim\limits_{t \to \infty} D'(t) = 1$; $\lim\limits_{t \to -\infty} D'(t) = -1$.

26. In economic theory, the **rate of growth** (as opposed to rate of change) of a function f is defined to be the ratio

$$\frac{f'(t)}{f(t)} = \frac{d}{dt}[\ln(f(t))] = \frac{\text{marginal function}}{\text{total function}}.$$

Show, according to this definition, that if the rate of growth of consumption C is a, and if the rate of growth of population P is b, then the rate of growth of per capita consumption, $R = \dfrac{C}{P}$, is $a - b$.

27. The money supply in the economy is defined by the equation $M = C + D$ where C is the cash on deposit in banks and D is the total of all time and demand deposits. If m, c, and d are the respective rates of growth of M, C, and D, show that

$$m = \frac{cC}{C + D} + \frac{dD}{C + D}.$$

(See Exercise 26 for the definition of rate of growth.)

28. Let $f(x, y) = xe^{y^2}$, $\qquad x(r, \theta) = r \cos \theta$,

$y(r, \theta) = r \sin \theta$. Find

a. $\dfrac{\partial^2 f}{\partial r^2}$ $\qquad$ **b.** $\dfrac{\partial^2 f}{\partial \theta^2}$

29. Let $f(x, y) = x^2 + y^4$, $\qquad x(s, t) = s^2t$,

$y(s, t) = t^2 - s^2$. Find

a. $\dfrac{\partial^2 f}{\partial s^2}$ $\qquad$ **b.** $\dfrac{\partial^2 f}{\partial t^2}$

30. Find $\dfrac{d^2 f}{dt^2}$ for the function $f(x, y)$ in Exercise 11.

31. Find $\dfrac{\partial^2 f}{\partial r^2}$ for the function $f(x, y)$ in Exercise 14.

32. Find an expression for the mixed second order partial derivative $\dfrac{\partial^2 f}{\partial s \partial t}$ for the composite function

$$f(x, y) = f(x(s, t), y(s, t)).$$

33. Use the result of Exercise 32 to find $\dfrac{\partial^2 f}{\partial r \partial \theta}$ for the function f in Exercise 28.

34. Use the result of Exercise 32 to find $\dfrac{\partial^2 f}{\partial s \partial t}$ for the function f in Exercise 29.

35. By use of the Chain Rule, show that Laplace's equation

$$\frac{\partial^2 v}{\partial x^2} + \frac{\partial^2 v}{\partial y^2} + \frac{\partial^2 v}{\partial z^2} = 0$$

in cylindrical coordinates is

$$\frac{\partial^2 v}{\partial r^2} + \frac{1}{r} \cdot \frac{\partial v}{\partial r} + \frac{1}{r^2} \cdot \frac{\partial^2 v}{\partial \theta^2} + \frac{\partial^2 v}{\partial z^2} = 0.$$

36. Use the Chain Rule to show that Laplace's equation

$$\frac{\partial^2 v}{\partial x^2} + \frac{\partial^2 v}{\partial y^2} + \frac{\partial^2 v}{\partial z^2} = 0$$

in spherical coordinates is

$$\frac{\partial^2 v}{\partial \rho^2} + \frac{2}{\rho} \cdot \frac{\partial v}{\partial \rho} + \frac{1}{\rho^2} \cdot \frac{\partial^2 v}{\partial \phi^2} + \frac{\cot \phi}{\rho^2} \cdot \frac{\partial v}{\partial \phi}$$
$$+ \frac{\csc^2 \phi}{\rho^2} \cdot \frac{\partial^2 v}{\partial \theta^2} = 0.$$

17.7 DIRECTIONAL DERIVATIVES; THE GRADIENT

If f is a function of two variables defined in a neighborhood of a point (x_0, y_0), the partial derivatives $\dfrac{\partial f}{\partial x}(x_0, y_0)$ and $\dfrac{\partial f}{\partial y}(x_0, y_0)$ measure the rate of change of f in the directions of the x- and y-coordinate axes, respectively. But how might we calculate the rate of change of f in an arbitrary direction? The answer to this question is provided by the *directional derivative*.

Suppose that we wish to calculate the rate of change of f at (x_0, y_0) in the direction determined by the unit vector $\mathbf{u} = u_1\mathbf{i} + u_2\mathbf{j}$ (Figure 7.1). If t is a number for which $0 < t < 1$, the point $(x_0 + tu_1, y_0 + tu_2)$ lies on the line between the two points (x_0, y_0) and $(x_0 + u_1, y_0 + u_2)$, so the difference

$$f(x_0 + tu_1, y_0 + tu_2) - f(x_0, y_0)$$

represents the change in $f(x, y)$ in the direction of $\mathbf{u}$ over an interval of length $\sqrt{(tu_1)^2 + (tu_2)^2} = t\sqrt{u_1^2 + u_2^2} = t$, since $|\mathbf{u}| = \sqrt{u_1^2 + u_2^2} = 1$ (see Figure 7.2). The definition of the directional derivative is therefore the following.

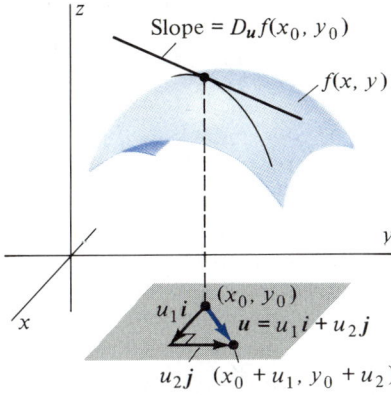

Figure 7.1 Directional derivative $D_\mathbf{u}f(x_0, y_0)$ is rate of change of f in the direction of the vector $\mathbf{u}$.

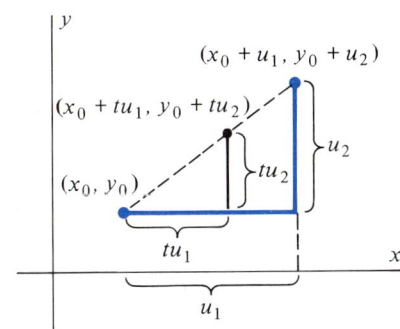

Figure 7.2 If $0 < t < 1$, the point $(x_0 + tu_1, y_0 + tu_2)$ lies between (x_0, y_0) and $(x_0 + u_1, y_0 + u_2)$.

DEFINITION 5

Let f be a function of two variables defined in a neighborhood of (x_0, y_0). Let $\mathbf{u} = u_1\mathbf{i} + u_2\mathbf{j}$ be a unit vector. The **directional derivative, $D_\mathbf{u}f(x_0, y_0)$,** in the direction of the unit vector $\mathbf{u}$, is the limit

$$D_\mathbf{u}f(x_0, y_0) = \lim_{t \to 0^+} \frac{f(x_0 + tu_1, y_0 + tu_2) - f(x_0, y_0)}{t}. \tag{1}$$

REMARK 1: Definition 5 has an obvious generalization to functions of more than two variables. If $\mathbf{u} = \langle u_1, u_2, \dots, u_n \rangle$ is a unit vector in $\mathbb{R}^n$ and f is a function of n variables, the directional derivative of f at $(x_1, x_2, \dots x_n)$ in the direction of $\mathbf{u}$ is the limit

$$D_\mathbf{u}f(x_1, x_2, \dots, x_n) \tag{2}$$

$$= \lim_{t \to 0^+} \frac{f(x_1 + tu_1, x_2 + tu_2, \dots, x_n + tu_n) - f(x_1, x_2, \dots, x_n)}{t}.$$

Unfortunately, equations (1) and (2) fail to provide a simple procedure for actually calculating the directional derivative when it exists. When f satisfies the hypotheses of the Approximation Theorem (Theorem 5), we can use this result to establish such a procedure. First, we write the numerator in (1) as

$$f(x_0 + tu_1, y_0 + tu_2) - f(x_0, y_0) = \frac{\partial f}{\partial x}(x_0, y_0)(tu_1) + \frac{\partial f}{\partial y}(x_0, y_0)(tu_2)$$
$$+ \epsilon_1(tu_1) + \epsilon_2(tu_2)$$

where $\epsilon_1 \to 0$ and $\epsilon_2 \to 0$ as $t \to 0$. Dividing both sides by $t \neq 0$ and applying $\lim_{t \to 0^+}$ shows that

$$D_u f(x_0, y_0) = \lim_{t \to 0^+} \frac{\frac{\partial f}{\partial x}(x_0, y_0)(tu_1) + \frac{\partial f}{\partial y}(x_0, y_0)(tu_2) + \epsilon_1(tu_1) + \epsilon_2(tu_2)}{t}$$

$$= \lim_{t \to 0^+} \left\{ \frac{\partial f}{\partial x}(x_0, y_0)u_1 + \frac{\partial f}{\partial y}(x_0, y_0)u_2 + \epsilon_1 u_1 + \epsilon_2 u_2 \right\}$$

$$= \frac{\partial f}{\partial x}(x_0, y_0)u_1 + \frac{\partial f}{\partial y}(x_0, y_0)u_2 + \lim_{t \to 0^+} \epsilon_1 u_1 + \lim_{t \to 0^+} \epsilon_2 u_2.$$

Since $\epsilon_1 \to 0$ and $\epsilon_2 \to 0$ as $t \to 0$, the last two terms on the right side of the above equation are zero. We have therefore established the following theorem.

THEOREM 7	If the function f and its first partial derivatives are continuous in a neighborhood of (x_0, y_0), the directional derivative $D_u f(x_0, y_0)$ in the direction of the unit vector $u = u_1 i + u_2 j$ is given by $$D_u f(x_0, y_0) = \frac{\partial f}{\partial x}(x_0, y_0)u_1 + \frac{\partial f}{\partial y}(x_0, y_0)u_2. \qquad (3)$$

REMARK 2: The corresponding result for functions of n variables, $n \geq 3$, is that the directional derivative in (2) may be expressed as

$$D_u f(x_1, x_2, \ldots, x_n) = \frac{\partial f}{\partial x_1}(x_1, \ldots, x_n)u_1 \qquad (4)$$

$$+ \frac{\partial f}{\partial x_2}(x_1, x_2, \ldots, x_n)u_2 + \cdots$$

$$+ \frac{\partial f}{\partial x_n}(x_1, x_2, \ldots, x_n)u_n$$

when f and its first order partial derivatives are continuous in a neighborhood of $(x_1, x_2, \ldots, x_n)$ in $\mathbb{R}^n$.

REMARK 3: It is important to note that the vector u in expressions (1) through (4) is a *unit* vector. If you wish to calculate the directional derivative of f in the direction of a vector w with $|w| \neq 1$, you must first obtain the unit vector $u = \frac{w}{|w|}$ in the direction of w.

REMARK 4: Note that if $u = i$ (so that $u_1 = 1$ and $u_2 = 0$), and if the partial derivative $\frac{\partial f}{\partial x}(x, y)$ exists, then the directional derivative $D_i f(x, y)$ is just $\frac{\partial f}{\partial x}(x, y)$.

Similarly, if $u = j$ and $\dfrac{\partial f}{\partial y}(x, y)$ exists, then $D_jf(x, y)$ is equal to $\dfrac{\partial f}{\partial y}(x, y)$. This is most easily seen from the form for $D_uf(x, y)$ in Theorem 7. However, $D_uf(x, y)$ may exist, as defined in Definition 5, for all u even when one or both of the partial derivatives do not exist (see Exercise 41).

Example 1

Find the directional derivative $D_uf(2, 1)$ where $f(x, y) = x^2e^{3y}$ and $u = \dfrac{1}{\sqrt{5}}i + \dfrac{2}{\sqrt{5}}j$.

Solution: The partial derivatives are

$$\frac{\partial f}{\partial x}(2, 1) = 2xe^{3y}\bigg|_{(2,1)} = 4e^3$$

and

$$\frac{\partial f}{\partial y}(2, 1) = 3x^2e^{3y}\bigg|_{(2,1)} = 12e^3.$$

Since

$$|u| = \sqrt{\frac{1}{5} + \frac{4}{5}} = 1,$$

we have from (3) that

$$D_uf(2, 1) = (4e^3)\left(\frac{1}{\sqrt{5}}\right) + 12e^3\left(\frac{2}{\sqrt{5}}\right)$$

$$= \frac{28e^3}{\sqrt{5}} \approx 251.5. \qquad \diamond$$

Example 2

Find the directional derivative of the function $f = e^x \cos y + xz$ at the point $(1, \pi, -1)$ in the direction of the vector $w = i - 3j + 4k$.

Solution: The partial derivatives of f are

$$\frac{\partial f}{\partial x}(1, \pi, -1) = e^x \cos y + z\bigg|_{(1,\pi,-1)} = -e - 1,$$

$$\frac{\partial f}{\partial y}(1, \pi, -1) = -e^x \sin y\bigg|_{(1,\pi,-1)} = 0,$$

and

$$\frac{\partial f}{\partial z}(1, \pi, -1) = x\bigg|_{(1,\pi,-1)} = 1.$$

Since $|w| = \sqrt{1^2 + 3^2 + 4^2} = \sqrt{26}$, a *unit* vector in the direction of w is $u = \frac{1}{\sqrt{26}}(i - 3j + 4k)$, so $u_1 = \frac{1}{\sqrt{26}}$, $u_2 = \frac{-3}{\sqrt{26}}$, and $u_3 = \frac{4}{\sqrt{26}}$. According to (4),

$$D_u f(1, \pi, -1) = (-e - 1)\left(\frac{1}{\sqrt{26}}\right) + (0)\left(\frac{-3}{\sqrt{26}}\right) + (1)\left(\frac{4}{\sqrt{26}}\right)$$

$$= \frac{-e + 3}{\sqrt{26}}$$

$$\approx 0.055. \qquad \diamond$$

If $u = u_1 i + u_2 j$ is a unit vector and θ is the angle formed between u and the positive x-axis, then

$$u_1 = \frac{u_1}{|u|} = \cos \theta, \qquad \text{and} \qquad u_2 = \frac{u_2}{|u|} = \sin \theta.$$

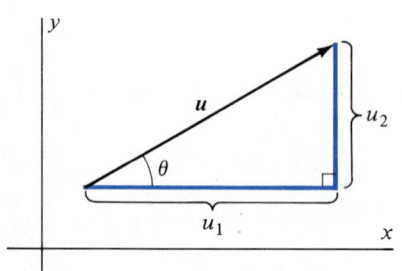

Figure 7.3 If $u = u_1 i + u_2 j$ with $|u| = 1$, then $u_1 = \cos \theta$, $u_2 = \sin \theta$.

(See Figure 7.3.) Using these equations, we may write the directional derivative $D_u f(x_0, y_0)$ in (3) in the form

$$D_u f(x_0, y_0) = \frac{\partial f}{\partial x}(x_0, y_0) \cos \theta + \frac{\partial f}{\partial y}(x_0, y_0) \sin \theta. \qquad (5)$$

Equation (5) makes explicit the fact that when f and its first partial derivatives are continuous, the directional derivative depends only on the partial derivatives and the direction of the unit vector u.

Example 3

Let $f(x, y) = xy - y^3$. Find the unit vector u for which the directional derivative $D_u f(2, 1)$ is a maximum.

Solution: We begin by calculating the partial derivatives for f:

$$\frac{\partial f}{\partial x}(2, 1) = y \bigg|_{(2,1)} = 1,$$

$$\frac{\partial f}{\partial y}(2, 1) = x - 3y^2 \bigg|_{(2,1)} = -1.$$

According to (5), the directional derivative $D_u f(2, 1)$ is

$$D_u f(2, 1) = (1) \cos \theta + (-1) \sin \theta$$
$$= \cos \theta - \sin \theta.$$

We must therefore find the value of θ for which the function

$$g(\theta) = \cos \theta - \sin \theta$$

is a maximum. To do so we set

$$g'(\theta) = -\sin \theta - \cos \theta = 0$$

and obtain the equation

$$\sin \theta = -\cos \theta, \qquad \text{or} \qquad \tan \theta = -1,$$

which has solutions $\theta = \dfrac{3\pi}{4}$ and $\theta = \dfrac{7\pi}{4}$ for $0 \le \theta \le 2\pi$.

Since

$$g''\left(\frac{3\pi}{4}\right) = -\cos\left(\frac{3\pi}{4}\right) + \sin\left(\frac{3\pi}{4}\right) = \sqrt{2} > 0$$

and

$$g''\left(\frac{7\pi}{4}\right) = -\cos\left(\frac{7\pi}{4}\right) + \sin\left(\frac{7\pi}{4}\right) = -\sqrt{2} < 0,$$

the angle $\theta = \dfrac{7\pi}{4}$ corresponds to the maximum. For this angle, $\boldsymbol{u} = (\cos\theta)\boldsymbol{i} + (\sin\theta)\boldsymbol{j} = \dfrac{\sqrt{2}}{2}\boldsymbol{i} - \dfrac{\sqrt{2}}{2}\boldsymbol{j}$ is the required unit vector.   ◇

The Gradient

The form of the directional derivative given by Theorem 7 can be written as a dot product:

$$D_{\boldsymbol{u}}f(x_0, y_0) = \frac{\partial f}{\partial x}(x_0, y_0)u_1 + \frac{\partial f}{\partial y}(x_0, y_0)u_2 \tag{6}$$

$$= \left[\frac{\partial f}{\partial x}(x_0, y_0)\boldsymbol{i} + \frac{\partial f}{\partial y}(x_0, y_0)\boldsymbol{j}\right] \cdot [u_1\boldsymbol{i} + u_2\boldsymbol{j}].$$

The second factor in this dot product is just the unit vector $\boldsymbol{u} = u_1\boldsymbol{i} + u_2\boldsymbol{j}$. The first factor is called the *gradient* of f at (x_0, y_0). It is usually written as

$$\nabla f(x_0, y_0) = \frac{\partial f}{\partial x}(x_0, y_0)\boldsymbol{i} + \frac{\partial f}{\partial y}(x_0, y_0)\boldsymbol{j} \tag{7}$$

or just

$$\nabla f = \frac{\partial f}{\partial x}\boldsymbol{i} + \frac{\partial f}{\partial y}\boldsymbol{j}.$$

For functions of three variables, the gradient is

$$\nabla f = \frac{\partial f}{\partial x}\boldsymbol{i} + \frac{\partial f}{\partial y}\boldsymbol{j} + \frac{\partial f}{\partial z}\boldsymbol{k}.$$

Thus, the gradient ∇f is a *vector* whose components are the partial derivatives of f.

Example 4

For $f(x, y) = x \cos y$, the gradient is

$$\nabla f(x, y) = \left[\frac{\partial}{\partial x}(x \cos y)\right]\boldsymbol{i} + \left[\frac{\partial}{\partial y}(x \cos y)\right]\boldsymbol{j}$$

$$= \cos y\boldsymbol{i} - x \sin y\boldsymbol{j}$$

and the vector given by the gradient at $\left(2, \dfrac{\pi}{4}\right)$ is

$$\nabla f(2, \pi/4) = \frac{\sqrt{2}}{2}\boldsymbol{i} - \sqrt{2}\boldsymbol{j}.$$ ◇

Example 5

For $f(x, y, z) = \sqrt{x}\, e^y\, \text{Tan}^{-1}\, z$, the gradient is

$$\nabla f(x, y, z) = \frac{e^y\, \text{Tan}^{-1}\, z}{2\sqrt{x}}\boldsymbol{i} + \sqrt{x}\, e^y\, \text{Tan}^{-1}\, z\boldsymbol{j} + \frac{\sqrt{x}\, e^y}{1 + z^2}\boldsymbol{k},$$

and the vector given by the gradient at $(4, 0, 1)$ is

$$\nabla f(4, 0, 1) = \frac{\pi}{16}\boldsymbol{i} + \frac{\pi}{2}\boldsymbol{j} + \boldsymbol{k}.$$

◇

Vector Notation and Gradients

Using vector notation, we may express the directional derivative as simply

$$D_{\boldsymbol{u}}f(\boldsymbol{x}) = \nabla f(\boldsymbol{x}) \cdot \boldsymbol{u}. \tag{8}$$

Equation (8) provides insight into the geometry associated with $\nabla f(\boldsymbol{x})$. From Theorem 4, Chapter 16, we know that

$$\begin{aligned}
\nabla f(\boldsymbol{x}) \cdot \boldsymbol{u} &= |\nabla f(\boldsymbol{x})| \cdot |\boldsymbol{u}| \cdot \cos\theta \\
&= |\nabla f(\boldsymbol{x})| \cos\theta \qquad (|\boldsymbol{u}| = 1)
\end{aligned} \tag{9}$$

where θ is the angle between the vectors $\nabla f(\boldsymbol{x})$ and $\boldsymbol{u}$. Combining equations (8) and (9) gives

$$D_{\boldsymbol{u}}f(\boldsymbol{x}) = |\nabla f(\boldsymbol{x})| \cos\theta. \tag{10}$$

Thus, since $-1 \le \cos\theta \le 1$ for all θ, equation (10) shows that

(i) $-|\nabla f(\boldsymbol{x})| \le D_{\boldsymbol{u}}f(\boldsymbol{x}) \le |\nabla f(\boldsymbol{x})|$, and
(ii) $D_{\boldsymbol{u}}f(\boldsymbol{x})$ assumes its maximum value when $\cos\theta = 1$, that is, when $\theta = 0$.

Since the case $\theta = 0$ occurs precisely when $\nabla f(\boldsymbol{x})$ and the direction vector $\boldsymbol{u}$ point in the same direction, conclusion (ii) says that $\nabla f(\boldsymbol{x})$ points in the direction of the maximum value of $D_{\boldsymbol{u}}f(\boldsymbol{x})$. That is,

$$\nabla f(\boldsymbol{x}) \text{ points in the direction of most rapid increase for } f \text{ at } \boldsymbol{x} \tag{11}$$

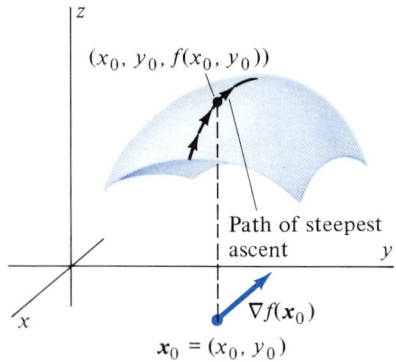

Figure 7.4 $\nabla f(\boldsymbol{x})$ points in the direction of most rapid increase of the function f at $\boldsymbol{x}$.

(assuming, of course, that the hypotheses of Theorem 7 hold).

This observation has several important applications. If $\boldsymbol{x} = (x, y)$ is a point in the domain of the function f, then $\nabla f(x, y)$ points in the direction in which a path through $(x, y, f(x, y))$ on the graph will rise most rapidly (see Figure 7.4). A path on a surface having the property that the tangent at each point of the path is parallel to the gradient of the function defining the surface is called a **path of steepest ascent.** Of obvious relevance to mountain climbers, this concept is also used to develop procedures for approximating extrema for complicated functions of several variables. (It should be geometrically obvious that proceeding in the opposite direction, $-\nabla f(x, y)$, leads to the **path of steepest descent.** In Exercise 33, you are asked to prove this analytically.)

A second application of observation (11) concerns particles moving so as to maximize an attribute of the medium through which they are moving. Examples of this are insects flying toward a light source (maximizing the intensity of light), heat-seeking missiles (maximizing temperature), and sharks in water (maximizing blood concentration). In such situations, if $A(\boldsymbol{x})$ is the value of the attribute at location x, the particle will move so that its tangent vector points in the direction of $\nabla A(x)$ (see Example 7).

Example 6

For the function $f(x, y) = 9 - \dfrac{x^2 + y^2}{4}$, the gradient at $x = (x, y)$ is

$$\nabla f(x) = -\frac{x}{2}i - \frac{y}{2}j$$

$$= -\frac{1}{2}(xi + yj)$$

$$= -\frac{1}{2}x.$$

[handwritten: $\dfrac{(2x)4' - 0}{16}$ $\dfrac{8x}{16} = \dfrac{x}{2}$]

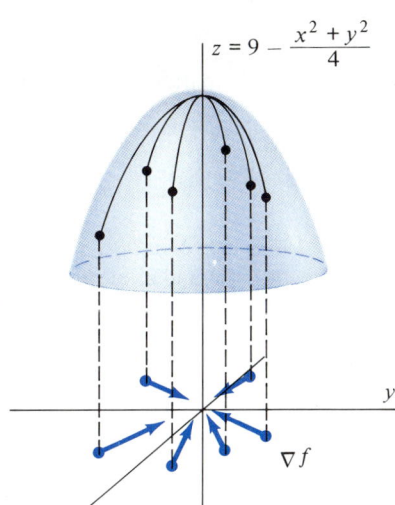

$z = 9 - \dfrac{x^2 + y^2}{4}$

∇f

x

Figure 7.5 $\quad \nabla f(x) = -\dfrac{1}{2}x \quad$ for

$f(x, y) = 9 - \dfrac{x^2 + y^2}{4}.$

Since $x = xi + yj$ is the position vector of the point (x, y), the vector $\nabla f(x) = -\dfrac{1}{2}x$

points toward the origin for all $(x, y) \neq (0, 0)$. In particular,

if $x = (2, 2),$ $\nabla f = -i - j,$

if $x = (1, 4),$ $\nabla f = -\dfrac{1}{2}i - 2j,$

if $x = (2, -1),$ $\nabla f = -i + \dfrac{1}{2}j.$

This should not be surprising, since the graph of f is a paraboloid. At any point on this surface, the z-coordinate is increased most rapidly by moving directly toward the z-axis (see Figure 7.5). ◇

Example 7

The temperature distribution across the surface of a rectangular plate is given by the function $T(x, y) = 100 - x^2 - 2y^2$. Find the path followed by a heat-seeking particle placed on the plate at the point $(4, 2)$.

Solution:

Let the path followed by the particle have parameterization

$$r(t) = x(t)i + y(t)j.$$

Assuming the components x and y to be differentiable functions of t, we recall from Chapter 17 that the tangent vector $T(t)$ at each point along the path is

$$T(t) = x'(t)i + y'(t)j.$$

Since the particle is heat seeking, this tangent should point in the same direction as the gradient

$$\nabla T(x, y) = -2xi - 4yj.$$

This will be the case when

$$x'(t) = -2x(t) \qquad \text{and} \qquad y'(t) = -4y(t).$$

These are the familiar differential equations for exponential decay. Their solutions are

$$x(t) = x(0)e^{-2t}, \quad \text{and} \quad y(t) = y(0)e^{-4t}.$$

Since the particle begins at (4, 2), we have $x(0) = 4$ and $y(0) = 2$. Thus

$$x(t) = 4e^{-2t}, \quad y(t) = 2e^{-4t}$$

are the components of a parameterization of the path. We may eliminate the parameter t by noting that

$$y(t) = 2e^{-4t} = \frac{1}{8}(4e^{-2t})^2 = \frac{1}{8}[x(t)]^2.$$

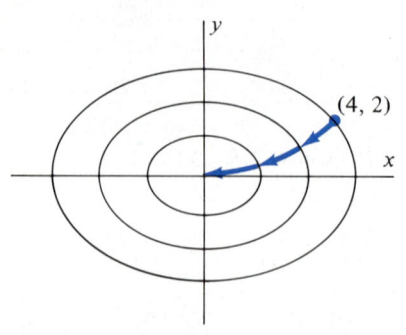

Figure 7.6 Heat-seeking particle approaches origin along parabolic path.

The particle therefore approaches the origin along the parabola $y = \frac{1}{8}x^2$ (Figure 7.6). ◇

Level Curves and Surfaces

Let $r(t) = x(t)i + y(t)j + z(t)k$ be a parameterization for a curve in $\mathbb{R}^3$, and let $w = f(x, y, z)$ be a function of three variables. Using the gradient and assuming that all necessary derivatives exist, we may write the Chain Rule (Theorem 6)

$$\frac{d}{dt}f(x, y, z) = \frac{\partial f}{\partial x} \cdot \frac{dx}{dt} + \frac{\partial f}{\partial y} \cdot \frac{dy}{dt} + \frac{\partial f}{\partial z} \cdot \frac{dz}{dt}$$

as

$$\frac{d}{dt}f(r(t)) = \nabla f(r(t)) \cdot r'(t). \tag{12}$$

Equation (12) says that the derivative of the composite function $f(r(t))$ is the dot product of the gradient $\nabla f(r(t))$ with the tangent vector $r'(t) = x'(t)i + y'(t)j + z'(t)k$ for each t. Equation (12) also holds for curves $r(t) = x(t)i + y(t)j$ in the plane composed with functions $z = f(x, y)$ of two variables, as does Theorem 6. (Of course, the beauty of the vector notation is that we need not distinguish between these two cases in writing equation (12).)

Equation (12) reveals an important relationship between gradients and level curves. To see this, suppose that f is a function of two variables for which the equation

$$f(x, y) = k, \quad k \text{ constant} \tag{13}$$

determines a curve C in the xy-plane with parameterization

$$r(t) = x(t)i + y(t)j.$$

Then along this level curve the composite function

$$f(r(t)) = f(x(t), y(t)) \tag{14}$$

is constant, according to equation (13).

Now let $x_0 = r(t_0) = (x(t_0), y(t_0))$ be a point on C. If the functions f, x, and y satisfy the hypotheses of Theorem 6 in a neighborhood of x_0, we may differentiate

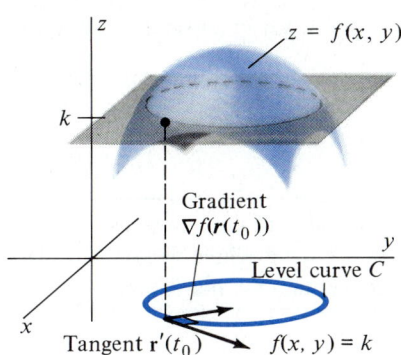

Figure 7.7 $\nabla f(\boldsymbol{x}_0)$ is orthogonal to the level curve through $\boldsymbol{x}_0$. (See **Plate 12.**)

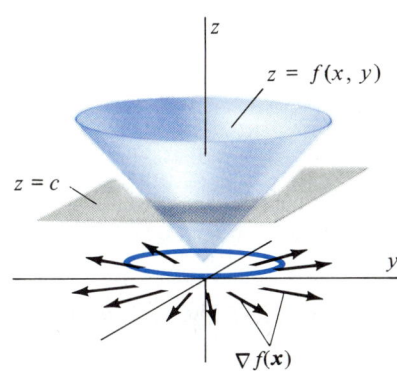

Figure 7.8 For $f(x, y) = \sqrt{x^2 + y^2}$ level curves are circles; $\nabla f(\boldsymbol{x})$ is parallel to a radius for the circle.

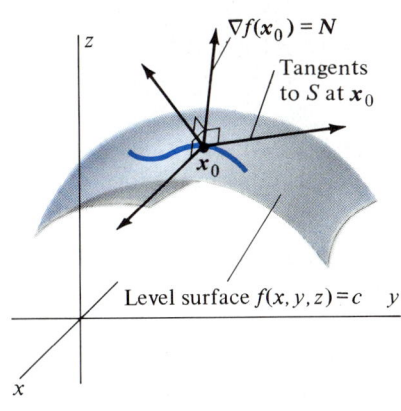

Figure 7.9 $\nabla f(\boldsymbol{x}_0)$ is orthogonal to the level surface for f containing $\boldsymbol{x}_0$.

both sides of equation (14), using equation (12), to conclude that

$$\frac{d}{dt} f(\boldsymbol{r}(t)) \bigg|_{t=t_0} = \nabla f(\boldsymbol{r}(t_0)) \cdot \boldsymbol{r}'(t_0) = 0. \tag{15}$$

Since $\boldsymbol{r}'(t_0)$ is a vector tangent to the level curve C, equation (15) shows that *the gradient $\nabla f(\boldsymbol{r}(t_0))$ is orthogonal to the level curve C at $\boldsymbol{x}_0$* (see Figure 7.7). That is:

> Let f be a function of two variables. If f and its first partial derivatives are continuous, then at each point in the domain of f the gradient vector, if nonzero, is orthogonal to the level curve through that point. (16)

Example 8

The graph of the function $z = \sqrt{x^2 + y^2}$ is a cone. The level curves associated with this graph are the circles

$$\sqrt{x^2 + y^2} = k, \qquad \text{or} \qquad x^2 + y^2 = k^2.$$

The gradient for this function is

$$\nabla f(x, y) = \frac{x}{\sqrt{x^2 + y^2}} \boldsymbol{i} + \frac{y}{\sqrt{x^2 + y^2}} \boldsymbol{j} = \frac{x\boldsymbol{i} + y\boldsymbol{j}}{\sqrt{x^2 + y^2}}.$$

In vector notation, the gradient may be written as

$$\nabla f(\boldsymbol{x}) = \frac{\boldsymbol{x}}{|\boldsymbol{x}|}.$$

This shows that the gradient vectors are parallel to radius vectors for the circular level curves and, therefore, orthogonal to the level curves (Figure 7.8). ◇

A similar relationship exists between gradients for functions of three variables and level surfaces. Suppose that

$$f(x, y, z) = k$$

is the equation of a level surface S for the function f, and suppose that the point $\boldsymbol{x}_0$ lies on S. Choose a curve C on the surface S with parameterization

$$\boldsymbol{r}(t) = x(t)\boldsymbol{i} + y(t)\boldsymbol{j} + z(t)\boldsymbol{k}$$

for which $\boldsymbol{r}(t_0) = \boldsymbol{x}_0$ (see Figure 7.9). Then, along this curve

$$f(\boldsymbol{r}(t)) = f(x(t), y(t), z(t)) = k, \tag{17}$$

that is, the composite function $f \circ \boldsymbol{r}$ is constant. As before, if the function f and the component functions x, y, and z satisfy the hypotheses of Theorem 6 in a neighborhood of $\boldsymbol{x}_0 = \boldsymbol{r}(t_0)$, we may differentiate both sides of equation (17), using the Chain Rule, to conclude that

$$\frac{d}{dt} f(\boldsymbol{r}(t)) \bigg|_{t=t_0} = \nabla f(\boldsymbol{r}(t_0)) \cdot \boldsymbol{r}'(t_0) = 0. \tag{18}$$

Equation (18) shows that the gradient $\nabla f(\boldsymbol{x}_0)$ is orthogonal to the tangent $\boldsymbol{r}'(t_0)$ to the curve C through $\boldsymbol{x}_0$ on the surface S. Since C was an *arbitrary* curve on S, it follows that $\nabla f(x_0)$ is orthogonal to *every* tangent to S at x_0. In particular, $\nabla f(x_0)$ is orthogonal to the tangent plane determined by $\dfrac{\partial f}{\partial x}(x_0)$ and $\dfrac{\partial f}{\partial y}(x_0)$ and therefore to the level surface S itself.

> Let f be a function of three variables. If f and its first partial derivatives are continuous, then at each point in the domain of f the gradient vector, if nonzero, is orthogonal to the level surface containing that point. $\qquad$ (19)

Exercise Set 17.7

In Exercises 1–9, find the gradient of the given function at the point P.

1. $f(x, y) = x^2 y$, $\quad P = (3, 1)$

2. $f(x, y) = xy^2 - ye^x$, $\quad P = (0, 2)$

3. $f(x, y) = x \cos(y - x)$, $\quad P = (\pi/2, \pi/4)$

4. $f(x, y) = \dfrac{2x}{y - x}$, $\quad P = (2, 1)$

5. $f(x, y, z) = x^2 y + xz^2$, $\quad P = (1, 1, 2)$

6. $f(x, y, z) = xe^{yz}$, $\quad P = (2, 0, 1)$

7. $f(x, y, z) = x^2 y + xz^3 - y^2 z$, $\quad P = (1, 2, 1)$

8. $f(x, y, z) = e^x \cos y - e^y \sin z$, $\quad P = (0, \pi/4, \pi/3)$

9. $f(x, y, z) = x\sqrt{y} \cosh z$, $\quad P = (2, 4, 1)$

In Exercises 10–19, find the directional derivative of the given function at the given point in the direction of the given vector.

10. $f(x, y) = 3x^2 - y^2$, $\quad P = (1, 2)$, $\quad w = i + j$

11. $f(x, y) = x^2 y^2$, $\quad P = (-2, 3)$, $\quad w = i - j$

12. $f(x, y) = \sin(y - x)$, $\quad P = (\pi/2, \pi/4)$, $\quad w = i + 2j$

13. $f(x, y) = \dfrac{x}{x + y}$, $\quad P = (1, 2)$, $\quad w = \sqrt{3}i + j$

14. $f(x, y) = xe^{y^2} - y$, $\quad P = (1, 3)$, $\quad w = \sqrt{2}i + \sqrt{2}j$

15. $f(x, y, z) = xy + xz + yz$, $\quad P = (1, 2, 1)$, $w = i + j - k$

16. $f(x, y, z) = \cosh x - \sinh y + \mathrm{Tan}^{-1} z$, $P = (-1, 1, 3)$, $\quad w = i - j - 2k$

17. $f(x, y, z) = xe^{yz}$, $\quad P = (2, 0, 1)$, $w = \sqrt{3}i + \sqrt{3}j - \sqrt{5}k$

18. $f(x, y, z) = xy^2 - x^2 z + (yz)^3$, $\quad P = (1, 0, 1)$, $w = i + 2j + k$

19. $f(x, y, z) = x \cos y - y \sinh z$, $\quad P = (6, \pi/4, -1)$, $w = i - 2j - 2k$

In Exercises 20–24, write vector equations for the normal and tangent lines to the given curves at the given points.

20. $2x^2 + 3y^2 = 11$, $\quad P = (2, 1)$

21. $4x^2 - y^2 = 7$, $\quad P = (2, 3)$

22. $3x^2 - y^4 + x = 13$, $\quad P = (2, 1)$

23. $\sqrt{x} + \sqrt{y} = 4$, $\quad P = (4, 4)$

24. $6x^2 - 4y^2 = 18$, $\quad P = (3, -3)$

25. For $f(x, y) = xy^2 + ye^x$, find the directional derivative at $(0, 1)$ in the direction of most rapid increase of f.

26. For $f(x, y) = x^2 + 2y^3$, find the directional derivative at $(2, 3)$ in the direction toward the origin.

27. For $f(x, y)$ as in Exercise 25, find $D_u f(0, 1)$ in the direction of the origin.

28. For $f(x, y) = \dfrac{x}{x + y}$, find $D_u f(1, 1)$ in the direction of the point $(2, 3)$.

29. For f as in Exercise 28, find $D_u f(1, 1)$ in the direction of most rapid increase of f.

30. Use the gradient to find the point(s) on the graph of the hyperbola $3x^2 - 2y^2 - 6x + 8y = 3$ where the tangent is horizontal.

31. Use the gradient to find the point(s) on the graph of the ellipse $x^2 - 6x + 2y^2 - 4y = -7$ where the tangent is **a.** horizontal, **b.** vertical.

32. Use equation (8) to show that the directional derivative of f in the direction of the unit vector u is the component of ∇f in the direction of u.

33. Use equation (8) to show that the negative gradient, $-\nabla f(x)$, points in the direction of most rapid decrease of the function f at x.

34. The temperature distribution in a room is given by the function $T(x, y, z) = 30 - (x^2 + 2y^2 + 3z^2)$. An insect flies so as to experience the most rapid decrease in temperature. In what direction does it move when located at point (2, 1, 1)?

35. Show that $|\nabla f(x_0)|$ is the maximum value of $D_u f(x_0)$ when f and its first partials are continuous at x_0.

36. Show that if $\cos \alpha$, $\cos \beta$, and $\cos \gamma$ are the direction cosines for the unit vector u, then $D_u f = \dfrac{\partial f}{\partial x} \cos \alpha i + \dfrac{\partial f}{\partial y} \cos \beta j + \dfrac{\partial f}{\partial z} \cos \gamma k$.

37. Find the direction of steepest ascent at the point above (4, 2) on the graph of $z = e^{x^2 + y^2}$.

38. Find the direction of steepest ascent at the point above (1, 1) on the graph of $z = xe^y + ye^x$.

39. Find the direction of steepest *descent* at the point above (1, 2) on the graph of $z = 4x^2 - 2y^2$.

40. The temperature distribution on a metal plate is $T(x, y) = 200 - (x^2 + 4y^2)$. Find the path followed by a heat-seeking particle placed at the point (0, 2).

41. Show that if the partial derivative $\dfrac{\partial f}{\partial x}(x_0, y_0)$ exists, then $D_i f(x_0, y_0)$ exists and $D_i f(x_0, y_0) = \dfrac{\partial f}{\partial x}(x_0, y_0)$. Give an example to show that $D_u f(x_0, y_0)$ can exist for all u while $\dfrac{\partial f}{\partial x}(x_0, y_0)$ (and, therefore, $\nabla f(x_0, y_0)$) fails to exist.

17.8 CONSTRAINED EXTREMA: THE METHOD OF LAGRANGE MULTIPLIERS

As an application of the theory of the gradient, we discuss a method due to the French mathematician Joseph L. Lagrange (1736–1813) for finding extrema of functions of several variables subject to constraints. Examples of this kind of problem are the following:

Example 1

Find the maximum and minimum values of the function $f(x, y) = 2x^2 + 4y^2$, given that $x^2 + y^2 = 1$.

Example 2

Find the point(s) on the hyperboloid $z = (y + 1)^2 - (x - 2)^2 + 1$ nearest the point (2, −1, 2).

Example 3

A cylindrical tin can, with a top and a bottom, is to be manufactured using 100 cm^2 of tin, ignoring waste. What dimensions produce the can of maximum volume?

Each of these examples involves finding the extreme values of a function

$$w = f(x), \qquad x \in \mathbb{R}^2 \text{ or } \mathbb{R}^3 \qquad \text{(function to be optimized)} \qquad (1)$$

subject to a **constraint** of the form

$$g(x) = 0 \qquad \text{(constraint equation).} \qquad (2)$$

In Example 1 the function to be optimized is $f(x, y) = 2x^2 + 4y^2$, and the constraint is $x^2 + y^2 = 1$. The latter equation can be put in the form of equation (2) by writing it as

$$g(x, y) = x^2 + y^2 - 1 = 0.$$

In Example 2, we are to minimize the distance function $D(x, y, z) = \sqrt{(x - 2)^2 + (y + 1)^2 + (z - 2)^2}$ subject to the constraint that (x, y, z) lie on the hyperboloid. We can write this constraint in the form of equation (2) as

$$g(x, y, z) = z - (y + 1)^2 + (x - 2)^2 - 1 = 0.$$

In Example 3, the dimensions of the tin can are r (radius) and h (height), so we need to maximize the volume $V = \pi r^2 h$ subject to the constraint that total surface area equal 100 cm^2. This constraint may be expressed as the equation

$$g(r, h) = 2\pi r^2 + 2\pi r h - 100 = 0.$$

Each of these examples illustrates the important difference between finding *relative* extrema for functions f of several variables and finding *extrema in the presence of constraints,* or *constrained extrema.* The constraint equation $g(x) = 0$ restricts the points x that may be considered to just those satisfying this constraint. Constrained extrema in general will not correspond to relative extrema.

Figure 8.1 illustrates the geometry associated with the constrained extrema problem of Example 1. Notice how the constraint $x^2 + y^2 = 1$ restricts the graph of f to

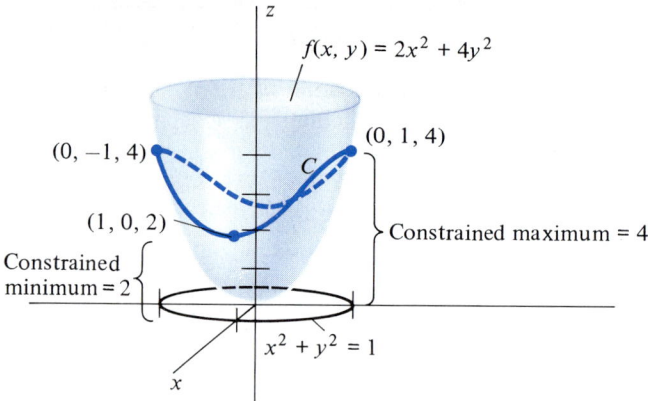

Figure 8.1 Constraint $g(x, y) = x^2 + y^2 - 1 = 0$ restricts graph of $f(x, y) = 2x^2 + 4y^2$ to the curve C.

just the curve C above the unit circle. Note also that neither the constrained maxima nor the constrained minima correspond to relative extrema for the function f.

It is sometimes possible to solve a problem involving a constraint by solving the constraint equation (2) for one of the independent variables and then substituting for this variable in equation (1). The method due to Lagrange is more general, in that it does not depend on our ability to solve the constraint equation for any particular variable. It is based on the following theorem.

THEOREM 8

Let x denote a point (vector) in either $\mathbb{R}^2$ or $\mathbb{R}^3$, and let f and g be functions of either two or three variables. Assume that both f and g have continuous partial derivatives in a neighborhood of the point x_0. If x_0 maximizes or minimizes the function

$$w = f(x) \tag{3a}$$

subject to the constraint

$$g(x) = 0, \tag{3b}$$

and if $\nabla g(x_0) \neq 0$, then

$$\nabla f(x_0) = \lambda \nabla g(x_0) \qquad (4)$$

for some constant λ. In other words, $\nabla f(x_0)$ and $\nabla g(x_0)$ are parallel.

The method of Lagrange for finding relative extrema is now obvious—we simply check all points satisfying equation (4). Among the values $f(x_0)$ must lie the constrained extrema, if such extrema exist. We sketch a proof of Theorem 8 at the end of this section.

Theorem 8 establishes the **method of Lagrange multipliers** (for finding the extreme values of f subject to the constraint $g(x) = 0$):

1. Find all simultaneous solutions of the equations

   $$\nabla f(x_0) = \lambda \nabla g(x_0) \qquad (5)$$

   and

   $$g(x_0) = 0. \qquad (6)$$

   a. If $x = (x, y) \in \mathbb{R}^2$, equations (5) and (6) are equivalent to the three equations

   $$\begin{cases} \dfrac{\partial f}{\partial x}(x_0, y_0) = \lambda \dfrac{\partial g}{\partial x}(x_0, y_0), \\[2mm] \dfrac{\partial f}{\partial y}(x_0, y_0) = \lambda \dfrac{\partial g}{\partial y}(x_0, y_0), \\[2mm] g(x_0, y_0) = 0. \end{cases} \qquad (7)$$

   b. If $x = (x, y, z) \in \mathbb{R}^3$, equations (5) and (6) are equivalent to the four equations

   $$\begin{cases} \dfrac{\partial f}{\partial x}(x_0, y_0, z_0) = \lambda \dfrac{\partial g}{\partial x}(x_0, y_0, z_0), \\[2mm] \dfrac{\partial f}{\partial y}(x_0, y_0, z_0) = \lambda \dfrac{\partial g}{\partial y}(x_0, y_0, z_0), \\[2mm] \dfrac{\partial f}{\partial z}(x_0, y_0, z_0) = \lambda \dfrac{\partial g}{\partial z}(x_0, y_0, z_0), \\[2mm] g(x_0, y_0, z_0) = 0. \end{cases} \qquad (8)$$

2. Calculate $f(x_0)$ for all x_0 obtained in step 1.
3. On geometric, analytic, or physical grounds, determine which of the numbers obtained in step 2 correspond to constrained extrema.

Solution to Example 1: We wish to find the extreme values of

$$f(x, y) = 2x^2 + 4y^2$$

subject to

$$g(x, y) = x^2 + y^2 - 1 = 0.$$

The three equations corresponding to equations (7) are:

$$4x = 2\lambda x \qquad\qquad (f_x = \lambda g_x), \qquad\qquad\qquad (9)$$

$$8y = 2\lambda y \qquad\qquad (f_y = \lambda g_y), \qquad\qquad\qquad (10)$$

and

$$x^2 + y^2 - 1 = 0 \qquad (g = 0). \qquad\qquad\qquad (11)$$

To solve this system of three equations, we begin with equation (9). If $x = 0$ equation (9) is satisfied, equation (11) then becomes simply $y^2 - 1 = 0$, so $y = \pm 1$. We therefore obtain the two points $(0, 1)$ and $(0, -1)$ that must be checked for extrema.

On the other hand, if $x \neq 0$ in (9), we may divide both sides of (9) by x to obtain $4 = 2\lambda$, so $\lambda = 2$. Substituting this value of λ into (10) then gives $8y = 4y$, so y must equal zero. With $y = 0$, equation (11) becomes $x^2 - 1 = 0$, so $x = \pm 1$. We have therefore obtained two additional points, $(1, 0)$ and $(-1, 0)$.

Finally, we simply calculate the value of $f(x, y)$ for each of the four points $(0, 1)$, $(0, -1)$, $(1, 0)$, and $(-1, 0)$. We find

$$f(0, 1) = 4 \qquad \text{(maximum)},$$
$$f(0, -1) = 4 \qquad \text{(maximum)},$$
$$f(1, 0) = 2 \qquad \text{(minimum)},$$
$$f(-1, 0) = 2 \qquad \text{(minimum)}.$$

As Figure 8.1 illustrates, the constrained maximum is 4, occurring at $(0, 1)$ and $(0, -1)$, and the constrained minimum is $2 = f(1, 0) = f(-1, 0)$. ◇

REMARK 1: In reading these examples, you will notice that the method by which we solve the systems of equations (7) and (8) varies from problem to problem. The great versatility of the method of Lagrange multipliers is that it applies to a wide variety of functions and constraints. However, the resulting systems of equations will therefore be of various types, and often most or all of the resulting equations will be nonlinear. Your success in using this method will depend on your ability to solve these various sorts of systems of equations.

REMARK 2: Note that although we obtained the value $\lambda = 2$ at one point in the solution of Example 1, the value of λ did not appear in the solution. This will always be the case, since the Lagrange multiplier λ is an ''artificial'' variable that is introduced merely as a device by which we can solve for the ''real'' variables in the problem.

The solution of Example 2 is similar to that of Example 1, and we leave it to you as Exercise 15.

Solution to Example 3: Since the volume of the cylinder is $V(r, h) = \pi r^2 h$, the areas of the top and bottom are each πr^2, and the lateral surface area is $2\pi r h$, we must maximize the function

$$V(r, h) = \pi r^2 h$$

subject to the side condition

$$g(r, h) = 2\pi r^2 + 2\pi r h - 100 = 0.$$

The equations corresponding to equations (7) are therefore

$$2\pi rh = \lambda(4\pi r + 2\pi h) \qquad (V_r = \lambda g_r), \tag{12}$$

$$\pi r^2 = \lambda(2\pi r) \qquad (V_h = \lambda g_h), \tag{13}$$

and

$$2\pi r^2 + 2\pi rh - 100 = 0 \qquad (g = 0). \tag{14}$$

Solving equation (12) for λ gives

$$\lambda = \frac{rh}{2r + h}, \tag{15}$$

while solving equation (13) for λ gives

$$\lambda = \frac{r}{2}. \tag{16}$$

Equating the right sides of (15) and (16) we find

$$\frac{rh}{2r + h} = \frac{r}{2}, \qquad \text{so} \qquad 2r + h = 2h,$$

or $h = 2r$. Substituting $h = 2r$ for h in (14) then gives

$$2\pi r^2 + 2\pi r(2r) - 100 = 0,$$

which has solution $r = \dfrac{10}{\sqrt{6\pi}}$. Since it is clear from the geometry of the situation that at least one pair (r, h) must produce a maximum volume, we conclude that the maximum volume of

$$V = \pi \left(\frac{10}{\sqrt{6\pi}}\right)^2 \left(\frac{20}{\sqrt{6\pi}}\right) = \frac{1000}{3\sqrt{6\pi}} \text{ cm}^3$$

corresponds to the dimensions $r = \dfrac{10}{\sqrt{6\pi}}$ cm and $h = 2r = \dfrac{20}{\sqrt{6\pi}}$ cm. ◇

Example 4

A builder wishes to design a rectangular house containing V cubic meters of heated space, so as to minimize heating costs. One wall of the building is to face south. The annual heating costs are estimated to be $4 per square meter of floor space, $3 per square meter for all exterior wall not facing south, and $2 per square meter for exterior wall space facing south. What dimensions will produce the most energy-efficient building?

Solution: As in Figure 8.2, we let x denote the length of the wall facing south, and we refer to this dimension as the width of the house. We let y be the depth, and we let z be the height of the heated portion of the house. Since the area of the floor is xy, the area of each side wall is yz, and the area of each front and rear wall is xz, the

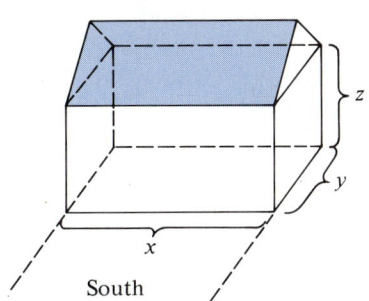

South

Figure 8.2

annual heating cost is

$$C(x, y, z) = 4xy + 3(2yz) + 3xz + 2xz$$

Cost for south wall

Cost for rear (north) wall

Costs for side walls (both)

Costs for roof (floor space)

$$= 4xy + 6yz + 5xz.$$

The constraint is that the volume of the building is to equal the constant V. That is,

$$g(x, y, z) = xyz - V = 0.$$

Applying equations (8), we find

$$4y + 5z = \lambda yz \qquad (C_x = \lambda g_x), \tag{17}$$

$$4x + 6z = \lambda xz \qquad (C_y = \lambda g_y), \tag{18}$$

$$5x + 6y = \lambda xy \qquad (C_z = \lambda g_z), \tag{19}$$

and

$$xyz - V = 0 \qquad (g = 0). \tag{20}$$

This time it is helpful to begin by multiplying both sides of equation (17) by x, both sides of (18) by y, and both sides of (19) by z. The result is the equations

$$4xy + 5xz = \lambda xyz, \tag{21}$$

$$4xy + 6yz = \lambda xyz, \tag{22}$$

and

$$5xz + 6yz = \lambda xyz. \tag{23}$$

Equating the left sides of (21) and (22) then gives $4xy + 5xz = 4xy + 6yz$, so $y = \dfrac{5}{6}x$. Similarly, equating the left sides of (22) and (23) gives $4xy + 6yz = 5xz + 6yz$, or $z = \dfrac{4y}{5} = \dfrac{2x}{3}$. With these substitutions (20) becomes

$$x\left(\frac{5}{6}x\right)\left(\frac{2}{3}x\right) = \frac{5}{9}x^3 = V.$$

so

$$x = \sqrt[3]{\frac{9V}{5}}, \qquad y = \frac{5}{6}\sqrt[3]{\frac{9V}{5}}, \qquad \text{and} \qquad z = \frac{2}{3}\sqrt[3]{\frac{9V}{5}}. \qquad \diamond$$

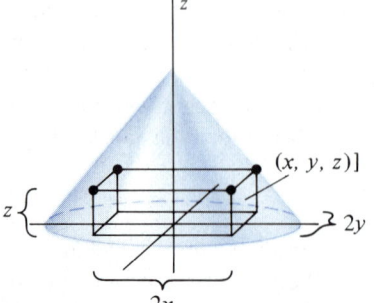

Figure 8.3 Box inscribed within a cone.

Example 5

A rectangular box is to be inscribed in the cone $z = 9 - \sqrt{x^2 + y^2}$, $z \geq 0$ (see Figure 8.3). Find the dimensions for the box that maximize its volume.

Solution: It is clear that to achieve maximum volume we should position the box with one face lying in the xy-plane. If (x, y, z) denotes one corner of the box lying in the first octant, x, y, $z > 0$, the dimensions of the box are

$$\text{length} = 2x, \qquad \text{width} = 2y, \qquad \text{and} \qquad \text{height} = z.$$

The problem is therefore to maximize the function

$$V(x, y, z) = 4xyz$$

subject to the constraint

$$g(x, y, z) = z + \sqrt{x^2 + y^2} - 9 = 0.$$

Applying the Lagrange criterion (8), we find

$$4yz = \frac{\lambda x}{\sqrt{x^2 + y^2}} \qquad (f_x = \lambda g_x), \tag{24}$$

$$4xz = \frac{\lambda y}{\sqrt{x^2 + y^2}} \qquad (f_y = \lambda g_y), \tag{25}$$

$$4xy = \lambda, \tag{26}$$

and

$$z + \sqrt{x^2 + y^2} - 9 = 0. \tag{27}$$

Multiplying (24) through by x, (25) by y, and (26) by z gives

$$4xyz = \frac{\lambda x^2}{\sqrt{x^2 + y^2}}; \qquad 4xyz = \frac{\lambda y^2}{\sqrt{x^2 + y^2}}; \qquad 4xyz = \lambda z.$$

Equating the right-hand sides of these equations gives

$$\frac{\lambda x^2}{\sqrt{x^2 + y^2}} = \frac{\lambda y^2}{\sqrt{x^2 + y^2}} \tag{28}$$

and

$$\lambda z = \frac{\lambda x^2}{\sqrt{x^2 + y^2}}. \tag{29}$$

From (28) it follows that $y = x$, since neither x nor y can be negative. With $y = x$, (29) gives $z = \dfrac{x^2}{\sqrt{2x^2}} = \dfrac{x}{\sqrt{2}}$. Substituting these expressions for y and z in (27) then gives

$$\frac{x}{\sqrt{2}} + \sqrt{x^2 + x^2} - 9 = \left(\frac{1}{\sqrt{2}} + \sqrt{2}\right)x - 9 = 0,$$

so

$$x = \frac{9}{\dfrac{1}{\sqrt{2}} + \sqrt{2}} = \frac{9\sqrt{2}}{1 + 2} = 3\sqrt{2}.$$

Thus, $y = 3\sqrt{2}$ and $z = 3$. The maximum volume is therefore

$$V = 4(3\sqrt{2})(3\sqrt{2})(3) = 216. \qquad \diamondsuit$$

Proof of Theorem 8 (Sketch): Let's first consider the planar case $x = (x, y) \in \mathbb{R}^2$. Then the constraint equation $g(x) = g(x, y) = 0$ determines a curve C in the xy-plane on which the point x_0 must lie. This curve C is a *level curve* for the function g. Thus, by statement (16), Section 17.7, if $\nabla g(x_0) \neq 0$ then $\nabla g(x_0)$ is orthogonal to C at x_0. We will complete the proof for the planar case by showing that $\nabla f(x_0)$ is also orthogonal to C at x_0.

Let $r(t) = x(t)i + y(t)j$ be a parameterization for C so that $r(t_0) = x_0$ and so that $r'(t_0) \neq 0$.* Since $x_0 = r(t_0)$ maximizes $f(x)$ on C, the scalar function $f \circ r$ has a relative extremum at $t = t_0$. Thus, by the Chain Rule,

$$(f \circ r)'(t_0) = \nabla f(r(t_0)) \cdot r'(t_0) = 0. \tag{30}$$

This shows that $\nabla f(r(t_0)) = \nabla f(x_0)$ is orthogonal to $r'(t_0)$. Since $r'(t_0)$ is *tangent* to C at x_0, it follows that $\nabla f(x_0)$ is orthogonal to C at x_0.

Since $\nabla f(x_0)$, $\nabla g(x_0)$, and C all lie in the xy-plane, we may now conclude, since $\nabla f(x_0)$ and $\nabla g(x_0)$ are both orthogonal to C at x_0, that $\nabla f(x_0)$ and $\nabla g(x_0)$ are parallel. That is, $\nabla f(x_0) = \lambda \nabla g(x_0)$ for some constant λ.

The space case $x \in \mathbb{R}^3$ is similar to the planar case. However, the constraint equation $g(x) = 0$ now determines a surface S in $\mathbb{R}^3$ that is a level surface for the function $u = g(x)$. By statement (19), Section 17.7, if $\nabla g(x_0) \neq 0$ then $\nabla g(x_0)$ is orthogonal to S at x_0.

As before, we will show that $\nabla f(x_0)$ is also orthogonal to S at x_0. We begin by letting C be an arbitrary curve on S with parameterization $r(t) = x(t)i + y(t)j + z(t)k$ for which $r(t_0) = x_0$ and $r'(t_0) \neq 0$. Just as in the plane case, the scalar function $f \circ r$ has a relative extremum at t_0, so equation (30) again holds. Since $r'(t_0)$ is tangent to C at x_0, it is also tangent to S at x_0. This shows that $\nabla f(x_0)$ is orthogonal to *a* tangent at x_0. But since the curve C was arbitrary, so is its tangent $r'(t_0)$. It follows that $\nabla f(x_0)$ is orthogonal to *all* tangents to S at x_0, and therefore that $\nabla f(x_0)$ is orthogonal to S. Since two vectors orthogonal to a (smooth) surface at a common point must be parallel, it follows that $\nabla f(x_0) = \lambda \nabla g(x_0)$ for some constant λ. ◆

Exercise Set 17.8

In Exercises 1–14, find the maximum and minimum values of the given function subject to the given constraint.

1. $f(x, y) = 2x^2 + 4y^2$ subject to $x^2 + y^2 = 1$

2. $f(x, y) = x^2 + y$ subject to $x^2 + y^2 = 9$

3. $f(x, y) = x^3 - y^3$ subject to $x - y = 2$

4. $f(x, y) = xy$ subject to $x^2 + y^2 - 4y = 5$

5. $f(x, y) = xy$ subject to $x^2 + y^2 = 1$

6. $f(x, y) = y - x$ subject to $x^2 + y^2 = 2$

7. $f(x, y) = x^2 + 4x + 4y^2$ subject to $x^2 + 2y^2 = 4$

8. $f(x, y, z) = x + y + z$ subject to $x^2 + y^2 + z^2 = 4$

9. $f(x, y) = x^2 - 4x + 4y^2$ subject to $x^2 + y^2 = 1$

10. $f(x, y, z) = xyz$ subject to $2x^2 + y^2 + 4z^2 = 9$

11. $f(x, y, z) = x + 2y - z$ subject to $x^2 + y^2 + z^2 = 1$

12. $f(x, y, z) = \sqrt{xyz}$ subject to $x + y + z = 4$

13. $f(x, y, z) = x + y + z$ subject to $x^2 + y^2 + z^2 = 12$

---

*$r(t)$ can always be chosen so that $r'(t_0) \neq 0$ if $\nabla g(x_0) \neq 0$. This follows from the Implicit Function Theorem, a result usually discussed in courses on advanced calculus. Actually, the Implicit Function Theorem underlies several of the seemingly obvious statements being made here, which is why we refer to this argument only as a sketch of a proof.

14. $f(x, y, z) = x^2 + 2y^2 + 4z^2$ subject to $x^2 + y^2 + z^2 = 1$

15. Find the solution of Example 2.

16. Find the point on the ellipsoid $9x^2 + 36y^2 + 4z^2 = 36$ nearest the origin.

17. Find the point on the circle $(x - 3)^2 + (y + 2)^2 = 9$ nearest the origin.

18. Find the point on the sphere $x^2 + y^2 + z^2 = 1$ furthest from the point $(3, 2, 1)$.

19. Find the point on the ellipsoid $(x - 1)^2 + \dfrac{(y - 2)^2}{4} + \dfrac{(z - 1)^2}{9} = 1$ nearest the point $(1, -1, 1)$.

20. Find the point on the cone $z = \sqrt{x^2 + y^2}$ nearest the point $(3, 1, 0)$.

21. Find the point(s) on the hyperboloid $z = (y + 1)^2 - (x - 2)^2 + 1$ nearest the point $(2, -1, 1)$.

22. The plane $2y - 3z = 8$ intersects the cone $z^2 = 4x^2 + 4y^2$ in an ellipse. Find the highest and lowest points of intersection.

23. Find the dimensions of the rectangular box of maximum volume that can be inscribed in a sphere of radius r.

24. Find the dimensions of the rectangular box of maximum volume which can be inscribed in the ellipsoid $x^2 + 4y^2 + 2z^2 = 8$.

25. Find the dimensions for the cylindrical jar, with volume 2 liters, that has minimum exterior surface area. (Assume that the jar has a lid and that the thickness of its walls is negligible.)

26–31. Rework Exercises 23–28 of Section 17.4 using the method of Lagrange multipliers.

17.9 RECONSTRUCTING A FUNCTION FROM ITS GRADIENT

In several ways, the gradient has emerged as the analogue of the derivative for functions of a single variable. It is therefore reasonable to ask whether the concept of antidifferentiation makes sense for gradients and, if so, whether this concept is useful. The goal of this section is to define what we mean by the "antiderivative" of a gradient (we don't really use this particular terminology), to show how such functions can be found, and to prove a theorem indicating when such functions exist. We will restrict our discussion to functions of two variables, although analogous results may be developed in more general settings. The results of this section will be used extensively in Chapter 19.

Potential Functions

We begin by recalling that a gradient is a vector-valued function

$$\nabla f(x, y) = \frac{\partial f}{\partial x}(x, y)\boldsymbol{i} + \frac{\partial f}{\partial y}(x, y)\boldsymbol{j}.$$

The antidifferentiation question for gradients is therefore the following: Given a vector-valued function of the form

$$\boldsymbol{F}(x, y) = M(x, y)\boldsymbol{i} + N(x, y)\boldsymbol{j}$$

is there a function f of two variables for which

$$\boldsymbol{F}(x, y) = \nabla f(x, y)?$$

We prefer not to refer to f as the antiderivative for the vector-valued function $\boldsymbol{F}$. Instead, we call it a *potential function*, or simply a *potential*. (The reason for this particular terminology has to do with the roles played by such functions in mechanics and in the theory of electricity and magnetism.)

Given the vector-valued function $\boldsymbol{F}$, the task of finding a function f for which $\boldsymbol{F}(x, y) = \nabla f(x, y)$ is therefore referred to as finding a potential for $\boldsymbol{F}$. This is also what we mean by "reconstructing the function f from its gradient." As Example 1 shows, potentials are not unique.

DEFINITION 6

The function f of two variables is called a **potential** for the vector-valued function

$$F(x, y) = M(x, y)\mathbf{i} + N(x, y)\mathbf{j}$$

in a rectangle $Q \subseteq \mathbb{R}^2$ if $\dfrac{\partial f}{\partial x}(x, y)$ and $\dfrac{\partial f}{\partial y}(x, y)$ exist and if

$$F(x, y) = \nabla f(x, y)$$

for all (x, y) in Q.

Example 1

If $f(x, y) = xe^{2y}$, then

$$\nabla f(x, y) = e^{2y}\mathbf{i} + 2xe^{2y}\mathbf{j}.$$

Thus, the function

$$F(x, y) = e^{2y}\mathbf{i} + 2xe^{2y}\mathbf{j}$$

has the function $f(x, y) = xe^{2y}$ as a potential. However, the function $g(x, y) = xe^{2y} + C$ is also a potential for F, since

$$\nabla g(x, y) = e^{2y}\mathbf{i} + 2xe^{2y}\mathbf{j} = F(x, y).$$

Thus, if f is a potential for F, so is $f + C$ for any constant C. ◇

Finding Potentials

If a potential f for F exists, how is it found? According to Definition 6 the following must be true.

> If $F(x, y) = M(x, y)\mathbf{i} + N(x, y)\mathbf{j} = \nabla f(x, y)$, then   (1)
>
> $$M(x, y) = \frac{\partial f}{\partial x}(x, y)$$
>
> and
>
> $$N(x, y) = \frac{\partial f}{\partial y}(x, y).$$ (2)

Equations (1) and (2) are the keys to finding f. Beginning with equation (1) and integrating "partially with respect to x" we find

$$f(x, y) = \int \left(\frac{\partial f}{\partial x}(x, y)\right) dx = \int M(x, y)\, dx = G_1(x, y) + h_1(y) + C_1. \quad (3)$$

By integrating "partially with respect to x" we mean treating y as a constant and integrating $M(x, y)$ as a function of x alone—just the reverse of partial differentiation. We must remember, however, that functions of y alone vanish entirely when differentiated partially with respect to x, so we must allow for the appearance of an entire function $h_1(y)$ of y alone when integrating partially with respect to x.

Applying the same idea to equation (2), we integrate partially with respect to y

to obtain an equation of the form

$$f(x, y) = \int \left(\frac{\partial f}{\partial y}(x, y) \right) dy = \int N(x, y) \, dy = G_2(x, y) + h_2(x) + C_2, \qquad (4)$$

where $h_2(x)$ is a function of x alone.

If we are fortunate, we can next equate the right-hand sides of equations (3) and (4) and determine the functions G_1, G_2, h_1, and h_2. The potential f can then be obtained from either equation (3) or equation (4).

Example 2

Let **F** be the vector-valued function

$$F(x, y) = (3x^2 + 2y^2)i + 4xyj.$$

Find a function f for which $F(x, y) = \nabla f(x, y)$ for all (x, y).

Strategy

Identify M and N.

Solution

Here $F(x, y) = M(x, y)i + N(x, y)j$ with

$$M(x, y) = 3x^2 + 2y^2; \qquad N(x, y) = 4xy.$$

Integrate M partially with respect to x to find an expression for $f(x, y)$.

Partially integrating M with respect to x according to equation (3) gives

$$f(x, y) = \int M(x, y) \, dx \qquad (5)$$

$$= \int (3x^2 + 2y^2) \, dx$$

$$= x^3 + 2xy^2 + h_1(y) + C_1$$

Integrate N partially with respect to y to obtain a second expression for $f(x, y)$.

where h_1 is a function of y alone. Then, partially integrating N with respect to y, we find

$$f(x, y) = \int N(x, y) \, dy \qquad (6)$$

$$= \int 4xy \, dy$$

$$= 2xy^2 + h_2(x) + C_2$$

where h_2 is a function of x alone.

Equate the two expressions for $f(x, y)$ and attempt to identify the unknown functions h_1 and h_2.

Equating these two expressions for $f(x, y)$ gives

$$x^3 + 2xy^2 + h_1(y) + C_1 = 2xy^2 + h_2(x) + C_2.$$

This equation is true if $h_1(y) = 0$, $h_2(x) = x^3$, and $C_1 = C_2 = C$. Both (5) and (6) then give the desired potential as

Read $f(x, y)$ from either expression.

$$f(x, y) = x^3 + 2xy^2 + C.$$

Partial differentiation verifies that

$$\nabla f(x, y) = F(x, y). \qquad \diamondsuit$$

Example 3

Find a potential for the function

$$F(x, y) = (ye^{xy} - 2x \sin x^2)i + \left(\frac{1}{\sqrt{y}} + xe^{xy}\right)j.$$

Solution: Here $F(x, y) = M(x, y)i + N(x, y)j$ with

$$M(x, y) = ye^{xy} - 2x \sin x^2; \qquad N(x, y) = \frac{1}{\sqrt{y}} + xe^{xy}.$$

Integrating M partially with respect to x gives

$$f(x, y) = \int M(x, y) \, dx = \int (ye^{xy} - 2x \sin x^2) \, dx \tag{7}$$

$$= e^{xy} + \cos x^2 + h_1(y) + C_1$$

where h_1 is a function of y alone. Integrating N partially with respect to y gives

$$f(x, y) = \int N(x, y) \, dy = \int \left(\frac{1}{\sqrt{y}} + xe^{xy}\right) dy \tag{8}$$

$$= 2\sqrt{y} + e^{xy} + h_2(x) + C_2$$

where h_2 is a function of x alone. Equating the right-hand sides of equations (7) and (8) then shows that

$$h_1(y) = 2\sqrt{y}, \qquad h_2(x) = \cos x^2, \qquad \text{and} \qquad C_1 = C_2 = C. \tag{9}$$

With the information in line (9), either of equations (7) and (8) gives that

$$f(x, y) = e^{xy} + \cos x^2 + 2\sqrt{y} + C.$$

Partial differentiation then verifies that $\nabla f(x, y) = F(x, y)$. $\diamond$

Example 4

Show that the function

$$F(x, y) = yi - xj$$

has no potential. That is, F cannot be the gradient of a function f of two variables.

Solution: On the contrary, let us assume that there does exist a function f with $F(x, y) = \nabla f(x, y)$. Since

$$M(x, y) = y, \qquad \text{and} \qquad N(x, y) = -x,$$

we have

$$f(x, y) = \int M(x, y) \, dx = \int y \, dx = xy + h(y) + C_1 \tag{10}$$

where h is a function of y alone. Furthermore, since we are assuming that $\dfrac{\partial f}{\partial y}(x, y)$ exists, h must be a differentiable function of y. We may therefore differentiate both sides of (10) with respect to y. Doing so, and using equation (2), gives

$$\frac{\partial f}{\partial y}(x, y) = x + h'(y) = -x = N(x, y).$$

Thus,

$$h'(y) = -2x. \tag{11}$$

But since h must be a function of y alone, so is h'. Thus, since h' cannot involve x, $\frac{\partial}{\partial x}h' = 0$. This shows that the result of differentiating both sides of equation (11) with respect to x is the contradictory statement $0 = -2$. The assumption that $F(x, y) = y\boldsymbol{i} - x\boldsymbol{j}$ has a potential therefore leads to a contradiction, and is therefore false. $F(x, y) = y\boldsymbol{i} - x\boldsymbol{j}$ has no potential. ◇

The result of Example 4 leaves us in need of information as to when a given vector function F has a potential. The information is supplied by the following theorem, whose proof employs the ideas of Examples 2 through 4.

THEOREM 9

Let M and N have continuous first partial derivatives in an open rectangle R in the xy-plane. Then the vector-valued function

$$F(x, y) = M(x, y)\boldsymbol{i} + N(x, y)\boldsymbol{j}$$

has a potential in R if and only if

$$\frac{\partial M}{\partial y}(x, y) = \frac{\partial N}{\partial x}(x, y) \tag{12}$$

for all (x, y) in R.

Before giving the proof, we note that condition (12) is satisfied for F in Example 2, since

$$\frac{\partial M}{\partial y}(x, y) = \frac{\partial}{\partial y}(3x^2 + 2y^2) = 4y = \frac{\partial N}{\partial x}(x, y).$$

Also, equation (12) holds for F in Example 3:

$$\frac{\partial M}{\partial y}(x, y) = \frac{\partial}{\partial y}(ye^{xy} - 2x \sin x^2)$$

$$= e^{xy} + xye^{xy}$$

$$= \frac{\partial}{\partial x}\left(\frac{1}{\sqrt{y}} + xe^{xy}\right)$$

$$= \frac{\partial N}{\partial x}(x, y).$$

However, condition (12) fails for F in Example 4 since

$$\frac{\partial M}{\partial y} = \frac{\partial}{\partial y}(y) = 1 \neq -1 = \frac{\partial}{\partial x}(-x) = \frac{\partial N}{\partial x}.$$

Proof of Theorem 9: We first assume that F has a potential f and show that equation (12) holds. This means that

$$F(x, y) = \nabla f(x, y)$$

where $M(x, y) = \dfrac{\partial f}{\partial x}(x, y)$ and $N(x, y) = \dfrac{\partial f}{\partial y}(x, y)$. Since both M and N are assumed to have continuous first partial derivatives in R, it follows that

$$\frac{\partial^2 f}{\partial y \partial x} = \frac{\partial}{\partial y}\left(\frac{\partial f}{\partial x}\right) = \frac{\partial M}{\partial y}(x, y) \tag{13}$$

and

$$\frac{\partial^2 f}{\partial x \partial y} = \frac{\partial}{\partial x}\left(\frac{\partial f}{\partial y}\right) = \frac{\partial N}{\partial x}(x, y). \tag{14}$$

Moreover, by Theorem 2 the mixed partial derivatives on the left-hand sides of equations (13) and (14) are equal. Thus, the right-hand sides of these equations are equal, and condition (12) is obtained.

The remaining part of the proof involves assuming that condition (12) holds and actually finding (at least formally) a potential for F. Taking a cue from Examples 2 through 4 we begin by fixing a point (x_0, y_0) in R and defining the function f as

$$f(x, y) = \int_{x_0}^{x} M(t, y_0)\, dt + \int_{y_0}^{y} N(x, s)\, ds. \tag{15}$$

Figure 9.1 may help you understand how f has been "pulled out of the hat." Since we want to construct f in such a way that $\dfrac{\partial f}{\partial x}(x, y) = M(x, y)$ and $\dfrac{\partial f}{\partial y}(x, y) =$

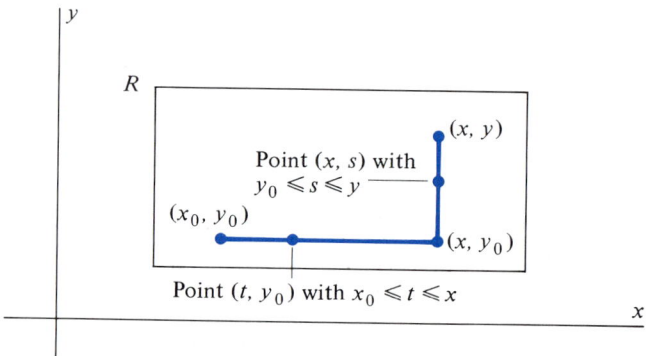

Figure 9.1 Domain of $M(x, y)$ and $N(x, y)$ in Theorem 9.

$N(x, y)$, the idea is to integrate M along a path involving only change in x and to integrate N along a path involving only change in y. The reason we require R to be a rectangle is simply to ensure that such a path lies entirely within R.

Since M and N are continuous in R, and since R is a rectangle, $f(x, y)$ is defined for all $(x, y) \in R$. Moreover, by the Fundamental Theorem of Calculus it follows that

$$\frac{\partial f}{\partial y}(x, y) = \frac{\partial}{\partial y}\left\{\int_{x_0}^{x} M(t, y_0)\, dt + \int_{y_0}^{y} N(x, s)\, ds\right\} \tag{16}$$

$$= \frac{\partial}{\partial y} \int_{y_0}^{y} N(x, s) \, ds$$

$$= N(x, y)$$

since the first integral inside the braces does not involve y. We also want to calculate $\frac{\partial f}{\partial x}(x, y)$, but this is a bit harder. We find that

$$\frac{\partial f}{\partial x}(x, y) = \frac{\partial}{\partial x} \left\{ \int_{x_0}^{x} M(t, y_0) \, dt + \int_{y_0}^{y} N(x, s) \, ds \right\} \qquad (17)$$

$$= M(x, y_0) + \frac{\partial}{\partial x} \int_{y_0}^{y} N(x, s) \, ds.$$

In courses on advanced calculus it is proved that

$$\frac{\partial}{\partial x} \int_{y_0}^{y} N(x, s) \, ds = \int_{y_0}^{y} \frac{\partial}{\partial x} N(x, s) \, ds. \qquad (18)$$

That is, we may pass the partial differentiation with respect to x under the y-integral sign in the integral on the right-hand side of equation (17). Using (18) we return to (17) and use condition (15) to find

$$\frac{\partial f}{\partial x}(x, y) = M(x, y_0) + \int_{y_0}^{y} \frac{\partial}{\partial x} N(x, s) \, ds \qquad (19)$$

$$= M(x, y_0) + \int_{y_0}^{y} \frac{\partial}{\partial s} M(x, s) \, ds \qquad \left(\frac{\partial N}{\partial x}(x, s) = \frac{\partial M}{\partial s}(x, s) \right)$$

$$= M(x, y_0) + M(x, s) \Big]_{s=y_0}^{s=y}$$

$$= M(x, y_0) + [M(x, y) - M(x, y_0)]$$

$$= M(x, y).$$

Reviewing equations (16) and (19), we find that the function f defined in line (15) has the properties

$$\frac{\partial f}{\partial x}(x, y) = M(x, y); \qquad \frac{\partial f}{\partial y}(x, y) = N(x, y).$$

That is, $\nabla f = F$, so F indeed has a potential. ◆

Exercise Set 17.9

In Exercises 1–12, determine whether the given vector function has a potential. If so, find a potential for F.

1. $F(x, y) = i - j$

2. $F(x, y) = 6yi + 6xj$

3. $F(x, y) = \pi yi - \pi xj$

4. $F(x, y) = 3xi + 4yj$

5. $F(x, y) = (x^2 - y)i + (x + y^2)j$

6. $F(x, y) = \sin yi + x \cos yj$

7. $F(x, y) = (3x^2 \cos y - 1)i + (2y - x^3 \sin y)j$

8. $F(x, y) = (2xy^3 - 2x)i + (3x^2y^2 + 1)j$

9. $F(x, y) = \left(2xe^{xy} + x^2 \, ye^{xy} \right)i + \left(x^3e^{xy} - \frac{1}{2\sqrt{y}} \right)j$

10. $F(x, y) = (x^3 - y \sin x)i + (y^3 - \cos x)j$

11. $F(x, y) = (2xy^2 - y \sin x)i + (2x^2y + \cos x)j$

12. $F(x, y) = 2x \, \mathrm{Tan}^{-1} \, yi + \frac{x^2}{1 + y^2}j$

SUMMARY OUTLINE OF CHAPTER 17

◆ A **level curve** for the function $z = f(x, y)$ is the graph of the equation $f(x, y) = c$ in the xy-plane. A **level surface** for the function $w = f(x, y, z)$ is the graph of an equation $f(x, y, z) = c$ in space. (page 753)

◆ If $x \in \mathbb{R}^n$ and $f \colon \mathbb{R}^n \to \mathbb{R}$, $\lim_{x \to x_0} f(x) = L$ means $|f(x) - L| \to 0$ as $x \to x_0$. (page 757)

◆ The function $f \colon \mathbb{R}^n \to \mathbb{R}$ is **continuous** at x_0 if $\lim_{x \to x_0} f(x) = f(x_0)$. (page 759)

◆ The **partial derivatives** of f are the limits (page 761)

$$\frac{\partial f}{\partial x}(x, y, z) = \lim_{h \to 0} \frac{f(x + h, y, z) - f(x, y, z)}{h},$$

$$\frac{\partial f}{\partial y}(x, y, z) = \lim_{h \to 0} \frac{f(x, y + h, z) - f(x, y, z)}{h},$$

and

$$\frac{\partial f}{\partial z}(x, y, z) = \lim_{h \to 0} \frac{f(x, y, z + h) - f(x, y, z)}{h}.$$

◆ **Theorem:** $\frac{\partial^2 f}{\partial y \partial x}(x, y) = \frac{\partial^2 f}{\partial x \partial y}(x, y)$ if f, $\frac{\partial f}{\partial x}$, $\frac{\partial f}{\partial y}$, $\frac{\partial^2 f}{\partial x \partial y}$, and $\frac{\partial^2 f}{\partial y \partial x}$ are continuous. (page 767)

◆ The plane tangent to the graph of $z = f(x, y)$ at (x_0, y_0, z_0) has equation (page 770)

$$\frac{\partial f}{\partial x}(x_0, y_0)(x - x_0) + \frac{\partial f}{\partial y}(x_0, y_0)(y - y_0) - (z - z_0) = 0.$$

◆ **Theorem:** If f has a relative extremum at (x_0, y_0) then either (page 775)

(i) $\frac{\partial f}{\partial x}(x_0, y_0) = \frac{\partial f}{\partial y}(x_0, y_0) = 0$, or

(ii) one or both of $\frac{\partial f}{\partial x}(x_0, y_0)$ and $\frac{\partial f}{\partial y}(x_0, y_0)$ fail to exist.

◆ **Theorem:** If $\frac{\partial f}{\partial x}(x_0, y_0) = \frac{\partial f}{\partial y}(x_0, y_0) = 0$ and if $A = \frac{\partial^2 f}{\partial x^2}(x_0, y_0)$, $B = \frac{\partial^2 f}{\partial y \partial x}(x_0, y_0)$, $C = \frac{\partial^2 f}{\partial y^2}(x_0, y_0)$, and $D = B^2 - AC$, then (page 778)

(i) If $D < 0$, and $A < 0$, $f(x_0, y_0)$ is a relative maximum.
(ii) If $D < 0$ and $A > 0$, $f(x_0, y_0)$ is a relative minimum.
(iii) If $D > 0$, (x_0, y_0) is a saddle point.
(iv) If $D = 0$ there is no conclusion.

◆ **Theorem:** (Linear Approximation) If f and its first partials are continuous, then (page 784)

$$f(x_0 + \Delta x, y_0 + \Delta y) = f(x_0, y_0) + \frac{\partial f}{\partial x}(x_0, y_0)\,\Delta x + \frac{\partial f}{\partial y}(x_0, y_0)\,\Delta y + \epsilon_1 \Delta x + \epsilon_2 \Delta y$$

where $\epsilon_1 \to 0$ and $\epsilon_2 \to 0$ as $\Delta x \to 0$ and $\Delta y \to 0$.

◆ **Chain Rule:** If $w = f(x(t), y(t))$ then (page 792)

$$\frac{dw}{dt} = \frac{\partial f}{\partial x} \cdot \frac{dx}{dt} + \frac{\partial f}{\partial y} \cdot \frac{dy}{dt}.$$

◆ The **directional derivative** $D_u f(x_0, y_0)$ in the direction of the unit vector $u = u_1 i + u_2 j$ is (page 799)

$$D_u f(x_0, y_0) = \lim_{t \to 0^+} \frac{f(x_0 + tu_1, y_0 + tu_2) - f(x_0, y_0)}{t}.$$

◆ The **gradient** $\nabla f(x_0, y_0)$ of the function f at (x_0, y_0) is the **vector**   (page 803)

$$\nabla f(x_0, y_0) = \frac{\partial f}{\partial x}(x_0, y_0)\mathbf{i} + \frac{\partial f}{\partial y}(x_0, y_0)\mathbf{j}$$

or

$$\nabla f(x_0, y_0, z_0) = \frac{\partial f}{\partial x}(x_0, y_0, z_0)\mathbf{i} + \frac{\partial f}{\partial y}(x_0, y_0, z_0)\mathbf{j} + \frac{\partial f}{\partial z}(x_0, y_0, z_0)\mathbf{k}.$$

◆ If $\dfrac{\partial f}{\partial x}(x, y)$ and $\dfrac{\partial f}{\partial y}(x, y)$ are continuous, then   (page 804)

$$D_{\mathbf{u}} f(x_0, y_0) = \frac{\partial f}{\partial x}(x_0, y_0)u_1 + \frac{\partial f}{\partial x}(x_0, y_0)u_2$$

$$= \nabla f(x_0, y_0) \cdot \mathbf{u}.$$

◆ $\nabla f(x_0)$ points in the direction of most rapid increase of the function f at x_0.   (page 804)

◆ $\nabla f(x_0)$ is orthogonal to the level curve for $z = f(x, y)$ at $x_0 = (x_0, y_0)$.   (page 807)

◆ $\nabla f(x_0)$ is orthogonal to the level surface for $z = f(x, y, z)$ at $x_0 = (x_0, y_0, z_0)$.   (page 808)

◆ The extreme values of the function $w = f(x)$ subject to the constraint $g(x) = 0$ occur at points x_0 where $\nabla f(x_0) = \lambda \nabla g(x_0)$, where λ is constant.   (page 810)

◆ The vector-valued function $\mathbf{F}(x, y) = M(x, y)\mathbf{i} + N(x, y)\mathbf{j}$ is said to have the **potential** f if $\mathbf{F}(x, y) = \nabla f(x, y)$.   (page 817)

◆ *Theorem:* If M, N and their first partial derivatives are continuous in a rectangle R, then   (page 821)

$$\mathbf{F}(x, y) = M(x, y)\mathbf{i} + N(x, y)\mathbf{j}$$

has a potential in R if and only if $\dfrac{\partial M}{\partial y} = \dfrac{\partial N}{\partial x}$ for all $(x, y) \in R$.

REVIEW EXERCISES—CHAPTER 17

1. Does $\displaystyle\lim_{(x,y)\to(0,0)} \frac{x^2 - y^2}{x^2 + y^2}$ exist? If so, find it.

2. Find $\displaystyle\lim_{(x,y)\to(0,0)} \frac{2x^2 y^3}{(x^2 + y^2)^2}$.

3. Sketch level curves for the function $f(x, y) = \dfrac{x}{x^2 + y^2}$ corresponding to levels $z = -2, -1, 1, 2, 4$.

4. Find $\dfrac{df}{dt}$ if $f(x, y) = x \operatorname{Tan}^{-1}\left(\dfrac{y}{x}\right)$, $x = 1 + t^2$, and $y = 1 - t$.

5. Find a vector normal to the curve $x^3 - 2xy^2 + y + 5 = 0$ at the point $(1, 2)$.

6. Find an equation for the plane tangent to the graph of the equation $x^3 + y^3 - 6xy + z = 0$ at the point $(2, 2, 8)$.

7. Find a vector normal to the surface $e^x \cos y - z = 4$ at the point $(0, \pi, -5)$.

8. What is the z-coordinate of the point $P = (1, 3, z)$ if P lies on the plane tangent to the ellipsoid $4x^2 + y^2 + 9z^2 = 17$ at the point $(1, 2, 1)$?

9. Find both first order partial derivatives for the function $f(x, y) = x \sin \sqrt{x^2 + y^2}$.

10. Find an equation for the plane tangent to the graph of $x^3 + y^3 + xz^2 + z^3 - 9 = 0$ at the point $(2, 1, -2)$.

11. Let $f(x, y, z) = xy^3 + x^2\sqrt{y^2 + z^2}$. Find the directional derivative of f in the direction of the vector $2\mathbf{i} + \mathbf{j} + 2\mathbf{k}$ at the point $(2, 3, 4)$.

12. Show that $f(x, y, z) = (ax + by + cz)^3$ satisfies the partial differential equation

$$x\frac{\partial f}{\partial x} + y\frac{\partial f}{\partial y} + z\frac{\partial f}{\partial z} = 3f.$$

13. Find $\dfrac{\partial f}{\partial r}$ and $\dfrac{\partial f}{\partial s}$ if $f(x, y) = 3x^3 + 2xy^2 - y^2$, $x = 2r + 5s$, and $y = r - 2s^2$.

14. Suppose that $w = f(x, y)$ has partial derivatives with respect to both variables. Show that the function $z = f(x - y, y - x)$ satisfies the partial differential equation

$$\frac{\partial z}{\partial x} + \frac{\partial z}{\partial y} = 0.$$

15. Find an equation for the line tangent to the graph of $x \cos \pi y + x^2 e^y = 6$ at the point $(2, 0)$.

16. Find the directional derivative $D_u f$ of the function $f(x, y) = y^2 \operatorname{Sin}^{-1} x + y e^x$ at the point $(0, 1)$ in the direction of the vector $w = i + \sqrt{e} j$.

17. Let $f(x, y)$ and $g(x, y)$ have partial derivatives with respect to both variables. Show that $\nabla(\alpha f + \beta g) = \alpha \nabla f + \beta \nabla g$ where α and β are constants.

18. Find the maximum value of the directional derivative for the function $w = xy^3 + z \cos y - \ln(x^2 + y)$ at the point $(1, 0, 4)$.

19. Find a vector pointing in the direction of most rapid decrease for the function $f(x, y, z) = xyz - z \operatorname{Tan}^{-1}(y/x)$ at the point $(1, 2, 3)$.

20. Find an equation for the plane tangent to the surface $z^2 x - 2zy + e^{xy} = 9$ at the point $(2, 0, 2)$.

21. Show that the function $f(x, t) = e^{x+ct}$ is a solution of the wave equation

$$\frac{\partial^2 f}{\partial t^2} = c^2 \frac{\partial^2 f}{\partial x^2}.$$

22. Find the maximum and minimum values of the function $f(x, y) = 4x^2 + xy + 2y^2$ on the square $S = \{(x, y) \mid -1 \le x \le 1, -1 \le y \le 1\}$.

23. Use the method of Lagrange multipliers to find the rectangular box of largest volume that can be inscribed in the ellipsoid $x^2 + \dfrac{y^2}{9} + \dfrac{z^2}{4} = 1$.

Find and classify all relative extrema.

24. $f(x, y) = 4y^2 - 2x^2$

25. $f(x, y) = 3x^2 + xy - 6y^2$

26. $f(x, y) = x^2 y + xy^2 + 4x + 4y$

27. $f(x, y) = e^{x^2 - 4xy}$

28. $f(x, y) = \ln(1 + x^2 + y^2)$

29. $f(x, y) = e^{1 + x^2 - y^2}$

30. $f(x, y) = 6x^2 - 2x - 3xy + y^2 + 5y + 5$

31. The directional derivative of $w = f(x, y)$ at P in the direction of the vector $u_1 = j$ is $D_{u_1} f(P) = 3$, and the directional derivative of f at P in the direction of $u_2 = 3i + 4j$ is $D_{u_2} f(P) = 3$ also. Find

   a. $\dfrac{\partial f}{\partial x}(P)$,   **b.** $\dfrac{\partial f}{\partial y}(P)$,   and   **c.** $\nabla f(P)$.

32. The directional derivative of f at (x_0, y_0) in the direction of the unit vector u is 6. What is
  a. $D_{2u} f(x_0, y_0)$?
  b. $D_{-u} f(x_0, y_0)$?

33. Find a function f for which $\nabla f(x, y) = 2xe^y i + (x^2 e^y - \sin y) j$.

34. Find a potential for the vector function $F(x, y) = (ye^x + e^y) i + (1 + e^x + xe^y) j$.

35. When three resistors r_1, r_2, and r_3 are connected in parallel, the net resistance R is determined by the equation

$$\frac{1}{R} = \frac{1}{r_1} + \frac{1}{r_2} + \frac{1}{r_3}.$$

Suppose $r_1 = 10$ ohms, $r_2 = 20$ ohms, and $r_3 = 25$ ohms. Approximate the change in the value of R that results from each of r_1, r_2, and r_3 being increased by 10%.

36. Is the function

$$f(x, y) = \begin{cases} \dfrac{6xy}{x^2 + y^2}, & (x, y) \ne (0, 0) \\ 0, & (x, y) = (0, 0) \end{cases}$$

differentiable at $(0, 0)$? Why or why not?

37. Show that the function

$$w = e^{x-y} + \cos(y - z) + \sqrt{z - x}$$

satisfies the partial differential equation

$$\frac{\partial w}{\partial x} + \frac{\partial w}{\partial y} + \frac{\partial w}{\partial z} = 0.$$

38. An airplane flying due north at a speed of 200 km/h and an altitude of 2 km passes directly over an automobile travelling due east along a straight highway at a speed of 100 km/h. At what rate are the plane and the automobile moving apart after 6 minutes?

39. Find the maximum and minimum values of the function

$$f(x, y) = 3x - 2y + 5$$

on or inside the ellipse $\dfrac{x^2}{4} + \dfrac{y^2}{9} = 1$.

40. A closed rectangular box is to contain 1000 cm³. If the material for the top and bottom costs 2¢/cm² and the material for the sides costs 3¢/cm², find the dimensions which minimize cost.

41. Let u and v be distinct unit vectors. Is $D_{u+v} f(x_0, y_0) = D_u f(x_0, y_0) + D_v f(x_0, y_0)$? Why or why not?

42. Find an equation for the plane through the point $(1, 2, 1)$ with positive x-, y-, and z-intercepts that bounds the solid of least volume in the first octant.

43. Find the point on the surface

$$x^2 + y^2 + z^2 = 16$$

where the function $f(x, y, z) = x + 2y - 3z + 1$ is a maximum.

44. Suppose that the level surfaces $f(x, y, z) = K$ and $g(x, y, z) = L$ intersect in a curve C. Let x_0 maximize $h(x, y, z)$ on C, and assume that f, g, and h have continuous partial derivatives. Show that $\nabla f(x_0)$, $\nabla g(x_0)$, and $\nabla h(x_0)$ lie in a common plane when based at x_0. Conclude that $\nabla f(x_0) = \lambda \nabla g(x_0) + \mu \nabla h(x_0)$ for some constants λ and μ.

45. Prove that if f is differentiable for all (x, y) in a closed and bounded disc D, then f is bounded on D. (This means that there exists a number M with $|f(x, y)| \le M$ for all $(x, y) \in D$.)

46. The function $f(x, y, z)$ is called homogeneous of degree n if $f(tx, ty, tz) = t^n f(x, y, z)$ for all x, y, z, and t. Show that if n is an integer and f has continuous partial derivatives, then

$$x\frac{\partial f}{\partial x} + y\frac{\partial f}{\partial y} + z\frac{\partial f}{\partial z} = nf$$

when f is homogeneous of degree n.

47. For the function $f(x, y) = \sqrt{x^2 + y^2}$, show that all directional derivatives exist at $(0, 0)$, but that neither partial derivative exists at $(0, 0)$.

48. Show that the spheres with equations $x^2 + (y - 1)^2 + z^2 = 2$ and $x^2 + (y + 1)^2 + z^2 = 2$ intersect in a circle. Then show that at each point on the circle of intersection the tangent planes to the two circles are orthogonal.

49. Use the method of Lagrange multipliers to find the point on the plane $x - 3y + 2z = 6$ nearest the origin.

50. Find a vector in the direction of most rapid increase of the function $f(x, y, z) = e^{x-z} \operatorname{Tan}^{-1}(y + z)$ at the point $(3, 1, 2)$.

51. Find an equation for the plane tangent to the surface $x^2 + 4y^2 + z^2 = 12$ at the point $(2, 1, 2)$.

52. Find $\dfrac{d^2 f}{dt^2}$ if $f(x, y) = e^{y^2 - x^2}$, $x(t) = 4 - t^2$, $y(t) = t \cos \pi t$.

53. Find the minimum value of the function $f(x, y) = 3y^2 - xy + x^2$ subject to the constraint that $x^2 + 3y^2 = 3$.

54. Find the maximum volume for a rectangular solid inscribed in the ellipsoid

$$\frac{x^2}{4} + \frac{y^2}{9} + \frac{z^2}{4} = 1.$$

55. Show that $u = f(xy)$ satisfies the partial differential equation

$$x\frac{\partial u}{\partial x} = y\frac{\partial u}{\partial y}.$$

55. Show that $u = f(x + 2y^2)$ satisfies the partial differential equation

$$\frac{\partial u}{\partial y} = 4y\frac{\partial u}{\partial x}.$$

57. Show that $u = f(t - x)$ is a solution of the partial differential equation

$$\frac{\partial^2 u}{\partial x^2} + 2\frac{\partial^2 u}{\partial x \partial t} + \frac{\partial^2 u}{\partial t^2} = 0.$$

58. Show that $u = f(x^2 + y^2)$ satisfies the partial differential equation

$$\frac{\partial u}{\partial y} = \frac{y}{x} \cdot \frac{\partial u}{\partial x}.$$

59. Show that if f is any differentiable function, the composite function $u(x, t) = f(bx - t)$ is a solution of the partial differential equation

$$\frac{\partial u}{\partial x} = -b\frac{\partial u}{\partial t}.$$

Chapter 18
Double and Triple Integrals

The goal of this chapter is to extend the theory of the Riemann integral to functions of two and three variables. You will find that most of what we do here is a straightforward generalization of the theory of integration for functions of a single variable.

18.1 THE DOUBLE INTEGRAL OVER A RECTANGLE

The original motivation for the definite integral was the problem of calculating the area of a region R bounded by the graph of the continuous nonnegative function $y = f(x)$ and the x-axis for $a \leq x \leq b$ (Figure 1.1). We did so by partitioning the

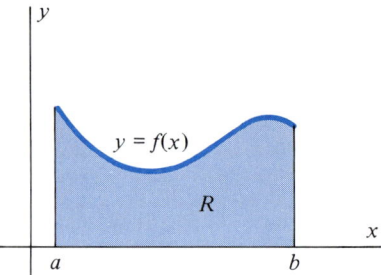

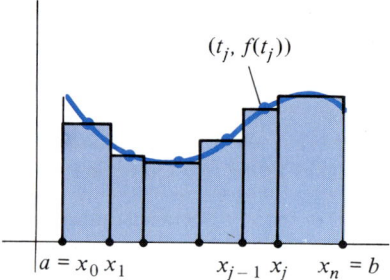

Figure 1.1 Region bounded by continuous nonnegative function $f(x)$.

Figure 1.2 Riemann sum approximates region R by rectangles.

interval $[a, b]$ into subintervals of length $\Delta x_j = x_j - x_{j-1}$. After choosing one number t_j arbitrarily in each interval, we formed the approximating Riemann sum

$$S_n = \sum_{j=1}^{n} f(t_j)\,\Delta x_j$$

representing the sum of the areas of the rectangles illustrated in Figure 1.2. We then proved that the limit of this Riemann sum, as $n \to \infty$ and as the norm of the partition $\|P_n\| \to 0$, is the desired area. This led to the definition of the definite integral

$$\int_a^b f(x)\,dx = \lim_{n \to \infty} \sum_{j=1}^{n} f(t_j)\,\Delta x_j.$$

For functions of two variables, the primary motivation leading to the corresponding integral (called the *double* integral) will be the calculation of volume rather than area. This is because a region Q in the domain of $z = f(x, y)$ and the graph of f over

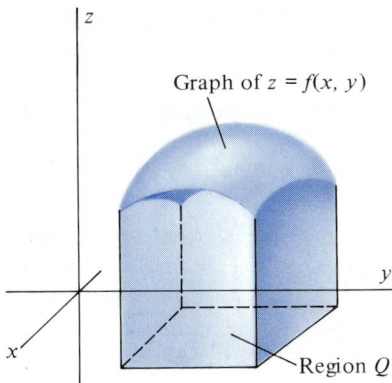

Figure 1.3 Graph of $z = f(x, y)$ over the region Q bounds a solid in xyz space.

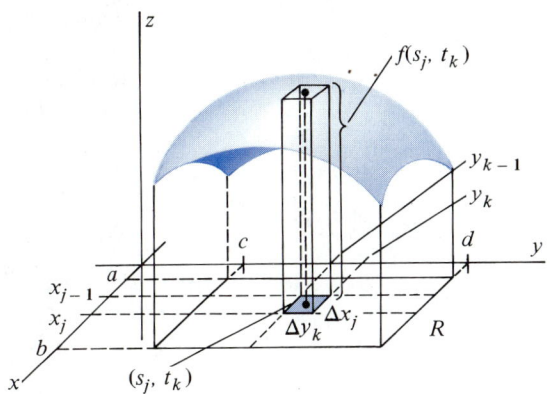

Figure 1.4 Volume of solid bounded by graph of f over R is approximated using rectangular prisms of volume $\Delta V_{jk} = f(s_j, t_k)\Delta x_j \Delta y_k$.

Q bound a solid in space (see Figure 1.3). We will approximate the volume of this solid by rectangular prisms (Figure 1.4), and the entire development will be very much analogous to the one-variable case.

Developing the Double Integral Over a Rectangle

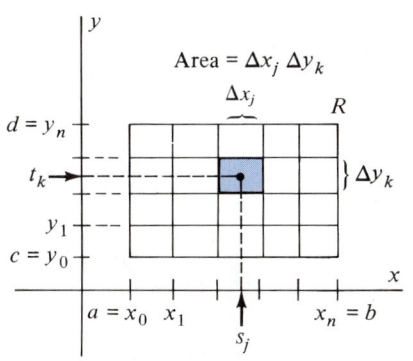

Figure 1.5 Rectangle R in xy-plane. Area of rectangle R_{jk} is $\Delta A = \Delta x_j \Delta y_k$.

We begin with the problem of calculating the volume V of a solid bounded above by the graph of the continuous nonnegative function $z = f(x, y)$, below by the rectangle $R = \{(x, y) \mid a \le x \le b, c \le y \le d\}$ in the xy-plane, and on four sides by the vertical planes $x = a$, $x = b$, $y = c$, and $y = d$ (see Figures 1.4 and 1.5).

Using the same terminology as in the one-variable case, we let $P_1 = \{a = x_0, x_1, x_2, \ldots, x_n = b\}$ be a partition of the interval $[a, b]$, and we let $P_2 = \{c = y_0, y_1, y_2, \ldots, y_m = d\}$ be a partition of the interval $[c, d]$. Also, we let

$$\Delta x_j = x_j - x_{j-1}, \qquad j = 1, 2, \ldots, n$$

and

$$\Delta y_k = y_k - y_{k-1}, \qquad k = 1, 2, \ldots, m.$$

As Figure 1.5 illustrates, these partitions determine a grid dividing the region R into rectangles R_{jk} of area $\Delta A_{jk} = \Delta x_j \Delta y_k$ for $j = 1, 2, \ldots, n$ and $k = 1, 2, \ldots, m$.

We refer to this grid as the *partition P* induced on R by the partitions P_1 and P_2, and we define the norm $\|P\|$ of this partition to be the larger of the norms $\|P_1\|$ and $\|P_2\|$ of the partitions P_1 and P_2. That is,

$$\|P\| = \max\{\|P_1\|, \|P_2\|\}$$
$$= \max\{\Delta x_1, \Delta x_2, \ldots, \Delta x_n, \Delta y_1, \Delta y_2, \ldots, \Delta y_m\}.$$

Our idea is now to approximate the volume of the region above the rectangle R_{jk} and below the graph of f with the volume of the rectangular prism with base of area $\Delta A_{jk} = \Delta x_j \Delta y_k$. For the height of this prism we use the function value $f(s_j, t_k)$, where the "test point" (s_j, t_k) is chosen arbitrarily in the rectangle R_{jk}. This leads to the approximation

$$S_{n,m} = \sum_{j=1}^{n} \sum_{k=1}^{m} f(s_j, t_k) \, \Delta A_{jk}, \qquad \Delta A_{jk} = \Delta x_j \Delta y_k, \tag{1}$$

which we refer to as a *Riemann double sum* for the function f on the rectangle R. Just as in the one-variable case, we now wish to ask whether the limit of such Riemann sums, as the sizes of the rectangles R_{jk} become very small, exists. Here is a formal definition of what we mean by the limit of the Riemann sums in equation (1).

DEFINITION 1

We say that the number S is the *limit* of the Riemann sum in equation (1) as $\|P\| \to 0$ (and, therefore, as $n \to \infty$ and $m \to \infty$), written

$$S = \lim_{\|P\| \to 0} \sum_{j=1}^{n} \sum_{k=1}^{m} f(s_j, t_k)\, \Delta A_{jk},$$

if, given any number $\epsilon > 0$, there exists a corresponding number $\delta > 0$ so that

$$\text{if } 0 < \|P\| < \delta, \text{ then } \left| S - \sum_{j=1}^{n} \sum_{k=1}^{m} f(s_j, t_k)\, \Delta A_{jk} \right| < \epsilon$$

regardless of how the test points (s_j, t_k) are chosen in R_{jk}.

Definition 1 simply says that the limit of the Riemann sums in equation (1) is the number S if these sums approach the number S as the number of rectangles R_{jk} in the partition P becomes infinite *and* as both dimensions (Δx_j and Δy_k) of each rectangle R_{jk} approach zero.

As in the one-variable case, the Riemann double sum in equation (1) will have a limit S when the function f is *continuous* on the rectangle R. We state this result without proof.

THEOREM 1

Let f be a function of two variables defined on the rectangle R. If f is continuous on R and P denotes a partition of R as described above, the limit

$$S = \lim_{\|P\| \to 0} \sum_{j=1}^{n} \sum_{k=1}^{m} f(s_j, t_k)\, \Delta A_{jk}$$

exists.

As before, we refer to the limit of the Riemann sum in equation (1), when it exists, as the Riemann integral of f on R. Because we are now in the two-variable case, we refer to this integral as a *double integral*.

DEFINITION 2

Let f be a continuous function of two variables on the rectangle $R = \{(x, y) \mid a \le x \le b,\ c \le y \le d\}$. The **double integral of f over the rectangle R** is the number

$$\iint_R f(x, y)\, dA = \lim_{\|P\| \to 0} \sum_{j=1}^{n} \sum_{k=1}^{m} f(s_j, t_k)\, \Delta A_{jk}, \qquad \Delta A_{jk} = \Delta x_j \Delta y_k$$

where P denotes a partition of the rectangle R and (s_j, t_k) is an arbitrary point in the rectangle $R_{jk} = \{(x, y) \mid x_{j-1} \le x \le x_j,\ y_{k-1} \le y \le y_k\}$.

In the symbol

$$\iint\limits_{R} f(x, y)\, dA,$$

we use two integral signs to indicate that it represents the result of a double limit process—the x-interval $[a, b]$ and the y-interval $[c, d]$ have both been partitioned into increasingly small subintervals. The subscript R denotes the rectangle over which the integral is evaluated. For now, the symbol dA (which may also be written $dx\, dy$) indicates that the Riemann sum has been obtained by partitioning R into rectangles of area $\Delta A_{jk} = \Delta x_j \Delta y_k$. As before, the function f is referred to as the **integrand.**

According to the development that led to the Riemann sum, we conclude that the volume V of the solid bounded above by the graph of the continuous nonnegative function f and below by the rectangle R in the xy-plane is

$$V = \iint\limits_{R} f(x, y)\, dA. \tag{2}$$

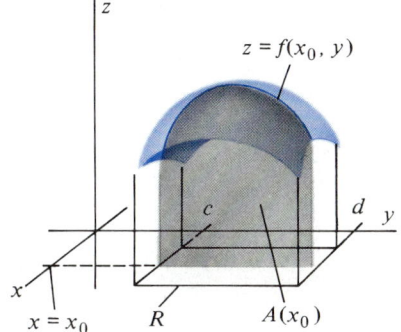

Figure 1.6 Area of cross section at x_0 is

$$A(x_0) = \int_c^d f(x_0, y)\, dy.$$

But how do we evaluate the integral in (2)? Ideally, there would be a result analogous to the Fundamental Theorem of Calculus that would enable us to do so.

Actually, the answer to this question is straightforward. Recall from Chapter 7 that the volume V is given by the definite integral

$$V = \int_a^b A(x)\, dx, \qquad a < b \tag{3}$$

when $A(x)$ is the area of the cross section taken perpendicular to the x-axis. Now if $x_0 \in [a, b]$ is fixed, the area of this cross section is just

$$A(x_0) = \int_c^d f(x_0, y)\, dy, \qquad c < d, \tag{4}$$

since the cross section is bounded above by the continuous function $g(y) = f(x_0, y)$ (see Figure 1.6).

Combining equations (3) and (4) we conclude that

$$V = \int_a^b \left\{ \int_c^d f(x, y)\, dy \right\} dx, \qquad a < b, \qquad c < d. \tag{5}$$

The meaning of equation (5) is that the volume V is calculated by first integrating f with respect to y (treating x as a constant) from c to d, and then integrating the resulting function of x from a to b. As Figure 1.7 illustrates, we could have begun by fixing y_0 and obtaining the area of the cross section perpendicular to the y-axis at $y = y_0$ as

$$A(y_0) = \int_a^b f(x, y_0)\, dx, \qquad a < b.$$

The resulting calculation for volume is

$$V = \int_c^d \left\{ \int_a^b f(x, y)\, dx \right\} dy, \qquad a < b, \quad c < d. \tag{6}$$

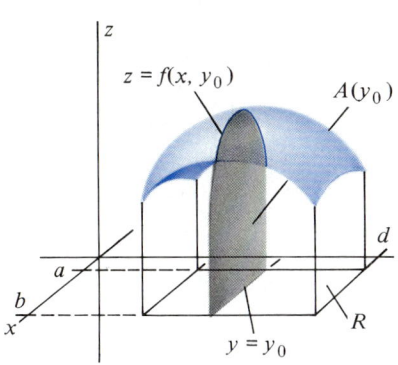

Figure 1.7 Area of cross section at y_0 is

$$A(y_0) = \int_a^b f(x, y_0)\, dx.$$

The integrals in (5) and (6) are called **iterated integrals,** because they involve the composition of two integrations, each with respect to a single variable. We usually omit the braces and simply write

$$\int_a^b \int_c^d f(x, y) \, dy \, dx = \int_a^b \left\{ \int_c^d f(x, y) \, dy \right\} dx \tag{7}$$

and

$$\int_c^d \int_a^b f(x, y) \, dx \, dy = \int_c^d \left\{ \int_a^b f(x, y) \, dx \right\} dy. \tag{8}$$

It is important to note that iterated integrals are interpreted "from inside out," meaning just what is specified by equations (7) and (8).

Example 1

Evaluate the iterated integral

$$\int_1^2 \int_1^3 (x^2 + 2xy) \, dy \, dx$$

and interpret the result geometrically.

Solution: The integral is evaluated as in equation (7):

$$\int_1^2 \int_1^3 (x^2 + 2xy) \, dy \, dx = \int_1^2 \left\{ \int_1^3 (x^2 + 2xy) \, dy \right\} dx$$

$$= \int_1^2 \left\{ x^2 y + xy^2 \Big]_{y=1}^{y=3} \right\} dx$$

$$= \int_1^2 [(3x^2 + 9x) - (x^2 + x)] \, dx$$

$$= \int_1^2 (2x^2 + 8x) \, dx$$

$$= \frac{2}{3}x^3 + 4x^2 \Big]_1^2$$

$$= \left[\frac{2}{3}(8) + 4(4) \right] - \left[\frac{2}{3}(1) + 4(1) \right]$$

$$= \frac{50}{3}.$$

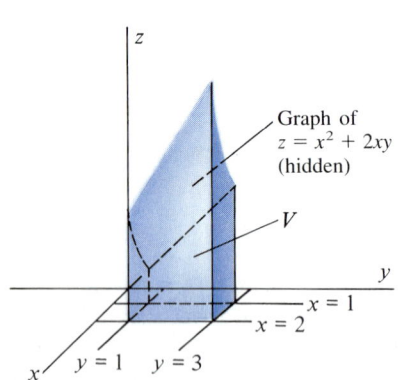

Figure 1.8 Volume of V is $\int_1^2 \int_1^3 (x^2 + 2xy) \, dy \, dx$. (See **Plate 13.**)

Since $f(x, y) = x^2 + 2xy$ is nonnegative for (x, y) in the rectangle $R = \{(x, y) \mid 1 \le x \le 2, 1 \le y \le 3\}$, the number $\frac{50}{3}$ is the volume of the solid bounded above by the graph of $f(x, y) = x^2 + 2xy$ and below by the xy-plane over the rectangle R (Figure 1.8). This volume may also be calculated using (8) as

$$\int_1^3 \int_1^2 (x^2 + 2xy) \, dx \, dy = \int_1^3 \left\{ \int_1^2 (x^2 + 2xy) \, dx \right\} dy$$

$$= \int_1^3 \left\{ \frac{x^3}{3} + x^2 y \Big]_{x=1}^{x=2} \right\} dy$$

$$= \int_1^3 \left[\left(\frac{8}{3} + 4y \right) - \left(\frac{1}{3} + y \right) \right] dy$$

$$= \int_1^3 \left(\frac{7}{3} + 3y \right) dy$$

$$= \frac{7y}{3} + \frac{3y^2}{2} \Big]_1^3$$

$$= \frac{50}{3}. \qquad \diamond$$

Example 2

Calculate the volume of the solid over the square $R = \{(x, y) \mid -1 \leq x \leq 1, -1 \leq y \leq 1\}$ bounded above by the graph of $f(x, y) = 8 - x^2 - y^2$ and below by the xy-plane.

Solution: The desired volume may be calculated as

$$\iint_R (8 - x^2 - y^2) \, dA = \int_{-1}^1 \left\{ \int_{-1}^1 (8 - x^2 - y^2) \, dx \right\} dy$$

$$= \int_{-1}^1 \left\{ 8x - \frac{1}{3}x^3 - xy^2 \right]_{x=-1}^{x=1} \right\} dy$$

$$= \int_{-1}^1 \left(\frac{46}{3} - 2y^2 \right) dy$$

$$= \frac{46y}{3} - \frac{2y^3}{3} \Big]_{y=-1}^{y=1}$$

$$= \frac{88}{3}.$$

(See Figure 1.9.) $\qquad \diamond$

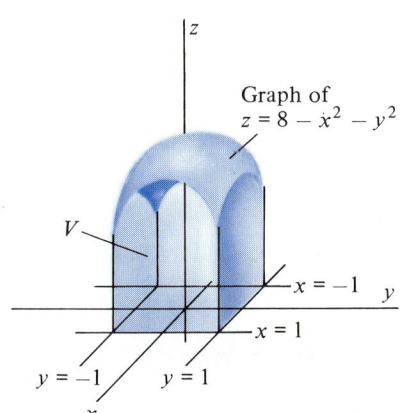

Figure 1.9 Volume of V is
$$\int_{-1}^1 \int_{-1}^1 (8 - x^2 - y^2) \, dx \, dy.$$

Finally, we note that the definition of the double integral in Definition 2 does not depend on the function f being nonnegative. The double integral $\iint_R f(x, y) \, dA$ is therefore defined for any continuous function f, regardless of its sign on the rectangle R.

When $f(x, y) \geq 0$ for all (x, y) in R, we may combine statement (2) with statements (5) through (8) to obtain the equations

$$\iint_R f(x, y) \, dA = \int_a^b \int_c^d f(x, y) \, dy \, dx = \int_a^b \left\{ \int_c^d f(x, y) \, dy \right\} dx \qquad (9)$$

and

$$\iint_R f(x, y) \, dA = \int_c^d \int_a^b f(x, y) \, dx \, dy = \int_c^d \left\{ \int_a^b f(x, y) \, dx \right\} dy. \qquad (10)$$

Equations (9) and (10) are true regardless of the sign of $f(x, y)$, although we shall omit the details involved in proving the more general case (see Exercise 36 for a sketch of a proof of this statement). However, you must keep in mind that the integrals in (9) and (10) correspond to the volume of the region between the rectangle R and the graph of $z = f(x, y)$ *only when* $f(x, y) \geq 0$ for all (x, y) in R, $a < b$, and $c < d$. We shall have more to say about properties of the double integral in the next section.

Exercise Set 18.1

In Exercises 1–14, evaluate the iterated integral.

1. $\int_0^1 \int_0^2 xy \, dx \, dy$

2. $\int_{-1}^1 \int_1^3 (y - x) \, dy \, dx$

3. $\int_1^3 \int_1^2 (4 + x - y) \, dx \, dy$

4. $\int_0^2 \int_{-1}^1 (x + y)^2 \, dx \, dy$

5. $\int_0^1 \int_0^{\pi/2} x \sin y \, dy \, dx$

6. $\int_0^1 \int_0^1 y e^{x-y^2} \, dy \, dx$

7. $\int_0^2 \int_1^e y^2 \ln x \, dx \, dy$

8. $\int_0^9 \int_1^4 \sqrt{\frac{y}{x}} \, dx \, dy$

9. $\int_0^1 \int_0^1 x \cosh y \, dx \, dy$

10. $\int_0^2 \int_0^2 xy e^{xy^2} \, dy \, dx$

11. $\int_0^1 \int_0^{\pi/2} xy \sin x \, dx \, dy$

12. $\int_0^\pi \int_0^1 \sinh x \cosh(\pi - y) \, dx \, dy$

13. $\int_0^1 \int_0^4 \frac{\sqrt{y}}{1 + x^2} \, dy \, dx$

14. $\int_0^{\pi/4} \int_0^{\pi/4} \tan x \sec^2 y \, dy \, dx$

In Exercises 15–21, evaluate the double integral over the rectangle R.

15. $\iint_R (x + y^2) \, dA$, $\quad R = \{(x, y) \mid 0 \leq x \leq 1, 0 \leq y \leq 1\}$

16. $\iint_R (x^2 + y^2) \, dA$, $\quad R = \{(x, y) \mid 0 \leq x \leq a, 0 \leq y \leq b\}$

17. $\iint_R x \cos y \, dA$, $\quad R = \{(x, y) \mid 0 \leq x \leq 4, 0 \leq y \leq \pi/2\}$

18. $\iint_R \frac{xy}{\sqrt{x^2 + y^2}} \, dA$, $\quad R = \{(x, y) \mid 1 \leq x \leq 2, 1 \leq y \leq 2\}$

19. $\iint_R xy \sec^2(xy^2) \, dA$, $\quad R = \{(x, y) \mid 0 \leq x \leq \pi/4, 0 \leq y \leq 1\}$

20. $\iint_R \frac{1}{\sqrt{x + y}} \, dA$, $\quad R = \{(x, y) \mid 4 \leq x \leq 8, 0 \leq y \leq 4\}$

21. $\iint_R y \cos(x + y) \, dA$, $\quad R = \{(x, y) \mid 0 \leq x \leq \pi/4, 0 \leq y \leq \pi/4\}$

In Exercises 22–27, use a double integral to calculate the volume of the solid bounded above by the graph of f and below by the rectangle R in the xy-plane.

22. $f(x, y) = 16 - 4x - 2y$, $\quad R = \{(x, y) \mid 0 \leq x \leq 2, 0 \leq y \leq 1\}$

23. $f(x, y) = x$, $\quad R = \{(x, y) \mid 0 \leq x \leq 2, 0 \leq y \leq 3\}$

24. $f(x, y) = 9 - x^2 - y^2$, $\quad R = \{(x, y) \mid -1 \leq x \leq 1, -1 \leq y \leq 1\}$

25. $f(x, y) = x \sin y$, $\quad R = \{(x, y) \mid 0 \leq x \leq 1, 0 \leq y \leq \pi/2\}$

26. $f(x, y) = \frac{\sqrt{x}}{1 + y^2}$, $\quad R = \{(x, y) \mid 0 \leq x \leq 4, 0 \leq y \leq 1\}$

27. $f(x, y) = \frac{e^{\sqrt{x}}}{\sqrt{xy}}$, $\quad R = \{(x, y) \mid 1 \leq x \leq 4, 1 \leq y \leq 9\}$

28. Show that if f is continuous on the rectangle R, then

$$\iint_R cf(x, y) \, dA = c \iint_R f(x, y) \, dA$$

for any constant c.

29. Show that if f and g are continuous on the rectangle R then

$$\iint_R [f(x, y) + g(x, y)] \, dA = \iint_R f(x, y) \, dA$$
$$+ \iint_R g(x, y) \, dA.$$

30. Show that if $f(x, y) \leq 0$ for all $(x, y) \in R = \{(x, y) \mid a \leq x \leq b, c \leq y \leq d\}$ and if V is the volume of the solid bounded by R and the graph of $z = f(x, y)$, then

$$V = -\iint_R f(x, y) \, dA.$$

31. Show that if f is continuous for $(x, y) \in R = \{(x, y) \mid a \leq x \leq b, c \leq y \leq d\}$ then

$$\int_a^b \int_c^d f(x, y) \, dy \, dx = -\int_b^a \int_c^d f(x, y) \, dy \, dx$$

$$= -\int_a^b \int_d^c f(x, y) \, dy \, dx$$

$$= \int_b^a \int_d^c f(x, y) \, dy \, dx.$$

32. *(Computer)* Program 8 in Appendix I is a BASIC computer program for calculating Riemann sums for double integrals. Use Program 8 to approximate the volume of the solid bounded above by the graph of $z = \sin \sqrt{xy}$ and below by the rectangle

$$R = \{(x, y) \mid 0 \leq x \leq 1, 0 \leq y \leq 1\}.$$

33. *(Computer)* Use Program 8 to approximate the double integral

$$\iint_R e^{x^2 + y^2} \, dA \qquad \text{where } R \text{ is the rectangle}$$

$$R = \{(x, y) \mid 0 \leq x \leq 1, 0 \leq y \leq 2\}.$$

34. *(Computer)* Use Program 8 to approximate the iterated integral

$$\int_0^1 \int_1^2 \sin(x^2 + y^2) \, dx \, dy.$$

35. *(Computer)* Use Program 8 to approximate the iterated integral

$$\int_1^3 \int_2^4 \sqrt{x^3 + y^3} \, dx \, dy.$$

36. The purpose of this exercise is to enable you to sketch a proof of the fact that

$$\iint_R f(x, y) \, dA = \int_a^b \int_c^d f(x, y) \, dy \, dx$$

where f is continuous and $R = \{(x, y) \mid a \leq x \leq b, c \leq y \leq d\}$.

a. Begin with the partitions $a = x_0 < x_1 < \cdots < x_n = b$

and $c = y_0 < y_1 < \cdots < y_m = d$ and the Riemann sum

$$S_{n, m} = \sum_{j=1}^n \sum_{k=1}^m f(s_j, t_k) \, \Delta y_k \Delta x_j$$

$$= \sum_{j=1}^n \left\{ \sum_{k=1}^m f(s_j, t_k) \, \Delta y_k \right\} \Delta x_j.$$

b. With j fixed, pick t_k in each interval $[y_{k-1}, y_k]$ to be the number for which

$$f(s_j, t_k) \, \Delta y_k = \int_{y_{k-1}}^{y_k} f(s_j, t) \, dt.$$

Why can this be done?

c. Conclude that

$$\sum_{k=1}^m f(s_j, t_k) \, \Delta y_k = \sum_{k=1}^m \int_{y_{k-1}}^{y_k} f(s_j, y) \, dy = \int_c^d f(s_j, y) \, dy.$$

Why is this true?

d. From (a) and (c) conclude that

$$S_{n, m} = \sum_{j=1}^n \left\{ \int_c^d f(s_j, y) \, dy \right\} \Delta x_j.$$

e. Note that $S_{n, m}$ in (d) has the form $S_{n, m} = \sum_{j=1}^n G(s_j) \, \Delta x_j$ where

$$G(s_j) = \int_c^d f(s_j, t) \, dt. \qquad \text{Conclude that}$$

$$\lim_{\|P\| \to 0} S_{n, m} = \lim_{n \to \infty} \sum_{j=1}^n G(s_j) \, \Delta x_j = \int_a^b G(x) \, dx$$

$$= \int_a^b \left\{ \int_c^d f(x, y) \, dy \right\} dx.$$

37. Show that if f is continuous on the interval $[a, b]$, and if g is continuous on the interval $[c, d]$, then

$$\iint_R f(x)g(y) \, dA = \left[\int_a^b f(x) \, dx \right] \cdot \left[\int_c^d g(y) \, dy \right]$$

where $R = \{(x, y) \mid a \leq x \leq b, c \leq y \leq d\}$.

18.2 DOUBLE INTEGRALS OVER MORE GENERAL REGIONS

Double integrals may be defined over regions in the plane more general than just rectangles. In particular, let Q be a region in the plane that is bounded (meaning it is contained in some finite rectangle) and whose boundary is a piecewise smooth curve that does not cross itself. A rectangle is certainly an example of such a region. But so is a pentagon, a triangle, an ellipse, or even a more general region such as that in Figure 2.1.

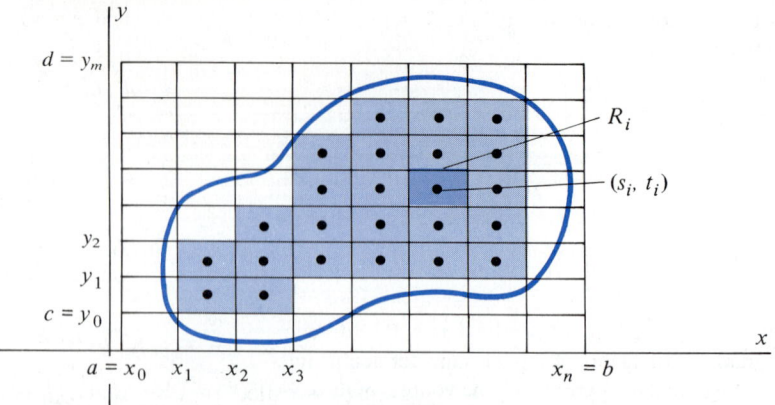

Figure 2.1 Partitioning a more general region Q.

Now suppose f is a continuous function of two variables defined on Q and that Q is entirely contained within a rectangle $R = \{(x, y) \mid a \le x \le b, c \le y \le d\}$. We construct a grid over R as before, which partitions R into smaller rectangles. We again denote the norm of this partition P by $\|P\|$.

Now some of these smaller rectangles will lie entirely within Q and some will not. Let $R_1, R_2, \ldots, R_m$ be a list of all such rectangles lying entirely within Q. For each such rectangle R_i, let (s_i, t_i) be a point in R_i. Finally, form the sum

$$S_m = \sum_{i=1}^{m} f(s_i, t_i)\, \Delta A_i, \qquad \Delta A_i = \text{area of } R_i. \tag{1}$$

Under the conditions stated on Q and on f, as $m \to \infty$ and as $\|P\| \to 0$ this Riemann sum will converge to the number we define to be the double integral of f over the region Q:

$$\iint\limits_{Q} f(x, y)\, dA = \lim_{\|P\| \to 0} \sum_{i=1}^{m} f(s_i, t_i)\, \Delta A_i. \tag{2}$$

Statement (2) is actually a definition of the double integral in this more general setting, although we prefer not to state it formally as such since this would require repeating nearly all the preceding statements. Before looking at ways in which such integrals may be evaluated, we wish to emphasize two points concerning this definition.

(i) Although only one summation sign appears in (2), this definition agrees with Definition 2 when Q is a rectangle. The difference is only that we have used a two-dimensional counting procedure (rows by columns) in stating Definition 2. In statement (2) we have simply listed all included rectangles $R_1, R_2, \ldots, R_m$ using a single index.

(ii) When $f(x, y) \ge 0$ for all $(x, y) \in Q$, the double integral (2) gives the volume V of the solid bounded by the graph of $z = f(x, y)$ and the region Q. The reason for this is the same as for the simpler case of the double integral over a rectangle.

Regular Regions

There are two special types of regions for which the double integral in (2) can be evaluated as an iterated integral: *x*-simple and *y*-simple regions.

DEFINITION 3

A region Q in the *xy*-plane is called **y-simple** if there exist continuous functions g_1 and g_2 so that

$$Q = \{(x, y) \mid a \leq x \leq b, \ g_1(x) \leq y \leq g_2(x)\}.$$

The region Q is called **x-simple** if there exist continuous functions h_1 and h_2 so that

$$Q = \{(x, y) \mid c \leq y \leq d, \ h_1(y) \leq x \leq h_2(y)\}.$$

The region Q is called **regular** if it is both *x*-simple and *y*-simple.

Figure 2.2 gives two illustrations of *y*-simple regions. The condition that $g_1(x) \leq y \leq g_2(x)$ for all $x \in [a, b]$ simply means that *the vertical line segment*

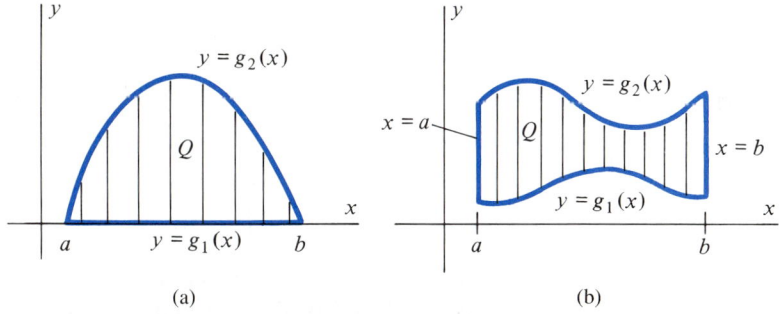

(a)   (b)

Figure 2.2   Two *y*-simple regions: Vertical lines intersect the boundary of Q at most twice.

connecting $(x, g_1(x))$ and $(x, g_2(x))$ lies entirely within the region Q. Another way to say this is that lines parallel to the *y*-axis intersect the boundary of Q at most twice.

Figure 2.3 shows two *x*-simple regions. This condition means that lines parallel to the *x*-axis intersect the boundary of Q at most twice. Note also that Figures 2.2(a)

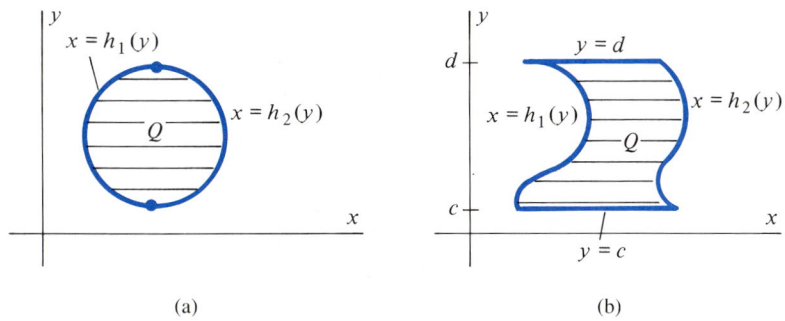

(a)   (b)

Figure 2.3   Two *x*-simple regions: Horizontal lines intersect the boundary of Q at most twice.

and 2.3(a) are *regular* (both x-simple and y-simple). However, Figure 2.2(b) is not x-simple, and Figure 2.3(b) is not y-simple.

The following theorem shows how double integrals over x-simple or y-simple regions may be evaluated as iterated integrals.

THEOREM 2

Let f be a continuous function of two variables on the region Q.

(i) If $Q = \{(x, y) \mid a \le x \le b, g_1(x) \le y \le g_2(x)\}$ is y-simple, then

$$\iint\limits_{Q} f(x, y) \, dA = \int_a^b \int_{g_1(x)}^{g_2(x)} f(x, y) \, dy \, dx = \int_a^b \left\{ \int_{g_1(x)}^{g_2(x)} f(x, y) \, dy \right\} dx. \qquad (3)$$

(ii) If $Q = \{(x, y) \mid c \le y \le d, h_1(y) \le x \le h_2(y)\}$ is x-simple, then

$$\iint\limits_{Q} f(x, y) \, dA = \int_c^d \int_{h_1(y)}^{h_2(y)} f(x, y) \, dx \, dy = \int_c^d \left\{ \int_{h_1(y)}^{h_2(y)} f(x, y) \, dx \right\} dy. \qquad (4)$$

Equation (3) says that if Q is y-simple, we first integrate f as a function of y alone, between the limits $g_1(x)$ and $g_2(x)$. The resulting function of x alone is then integrated over the constant limits $a \le x \le b$. Equation (4) says that if Q is x-simple, we first integrate f as a function of x alone, between the limits $h_1(y)$ and $h_2(y)$.

We shall not prove Theorem 2. However, Figure 2.4 embodies an argument for the validity of equation (4) when f is nonnegative on Q and Q is x-simple: For y-fixed, the integral $\int_{h_1(y)}^{h_2(y)} f(x, y) \, dx = A(y)$ gives the area of the cross section of the solid bounded by the graph of f and the region Q. The familiar formula for the volume V of a solid with known cross section then gives

$$V = \int_c^d A(y) \, dy = \int_c^d \int_{h_1(y)}^{h_2(y)} f(x, y) \, dx \, dy. \qquad (5)$$

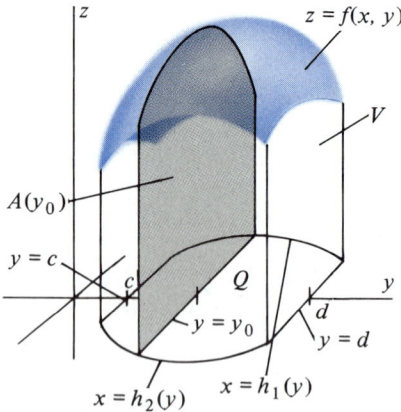

Figure 2.4 If Q is x-simple, area of cross section perpendicular to the y-axis at y_0 is

$$A(y_0) = \int_{h_1(y_0)}^{h_2(y_0)} f(x, y_0) \, dx.$$

Since the left side of equation (5) corresponds to the double integral $\iint\limits_{Q} f(x, y)\, dA$, we obtain equation (4).

In using Theorem 2 to evaluate double integrals, it is very important to first sketch the region Q to determine whether it is x-simple or y-simple. This determination will often dictate the order of integration in the iterated integral. (Of course, if Q is a regular region, you may proceed in either order. Just be careful that the limits of integration correspond to the chosen order of integration.)

Example 1

Evaluate the double integral $\iint\limits_{Q} (2xy + y^2)\, dA$ where Q is the triangle with vertices $(0, 0)$, $(1, 0)$, and $(1, 2)$.

Solution: The triangular region Q is sketched in Figure 2.5. Since Q is y-simple, we find the equation of the line segment joining $(0, 0)$ and $(1, 2)$. It is simply $y = 2x$. A vertical line segment through Q extends from $y = g_1(x) = 0$ to $y = g_2(x) = 2x$. That is, the limits of integration are determined by the inequalities

$$0 \leq x \leq 1 \qquad \text{and} \qquad 0 \leq y \leq 2x. \tag{6}$$

According to equation (3), the double integral is evaluated as

$$\iint\limits_{Q} (2xy + y^2)\, dA = \int_0^1 \int_0^{2x} (2xy + y^2)\, dy\, dx$$

$$= \int_0^1 \left\{ xy^2 + \frac{1}{3}y^3 \right]_{y=0}^{y=2x} \right\} dx$$

$$= \int_0^1 \frac{20x^3}{3}\, dx$$

$$= \frac{5}{3}x^4 \bigg]_0^1$$

$$= \frac{5}{3}.$$

Now the region Q is also x-simple. However, to use equation (4) we must first write the equation for the line segment joining $(0, 0)$ and $(1, 2)$ as a function of y. Solving $y = 2x$ for x gives $x = h_1(y) = y/2$ as the left boundary. As Figure 2.6

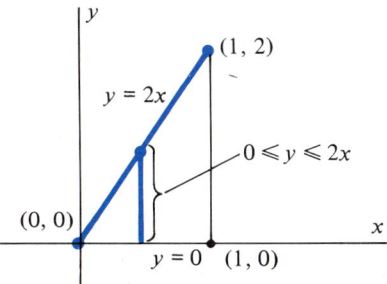

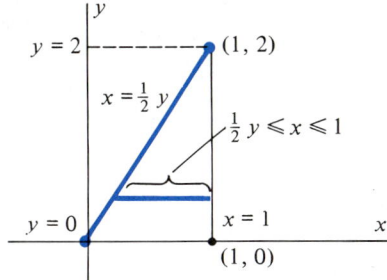

Figure 2.5 Treating the triangle Q as y-simple.

Figure 2.6 Treating the triangle Q as x-simple.

illustrates, $h_2(y) = 1$ is the right boundary of Q. The limits of integration are therefore

$$0 \leq y \leq 2 \quad \text{and} \quad \frac{y}{2} \leq x \leq 1. \tag{7}$$

Equation (4) then gives

$$
\begin{aligned}
\iint\limits_Q (2xy + y^2) \, dA &= \int_0^2 \int_{y/2}^1 (2xy + y^2) \, dx \, dy \\
&= \int_0^2 \left\{ x^2 y + xy^2 \Big]_{x=y/2}^{x=1} \right\} dy \\
&= \int_0^2 \left(-\frac{3}{4} y^3 + y^2 + y \right) dy \\
&= \left[-\frac{3}{16} y^4 + \frac{1}{3} y^3 + \frac{1}{2} y^2 \right]_0^2 \\
&= \frac{5}{3}. \quad\quad\quad \diamondsuit
\end{aligned}
$$

REMARK: Notice that the order of integration is entirely determined by our choice of whether we describe Q as y-simple, using inequalities (6), or as x-simple, using inequalities (7). Whichever integration is performed last must involve only constant limits of integration, since your answer cannot contain variables.

Example 2

Evaluate the double integral

$$\iint\limits_Q 4xy \, dA$$

where Q is the region bounded by the graphs of the equations $y = x + 1$ and $x = 1 - y^2$.

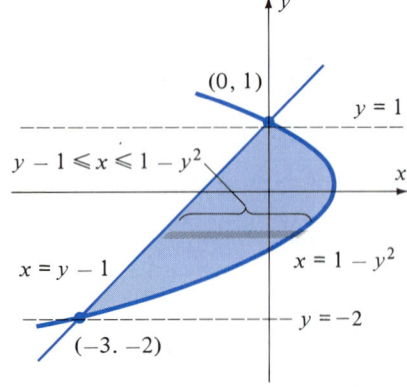

Figure 2.7 Region bounded by graphs of $y = x + 1$ and $x = 1 - y^2$ is x-simple.

Solution: The region Q is sketched in Figure 2.7. This region, like that in Example 1, is both x-simple and y-simple. However, we shall choose to work with Q as an x-simple region, since every horizontal line through Q originates on the line $y = x + 1$ and terminates on the parabola $x = 1 - y^2$. (Viewing Q as y-simple would be much messier—some vertical lines terminate on the line and others terminate on the parabola.)

Solving the equation $y = x + 1$ for x gives $x = h_1(y) = y - 1$ as the left boundary of Q. The right boundary is $h_2(y) = 1 - y^2$. The limits of integration are therefore

$$-2 \leq y \leq 1 \quad \text{and} \quad y - 1 \leq x \leq 1 - y^2.$$

The double integral is evaluated as

$$
\begin{aligned}
\iint\limits_Q 4xy \, dA &= \int_{-2}^1 \int_{y-1}^{1-y^2} 4xy \, dx \, dy \\
&= \int_{-2}^1 \left\{ 2x^2 y \Big]_{x=y-1}^{x=1-y^2} \right\} dy
\end{aligned}
$$

$$= \int_{-2}^{1} (2y^5 - 6y^3 + 4y^2)\, dy$$

$$= \frac{1}{3}y^6 - \frac{3}{2}y^4 + \frac{4}{3}y^3 \Big]_{-2}^{1}$$

$$= \frac{27}{2}.$$

◇

Example 3

Find the volume V of the solid bounded above by the graph of $f(x, y) = 2x^2y$ and below by the region Q in the xy plane bounded by the semicircle $y = \sqrt{1 - x^2}$ and the lines $y = 2$, $x = -1$, and $x = 1$.

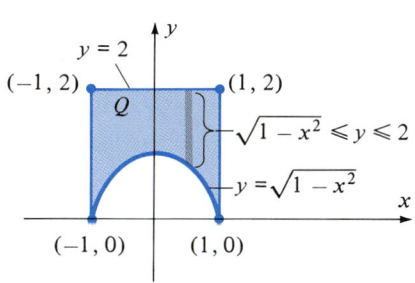

Figure 2.8 Region bounded by $y = \sqrt{1 - x^2}$ and $y = 2$ is y-simple, but not x-simple.

Solution: The region Q is sketched in Figure 2.8. Since $f(x, y) \geq 0$ for all $(x, y) \in Q$, V is given by the double integral $\iint\limits_{Q} 2x^2y\, dA$. As to the order of integration, we note that Q is y-simple, but not x-simple. We therefore integrate first with respect to y. The limits of integration are

$$-1 \leq x \leq 1 \qquad \text{and} \qquad \sqrt{1 - x^2} \leq y \leq 2.$$

The volume is

$$V = \iint\limits_{Q} 2x^2y\, dA = \int_{-1}^{1} \int_{\sqrt{1-x^2}}^{2} 2x^2y\, dy\, dx$$

$$= \int_{-1}^{1} \left\{ x^2 y^2 \Big]_{y=\sqrt{1-x^2}}^{y=2} \right\} dx$$

$$= \int_{-1}^{1} [4x^2 - x^2(1 - x^2)]\, dx$$

$$= \int_{-1}^{1} (3x^2 + x^4)\, dx$$

$$= x^3 + \frac{1}{5}x^5 \Big]_{-1}^{1}$$

$$= \frac{12}{5}.$$

◇

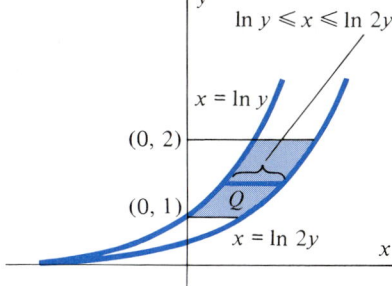

Figure 2.9 Region Q in Example 4.

Example 4

Find the volume V of the solid bounded by the graph of $f(x, y) = e^{x+y^2}$ and the region

$$Q = \{(x, y) \mid \ln y \leq x \leq \ln 2y,\ 1 \leq y \leq 2\}$$

in the xy plane.

Solution: The region Q is sketched in Figure 2.9. Although Q is both x-simple and y-simple, this time the order of integration is determined by the integrand $f(x, y) = e^{x+y^2}$. We must attempt to integrate first with respect to x, since there is no hope of finding an antiderivative with respect to y. Treating Q as an x-simple region, we use the limits

$$\ln y \le x \le \ln 2y \qquad \text{and} \qquad 1 \le y \le 2$$

and obtain

$$V = \iint_Q e^{x+y^2} \, dA = \int_1^2 \int_{\ln y}^{\ln 2y} e^{x+y^2} \, dx \, dy$$

$$= \int_1^2 \left\{ e^{x+y^2} \Big]_{x=\ln y}^{x=\ln 2y} \right\} dy$$

$$= \int_1^2 [(e^{\ln 2y}) e^{y^2} - (e^{\ln y}) e^{y^2}] \, dy$$

$$= \int_1^2 (2y e^{y^2} - y e^{y^2}) \, dy$$

$$= \int_1^2 y e^{y^2} \, dy$$

$$= \frac{1}{2} e^{y^2} \Big]_1^2$$

$$= \frac{e}{2}(e^3 - 1)$$

$$\approx 25.94. \qquad \diamondsuit$$

Finding Areas by Double Integration

The fact that volumes of solids may be calculated by use of double integrals means that areas of regions in the plane may also be calculated in this way. As Figure 2.10

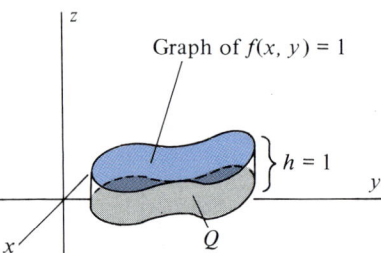

Figure 2.10 Area of $Q = \displaystyle\iint_Q 1 \, dA$.

illustrates, the area of Q is the same as the volume of the cylinder with base Q and uniform height $h = 1$. To calculate the area of Q, we simply evaluate the integral

$$\text{Area of } Q = \iint_Q 1 \, dA. \qquad (8)$$

Example 5

Use a double integral to find the area of the region Q lying inside the circle $x^2 + y^2 = 4$ and above the line $y = 1$.

Solution: The points on the circle $x^2 + y^2 = 4$ with y-coordinate equal to 1 have x-coordinates $x = \pm\sqrt{4 - 1} = \pm\sqrt{3}$. The region may therefore be described by the inequalities

$$-\sqrt{3} \le x \le \sqrt{3}, \qquad 1 \le y \le \sqrt{4-x^2}.$$

(See Figure 2.11.) According to (8),

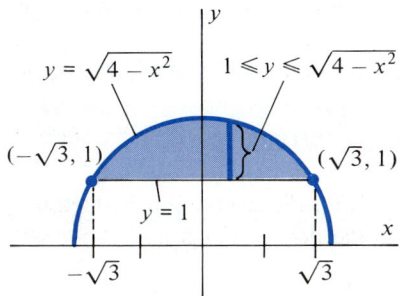

Figure 2.11 Region inside circle $x^2 + y^2 = 4$ and above $y = 1$.

$$\text{Area of } Q = \iint_Q 1 \; dA$$

$$= \int_{-\sqrt{3}}^{\sqrt{3}} \int_1^{\sqrt{4-x^2}} 1 \cdot dy \; dx$$

$$= \int_{-\sqrt{3}}^{\sqrt{3}} \left\{ \Big[y \Big]_{y=1}^{y=\sqrt{4-x^2}} \right\} dx$$

$$= \int_{-\sqrt{3}}^{\sqrt{3}} (\sqrt{4-x^2} - 1) \; dx$$

$$= 2\int_0^{\sqrt{3}} \sqrt{4-x^2} \; dx - 2\sqrt{3}.$$

A straightforward application of the method of trigonometric substitutions shows that

$$\int \sqrt{4-x^2} \; dx = 2 \; \text{Sin}^{-1} \left(\frac{x}{2} \right) + \frac{x\sqrt{4-x^2}}{2} + C.$$

The desired area is therefore

$$2\left[2 \; \text{Sin}^{-1} \left(\frac{\sqrt{3}}{2} \right) + \frac{\sqrt{3}}{2} \right] - 2\sqrt{3} \approx 2.457. \qquad \diamond$$

Interchanging the Order of Integration

There are occasions on which we need to interchange the order of integration in an iterated integral because an antiderivative cannot be found with respect to the "inside" variable. For example, in the iterated integral

$$\int_0^1 \int_{y^2}^1 ye^{x^2} \; dx \; dy \qquad\qquad (9)$$

we cannot find a (formal) antiderivative for the integrand ye^{x^2} with respect to x.

However, such integrals can often be evaluated by reversing the order of integration. In particular, the antiderivative of ye^{x^2} with respect to y is easily seen to be the function $\frac{1}{2}y^2e^{x^2}$. But if we are going to reverse the order in which the integrations are performed, we must also determine correct limits of integration corresponding to this new order of integration. We do so as follows:

To reverse the order of integration in the iterated integral

$$\int_c^d \int_{h_1(y)}^{h_2(y)} f(x, y)\, dx\, dy:$$

(i) Identify the region Q for which the iterated integral can be written as the double integral

$$\int_c^d \int_{h_1(y)}^{h_2(y)} f(x, y)\, dx\, dy = \iint_Q f(x, y)\, dA.$$

(ii) Find constants a and b, and continuous functions g_1 and g_2, so that the region Q can be expressed as

$$Q = \{(x, y) \mid a \le x \le b,\ g_1(x) \le y \le g_2(x)\}.$$

(iii) Rewrite the iterated integral as

$$\int_c^d \int_{h_1(y)}^{h_2(y)} f(x, y)\, dx\, dy = \iint_Q f(x, y)\, dA$$

$$= \int_a^b \int_{g_1(x)}^{g_2(x)} f(x, y)\, dy\, dx.$$

Obviously, this procedure can be applied only when Q is a regular region. Also, though the procedure is stated for reversing the order of integration from $dx\, dy$ to $dy\, dx$, the procedure for changing from order $dy\, dx$ to order $dx\, dy$ is analogous. Finally, there is no guarantee that the resulting integral can be more easily evaluated than the original one.

Example 6

Use the procedure for reversing order of integration to evaluate the iterated integral

$$\int_0^1 \int_{y^2}^1 ye^{x^2}\, dx\, dy.$$

Solution: From the given limits of integration, the region Q is described by the inequalities

$$y^2 \le x \le 1, \qquad 0 \le y \le 1.$$

That is, Q is the region bounded between the graphs of $x = y^2$ and $x = 1$ for $0 \le y \le 1$. From Figure 2.12 (which must always be sketched when applying this method), we can see that Q is regular and can also be described by the inequalities

$$0 \le x \le 1, \qquad 0 \le y \le \sqrt{x} \qquad \text{(Figure 2.13)}.$$

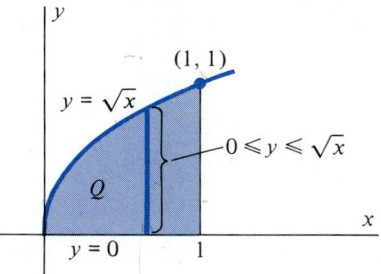

Figure 2.12 $Q = \{(x, y) \mid y^2 \leq x \leq 1,\ 0 \leq y \leq 1\}$.

Figure 2.13 $Q = \{(x, y) \mid 0 \leq x \leq 1,\ 0 \leq y \leq \sqrt{x}\}$.

Beginning with the given integral, we therefore reverse the order of integration as follows:

$$\int_0^1 \int_{y^2}^1 y e^{x^2}\, dx\, dy = \iint_Q y e^{x^2}\, dA = \int_0^1 \int_0^{\sqrt{x}} y e^{x^2}\, dy\, dx$$

$$= \int_0^1 \left\{ \frac{1}{2} y^2 e^{x^2} \right]_{y=0}^{y=\sqrt{x}} \right\}\, dx$$

$$= \int_0^1 \frac{1}{2} x e^{x^2}\, dx$$

$$= \frac{1}{4} e^{x^2} \Big]_0^1$$

$$= \frac{1}{4}(e - 1)$$

$$\approx 0.43. \qquad \diamond$$

REMARK: It is important to note that we cannot simply interchange limits of integration when we interchange the order of integration. There is no alternative to sketching the region Q and working out the new limits of integration from knowledge of the boundary of Q.

Example 7

Evaluate the iterated integral

$$\int_0^1 \int_0^{\sqrt{1-x}} xy^2\, dy\, dx$$

$0 \leq x \leq 1$
$0 \leq y \leq \sqrt{1-x}$

by first reversing the order of integration.

Solution: The given limits of integration are

$$0 \leq x \leq 1, \qquad 0 \leq y \leq \sqrt{1-x},$$

which describe the region Q bounded above by the graph of $y = \sqrt{1-x}$ and below by the x-axis for $0 \leq x \leq 1$ (see Figure 2.14). By solving the equation $y = \sqrt{1-x}$ for x, we find that this region may also be described by the inequalities

$$0 \leq x \leq 1 - y^2, \qquad 0 \leq y \leq 1.$$

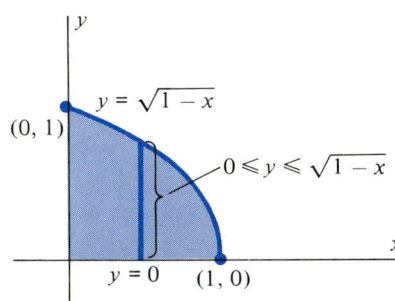

Figure 2.14 $Q = \{(x, y) \mid 0 \leq x \leq 1,\ 0 \leq y \leq \sqrt{1-x}\}$.

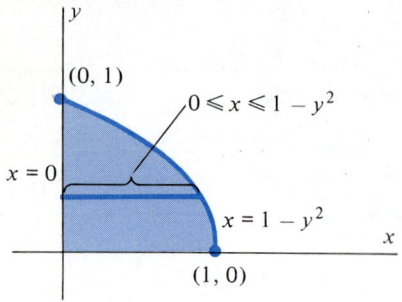

Figure 2.15 $Q = \{(x, y) \mid 0 \le x \le 1 - y^2, 0 \le y \le 1\}$.

(See Figure 2.15.) We may therefore evaluate the iterated integral as

$$\int_0^1 \int_0^{\sqrt{1-x}} xy^2 \, dy \, dx = \iint_Q xy^2 \, dA = \int_0^1 \int_0^{1-y^2} xy^2 \, dx \, dy$$

$$= \int_0^1 \left\{ \frac{1}{2} x^2 y^2 \right]_{x=0}^{x=1-y^2} \right\} dy$$

$$= \int_0^1 \frac{1}{2} (1 - y^2)^2 y^2 \, dy$$

$$= \int_0^1 \left(\frac{1}{2} y^2 - y^4 + \frac{1}{2} y^6 \right) dy$$

$$= \frac{1}{6} y^3 - \frac{1}{5} y^5 + \frac{1}{14} y^7 \bigg]_0^1$$

$$= \frac{4}{105} . \qquad \diamond$$

Properties of Double Integrals

As defined in this section, the double integral satisfies the following properties:

(i) $\displaystyle \iint_Q [f(x, y) + g(x, y)] \, dA = \iint_Q f(x, y) \, dA + \iint_Q g(x, y) \, dA,$

(ii) $\displaystyle \iint_Q cf(x, y) \, dA = c \iint_Q f(x, y) \, dA, \qquad c = \text{constant},$

(iii) $\displaystyle \iint_{Q_1 \cup Q_2} f(x, y) \, dA = \iint_{Q_1} f(x, y) \, dA + \iint_{Q_2} f(x, y) \, dA.$

In (iii) we mean that $Q = Q_1 \cup Q_2$ is the union of two nonoverlapping regions Q_1 and Q_2, each satisfying the properties required of Q in this section. Property (iii) extends to unions of finitely many nonoverlapping regions $Q_1, Q_2, \ldots, Q_n$ of this type.

We shall not prove properties (i) through (iii). The proofs are analogous to those in the one variable case.

Exercise Set 18.2

In Exercises 1–12, sketch the region Q determined by the limits of integration and evaluate the iterated integral.

1. $\displaystyle \int_0^1 \int_0^x (y^2 - x^2) \, dy \, dx$

2. $\displaystyle \int_0^1 \int_0^{1-x} 2xy \, dy \, dx$

3. $\displaystyle \int_{-1}^0 \int_{-1}^{y+1} (xy - x) \, dx \, dy$

4. $\displaystyle \int_0^1 \int_0^y xye^{x^2} \, dx \, dy$

5. $\displaystyle \int_{-1}^1 \int_0^{1-x^2} xy \, dy \, dx$

6. $\displaystyle \int_0^1 \int_{-x}^{\sqrt{x}} \frac{y}{1+x} \, dy \, dx$

7. $\displaystyle \int_0^1 \int_{x^3}^{x^2} x \, dy \, dx$

8. $\displaystyle \int_0^{\sqrt{\pi/2}} \int_0^{\sqrt{y}} x \sin y^2 \, dx \, dy$

9. $\displaystyle \int_0^{\pi/2} \int_0^{\sin x} 2y \cos x \, dy \, dx$

10. $\displaystyle \int_1^e \int_0^{\ln y} e^{x+y} \, dx \, dy$

11. $\displaystyle \int_0^1 \int_0^{y^2} e^{x/y} \, dx \, dy$

12. $\displaystyle \int_0^{\pi/2} \int_0^{\cos y} x \sin y \, dx \, dy$

In Exercises 13–22, evaluate the double integral.

13. $\displaystyle \iint_Q \sqrt{x} \, y \, dA, \qquad Q = \{(x, y) \mid 0 \le x \le y^2, 0 \le y \le 1\}$

14. $\displaystyle \iint_Q x \cos \pi y \, dA, \qquad Q = \{(x, y) \mid 0 \le x \le 1, 0 \le y \le x\}$

15. $\displaystyle\iint_Q (x + 2)\sqrt{1 + e^y}\, dA,$

$Q = \{(x, y) \mid 0 \le x \le e^{y/2},\ 0 \le y \le 1\}$

16. $\displaystyle\iint_Q \frac{x^2}{1 + y}\, dA,\quad Q = \{(x, y) \mid 0 \le x \le 1,\ 0 \le y \le e^x - 1\}$

17. $\displaystyle\iint_Q xy\, dA,\qquad Q = \{(x, y) \mid y \le x \le \sqrt{y},\ 0 \le y \le 1\}$

18. $\displaystyle\iint_Q (x^2 + y^2)\, dA,$

$Q = \{(x, y) \mid -2 \le x \le 2,\ -3 \le y \le 1 - x^2\}$

19. $\displaystyle\iint_Q y\, dA,\qquad Q = \{(x, y) \mid -1 \le x \le 1,\ e^x \le y \le e\}$

20. $\displaystyle\iint_Q ye^x\, dA,\ Q = \{(x, y) \mid -\ln y \le x \le \ln y,\ 1 \le y \le 2\}$

21. $\displaystyle\iint_Q x\, dA,\qquad Q = \{(x, y) \mid 0 \le x \le \sqrt{1 - y^2},\ 0 \le y \le 1\}$

22. $\displaystyle\iint_Q xy\, dA,\qquad Q = \{(x, y) \mid 0 \le x^2 + y^2 \le 1\}$

In Exercises 23–28, sketch the region Q determined by the limits of integration, interchange the order of integration, and evaluate the given integral, where possible.

23. $\displaystyle\int_{-1}^{1}\int_0^{x+1} (x + y)\, dy\, dx$

24. $\displaystyle\int_0^1 \int_{x^2}^1 xe^{y^2}\, dy\, dx$

25. $\displaystyle\int_0^1 \int_0^y xy^2\, dx\, dy$

26. $\displaystyle\int_{-2}^0 \int_{x^2}^4 xe^{y^2}\, dy\, dx$

27. $\displaystyle\int_1^e \int_0^{\ln x} f(x, y)\, dy\, dx$

28. $\displaystyle\int_0^1 \int_0^{\operatorname{Sin}^{-1} x} f(x, y)\, dy\, dx$

In Exercises 29–33, use a double integral to find the area of Q.

29. Q is the region bounded by the graphs of $y = 4 - x^2$ and the line $y = x + 2$.

30. Q is the region bounded by the graphs of $y = x^2$ and $y = x^3$.

31. Q is the region bounded by the graphs of $y = \sin x$ and $y = \cos x$ for $0 \le x \le \pi/4$.

32. Q is the region bounded by the graphs of $y = x^3$ and $y = 4x$.

33. Q is the region bounded by the graphs of $y = \sqrt{x}$ and $y = x^2$.

34. Use a double integral to find the volume of the tetrahedron with vertices $(1, 0, 0)$, $(0, 1, 0)$, $(0, 0, 0)$, and $(0, 0, 1)$.

35. Find the area of the ellipse $\dfrac{x^2}{4} + \dfrac{y^2}{9} = 1$ by use of a double integral.

36. Use a double integral to establish the formula for the volume of a right circular cone of radius r and height h.

37. Find the volume of the solid bounded by the paraboloid $z = \dfrac{1}{4}(x^2 + y^2)$ and the paraboloid $z = 5 - x^2 - y^2$.

38. Use a double integral to find the volume of the sphere $x^2 + y^2 + z^2 = 4$ lying above the plane $z = 1$.

39. Find the volume of the portion of the solid bounded by the plane $z = x$, the cylinder $x^2 + y^2 = 4$, and the plane $z = 0$.

40. Find the volume of the solid bounded by the coordinate planes and the plane $6x + 3y + 2z = 6$.

41. Sketch the region in the first octant common to the two cylinders $z^2 = 1 - x^2$ and $z^2 = 1 - y^2$. Find its volume.

18.3 DOUBLE INTEGRALS IN POLAR COORDINATES

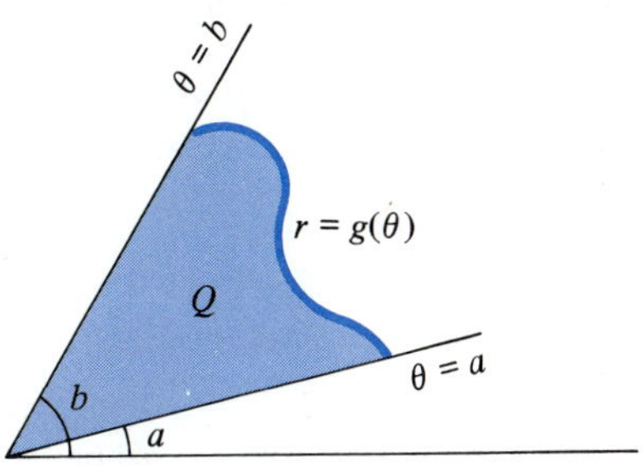

Figure 3.1 The region Q, bounded by the graph of $r = g(\theta)$ and the rays $\theta = a$ and $\theta = b$.

So far, we have defined double integrals only in the case where Q is a region in the xy- (Cartesian) coordinate plane and f is a function written in terms of Cartesian coordinates. We can also define the double integral when the region Q and the function f are described in polar coordinates.

Figure 3.1 shows a region Q bounded by the rays $\theta = a$ and $\theta = b$ and the graph of the function $r = g(\theta)$.

As Figure 3.2 illustrates, the circular arcs $r = r_j$ and the rays $\theta = \theta_k$ partition the region Q into wedge-shaped subregions R_{jk}. To determine the area of R_{jk} we use two facts (see Figure 3.3):

(i) The region R_{jk} lies between the concentric circles $r = r_{j-1}$ and $r = r_j$. The area of the entire annulus lying between these circles is therefore

$$\text{Area of annulus} = \pi r_j^2 - \pi r_{j-1}^2.$$

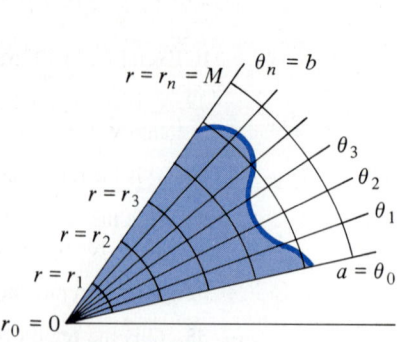

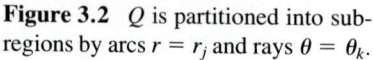

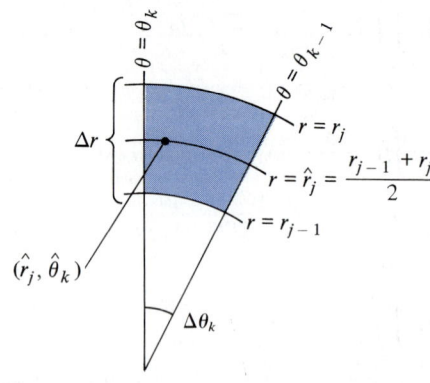

Figure 3.2 Q is partitioned into sub-regions by arcs $r = r_j$ and rays $\theta = \theta_k$.

Figure 3.3 Area of region R_{jk} is $\Delta A_{jk} = \hat{r}_j \, \Delta r_j \, \Delta \theta_k$.

(ii) The angle $\Delta \theta_k$, measured in radians, determines the proportion of the entire annulus lying in the region R_{jk}. That is,

$$\text{Proportion of annulus in } R_{jk} = \frac{\Delta \theta_k}{2\pi}.$$

From statements (i) and (ii), you can see that the area of the region R_{jk} is

$$\Delta A_{jk} = (\pi r_j^2 - \pi r_{j-1}^2)\left(\frac{\Delta \theta_k}{2\pi}\right) \tag{1}$$

$$= \frac{1}{2}(r_j^2 - r_{j-1}^2) \, \Delta \theta_k.$$

Since

$$r_j^2 - r_{j-1}^2 = (r_j + r_{j-1})(r_j - r_{j-1})$$

$$= 2\left(\frac{r_j + r_{j-1}}{2}\right)(r_j - r_{j-1}),$$

we may write

$$r_j^2 - r_{j-1}^2 = 2\hat{r}_j \, \Delta r_j \tag{2}$$

where $\hat{r}_j = \dfrac{r_j + r_{j-1}}{2}$ and $\Delta r_j = r_j - r_{j-1}$. It is important to note that the number $\hat{r}_j$ is simply the average of the radii r_{j-1} and r_j (see Figure 3.3). Combining (1) and (2), we may now write the area of the region R_{jk} as

$$\Delta A_{jk} = \hat{r}_j \, \Delta r_j \, \Delta \theta_k. \tag{3}$$

Now let f be a continuous function defined on the region Q, and let $\hat{\theta}_k$ be any number with $\theta_{k-1} \le \hat{\theta}_k \le \theta_k$. Then the number $f(\hat{r}_j, \hat{\theta}_k)$ is the value of the function f at the point $(\hat{r}_j, \hat{\theta}_k)$ in the region R_{jk}. If $f(r, \theta) \ge 0$ for all $(r, \theta) \in Q$, then the product $f(\hat{r}_j, \hat{\theta}_k) \, \Delta A_{jk}$ approximates the volume of the solid bounded by the graph of f over R_{jk}, and the sum

$$\sum_{j=1}^{n} \sum_{k=1}^{m} f(\hat{r}_j, \hat{\theta}_k) \, \Delta A_{jk} = \sum_{j=1}^{n} \sum_{k=1}^{m} f(\hat{r}_j, \hat{\theta}_k)\hat{r}_j \, \Delta r_j \, \Delta \theta_k \tag{4}$$

approximates the volume of the solid bounded by the graph of f over the entire region Q. (In writing the sums in line (4) we mean that the sums are taken only over those regions R_{jk} that lie entirely within Q.) We therefore define the double integral of $f(r, \theta)$ over Q as

$$\iint\limits_{Q} f(r, \theta) \, dA = \lim_{\|P\| \to 0} \sum_{j=1}^{n} \sum_{k=1}^{m} f(\hat{r}_j, \hat{\theta}_k)\hat{r}_j \, \Delta r_j \, \Delta \theta_k \tag{5}$$

where $\|P\|$ is an appropriately defined norm for the partition P which we have described above.

The integral in (5) is defined whenever f is continuous on Q, regardless of the sign of $f(r, \theta)$. However, as for the double integral in Cartesian coordinates, if $f(r, \theta) \geq 0$ for all $(r, \theta) \in Q$, we have

$$\iint\limits_{Q} f(r, \theta) \, dA = \text{volume of solid bounded above by the graph of} \tag{6}$$
$$f \text{ and below by } Q, \text{ and}$$

$$\iint\limits_{Q} 1 \cdot dA = \text{area of the region } Q. \tag{7}$$

Of course, the integral defined by equation (5) is of little value unless we know how to evaluate the limit of the approximating sums. The following theorem, analogous to Theorem 2, shows how this may be done. The idea behind its proof is contained in equation (4).

THEOREM 3

Let Q be the region bounded by the graph of the continuous function $r = g(\theta)$ and the rays $\theta = a$ and $\theta = b$, as in Figure 3.1. Let f be a continuous function defined on the region Q. Then

$$\iint\limits_{Q} f(r, \theta) \, dA = \int_{a}^{b} \int_{0}^{g(\theta)} f(r, \theta)r \, dr \, d\theta. \tag{8}$$

Theorem 3 says that the double integral in (5) may be evaluated as an iterated integral, where we integrate first with respect to r alone, for $0 \leq r \leq g(\theta)$, and then integrate the resulting function of θ over the limits $a \leq \theta \leq b$. However, *it is important to note that the integrand in the iterated integral is the product $f(r, \theta)r$, not just $f(r, \theta)$*. The reason for this can be seen from equation (3): The area of the region R_{jk} is $\Delta A_{jk} = \hat{r}_j \, \Delta r_j \, \Delta \theta_k$. In the Cartesian-coordinate case the rectangles R_{jk} used in approximating the area of Q had area $\Delta A_{jk} = \Delta x_j \, \Delta y_k$. The extra factor r appearing in the iterated polar double integral appears because of this formula for the area of the approximating region R_{jk}.

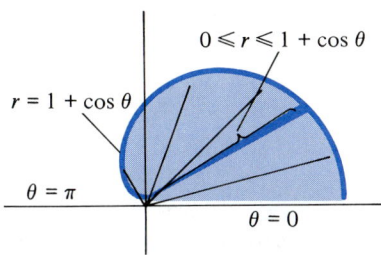

Figure 3.4 Region bounded by graph of $r = 1 + \cos \theta$ for $0 \leq \theta \leq \pi$.

Example 1

Evaluate the double integral $\iint\limits_{Q} r \sin \theta \, dA$ where Q is the region inside the upper half of the cardioid $r = 1 + \cos \theta$ (see Figure 3.4).

Solution: Here $f(r, \theta) = r \sin \theta$, and the region Q may be described by the inequalities

$$0 \le r \le 1 + \cos \theta, \qquad 0 \le \theta \le \pi.$$

Thus, $g(\theta) = 1 + \cos \theta$ in equation (8), and

$$\iint_Q r \sin \theta \, dA = \int_0^\pi \int_0^{1+\cos\theta} r \sin \theta \cdot r \, dr \, d\theta \qquad \text{note extra factor } r$$

$$= \int_0^\pi \int_0^{1+\cos\theta} r^2 \sin \theta \, dr \, d\theta$$

$$= \int_0^\pi \left\{ \frac{1}{3} r^3 \sin \theta \right]_{r=0}^{r=1+\cos\theta} \right\} d\theta$$

$$= \int_0^\pi \frac{1}{3} (1 + \cos \theta)^3 \sin \theta \, d\theta$$

$$= -\frac{1}{12} (1 + \cos \theta)^4 \Big]_{\theta=0}^{\theta=\pi}$$

$$= -\frac{1}{12} (0 - 2^4) = \frac{4}{3}.$$

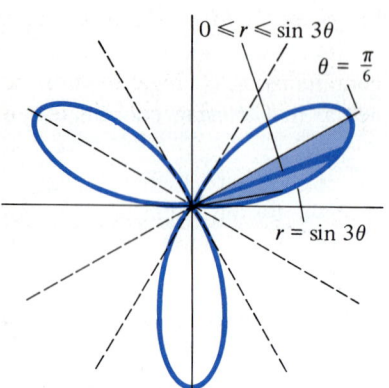

$0 \le r \le \sin 3\theta$

$\theta = \dfrac{\pi}{6}$

$r = \sin 3\theta$

Figure 3.5 Region enclosed by three-leaved rose $r = \sin 3\theta$.

Example 2

Find the area of the region enclosed by the three-leaved rose $r = \sin 3\theta$ (see Figure 3.5).

Solution: As illustrated in Figure 3.5, one sixth of the entire region Q lies between the rays $\theta = 0$ and $\theta = \pi/6$. Using the symmetry of the region, equation (7), and equation (8), we find that

$$\text{Area of } Q = \iint_Q 1 \, dA$$

$$= 6 \int_0^{\pi/6} \int_0^{\sin 3\theta} 1 \cdot r \, dr \, d\theta \qquad \text{note extra factor } r$$

$$= 6 \int_0^{\pi/6} \left\{ \frac{r^2}{2} \right]_{r=0}^{r=\sin 3\theta} \right\} d\theta$$

$$= 3 \int_0^{\pi/6} \sin^2 3\theta \, d\theta$$

$$= \frac{3}{2} \int_0^{\pi/6} (1 - \cos 6\theta) \, d\theta \qquad \left(\sin^2 3\theta = \frac{1}{2} (1 - \cos 6\theta) \right)$$

$$= \frac{3}{2} \left[\theta - \frac{1}{6} \sin 6\theta \right]_{\theta=0}^{\theta=\pi/6}$$

$$= \frac{\pi}{4}.$$

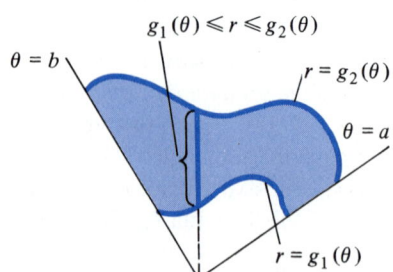

$g_1(\theta) \le r \le g_2(\theta)$

$\theta = b$

$r = g_2(\theta)$

$\theta = a$

$r = g_1(\theta)$

Figure 3.6 Region Q bounded by two polar curves, $r = g_1(\theta)$ and $r = g_2(\theta)$.

Figure 3.6 shows a region Q that is **r-simple**: it lies *between* the graphs of the polar equations $r = g_1(\theta)$ and $r = g_2(\theta)$ for $a \le \theta \le b$. That is,

$$Q = \{(r, \theta) \mid g_1(\theta) \le r \le g_2(\theta), \ a \le \theta \le b\}.$$

If we let

$$Q_1 = \{(r, \theta) \mid 0 \le r \le g_1(\theta), \, a \le \theta \le b\},$$

and

$$Q_2 = \{(r, \theta) \mid 0 \le r \le g_2(\theta), \, a \le \theta \le b\},$$

then $Q = Q_2 - Q_1$. Accordingly, we have

$$\iint_Q f(r, \theta) \, dA = \iint_{Q_2} f(r, \theta) \, dA - \iint_{Q_1} f(r, \theta) \, dA$$

$$= \int_a^b \int_0^{g_2(\theta)} f(r, \theta) \, r \, dr \, d\theta - \int_a^b \int_0^{g_1(\theta)} f(r, \theta) r \, dr \, d\theta$$

$$= \int_a^b \left\{ \int_0^{g_2(\theta)} f(r, \theta) r \, dr - \int_0^{g_1(\theta)} f(r, \theta) r \, dr \right\} d\theta$$

$$= \int_a^b \int_{g_1(\theta)}^{g_2(\theta)} f(r, \theta) r \, dr \, d\theta.$$

Thus, if Q is described by the polar inequalities

$$g_1(\theta) \le r \le g_2(\theta), \qquad a \le \theta \le b,$$

then

$$\iint_Q f(r, \theta) \, dA = \int_a^b \int_{g_1(\theta)}^{g_2(\theta)} f(r, \theta) r \, dr \, d\theta. \tag{9}$$

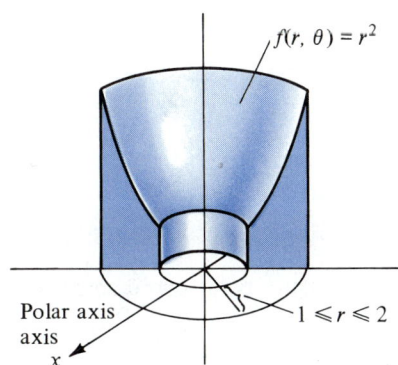

Figure 3.7 One half of the bearing sleeve of Example 3. (See **Plate 14**.)

Example 3

A bearing sleeve has the shape of the solid bounded above by the graph of $f(r, \theta) = r^2$ and below by the xy-plane for $1 \le r \le 2$. Find the volume of the sleeve.

Solution: The solid is illustrated in Figure 3.7. The region Q over which the solid is defined is described by the inequalities

$$1 \le r \le 2, \qquad 0 \le \theta \le 2\pi.$$

The volume is therefore

$$\iint_Q f(r, \theta) \, dA = \int_0^{2\pi} \int_1^2 r^2 \cdot r \, dr \, d\theta$$

$$= \int_0^{2\pi} \int_1^2 r^3 \, dr \, d\theta$$

$$= \int_0^{2\pi} \left\{ \frac{1}{4} r^4 \right]_{r=1}^{r=2} \right\} d\theta$$

$$= \int_0^{2\pi} \frac{15}{4} \, d\theta$$

$$= \frac{15\pi}{2}. \qquad \diamond$$

Changing From Cartesian to Polar Coordinates

Often, a double integral in Cartesian coordinates is more easily evaluated by first changing to polar coordinates. For example, in Exercise 38 in the last section you were asked to find the volume of the sphere $x^2 + y^2 + z^2 = 4$ lying above the plane $z = 1$. This led to a double integral equivalent to

$$V = \int_{-\sqrt{3}}^{\sqrt{3}} \int_{-\sqrt{3-x^2}}^{\sqrt{3-x^2}} (\sqrt{4 - (x^2 + y^2)} - 1) \, dy \, dx. \tag{10}$$

(See Figure 3.8.)

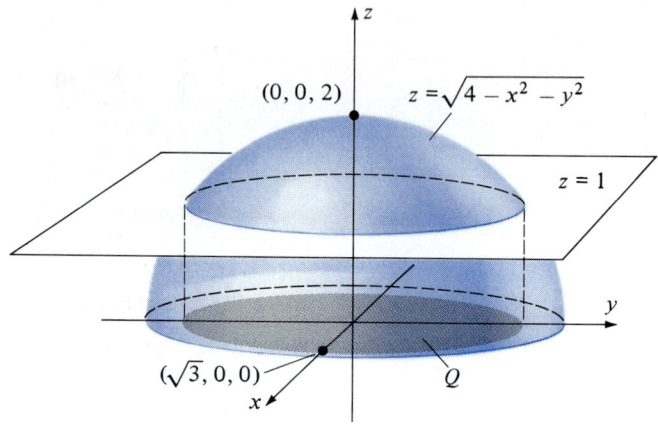

Figure 3.8 $Q = \{(x, y) \mid x^2 + y^2 \le 3\}$ may be described in polar coordinates as

$$Q = \{(r, \theta) \mid 0 \le r \le \sqrt{3}, \, 0 \le \theta \le 2\pi\}.$$

We can calculate this volume much more simply using polar coordinates, together with the equations

$$x = r \cos \theta \tag{11a}$$

$$y = r \sin \theta \tag{11b}$$

for changing from Cartesian coordinates to polar coordinates.

The region

$$Q = \{(x, y) \mid -\sqrt{3} \le x \le \sqrt{3}, \, -\sqrt{3 - x^2} \le y \le \sqrt{3 - x^2}\}$$

can be described in polar coordinates as

$$Q = \{(r, \theta) \mid 0 \le r \le \sqrt{3}, \, 0 \le \theta \le 2\pi\},$$

and the function $f(x, y) = \sqrt{4 - (x^2 + y^2)}$ in polar coordinates, obtained by using equations (11a) and (11b), is simply

$$f(r, \theta) = \sqrt{4 - (r^2 \cos^2 \theta + r^2 \sin^2 \theta)} = \sqrt{4 - r^2}.$$

The desired volume is then obtained using equations (6) and (8):

$$V = \iint_Q f(r \cos \theta, r \sin \theta) \, dA = \int_0^{2\pi} \int_0^{\sqrt{3}} (\sqrt{4 - r^2} - 1) r \, dr \, d\theta \qquad \text{note extra } r$$

$$= \int_0^{2\pi} \left\{ -\frac{1}{3}(4 - r^2)^{3/2} - \frac{r^2}{2} \right\}_{r=0}^{r=\sqrt{3}} d\theta$$

$$= \int_0^{2\pi} \left[-\frac{1}{3}(1 - 4^{3/2}) - \frac{3}{2} \right] d\theta$$

$$= \int_0^{2\pi} \frac{5}{6} \, d\theta = \frac{5\pi}{3}.$$

This example is typical of the following more general problem: Given an iterated integral in Cartesian coordinates, how can we evaluate the integral using an iterated integral in polar coordinates (without, of course, changing the value of the integral)? The answer is the following:

To express the iterated integral

$$\int_c^d \int_{h_1(y)}^{h_2(y)} f(x, y) \, dx \, dy \qquad \text{(or equivalent)}$$

in polar coordinates,

(i) express the region $Q = \{(x, y) \mid h_1(y) \le x \le h_2(y), \, c \le y \le d\}$ in polar coordinates as

$$Q = \{(r, \theta) \mid g_1(\theta) \le r \le g_2(\theta), \, a \le \theta \le b\}, \text{ and}$$

(ii) using the substitutions $x = r \cos \theta$ and $y = r \sin \theta$, replace the integrand

$$f(x, y) \qquad \text{by} \qquad f(r \cos \theta, \, r \sin \theta) \cdot r.$$

The result of steps (i) and (ii) is the iterated integral

$$\int_a^b \int_{g_1(\theta)}^{g_2(\theta)} f(r, \theta) r \, dr \, d\theta.$$

Another way to write this is simply that

$$\iint_Q f(x, y) \, dx \, dy = \iint_Q f(r \cos \theta, \, r \sin \theta) r \, dr \, d\theta. \qquad (12)$$

Equation (12) is referred to as a **change of variables** formula.

We shall not prove statement (12). However, it may be justified by comparing Theorem 2, which expresses $\iint_Q f \, dA$ as an iterated integral in Cartesian coordinates, with equation (9), which gives $\iint_Q f \, dA$ as an iterated integral in polar coordinates. We may paraphrase equation (12) by saying that in changing from Cartesian to polar coordinates we replace the *element of area*

$$dA = dx \, dy$$

in rectangular coordinates with the element of area

$$dA = r \, dr \, d\theta \qquad (13)$$

in polar coordinates. It is very important to note the extra factor r in (13), and to remember that, in the integral on the right side of equation (12), Q must be described using polar coordinates.

Example 4

Evaluate the integral

$$\int_{-2}^{2} \int_{-\sqrt{4-x^2}}^{\sqrt{4-x^2}} e^{x^2+y^2} \, dy \, dx$$

by first changing to polar coordinates.

Solution: From the limits of integration, we can see that the integral is being evaluated over the disc-shaped region

$$Q = \{(x, y) \mid -2 \leq x \leq 2, \ -\sqrt{4 - x^2} \leq y \leq \sqrt{4 - x^2}\}.$$

(See Figure 3.9.) This region can be described in polar coordinates by the inequalities

$$0 \leq r \leq 2, \qquad 0 \leq \theta \leq 2\pi \qquad \text{(Figure 3.10)}.$$

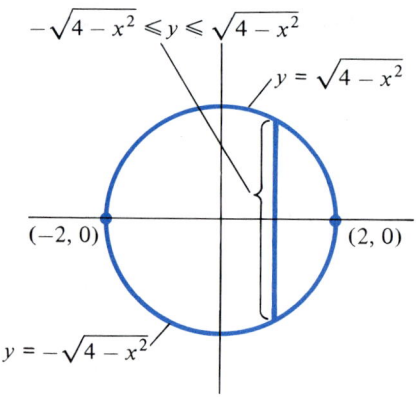

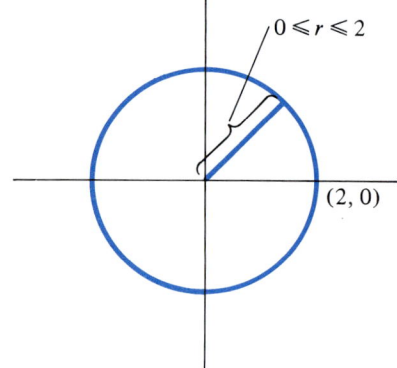

Figure 3.9 $Q: -2 \leq x \leq 2$
$-\sqrt{4 - x^2} \leq y \leq \sqrt{4 - x^2}.$

Figure 3.10 $Q: 0 \leq r \leq 2$
$0 \leq \theta \leq 2\pi.$

With $x = r \cos \theta$ and $y = r \sin \theta$, the function $f(x, y) = e^{x^2+y^2}$ becomes

$$f(r \cos \theta, r \sin \theta) = e^{r^2 \cos^2 \theta + r^2 \sin^2 \theta}$$
$$= e^{r^2}.$$

Using (12), we obtain

$$\int_{-2}^{2} \int_{-\sqrt{4-x^2}}^{\sqrt{4-x^2}} e^{x^2+y^2} \, dy \, dx = \int_{0}^{2\pi} \int_{0}^{2} e^{r^2} r \, dr \, d\theta \qquad \text{note the extra factor } r$$

$$= \int_{0}^{2\pi} \left\{ \frac{1}{2} e^{r^2} \right]_{r=0}^{r=2} \right\} d\theta$$

$$= \int_{0}^{2\pi} \frac{1}{2}(e^4 - 1) \, d\theta$$

$$= \pi(e^4 - 1).$$

REMARK: It is important to note that the iterated integral in Example 4 could not be evaluated in rectangular coordinates, since we would not be able to find an antiderivative for $e^{x^2+y^2}$ with respect to either x or y. Thus, it is sometimes essential that we change to polar coordinates, if possible. The next example involves a volume that could be calculated using Cartesian coordinates, but the calculation in polar coordinates is much easier. It is typical of the kind of problem we will encounter in Chapter 19.

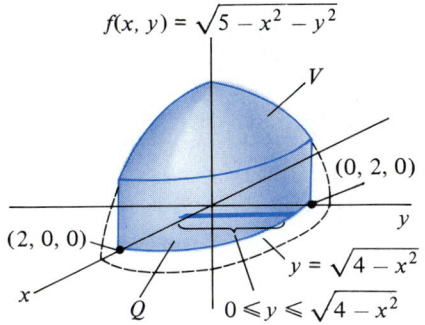

Figure 3.11 Solid V over region Q in Example 5.

Example 5

Interpret the iterated integral

$$\int_0^2 \int_0^{\sqrt{4-x^2}} \sqrt{5 - x^2 - y^2} \, dy \, dx$$

geometrically, and evaluate the integral by first changing to polar coordinates.

Solution: Since the integrand is nonnegative, the integral may be interpreted as the volume of the solid bounded above by the graph of $f(x, y) = \sqrt{5 - x^2 - y^2}$ and below by the quarter disc $Q = \{(x, y) \mid 0 \le x \le 2, 0 \le y \le \sqrt{4 - x^2}\}$ of radius 2 (Figure 3.11). In polar coordinates, Q is determined by the inequalities

$$0 \le r \le 2, \qquad 0 \le \theta \le \frac{\pi}{2}.$$

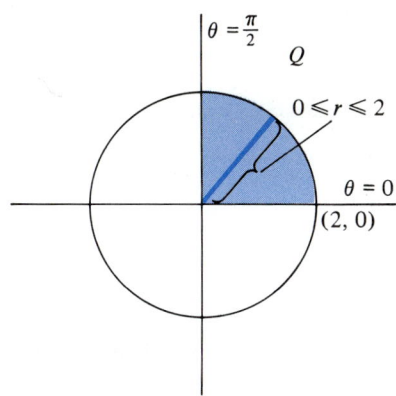

Figure 3.12 Region Q in Example 5: $0 \le r \le 2$, $0 \le \theta \le \pi/2$.

(See Figure 3.12.) Thus

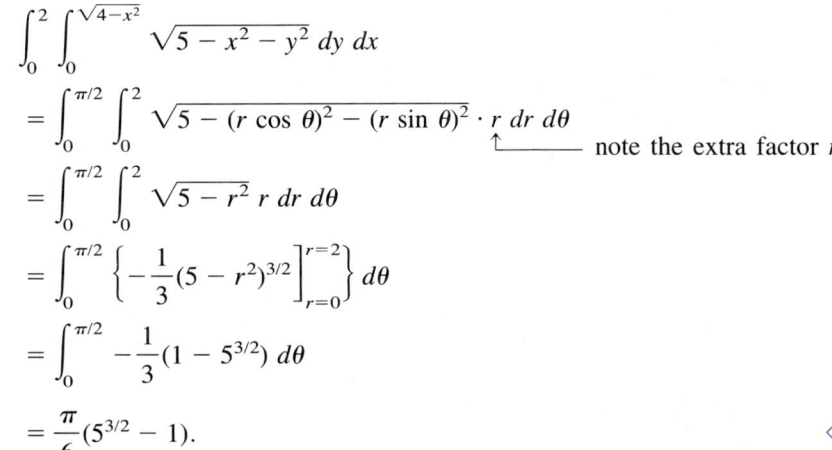

$$\int_0^2 \int_0^{\sqrt{4-x^2}} \sqrt{5 - x^2 - y^2} \, dy \, dx$$

$$= \int_0^{\pi/2} \int_0^2 \sqrt{5 - (r \cos \theta)^2 - (r \sin \theta)^2} \cdot r \, dr \, d\theta \qquad \text{note the extra factor } r$$

$$= \int_0^{\pi/2} \int_0^2 \sqrt{5 - r^2} \, r \, dr \, d\theta$$

$$= \int_0^{\pi/2} \left\{ -\frac{1}{3}(5 - r^2)^{3/2} \right]_{r=0}^{r=2} \right\} d\theta$$

$$= \int_0^{\pi/2} -\frac{1}{3}(1 - 5^{3/2}) \, d\theta$$

$$= \frac{\pi}{6}(5^{3/2} - 1). \qquad \diamond$$

Exercise Set 18.3

In Exercises 1–7, sketch the region Q and evaluate the double integral $\iint_Q f(r, \theta) \, dA$ for the given function over the given region Q.

1. $f(r, \theta) = r$, $\quad Q = \{(r, \theta) \mid 0 \le r \le 1, 0 \le \theta \le 2\pi\}$

2. $f(r, \theta) = 1 - r$, $\quad Q = \{(r, \theta) \mid 0 \le r \le 2, 0 \le \theta \le \pi\}$

3. $f(x, y) = e^{x^2+y^2}$, $\quad Q = \{(r, \theta) \mid r \le a\}$

4. $f(x, y) = e^{x^2+y^2}$, $\quad Q = \{(r, \theta) \mid 1 \le r \le 2, 0 \le \theta \le \pi/4\}$

5. $f(r, \theta) = 3r^2$, $\quad Q = \{(r, \theta) \mid 1 \le r \le 2, 0 \le \theta \le \pi/4\}$

6. $f(r, \theta) = 2 - r$, $\quad Q = \{(r, \theta) \mid 0 \le r \le (1 + \cos \theta), 0 \le \theta \le \pi/2\}$

7. $f(r, \theta) = r,$ $Q = \{(r, \theta) \mid 0 \le r \le \sin 3\theta,$
$\qquad\qquad\qquad\qquad 0 \le \theta \le \pi/3\}$

In Exercises 8–14, find the area of the region Q by use of a double integral.

8. Q is the region enclosed by the circle $r = 2 \cos \theta$.

9. Q is the region enclosed by the three-leaved rose $r = \sin 3\theta$.

10. Q is the region enclosed by the cardioid $r = a(1 + \cos \theta)$.

11. Q is the region inside the cardioid $r = 1 + \cos \theta$ and outside the circle $r = 1$.

12. Q is the region bounded by the graph of $r^2 = a^2 \sin 2\theta$.

13. Q is the region enclosed by the graph of $r = 2 - \sin \theta$.

14. Q is the region enclosed by the graph of the lemniscate $r^2 = 4 \cos^2 \theta$.

In Exercises 15–22, change the iterated integral from Cartesian to polar coordinates. Then evaluate the resulting integral.

15. $\int_0^1 \int_0^{\sqrt{1-x^2}} 2 \, dy \, dx$

16. $\int_{-2}^{\sqrt{2}} \int_x^{\sqrt{4-x^2}} 1 \, dy \, dx$

17. $\int_0^1 \int_0^{\sqrt{1-y^2}} e^{x^2+y^2} \, dx \, dy$

18. $\int_0^2 \int_{-\sqrt{2y-y^2}}^{\sqrt{2y-y^2}} \sqrt{x^2+y^2} \, dx \, dy$

19. $\int_0^2 \int_0^{\sqrt{2x-x^2}} \frac{1}{\sqrt{x^2+y^2}} \, dy \, dx$

20. $\int_0^1 \int_0^{\sqrt{4-x^2}} (x^2+y^2)^{3/2} \, dy \, dx$

21. $\int_{-2}^2 \int_0^{\sqrt{4-x^2}} \frac{x}{\sqrt{x^2+y^2}} \, dy \, dx$

22. $\int_0^1 \int_0^{\sqrt{1-y^2}} (x^2+y^2)^{3/2} \, dx \, dy$

23. Find the volume of the portion of the cylinder $x^2 + (y-1)^2 = 4$ bounded above by the plane $z = x + 4$ and below by the xy-plane.

24. Find the area of the region outside the spiral $r = \theta$ and inside the spiral $r = 2\theta$ for $0 \le \theta \le 2\pi$.

25. Find the volume of the region lying inside the sphere $x^2 + y^2 + z^2 = 4$ and outside the cylinder $x^2 + y^2 = 1$.

26. Find the volume of the region lying inside both the cone $z^2 = x^2 + y^2$ and the sphere $x^2 + y^2 + z^2 = 2$.

27. Find the volume of the solid bounded above by the plane $z = y + 2$ and below by the region inside the cardioid $r = 1 + \cos \theta$.

28. Use a double integral in polar coordinates to obtain the formula for the volume of a right circular cylinder of radius r and height h.

29. Find the volume of the solid bounded above by the graph of $z = 9 - x^2 - y^2$ and below by the graph of $z = 1 + x^2 + y^2$.

30. A hole 2 cm in diameter is drilled through the center of a spherical bearing of radius 3 cm. Find the volume of the remaining solid.

31. Change the order of integration in

$$\int_0^{\pi/2} \int_0^{\sin \theta} \sin \theta \, dr \, d\theta$$

and evaluate the integral.

32. Find the volume of the solid inside both the ellipsoid $z^2 + 4r^2 = 4$ and the cylinder $r = \sin \theta$.

33. Find the volume of the solid bounded by the cone $z = \sqrt{x^2 + y^2}$, the cylinder $x^2 + y^2 = 4$, and the xy-plane.

18.4 CALCULATING MASS AND CENTERS OF MASS

In this section we use the double integral to calculate mass and centroids for lamina (thin flat objects) lying in the plane. The difference between the discussion of Chapter 7 and what we do here is that we previously had assumed the density ρ of the material to be constant. Here we will treat the more general case of a variable density ρ. We assume the density function ρ to be continuous throughout the planar region Q that describes the lamina.

More specifically, let Q be a region in the plane (which we think of as the base of the lamina) and let $\rho(x, y)$ be the **mass per unit area** of the lamina at point (x, y). This is what we mean by density. (Thus, $\rho(x, y)$ is affected both by the thickness of the material and by its mass per unit volume. We will not be concerned about these two factors individually.) If R_j is a rectangle in Q of area ΔA, and if (s_j, t_j) is a point in R_j, then the product

$$\Delta M_j = \rho(s_j, t_j)\, \Delta A \qquad \text{(mass = mass per unit area times area)}$$

is an approximation to the mass of the lamina over the rectangle R_j. If, as in Section 18.2, the region Q is partitioned by a rectangular grid* and $R_1, R_2, \ldots, R_n$ is a list of all rectangles lying within Q, then

$$M \approx \sum_{j=1}^{n} \Delta M_j = \sum_{j=1}^{n} \rho(s_j, t_j)\, \Delta A \tag{1}$$

is an approximation to the mass of the lamina over Q. Since, as $n \to \infty$, the union of the rectangles R_j provides an increasingly accurate approximation to the region Q, we obtain M, the mass of Q, as the double integral

$$\boxed{M = \iint\limits_{Q} \rho(x, y)\, dA.} \tag{2}$$

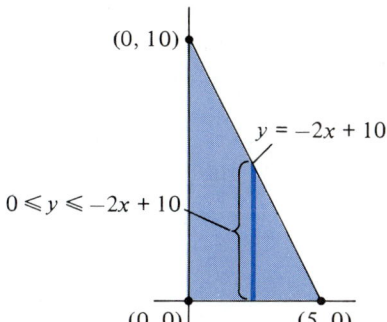

Figure 4.1 The triangular region in Example 1.

Example 1

A thin plate has the shape of a right triangle with legs of length 5 cm and 10 cm. The density, in terms of mass per unit area, at each point on the plate is proportional to the square of the distance from the vertex corresponding to the right angle. What is the mass of the plate?

Solution: With the triangle positioned as in Figure 4.1, the region it occupies may be described by the inequalities

$$0 \le x \le 5, \qquad 0 \le y \le -2x + 10.$$

Since the right angle is at the origin, the density function is

$$\rho(x, y) = \lambda(x^2 + y^2)$$

where λ is constant. By (2), the mass is

$$
\begin{aligned}
M = \iint\limits_{Q} \lambda(x^2 + y^2)\, dA &= \int_0^5 \int_0^{-2x+10} \lambda(x^2 + y^2)\, dy\, dx \\
&= \int_0^5 \lambda \left\{ x^2 y + \frac{1}{3} y^3 \right\}_{y=0}^{y=-2x+10} \, dx \\
&= \int_0^5 \lambda \left(-\frac{14}{3} x^3 + 50x^2 - 200x + \frac{1000}{3} \right) dx \\
&= \lambda \left[-\frac{7}{6} x^4 + \frac{50}{3} x^3 - 100x^2 + \frac{1000}{3} x \right]_0^5 \\
&= \frac{3125\lambda}{6}. \qquad \Diamond
\end{aligned}
$$

---

*As we did in Chapter 7, we shall henceforth use *regular* grids to partition a region Q into rectangles of equal dimensions Δx and Δy, and areas $\Delta A = \Delta x\, \Delta y$.

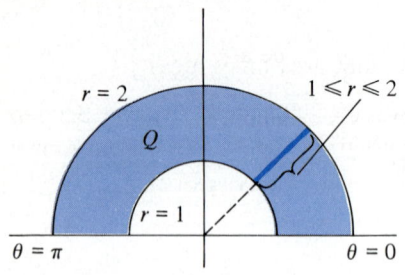

Figure 4.2 Annular region of Example 2.

Example 2

A machine part has the shape of a half annulus—the region between two concentric semicircles of radius $r = 1$ and $r = 2$ (see Figure 4.2). Find the mass of the part if the density at each point is proportional to the distance from that point to the common center of the two semicircles.

Solution: The density function can be written as

$$\rho(x, y) = \lambda \sqrt{x^2 + y^2}$$

where λ is constant. However, it is most convenient to describe the region Q in polar coordinates using the inequalities

$$0 \le \theta \le \pi, \qquad 1 \le r \le 2,$$

in which case the density function is $\rho(r, \theta) = \lambda r$.

The mass is therefore

$$M = \iint_Q \lambda r \, dA = \int_0^\pi \int_1^2 \lambda r \cdot r \, dr \, d\theta$$

$$= \int_0^\pi \int_1^2 \lambda r^2 \, dr \, d\theta = \int_0^\pi \left\{ \frac{\lambda}{3} r^3 \right\}_{r=1}^{r=2} d\theta$$

$$= \int_0^\pi \frac{7\lambda}{3} \, d\theta = \frac{7\lambda\pi}{3}.$$
◇

Centers of Mass

In Chapter 7 we determined that the x-coordinate of the *centroid* of the lamina with shape Q and constant density ρ is

$$\bar{x} = \frac{\displaystyle\int_a^b \rho x [f(x) - g(x)] \, dx}{\displaystyle\int_a^b \rho [f(x) - g(x)] \, dx}. \tag{3}$$

In writing equation (3), we assume that Q is a region that can be described by the inequalities

$$a \le x \le b, \qquad g(x) \le y \le f(x) \qquad \text{(Figure 4.3)}.$$

In other words, Q is y-simple.

To generalize to the case of a variable density function ρ, we begin with the integral in the numerator of $\bar{x}$ in (3), which we write as

$$\int_a^b \rho x [f(x) - g(x)] \, dx = \int_a^b \left\{ \rho x y \right\}_{y=g(x)}^{y=f(x)} dx \tag{4}$$

$$= \int_a^b \int_{g(x)}^{f(x)} \rho x \, dy \, dx.$$

The right-hand side of (4) allows us to define the first moment of mass for Q about the y-axis when ρ is nonconstant. We simply replace ρ by $\rho(x, y)$ and define

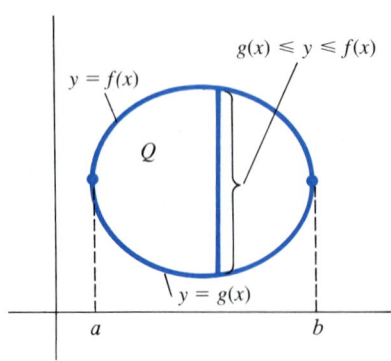

Figure 4.3

$$M_y = \int_a^b \int_{g(x)}^{f(x)} x\rho(x, y) \, dy \, dx = \iint_Q x\rho(x, y) \, dA.$$

Similarly, the first moment of mass for Q about the x-axis is defined to be

$$M_x = \iint\limits_{Q} y\rho(x, y) \, dA.$$

Finally, the denominator of $\bar{x}$ can be written

$$\int_a^b \rho[f(x) - g(x)] \, dx = \int_a^b \left\{ \rho y \big]_{y=g(x)}^{y=f(x)} \right\} dx$$

$$= \int_a^b \int_{g(x)}^{f(x)} \rho \, dy \, dx$$

$$= \iint\limits_{Q} \rho \, dy \, dx,$$

which is just the mass (area times density) of Q. We are therefore ready to generalize the concept of centroid to that of *center of mass*.

| DEFINITION 4 | Let Q be a region in the xy-plane that is either x-simple or y-simple, and let ρ be a continuous density function defined on Q. Then |

(i) The **mass** of Q is the number

$$M = \iint\limits_{Q} \rho(x, y) \, dA.$$

(ii) The **first moment of mass for Q about the y-axis** is the number

$$M_y = \iint\limits_{Q} x\rho(x, y) \, dA.$$

(iii) The **first moment of mass for Q about the x-axis** is the number

$$M_x = \iint\limits_{Q} y\rho(x, y) \, dA.$$

(iv) The **center of mass** of Q is the point $(\bar{x}, \bar{y})$, where

$$\bar{x} = \frac{M_y}{M}, \qquad \bar{y} = \frac{M_x}{M}.$$

Because of the way the concept of centroid is generalized by Definition 4, the physical interpretation of the center of mass is the same as before—the lamina associated with Q will "balance" at the point $(\bar{x}, \bar{y})$.

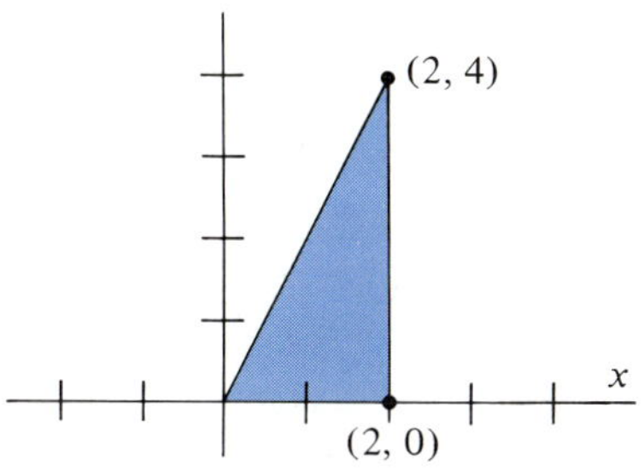

Example 3

A thin lamina has the shape of a triangle with vertices at $(0, 0)$, $(2, 0)$, and $(2, 4)$. The density function associated with the lamina has equation $\rho(x, y) = 4x + 2y + 2$. Find the mass and center of mass of the lamina (see Figure 4.4).

Solution: The region Q associated with the lamina may be described by the inequalities

$$0 \le x \le 2, \qquad 0 \le y \le 2x.$$

Figure 4.4

Thus, by Definition 4,

$$M = \iint_Q (4x + 2y + 2)\, dA = \int_0^2 \int_0^{2x} (4x + 2y + 2)\, dy\, dx$$

$$= \int_0^2 \left\{ 4xy + y^2 + 2y\right]_{y=0}^{y=2x} \right\} dx$$

$$= \int_0^2 (12x^2 + 4x)\, dx$$

$$= 4x^3 + 2x^2\big]_0^2$$

$$= 40,$$

$$M_y = \iint_Q x(4x + 2y + 2)\, dA = \int_0^2 \int_0^{2x} (4x^2 + 2xy + 2x)\, dy\, dx$$

$$= \int_0^2 \left\{ 4x^2y + xy^2 + 2xy\right]_{y=0}^{y=2x} \right\} dx$$

$$= \int_0^2 (12x^3 + 4x^2)\, dx = 3x^4 + \frac{4}{3}x^3\Big]_0^2$$

$$= \frac{176}{3},$$

and

$$M_x = \iint_Q y(4x + 2y + 2)\, dA = \int_0^2 \int_0^{2x} (4xy + 2y^2 + 2y)\, dy\, dx$$

$$= \int_0^2 \left\{ 2xy^2 + \frac{2}{3}y^3 + y^2\right]_{y=0}^{y=2x} \right\} dx$$

$$= \int_0^2 \left(\frac{40}{3}x^3 + 4x^2 \right) dx = \frac{10}{3}x^4 + \frac{4}{3}x^3\Big]_0^2$$

$$= \frac{192}{3}.$$

Thus,

$$\bar{x} = \frac{M_y}{M} = \frac{\left(\dfrac{176}{3}\right)}{40} = \frac{22}{15}$$

and

$$\bar{y} = \frac{M_x}{M} = \frac{\left(\dfrac{192}{3}\right)}{40} = \frac{8}{5}.$$

The center of mass is therefore $(\bar{x}, \bar{y}) = \left(\dfrac{22}{15}, \dfrac{8}{5} \right)$. ◇

Example 4

Find the center of mass for the lamina described in Example 2.

Solution: The density function in Example 2 is

$$\rho(x, y) = \lambda\sqrt{x^2 + y^2}$$

and the region Q is described in polar coordinates as

$$0 \le \theta \le \pi, \qquad 1 \le r \le 2.$$

We evaluate the integral for M_y in polar coordinates as

$$M_y = \iint\limits_Q x\rho(x, y) \, dA = \int_0^\pi \int_1^2 (r \cos \theta)\lambda r \cdot \underbrace{r \, dr \, d\theta}_{dA}$$

$$\rho = \lambda\sqrt{x^2 + y^2} = \lambda r$$
$$x = r \cos \theta$$

$$= \lambda \int_0^\pi \int_1^2 r^3 \cos \theta \, dr \, d\theta$$

$$= \lambda \int_0^\pi \left\{ \frac{1}{4}r^4 \cos \theta \Big]_{r=1}^{r=2} \right\} d\theta$$

$$= \lambda \int_0^\pi \frac{15}{4} \cos \theta \, d\theta$$

$$= \lambda\left(\frac{15}{4} \sin \theta \right)\Big]_{\theta=0}^{\theta=\pi}$$

$$= 0.$$

(This should not surprise you since both Q and ρ are symmetric with respect to the y-axis.) The integral for M_x is

$$M_x = \iint\limits_Q y\rho(x, y) \, dA = \int_0^\pi \int_1^2 (r \sin \theta) \cdot \lambda r \cdot \underbrace{r \, dr \, d\theta}_{dA}$$

$$\rho = \lambda\sqrt{x^2 + y^2} = \lambda r$$
$$y = r \sin \theta$$

$$= \lambda \int_0^\pi \int_1^2 r^3 \sin \theta \, dr \, d\theta$$

$$= \lambda \int_0^\pi \left\{ \frac{1}{4}r^4 \sin \theta \Big]_{r=1}^{r=2} \right\} d\theta$$

$$= \lambda \int_0^\pi \frac{15}{4} \sin \theta \, d\theta$$

$$= \lambda\left(-\frac{15}{4} \cos \theta \right)\Big]_{\theta=0}^{\theta=\pi}$$

$$= \frac{15\lambda}{2}.$$

Since we found in Example 2 that $M = \dfrac{7\lambda\pi}{3}$, we have

$$\bar{x} = \frac{M_y}{M} = 0$$

and

$$\bar{y} = \frac{M_x}{M} = \frac{\left(\dfrac{15\lambda}{2}\right)}{\left(\dfrac{7\lambda\pi}{3}\right)} = \frac{45}{14\pi} \approx 1.02.$$

The center of mass is therefore $\left(0, \dfrac{45}{14\pi}\right)$.

◇

Exercise Set 18.4

In Exercises 1–10, find the mass of a lamina with shape given by the region Q and density function ρ.

1. $\rho(x, y) = x + y$,
   $Q = \{(x, y) \mid 0 \le x \le 2, 0 \le y \le 1\}$

2. $\rho(x, y) = x^2 + y$,
   $Q = \{(x, y) \mid 0 \le x \le 2, 0 \le y \le x\}$

3. $\rho(x, y) = 6 + x$,
   $Q = \{(x, y) \mid -1 \le x \le 1, 0 \le y \le 1 - x^2\}$

4. $\rho(x, y) = xy$,
   $Q = \{(x, y) \mid 0 \le x \le 1 - y^2, 0 \le y \le 1\}$

5. $\rho(x, y) = \sin(x + y)$,
   $Q = \{(x, y) \mid 0 \le x \le \pi/4, 0 \le y \le \pi/4\}$

6. $\rho(x, y) = xy$, $Q = \{(x, y) \mid 0 \le x \le 1, 0 \le y \le 1\}$

7. $\rho(x, y) = x^2 + y^2$,
   $Q = \{(x, y) \mid -1 \le x \le 1, 0 \le y \le \sqrt{1 - x^2}\}$

8. $\rho(x, y) = \sqrt{x^2 + y^2}$,
   $Q = \{(r, \theta) \mid 0 \le r \le 2, 0 \le \theta \le \pi/2\}$

9. $\rho(x, y) = \dfrac{1}{\sqrt{x^2 + y^2}}$, Q is the annulus $1 \le r \le 2$

10. $\rho(x, y) = xy$,
    $Q = \{(r, \theta) \mid 0 \le r \le 1, 0 \le \theta \le \pi/2\}$

11. Find the center of mass of the lamina in Exercise 1.

12. Find the center of mass of the lamina in Exercise 3.

13. Find the center of mass of the lamina in Exercise 8.

14. A lamina has the shape of a triangle with vertices $(0, 0)$, $(0, 4)$, and $(1, 0)$. The density at each point (x, y) is $\rho(x, y) = y - x + 8$. Find the mass of the lamina.

15. Find the center of mass of the lamina in Exercise 14.

16. Find the centroid of the region bounded by the graph of $r = \sin 2\theta$ for $0 \le \theta \le \pi/2$. (Assume $\rho(x, y) = 1$ throughout the region.)

17. Find the centroid of the planar region bounded by the parabola $y = 4 - x^2$ and the x-axis.

18. Find the centroid of the region bounded by the graph of the function $f(x) = 1/x$ and the line $2x + 2y = 5$.

19. Find the centroid of the region obtained by connecting the points $(0, 0)$, $(4, 0)$, $(4, 4)$, $(2, 4)$, $(2, 1)$, $(0, 1)$, and $(0, 0)$ in order by line segments.

20. Find the centroid of the region bounded by the graph of $y = x^2$ and $y = x^3$.

21. Show that the centroid of the region
    $$R = \{(x, y) \mid 0 \le x \le a, 0 \le y \le \sqrt{a^2 - x^2}\}$$
    is $(\bar{x}, \bar{y}) = \left(\dfrac{4a}{3\pi}, \dfrac{4a}{3\pi}\right)$.

22. Find the centroid of the region bounded by the graph of $f(x) = \sinh x$ and the x-axis for $0 \le x \le 1$.

23. Find the centroid of the region bounded by the graph of $x = y(4 - y)$ and the y-axis.

24. A lamina has the shape of a circle of radius R. The density at any point P is proportional to the distance from the center. What is the mass?

25. A lamina has the shape of a right triangle with legs of length 2 and 6. The density at any point is proportional to the distance from the longer leg. What is the mass?

26. Find the distance from the vertex at the right angle to the center of mass for the triangle in Exercise 25.

27. Assume that the region in Example 1 is oriented with the 10 cm leg of the triangle along the x-axis, again with the right angle at the origin. Show that the mass calculated in this way is the same as that found in Example 1.

28. Find the center of mass of a lamina with density function $\rho(r, \theta) = \lambda(1 + \cos^2 \theta)$ over the half annulus $Q = \{(r, \theta) \mid 1 \le r \le 2, 0 \le \theta \le \pi\}$. Locate the center of mass on a sketch of the region Q. State (in words) how the mass of the lamina is distributed, and what relation this has to the location of the center of mass.

18.5 SURFACE AREA

As our final application of double integrals, we consider the problem of calculating the area of a surface in space. We shall restrict our considerations to surfaces that are graphs of functions of two variables, although the ideas discussed here can be extended to more general types of surfaces.

Let Q be a region in the plane which is either x-simple or y-simple, and let the surface S be the graph of the continuous function $z = f(x, y)$ on Q. We partition the region Q with a rectangular grid, and we denote by $R_1, R_2, \ldots, R_n$ the rectangles lying entirely within Q. This grid is constructed so that each of the rectangles R_m has dimensions Δx and Δy (see Figure 5.1). The grid on Q partitions the surface S into

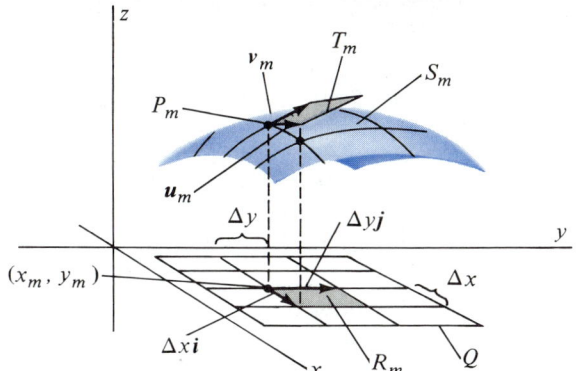

Figure 5.1 Surface is approximated by parallelograms pasted on above corners of approximating rectangles.

Figure 5.2 $\quad u_m = \Delta x i + \dfrac{\partial f}{\partial x} \Delta x k$

$$v_m = \Delta y j + \frac{\partial f}{\partial y} \Delta y k.$$

patches $S_1, S_2, \ldots, S_n$. Specifically, by the patch S_m we mean the portion of the surface S lying above the rectangle R_m.

As Figure 5.1 suggests, the patch S_m is nearly a parallelogram, but not quite, since S_m is part of the (generally) curved surface S. We therefore construct a parallelogram T_m which approximates the patch S_m in the following way. Let P_m be the point on the corner of S_m nearest the z-axis and let (x_m, y_m) be the vertex of the rectangle R_m directly beneath the point P_m. Then the vector

$$u_m = \Delta x \, i + \frac{\partial f}{\partial x}(x_m, y_m) \, \Delta x \, k, \tag{1}$$

when originating at point P_m, terminates at some point above the vertex $(x_m + \Delta x, y_m)$ of R_m. Moreover, u_m is tangent to S at P_m, by definition of the partial derivative $\dfrac{\partial f}{\partial x}(x_m, y_m)$. Similarly, the vector

$$v_m = \Delta y \, j + \frac{\partial f}{\partial y}(x_m, y_m) \, \Delta y \, k, \tag{2}$$

when originating at P_m, terminates above the vertex $(x_m, y_m + \Delta y)$ of R_m and is tangent to S at P_m (see Figure 5.2).

From the properties noted for the vectors u_m and v_m, we conclude that u_m and v_m determine a parallelogram T_m that

(i) lies directly above the rectangle R_m (and, hence, the patch S_m), and

(ii) lies tangent to the surface S at point P_m.

The idea is therefore to use the area of the parallelogram T_m as an approximation to the area of the patch S_m. From Section 16.5, we recall that the area of the parallelogram T_m is equal to the length of the cross product of the vectors u_m and v_m, or

$$\text{Area of } T_m = |u_m \times v_m|$$

$$= \left| \left(\Delta x \, i + \frac{\partial f}{\partial x}(x_m, y_m) \, \Delta x \, k \right) \times \left(\Delta y \, j + \frac{\partial f}{\partial y}(x_m, y_m) \, \Delta y \, k \right) \right|$$

$$= (\Delta x \Delta y) \left| \det \begin{bmatrix} i & j & k \\ 1 & 0 & \frac{\partial f}{\partial x}(x_m, y_m) \\ 0 & 1 & \frac{\partial f}{\partial y}(x_m, y_m) \end{bmatrix} \right|$$

$$= (\Delta x \Delta y) \left| -\frac{\partial f}{\partial x}(x_m, y_m) i - \frac{\partial f}{\partial y}(x_m, y_m) j + k \right|$$

$$= \sqrt{\left[\frac{\partial f}{\partial x}(x_m, y_m) \right]^2 + \left[\frac{\partial f}{\partial y}(x_m, y_m) \right]^2 + 1} \, \Delta x \Delta y.$$

Summing these approximations over all patches $S_1, S_2, \ldots, S_n$ gives the approximation to the area of S as

$$\text{Area of } S \approx \sum_{m=1}^{n} \sqrt{\left[\frac{\partial f}{\partial x}(x_m, y_m) \right]^2 + \left[\frac{\partial f}{\partial y}(x_m, y_m) \right]^2 + 1} \, \Delta x \Delta y.$$

If the partial derivatives $\dfrac{\partial f}{\partial x}$ and $\dfrac{\partial f}{\partial y}$ are continuous on Q, this Riemann sum converges to an integral as $n \to \infty$ and as $\Delta x \to 0$ and $\Delta y \to 0$. This integral provides our definition of surface area.

DEFINITION 5

Let Q be a region in the plane that is either x-simple or y-simple, and let S be the graph of the function $z = f(x, y)$ for $(x, y) \in Q$. If $\dfrac{\partial f}{\partial x}(x, y)$ and $\dfrac{\partial f}{\partial y}(x, y)$ are continuous on Q, the **area of the surface S** is defined to be

$$A_s = \iint_Q \sqrt{\left[\frac{\partial f}{\partial x}(x, y) \right]^2 + \left[\frac{\partial f}{\partial y}(x, y) \right]^2 + 1} \, dA. \tag{3}$$

Example 1

Find the area of the surface that is the graph of the equation $f(x, y) = x^2$ lying above the rectangle $Q = \{(x, y) \mid -1 \leq x \leq 1, -1 \leq y \leq 1\}$.

Solution: The surface is the "cylinder" sketched in Figure 5.3. Using the method of integration by trigonometric substitution, we may evaluate the surface area inte-

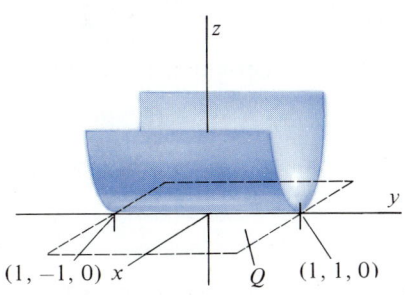

Figure 5.3 Cylinder $f(x, y) = x^2$ over the square

Q: $-1 \le x \le 1$, $-1 \le y \le 1$.

(See **Plate 15**.)

gral (3) as

$$A_s = \iint\limits_Q \sqrt{(2x)^2 + 1} \ dA$$

$$= \int_{-1}^{1} \int_{-1}^{1} \sqrt{4x^2 + 1} \ dx \ dy$$

$$= \int_{-1}^{1} \left\{ \frac{1}{2} x \sqrt{4x^2 + 1} + \frac{1}{4} \ln |2x + \sqrt{4x^2 + 1}| \right\}_{x=-1}^{x=1} \ dy$$

$$= \int_{-1}^{1} \left\{ \sqrt{5} + \frac{1}{4}[\ln(2 + \sqrt{5}) - \ln(\sqrt{5} - 2)] \right\} dy$$

$$= 2\sqrt{5} + \frac{1}{2}[\ln(2 + \sqrt{5}) - \ln(\sqrt{5} - 2)]$$

$$\approx 5.92. \qquad\qquad \diamondsuit$$

Example 2

Find the surface area of the portion of the paraboloid

$$z = 4 - x^2 - y^2$$

lying above the xy-plane.

Solution: The paraboloid intersects the xy-plane in the circle $4 - x^2 - y^2 = 0$, or $x^2 + y^2 = 4$. (See Figure 5.4.) In xy-coordinates, this region may be described by the inequalities

$$-2 \le x \le 2, \qquad -\sqrt{4 - x^2} \le y \le \sqrt{4 - x^2}.$$

With $f(x, y) = 4 - x^2 - y^2$, we have

$$\frac{\partial f}{\partial x}(x, y) = -2x, \qquad \text{and} \qquad \frac{\partial f}{\partial y}(x, y) = -2y.$$

By (3) the integral giving the surface area is

$$A_S = \iint\limits_Q \sqrt{4x^2 + 4y^2 + 1} \ dA = \int_{-2}^{2} \int_{-\sqrt{4-x^2}}^{\sqrt{4-x^2}} \sqrt{4x^2 + 4y^2 + 1} \ dy \ dx.$$

This integral is most easily evaluated in polar coordinates. With $x = r \cos \theta$ and $y = r \sin \theta$, and the region Q described as

$$0 \le \theta \le 2\pi, \qquad 0 \le r \le 2,$$

we have

$$A_s = \int_0^{2\pi} \int_0^{2} \sqrt{4(r \cos \theta)^2 + 4(r \sin \theta)^2 + 1} \ r \ dr \ d\theta$$

$$= \int_0^{2\pi} \int_0^{2} \sqrt{4r^2 + 1} \ r \ dr \ d\theta$$

$$= \int_0^{2\pi} \frac{1}{12}(4r^2 + 1)^{3/2} \Big]_{r=0}^{r=2} \ d\theta$$

$$= \frac{\pi}{6}(17^{3/2} - 1) \approx 11.5 \ \pi. \qquad\qquad \diamondsuit$$

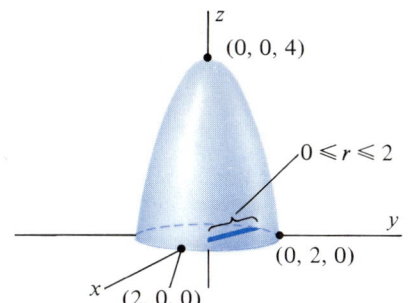

Figure 5.4 Paraboloid $z = 4 - x^2 - y^2$.

Example 3

Show that Definition 5 of surface area agrees with the usual definition of area when S is a flat surface over a rectangle in the plane.

Solution: If S is a flat surface in space, then S is a portion of a plane. Thus, we let

$$z = f(x, y) = Ax + By + C$$

be the equation for the plane, and we let

$$Q = \{(x, y) \mid a \le x \le b, \, c \le y \le d\}$$

be a rectangle in the plane.

As in Figure 5.1, the portion of the plane $Ax + By + C$ lying over the rectangle Q is a parallelogram. From (1) and (2) we see that the vectors

$$\boldsymbol{u} = \Delta x \, \boldsymbol{i} + \frac{\partial f}{\partial x} \Delta x \, \boldsymbol{k} = (b - a)\boldsymbol{i} + A(b - a)\boldsymbol{k}$$

and

$$\boldsymbol{v} = \Delta y \, \boldsymbol{j} + \frac{\partial f}{\partial y} \Delta y \, \boldsymbol{k} = (d - c)\boldsymbol{j} + B(d - c)\boldsymbol{k},$$

when positioned at the vertex $(a, c, f(a, c))$, form two adjacent sides of the parallelogram. By a standard calculation involving the cross product, the area of the parallelogram is

$$
\begin{aligned}
A_s &= |\boldsymbol{u} \times \boldsymbol{v}| \\
&= \left| \det \begin{bmatrix} \boldsymbol{i} & \boldsymbol{j} & \boldsymbol{k} \\ b - a & 0 & A(b - a) \\ 0 & d - c & B(d - c) \end{bmatrix} \right| \\
&= |-A(b - a)(d - c)\boldsymbol{i} - B(b - a)(d - c)\boldsymbol{j} + (b - a)(d - c)\boldsymbol{k}| \\
&= (b - a)(d - c)\sqrt{A^2 + B^2 + 1}.
\end{aligned}
$$

Using Definition 5 with $\dfrac{\partial f}{\partial x} = A$ and $\dfrac{\partial f}{\partial y} = B$ we find

$$A_s = \int_c^d \int_a^b \sqrt{A^2 + B^2 + 1} \, dx \, dy = (b - a)(d - c)\sqrt{A^2 + B^2 + 1}.$$

Thus, Definition 5 agrees with our usual concept of area for flat surfaces. ◇

The general notion of calculating surface area by double integration will be put to important use in the next chapter.

Exercise Set 18.5

In Exercises 1–10, find the surface area of the graph of $z = f(x, y)$ above the region Q in the plane.

1. $f(x, y) = x + y + 6$, $\quad Q = \{(x, y) \mid 0 \le x \le 1, 0 \le y \le 1\}$

2. $f(x, y) = 9 - x + 2y$, $\quad Q = \{(x, y) \mid 0 \le x^2 + y^2 \le 1\}$

3. $f(x, y) = 9 - x^2 - y^2$, $\quad Q = \{(x, y) \mid 0 \le x^2 + y^2 \le 3\}$

4. $f(x, y) = 4 + y^2$, $\quad Q = \{(x, y) \mid 0 \le x \le 1, 0 \le y \le 2\}$

5. $f(x, y) = 2 - x - y$, $Q = \{(x, y) \mid 0 \le x \le 2, 0 \le y \le 2 - x\}$

6. $f(x, y) = 3 + y^2$, $\quad Q = \{(x, y) \mid 0 \le x \le 2, 0 \le y \le 2\}$

7. $f(x, y) = \sqrt{x^2 + y^2}$, $Q = \{(x, y) \mid 1 \le x^2 + y^2 \le 4\}$

8. $f(x, y) = x + y^2$, $Q = \{(x, y) \mid 0 \le x \le 1, 0 \le y \le 2\}$

9. $f(x, y) = \sqrt{3}\, y - x^2$, $Q = \{(x, y) \mid 0 \le x \le 1, 0 \le y \le 1\}$

10. $f(x, y) = x^2 + y$, $Q = \{(x, y) \mid 0 \le x \le 1, 0 \le y \le x\}$

11. Find the surface area of the portion of the graph of $z = y + 2x^2$ over the triangular region with vertices $(0, 0)$, $(0, 1)$, and $(1, 1)$.

12. Find the area of the part of the plane $x + y + z = 4$ bounded by the cylinder $x^2 + y^2 = 4$.

13. Find the surface area of the portion of the paraboloid $z = 16 - x^2 - y^2$ lying between the planes $z = 4$ and $z = 9$.

14. Find the surface area of the part of the sphere $x^2 + y^2 + z^2 = 4$ lying above the plane $z = 1$.

15. Find the surface area of the part of the hemisphere $z = \sqrt{4 - x^2 - y^2}$ lying inside the cylinder $x^2 + y^2 = 1$.

16. Find the surface area of the part of the paraboloid $z = 4 - x^2 - y^2$ lying inside the cylinder $x^2 + y^2 = 1$.

17. Use Simpson's Rule to approximate the surface area of the portion of the paraboloid $z = 3 - x^2 - y^2 + 2y$ lying above the plane $z = 2y + 2$.

18. Develop the formula for the surface area of a sphere using a double integral.

19. What relationship exists between the formula for the surface area of a solid of revolution and Definition 5?

20. (A coordinate-free formula for surface area) Let u_m and v_m be the vectors in (1) and (2). Let ΔT_m be the area of the parallelogram tangent to S at P_m, over the rectangle R_m of area ΔA, as before.

   a. Show that $N_m = u_m \times v_m$ is normal to S at P_m.
   b. Show that $N \cdot k = \Delta x \Delta y = \Delta A$.
   c. Show that, also, $N \cdot k = |u_m \times v_m| \cos \theta$, where θ is the angle between N and k.
   d. Conclude from (b) and (c) that $\Delta T_m = |u_m \times v_m| = \dfrac{\Delta A}{\cos \theta} = \sec \theta\, \Delta A.$
   e. Conclude from (d) that $A_s = \displaystyle\iint_Q \sec \theta\, dA.$

18.6 TRIPLE INTEGRALS

In this section, we define the triple integral for a continuous function of three independent variables. As we did for double integrals, we shall first carry out this development for special types of regions Q (namely, boxes) and then indicate how the concept extends to more general regions.

The Triple Integral Over a Box

Let Q be the box-shaped region in $\mathbb{R}^3$ defined by the inequalities

$$a \le x \le b, \qquad c \le y \le d, \qquad p \le z \le q.$$

(See Figure 6.1.) Let f be a continuous function defined on Q. By constructing planes perpendicular to the x-axis at $x_0, x_1, \ldots, x_n$, planes perpendicular to the y-axis at $y_0, y_1, \ldots, y_m$, and planes perpendicular to the z-axis at $z_0, z_1, \ldots, z_\ell$, we partition the box Q into smaller rectangular boxes Q_{ijk}, each of which has volume $\Delta V_{ijk} = \Delta x_i \Delta y_j \Delta z_k$ (Figure 6.2).

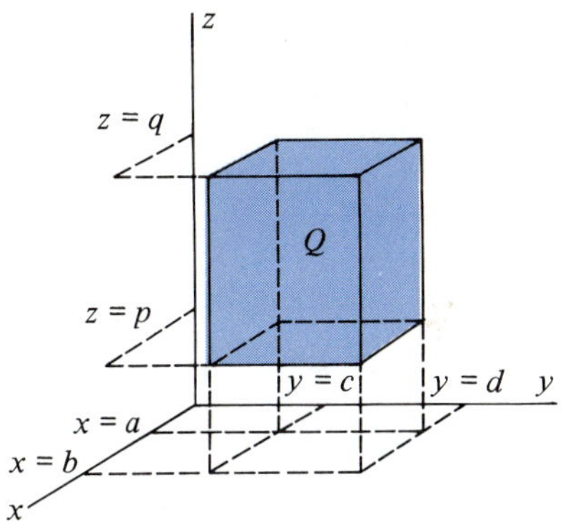

Next, we select one point (s_i, t_j, u_k) in each box Q_{ijk}, and we form the approximating sum

$$S_n = \sum_{i=1}^{n} \sum_{j=1}^{m} \sum_{k=1}^{\ell} f(s_i, t_j, u_k)\, \Delta V_{ijk}. \tag{1}$$

By analogy with the one- and two-variable cases, this approximating sum is called a Riemann sum for f on Q and we say that the set of rectangular boxes constitutes a *partition P* of Q. As in the one- and two-variable cases, if f is continuous on Q, this sum approaches a limit as $\|P\| \to 0$. This limit is defined to be the **triple integral** of f on the box Q:

Figure 6.1 The rectangular box Q: $a \le x \le b, c \le y \le d, p \le z \le q.$

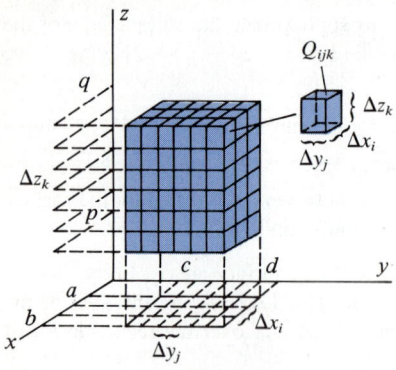

Figure 6.2 Partitioning the box into smaller boxes Q_{ijk} of volume $\Delta V_{ijk} = \Delta x_i \Delta y_j \Delta z_k$.

$$\iiint\limits_{Q} f(x, y, z)\, dV = \lim_{\|P\| \to 0} \sum_{i=1}^{n} \sum_{j=1}^{m} \sum_{k=1}^{p} f(s_i, t_j, u_k)\, \Delta V_{ijk}. \tag{2}$$

In the special case $f(x, y, z) \equiv 1$, we can give a geometric interpretation of the triple integral in (1). The terms in the approximating Riemann sum (1) are just

$$f(s_i, t_j, u_k)\, \Delta V_{ijk} = 1 \cdot \Delta V_{ijk} = \text{volume of } Q_{ijk}.$$

Since the union of the boxes Q_{ijk} is the box Q, it follows that

$$\iiint\limits_{Q} 1 \, dV = \text{Volume of } Q. \tag{3}$$

That is, the triple integral of the function $f(x, y, z) \equiv 1$ over the box Q is just the volume of Q.

As for double integrals, triple integrals over boxes can be evaluated as iterated integrals. In particular, with Q as above, we have

$$\iiint\limits_{Q} f(x, y, z)\, dV = \int_{p}^{q} \int_{c}^{d} \int_{a}^{b} f(x, y, z)\, dx\, dy\, dz. \tag{4}$$

In evaluating the iterated integral in (4), there is no reason why the first integration must be performed with respect to x. Since each of the variables ranges between constant limits, the order of integration can be any of the six possible orders xyz, xzy, yxz, yzx, zxy, or zyx.

We shall not prove equation (4), although it is easy to explain in the case $f(x, y, z) \equiv 1$. If v is any number in the z-interval $[p, q]$, the plane $z = v$ determines a rectangular cross section of Q of area

$$A(z) = \int_{c}^{d} \int_{a}^{b} 1 \cdot dx\, dy.$$

Thus,

$$\begin{aligned} \text{Volume of } Q &= \int_{p}^{q} A(z)\, dz \\ &= \int_{p}^{q} \left\{ \int_{c}^{d} \int_{a}^{b} 1 \cdot dx\, dy \right\} dz \\ &= \int_{p}^{q} \int_{c}^{d} \int_{a}^{b} 1 \cdot dx\, dy\, dz. \end{aligned} \tag{5}$$

Combining equations (3) and (5) results in equation (4) in this special case.

Example 1

Evaluate the triple integral $\displaystyle\iiint\limits_{Q} xe^y \cos z \, dV$ where Q is the box $\{(x, y, z) \mid 0 \le x \le 2,\ 0 \le y \le \ln 2,\ 0 \le z \le \pi/2\}$.

Solution: Using equation (4), we find

$$\iiint\limits_{Q} xe^y \cos z \, dV = \int_0^{\pi/2} \int_0^{\ln 2} \int_0^2 xe^y \cos z \, dx \, dy \, dz$$

$$= \int_0^{\pi/2} \int_0^{\ln 2} \left\{ \frac{x^2}{2} e^y \cos z \right]_{x=0}^{x=2} \right\} dy \, dz$$

$$= \int_0^{\pi/2} \int_0^{\ln 2} 2e^y \cos z \, dy \, dz$$

$$= \int_0^{\pi/2} \left\{ 2e^y \cos z \right]_{y=0}^{y=\ln 2} \right\} dz$$

$$= \int_0^{\pi/2} 2 \cos z \, dz$$

$$= 2 \sin z]_{z=0}^{z=\pi/2}$$

$$= 2. \qquad \diamondsuit$$

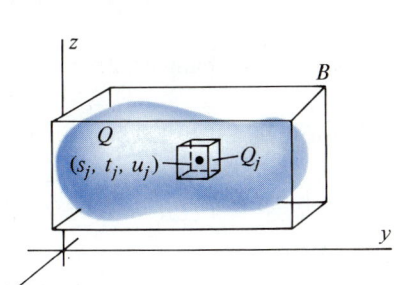

Figure 6.3 More general region Q enclosed by the box B.

Triple Integrals Over More General Regions

If Q is a bounded region in space (not necessarily a box), we define the triple integral of f over Q as follows. First, we find a box B containing the region Q. (See Figure 6.3.) Next, we partition the box B into smaller rectangular boxes, just as we did when Q itself was a box, and we let $Q_1, Q_2, \ldots, Q_n$ be a complete list of all such rectangular boxes *lying entirely within the region* Q. For each such box we select a point $(s_j, t_j, u_j) \in Q_j$ and we denote by ΔV_j the volume of Q_j. The limit of this sequence of approximating sums, when it exists, is called the **triple integral** of f over Q:

$$\iiint\limits_{Q} f(x, y, z) \, dV = \lim_{\|P\| \to 0} \sum_{j=1}^{n} f(s_j, t_j, u_j) \, \Delta V_j. \qquad (6)$$

The triple integral in (6) will exist whenever f is continuous on Q and Q is a "sufficiently nice" region in space. Rather than worry too much about the precise meaning of this last phrase, we shall state a theorem showing how triple integrals may be evaluated for regions of the type encountered in this text and in most applications. Before doing so, however, we need to make one observation concerning the approximating sum in equation (6). In the special case $f(x, y, z) \equiv 1$, the terms in the approximating sum are, as before, the volumes of the approximating boxes. The sum therefore approximates the volume of Q, and in the limit we obtain

$$\iiint\limits_{Q} 1 \cdot dV = \text{Volume of } Q \qquad (7)$$

just as in the case when Q itself is a box.

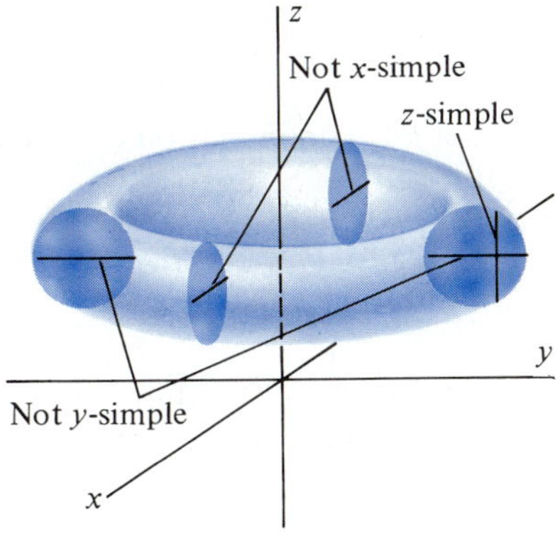

Figure 6.4 The torus is z-simple, but neither x-simple nor y-simple.

Evaluating Triple Integrals

We say that a region Q in $\mathbb{R}^3$ is z-**simple** if every vertical line (that is, a line parallel to the z-axis) intersects the boundary of the region at most twice. The notions of x-**simple** and y-**simple** regions in $\mathbb{R}^3$ are defined accordingly. Figure 6.4 illustrates the fact that a torus is an example of a region that is z-simple, but neither x-simple nor y-simple. We shall describe a certain type of z-simple region Q for which the triple integral $\iiint\limits_{Q} f(x, y, z)\, dV$ can be evaluated as an iterated integral. The same result applies to regions that are x-simple or y-simple by interchanging the roles of x and z or y and z.

Specifically, let Q be a z-simple region in $\mathbb{R}^3$ that can be described by inequalities of the form

$$g_1(x, y) \leq z \leq g_2(x, y), \qquad h_1(x) \leq y \leq h_2(x), \qquad a \leq x \leq b. \tag{8}$$

In the inequalities in (8), we assume that g_1, g_2, h_1, and h_2 are continuous functions of their arguments. Figure 6.5 shows such a region. The following theorem shows how triple integrals over such regions may be evaluated.

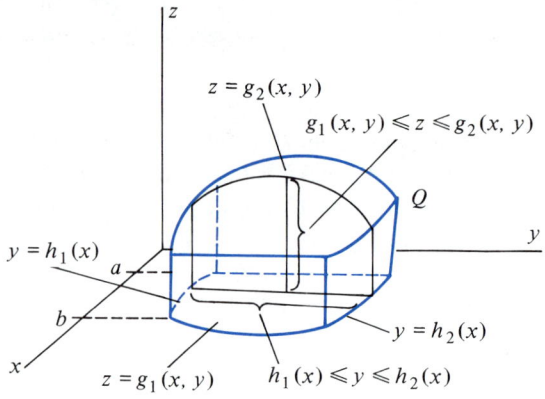

Figure 6.5 Region defined by inequalities (8).

THEOREM 4

Let Q be a region in $\mathbb{R}^3$ described by the inequalities

$$a \leq x \leq b, \qquad h_1(x) \leq y \leq h_2(x), \qquad g_1(x, y) \leq z \leq g_2(x, y)$$

where h_1, h_2, g_1, and g_2 are continuous functions. Let f be continuous on Q. Then

$$\iiint\limits_{Q} f(x, y, z)\, dV = \int_a^b \int_{h_1(x)}^{h_2(x)} \int_{g_1(x, y)}^{g_2(x, y)} f(x, y, z)\, dz\, dy\, dx.$$

A proof of Theorem 4 requires a deeper treatment of multiple integrals than we have given here and is left to more advanced courses. Before proceeding to apply this result, we need to emphasize two points.

REMARK 1: Theorem 4 remains true with the roles of the independent variables interchanged. For example, if the region Q is described by the inequalities

$$c \leq y \leq d, \qquad h_1(y) \leq z \leq h_2(y), \qquad g_1(y, z) \leq x \leq g_2(y, z),$$

then the iterated integral formula is

$$\iiint\limits_{Q} f(x, y, z)\, dV = \int_c^d \int_{h_1(y)}^{h_2(y)} \int_{g_1(y, z)}^{g_2(y, z)} f(x, y, z)\, dx\, dz\, dy.$$

REMARK 2: Theorem 4 may be paraphrased this way. To evaluate the triple integral

$$\iiint\limits_{Q} f(x, y, z) \, dV:$$

(i) Find the constants and/or functions that bound the region Q in each of the three directions corresponding to the coordinate axes. (Write these down!)

(ii) Evaluate $\iiint\limits_{Q} f(x, y, z) \, dV$ as an iterated integral, integrating first with respect to a variable whose bounds depend on the other one or two variables, integrating second with respect to a variable whose limits involve the remaining variable, and integrating last with respect to a variable whose limits involve only constants.

It is important to note that we cannot integrate over limits involving a variable for which an integration has already been performed. For example, the expressions

$$\int_{a}^{b} \int_{g_1(x, y)}^{g_2(x, y)} \int_{h_1(y)}^{h_2(y)} f(x, y, z) \, dz \, dy \, dx$$

and

$$\int_{h_1(x)}^{h_2(x)} \int_{g_1(x, z)}^{g_2(x, z)} \int_{a}^{b} f(x, y, z) \, dx \, dy \, dz$$

are nonsense.

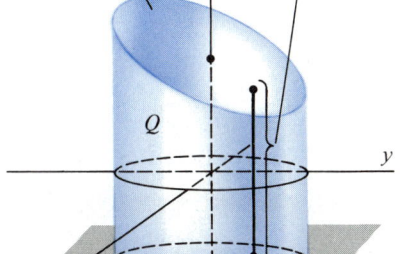

$z = 4 - x - y$

$-1 \le z \le 4 - x - y$

$z = -1$

Figure 6.6 Region of Example 2.

Example 2

Evaluate the triple integral

$$\iiint\limits_{Q} 2xy \, dV,$$

where Q is the region inside the cylinder $x^2 + y^2 = 1$ bounded by the planes $x + y + z = 4$ and $z = -1$.

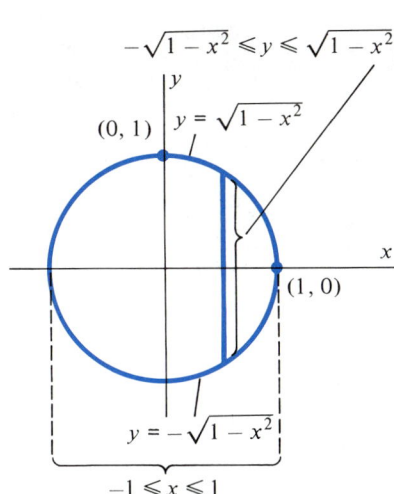

$-\sqrt{1 - x^2} \le y \le \sqrt{1 - x^2}$

$(0, 1)$ $y = \sqrt{1 - x^2}$

$(1, 0)$

$y = -\sqrt{1 - x^2}$

$-1 \le x \le 1$

Figure 6.7 Projection of Q into the xy-plane.

Solution: The region is sketched in Figure 6.6. Figure 6.7 shows the projection of Q into the xy-plane and illustrates how the inequalities involving x and y are obtained. The equation $x + y + z = 4$ gives $z = 4 - x - y$, so $-1 \le z \le 4 - x - y$. The region is therefore completely described by the inequalities

$$-1 \le x \le 1, \qquad -\sqrt{1 - x^2} \le y \le \sqrt{1 - x^2}, \qquad -1 \le z \le 4 - x - y.$$

Thus, according to Theorem 4,

$$\iiint\limits_{Q} 2xy \, dV = \int_{-1}^{1} \int_{-\sqrt{1-x^2}}^{\sqrt{1-x^2}} \int_{-1}^{4-x-y} 2xy \, dz \, dy \, dx$$

$$= \int_{-1}^{1} \int_{-\sqrt{4-x^2}}^{\sqrt{4-x^2}} \left\{ 2xyz \Big]_{z=-1}^{z=4-x-y} \right\} dy \, dx$$

$$= \int_{-1}^{1} \int_{-\sqrt{4-x^2}}^{\sqrt{4-x^2}} (10xy - 2x^2y - 2xy^2) \, dy \, dx$$

$$= \int_{-1}^{1} \left\{ 5xy^2 - x^2y^2 - \frac{2}{3}xy^3 \right]_{y=-\sqrt{4-x^2}}^{y=\sqrt{4-x^2}} \right\} dx$$

$$= \int_{-1}^{1} -\frac{4}{3}x(4-x^2)^{3/2} \, dx$$

$$= \frac{4}{15}(4-x^2)^{5/2} \Big]_{x=-1}^{x=1}$$

$$= 0.$$

(This result is explained by the fact that both the integrand and the region Q are symmetric with respect to the plane $y = x$.)   ◇

REMARK: The integral in Example 2 could also have been evaluated as

$$\iiint_Q 2xy \, dV = \int_{-1}^{1} \int_{-\sqrt{1-y^2}}^{\sqrt{1-y^2}} \int_{-1}^{4-x-y} 2xy \, dz \, dx \, dy.$$

Example 3

Evaluate the triple integral

$$\iiint_Q (x - y + z) \, dV$$

where Q is the tetrahedron with vertices $(0, 0, 0)$, $(1, 0, 0)$, $(0, 2, 0)$, and $(0, 0, 4)$.

Solution: The tetrahedron is sketched in Figure 6.8. Substituting the given points into the equation $z = Ax + By + C$ shows that the plane bounding Q above has equation $z = 4 - 4x - 2y$. As Figure 6.9 illustrates, the base of Q is a triangle in the xy plane bounded by the lines $x = 0$, $y = 0$, and $y = -2x + 2$. The region Q

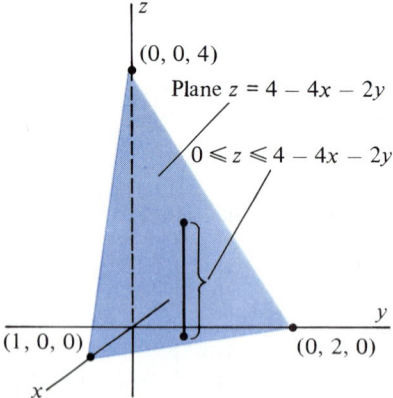

Figure 6.8 Tetrahedron of Example 3.

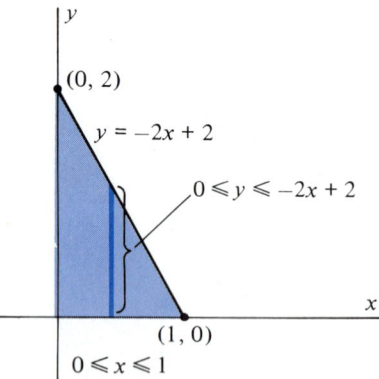

Figure 6.9 Projection of tetrahedron into the xy-plane.

may therefore be described by the inequalities

$$0 \le x \le 1, \qquad 0 \le y \le -2x + 2, \qquad 0 \le z \le 4 - 4x - 2y.$$

The integral is therefore

$$\iiint\limits_{Q} (x - y + z)\, dV = \int_0^1 \int_0^{2-2x} \int_0^{4-4x-2y} (x - y + z)\, dz\, dy\, dx$$

$$= \int_0^1 \int_0^{2-2x} \left\{ (x - y)z + \frac{1}{2}z^2 \right]_{z=0}^{z=4-4x-2y} \right\}\, dy\, dx$$

$$= \int_0^1 \int_0^{2-2x} (4x^2 - 12x + 10xy - 12y + 4y^2 + 8)\, dy\, dx$$

$$= \int_0^1 \left\{ (4x^2 - 12x + 8)y + 5xy^2 - 6y^2 + \frac{4}{3}y^3 \right]_{y=0}^{y=2-2x} \right\},$$

$$= \int_0^1 \left(\frac{4}{3}x^3 - 4x + \frac{8}{3} \right)\, dx$$

$$= \frac{1}{3}x^4 - 2x^2 + \frac{8}{3}x \Big]_0^1$$

$$= 1. \qquad \diamond$$

REMARK: The integral in Example 3 could also have been evaluated as

$$\int_0^2 \int_0^{1-y/2} \int_0^{4-4x-2y} (x - y + z)\, dz\, dx\, dy,$$

or

$$\int_0^1 \int_0^{4-4x} \int_0^{2-2x-1/2z} (x - y + z)\, dy\, dz\, dx,$$

or as one of three other iterated integrals (see Exercise 29).

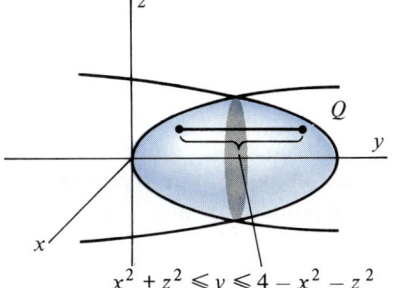

Figure 6.10 Intersecting paraboloids of Example 4. (See **Plate 16.**)

Example 4

Find the volume of the solid bounded by the paraboloids $y = 4 - x^2 - z^2$ and $y = x^2 + z^2$.

Solution: The region is sketched in Figure 6.10. From the description of the region we can see that

$$x^2 + z^2 \le y \le 4 - x^2 - z^2. \tag{9}$$

We therefore seek inequalities for the variables x and z. Equating the two expressions for y gives

$$4 - x^2 - z^2 = x^2 + z^2$$

or

$$x^2 + z^2 = 2.$$

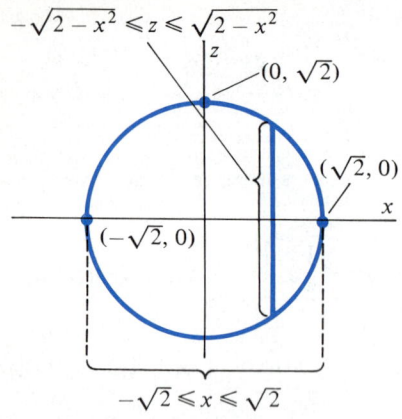

$-\sqrt{2-x^2} \leqslant z \leqslant \sqrt{2-x^2}$

$(0, \sqrt{2})$

$(\sqrt{2}, 0)$

$(-\sqrt{2}, 0)$

$-\sqrt{2} \leqslant x \leqslant \sqrt{2}$

Figure 6.11 Projection of Q into xz-plane.

This is the equation of the circle (in the plane $y = 2$) where the paraboloids intersect. The projection of this circle into the xz-plane is shown in Figure 6.11. From this sketch we see that the inequalities on x and on z are

$$-\sqrt{2} \leq x \leq \sqrt{2}, \qquad -\sqrt{2-x^2} \leq z \leq \sqrt{2-x^2}. \tag{10}$$

Using statement (7) and inequalities (9) and (10), we obtain the desired volume as

$$V = \iiint\limits_{Q} 1\ dV = \int_{-\sqrt{2}}^{\sqrt{2}} \int_{-\sqrt{2-x^2}}^{\sqrt{2-x^2}} \int_{x^2+z^2}^{4-x^2-z^2} 1\ dy\ dz\ dx$$

$$= \int_{-\sqrt{2}}^{\sqrt{2}} \int_{-\sqrt{2-x^2}}^{\sqrt{2-x^2}} \left\{ \Big[y \Big]_{y=x^2+z^2}^{y=4-x^2-z^2} \right\} dz\ dx$$

$$= \int_{-\sqrt{2}}^{\sqrt{2}} \int_{-\sqrt{2-x^2}}^{\sqrt{2-x^2}} (4 - 2x^2 - 2z^2)\ dz\ dx$$

$$= \int_{-\sqrt{2}}^{\sqrt{2}} \left\{ \Big[(4 - 2x^2)z - \frac{2}{3}z^3 \Big]_{z=-\sqrt{2-x^2}}^{z=\sqrt{2-x^2}} \right\} dx$$

$$= \int_{-\sqrt{2}}^{\sqrt{2}} \frac{8}{3}(2 - x^2)^{3/2}\ dx \tag{11}$$

$$= 4\pi.$$

(The integral in line (11) is evaluated by means of a trigonometric substitution.) ◇

REMARK: The triple integral in Example 4 could also have been evaluated as

$$\int_{-\sqrt{2}}^{\sqrt{2}} \int_{-\sqrt{2-z^2}}^{\sqrt{2-z^2}} \int_{x^2+z^2}^{4-x^2-z^2} 1\ dy\ dx\ dz.$$

Density

The triple integral may also be used to calculate the mass of a solid object, if we know the *density* of the material in units of mass per unit volume (such as gram/cm^3) as a continuous *density function* ρ. If Q is an object with a density function ρ of this type, we may approximate the mass of Q by partitioning Q into approximating boxes $Q_1, Q_2, \ldots, Q_n$, as before. If (s_j, t_j, u_j) is a point in Q_j, and if the volume ΔV_j of Q_j is small, then the quantity

$$M_j = \rho(s_j, t_j, u_j)\ \Delta V_j \qquad \text{(mass = density × volume)}$$

provides an approximation to the mass of the jth box Q_j. Summing these approximations over all boxes contained within Q gives the approximating sum

$$M \approx \sum_{j=1}^{n} M_j = \sum_{j=1}^{n} \rho(s_j, t_j, u_j)\ \Delta V_j.$$

Thus, the **mass** of Q is defined by the triple integral

$$M = \iiint\limits_{Q} \rho(x, y, z)\ dV. \tag{12}$$

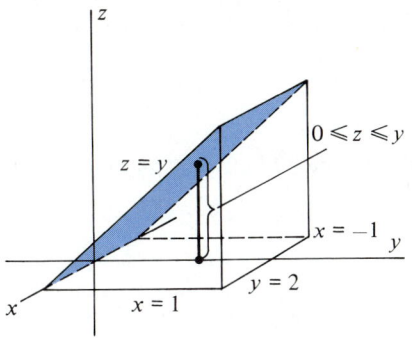

Figure 6.12 The wedge of Example 5.

That is, mass is the integral of the density function over the region Q, just as in the one- and two-variable cases.

Example 5

A small wedge has the shape of the region

$$Q = \{(x, y, z) \mid -1 \le x \le 1, 0 \le y \le 2, 0 \le z \le y\}.$$

Find its mass if the density at any point (x, y, z) is given by the density function $\rho(x, y, z) = 1 + x^2 + y^2$ gram/cm^3 and the dimensions for Q are in cm (Figure 6.12).

Solution: According to equation (12) and Theorem 4 the mass is

$$M = \iiint_Q (1 + x^2 + y^2) \, dV = \int_{-1}^{1} \int_{0}^{2} \int_{0}^{y} (1 + x^2 + y^2) \, dz \, dy \, dx$$

$$= \int_{-1}^{1} \int_{0}^{2} \left\{ (1 + x^2 + y^2)z \Big]_{z=0}^{z=y} \right\} dy \, dx$$

$$= \int_{-1}^{1} \int_{0}^{2} (y + x^2 y + y^3) \, dy \, dx$$

$$= \int_{-1}^{1} \left\{ \frac{1}{2} y^2 + \frac{1}{2} x^2 y^2 + \frac{1}{4} y^4 \Big]_{y=0}^{y=2} \right\} dx$$

$$= \int_{-1}^{1} (6 + 2x^2) \, dx$$

$$= 6x + \frac{2}{3} x^3 \Big]_{-1}^{1}$$

$$= \frac{40}{3} \text{ grams.} \qquad \diamond$$

Moments and Center of Mass

Let Q denote a solid in $\mathbb{R}^3$, and let ρ be a continuous function giving the density $\rho(x, y, z)$ of Q at each point (x, y, z). As before, let Q be partitioned into rectangular boxes $Q_1, Q_2, \ldots, Q_n$ and let (s_j, t_j, u_j) be a point in Q_j, $j = 1, 2, \ldots, n$. Since s_j is the distance from the point (s_j, t_j, u_j) to the yz-plane, the product $s_j \rho(s_j, t_j, u_j) \, \Delta V_j$ may be interpreted as an approximation to the product of the mass of Q_j and the length of its "lever arm" extending from the yz-plane. By analogy with our previous discussions on moments (see Section 19.4), we define *the first moment, M_{yz}, of the solid Q with respect to the yz-plane* to be

$$M_{yz} = \iiint_Q x\rho(x, y, z) \, dV = \lim_{n \to \infty} \sum_{j=1}^{n} s_j \rho(s_j, t_j, u_j) \, \Delta V_j.$$

Similarly, *the moments of Q with respect to the xz- and xy-planes* are

$$M_{xz} = \iiint_Q y\rho(x, y, z) \, dV$$

and

$$M_{xy} = \iiint\limits_{Q} z\rho(x, y, z) \, dV.$$

Finally, if M denotes the mass of Q, the **center of mass** of Q is the point $(\bar{x}, \bar{y}, \bar{z})$ where

$$\bar{x} = \frac{M_{yz}}{M}; \qquad \bar{y} = \frac{M_{xz}}{M}; \qquad \bar{z} = \frac{M_{xy}}{M}.$$

Example 6

Find the three first moments and the center of mass for the solid in Example 5.

Solution: With $\rho(x, y, z) = 1 + x^2 + y^2$ and Q as given in Example 5, we have

$$M_{yz} = \iiint\limits_{Q} x(1 + x^2 + y^2) \, dV = \int_{-1}^{1} \int_{0}^{2} \int_{0}^{y} (x + x^3 + xy^2) \, dz \, dy \, dx$$

$$= \int_{-1}^{1} \int_{0}^{2} \left\{ (x + x^3 + xy^2)z \Big]_{z=0}^{z=y} \right\} dy \, dx$$

$$= \int_{-1}^{1} \int_{0}^{2} [(x + x^3)y + xy^3] \, dy \, dx$$

$$= \int_{-1}^{1} \left\{ (x + x^3)\frac{y^2}{2} + \frac{1}{4}xy^4 \Big]_{y=0}^{y=2} \right\} dx$$

$$= \int_{-1}^{1} (6x + 2x^3) \, dx$$

$$= 3x^2 + \frac{1}{2}x^4 \Big]_{-1}^{1}$$

$$= 0.$$

Similar calculations show that

$$M_{xz} = \iiint\limits_{Q} y(1 + x^2 + y^2) \, dV$$

$$= \int_{-1}^{1} \int_{0}^{2} \int_{0}^{y} (y + x^2y + y^3) \, dz \, dy \, dx = \frac{896}{45}$$

and that

$$M_{xy} = \iiint\limits_{Q} z(1 + x^2 + y^2) \, dV$$

$$= \int_{-1}^{1} \int_{0}^{2} \int_{0}^{y} (z + x^2z + y^2z) \, dz \, dy \, dx = \frac{448}{45}.$$

Since we have previously calculated the mass to be $M = \dfrac{40}{3}$, the coordinates of the center of mass $(\bar{x}, \bar{y}, \bar{z})$ are

$$\bar{x} = \frac{M_{yz}}{M} = 0 \cdot \frac{3}{40} = 0,$$

$$\bar{y} = \frac{M_{xz}}{M} = \frac{896}{45} \cdot \frac{3}{40} = \frac{112}{75} \approx 1.49,$$

and

$$\bar{z} = \frac{M_{xy}}{M} = \frac{448}{45} \cdot \frac{3}{40} = \frac{56}{75} \approx 0.75. \qquad \diamond$$

Exercise Set 18.6

In Exercises 1–8, evaluate the iterated integral.

1. $\displaystyle\int_0^1 \int_0^1 \int_0^1 xyz \, dx \, dy \, dz$

2. $\displaystyle\int_0^2 \int_{-\pi/2}^{\pi/2} \int_1^2 x \cos y e^z \, dx \, dy \, dz$

3. $\displaystyle\int_0^1 \int_0^y \int_0^x 3 \, dz \, dx \, dy$

4. $\displaystyle\int_1^3 \int_0^{\pi/4} \int_0^x \cos(x + y) \, dy \, dx \, dz$

5. $\displaystyle\int_0^2 \int_0^x \int_0^{x+y} z \, dz \, dy \, dx$

6. $\displaystyle\int_{-1}^1 \int_{-\sqrt{1-x^2}}^{\sqrt{1-x^2}} \int_0^{\sqrt{1-x^2-y^2}} 1 \, dz \, dy \, dx$ (*Hint*: Use geometry.)

7. $\displaystyle\int_{-1}^1 \int_0^y \int_0^x ye^{x^2+y^2} \, dz \, dx \, dy$ **8.** $\displaystyle\int_0^2 \int_0^x \int_{x+y}^{x^2+y^2} 1 \, dz \, dy \, dx$

9. Interchange the order of integration in

$$\int_0^1 \int_0^{2x} \int_0^{x+y} f(x, y, z) \, dz \, dy \, dx$$

from order $dz \, dy \, dx$ to order $dz \, dx \, dy$.

10. Use a triple integral to find the volume of the tetrahedron with vertices $(0, 0, 0)$, $(1, 0, 0)$, $(1, 1, 0)$, and $(1, 1, 1)$.

11. Sketch the solid whose volume is given by the iterated integral

$$\int_0^2 \int_0^{2x} \int_0^{x+y} dz \, dy \, dx.$$

12. Sketch the solid whose volume is given by the iterated integral

$$\int_{-1}^1 \int_{-\sqrt{1-x^2}}^{\sqrt{1-x^2}} \int_{\sqrt{x^2+y^2}}^{2-\sqrt{x^2+y^2}} dz \, dy \, dx.$$

13. Find the volume of the region in Exercise 12.

14. Find the volume of the tetrahedron bounded by the plane $x + y + z = 1$ and the coordinate planes $x = 0$, $y = 0$, and $z = 0$.

15. Find the volume of the region lying above the xy-plane, inside the cylinder $x^2 + y^2 = 9$, and below the plane $z = y + 3$.

16. Sketch the solid whose volume is given by the integral

$$V = \int_0^2 \int_0^{\sqrt{2x-x^2}} \int_0^{2-x} dz \, dy \, dx$$

and find the volume.

17. Evaluate the integral $\displaystyle\iiint_Q (3x + xz) \, dV$ where Q is the region bounded by the cylinder $x^2 + z^2 = 9$, the plane $y + z = 3$, and the plane $y = 0$.

18. Interchange the order of integration in the integral

$$\int_{-2}^2 \int_0^{\sqrt{4-x^2}} \int_{-\sqrt{4-x^2-y^2}}^{\sqrt{4-x^2-y^2}} f(x, y, z) \, dz \, dy \, dx$$

from $dz \, dy \, dx$ to $dy \, dz \, dx$.

19. Find the volume of the region bounded by the paraboloids $x = y^2 + z^2$ and $x = 2 - y^2 - z^2$.

20. Find the volume of the solid bounded above by the paraboloid $z = 2 - x^2 - y^2$ and below by the plane $z = 2 - 2x$.

21. Let Q be the solid bounded by the cylinder $x^2 + y^2 = 9$ and the planes $z = 0$ and $x + z = 3$. Find the mass of Q if the density at each point (x, y, z) is given by the function $\rho(x, y, z) = z$.

22. Find the volume of the region common to the cylinders $x^2 + z^2 = 1$ and $y^2 + z^2 = 1$.

23. The density at each point of the box $Q = \{(x, y, z) \mid 0 \le x \le 1,\ 0 \le y \le 2,\ 0 \le z \le 2\}$ is proportional to the square of the distance from the origin. Find the mass.

24. Find the center of mass of the region in Exercise 14 if the density is constant.

25. Find the center of mass of the solid in Exercise 15 if the density is constant.

26. Find the center of mass of the solid in Exercise 21.

27. Find the center of mass of the part of the region enclosed by

the sphere $x^2 + y^2 + z^2 = 4$ lying in the first octant if the density is constant.

28. Find the center of mass of the cube $Q = \{(x, y, z) \mid 0 \le x \le 1,\ 0 \le y \le 1,\ 0 \le z \le 1\}$ if the density function is $\rho(x, y, z) = xyz$.

29. Find five other iterated integrals by which the triple integral of Example 3 may be evaluated.

30. Find the volume, mass, and center of gravity for the solid in Example 5 if the density is uniform $\rho(x, y, z) = 1$ gram/cm^3. Compare your answers with those in Examples 5 and 6, and account for the differences in physical terms.

18.7 TRIPLE INTEGRALS IN CYLINDRICAL AND SPHERICAL COORDINATES

The purpose of this section is to define the triple integral for functions written in cylindrical or spherical coordinates. In part, we want to know how to calculate volumes and masses for solids described in these coordinate systems. Another reason for studying these topics is that certain triple integrals, originally expressed in Cartesian coordinates, are more easily evaluated by changing to either cylindrical or spherical coordinates.

Triple Integrals in Cylindrical Coordinates

Recall the relationship between cylindrical and Cartesian coordinates: If a point P has cylindrical coordinates $P = (r, \theta, z)$ and Cartesian coordinates $P = (x, y, z)$, then (r, θ) are the polar coordinates for the point (x, y) in the xy-plane (Figure 7.1). Now suppose that $\mathbb{R}^3$ is coordinatized by cylindrical coordinates, that Q is a region in $\mathbb{R}^3$, and that f is a continuous function defined on Q. We shall define the triple integral of f over Q in the usual way—by partitioning the region Q, forming an approximating sum, and obtaining the limit of the approximating sum.

Suppose that the region Q lies within the "cylindrical box" determined by the inequalities

$$r_a \le r \le r_b, \qquad \theta_a \le \theta \le \theta_b, \qquad z_a \le z \le z_b.$$

(See Figure 7.1.) We divide the region containing Q into small cylindrical blocks

$$Q_{ijk} = \{(r, \theta, z) \mid r_{i-1} \le r \le r_i,\ \theta_{j-1} \le \theta \le \theta_j,\ z_{k-1} \le z \le z_k\}$$

as illustrated in Figure 7.2. According to equation (3), Section 19.3, the area of the base of block Q_{ijk} is

$$\Delta A_{ijk} = \hat{r}_i \Delta r_i \Delta \theta_j$$

where $\hat{r}_i = \dfrac{1}{2}(r_{i-1} + r_i)$. Since the block Q_{ijk} has height Δz_k, the volume of the block Q_{ijk} is

$$\Delta V_{ijk} = \hat{r}_i \Delta r_i \Delta \theta_j \Delta z_k. \tag{1}$$

Now let θ_j^* be any number in the interval $[\theta_{j-1}, \theta_j]$ and let z_k^* be any number in the interval $[z_{k-1}, z_k]$. Then, since $\hat{r}_i \in [r_{i-1}, r_i]$, the point $(\hat{r}_i, \theta_j^*, z_k^*)$ lies in the

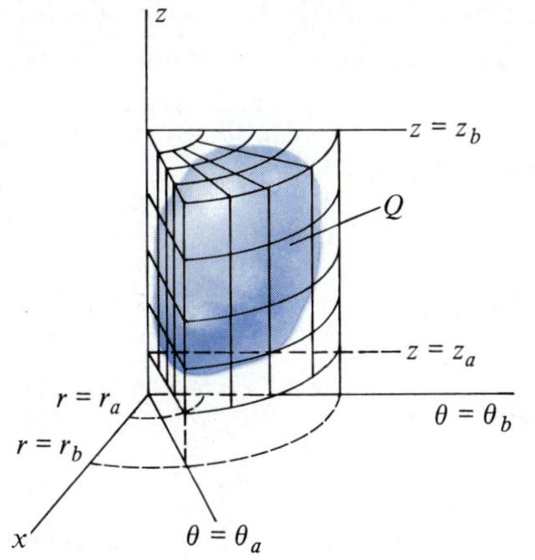

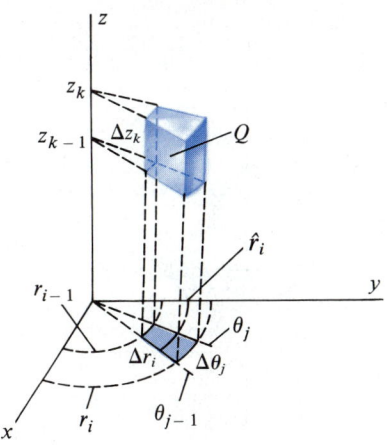

Figure 7.1 Partitioning the region Q using a grid specified in cylindrical coordinates.

Figure 7.2 Volume of circular wedge is

$$\Delta V_{ijk} = \hat{r}_i\,\Delta r_i\,\Delta\theta_j\,\Delta z_k$$

where $\hat{r}_i = \dfrac{r_{i-1} + r_i}{2}$.

block Q_{ijk}. The sum

$$\sum_i \sum_j \sum_k f(\hat{r}_i,\,\theta_j^*,\,z_k^*)\,\Delta V_{ijk} = \sum_i \sum_j \sum_k f(\hat{r}_i,\,\theta_j^*,\,z_k^*)\hat{r}_i\Delta r_i\Delta\theta_j\Delta z_k \qquad (2)$$

is the approximating Riemann sum for f over Q, where the sum is taken over all boxes Q_{ijk} lying entirely within Q. Its limit as the sizes of all blocks approach zero (written $\|P\| \to 0$ as before), which exists when f is continuous and Q is as described in the following theorem, is the triple integral of f over Q:

$$\iiint\limits_Q f(r,\,\theta,\,z)\,dV = \lim_{\|P\|\to 0} \sum_i \sum_j \sum_k f(\hat{r}_i,\,\theta_j^*,\,z_k^*)\hat{r}_i\Delta r_i\Delta\theta_j\Delta z_k. \qquad (3)$$

The following theorem shows how this triple integral may be evaluated for the types of regions encountered in this section and in most applications.

THEOREM 5

Let Q be a region in $\mathbb{R}^3$ of the form

$$Q = \{(r,\,\theta,\,z) \mid a \le \theta \le b,\, h_1(\theta) \le r \le h_2(\theta),\, g_1(r,\,\theta) \le z \le g_2(r,\,\theta)\}$$

where g_1, g_2, h_1, and h_2 are continuous functions. Let f be continuous on Q. Then

$$\iiint\limits_Q f(r,\,\theta,\,z)\,dV = \int_a^b \int_{h_1(\theta)}^{h_2(\theta)} \int_{g_1(r,\,\theta)}^{g_2(r,\,\theta)} f(r,\,\theta,\,z)\, r\, dz\, dr\, d\theta. \qquad (4)$$

REMARK 1: Note the extra factor r appearing in the iterated integral in (4). The reason for its appearance is the same as in Section 18.3: it can be considered a result of converting dV to cylindrical coordinates.

REMARK 2: When $f(r, \theta, z) \equiv 1$, it follows from equation (1) that the Riemann sum in (2) approximates the volume of Q. Thus

$$\text{Volume of } Q = \iiint\limits_{Q} dV \tag{5}$$

holds in cylindrical coordinates as well as in rectangular coordinates.

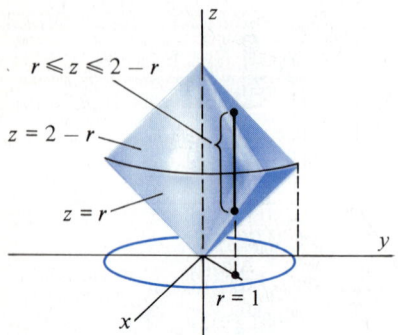

$r \leq z \leq 2 - r$

$z = 2 - r$

$z = r$

$r = 1$

Figure 7.3 Solid Q in Example 1.

Example 1

A solid "top" has the shape of the region

$$Q = \{(r, \theta, z) \mid 0 \leq r \leq 1, 0 \leq \theta \leq 2\pi, r \leq z \leq 2 - r\}$$

as illustrated in Figure 7.3. Find the mass of the object if the density at any point is proportional to the distance from the z-axis.

Solution: The density function described here is $\rho(r, \theta, z) = \lambda r$, where λ is the constant of proportionality. By equation (12), Section 18.6, and Theorem 5, we have

$$\text{Mass} = \iiint\limits_{Q} \rho(r, \theta, z) \, dV = \iiint\limits_{Q} \lambda r \, dV$$

$$= \int_{0}^{2\pi} \int_{0}^{1} \int_{r}^{2-r} \lambda r^2 \, dz \, dr \, d\theta \qquad \text{— note extra factor } r$$

$$= \int_{0}^{2\pi} \int_{0}^{1} \left\{ \lambda r^2 z \Big]_{z=r}^{z=2-r} \right\} dr \, d\theta$$

$$= \int_{0}^{2\pi} \int_{0}^{1} (2\lambda r^2 - 2\lambda r^3) \, dr \, d\theta$$

$$= \int_{0}^{2\pi} \left\{ \frac{2\lambda}{3} r^3 - \frac{\lambda}{2} r^4 \right]_{r=0}^{r=1} \right\} d\theta$$

$$= \int_{0}^{2\pi} \left(\frac{\lambda}{6} \right) d\theta$$

$$= \frac{\pi\lambda}{3}. \qquad \diamondsuit$$

Changing to Cylindrical Coordinates

You will sometimes find that a triple integral written in Cartesian coordinates is more easily evaluated by first changing to cylindrical coordinates. Doing so is analogous to changing a double integral from Cartesian to polar coordinates. Specifically, to write the iterated integral

$$\int_{a}^{b} \int_{h_1(x)}^{h_2(x)} \int_{g_1(x, y)}^{g_2(x, y)} f(x, y, z) \, dz \, dy \, dx$$

in cylindrical coordinates, we do the following:

(i) Express the region

$$Q = \{(x, y, z) \mid a \le x \le b, h_1(x) \le y \le h_2(x), g_1(x, y) \le z \le g_2(x, y)\}$$

in cylindrical coordinates as

$$Q = \{(r, \theta, z) \mid c \le \theta \le d, h_3(\theta) \le r \le h_4(\theta), g_3(r, \theta) \le z \le g_4(r, \theta)\}.$$

(ii) Using the substitutions $x = r \cos \theta$ and $y = r \sin \theta$, replace the integrand $f(x, y, z)$ by $f(r \cos \theta, r \sin \theta, z)r$. (Do not forget the extra factor r.)

(iii) Obtain the equation

$$\int_a^b \int_{h_1(x)}^{h_2(x)} \int_{g_1(x, y)}^{g_2(x, y)} f(x, y, z) \, dz \, dy \, dx \qquad (6)$$
$$= \int_c^d \int_{h_3(\theta)}^{h_4(\theta)} \int_{g_3(r, \theta)}^{g_4(r, \theta)} f(r \cos \theta, r \sin \theta, z) \, r \, dz \, dr \, d\theta.$$

Equation (6) is verified by comparing Theorems 4 and 5, each of which expresses the triple integral $\iiint_Q f \, dV$ as one of the two iterated integrals in (6). One way to paraphrase equation (6) is to say that *in changing from Cartesian coordinates to cylindrical coordinates, the* **volume element**

$$dV = dz \, dy \, dx$$

in Cartesian coordinates is replaced by the volume element

$$dV = r \, dz \, dr \, d\theta \qquad (7)$$

in cylindrical coordinates. It is very important to note the extra factor r that appears in the integrand on the right-hand side of (6) and in equation (7).

Example 2

Calculate the volume of the ellipsoid

$$4x^2 + 4y^2 + z^2 = 4.$$

Solution: The ellipsoid is sketched in Figure 7.4. Since the ellipsoid is symmetric with respect to the xy-plane, we may calculate the volume as twice the volume of the region Q lying above the xy-plane. Since this region is described by the inequalities

$$0 \le z \le 2\sqrt{1 - x^2 - y^2}, \qquad -\sqrt{1 - x^2} \le y \le \sqrt{1 - x^2}, \qquad -1 \le x \le 1,$$

the volume is given by the iterated integral

$$V = 2 \int_{-1}^1 \int_{-\sqrt{1-x^2}}^{\sqrt{1-x^2}} \int_0^{2\sqrt{1-x^2-y^2}} 1 \, dz \, dy \, dx. \qquad (8)$$

Clearly, this integral will be difficult to evaluate, so we try switching to cylindrical coordinates. In cylindrical coordinates the region Q is described by the inequalities

$$0 \le r \le 1, \qquad 0 \le \theta \le 2\pi, \qquad 0 \le z \le 2\sqrt{1 - r^2}$$

since $2\sqrt{1 - x^2 - y^2} = 2\sqrt{1 - (x^2 + y^2)} = 2\sqrt{1 - r^2}$. Using equation (6) we

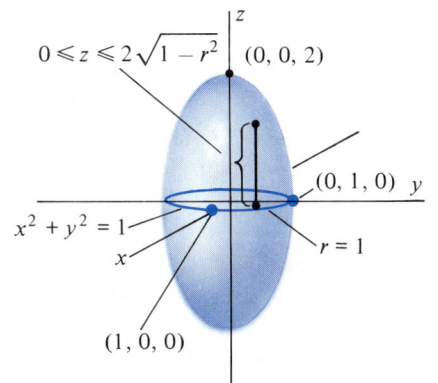

$0 \le z \le 2\sqrt{1 - r^2}$ $(0, 0, 2)$

$(0, 1, 0)$ y

$x^2 + y^2 = 1$

x $r = 1$

$(1, 0, 0)$

Figure 7.4 Ellipsoid

$4x^2 + 4y^2 + z^2 = 4$

in Example 2.

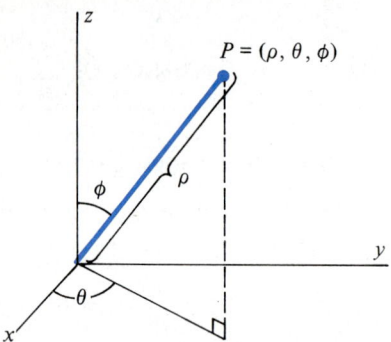

Figure 7.5 Spherical coordinates (ρ, θ, ϕ) for P.

may rewrite the integral (8) as

$$V = 2 \int_0^{2\pi} \int_0^1 \int_0^{2\sqrt{1-r^2}} r\, dz\, dr\, d\theta \quad \text{note the extra factor } r$$

$$= 2 \int_0^{2\pi} \int_0^1 \left\{ zr \right]_{z=0}^{z=2\sqrt{1-r^2}} \right\} dr\, d\theta$$

$$= 2 \int_0^{2\pi} \int_0^1 2r\sqrt{1-r^2}\, dr\, d\theta$$

$$= 2 \int_0^{2\pi} \left\{ -\frac{2}{3}(1-r^2)^{3/2} \right]_{r=0}^{r=1} \right\} d\theta$$

$$= \frac{8\pi}{3}. \qquad \diamond$$

Spherical Coordinates

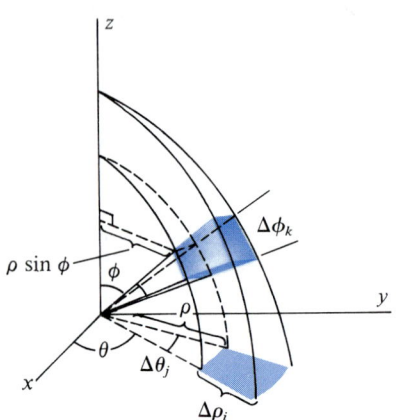

Figure 7.6 The spherical block.

Recall that the point P has spherical coordinates $P = (\rho, \theta, \phi)$ if $\rho \geq 0$ is its distance from the origin, θ is its rotation angle (as in cylindrical coordinates), and ϕ is its angle of inclination from the vertical (see Figure 7.5). Figure 7.6 illustrates the ''spherical block'' determined by the inequalities

$$\rho_{i-1} \leq \rho \leq \rho_i, \qquad \theta_{j-1} \leq \theta \leq \theta_j, \qquad \phi_{k-1} \leq \phi \leq \phi_k \qquad (9)$$

where $\Delta\rho_i = \rho_i - \rho_{i-1}$, $\Delta\theta_j = \theta_j - \theta_{j-1}$, and $\Delta\phi_k = \phi_k - \phi_{k-1}$. A fact, which we shall not prove, is that the volume of this spherical block is given by

$$\Delta V_{ijk} = \rho_i^{*2} \sin \phi_k^* \, \Delta\rho_i \Delta\theta_j \Delta\phi_k$$

for some $\rho_i^* \in [\rho_{i-1}, \rho_i]$ and $\phi_k^* \in [\phi_{k-1}, \phi_k]$ (see Exercise 31).

In order to define the integral of a continuous function f defined on Q, we select one point $(\rho_i^*, \theta_j^*, \phi_k^*)$ in each block Q_{ijk}, form the approximating sum, and evaluate the limit of this sum as the sizes of the blocks approach zero. When it exists, it is the triple integral of f over Q:

$$\iiint_Q f(\rho, \theta, \phi)\, dV$$

$$= \lim_{n \to \infty} \sum_i \sum_j \sum_k f(\rho_i^*, \theta_j^*, \phi_k^*)\rho_i^{*2} \sin \phi_k^* \, \Delta\rho_i \Delta\theta_j \Delta\phi_k.$$

As with other types of triple integrals, the triple integral in spherical coordinates may be evaluated as an iterated integral in certain cases.

THEOREM 6

Let Q be a region in $\mathbb{R}^3$ of the form

$$Q = \{(\rho, \theta, \phi) \mid a \leq \theta \leq b, h_1(\theta) \leq \phi \leq h_2(\theta), g_1(\theta, \phi) \leq \rho \leq g_2(\theta, \phi)\}$$

where h_1, h_2, g_1, and g_2 are continuous functions. Let f be continuous on Q. Then,

$$\iiint_Q f(\rho, \theta, \phi)\, dV = \int_a^b \int_{h_1(\theta)}^{h_2(\theta)} \int_{g_1(\theta, \phi)}^{g_2(\theta, \phi)} f(\rho, \theta, \phi)\, \rho^2 \sin \phi\, d\rho\, d\phi\, d\theta.$$

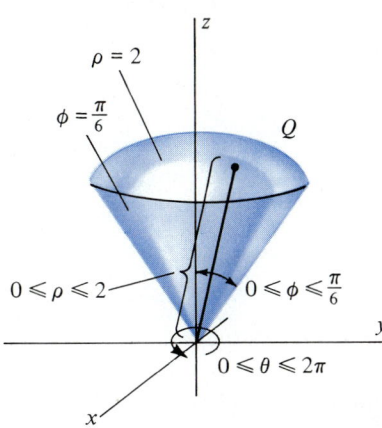

Figure 7.7 Region Q of Example 3.

REMARK: Note the extra factor $\rho^2 \sin \phi$ in the integrand in the iterated integral.

Example 3

Find the mass of the solid occupying the region

$$Q = \{(\rho, \theta, \phi) \mid 0 \le \rho \le 2,\ 0 \le \theta \le 2\pi,\ 0 \le \phi \le \pi/6\}$$

if the density at any point is proportional to its distance from the origin.

Solution: The region Q is the "ice cream cone" region sketched in Figure 7.7. As before, we obtain the mass of the solid by integrating the density function $D(\rho, \theta, \phi) = \lambda\rho$. (Note that we are now using ρ for radial distance, not density.) Thus, by Theorem 6

$$\text{Mass} = \iiint\limits_{Q} \lambda\rho\, dV = \int_0^{2\pi} \int_0^{\pi/6} \int_0^2 (\lambda\rho)(\rho^2 \sin \phi)\, d\rho\, d\phi\, d\theta$$

$$\text{note extra factor } \rho^2 \sin \phi$$

$$= \int_0^{2\pi} \int_0^{\pi/6} \int_0^2 \lambda\rho^3 \sin \phi\, d\rho\, d\phi\, d\theta$$

$$= \int_0^{2\pi} \int_0^{\pi/6} \left\{ \frac{\lambda}{4}\rho^4 \sin \phi \right]_{\rho=0}^{\rho=2} \right\} d\phi\, d\theta$$

$$= \int_0^{2\pi} \int_0^{\pi/6} 4\lambda \sin \phi\, d\phi\, d\theta$$

$$= \int_0^{2\pi} \left\{ -4\lambda \cos \phi \right]_{\phi=0}^{\phi=\pi/6} \right\} d\theta$$

$$= \int_0^{2\pi} 4\lambda\left(1 - \frac{\sqrt{3}}{2}\right) d\theta$$

$$= 8\lambda\pi\left(1 - \frac{\sqrt{3}}{2}\right). \qquad \diamondsuit$$

Example 4

Obtain the formula for the volume of a sphere of radius a using spherical coordinates.

Solution: The sphere of radius a is described by the spherical inequalities

$$0 \le \rho \le a, \qquad 0 \le \phi \le \pi, \qquad 0 \le \theta \le 2\pi \qquad \text{(Figure 7.8).}$$

Thus

$$\text{Volume} = \iiint\limits_{Q} 1\, dV = \int_0^{2\pi} \int_0^{\pi} \int_0^a \rho^2 \sin \phi\, d\rho\, d\phi\, d\theta$$

$$\text{note extra factor } \rho^2 \sin \phi$$

$$= \int_0^{2\pi} \int_0^{\pi} \left\{ \frac{1}{3}\rho^3 \sin \phi \right]_{\rho=0}^{\rho=a} \right\} d\phi\, d\theta$$

$$= \int_0^{2\pi} \int_0^{\pi} \frac{a^3}{3} \sin \phi\, d\phi\, d\theta$$

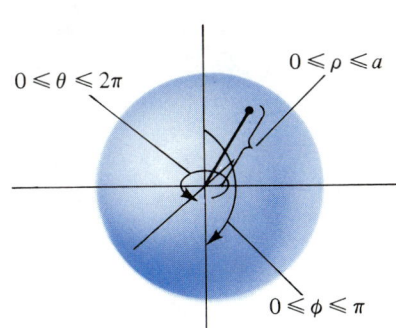

Figure 7.8 Sphere is determined by the inequalities

$0 \le \rho \le a$,
$0 \le \phi \le \pi$, $0 \le \theta \le 2\pi$.

$$= \int_0^{2\pi} \left\{ -\frac{a^3}{3} \cos \phi \right]_{\phi=0}^{\phi=\pi} \right\} d\theta$$

$$= \int_0^{2\pi} \frac{2a^3}{3} \, d\theta$$

$$= \frac{4}{3} \cdot \pi a^3.$$

Changing From Cartesian to Spherical Coordinates

If we encounter a function f written in Cartesian coordinates for which we wish to calculate a triple integral over a spherical region, it is often useful to rewrite the function in spherical coordinates using the substitutions

$$x = \rho \sin \phi \cos \theta, \qquad y = \rho \sin \phi \sin \theta, \qquad z = \rho \cos \phi.$$

(See Section 15.6 for a review of how these equations are obtained.) We shall not write out a general formula for converting triple integrals from Cartesian coordinates to spherical coordinates, since the situations in which such changes are possible are almost exclusively limited to functions involving the expression

$$x^2 + y^2 + z^2 = \rho^2 \sin^2 \phi \cos^2 \theta + \rho^2 \sin^2 \phi \sin^2 \theta + \rho^2 \cos^2 \phi$$
$$= \rho^2 \sin^2 \phi (\cos^2 \theta + \sin^2 \theta) + \rho^2 \cos^2 \phi$$
$$= \rho^2.$$

Example 5

A hemispherical solid is described in Cartesian coordinates as

$$Q = \{(x, y, z) \mid 0 \le x^2 + y^2 \le 1, 0 \le z \le \sqrt{1 - x^2 - y^2}\}.$$

Find the mass of the object if the density at the point (x, y, z) is given by the function

$$D(x, y, z) = e^{-(x^2+y^2+z^2)^{3/2}}.$$

Solution: The hemisphere may be described by the inequalities

$$0 \le \rho \le 1, \qquad 0 \le \phi \le \pi/2, \qquad 0 \le \theta \le 2\pi,$$

and the density function, in spherical coordinates, is

$$D(\rho, \theta, \phi) = e^{-\rho^3}.$$

Thus,

$$\text{Mass} = \int_0^{2\pi} \int_0^{\pi/2} \int_0^1 e^{-\rho^3} \underbrace{\rho^2 \sin \phi} \, d\rho \, d\phi \, d\theta$$

note extra factor $\rho^2 \sin \phi$

$$= \int_0^{2\pi} \int_0^{\pi/2} \left\{ -\frac{1}{3} e^{-\rho^3} \sin \phi \right]_{\rho=0}^{\rho=1} \right\} d\phi \, d\theta$$

$$= \int_0^{2\pi} \int_0^{\pi/2} \frac{1}{3} \left(1 - \frac{1}{e} \right) \sin \phi \, d\phi \, d\theta$$

$$= \int_0^{2\pi} \left\{ \left(-\frac{e-1}{3e} \right) \cos \phi \Big]_{\phi=0}^{\phi=\pi/2} \right\} d\theta$$

$$= \int_0^{2\pi} \left(\frac{e-1}{3e} \right) d\theta$$

$$= \frac{2\pi(e-1)}{3e}. \qquad \Diamond$$

Exercise Set 18.7

In Exercises 1–8, use cylindrical coordinates.

1. Find the volume of the solid bounded by the graphs of $z = x^2 + y^2$ and $z = 9$.

2. Find the volume of the solid bounded by the graphs of the equations $z = 9 - x^2 - y^2$, $x^2 + y^2 = 3$, and $z = 9$.

3. Find the center of mass of a solid having the shape of the solid in Exercise 1 if the density is constant.

4. Find the volume of the region lying inside both the cylinder $x^2 - 2x + y^2 = 0$ and the sphere $x^2 + y^2 + z^2 = 4$.

5. Find the volume of the solid remaining when the region inside the cylinder $r = 3 \sin \theta$ is removed from the solid region bounded by the sphere $x^2 + y^2 + z^2 = 9$.

6. A solid is bounded by the cylinder $y^2 + z^2 = 4$ and the planes $x = 0$ and $x = 4$. Find its mass if the density at each point is proportional to the distance of that point from the central axis of the cylinder.

7. Find the volume of the solid bounded above and below by the cone $z^2 = 2x^2 + 2y^2$ and on the sides by the cylinder $x^2 + y^2 - 4y = 0$.

8. Find the volume of the solid bounded above by the plane $z = x$ and below by the paraboloid $z = x^2 + y^2$.

In Exercises 9–12, evaluate the iterated integral by first changing to cylindrical coordinates.

9. $\int_0^2 \int_{-\sqrt{2x-x^2}}^{\sqrt{2x-x^2}} \int_0^{\sqrt{x^2+y^2}} 1 \, dz \, dy \, dx$

10. $\int_{-1}^1 \int_{-\sqrt{1-x^2}}^{\sqrt{1-x^2}} \int_{\sqrt{x^2+y^2}}^1 x^2 \, dz \, dy \, dx$

11. $\int_{-2}^2 \int_{-\sqrt{4-x^2}}^{\sqrt{4-x^2}} \int_0^{y+2} xy \, dz \, dy \, dx$

12. $\int_{-1}^1 \int_0^{\sqrt{1-x^2}} \int_{x^2+y^2}^1 z \, dz \, dy \, dx$

In Exercises 13–18, use spherical coordinates.

13. Find the volume of the solid bounded above by the sphere $x^2 + y^2 + z^2 - 2z = 0$ and below by the cone $z = \sqrt{x^2 + y^2}$.

14. Find the mass of a sphere of radius $r = 2$ if the density at each point is proportional to the square of the distance from the center.

15. Find the center of mass of the quarter sphere

$$Q = \{(\rho, \theta, \phi) \mid 0 \le \rho \le 1, 0 \le \phi \le \pi/2, 0 \le \theta \le \pi\}$$

if the density is constant.

16. Find the mass of the solid lying between the spheres $x^2 + y^2 + z^2 = 1$ and $x^2 + y^2 + z^2 = 4$ if the density at each point is proportional to the reciprocal of the distance from the center of the spheres.

17. Find the volume of the solid remaining when the region lying inside the cone $z^2 = x^2 + y^2$ is removed from the region bounded by the sphere $x^2 + y^2 + z^2 = 4$.

18. Use spherical coordinates to obtain the formula for the volume of a right circular cone of radius r and height h.

In Exercises 19 and 20, evaluate the iterated integral by first changing to spherical coordinates.

19. $\int_{-1}^1 \int_{-\sqrt{1-x^2}}^{\sqrt{1-x^2}} \int_0^{\sqrt{1-x^2-y^2}} 3 \, dz \, dy \, dx$

20. $\int_{-2}^2 \int_{-\sqrt{4-x^2}}^{\sqrt{4-x^2}} \int_{-\sqrt{4-x^2-y^2}}^{\sqrt{4-x^2-y^2}} (x^2 + y^2) \, dz \, dy \, dx$

21. Two circular cylinders of radius R meet at right angles. Use triple integration to find the volume of the solid common to both cylinders.

22. Find the volume of the region lying inside the sphere $x^2 + y^2 + z^2 = 4$ and outside the cylinder $x^2 + z^2 = 1$.

23. Evaluate the triple integral

$$\iiint_Q \frac{z}{(x^2 + y^2)^{3/2}} \, dV$$

where Q is the region

$$Q = \{(x, y, z) \mid 1 \le x^2 + y^2 \le 3, 0 \le z \le 3\}.$$

24. Evaluate the triple integral $\iiint\limits_{Q} \cos \pi y \cdot \sqrt{x^2 + z^2}\, dV$

where Q is the region bounded by the cylinders $x^2 + z^2 = 1$ and $x^2 + z^2 = 4$, and the planes $y = -1$ and $y = 2$.

25. Evaluate the triple integral $\iiint\limits_{Q} \sqrt{\dfrac{x}{y^2 + z^2}}\, dV$ where Q is

the region bounded by the cone $y^2 + z^2 = x^2$, the cylinder $y^2 + z^2 = 4$, and the planes $x = 0$ and $x = 2$.

26. Find the volume of the solid that remains when the cone $3y^2 = x^2 + z^2$ is removed from the sphere $x^2 + y^2 + z^2 = 4$.

27. Find the volume of the smaller of the two parts of the sphere $\rho = 4$ determined by the plane $y = 2$.

28. Evaluate the integral $\iiint\limits_{Q} (x^2 + y^2)\, dV$ where Q is the

sphere $x^2 + y^2 + z^2 \leq 4$.

29. Find the volume of the solid lying between the spheres $x^2 + y^2 + z^2 = 1$ and $x^2 + y^2 + z^2 = 9$ and inside the cone $y^2 = x^2 + z^2$.

30. Find the volume of the solid bounded below by the graph of $x^2 + y^2 + z^2 + 2z = 0$ and above by the graph of $z^2 = x^4 + 2x^2y^2 + y^4$.

31. Obtain the approximation $\Delta V_{ijk} \approx \rho_i^2 \sin \phi_k\, \Delta\rho_i \Delta\theta_j \Delta\phi_k$ for the spherical block in Figure 7.9 as follows.

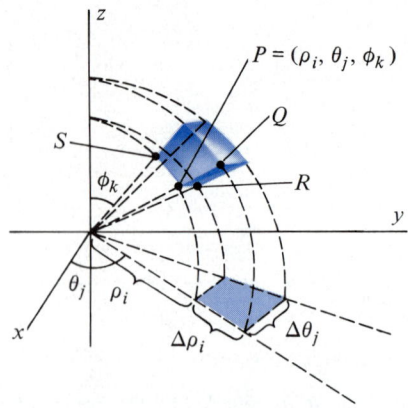

Figure 7.9 Spherical block of volume ΔV_{ijk}.

a. Observe that the spherical block is nearly a parallelepiped. We shall therefore approximate its volume by the product (length of PQ) · (length of PS) · (length of PR).

b. Show that the length of $PQ = \Delta\rho_i$.

c. Show that the length of the arc PS is $\rho_i \Delta\phi_k$. (*Hint:* It lies on a circle of radius ρ_i.)

d. Show that the length of the arc PR is $\rho_i \sin \phi_k\, \Delta\theta_j$. (*Hint:* It lies on a horizontal circle of radius $\rho_i \sin \phi_k$. Why?)

SUMMARY OUTLINE OF CHAPTER 18

◆ If f is continuous on the planar region $Q = \{(x, y) \mid a \leq x \leq b,\, g_1(x) \leq y \leq g_2(x)\}$ then $\qquad$ (page 838)

(i) $\displaystyle\iint\limits_{Q} f(x, y)\, dA = \int_a^b \int_{g_1(x)}^{g_2(x)} f(x, y)\, dy\, dx.$

(ii) Area of $Q = \displaystyle\iint\limits_{Q} 1\, dA.$

◆ If f is continuous on the region Q described in polar coordinates as $\qquad$ (page 849)

$Q = \{(r, \theta) \mid g_1(\theta) \leq r \leq g_2(\theta),\, a \leq \theta \leq b\},$

then

(i) $\displaystyle\iint\limits_{Q} f(r, \theta)\, dA = \int_a^b \int_{g_1(\theta)}^{g_2(\theta)} f(r, \theta)\, r\, dr\, d\theta.$

(ii) Area of $Q = \displaystyle\int_a^b \int_{g_1(\theta)}^{g_2(\theta)} r\, dr\, d\theta$ if $g_1(\theta) \geq 0$ for $a \leq \theta \leq b.$

◆ In changing an iterated double integral from Cartesian to polar coordinates, the differential for area $dA = dx\, dy$ is $\qquad$ (page 853) replaced by the polar differential for area $dA = r\, dr\, d\theta.$

◆ If a thin lamina with density ρ has the shape of the planar region Q, then $\qquad$ (page 859)

(i) Mass $= M = \displaystyle\iint\limits_{Q} \rho(x, y)\, dA.$

(ii) Center of Mass $= (\bar{x}, \bar{y})$ where $\bar{x} = \dfrac{M_y}{M}$, $\bar{y} = \dfrac{M_x}{M}$,

$$M_y = \iint\limits_{Q} x\rho(x, y)\, dA,$$

and

$$M_x = \iint\limits_{Q} y\rho(x, y)\, dA.$$

◆ The **surface area** of the graph of $z = f(x, y)$ over the region Q is   (page 864)

$$A_S = \iint\limits_{Q} \sqrt{\left[\frac{\partial f}{\partial x}(x, y)\right]^2 + \left[\frac{\partial f}{\partial y}(x, y)\right]^2 + 1}\; dA.$$

◆ If the region Q in $\mathbb{R}^3$ is described by the inequalities $a \leq x \leq b$, $h_1(x) \leq y \leq h_2(x)$, $g_1(x, y) \leq z \leq g_2(x, y)$,   (page 868) then

(i) $\displaystyle\iiint\limits_{Q} f(x, y, z)\, dV = \int_a^b \int_{h_1(x)}^{h_2(x)} \int_{g_1(x, y)}^{g_2(x, y)} f(x, y, z)\, dz\, dy\, dx.$

(ii) Volume of $Q = \displaystyle\iiint\limits_{Q} 1\, dV.$

◆ If a solid has the shape of the region Q in $\mathbb{R}^3$ and density function ρ, then   (page 874)

(i) Mass $= M = \displaystyle\iiint\limits_{Q} \rho(x, y, z)\, dV.$

(ii) Center of mass $= (\bar{x}, \bar{y}, \bar{z})$ where $\bar{x} = \dfrac{M_{yz}}{M}$, $\bar{y} = \dfrac{M_{xz}}{M}$, $\bar{z} = \dfrac{M_{xy}}{M}$,

$$M_{yz} = \iiint\limits_{Q} x\rho(x, y, z)\, dV,$$

$$M_{xz} = \iiint\limits_{Q} y\rho(x, y, z)\, dV,$$

$$M_{xy} = \iiint\limits_{Q} z\rho(x, y, z)\, dV.$$

◆ If the region Q is described in cylindrical coordinates by the inequalities $a \leq \theta \leq b$, $h_1(\theta) \leq r \leq h_2(\theta)$,   (page 879) $g_1(r, \theta) \leq z \leq g_2(r, \theta)$, then

$$\iiint\limits_{Q} f(r, \theta, z)\, dV = \int_a^b \int_{h_1(\theta)}^{h_2(\theta)} \int_{g_1(r, \theta)}^{g_2(r, \theta)} f(r, \theta, z) r\, dz\, dr\, d\theta.$$

◆ In changing an iterated triple integral from Cartesian coordinates to cylindrical coordinates, the Cartesian differential   (page 881) for volume $dV = dx\, dy\, dz$ is replaced by the cylindrical differential for volume $dV = r\, dz\, dr\, d\theta$.

◆ If the region Q in $\mathbb{R}^3$ is described in spherical coordinates by the inequalities $a \leq \theta \leq b$, $h_1(\theta) \leq \phi \leq h_2(\theta)$,   (page 882) $g_1(\theta, \phi) \leq \rho \leq g_2(\theta, \phi)$, then

$$\iiint\limits_{Q} f(\rho, \theta, \phi)\, dV = \int_a^b \int_{h_1(\theta)}^{h_2(\theta)} \int_{g_1(\theta, \phi)}^{g_2(\theta, \phi)} f(\rho, \theta, \phi)\rho^2 \sin\phi\, d\rho\, d\phi\, d\theta.$$

◆ In changing an iterated triple integral from Cartesian to spherical coordinates, the Cartesian differential for volume   (page 884) $dV = dx\, dy\, dz$ is replaced by the spherical differential for volume $dV = \rho^2 \sin\phi\, d\rho\, d\phi\, d\theta.$

REVIEW EXERCISES—CHAPTER 18

1. Find the volume of the solid bounded above by the graph of $z = xy^2$ and below by the triangle with vertices $(0, 0)$, $(2, 0)$, and $(0, 1)$.

2. Evaluate the double integral $\iint\limits_Q \dfrac{x}{xy + 2} \, dA$ where Q is the rectangle

   $Q = \{(x, y) \mid 0 \le x \le 1, 0 \le y \le 1\}$.

3. Evaluate $\iint\limits_Q e^{-x^2/2} \, dy \, dx$ where Q is the region

   $Q = \{(x, y) \mid 0 \le x \le 1, 0 \le y \le 2x\}$.

4. Evaluate $\iint\limits_Q (x - 4y) \, dA$ where Q is the rectangle

   $Q = \{(x, y) \mid -1 \le x \le 1, 0 \le y \le 2\}$.

5. Find the volume of the solid in the first octant bounded by the graphs of $z = y^2$, $y = x$, $z = 0$, and $y = 4$.

6. Change the order of integration:

   $$\int_{-1}^{2} \int_{x^2-2}^{x} f(x, y) \, dy \, dx.$$

7. Find the surface area of that part of the sphere $x^2 + y^2 + z^2 = 4$ lying inside the cylinder $x^2 + y^2 - 2x = 0$.

8. Find the center of mass of a solid having the shape of the region bounded above by the paraboloid $4z = 4 - (x^2 + y^2)$ and below by the xy-plane if the density of the material is uniform.

9. Find the volume of the solid bounded by the cylinder $r = 2 \cos \theta$, the paraboloid $z = 2r^2$ and the plane $z = 0$ (use cylindrical coordinates).

10. Rewrite the integral

    $$\int_{-\pi/2}^{\pi/2} \int_{0}^{2\cos\theta} \int_{-\sqrt{4-r^2}}^{\sqrt{4-r^2}} r \, dz \, dr \, d\theta$$

    in rectangular coordinates.

11. Find the volume of the solid bounded by the graph of $\dfrac{x}{a} + \dfrac{y}{b} + \dfrac{z}{c} = 1$ $(a > 0, b > 0, c > 0)$ and the three coordinate planes.

12. Sketch the region Q corresponding to the iterated integral, reverse the order of integration, and evaluate the resulting integral:

    $$\int_{-2}^{0} \int_{-\sqrt{x+2}}^{\sqrt{x+2}} y^2 \, dy \, dx.$$

13. Calculate the volume of the solid lying inside the cylinder $x^2 + y^2 = 4$ and between the planes $y + z = 9$ and $z = 0$.

14. Use a double integral to calculate the area of the region lying between the graph of $\sqrt{x} + \sqrt{y} = 1$ and the graph of $x + y = 1$.

15. Find the center of mass of a solid bounded by the cylinder $r = 2$, the cone $z = r$, and the plane $z = 0$ if the density of the material is uniform.

16. Calculate the volume of the ellipse

    $$\frac{x^2}{4} + \frac{y^2}{9} + \frac{z^2}{4} = 1.$$

17. Find the volume of the region common to the sphere $r^2 + z^2 = a^2$ and the cylinder $r = a \cos \theta$.

18. Find the surface area of the paraboloid $z = x^2 + y^2$ lying between the planes $z = 1$ and $z = 9$.

19. Find the area of the region cut from the plane $z = 4y$ by the cylinder $x^2 + y^2 = 4$.

20. Find the volumes of the two regions cut from the sphere $\rho = 4$ by the plane $x = 2$.

21. Find the volume of the region common to the sphere $x^2 + y^2 + z^2 = 16$ and the cylinder $x^2 + z^2 = 4$.

22. Use a double integral to find the area enclosed by the lemniscate $r^2 = 2 \cos 2\theta$.

23. A solid is bounded below by the region bounded by the graphs of $y = x$ and $y = x^2 - 2$. It is bounded above by the plane $z - x + 2y = 10$. Find its volume.

24. Find $\iint\limits_Q \dfrac{\sin x}{x} \, dA$ where Q is the triangle with vertices

    $(0, 0)$, $(2, 0)$ and $(2, 2)$.

25. Find the centroid of the half disc $r = 4$, $0 \le \theta \le \pi$.

26. Find the mass of a right circular cone of radius r and height h if the density at each point is proportional to the distance of that point from the vertex.

27. Find $\iiint\limits_Q e^{(x^2+y^2+z^2)^{3/2}} \, dV$ where Q is the half sphere

    $Q = \{(r, \theta, \phi) \mid 0 \le r \le 1, 0 \le \phi \le \pi/2, 0 \le \theta \le 2\pi\}$.

28. Show that if a thin lamina of constant density has the shape of the region Q described in polar coordinates, then the coordinates $(\bar{x}, \bar{y})$ of the center of mass may be calculated by the formulas

    $$\bar{x} = \frac{1}{\text{area}} \iint\limits_Q r^2 \cos \theta \, dr \, d\theta;$$

    $$\bar{y} = \frac{1}{\text{area}} \iint\limits_Q r^2 \sin \theta \, dr \, d\theta.$$

29. Use the result of Exercise 28 to calculate the center of mass

for a thin lamina of constant density whose shape is the cardioid $r = 1 + \cos \theta$.

30. Find the volume of the solid bounded above by the plane $z - x = 2$ and below by the paraboloid $z = x^2 + y^2$.

31. Find the volume of the region bounded by the paraboloids

$$z = 4 - x^2 + 2x - y^2 - 4y$$

and

$$z = x^2 - 2x + y^2 + 4y + 5.$$

32. Find the volume of the solid bounded above by the cylinder $z = 4 - x^2$ and below by the paraboloid $3x^2 + y^2 = z$.

33. A solid corresponds to the region bounded by the cylinder $x^2 + y^2 = 4$ and the planes $z = 0$ and $z = 4$. Calculate the mass of the solid if the density at each point is proportional to the distance from the xy-plane.

34. Evaluate the integral $\displaystyle\iint_Q \cos \sqrt{x^2 + y^2} \, dA$ where Q is the disc

$$Q = \{(x, y) \mid 0 \le x^2 + y^2 \le 4\}.$$

35. Evaluate the integral $\displaystyle\iint_Q \frac{1}{\sqrt{1 + x^2 + y^2}} \, dA$ where Q is the quarter circle

$$Q = \{(x, y) \mid 0 \le x \le 1, 0 \le y \le \sqrt{1 - x^2}\}.$$

36. Find the volume of the solid bounded above by the graph of $z = 4 - r$ and below by the region in the $r\theta$-plane bounded by the graph of $r = 3 \sin \theta$.

37. Find the volume of the solid bounded above by the cone $z^2 = x^2 + y^2$, on the sides by the cylinder $x^2 + y^2 - 4y = 0$, and below by the xy-plane.

38. Find the volume of the solid inside the cylinder $x^2 - 4x + y^2 = 0$ lying above the xy-plane and below the plane $z - x = 4$.

39. *(Computer)* Use a modification of Program 8 in Appendix I to approximate

$$\int_0^2 \int_1^3 \sin^2 \sqrt{x - y^2} \, dy \, dx.$$

40. Find the area of the region bounded by the graphs of $y = \sqrt{x + 2}$ and $x - 3y + 2 = 0$ by double integration.

41. Find the volume of the solid bounded by the cone $z^2 = x^2 + y^2$ and the cylinder $x^2 + y^2 = 9$.

42. Find the centroid of the region bounded by the graph of $r = \cos 2\theta$.

43. Find the surface area of the part of the sphere $x^2 + y^2 + z^2 = 4$ lying outside the cylinder $x^2 + y^2 = 1$.

Chapter 19
Topics in Vector Analysis

The goal of this chapter is to develop a theory of integral calculus for vector-valued functions defined in the plane or in space. These are functions of the form $w = F(r)$ where both w and r are vectors.

19.1 VECTOR FIELDS

To avoid confusion between functions of this type and vector-valued functions of a single real variable, we shall use the term *vector fields*.

DEFINITION 1

Let Q be a subset of $\mathbb{R}^3$. A **vector field** on Q is a function

$$F(x, y, z) = M(x, y, z)i + N(x, y, z)j + P(x, y, z)k \tag{1}$$

that assigns a vector $w = F(x, y, z)$ to each point (x, y, z) in Q.
  If Q is a subset of $\mathbb{R}^2$, a vector field on Q is a function of the form

$$F(x, y) = M(x, y)i + N(x, y)j. \tag{2}$$

The real-valued functions M, N, and P in equation (1) are referred to as the **component** functions of the vector field F. We say that a vector field is **continuous** if each of its component functions is continuous. Using the position vector

$$r = xi + yj + zk, \qquad (x, y, z) \in Q,$$

we may write the vector field in equation (1) as simply $w = F(r)$. Functions of the form $w = f(x, y, z)$, which assign real numbers to vectors, will now be referred to as **scalar functions** or scalar fields.
  The following examples indicate the types of vector fields and scalar functions that we shall study in this chapter.

Example 1

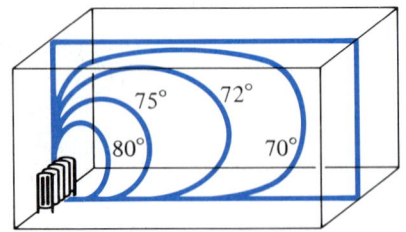

Figure 1.1 Isothermal curves in a cross section of a room with a single heat source.

Figure 1.1 represents a room with a single radiator. The function T, which gives the temperature $T(x, y, z)$ at the point in the room with coordinates (x, y, z), is an example of a *scalar* function. The curves drawn in a cross section of the room are locations of points of equal temperature and are called **isothermal curves.**
  Since warmer air rises and cooler air falls, the presence of a single heat source in a room causes air to flow about the room in **convection currents.** If the vector

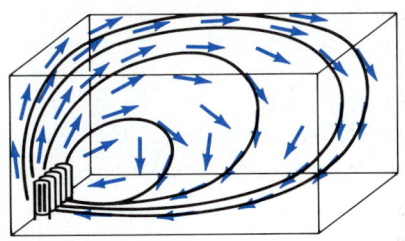

Figure 1.2 Tangents $F(x, y, z)$ to convection currents determine a vector field in the room.

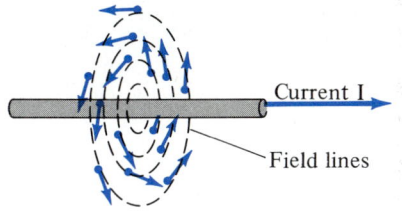

Current I

Field lines

Figure 1.3 Tangents to magnetic field lines form a (magnetic) vector field in space.

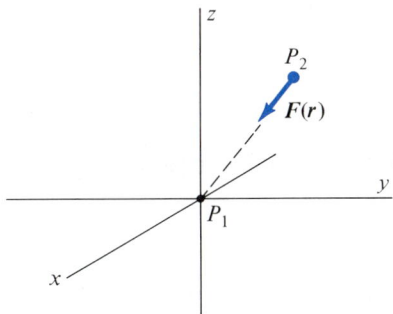

Figure 1.4 Gravitational force vector $F(r)$ acting on particle P_2.

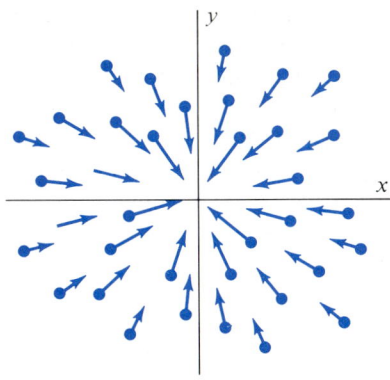

Figure 1.5 Gravitational force field (drawing represents xy-plane only).

$F(x, y, z)$ represents the velocity vector of the convection current passing through the point (x, y, z), the function $w = F(x, y, z)$ is an example of a vector field defined throughout the room (Figure 1.2). ◇

Example 2

An electric current of magnitude I flowing through a thin wire induces a **magnetic field** around the wire. (The statement of Ampère's Law, giving the relationship between the electric current I and its magnetic effect, involves the idea of a *line integral*, introduced in Section 19.2.) The set of vectors $F(x, y, z)$, giving the direction and intensity of the magnetic field at location (x, y, z), determines a vector field in the space surrounding the wire (Figure 1.3). ◇

Example 3

An example of a force field is a **gravitational field.** According to Newton's Law of Gravitation, the magnitude of the force of attraction exerted on a particle P_2 of mass m_2 by a particle P_1 of mass m_1 is given by the equation

$$F = \frac{Gm_1m_2}{r^2} \tag{3}$$

where r is the distance between the particles and G is a constant. If we assume the particle P_1 to be located at the origin in xyz-space, then the force in (3) is a function of the coordinates (x, y, z) giving the location of the particle P_2. That is,

$$F(x, y, z) = \frac{Gm_1m_2}{x^2 + y^2 + z^2}. \tag{4}$$

Since the force whose magnitude is given by (4) acts toward the origin, it acts in the direction opposite that of the position vector $r = xi + yj + zk$ of P_2. We may therefore write the force vector $F(x, y, z)$ as

$$F(x, y, z) = \left(\frac{Gm_1m_2}{(x^2 + y^2 + z^2)}\right)\left(-\frac{xi + yj + zk}{\sqrt{x^2 + y^2 + z^2}}\right) \tag{5}$$

$$= \left(\frac{-Gm_1m_2}{(x^2 + y^2 + z^2)^{3/2}}\right)(xi + yj + zk).$$

Using the position vector $r = xi + yj + zk$, we may write the force field in (5) as

$$F(r) = -\frac{Gm_1m_2}{|r|^3}r, \qquad r \neq 0 \qquad \text{(Figure 1.4).} \tag{6}$$

Since the vector F in (5) and (6) is defined for all position vectors $r \neq 0$, either equation defines a vector field at all points of $\mathbb{R}^3$ except the origin (Figure 1.5). ◇

The gravitational field of Example 3 is an example of a **central force field,** since the force vector at each point points toward the origin. More generally, a central force field in space has the form

$$F(x, y, z) = f(x, y, z)(xi + yj + zk)$$

where f is a real-valued function of three variables.

Example 4

Another example of a central force field is obtained by letting $\boldsymbol{F}(x, y, z)$ be the vector representing the force exerted on a particle P_2 with electric charge q_2 at location (x, y, z) by a particle P_1 with electric charge q_1 located at the origin. According to **Coulomb's Law,** this force vector is

$$\boldsymbol{F}(\boldsymbol{r}) = \frac{kq_1q_2}{|\boldsymbol{r}|^3}\boldsymbol{r}, \qquad \boldsymbol{r} \neq \boldsymbol{0} \tag{7}$$

where $\boldsymbol{r}$ is the position vector $\boldsymbol{r} = x\boldsymbol{i} + y\boldsymbol{j} + z\boldsymbol{k}$ of P_2, and k is a constant that depends on the choice of units for $\boldsymbol{r}$, q_1, and q_2. Figure 1.6 represents the **electric force field $\boldsymbol{F}$** (for positive charges P_2) due to a positive charge at the origin. Figure 1.7 represents the force field $\boldsymbol{F}$ (for positive charges P_2) due to a negative charge at

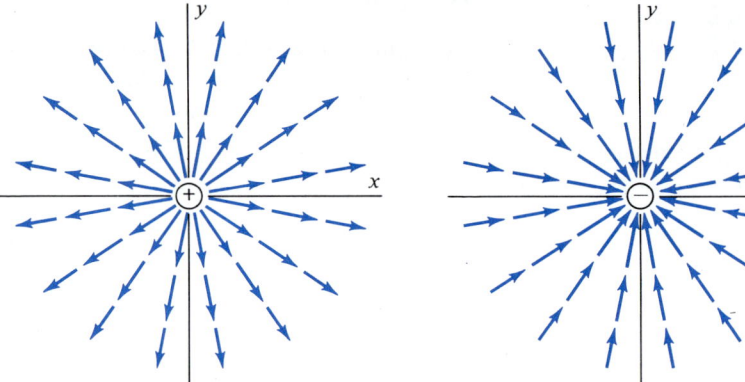

Figure 1.6 Electric force field due to a positive charge at the origin.

Figure 1.7 Electric field due to a negative charge at the origin.

the origin. (Although the force field is defined in space, we show only the force vectors lying in the xy-plane in these figures.) ◇

If charge is considered to be analogous to mass, it is clear from comparison of equations (6) and (7) that the gravitational and electric force fields have the same form; physicists say that both forces obey an *inverse square law*. The major difference is that the direction of $\boldsymbol{F}(\boldsymbol{r})$ may be either toward or away from the origin for the electric field, but it can only be toward the origin for the gravitational field (since all masses are positive).

Electric force fields satisfy the **principle of superposition,** which states that the force acting on a charge at a point P due to two separate charges is the vector sum of the individual forces acting on the charge at P. Figure 1.8 represents the electric field, acting on a positive charge, that results from a positive charge at $(-1, 0)$ and a negative charge of equal magnitude at $(1, 0)$. Note that while the individual force fields in Figure 1.6 and 1.7 are central, their sum is not.

Example 5

If f is a differentiable scalar function of three variables, the gradient function

$$\boldsymbol{F}(x, y, z) = \nabla f(x, y, z) = f_x(x, y, z)\boldsymbol{i} + f_y(x, y, z)\boldsymbol{j} + f_z(x, y, z)\boldsymbol{k} \tag{8}$$

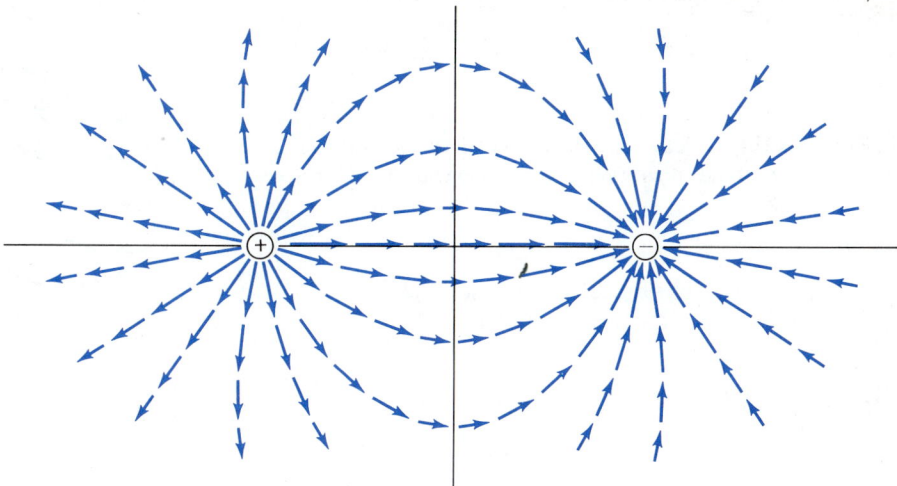

Figure 1.8 Electric force field due to two opposite charges is the vector sum of the individual central force fields.

defines a vector field on $\mathbb{R}^3$. If f is a differentiable scalar function of two variables, the corresponding gradient field on $\mathbb{R}^2$ is

$$F(x, y) = \nabla f(x, y) = f_x(x, y)i + f_y(x, y)j. \tag{9}$$

For example, the scalar function $f(x, y, z) = xe^{2y} \sin z$ yields the gradient vector field

$$F(x, y, z) = e^{2y} \sin zi + 2xe^{2y} \sin zj + xe^{2y} \cos zk. \qquad \diamond$$

As in Section 17.9, we will refer to the function f in equation (8) as a **scalar potential** for the vector field F. Although every differentiable scalar function yields a gradient vector field, a given vector field F need not be the gradient of a scalar potential. Those vector fields that can be written as gradients of scalar potentials are called **conservative vector fields.**

There are many other familiar examples of vector fields. The flow of air currents around a moving automobile or airplane determines a **velocity field.** The study of such vector fields is called *aerodynamics*. Similarly, the velocity field determined by the flow of water through a container, such as a pipe or a dam, is the subject of *hydrodynamics*.

Exercise Set 19.1

In Exercises 1–10, sketch enough vectors in the given vector field to get a sense of the nature of the vector field.

1. $F(x, y) = xi - yj$

2. $F(x, y) = -yi + xj$

3. $F(x, y) = 3i$

4. $F(x, y) = -xi - yj$

5. $F(x, y) = \dfrac{x}{\sqrt{x^2 + y^2}}i + \dfrac{y}{\sqrt{x^2 + y^2}}j$

6. $F(x, y) = \dfrac{x}{\sqrt{x^2 + y^2}}i - \dfrac{y}{\sqrt{x^2 + y^2}}j$

7. $F(x, y, z) = j + k$

8. $F(x, y, z) = xi + yj$

9. $F(x, y, z) = 2xi + 2yj + 2zk$

10. $F(x, y, z) = i - j + k$

11. Which of the vector fields in Exercises 1–10 are central?

Find the gradient vector field $F = \nabla f$ for each of the functions in Exercises 12–17.

12. $f(x, y) = xy$

13. $f(x, y) = x^2 - y^2$

14. $f(x, y) = x \tan xy$

15. $f(x, y, z) = \sqrt{x^2 + y^2 + z^2}$

16. $f(x, y, z) = x \ln(y^2 + z^2)$

17. $f(x, y, z) = ze^{x-y}$

In Exercises 18–23, determine whether the given vector field is a gradient vector field. If so, find a potential ϕ with $F = \nabla\phi$ (see Section 18.9).

18. $F(x, y) = \sin y\, i + \cos y\, j$

19. $F(x, y) = (\tan xy - xy \sec^2 xy)i + x^2 \sec^2 xy\, j$

20. $F(x, y) = e^{\sqrt{xy}}i - \sqrt{x}e^{\sqrt{xy}}j$

21. $F(x, y) = \left[\ln(y - x) + \dfrac{x}{y - x}\right]i + \left(\dfrac{x}{y - x}\right)j$

22. $F(x, y, z) = x^2 i + x^2 \sin zj - \cos yz\, k$

23. $F(x, y, z) = yze^{xyz}i + xze^{xyz}j + xye^{xyz}k$

24. The **flow lines** for a vector field F are curves $\alpha(t)$ tangent to the vector field. That is, $\alpha'(t) = F(\alpha(t))$ for every t and every flow line. (An example of flow lines associated with a vector field in the plane is shown in Figure 1.9.) Sketch several flow lines for each of the vector fields in Exercises 1–10.

25. Imagine water flowing through a pipe as in Figure 1.10. At each point (x, y, z) within the pipe let $F(x, y, z)$ be the velocity vector for the water at that point. This determines a vector field within the pipe. Why are the vectors in the narrow part of the pipe larger in magnitude?

26. Show that the central force field $F(r) = \left(\dfrac{k}{|r|^3}\right)r$ is the gradient of the scalar field $f(r) = \dfrac{k}{|r|}$. (r denotes the position vector $r = xi + yj + zk$ associated with the point (x, y, z).)

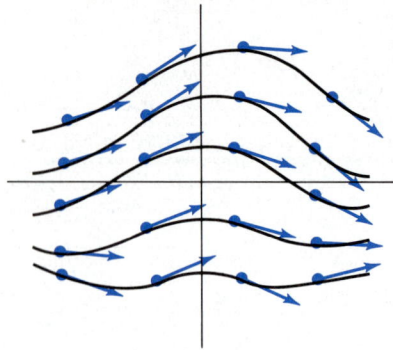

Figure 1.9 Flow lines for a vector field.

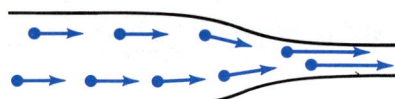

Figure 1.10 Velocity field in a water pipe.

27. Let $F(x, y, z)$ be the electric field describing the force on a particle with charge $q = +1$ at location (x, y, z) due to the combined effect of a charge $q_1 = +3$ at $(-1, 0, 0)$ and a charge $q_2 = -1$ at $(1, 0, 0)$.
  a. Find the force vector $F(0, 0, 0)$.
  b. Find the force vector $F(-2, 0, 0)$.
  c. Sketch enough vectors $F(x, y, 0)$ to get an idea of the force field in the xy-plane.
  d. Sketch several of the flow lines for the vector field (see Exercise 24).

19.2 WORK AND LINE INTEGRALS

When a force of magnitude F moves an object d units along a line, we define the **work** done on the object by the force as the product $W = Fd$. That is,

$$\text{Work} = (\text{force}) \times (\text{distance}) \tag{1}$$

for constant forces applied along a line.

In Chapter 7 we applied the theory of the definite integral to calculate the work done by a continuously varying force f in moving an object from location $x = a$ to location $x = b$ along a line:

$$W = \lim_{n \to \infty} \sum_{j=1}^{n} f(t_j)\, \Delta x = \int_a^b f(x)\, dx,$$

which agrees with (1) when f is a constant force (Figures 2.1 and 2.2).

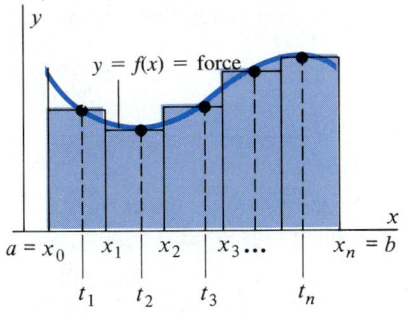

Figure 2.1 $W \approx \sum\limits_{j=1}^{n} f(t_j) \, \Delta x.$

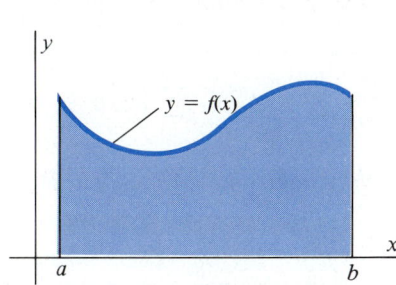

Figure 2.2 $W = \int_a^b f(x) \, dx.$

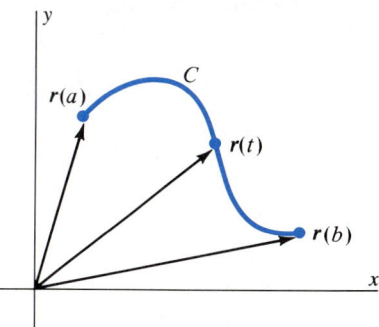

Figure 2.3 $C: \quad r(t) = x(t)\mathbf{i} + y(t)\mathbf{j} + z(t)\mathbf{k}.$

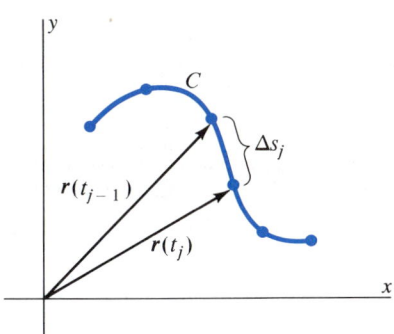

Figure 2.4 $\Delta s_j = \int_{t_{j-1}}^{t_j} |r'(t)| \, dt.$

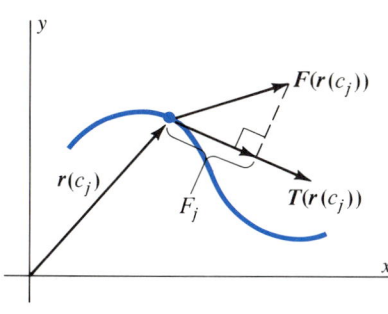

Figure 2.5 F_j is the component of the force field in the direction of the unit tangent at $r(c_j)$.

In this section we wish to generalize these concepts to calculate the work done on a particle by a force that moves the particle along a curve C lying in the plane or in space. Examples of this situation include small charged particles moving in electric fields, objects moving in gravitational fields, and others suggested by the discussion of Section 19.1.

To formulate the problem more precisely, assume that $\mathbf{F}$ is a vector field defined on an open box D of $\mathbb{R}^3$ that determines a force vector $\mathbf{F}(r)$ for each position vector $r = x\mathbf{i} + y\mathbf{j} + z\mathbf{k}$ in D. Also, let C be a curve lying within D that is parameterized by the vector function

$$C: \quad r(t) = x(t)\mathbf{i} + y(t)\mathbf{j} + z(t)\mathbf{k}, \qquad a \le t \le b \qquad \text{(see Figure 2.3).} \qquad (2)$$

If we partition the parameter interval $[a, b]$ into n subintervals of equal length $\Delta t = \dfrac{b - a}{n}$ with endpoints $a = t_0 < t_1 < t_2 < \cdots < t_n = b$, the position vectors $r(t_0), r(t_1), \ldots, r(t_n)$ divide the curve C into n arcs. If the curve C is smooth (meaning that $r'(t)$ is continuous for $t \in [a, b]$), equation (9) of Section 16.3 gives the length of the jth arc as

$$\Delta s_j = \int_{t_{j-1}}^{t_j} |r'(t)| \, dt \qquad \text{(see Figure 2.4).}$$

Under this assumption, the Mean Value Theorem for Integrals guarantees the existence of a number c_j in the interval $[t_{j-1}, t_j]$ for which

$$\Delta s_j = \int_{t_{j-1}}^{t_j} |r'(t)| \, dt = |r'(c_j)| \cdot (t_j - t_{j-1}) \qquad (3)$$

$$= |r'(c_j)| \, \Delta t.$$

To approximate the work ΔW_j done by the force field in moving a particle along the jth arc, we assume that the magnitude F_j of the force acting throughout the jth arc is the tangential component of $\mathbf{F}(r)$ at the point with position vector $r(c_j)$. That is, we use

$$F_j = \operatorname{comp}_{T(r(c_j))} \mathbf{F}(r(c_j)) = \mathbf{F}(r(c_j)) \cdot T(r(c_j)) \qquad (4)$$

where $T(r(c_j))$ is the unit tangent to C at $r(c_j)$ (see Figure 2.5).

Multiplying (3) and (4) and summing over all n arcs, we obtain the approximation

$$W \approx \sum_{j=1}^{n} F_j \, \Delta s_j = \sum_{j=1}^{n} \boldsymbol{F}(\boldsymbol{r}(c_j)) \cdot \boldsymbol{T}(\boldsymbol{r}(c_j)) |\boldsymbol{r}'(c_j)| \, \Delta t$$

for the work done by the force field $\boldsymbol{F}$ in moving the particle over the curve C. We therefore define the work W as the limit as $n \to \infty$ of this approximating sum:

$$W = \lim_{n \to \infty} \sum_{j=1}^{n} \boldsymbol{F}(\boldsymbol{r}(c_j)) \cdot \boldsymbol{T}(\boldsymbol{r}(c_j)) |\boldsymbol{r}'(c_j)| \, \Delta t \tag{5}$$

$$= \int_a^b \boldsymbol{F}(\boldsymbol{r}(t)) \cdot \boldsymbol{T}(\boldsymbol{r}(t)) |\boldsymbol{r}'(t)| \, dt.$$

Equation (5) may be simplified somewhat by recalling that, if $|\boldsymbol{r}'(t)| \neq 0,$* the unit tangent $\boldsymbol{T}(\boldsymbol{r}(t))$ to C at $\boldsymbol{r}(t)$ is given by

$$\boldsymbol{T}(\boldsymbol{r}(t)) = \frac{\boldsymbol{r}'(t)}{|\boldsymbol{r}'(t)|}. \tag{6}$$

Combining equations (5) and (6) gives the desired definition of work done by a (continuous) force field in moving a particle along a smooth curve C.

DEFINITION 2

Let $\boldsymbol{F}$ be a continuous force field defined in some open box containing the smooth curve

$$C: \quad \boldsymbol{r}(t) = x(t)\boldsymbol{i} + y(t)\boldsymbol{j} + z(t)\boldsymbol{k}, \qquad a \leq t \leq b, \qquad |\boldsymbol{r}'(t)| \neq 0.$$

The **work done by the force field $\boldsymbol{F}$** in moving a particle along C from $\boldsymbol{r}(a)$ to $\boldsymbol{r}(b)$ is given by the definite integral

$$W = \int_a^b \boldsymbol{F}(\boldsymbol{r}(t)) \cdot \boldsymbol{r}'(t) \, dt. \tag{7}$$

REMARK: For the curve C as in Definition 2,

$$\boldsymbol{r}'(t) = \frac{d}{dt}(\boldsymbol{r}(t)) = \frac{dx}{dt}\boldsymbol{i} + \frac{dy}{dt}\boldsymbol{j} + \frac{dz}{dt}\boldsymbol{k},$$

which suggests the notation

$$d\boldsymbol{r} = \boldsymbol{r}'(t) \, dt. \tag{8}$$

Using (8), we sometimes abbreviate the integral in (7) as

$$W = \int_C \boldsymbol{F} \cdot d\boldsymbol{r}. \tag{9}$$

In using equation (9), it is important to remember that one must first find a smooth

---

*The requirement that $|\boldsymbol{r}'(t)| \neq 0$ is not as restrictive as it might seem, since in most applications we are free to choose the particular parameterization used to describe the curve C.

parameterization $r(t)$ for C and then proceed as in equation (7). It is a fact, which we shall not prove, that the value of the integral in (9) does not depend on the particular parameterization chosen for C.

Example 1

Find the work done by the force field

$$F(x, y, z) = 2x\mathbf{i} + 3y\mathbf{j} + z\mathbf{k} \qquad \text{(see Figures 2.6 and 2.7)}$$

in moving a particle along the circular helix

$$C: \quad r(t) = \cos t\mathbf{i} + \sin t\mathbf{j} + t\mathbf{k}$$

from point $r(0) = \mathbf{i}$ to point $r(\pi) = -\mathbf{i} + \pi\mathbf{k}$.

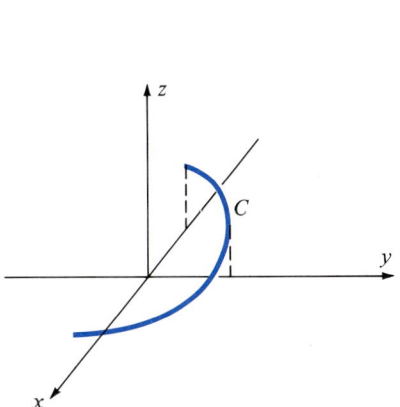

Figure 2.6 The helix C: $r(t) = \cos t\mathbf{i} + \sin t\mathbf{j} + t\mathbf{k}$.

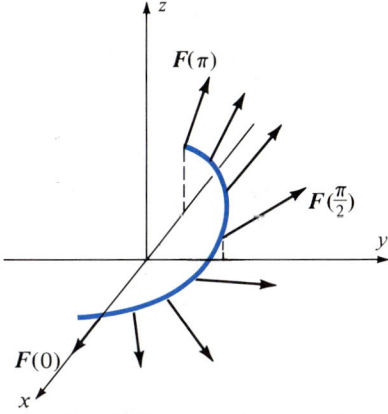

Figure 2.7 Selected vectors in the force field $F(x, y, z) = 2x\mathbf{i} + 3y\mathbf{j} + z\mathbf{k}$ for which (x, y, z) lies on C.

Solution: The parameterization given for C is

$$x(t) = \cos t; \qquad y(t) = \sin t; \qquad z(t) = t,$$

so

$$F(r(t)) = 2 \cos t\mathbf{i} + 3 \sin t\mathbf{j} + t\mathbf{k}.$$

Also,

$$r'(t) = -\sin t\mathbf{i} + \cos t\mathbf{j} + \mathbf{k}.$$

Thus, by Definition 2,

$$W = \int_0^\pi [2 \cos t\mathbf{i} + 3 \sin t\mathbf{j} + t\mathbf{k}] \cdot [-\sin t\mathbf{i} + \cos t\mathbf{j} + \mathbf{k}] \, dt$$

$$= \int_0^\pi (-2 \cos t \sin t + 3 \sin t \cos t + t) \, dt = \int_0^\pi (\sin t \cos t + t) \, dt$$

$$= \frac{1}{2} \sin^2 t + \frac{1}{2} t^2 \Big]_0^\pi = \frac{\pi^2}{2}. \qquad \diamond$$

Example 2

Find the work done by the force field

$$F(x, y) = 3y\mathbf{i} - x^2\mathbf{j}$$

in moving a particle along the plane curve $y = \sqrt{x}$ from the point $(1, 1)$ to the point $(4, 2)$.

Solution: Here we must first find a parameterization for the arc C of the graph of $y = \sqrt{x}$ from $(1, 1)$ to $(4, 2)$. We do so by setting $x(t) = t$, $y(t) = \sqrt{t}$. Then

$$\mathbf{r}(t) = t\mathbf{i} + \sqrt{t}\,\mathbf{j}, \qquad 1 \le t \le 4,$$

$$\mathbf{r}'(t) = \mathbf{i} + \frac{1}{2\sqrt{t}}\mathbf{j},$$

and

$$F(\mathbf{r}(t)) = 3\sqrt{t}\,\mathbf{i} - t^2\mathbf{j}.$$

Thus, by (7)

$$W = \int_1^4 [3\sqrt{t}\,\mathbf{i} - t^2\mathbf{j}] \cdot \left[\mathbf{i} + \frac{1}{2\sqrt{t}}\mathbf{j}\right] dt$$

$$= \int_1^4 \left(3t^{1/2} - \frac{1}{2}t^{3/2}\right) dt$$

$$= 2t^{3/2} - \frac{1}{5}t^{5/2}\Big]_1^4$$

$$= \frac{39}{5}. \qquad \diamond$$

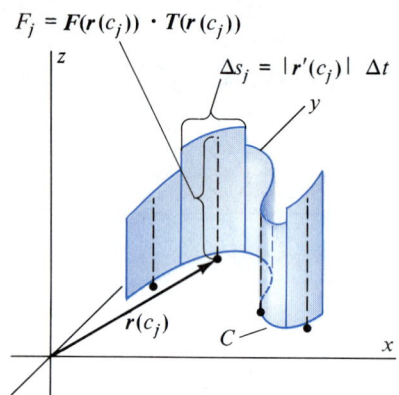

$F_j = F(\mathbf{r}(c_j)) \cdot T(\mathbf{r}(c_j))$

$\Delta s_j = |\mathbf{r}'(c_j)|\,\Delta t$

$\mathbf{r}(c_j)$

C

Figure 2.8 $F(\mathbf{r}(c_j)) \cdot T(\mathbf{r}(c_j))|\mathbf{r}'(c_j)|\,\Delta t$ in (5) is surface area of a curved slab over C.

Example 3

Show that the result of Example 2 remains unchanged if the parameterization $x(t) = t^2$, $y(t) = t$, $1 \le t \le 2$, is used for C.

Solution: Here $\mathbf{r}(t) = t^2\mathbf{i} + t\mathbf{j}$, $\mathbf{r}'(t) = 2t\mathbf{i} + \mathbf{j}$, and $F(\mathbf{r}(t)) = 3t\mathbf{i} - t^4\mathbf{j}$. Thus,

$$W = \int_1^2 [3t\mathbf{i} - t^4\mathbf{j}] \cdot [2t\mathbf{i} + \mathbf{j}]\,dt$$

$$= \int_1^2 (6t^2 - t^4)\,dt = 2t^3 - \frac{1}{5}t^5\Big]_1^2 = \frac{39}{5}.$$

This result illustrates our earlier remark that the value of the integral in (7) depends only on F and on C, not on the particular parameterization chosen for C. $\diamond$

Figure 2.8 gives a geometric interpretation of the approximating sum for the integral in equation (5). As Figure 2.9 illustrates, the integral in equation (7) may be interpreted as surface areas of curved vertical "slabs" of variable height $F(\mathbf{r}(t)) \cdot T(\mathbf{r}(t))$.

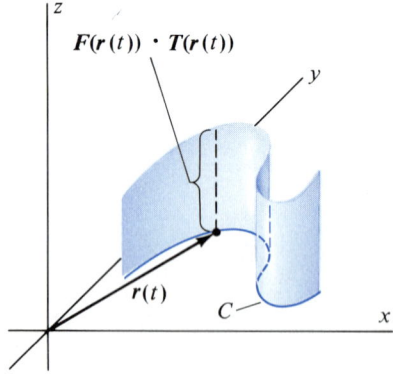

$F(\mathbf{r}(t)) \cdot T(\mathbf{r}(t))$

$\mathbf{r}(t)$

C

Figure 2.9 $\displaystyle\int_C F \cdot d\mathbf{r}$ is a surface area.

Line Integrals

The existence of the integral in Definition 2 depends only on the mathematical properties of the force field F and the curve C. We may therefore generalize Definition 2 to apply to any vector field F, not necessarily a force field. The result is called a *line integral*.

DEFINITION 3

Let F be a continuous vector field in some open box D and let C be a smooth curve lying within D with parameterization $C = \{r(t) \mid a \le t \le b\}$. The **line integral** of F over C is defined as

$$\int_C F \cdot dr = \int_a^b F(r(t)) \cdot r'(t)\, dt. \tag{10}$$

Definitions 2 and 3 are the same except that Definition 2 refers to the more specific situation of calculating work when F is a force field.

The line integral in (10) may be written in various other ways. When the vector field F has the form

$$F(x, y, z) = M(x, y, z)i + N(x, y, z)j + P(x, y, z)k,$$

we use the notation

$$dr = dx\, i + dy\, j + dz\, k$$

to write $F \cdot dr$ as

$$F \cdot dr = M(x, y, z)\, dx + N(x, y, z)\, dy + P(x, y, z)\, dz.$$

If the parameterization for C has the form

$$r(t) = x(t)i + y(t)j + z(t)k,$$

then

$$r'(t) = x'(t)i + y'(t)j + z'(t)k.$$

With this notation, the line integral in (10) can be written

$$\int_C F \cdot dr = \int_a^b \{M(x(t), y(t), z(t))x'(t) \\ + N(x(t), y(t), z(t))y'(t) \\ + P(x(t), y(t), z(t))z'(t)\}\, dt. \tag{11}$$

Equation (11) may be abbreviated simply as

$$\int_C F \cdot dr = \int_C M(x, y, z)\, dx + N(x, y, z)\, dy + P(x, y, z)\, dz. \tag{12}$$

When the vector field F and the curve C are defined in the plane by

$$F(x, y) = M(x, y)i + N(x, y)j$$

and

$$r(t) = x(t)i + y(t)j, \qquad a \le t \le b,$$

the line integral in Definition 3 takes the form

$$\int_C F \cdot dr = \int_a^b [M(x(t), y(t))x'(t) + N(x(t), y(t))y'(t)]\, dt \tag{13}$$

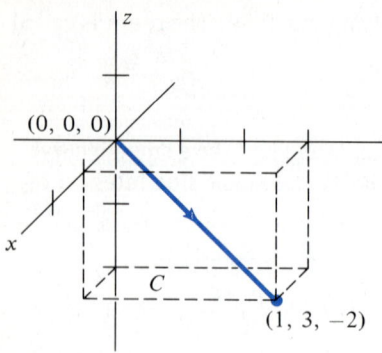

Figure 2.10 Path C from $(0, 0, 0)$ to $(1, 3, -2)$.

or

$$\int_C \boldsymbol{F} \cdot d\boldsymbol{r} = \int_C M(x, y)\,dx + N(x, y)\,dy. \tag{14}$$

Example 4

Evaluate the line integral $\displaystyle\int_C \boldsymbol{F} \cdot d\boldsymbol{r}$ where $\boldsymbol{F}$ is the vector field

$$\boldsymbol{F}(x, y, z) = x^2\boldsymbol{i} + xy\boldsymbol{j} + xz\boldsymbol{k}$$

and C is the line segment from $(0, 0, 0)$ to $(1, 3, -2)$ (see Figure 2.10).

Solution: Here $\boldsymbol{F} = M\boldsymbol{i} + N\boldsymbol{j} + P\boldsymbol{k}$ with

$$M(x, y, z) = x^2, \qquad N(x, y, z) = xy, \qquad P(x, y, z) = xz. \tag{15}$$

The line segment C from $(0, 0, 0)$ to $(1, 3, -2)$ may be parameterized with

$$x(t) = t, \qquad y(t) = 3t, \qquad z(t) = -2t \tag{16}$$
$$\text{for} \qquad 0 \le t \le 1 \qquad \text{(Figure 2.10).}$$

Thus,

$$x'(t) = 1, \qquad y'(t) = 3, \qquad z'(t) = -2. \tag{17}$$

Combining equations (15) and (16) shows that

$$M(x(t), y(t), z(t)) = t^2; \qquad N(x(t), y(t), z(t)) = 3t^2; \tag{18}$$
$$P(x(t), y(t), z(t)) = -2t^2.$$

Using equations (11), (17), and (18), we find

$$\int_C \boldsymbol{F} \cdot d\boldsymbol{r} = \int_0^1 [(t^2)(1) + (3t^2)(3) + (-2t^2)(-2)]\,dt$$
$$= \int_0^1 14t^2\,dt = \frac{14}{3}t^3\Big]_0^1 = \frac{14}{3}. \qquad \diamond$$

Example 5

Evaluate the line integral

$$\int_C xy\,dx - 2y^2\,dy$$

where C is the arc of the unit circle from $(1, 0)$ to $(0, 1)$ traversed counterclockwise (Figure 2.11).

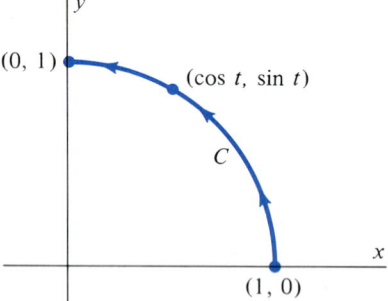

Figure 2.11 Path C from $(1, 0)$ to $(0, 1)$.

Solution: The line integral has the form $\boldsymbol{F} = M\boldsymbol{i} + N\boldsymbol{j}$ where

$$M(x, y) = xy, \qquad N(x, y) = -2y^2.$$

A parameterization for the curve C is given by the equations

$$x(t) = \cos t, \qquad y(t) = \sin t, \qquad 0 \le t \le \pi/2 \qquad \text{(Figure 2.11).}$$

Thus

$$x'(t) = -\sin t, \qquad y'(t) = \cos t.$$

Equating the right sides of equations (13) and (14) and using the above functions we obtain

$$\int_C xy \, dx - 2y^2 \, dy = \int_0^{\pi/2} [(\cos t)(\sin t)(-\sin t) - 2(\sin^2 t)(\cos t)] \, dt.$$

$$= \int_0^{\pi/2} -3 \sin^2 t \cos t \, dt$$

$$= -\sin^3 t]_0^{\pi/2}$$

$$= -1. \qquad \diamondsuit$$

REMARK: It is important to note that the value of the line integral in Definition 3 depends on the *direction* in which the curve C is traced out by the parameterization $r(t)$. *Reversing the direction along C changes the sign of the line integral.* The reason for this can be most easily seen in line (5): Reversing the direction in which C is traversed changes the sign of the unit tangent vector $T(r(c_j))$.

To illustrate this remark, we evaluate the line integral in Example 5 over the same curve C, but traversed in the opposite direction, from $(0, 1)$ to $(1, 0)$. (We will refer to the curve C with its new orientation as $-C$). A parameterization for $-C$ is

$$x(t) = \cos\left(\frac{\pi}{2} - t\right), \qquad y(t) = \sin\left(\frac{\pi}{2} - t\right), \qquad 0 \le t \le \frac{\pi}{2},$$

so

$$x'(t) = \sin\left(\frac{\pi}{2} - t\right), \qquad y'(t) = -\cos\left(\frac{\pi}{2} - t\right),$$

and

$$\int_{-C} xy \, dx - 2y^2 \, dy = \int_0^{\pi/2} \left[\cos\left(\frac{\pi}{2} - t\right) \sin^2\left(\frac{\pi}{2} - t\right) \right.$$

$$\left. + 2 \sin^2\left(\frac{\pi}{2} - t\right) \cos\left(\frac{\pi}{2} - t\right) \right] dt$$

$$= \int_0^{\pi/2} 3 \sin^2\left(\frac{\pi}{2} - t\right) \cos\left(\frac{\pi}{2} - t\right) dt$$

$$= -\sin^3\left(\frac{\pi}{2} - t\right) \Big]_0^{\pi/2}$$

$$= 1.$$

This is the negative of the result obtained in Example 5. We summarize this remark by writing

$$\int_{-C} F \cdot dr = -\int_C F \cdot dr.$$

When the curve C is a line segment parallel to the x-axis, the line integral in (11) and (12) reduces to

$$\int_C F \cdot dr = \int_C M(x, y, z) \, dx = \int_a^b M(x(t), y(t), z(t))x'(t) \, dt.$$

Such integrals are called **line integrals with respect to x.** Line integrals with respect to y and with respect to z are defined similarly. Using these definitions we may write the line integral in (12) as

$$\int_C \boldsymbol{F} \cdot d\boldsymbol{r} = \int_C M(x, y, z) \, dx + \int_C N(x, y, z) \, dy + \int_C P(x, y, z) \, dz.$$

Piecewise Smooth Curves

Up to this point we have defined the line integral only for smooth curves C and continuous vector fields $\boldsymbol{F}$. (Recall that the curve $C = \{r(t) \mid a \le t \le b\}$ is *smooth* if $r'(t)$ exists and is continuous for all t in the interval $[a, b]$.) The reason why C has been taken to be smooth is that the integrand in the definite integral in line (10) involves the derivative, $r'(t)$. Thus, when $\boldsymbol{F}$ and r' are continuous, the theory developed in Chapter 6 guarantees that the definite integral in equation (10) exists.

However, this integral exists under slightly less restrictive conditions. In particular, it was stated in Chapter 6 that the definite integral exists when the integrand is merely piecewise continuous. Thus, we need only require that r' be piecewise continuous for the line integral in Definition 3 to exist. Curves of this type are called *piecewise smooth*. We shall refer to such curves as *paths* (see Figures 2.12 and 2.13).

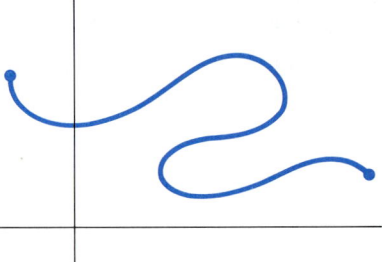

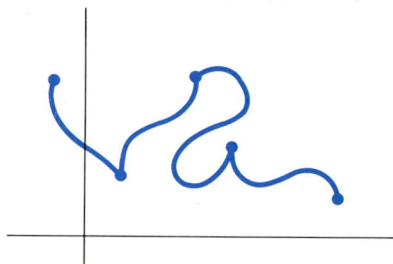

Figure 2.12 A smooth curve (r' continuous).

Figure 2.13 A piecewise smooth curve, or *path* (r' piecewise continuous).

DEFINITION 4

A curve $C = \{r(t) \mid a \le t \le b\}$ is **piecewise smooth** if there exist numbers $a = t_0 < t_1 < t_2 < \cdots < t_n = b$ so that r is continuous on $[a, b]$ and so that $r'(t)$ exists and is continuous on each of the subintervals $(t_0, t_1), (t_1, t_2), \ldots, (t_{n-1}, t_n)$. A piecewise smooth curve is called a **path**.

REMARK: For piecewise smooth arcs, we evaluate the line integral by first integrating over each of the subarcs on which r' is continuous and then summing the results:

$$\int_C \boldsymbol{F} \cdot d\boldsymbol{r} = \int_a^b \boldsymbol{F}(r(t)) \cdot r'(t) \, dt \tag{19}$$

$$= \int_{t_0}^{t_1} \boldsymbol{F}(r(t)) \cdot r'(t) \, dt + \cdots + \int_{t_{n-1}}^{t_n} \boldsymbol{F}(r(t)) \cdot r'(t) \, dt$$

$$= \int_{C_1} \boldsymbol{F} \cdot d\boldsymbol{r} + \int_{C_2} \boldsymbol{F} \cdot d\boldsymbol{r} + \cdots + \int_{C_n} \boldsymbol{F} \cdot d\boldsymbol{r}$$

where C_j denotes the arc $C_j = \{r(t) \mid t_{j-1} \le t \le t_j\}$.

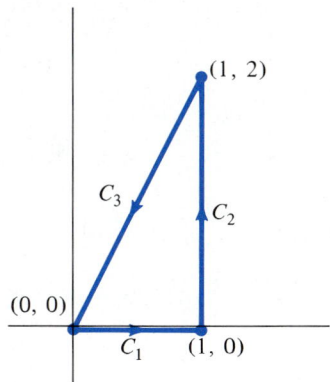

Figure 2.14

Example 6

Evaluate the line integral

$$\int_C (x + y)\, dx + xy\, dy$$

over the path C consisting of the line segments

C_1: from $(0, 0)$ to $(1, 0)$,

C_2: from $(1, 0)$ to $(1, 2)$,

C_3: from $(1, 2)$ to $(0, 0)$,

(See Figure 2.14.)

Solution: Here the vector field is $\mathbf{F}(x, y) = M(x, y)\mathbf{i} + N(x, y)\mathbf{j}$ with

$$M(x, y) = x + y, \qquad N(x, y) = xy.$$

A parameterization for C_1 is

$$C_1: \quad \left. \begin{array}{l} x(t) = t \\ y(t) = 0 \end{array} \right\} 0 \leq t \leq 1.$$

On this arc, $x'(t) = 1$ and $y'(t) = 0$. The line integral over C_1 is therefore

$$\int_{C_1} (x + y)\, dx + xy\, dy = \int_0^1 [(t + 0)(1) + (t \cdot 0)(0)]\, dt$$

$$= \int_0^1 t\, dt = \frac{1}{2}.$$

A parameterization for C_2 is

$$C_2: \quad \left. \begin{array}{l} x(t) = 1 \\ y(t) = t \end{array} \right\} 0 \leq t \leq 2.$$

On this arc, $x'(t) = 0$ and $y'(t) = 1$. Thus,

$$\int_{C_2} (x + y)\, dx + xy\, dy = \int_0^2 [(1 + t)(0) + (1 \cdot t)(1)]\, dt$$

$$= \int_0^2 t\, dt = 2.$$

Finally, a parameterization for C_3 is

$$C_3: \quad \left. \begin{array}{l} x(t) = 1 - t \\ y(t) = 2 - 2t \end{array} \right\} 0 \leq t \leq 1.$$

On this arc, $x'(t) = -1$ and $y'(t) = -2$. Thus, by (13),

$$\int_{C_3} (x + y)\, dx + xy\, dy = \int_0^1 \{[(1 - t) + (2 - 2t)](-1)$$

$$+ (1 - t)(2 - 2t)(-2)\}\, dt$$

$$= \int_0^1 (-4t^2 + 11t - 7)\, dt = -\frac{17}{6}.$$

Applying equation (19), we conclude

$$\int_C (x + y) \, dx + xy \, dy = \int_{C_1} (x + y) \, dx + xy \, dy$$

$$+ \int_{C_2} (x + y) \, dx + xy \, dy$$

$$+ \int_{C_3} (x + y) \, dx + xy \, dy$$

$$= \frac{1}{2} + 2 - \frac{17}{6}$$

$$= -\frac{1}{3}.$$ ◇

Line Integrals with Respect to Arc Length

Thus far we have defined line integrals only for vector fields. Another type of line integral, called a **line integral with respect to arc length,** concerns a scalar valued function, say f on $\mathbb{R}^2$. Let C be a path with parameterization

$$C = \{r(t) = x(t)i + y(t)j \mid a \le t \le b\}.$$

The usual partitioning of the interval $[a, b]$ produces arcs $C_1, C_2, \ldots, C_n$ whose lengths Δs_j, as before, are

$$\Delta s_j = \sqrt{[x'(c_j)]^2 + [y'(c_j)]^2} \, \Delta t, \qquad j = 1, 2, \ldots, n,$$

where c_j is a number in the jth subinterval of $[a, b]$. If the function f is continuous on an open rectangle containing the path C, the Riemann sum

$$\sum_{j=1}^{n} f(x(c_j), y(c_j)) \, \Delta s_j \tag{20}$$

$$= \sum_{j=1}^{n} f(x(c_j), y(c_j)) \sqrt{[x'(c_j)]^2 + [y'(c_j)]^2} \, \Delta t$$

converges to the definite integral

$$\int_C f(x, y) \, ds = \int_a^b f(x(t), y(t)) \sqrt{[x'(t)]^2 + [y'(t)]^2} \, dt. \tag{21}$$

The integral in equation (21) is called the *line integral of f over C with respect to arc length*. The difference between the integral in (21) and the line integral in Definition 3 is that (21) is an integral of a scalar function f with respect to change in arc length, while the integral in (10) is an integral of the dot product of a vector field and the unit tangent with respect to change in the parameter t. By using the differential for arc length

$$ds = \sqrt{[x'(t)]^2 + [y'(t)]^2} \, dt = |r'(t)| \, dt,$$

we may interpret the line integral in Definition 3 as a special case of the integral in (21) with $f(x(t), y(t)) = F(r(t)) \cdot T(r(t))$. That is,

$$\int_C F \cdot dr = \int_C F \cdot T \, ds. \tag{22}$$

This can be seen by comparing the approximating sums in lines (5) and (20).

Although we shall work almost exclusively in what follows with line integrals as defined in Definition 3, equation (22) will be used in Section 19.4 to interpret certain statements about line integrals.

Example 7

Find $\int_C xy^3 \, ds$ where C is the quarter circle

$$C = \left\{ \cos t\boldsymbol{i} + \sin t\boldsymbol{j} \mid 0 \le t \le \frac{\pi}{2} \right\}.$$

Solution: Here $x(t) = \cos t$, $y(t) = \sin t$, $x'(t) = -\sin t$, and $y'(t) = \cos t$. Thus, by (21)

$$\int_C xy^3 \, ds = \int_0^{\pi/2} (\cos t)(\sin^3 t)\sqrt{[-\sin t]^2 + [\cos t]^2} \, dt$$

$$= \int_0^{\pi/2} \sin^3 t \cos t \, dt$$

$$= \frac{1}{4} \sin^4 t \Big]_0^{\pi/2}$$

$$= \frac{1}{4}. \qquad \diamond$$

It is important to note that, unlike line integrals of the form $\int_C \boldsymbol{F} \cdot d\boldsymbol{r}$, line integrals with respect to arc length are independent of the direction along which the curve C is traversed (see Exercise 28).

An application of line integrals with respect to arc length concerns finding the mass M of a thin wire of variable density whose shape is given by the curve $C = \{\boldsymbol{r}(t) = x(t)\boldsymbol{i} + y(t)\boldsymbol{j} \mid a \le t \le b\}$. If the mass density (in units of mass per unit length) is given by the continuous function f, the expression

$$\Delta m_j = f(x(c_j), y(c_j)) \, \Delta s_j$$

approximates the mass of a section of the wire of length Δs_j, one point of which has coordinates $(x(c_j), y(c_j))$. The sum in equation (20) therefore approximates the total mass of the wire, which is given precisely by the line integral with respect to arc length in equation (21).

For curves in space, line integrals with respect to arc length are defined in an entirely analogous manner. That is, if f is continuous in an open box containing the piecewise smooth curve

$$C = \{\boldsymbol{r}(t) = x(t)\boldsymbol{i} + y(t)\boldsymbol{j} + z(t)\boldsymbol{k} \mid a \le t \le b\},$$

then

$$\int_C f(x, y, z) \, ds$$

$$= \int_a^b f(x(t), y(t), z(t))\sqrt{[x'(t)]^2 + [y'(t)]^2 + [z'(t)]^2} \, dt.$$

Exercise Set 19.2

In Exercises 1–8, calculate the work done by the force field F in moving a particle along the specified path C.

1. $F(x, y) = x\mathbf{i} + y\mathbf{j}$, C: $\mathbf{r}(t) = t^2\mathbf{i} + (3 + t)\mathbf{j}$, $0 \leq t \leq 2$

2. $F(x, y) = xy\mathbf{i} + x^2\mathbf{j}$,
   C: $\mathbf{r}(t) = \cos t\mathbf{i} + \sin t\mathbf{j}$, $0 \leq t \leq \pi$

3. $F(x, y) = x^2\mathbf{i} + y^2\mathbf{j}$,
   C: $\mathbf{r}(t) = \sqrt{t}\mathbf{i} + 3t\mathbf{j}$, $1 \leq t \leq 4$

4. $F(x, y) = -y\mathbf{i} + x\mathbf{j}$,
   C: $\mathbf{r}(t) = (t + 1)^2\mathbf{i} + (t - 1)^2\mathbf{j}$, $0 \leq t \leq 2$

5. $F(x, y) = (x^2 + y^2)\mathbf{i} + xy\mathbf{j}$,
   C: $\mathbf{r}(t) + t\mathbf{i} + t^2\mathbf{j}$, $0 \leq t \leq 2$

6. $F(x, y) = x^2y\mathbf{i} + xy^2\mathbf{j}$,
   C: $\mathbf{r}(t) = t\mathbf{i} + 2t^2\mathbf{j}$, $0 \leq t \leq 1$

7. $F(x, y, z) = xy\mathbf{i} + xz\mathbf{j} + yz\mathbf{k}$,
   C: $\mathbf{r}(t) = \cos t\mathbf{i} + \sin t\mathbf{j} + t\mathbf{k}$, $0 \leq t \leq \pi$

8. $F(x, y, z) = x^2\mathbf{i} + y^2\mathbf{j} + z^2\mathbf{k}$,
   C: $\mathbf{r}(t) = t\mathbf{i} + 3t^2\mathbf{j} - 2t\mathbf{k}$; $0 \leq t \leq 2$

In Exercises 9–15, evaluate the line integral of the vector field F over the indicated path C.

9. $F(x, y) = y\mathbf{i} + 2x\mathbf{j}$, C is the line segment from $(0, 0)$ to $(4, 2)$.

10. $F(x, y) = -x^2\mathbf{i} + y^2\mathbf{j}$, C is the upper unit semicircle from $(1, 0)$ to $(-1, 0)$.

11. $F(x, y) = (y - x)\mathbf{i} + xy\mathbf{j}$, C is the unit circle traversed counterclockwise.

12. $F(x, y) = xy^2\mathbf{i} + (x + y)\mathbf{j}$, C is the triangular path from $(0, 0)$ to $(4, 0)$ to $(4, 2)$ to $(0, 0)$.

13. $F(x, y) = (x + y)\mathbf{i} + (y^2 - x^2)\mathbf{j}$, C is the triangle with vertices $(-1, 0)$, $(1, -4)$, $(0, 2)$ traversed counterclockwise.

14. $F(x, y, z) = xy\mathbf{i} + y\mathbf{j} + xz\mathbf{k}$, C is the line segment from $(0, 0, 0)$ to $(2, 4, -6)$.

15. $F(x, y, z) = y\mathbf{i} - x\mathbf{j} + 2z\mathbf{k}$, C is the arc of the circular helix $\mathbf{r}(t) = \cos t\mathbf{i} + \sin t\mathbf{j} + t\mathbf{k}$ from $(1, 0, 0)$ to $(0, 1, \pi/2)$.

16. Evaluate $\int_C xy \, dx$ where C is the arc of the unit circle from $(0, 1)$ to $(-1, 0)$ traversed counterclockwise.

17. Evaluate $\int_C (x^2 + y^2) \, dy$ where C is the path in Exercise 16.

18. Evaluate $\int_C F \cdot d\mathbf{r}$ where $F(x, y) = (x^2 + y^2)\mathbf{i} + 2xy\mathbf{j}$ and C is the arc of the circle $x^2 + y^2 = 1$ from $(1, 0)$ to $(0, 1)$, followed by the line segment from $(0, 1)$ to $(-1, 1)$ (Figure 2.15).

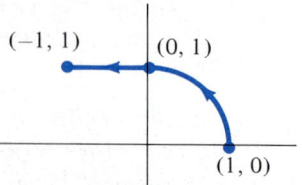

Figure 2.15

19. Evaluate $\int_C (x^2 + y^2 + z^2) \, dy$ along the line segment from $(0, 1, 0)$ to $(-1, 2, 1)$.

20. Find the work done on a particle by the force field $F(x, y) = (x - y)\mathbf{i} + xy\mathbf{j}$ in moving a particle counterclockwise around the ellipse $9x^2 + 4y^2 = 36$ from $(2, 0)$ to $(-2, 0)$.

21. Evaluate $\int_C xy \, dx + xy^2 \, dy$ where C is the curve in Exercise 18.

22. Find the work done by the force field $F(x, y) = (x\mathbf{i} - y\mathbf{j})$ in moving a particle along the path in Figure 2.16.

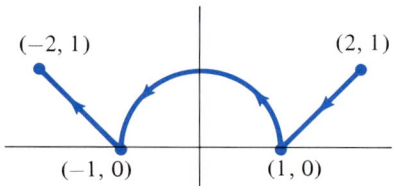

Figure 2.16

23. Evaluate $\int_C (x + y) \, dx + (x^2 - y^2) \, dy$ where C is the arc of the parabola $y = 2x^2$ from $(0, 0)$ to $(2, 8)$.

24. Evaluate $\int_C (x - y) \, dx + xy \, dy$ where C is the path consisting of the line segments from $(0, 1)$ to $(2, -1)$, from $(2, -1)$ to $(2, 5)$, and from $(2, 5)$ to $(4, 1)$.

25. Evaluate $\int_C x^2y \, dx + xy^2 \, dy$ where C is the path from $(0, 2)$ to $(4, 1)$ along the three line segments parallel to the x- and y-axes.

26. Let F be the central force field $F(\mathbf{r}) = \dfrac{2}{|\mathbf{r}|^3}\mathbf{r}$, $\mathbf{r} \neq \mathbf{0}$. Find the work done by F in moving a particle along the upper half of the unit circle from $(1, 0)$ to $(-1, 0)$.

27. Show that Definition 2 agrees with the definition of work when a force $f(x)$ is applied to a particle moving along the x-axis.

28. Show that the line integral with respect to arc length in line (21) is independent of the direction along which C is traversed.

Evaluate each of the following line integrals with respect to arc length.

29. $\displaystyle\int_C (x + 2y^2)\, ds$ where $C = \{r(t) = 2t i - 4t j \mid 0 \le t \le 1\}$

30. $\displaystyle\int_C x e^y\, ds$ where $C = \{r(t) = \cos t i + \sin t j \mid 0 \le t \le \pi\}$

31. $\displaystyle\int_C (8x + y - 3)\, ds$

where $C = \{r(t) = t i + 3j + t^2 k \mid 0 \le t \le 1\}$

32. Give an argument to support the conclusion that if the

curve C describes the shape of a thin wire of mass density $\rho(x, y, z)$ then the total mass of the wire is

$$M = \int_C \rho\, ds$$

$$= \int_a^b \rho(x(t), y(t), z(t))\sqrt{[x'(t)]^2 + [y'(t)]^2 + [z'(t)]^2}\, dt.$$

33. Refer to Exercise 32. Find the total mass of a wire with shape

$$C: \quad r(t) = \cos t i - \sin t j + 2k, \qquad 0 \le t \le \pi$$

if the density is $\rho(x, y, z) = x^2 + y^2 + z^2$.

34. Refer to Exercise 32. Find the mass of a wire of constant density $\rho(x, y, z) = k$ shaped like the helix $r(t) = 4\cos t i + 4\sin t j + 3t k$ for $0 \le t \le 2\pi$.

19.3 LINE INTEGRALS: INDEPENDENCE OF PATH

The Fundamental Theorem of Calculus allows us to evaluate definite integrals of functions of a single variable by means of an antiderivative. In this section, we develop a similar theorem for line integrals. To do so requires the concept of a *scalar potential* ϕ for F. Recall from Section 17.9 that the differentiable scalar function ϕ is a scalar potential for the vector function F if $\nabla\phi = F$.

Our Fundamental Theorem for line integrals is the following.

THEOREM 1

Let $C\{r(t) \mid a \le t \le b\}$ be a piecewise smooth curve in an open rectangle D. Let F be a vector field on D and let ϕ be a differentiable scalar function for which $F(r) = \nabla\phi(r)$ for all r in D. Then

$$\int_C F \cdot dr = \phi(r(b)) - \phi(r(a)).$$

Theorem 1 says this. If the vector field F is conservative, meaning that it is the gradient of some scalar function ϕ, then the line integral of F over C from $A = r(a)$ to $B = r(b)$ depends only on the values of the scalar potential ϕ at A and at B. Thus, if the vector field F is *conservative*, we can determine the value of the line integral by finding a scalar potential ϕ for F. Before proving Theorem 1, we present two examples.

Example 1

Find $\displaystyle\int_C F \cdot dr$ where F is the vector field $F(x, y) = 2xy i + x^2 j$ and C is the line segment from $(-1, 2)$ to $(4, -3)$.

Solution: Here $\phi(x, y) = x^2 y$ is a potential for F since

$$\nabla\phi(x, y) = 2xy i + x^2 j = F(x, y).$$

Thus, by Theorem 1,

$$\int_C F(r) \cdot dr = \phi(4, -3) - \phi(-1, 2) = 4^2(-3) - (-1)^2 2 = -50. \qquad \diamond$$

Example 2

Evaluate the line integral

$$\int_C 2xye^{x^2} \, dx + e^{x^2} \, dy$$

over a path C from $(0, 1)$ to $(1, 3)$.

Solution: This integral has the form $\int_C F \cdot dr$ where

$$F(x, y) = 2xye^{x^2} i + e^{x^2} j$$

and $dr = dx \, i + dy \, j$. A potential for F is

$$\phi(x, y) = ye^{x^2}$$

since

$$\nabla \phi(x, y) = 2xye^{x^2} i + e^{x^2} j = F(x, y).$$

Thus, by Theorem 1

$$\int_C 2xye^{x^2} \, dx + e^{x^2} \, dy = \phi(1, 3) - \phi(0, 1)$$

$$= 3e - e^0$$

$$= 3e - 1. \qquad \diamond$$

Proof of Theorem 1: We will need to make use of the Chain Rule for vector functions: If x and y are differentiable component functions, then

$$\frac{d}{dt} \phi(x(t), y(t)) = \phi_x(x(t), y(t))x'(t) + \phi_y(x(t), y(t))y'(t)$$

$$= \nabla \phi(x(t), y(t)) \cdot [x'(t)i + y'(t)j].$$

Using the position vector notation, this can be written

$$\frac{d}{dt} \phi(r(t)) = \nabla \phi(r(t)) \cdot r'(t). \qquad (1)$$

Assume first that C is a smooth curve. Then, under the hypotheses of Theorem 1, equation (1) holds. Thus,

$$\int_C F \cdot dr = \int_a^b F(r(t)) \cdot r'(t) \, dt$$

$$= \int_a^b \nabla \phi(r(t)) \cdot r'(t) \, dt$$

$$= \int_a^b \frac{d}{dt} [\phi(r(t))] \, dt \qquad \text{(equation (1))}$$

$$= \phi(r(t))]_a^b$$

$$= \phi(r(b)) - \phi(r(a)).$$

The extension to the case C piecewise smooth is easy and is left as an exercise. ◆

In trying to determine whether the vector field F is conservative, it is helpful to recall Theorem 9, Section 17.9: If the functions M and N are continuously differentiable in an open rectangle D, *the vector function*

$$F(x, y) = M(x, y)i + N(x, y)j$$

is conservative if and only if

$$M_y(x, y) = N_x(x, y) \tag{2}$$

for all (x, y) *in* D. Equation (2) provides a quick check of whether Theorem 1 applies for a vector field F in the plane. When equation (2) holds, the method of Section 17.9 can sometimes be applied to find the potential ϕ.

Example 3

Find the work done by the vector field

$$F(x, y) = (e^y + 3x^2y^2)i + (xe^y + 2x^3y + 2)j$$

in moving a particle from the point $(0, 0)$ to the point $(3, 2)$ in the plane.

Solution: Here $F(x, y) = M(x, y)i + N(x, y)j$ where

$$M(x, y) = e^y + 3x^2y^2, \qquad N(x, y) = xe^y + 2x^3y + 2.$$

Then

$$M_y(x, y) = e^y + 6x^2y = N_x(x, y),$$

so the vector field F is conservative. To find a potential ϕ with

$$\nabla\phi(x, y) = \phi_x(x, y)i + \phi_y(x, y)j \tag{3}$$
$$= M(x, y)i + N(x, y)j = F(x, y),$$

we first equate i-components and integrate partially with respect to x to find that

$$\phi(x, y) = \int (e^y + 3x^2y^2)\, dx = xe^y + x^3y^2 + f(y) + C \tag{4}$$

where f is a function of y alone. Equating j-components in (3) and integrating partially with respect to y shows that

$$\phi(x, y) = \int (xe^y + 2x^3y + 2)\, dy = xe^y + x^3y^2 + 2y + g(x) + C \tag{5}$$

where g is a function of x alone.

Equating the expressions for ϕ in lines (4) and (5) shows that a scalar potential for F is

$$\phi(x, y) = xe^y + x^3y^2 + 2y.$$

We may therefore apply Theorem 1. The desired work is

$$W = \int_C F \cdot dr = \phi(3, 2) - \phi(0, 0)$$
$$= (3e^2 + 3^3 \cdot 2^2 + 2 \cdot 2) - 0$$
$$= 3e^2 + 112.$$

◇

Whether or not an explicit potential ϕ can actually be found, Theorem 1 guarantees that the value of a line integral for a conservative vector field depends only on the endpoints of the path C and not on the path itself. (Physicists paraphrase this statement by saying that "work is a function of position, not of path.") The terminology we choose to use is that the line integral of a conservative force field is **independent of path.** The following theorem characterizes this notion.

THEOREM 2	Let C be a piecewise smooth curve in an open rectangle D. The line integral $$\int_C \boldsymbol{F} \cdot d\boldsymbol{r}$$ is independent of path if and only if the vector field $\boldsymbol{F}$ is conservative in D.

The "if" part of this theorem follows directly from Theorem 1. The "only if" part is an important observation in more advanced courses in mathematics but will not be used here in a direct way. The proof is omitted.

The following corollary is a useful formulation of Theorem 2 and the statement concerning equation (2) for vector fields in the plane.

COROLLARY 1	Let D be an open rectangle in the plane and let C be any path from point A to point B lying within D. Let M and N be continuously differentiable in D. The line integral $$\int_C M(x, y)\, dx + N(x, y)\, dy \qquad (6)$$ is independent of path if and only if $$M_y(x, y) = N_x(x, y) \qquad (7)$$ for all (x, y) in D.

Example 4

Evaluate the line integral

$$\int_C (\cos xy - xy \sin xy)\, dx - x^2 \sin xy\, dy$$

where C is the upper unit semicircle from $(1, 0)$ to $(-1, 0)$.

Solution: This integral has the form (6) with

$$M(x, y) = \cos xy - xy \sin xy, \qquad N(x, y) = -x^2 \sin xy,$$

so

$$M_y(x, y) = -2x \sin xy - x^2 y \cos xy = N_x(x, y).$$

Thus, equation (7) holds, so the integral is independent of path. However, instead of applying the method of Example 3 to actually find the potential ϕ, we demonstrate how to apply Corollary 1 in evaluating this line integral.

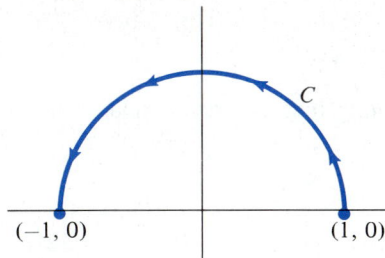

Figure 3.1 Path C from $(1, 0)$ to $(-1, 0)$.

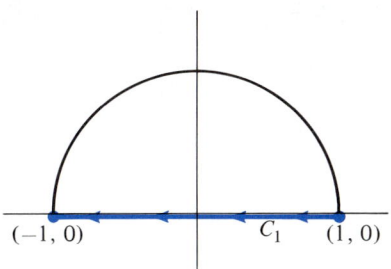

Figure 3.2 Path C_1 from $(1, 0)$ to $(-1, 0)$.

Since the integral is independent of path, we may choose a path that makes the line integral easier to compute than does C. In this case, we choose the path C_1 consisting of the (straight) line segment from $(1, 0)$ to $(-1, 0)$ (see Figures 3.1 and 3.2). A parameterization for C_1 is

$$C_1: \left. \begin{matrix} x(t) = 1 - t \\ y(t) = 0 \end{matrix} \right\} \quad 0 \le t \le 2.$$

Thus, $x'(t) = -1$ and $y'(t) = 0$. Calculating the line integral according to Definition 2 gives

$$\int_C (\cos xy - xy \sin xy)\, dx - x^2 \sin xy\, dy$$

$$= \int_0^2 \{[\cos(0) - 0 \cdot \sin(0)](-1) - (1 - t)^2 \cdot \sin(0)(0)\}\, dt$$

$$= \int_0^2 (-1)\, dt$$

$$= -2.$$

Now we note that a scalar potential for this vector field is $\phi(x, y) = x \cos xy$. With this information the line integral could have been evaluated using Theorem 1 as

$$\phi(-1, 0) - \phi(1, 0) = (-1)\cos(0) - (1)\cos(0) = -2. \qquad \diamondsuit$$

REMARK: Since a line integral over a path from point A to point B for a conservative force field $\boldsymbol{F}$ depends only on the points A and B, the notation

$$\int_C \boldsymbol{F} \cdot d\boldsymbol{r} = \int_A^B \boldsymbol{F} \cdot d\boldsymbol{r}$$

is sometimes used. If $\boldsymbol{F} = \boldsymbol{\nabla}\phi$, Theorem 1 is stated in this notation as

$$\int_A^B \boldsymbol{F} \cdot d\boldsymbol{r} = \phi(B) - \phi(A).$$

In using this notation, one must exercise caution to ensure that the hypotheses of Theorem 1 are true. Otherwise, this notation is meaningless.

Example 5

In the above notation, the result of Example 4 can be stated as

$$\int_{(1,\, 0)}^{(-1,\, 0)} (\cos xy - xy \sin xy)\, dx - x^2 \sin xy\, dy = -2. \qquad \diamondsuit$$

Example 6

Although all of the preceding examples have dealt with paths in the plane, Theorems 1 and 2 apply also to paths and vector fields in space. The line integral

$$\int_C yz\, dx + xz\, dy + xy\, dz$$

over the helical path

$$C: \quad r(t) = \cos t i + \sin t j + t k, \qquad 0 \le t \le \pi/4$$

may be handled by means of Theorem 1 by noting that the vector field

$$F(x, y, z) = yz i + xz j + xy k$$

is the gradient of the scalar potential

$$\phi(x, y, z) = xyz.$$

Since the endpoints of C are $A = r(0) = (1, 0, 0)$ and $B = r(\pi/4) = (\sqrt{2}/2, \sqrt{2}/2, \pi/4)$, we have

$$\int_C yz \, dx + xz \, dy + xy \, dz = \int_{(1, 0, 0)}^{(\sqrt{2}/2, \sqrt{2}/2, \pi/4)} yz \, dx + xz \, dy + xy \, dz$$

$$= \phi\left(\frac{\sqrt{2}}{2}, \frac{\sqrt{2}}{2}, \frac{\pi}{4}\right) - \phi(1, 0, 0)$$

$$= \frac{\pi}{8}. \qquad \diamond$$

Line Integrals Over Closed Paths

The path (piecewise smooth curve) $C = \{r(t) \mid a \le t \le b\}$ is called **closed** if $r(a) = r(b)$. Thus, circles, ellipses, triangles, squares, and rectangles are all examples of closed paths in the plane. There is a very simple situation governing line integrals over closed paths. *If the vector field F is conservative, the line integral $\int_C F \cdot dr$ over a closed path is zero.* This statement follows directly from Theorem 1: Since a closed path $C = \{r(t) \mid a \le t \le b\}$ satisfies the equation $r(a) = r(b)$,

$$\int_C F \cdot dr = \phi(r(b)) - \phi(r(a)) = 0.$$

This fact is well known in physics. In terms of work, it says that the work done by a conservative force field (meaning that no energy is lost in the process) in moving a particle around a closed path is zero.

Conservation of Energy

Let's take the physics one step further. Let F be a conservative force field. If a particle of mass m moves in the vector field F with velocity $v(t) = r'(t)$, we define the **kinetic energy** of the particle at time t as

$$K(t) = \frac{1}{2}m|v(t)|^2. \tag{8}$$

Also, the change in **potential energy** U for the particle as it is moved from point A to point B is defined as the negative of the work done by the force field F in moving it from A to B:

$$U(B) - U(A) = -\int_A^B F \cdot dr. \tag{9}$$

Now if $a(t)$ denotes the acceleration of the particle at time t, Newton's second law of motion states that

$$F(r(t)) = ma(t) = mr''(t),$$

so we can rewrite the integral in Definition 2 as

$$\boldsymbol{F}(\boldsymbol{r}(t)) \cdot \boldsymbol{r}'(t) = m\boldsymbol{r}''(t) \cdot \boldsymbol{r}'(t) \tag{10}$$

$$= \frac{1}{2}m[2\boldsymbol{r}''(t) \cdot \boldsymbol{r}'(t)]$$

$$= \frac{1}{2}m \frac{d}{dt}[\boldsymbol{r}'(t) \cdot \boldsymbol{r}'(t)]$$

$$= \frac{d}{dt}\left[\frac{1}{2}m|\boldsymbol{v}(t)|^2\right].$$

Using (8) and (10) and the fact that $\boldsymbol{F}$ is conservative, we find that

$$\text{Work} = \int_A^B \boldsymbol{F} \cdot d\boldsymbol{r} = \int_a^b \boldsymbol{F}(\boldsymbol{r}(t)) \cdot \boldsymbol{r}'(t) \, dt \tag{11}$$

$$= \int_a^b \frac{d}{dt}\left[\frac{1}{2}m|\boldsymbol{v}(t)|^2\right] dt$$

$$= \frac{1}{2}m|\boldsymbol{v}(b)|^2 - \frac{1}{2}m|\boldsymbol{v}(a)|^2$$

$$= K(B) - K(A)$$

where $A = \boldsymbol{r}(a)$, $B = \boldsymbol{r}(b)$. That is,

$$\boxed{\text{Work} = K(B) - K(A).} \tag{12}$$

Equation (12) may be interpreted as the principle that *in a conservative system, the work done on an object (by a force) equals the change in its kinetic energy.* Finally, combining equations (9) and (11) gives the equation

$$U(A) - U(B) = K(B) - K(A)$$

or

$$\boxed{K(A) + U(A) = K(B) + U(B).} \tag{13}$$

Equation (13) is the famous **principle of conservation of energy:** In a conservative system, the sum of kinetic and potential energy of a moving particle remains constant from point to point.

Exercise Set 19.3

In Exercises 1–10, evaluate the line integral by verifying that the vector field is conservative and applying Theorem 1.

1. $\int_C \boldsymbol{F} \cdot d\boldsymbol{r}$, $\boldsymbol{F}(x, y) = y\boldsymbol{i} + x\boldsymbol{j}$, C is a path from $(0, 0)$ to $(3, 1)$.

2. $\int_C \boldsymbol{F} \cdot d\boldsymbol{r}$, $\boldsymbol{F}(x, y) = 3x^2y^2\boldsymbol{i} + 2x^3y\boldsymbol{j}$, C is a path from $(-3, 1)$ to $(2, 2)$.

3. $\int_C \boldsymbol{F} \cdot d\boldsymbol{r}$, $\boldsymbol{F}(x, y) = ye^{xy}\boldsymbol{i} + xe^{xy}\boldsymbol{j}$, C is a path from $(0, 0)$ to $(1, 2)$.

4. $\int_C \boldsymbol{F} \cdot d\boldsymbol{r}$, $\boldsymbol{F}(x, y, z) = yze^{xyz}\boldsymbol{i} + xze^{xyz}\boldsymbol{j} + xye^{xyz}\boldsymbol{k}$, C is a path from $(0, 0, 0)$ to $(1, 0, 1)$.

5. $\int_C \boldsymbol{F} \cdot d\boldsymbol{r}$, $\boldsymbol{F}(x, y) = 2x\boldsymbol{i} + 2y\boldsymbol{j}$, C is the upper unit semicircle traversed counterclockwise.

6. $\int_{(1, 1)}^{(2, 3)} (1 + 2xy^2) \, dx + 2x^2y \, dy$

7. $\int_{(0, 0)}^{(2, 1)} e^{y^2} \, dx + 2xye^{y^2} \, dy$

8. $\displaystyle\int_{(0,\,0)}^{(\pi/2,\,1)} y \sin xy \, dx + x \sin xy \, dy$

9. $\displaystyle\int_{(0,\,0)}^{(1,\,\pi/4)} e^x \sin y \, dx + e^x \cos y \, dy$

10. $\displaystyle\int_{(0,\,0,\,0)}^{(0,\,\pi/4,\,\pi/4)} e^x \sin y \cos z \, dx$
$+ \, e^x \cos y \cos z \, dy - e^x \sin y \sin z \, dz$

11. Let $F(x, y)$ define a conservative vector field throughout the plane. Let C_1 be the path along the upper semicircle from $(1, 0)$ to $(-1, 0)$. Let C_2 be the path along the x-axis from $(-1, 0)$ to $(1, 0)$. If $\displaystyle\int_{C_1} F \cdot dr = a$, what is $\displaystyle\int_{C_2} F \cdot dr$? Why?

12. Evaluate $\displaystyle\int_C \sin y \, dx - x \cos y \, dy$ where C is the curve with parameterization C: $\quad r(t) = \cos t i + \sin t j, \quad 0 \le t \le \pi$.

13. Use Theorem 2 to prove that the vector field F is conservative in the open set D if and only if $\displaystyle\int_C F \cdot dr = 0$ for every closed path in D.

14. Use Theorem 2 and Exercise 13 to prove that the line integral $\displaystyle\int_C F \cdot dr$ is independent of path in the open set D if and only if $\displaystyle\int_C F \cdot dr = 0$ for every closed path in D.

15. Let $F(x, y) = 2xye^{x^2y}i + x^2e^{x^2y}j$. Let C_1 be the path from $A = (-1, 0)$ to $B = (1, 0)$ clockwise along the unit circle, and let C_2 be the path consisting of the line segment from A to B. Use the method of Section 19.2 to evaluate each of the

line integrals $\displaystyle\int_{C_1} F \cdot dr$ and $\displaystyle\int_{C_2} F \cdot dr$. Are they equal? Why?

16. Repeat Exercise 15 for the vector field $F(x, y) = x^3y^2i + x^3yj$.

17. Show that, "in a conservative force field, the force is equal to the negative gradient of the potential."

18. Find a force field F so that $\displaystyle\int_{(0,\,0,\,0)}^{(x,\,y,\,z)} F \cdot dr = xyz$ for all $(x, y, z) \in \mathbb{R}^3$, independent of path.

19. Are line integrals of $F(x, y) = (x - y)i + (x + y)j$ independent of path in $\mathbb{R}^2$? Why or why not?

20. a. Show that
$$\int_C \frac{1}{x^2 + y^2}(x \, dy - y \, dx) = 2\pi$$
where C is the unit circle oriented counterclockwise.

  b. Show that the vector field
$$F(x, y) = \frac{x}{x^2 + y^2}i - \frac{y}{x^2 + y^2}j$$
  is not conservative.

  c. Show that, for
$$M(x, y) = \frac{x}{x^2 + y^2} \quad \text{and} \quad N(x, y) = \frac{-y}{x^2 + y^2},$$
$$M_y(x, y) = N_x(x, y), \text{ with } x \ne 0 \text{ and } y \ne 0.$$

  d. Explain why parts (a) through (c) do not contradict the statement involving equation (2).

21. Prove Theorem 1 in the case C piecewise smooth.

19.4 GREEN'S THEOREM

The Fundamental Theorem of Calculus states that if $F'(x) = f(x)$ for all x in $[a, b]$, the definite integral $\displaystyle\int_a^b f(x) \, dx$ may be evaluated by means of the formula

$$\int_a^b f(x) \, dx = F(b) - F(a). \tag{1}$$

In other words, the value of the integral in (1) is completely determined by the associated antiderivative at the endpoints of the interval $[a, b]$. In this section, we establish a similar result for certain functions of two variables: The value of a double integral of such a function over a region R in the plane is the same as an associated line integral taken around the boundary of R. This remarkable result is due to George Green, an English mathematician and physicist (1793–1841).

The statement of Green's Theorem involves the concept of **simple closed curves.** A closed curve $C = \{r(t) \mid a \le t \le b\}$ is called simple if $r(t_1) \ne r(t_2)$ for all numbers t_1 and t_2 in (a, b) with $t_1 \ne t_2$. (Remember, we must have $r(a) = r(b)$ if C

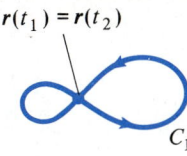

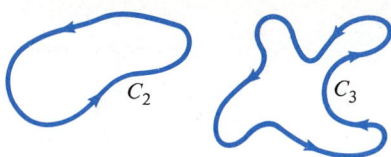

Figure 4.1 C_1 is not a simple closed curve.

Figure 4.2 C_2 and C_3 are simple closed curves.

is closed.) In other words, *a simple closed curve does not cross itself* (see Figures 4.1 and 4.2). Be careful not to confuse the notion of simple closed *curves* with our earlier concepts of vertically simple or horizontally simple *regions*.

THEOREM 3	Let C be a piecewise smooth simple closed curve in the plane, oriented counter-clockwise, and let Q be the region enclosed by C. If the functions $M(x, y)$ and $N(x, y)$ have continuous first partial derivatives in an open region containing Q, then
Green's Theorem	

$$\int_C M(x, y)\, dx + N(x, y)\, dy = \iint_Q \left(\frac{\partial N}{\partial x} - \frac{\partial M}{\partial y} \right) dA. \qquad (2)$$

Before discussing the proof of Green's Theorem, let's try to get a feel for what is going on by means of a few remarks and examples. We begin by expressing the line integral in (2) in vector form. If we write

$$\mathbf{F}(x, y) = M(x, y)\mathbf{i} + N(x, y)\mathbf{j}$$

and

$$C = \{\mathbf{r}(t) \mid a \le t \le b\},$$

then the conclusion (2) may be written as

$$\int_C \mathbf{F}(\mathbf{r}) \cdot d\mathbf{r} = \iint_Q \left(\frac{\partial N}{\partial x} - \frac{\partial M}{\partial y} \right) dA. \qquad (3)$$

Now if $\mathbf{F}$ is a conservative vector field, equation (3) is not surprising, since both integrals are zero. The line integral is zero since C is closed, and the double integral is zero since $\frac{\partial N}{\partial x} - \frac{\partial M}{\partial y} = 0$ if $\mathbf{F}$ is conservative. Thus, Green's Theorem is telling us something about evaluating line integrals for *nonconservative* vector fields. Figure 4.3 illustrates the result geometrically: the surface area associated with the line integral is numerically equal to the volume associated with the double integral.

Example 1

Use Green's Theorem to evaluate the line integral

$$\int_C xy^2\, dx + 2x^3y\, dy$$

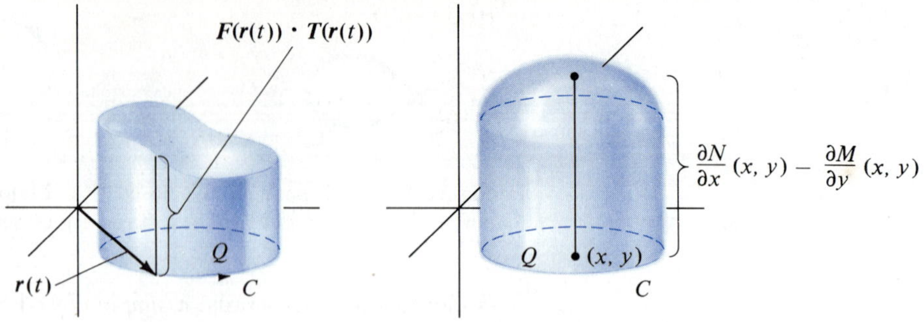

$$F(r(t)) \cdot T(r(t))$$

$$\frac{\partial N}{\partial x}(x, y) - \frac{\partial M}{\partial y}(x, y)$$

$$\int_C F(r) \cdot dr = \int_C M(x, y)\, dx + N(x, y)\, dy = \iint_Q \left(\frac{\partial N}{\partial x} - \frac{\partial M}{\partial y} \right) dA$$

Figure 4.3 Statement of Green's Theorem.

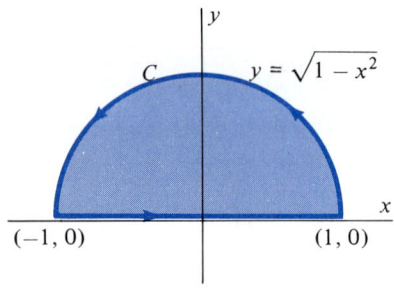

Figure 4.4 Region Q in Example 1.

where C is the path consisting of the upper unit semicircle traversed counterclockwise and the directed line segment from $(-1, 0)$ to $(1, 0)$ (see Figure 4.4).

Solution: Here $M(x, y) = xy^2$, $N(x, y) = 2x^3y$. The region Q enclosed by the path C may be described by the inequalities

$$Q: \quad -1 \le x \le 1; \quad 0 \le y \le \sqrt{1 - x^2}.$$

With $dA = dx\, dy$, Green's Theorem says that

$$\int_C xy^2\, dx + 2x^3y\, dy = \int_{-1}^{1} \int_{0}^{\sqrt{1-x^2}} \left[\frac{\partial}{\partial x}(2x^3y) - \frac{\partial}{\partial y}(xy^2) \right] dy\, dx$$

$$= \int_{-1}^{1} \int_{0}^{\sqrt{1-x^2}} (6x^2y - 2xy)\, dy\, dx$$

$$= \int_{-1}^{1} \{ [(3x^2 - x)y^2]_0^{\sqrt{1-x^2}} \}\, dx$$

$$= \int_{-1}^{1} (3x^2 - x)(1 - x^2)\, dx$$

$$= \int_{-1}^{1} (-3x^4 + x^3 + 3x^2 - x)\, dx$$

$$= \frac{4}{5}. \qquad \diamond$$

Example 2

Use Green's Theorem to evaluate the line integral

$$\int_C xy\, dx + e^y\, dy$$

where C is the path from $(0, 0)$ to $(1, 1)$ along the graph of $y = x^2$ and from $(1, 1)$ to $(0, 0)$ along the graph of $y = \sqrt{x}$.

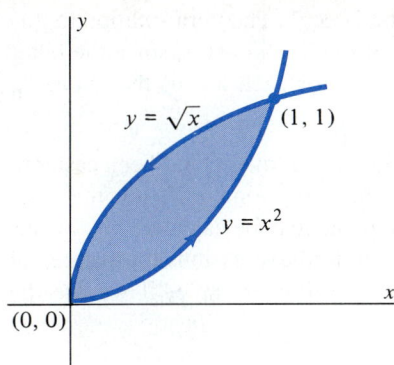

Figure 4.5 Region Q in Example 2.

Solution: Note in Figure 4.5 that C is oriented counterclockwise, as required. The region Q enclosed by C may be described by the inequalities

$$Q: \quad 0 \le x \le 1; \quad x^2 \le y \le \sqrt{x}.$$

Since $M(x, y) = xy$ and $N(x, y) = e^y$, an application of Green's Theorem gives

$$\int_C xy \, dx + e^y \, dy = \int_0^1 \int_{x^2}^{\sqrt{x}} \left[\frac{\partial}{\partial x}(e^y) - \frac{\partial}{\partial y}(xy) \right] dy \, dx$$

$$= \int_0^1 \int_{x^2}^{\sqrt{x}} -x \, dy \, dx$$

$$= \int_0^1 \{-xy]_{x^2}^{\sqrt{x}}\} \, dx$$

$$= \int_0^1 (-x^{3/2} + x^3) \, dx$$

$$= -\frac{3}{20}. \qquad \diamond$$

Example 3

Use Green's Theorem to evaluate the line integral

$$\int_C xy \, dx + (x + y) \, dy$$

where C is the path indicated in Figure 4.6.

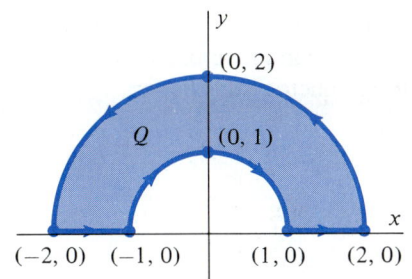

Figure 4.6

Solution: The region Q enclosed by C may be described in polar coordinates by the inequalities

$$Q: \quad 1 \le r \le 2; \quad 0 \le \theta \le \pi.$$

Since $M(x, y) = xy$ and $N(x, y) = x + y$, we have

$$\frac{\partial N}{\partial x} - \frac{\partial M}{\partial y} = \frac{\partial}{\partial x}(x + y) - \frac{\partial}{\partial y}(xy) = 1 - x. \qquad (4)$$

Since we are using polar coordinates, we set

$$x = r \cos \theta, \quad y = r \sin \theta, \quad \text{and} \quad dA = r \, dr \, d\theta. \qquad (5)$$

Combining (4) and (5) with the statement of Green's Theorem, we obtain

$$\int_C xy \, dx + (x + y) \, dy = \int_0^\pi \int_1^2 [1 - r \cos \theta] r \, dr \, d\theta$$

$$= \int_0^\pi \left[\frac{r^2}{2} - \frac{r^3}{3} \cos \theta \right]_{r=1}^{r=2} d\theta$$

$$= \int_0^\pi \left(\frac{3}{2} - \frac{7}{3} \cos \theta \right) d\theta$$

$$= \frac{3\theta}{2} - \frac{7}{3} \sin \theta \Big]_0^\pi$$

$$= \frac{3\pi}{2}. \qquad \diamond$$

There should be no doubt in your mind that the Green's Theorem solution to this problem is much easier than a direct evaluation of the line integral, since the latter would require four separate parameterizations, one for each arc of the curve C.

Calculating Area by Line Integrals

So far we have seen examples in which the double integral over Q has been easier to evaluate than the line integral over C in Green's Theorem. Sometimes it is the other way around. Since areas of certain regions in the plane may be calculated by means of a double integral, Green's Theorem enables us to find expressions for the area of Q (in Theorem 3) in terms of line integrals over C. For example, if we use the functions $M(x, y) = 0$ and $N(x, y) = x$, we obtain

$$\int_C x \, dy = \iint_Q \left[\frac{\partial}{\partial x}(x) - \frac{\partial}{\partial y}(0) \right] dA = \iint_Q 1 \, dA = \text{Area of } Q. \tag{6}$$

Similarly, by choosing $M(x, y) = y$ and $N(x, y) = 0$, we obtain

$$\int_C y \, dx = \iint_Q \left[\frac{\partial}{\partial x}(0) - \frac{\partial}{\partial y}(y) \right] dA = \iint_Q (-1) \, dA = -(\text{Area of } Q). \tag{7}$$

Subtracting the corresponding sides of equations (6) and (7) and dividing by 2 gives

$$\text{Area of } Q = \frac{1}{2} \int_C x \, dy - y \, dx. \tag{8}$$

Any of equations (6), (7), or (8) may be used in calculating the area of Q. In the following example, equation (8) provides an easy solution to a problem that otherwise requires the technique of integration by trigonometric substitution.

Example 4

Calculate the area A of the region enclosed by the ellipse

$$\frac{x^2}{a^2} + \frac{y^2}{b^2} = 1.$$

Solution: A parameterization for the ellipse with counterclockwise orientation is

$$x(t) = a \cos t, \qquad y(t) = b \sin t, \qquad 0 \le t \le 2\pi.$$

Thus

$$x'(t) = -a \sin t, \qquad \text{and} \qquad y'(t) = b \cos t.$$

By equation (8),

$$A = \frac{1}{2} \int_C x \, dy - y \, dx$$

$$= \frac{1}{2} \int_0^{2\pi} [(a \cos t)(b \cos t) - (b \sin t)(-a \sin t)] \, dt$$

$$= \frac{1}{2} \int_0^{2\pi} ab(\cos^2 t + \sin^2 t) \, dt$$

$$= \pi ab.$$

$\diamond$

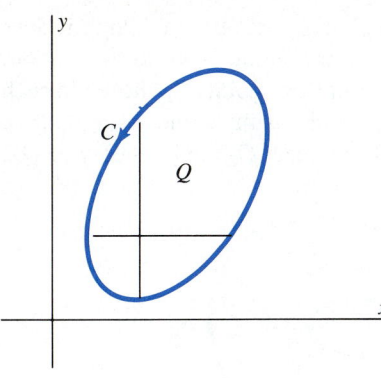

Figure 4.7 Q is both x-simple and y-simple.

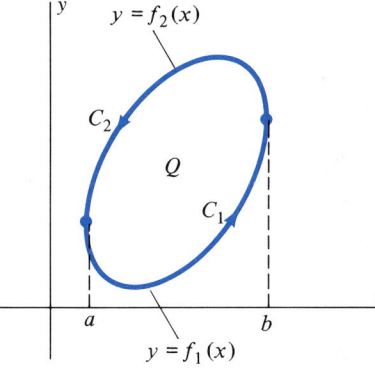

Figure 4.8 C_j is graph of $y = f_j(x)$, $j = 1, 2$.

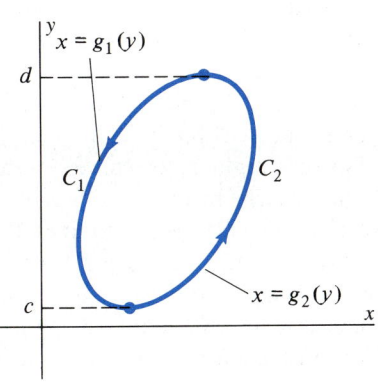

Figure 4.9 C_j is graph of $x = g_j(x)$, $j = 1, 2$.

Proof of Theorem 3: We will prove Theorem 3 only for regions which are both x-simple and y-simple and then comment on the idea by which the proof can be extended to more general regions.

Specifically, let us assume that Q is a region such as that in Figure 4.7 which can be described both as

$$Q = \{(x, y) \mid a \le x \le b, f_1(x) \le y \le f_2(x)\} \qquad \text{(Figure 4.8)} \qquad (9)$$

or as

$$Q = \{(x, y) \mid g_1(y) \le x \le g_2(y), c \le y \le d\} \qquad \text{(Figure 4.9)} \qquad (10)$$

where each of the functions f_1, f_2, g_1, and g_2 is continuous and piecewise differentiable.

We will prove Theorem 3 by showing both that

$$\int_C M(x, y)\, dx = -\iint_Q \frac{\partial M}{\partial y}\, dA \qquad (11)$$

and that

$$\int_C N(x, y)\, dy = \iint_Q \frac{\partial N}{\partial x}\, dA. \qquad (12)$$

Adding the corresponding sides of equations (11) and (12) then establishes equation (2), which is what we are trying to prove.

To prove (11), we first use Figure 4.8 and equation (9) to find that

$$\begin{aligned}
\int_C M(x, y)\, dx &= \int_{C_1} M(x, y)\, dx + \int_{C_2} M(x, y)\, dx \\
&= \int_a^b M(x, f_1(x))\, dx + \int_b^a M(x, f_2(x))\, dx \\
&= \int_a^b M(x, f_1(x))\, dx - \int_a^b M(x, f_2(x))\, dx.
\end{aligned}$$

Thus,

$$\int_C M(x, y)\, dx = \int_a^b [M(x, f_1(x)) - M(x, f_2(x))]\, dx. \qquad (13)$$

On the other hand, we also have

$$\begin{aligned}
\iint_Q \frac{\partial M}{\partial y}\, dA &= \int_a^b \int_{f_1(x)}^{f_2(x)} \frac{\partial}{\partial y} M(x, y)\, dy\, dx \qquad (14) \\
&= \int_a^b \{M(x, y)]_{y=f_1(x)}^{y=f_2(x)}\}\, dx \\
&= \int_a^b [M(x, f_2(x)) - M(x, f_1(x))]\, dx.
\end{aligned}$$

Comparing equations (13) and (14) shows that equation (11) holds. Similarly, using Figure 4.9 and equation (10) we can show that equation (12) holds. (This step is left to you as Exercise 23.) This completes the proof for regions Q of the stated type. ◆

Extending Green's Theorem to More General Regions

Figure 4.10 shows a region that is neither x-simple nor y-simple, although it does satisfy the hypotheses of Theorem 3. To extend the preceding proof to this region, we make a "cut" (that is, we draw a line) from point P_1 to point P_2, chosen in such a way that the resulting subregions Q_1 and Q_2 are both x-simple and y-simple (See Figure 4.11.) Thus, if C_1 denotes the boundary of Q_1, and C_2 the boundary of Q_2,

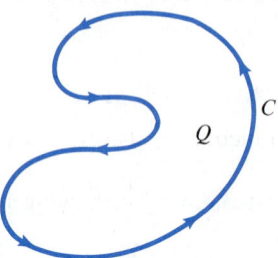

Figure 4.10 Q is neither x-simple nor y-simple.

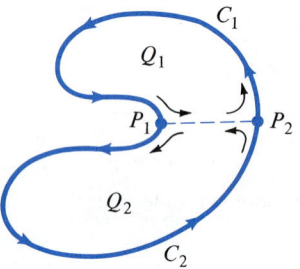

Figure 4.11 Q_1 and Q_2 are both x-simple and y-simple.

Green's Theorem on the individual subregions shows that

$$\int_{C_1} M(x, y)\, dx + N(x, y)\, dy = \iint_{Q_1} \left[\frac{\partial N}{\partial x} - \frac{\partial M}{\partial y} \right] dA \qquad (15)$$

and

$$\int_{C_2} M(x, y)\, dx + N(x, y)\, dy = \iint_{Q_2} \left[\frac{\partial N}{\partial x} - \frac{\partial M}{\partial y} \right] dA. \qquad (16)$$

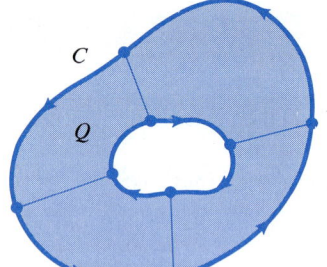

Figure 4.12 C consists of two arcs.

Adding the corresponding sides of these two equations then gives equation (2). The reason for this is that the contribution to the line integral in (15) resulting from the path from P_1 to P_2 is precisely the negative of the contribution of this same path to the line integral in (16), since the path is traversed in opposite directions by the two line integrals. Thus, the two line integrals taken over the path $P_1 P_2$ "cancel."

Green's Theorem may be extended to regions even more general than those in our statement of Theorem 3. In particular, Green's Theorem holds for regions Q containing "holes," such as that in Figure 4.12. For such regions, the curve C is the entire boundary—both the "inner" and "outer" curves. The meaning of "counter-clockwise" for such regions is that each arc of C is traversed so as to keep the region Q on the left-hand side, as indicated in Figure 4.12. The idea by which Green's Theorem is extended to include such regions is again to make "cuts" (see Figure 4.13) dividing Q into subregions without holes, to which Theorem 3 applies. As before, the contributions to the line integral along each cut path sum to zero, so equation (2) results from adding the corresponding sides of the equations holding in each region.

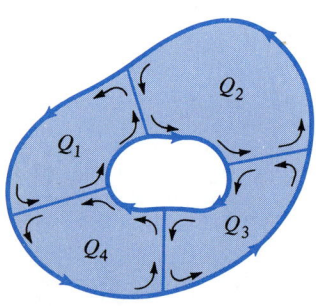

Figure 4.13 Cuts divide Q into regions without holes.

Other Formulations of Green's Theorem

There are two special forms in which Green's Theorem may be stated, each of which is a two-dimensional version of more general results yet to come. We will interpret each of these statements in terms of the flow of a thin layer of fluid, such as that very near the surface of the water in a swimming pool or in a cross section of a water pipe.

We begin by letting $F(x, y) = M(x, y)i + N(x, y)j$ be the vector field giving the velocity (speed and direction) of the fluid flow at each point (x, y) in the thin layer. Using equation (22), Section 19.2, we may write the vector form (3) of Green's Theorem as

$$\int_C F \cdot T \, ds = \int\int_Q \left(\frac{\partial N}{\partial x} - \frac{\partial M}{\partial y} \right) dA \tag{17}$$

where C is any smooth closed curve in the fluid layer. The integrand $F \cdot T$ in the line integral in (17) is just the component of the vector field (i.e., the flow) in the direction of the unit tangent T to the curve C at each point (see Figure 4.14). If the fluid has a tendency to circulate around the curve C (such as what you see when you pull the plug in your bathtub), we would intuitively expect the line integral in (17) to be large, since the velocity vector F and the tangent vector T should point rather consistently either in approximately the same directions (when the fluid rotates counterclockwise) or in opposite directions (when the fluid rotates clockwise), as illustrated in Figure 4.15. When the fluid has little tendency to rotate, such as occurs in a constant flow field, we expect the line integral in (17) to be small since the component of F on T will be of opposite sign on opposite sides of C (Figure 4.16). For these reasons we say that the line integral in (17) is a measure of the

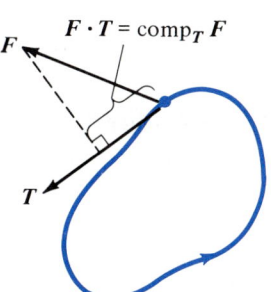

Figure 4.14 $F \cdot T$ is component of F in direction of T.

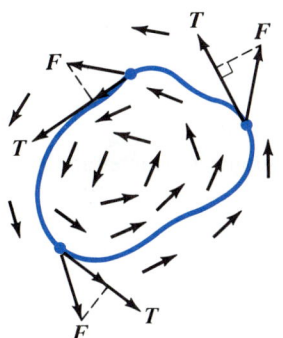

Figure 4.15 In a rotating fluid, $\left| \int_C F \cdot T \, ds \right|$ is large.

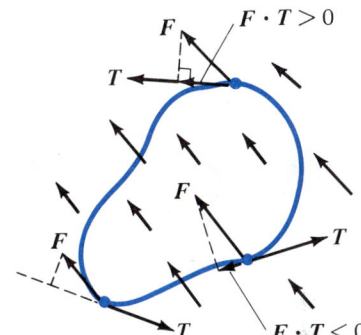

Figure 4.16 In a fluid with little rotation, $\left| \int_C F \cdot T \, ds \right|$ is small.

tendency of the fluid layer to circulate or to rotate. Vector fields in the plane for which this line integral is always zero are called **irrotational.**

Since the double integral in equation (17) has the same value as the line integral, it too measures the tendency of the fluid to rotate. However, it does so by integrating the scalar function $\dfrac{\partial N}{\partial x} - \dfrac{\partial M}{\partial x}$ over the region Q enclosed by C. For this reason we refer to this function as the **scalar curl** of the vector field F. That is,

$$\text{curl } F = \frac{\partial N}{\partial x} - \frac{\partial M}{\partial y}. \tag{18}$$

Using this definition, we may rewrite the formulation of Green's Theorem given

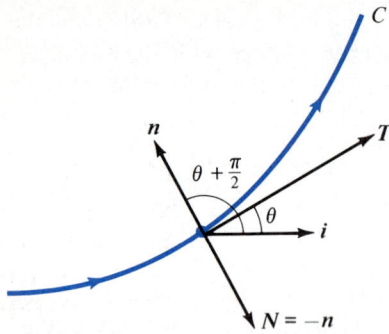

Figure 4.17 N is the outward unit normal.

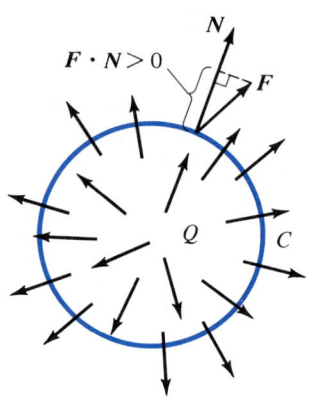

Figure 4.18 $\int_C F \cdot N \, ds > 0$ if net flow across C is outward.

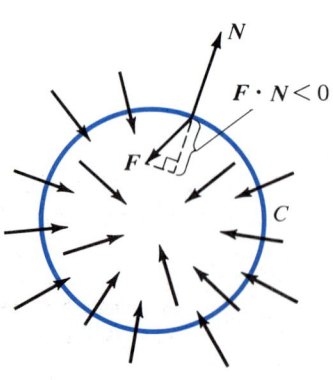

Figure 4.19 $\int_C F \cdot N \, ds < 0$ if net flow across C is inward.

in line (17) as

$$\int_C F \cdot T \, ds = \iint_Q \text{curl } F \, dA \qquad \text{(Circulation of } F \text{ around } C\text{).} \qquad (19)$$

Equation (19) is referred to as **Stokes' Theorem in the plane.** In Section 19.6 we shall study the generalizations of curl and of Stokes' Theorem to three dimensions.

A second vector formulation of Green's Theorem concerns the outward unit normal to the closed smooth simple curve C. To obtain this vector we let T be the unit tangent at a point on the curve C (oriented counterclockwise) and we let θ be the angle formed between T and the unit vector i (Figure 4.17). Then

$$T = \cos \theta i + \sin \theta j. \qquad (20)$$

The unit vector n obtained by rotating T 90° in the counterclockwise direction is the **unit normal**

$$\begin{aligned} n &= \cos(\theta + \pi/2)i + \sin(\theta + \pi/2)j \\ &= -\sin \theta i + \cos \theta j. \end{aligned}$$

However, this normal points inward for a closed curve C oriented counterclockwise, so the desired **outward unit normal** N is the vector

$$N = -n = \sin \theta i - \cos \theta j. \qquad (21)$$

Now recall from Section 17.3 that the unit tangent vector T may be written

$$T = \frac{dx}{ds}i + \frac{dy}{ds}j \qquad (22)$$

where s is the arc length parameter and $r(s) = x(s)i + y(s)j$ is a parameterization for C by arc length. From equations (20), (21), and (22) it follows that the outward unit normal N may be written

$$N = \frac{dy}{ds}i - \frac{dx}{ds}j. \qquad (23)$$

For the vector field $F(x, y) = M(x, y)i + N(x, y)j$, the expression (23) for N suggests that

$$\int_C F \cdot N \, ds = \int_C M(x, y) \, dy - N(x, y) \, dx, \qquad (24)$$

and Green's Theorem, applied to the right-hand side of equation (24) gives

$$\int_C M(x, y) \, dy - N(x, y) \, dx = \iint_Q \left(\frac{\partial M}{\partial x} + \frac{\partial N}{\partial y} \right) dA. \qquad (25)$$

Finally, combining (24) and (25), we obtain

$$\int_C F \cdot N \, ds = \iint_Q \left(\frac{\partial M}{\partial x} + \frac{\partial N}{\partial y} \right) dA. \qquad (26)$$

Equation (26) is referred to as the *Divergence Theorem in the plane* for reasons that we will now explain. Since the scalar quantity $F \cdot N$ in the line integral is the

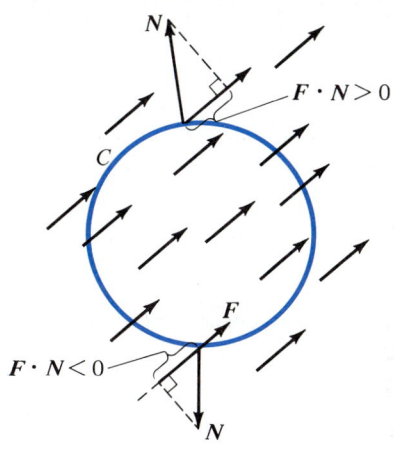

Figure 4.20 $\int_C \mathbf{F} \cdot \mathbf{N}\,ds = 0$ if net flow across C is zero.

component of the vector field (fluid flow) in the direction of the outward normal, the line integral in (26) measures the net flow of fluid outward across the boundary C of the region Q. The term **flux** is used to denote this net rate at which fluid flows across a boundary, and the line integral in (26) is called a **flux integral.** When the line integral is *positive,* more fluid is leaving Q than is entering. (Think of an open faucet somewhere inside Q.) When the line integral in (26) is *negative,* more fluid is flowing into Q than is leaving. (Think of an open drain somewhere inside Q.) When the line integral is small or zero, there is little or no net gain or loss of fluid in Q. (Think of fluid moving uniformly, or motionless fluid. See Figures 4.18 through 4.20.)

Since the double integral in (26) gives the same measure of net flow as does the line integral, the integrand $\dfrac{\partial M}{\partial x} + \dfrac{\partial N}{\partial y}$ is referred to as the **divergence** of the vector field $\mathbf{F}$, written div $\mathbf{F}$. That is

$$\text{div } \mathbf{F} = \frac{\partial M}{\partial x} + \frac{\partial N}{\partial y}. \tag{27}$$

Using (27), we may rewrite the Divergence Theorem in the plane as

$$\int_C \mathbf{F} \cdot \mathbf{N}\,ds = \iint_Q \text{div } \mathbf{F}\,dA. \qquad \text{(Flux of } \mathbf{F} \text{ across } C\text{).} \tag{28}$$

In Section 19.7 we will encounter generalizations of both the concept of divergence of a vector field and the Divergence Theorem to three dimensions, the primary motivation being fluid flow in space.

Exercise Set 19.4

(In all exercises the closed curves C are oriented counterclockwise.) In Exercises 1–10, use Green's Theorem to evaluate the given line integral.

1. $\int_C xy\,dx + (x + y)\,dy,$ C is the square with vertices $(0, 0)$, $(0, 1)$, $(1, 1)$, and $(1, 0)$.

2. $\int_C xy\,dx + (y - x)\,dy,$ C is the unit circle.

3. $\int_C x^2y^3\,dx + (x^2 + 1)\,dy,$ C is the square in Exercise 1.

4. $\int_C \sqrt{y}\,dx + \sqrt{x}\,dy,$ C is the rectangle with vertices $(1, 1)$, $(1, 2)$, $(4, 1)$, and $(4, 2)$.

5. $\int_C xy^2\,dy - x^2y\,dx,$ C is the quarter circular path from $(0, 0)$ to $(1, 0)$ to $(0, 1)$ to $(0, 0)$.

6. $\int_C e^x \tan y\,dx + e^x \sec^2 y\,dy,$ C is the triangle with vertices $(-1, 2)$, $(3, 0)$, and $(1, 6)$.

7. $\int_C (x^3 + y)\,dx + (y - x^2)\,dy,$ C is the boundary of the region bounded by the graphs of $y = x^2$ and $y = x^3$.

8. $\int_C (\text{Tan}^{-1} x + y^2)\,dx + (\ln^2 y - x^2)\,dy,$ C is the circle $x^2 + y^2 = 4$.

9. $\int_C (\cos^5 x + \sqrt{x})\,dx + \text{Tan}^{-1} y\,dy,$ C is the ellipse $4x^2 + y^2 = 1$.

10. $\int_C \sqrt{x}\,dx + \ln(x^2 + y^2)\,dy,$ C is the curve in Figure 4.6.

In Exercises 11–14, use Green's Thorem to find $\int_C \mathbf{F} \cdot d\mathbf{r}.$

11. $\mathbf{F}(x, y) = 2y\mathbf{i} - 3x\mathbf{j},$ C is the unit circle.

12. $\mathbf{F}(x, y) = xy\mathbf{i} + x^2\mathbf{j},$ C is the square with vertices $(0, 0)$, $(0, 2)$, $(2, 2)$, and $(2, 0)$.

13. $\mathbf{F}(x, y) = x^2(y^2 - x^2)\mathbf{i} + \dfrac{2}{3}x^3y\mathbf{j},$ C is the triangle with vertices $(1, 1)$, $(4, 1)$, and $(3, 5)$.

14. $F(x, y) = e^x \sin y\mathbf{i} + e^x \cos y\mathbf{j}$, C is the ellipse $x^2 + 9y^2 = 1$.

15. Find the work done by the force field $F(x, y) = (e^{x^2} + 2y)\mathbf{i} + (ye^y - x)\mathbf{j}$ in moving a particle once around the path indicated in Figure 4.6.

16. Use equation (8) to find the area of the region bounded by the graphs of $y = x$ and $y = x^2$.

17. Use equations (8) to find the area of the region bounded by the graphs of $y = x$ and $y = \sqrt{x}$.

18. Find the area of the region enclosed by the graph of the parametric equations $x(t) = \sin t \cos t$, $y(t) = \sin t$ for $0 \le t \le \pi$.

19. Find the area of the region bounded by the graph of the parametric equations $x(t) = a \cos^3 t$, $y(t) = a \sin^3 t$, $0 \le t \le 2\pi$.

20. Let Q be a region in the plane whose boundary is a simple closed piecewise smooth curve C. Let A be the area of Q.

Show that the coordinates of the centroid of Q are

$$\bar{x} = \frac{1}{2A}\int_C x^2\, dy, \qquad \bar{y} = -\frac{1}{2A}\int_C y^2\, dx.$$

21. Use Green's Theorem to prove that if F is a conservative vector field in the plane, then any line integral $\int_C F \cdot dr$ over a piecewise smooth curve C is independent of path.

22. Use Green's Theorem to prove that if F is a vector field in the plane for which $\int_A^B F \cdot dr$ is independent of path for all A and B, then $\int_C F \cdot dr = 0$ for every simple closed path C.

23. Complete the proof of Theorem 3 by showing that equation (12) holds.

24. Let $F(x, y) = M(x, y)\mathbf{i} + N(x, y)\mathbf{j}$ where M and N have continuous first partial derivatives in some open rectangle D. Show that F is conservative in D if and only if curl $F = 0$ for all $(x, y) \in D$.

19.5 SURFACE INTEGRALS

The primary motivation for the remainder of this chapter will be to extend the concept of flux, introduced in Section 19.4, to the flow of fluids in and out of regions in space. Since regions in space are bounded by surfaces rather than curves, we must begin by developing the concept of integrals over surfaces, or *surface integrals*. Although we are primarily interested in defining the integral of a *vector field F* over a surface S (Figure 5.1), we shall begin by defining the integral of a scalar function f over a surface S (Figure 5.2). This is because the development of

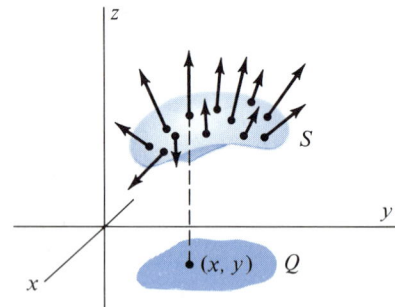

Figure 5.1 A vector field $F(x, y, z)$ assigns a vector to each point on the surface S.

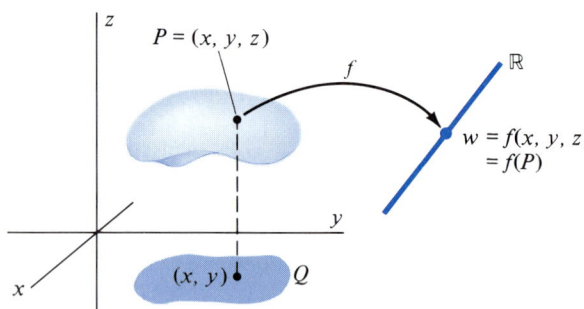

Figure 5.2 A scalar function $f(x, y, z)$ assigns a number to each point P on S.

the surface integral for scalar functions very closely parallels the discussion of surface area in Section 18.5.

By a surface S, we shall mean the graph of a continuous function of two variables, $z = g(x, y)$, over some regular region Q in the xy-plane. (Recall that the region Q is regular if it is both x-simple and y-simple.) Such a surface is illustrated in Figure 5.3. This is not the most general concept of surface, but techniques developed here will enable you to handle most surfaces encountered in practice.

Suppose that the continuous scalar function f is defined for all points $(x, y, g(x, y))$ on the surface S. Recall from Section 18.5 that a partition of the region Q into n rectangles of area $\Delta A_j = \Delta x \, \Delta y$ partitions "most" of the surface S into curvilinear "patches" S_j of area ΔS_j (Figure 5.4.) Proceeding in a purely

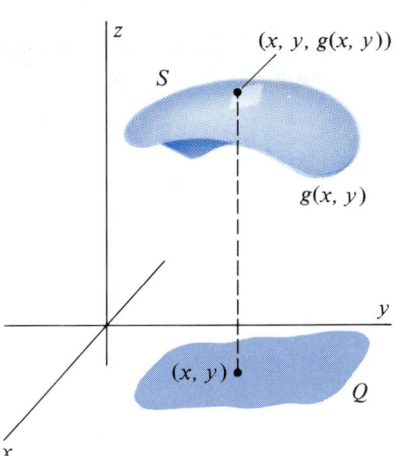

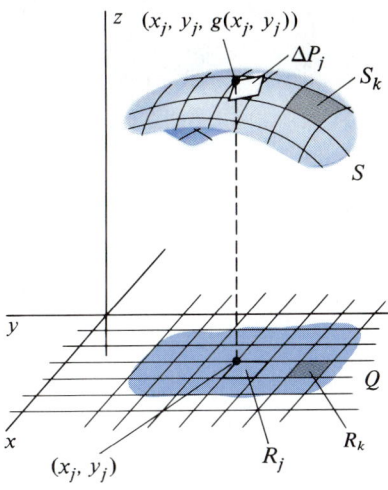

Figure 5.3 Surface S is the graph of a continuous function $z = g(x, y)$ over Q.

Figure 5.4 Parallelogram of area ΔP_j over rectangle R_j approximates the area ΔS_j of the "patch" S_j.

mathematical way, we form the product of the surface area ΔS_j of the patch S_j and the value $f(x_j, y_j, z_j)$ of the scalar function f at an arbitrary point (x_j, y_j, z_j) on the patch. Summing these products over all n patches gives the approximating sum

$$\sum_{j=1}^{n} f(x_j, y_j, z_j) \, \Delta S_j, \qquad (x_j, y_j, z_j) \in S_j.$$

The limit as the norm of the partition P approaches zero ($\|P\| \to 0$) of this approximating sum, when it exists, is defined as the surface integral of f over S:

$$\iint_S f(x, y, z) \, dS = \lim_{\|P\| \to 0} \sum_{j=1}^{n} f(x_j, y_j, z_j) \, \Delta S_j. \qquad (1)$$

We shall determine several ways in which this surface integral may be calculated. The first of these involves recalling that the area ΔS_j of the patch S_j may be approximated by the area ΔP_j of the parallelogram over R_j tangent to the surface S at one corner (Figure 5.4). That is, we have the approximation

$$\Delta S_j \approx \Delta P_j = \sqrt{\left(\frac{\partial g}{\partial x}\right)^2 + \left(\frac{\partial g}{\partial y}\right)^2 + 1} \; \Delta x \, \Delta y \qquad (2)$$

where the partial derivatives $\dfrac{\partial g}{\partial x}$ and $\dfrac{\partial g}{\partial y}$ are evaluated at one corner, (x_j, y_j), of the rectangle R_j. Using approximation (2) in equation (1) leads to the following definition.

DEFINITION 5

Let Q be a regular region in the plane and let g be a continuous function defined on some open rectangle containing Q on which both $\dfrac{\partial g}{\partial x}$ and $\dfrac{\partial g}{\partial y}$ are continuous. Let S be the graph of g over Q, and let f be a continuous scalar function defined on S. The **surface integral** of f over S is

$$\iint_S f(x, y, z) \, dS = \iint_Q f(x, y, g(x, y)) \sqrt{\left(\frac{\partial g}{\partial x}\right)^2 + \left(\frac{\partial g}{\partial y}\right)^2 + 1} \; dx \, dy. \qquad (3)$$

Example 1

Evaluate the surface integral

$$\iint_S (x^2 + y + z) \, dS$$

where S is the graph of the function $g(x, y) = \sqrt{3}\, y - x^2$ over the unit square $Q = \{(x, y) \mid 0 \leq x \leq 1, \, 0 \leq y \leq 1\}$.

Solution: The integrand is the function

$$f(x, y, z) = x^2 + y + z.$$

Also,

$$\frac{\partial g}{\partial x} = -2x, \qquad \text{and} \qquad \frac{\partial g}{\partial y} = \sqrt{3}.$$

Substituting into equation (3) gives

$$\iint_S (x^2 + y + z) \, dS = \iint_Q [x^2 + y + (\sqrt{3}\, y - x^2)] \sqrt{(-2x)^2 + (\sqrt{3})^2 + 1} \; dx \, dy$$

$$= \int_0^1 \int_0^1 (1 + \sqrt{3})y \sqrt{4x^2 + 4} \; dy \, dx$$

$$= (1 + \sqrt{3}) \int_0^1 \{y^2 \sqrt{x^2 + 1}\}_{y=0}^{y=1} \, dx$$

$$= (1 + \sqrt{3}) \int_0^1 \sqrt{x^2 + 1} \; dx \qquad \text{(trigonometric substitution)}$$

$$= (1 + \sqrt{3}) \left(\frac{\sqrt{2}}{2} + \frac{1}{2} \ln(1 + \sqrt{2}) \right)$$

$$\approx 3.14. \qquad \diamond$$

Example 2

Evaluate the surface integral

$$\iint_S (x^2 + y^2 + z)\, dS$$

where S is the portion of the graph of the function $z = 4 - x^2 - y^2$ bounded below by the xy-plane.

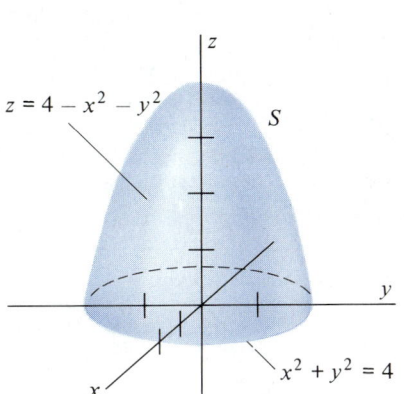

Figure 5.5 Surface S is graph of $z = 4 - x^2 - y^2$.

Solution: The integrand is $f(x, y, z) = x^2 + y^2 + z$, and the surface is the graph of the function $g(x, y) = 4 - x^2 - y^2$. Thus

$$\frac{\partial g}{\partial x} = -2x, \qquad \text{and} \qquad \frac{\partial g}{\partial y} = -2y.$$

The region Q in the xy-plane over which S is defined is

$$Q = \{(x, y) \mid x^2 + y^2 \le 4\} \qquad \text{(Figure 5.5)}.$$

According to Definition 5, the surface integral is

$$\iint_S (x^2 + y^2 + z)\, dS = \iint_Q [x^2 + y^2 + (4 - x^2 - y^2)]\sqrt{(-2x)^2 + (-2y)^2 + 1}\, dx\, dy$$

$$= \iint_Q 4\sqrt{4x^2 + 4y^2 + 1}\, dx\, dy.$$

Switching to polar coordinates allows us to evaluate this integral as

$$\iint_S (x^2 + y^2 + z)\, dS = \int_0^{2\pi} \int_0^2 4\sqrt{4r^2 + 1}\, r\, dr\, d\theta$$

$$= \int_0^{2\pi} \left\{ \left[\frac{1}{3}(4r^2 + 1)^{3/2} \right]_{r=0}^{r=2} \right\} d\theta$$

$$= \frac{2\pi}{3}(17^{3/2} - 1)$$

$$\approx 144.71. \qquad \diamondsuit$$

REMARK: For surfaces that are graphs of functions of the form $y = g(x, z)$ or $x = g(y, z)$, we may simply interchange roles among the variables x, y, and z as required. The following example presents one such situation. (Notice that the given surface cannot be described as the graph of a function of the form $z = g(x, y)$.)

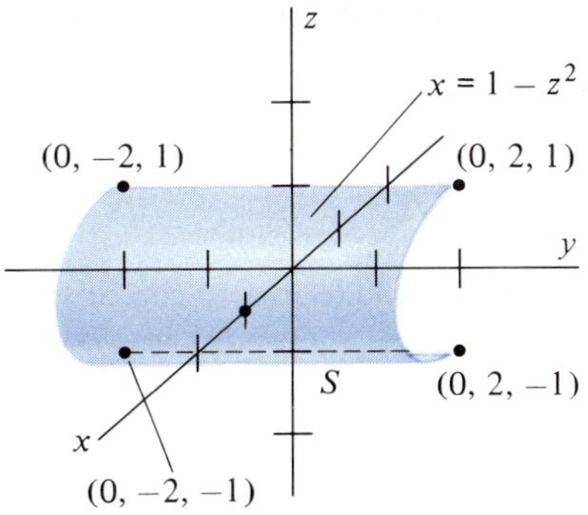

Example 3

Evaluate the surface integral

$$\iint_S yz\, dS$$

where S is the portion of the graph of $x = 1 - z^2$ bounded by the yz-plane and the planes $y = -2$ and $y = 2$ (Figure 5.6).

Figure 5.6 Surface S is graph of $x = 1 - z^2$.

Solution: In order to describe S as the graph of a function, we must take x as the dependent variable. Interchanging roles of x and z in Definition 5, we find that

$$f(x, y, z) = yz \qquad \text{and} \qquad g(y, z) = 1 - z^2.$$

Thus,

$$\frac{\partial g}{\partial z} = -2z \qquad \text{and} \qquad \frac{\partial g}{\partial y} = 0.$$

The region Q in the yz-plane over which S is defined is the rectangle

$$Q = \{(y, z) \mid -2 \le y \le 2, \, -1 \le z \le 1\}.$$

Definition 5 now gives

$$\iint_S yz \, dS = \iint_Q yz\sqrt{(-2z)^2 + 1} \, dz \, dy$$

$$= \int_{-2}^{2} \int_{-1}^{1} yz\sqrt{4z^2 + 1} \, dz \, dy$$

$$= \int_{-2}^{2} \left\{ \frac{y}{12} (4z^2 + 1)^{3/2} \right]_{z=-1}^{z=1} \right\} dy$$

$$= \int_{-2}^{2} \frac{y}{12} (5^{3/2} - 5^{3/2}) \, dy$$

$$= 0.$$

The result should not be surprising, since both f and the surface are symmetric with respect to both y and z. ◇

Calculating Mass

A straightforward application of surface integrals of the form (1) occurs in the calculation of the total mass of a hollow object (such as a basketball or a vase) whose mass density per unit surface area, $\rho(x, y, z)$, varies continuously over the surface of the object. That is, we assume that $\rho(x, y, z)$ gives the mass density (in grams/cm^2, say) of the object at location (x, y, z). Then if S_j is a small patch of area ΔS_j on the surface of the object, the total mass ΔM_j of S_j may be approximated by

$$\Delta M_j \approx \rho(x_j, y_j, z_j) \, \Delta S_j \qquad \left(\text{mass} = \frac{\text{mass}}{\text{unit area}} \times \text{area} \right) \tag{4}$$

where (x_j, y_j, z_j) is a point on S_j. Summing (4) over all patches in a partition covering S, we arrive at the approximation to total mass:

$$M \approx \sum_{j=1}^{n} \rho(x_j, y_j, z_j) \, \Delta S_j. \tag{5}$$

Comparing approximation (5) with equation (1), we conclude that in the limit as $\Delta S_j \to 0$, assuming appropriate conditions on the shape of the object and on the density function ρ,

$$M = \iint_S \rho(x, y, z) \, dS. \tag{6}$$

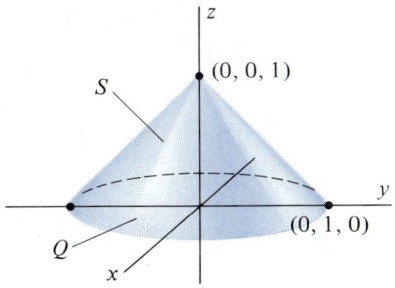

Figure 5.7

In other words, **the total mass of the object** is simply the integral, over the surface of the object, of the function giving the mass per unit area.

Example 4

Let S be the idealized cone described in cylindrical coordinates by the equation $z = 1 - r$, $0 \leq r \leq 1$. Find the mass of S if the density at any point is proportional to the square of the distance between the point and the origin (see Figure 5.7).

Solution: Here $\rho(x, y, z) = \lambda(x^2 + y^2 + z^2)$ is the mass density, where λ is a constant. We will work with ρ in cylindrical coordinates:

$$\rho(r, \theta, z) = \lambda(r^2 + z^2).$$

Also, the surface S is described by the function

$$z = g(r, \theta) = 1 - r.$$

Using the result of Exercise 25 of this section (for evaluating a surface integral in cylindrical coordinates) together with equation (6), we find that

$$M = \iint_S \rho(r, \theta, z) \, dS$$

$$= \iint_Q \rho(r, \theta, g(r, \theta)) \sqrt{r^2 + r^2\left(\frac{\partial g}{\partial r}\right)^2 + \left(\frac{\partial g}{\partial \theta}\right)^2} \, dr \, d\theta$$

$$= \int_0^{2\pi} \int_0^1 \lambda(r^2 + (1 - r)^2) \cdot \sqrt{r^2 + r^2(-1)^2 + 0^2} \, dr \, d\theta$$

$$= \int_0^{2\pi} \int_0^1 \sqrt{2}\lambda \cdot (2r^3 - 2r^2 + r) \, dr \, d\theta$$

$$= \frac{2\sqrt{2}\pi\lambda}{3}. \qquad \diamond$$

The concept of surface integral may be extended to various types of surfaces that do not fulfill the hypotheses of Definition 5. If a surface S is a union of a finite number of surfaces $S = S_1 \cup S_2 \cup \cdots \cup S_n$, each of which satisfies the hypotheses of Definition 5, we define the surface integral of $f(x, y, z)$ over S to be the sum of the individual surface integrals

$$\iint_S f(x, y, z) \, dS = \sum_{j=1}^n \iint_{S_j} f(x, y, z) \, dS_j, \qquad (7)$$

each of which is evaluated by use of Definition 5.

Example 5

Evaluate $\displaystyle\iint_S (x^2 + y^2 + z) \, dS$ where the surface S is the boundary of the region

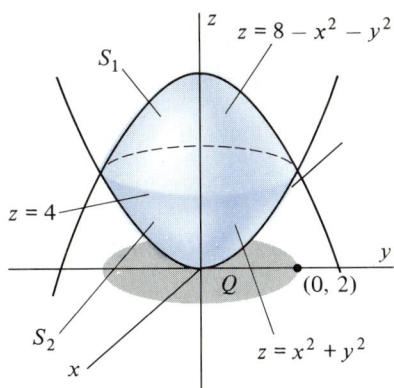

Figure 5.8 Surface S is not smooth along the curve $z = 4$.

bounded by the graphs of $g_1(x, y) = 8 - (x^2 + y^2)$ and $g_2(x, y) = x^2 + y^2$ (Figure 5.8).

Solution: The two graphs intersect above the circle $x^2 + y^2 = 4$. Let $Q = \{(x, y) \mid x^2 + y^2 \leq 4\}$. We may therefore describe the surface S as $S = S_1 \cup S_2$ where

$$S_1 = \{(x, y, 8 - (x^2 + y^2)) \mid (x, y) \in Q\},$$
$$S_2 = \{(x, y, x^2 + y^2) \mid (x, y) \in Q\}.$$

On the surface S_1, with $f(x, y, z) = x^2 + y^2 + z$ and $g_1(x, y) = 8 - (x^2 + y^2)$, Definition 5 gives

$$\iint\limits_{S_1} (x^2 + y^2 + z)\, dS = \iint\limits_{Q} [x^2 + y^2 + (8 - (x^2 + y^2))]\sqrt{(-2x)^2 + (-2y)^2 + 1}\, dx\, dy$$

$$= \int_0^{2\pi} \int_0^2 8\sqrt{4r^2 + 1} \cdot r\, dr\, d\theta \qquad \text{(switching to polar coordinates)}$$

$$= \frac{4\pi}{3}(17^{3/2} - 1).$$

On the surface S_2, with $f(x, y, z) = x^2 + y^2 + z$ and $g_2(x, y) = x^2 + y^2$, we obtain

$$\iint\limits_{S_2} (x^2 + y^2 + z)\, dS = \iint\limits_{Q} [x^2 + y^2 + (x^2 + y^2)]\sqrt{(2x)^2 + (2y)^2 + 1}\, dx\, dy$$

$$= \int_0^{2\pi} \int_0^2 2r^2\sqrt{4r^2 + 1}\, r\, dr\, d\theta \qquad \text{(let } u = 4r^2 + 1)$$

$$= \frac{\pi}{20}(17^{5/2} - 1) - \frac{\pi}{12}(17^{3/2} - 1).$$

Combining these results according to equation (7) gives

$$\iint\limits_{S} (x^2 + y^2 + z)\, dS = \frac{\pi}{20}(17^{5/2} - 1) + \frac{5\pi}{4}(17^{3/2} - 1)$$

$$\approx 458.3. \qquad \diamond$$

Definition 5 could not be applied directly to the entire surface S in Example 5 for two reasons. First, S is not the graph of a *function* $z = g(x, y)$, since there are two distinct points corresponding to some (x, y)-coordinates. Second, the surface S is not differentiable $\left(\text{meaning that } \dfrac{\partial g}{\partial x} \text{ and } \dfrac{\partial g}{\partial y} \text{ are not continuous}\right)$ along the "seam" where S_1 meets S_2.

Another difficulty that can arise in attempting to apply Definition 5 is that the integrand in the double integral has a singularity (i.e., becomes infinite) along the boundary of the region Q. In such cases we define $\iint\limits_{S} f(x, y, z)\, dS$ to be the corresponding improper integral if this integral exists. The following example is typical. When combined with the notion of equation (7), this idea allows us to evaluate surface integrals over spheres, an important aspect of many applications of surface integrals.

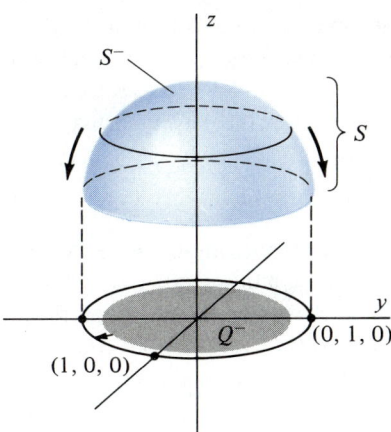

Figure 5.9 Surface integral over hemisphere $z = 1 + \sqrt{1 - x^2 - y^2}$ is evaluated as an improper integral.

Example 6

Evaluate the surface integral $\iint\limits_{S} z \, dS$ where S is the hemisphere $z = 1 + \sqrt{1 - x^2 - y^2}$ (Figure 5.9).

Solution: Here $g(x, y) = 1 + \sqrt{1 - x^2 - y^2}$ and $f(x, y, z) = z$. Equation (3) becomes

$$\iint\limits_{S} z \, dS = \iint\limits_{Q} [1 + \sqrt{1 - x^2 - y^2}]$$

$$\cdot \sqrt{\left(\frac{-x}{\sqrt{1 - x^2 - y^2}}\right)^2 + \left(\frac{-y}{\sqrt{1 - x^2 - y^2}}\right)^2 + 1} \; dx \, dy$$

$$= \iint\limits_{Q} \left[\frac{1}{\sqrt{1 - x^2 - y^2}} + 1\right] dx \, dy$$

where Q is the unit circle $x^2 + y^2 = 1$. However, this double integral is improper, since the denominator of the first term in the integrand vanishes as (x, y) approaches the boundary of Q. We evaluate this improper integral as suggested in Figure 5.9— by evaluating it over a disc Q^- of radius $\hat{r} < 1$ and then calculating the limit as $\hat{r}$ approaches 1. To do so we switch to polar coordinates and find that

$$\iint\limits_{S} z \, dS = \iint\limits_{Q} \left[\frac{1}{\sqrt{1 - r^2}} + 1\right] r \, dr \, d\theta$$

$$= \lim_{\hat{r} \to 1^-} \int_0^{\hat{r}} \int_0^{2\pi} \left[\frac{r}{\sqrt{1 - r^2}} + r\right] d\theta \, dr$$

$$= \lim_{\hat{r} \to 1^-} \int_0^{\hat{r}} 2\pi \left[\frac{r}{\sqrt{1 - r^2}} + r\right] dr$$

$$= \lim_{\hat{r} \to 1^-} 2\pi \left\{-\sqrt{1 - r^2} + \frac{1}{2} r^2\right\}\Big|_{r=0}^{r=\hat{r}}$$

$$= \lim_{\hat{r} \to 1^-} 2\pi \left\{-\sqrt{1 - \hat{r}^2} + \frac{1}{2} \hat{r}^2 - (-\sqrt{1})\right\}$$

$$= 3\pi. \qquad \qquad \diamond$$

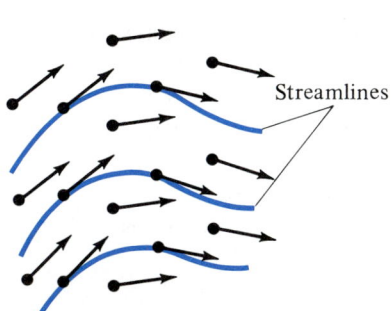

Figure 5.10 A velocity field and streamlines.

Flux Integrals

Returning now to our primary motivation concerning fluid flow, let us consider the motion of a fluid, such as water or a gas, through a region R in space. At each point (x, y, z) in the region, let $v(x, y, z)$ be the velocity vector for the motion of the fluid. That is, $v(x, y, z)$ is a vector pointing in the direction of motion whose length is the speed of the fluid at (x, y, z). We shall only consider the case of **steady state fluid motion.** This means that the velocity vector $v(x, y, z)$ does not change as time changes. Thus, $v(x, y, z)$ is a *vector field* defined on the region R. (The curves having the property of being tangent to the velocity field at each point are called the **streamlines** of the velocity field. See Figure 5.10.)

Now suppose that S is a permeable surface suspended in the fluid. (This means that the fluid can flow through the surface S, as coffee flows through a paper filter.)

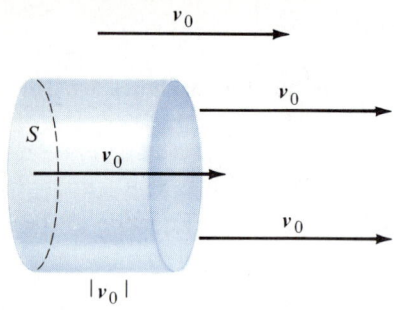

Figure 5.11 Volume of fluid flowing across S in unit time is $\Delta V = |v_0|\, \Delta S$.

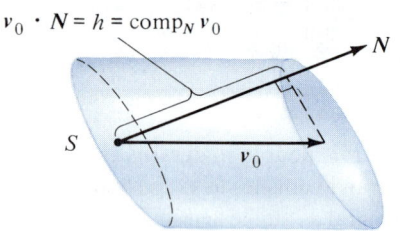

Figure 5.12 Volume of fluid crossing S in a unit of time is $\Delta V = (v_0 \cdot N)\, \Delta S$.

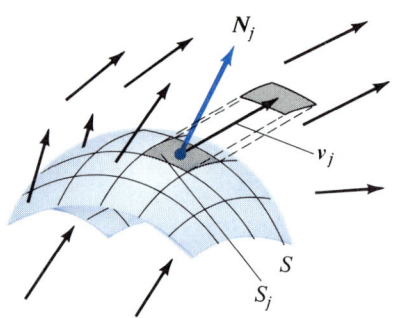

Figure 5.13 Volume of fluid crossing patch S_j is approximated by $\Delta V \approx (v_j \cdot N_j)\, \Delta S_j$.

For a steady state velocity field v, we define the **flux** of v across S as the rate at which mass is flowing across the surface S. Our interest here is in developing a formula by which flux can be calculated for a given surface and a given vector field.

We can simplify this problem somewhat by assuming that the density δ (mass per unit volume) is constant throughout the fluid. Since

$$\text{Mass} = (\text{density}) \times (\text{volume}) \tag{8}$$

we need only find the rate at which volume (i.e., fluid) is flowing across S. Multiplying by δ then gives the desired flux, according to equation (8).

The simplest case of this problem occurs when v is a constant vector field, $v(x, y, z) \equiv v_0$, and S is a flat surface perpendicular to v_0 (Figure 5.11). Since the speed at which fluid crosses S is $|v_0|$, in one unit of time the quantity

$$\Delta V = |v_0|\, \Delta S \tag{9}$$

crosses S, where ΔS is the area of S. If $v(x, y, z) = v_0$ is constant and S is flat, but not orthogonal to v_0, the situation is only slightly more complicated. Let N be a unit vector normal to S. (There are two possible choices for N. Take the one for which $0 \le \theta \le 90°$, where θ is the angle between N and v_0.) The volume of fluid that crosses S in a unit of time can be thought of as a prism with base of area ΔS and height $h = \text{comp}_N\, v_0 = v_0 \cdot N$, as illustrated in Figure 5.12. The desired volume is thus

$$\Delta V = (v_0 \cdot N)\, \Delta S. \tag{10}$$

The more general case of a curved surface S and a nonconstant velocity field v is handled in a familiar way. First, we assume that S has a normal vector $N(x, y, z)$ at each point and that $N(x, y, z)$ is continuous on S. Also, we assume that S can be recognized as having two sides and that $N(x, y, z)$ always remains on the same side of S. All these assumptions are met if S is the graph of a differentiable function $z = g(x, y)$ over a regular region Q in the plane.

By partitioning Q, as in Figure 5.4, we divide S into nonoverlapping patches $S_1, S_2, \ldots, S_n$. For each patch we choose one point $(x_j, y_j, z_j) \in S_j$, and we let ΔS_j denote the surface area of S_j. Also, we approximate $v(x, y, z)$ for every $(x, y, z) \in S_j$ by the vector $v_j = v(x_j, y_j, z_j)$ and we let $N_j = N(x_j, y_j, z_j)$. Using equation (10), we approximate the volume ΔV_j of fluid crossing the patch S_j in a unit of time as

$$\Delta V_j \approx (v_j \cdot N_j)\, \Delta S_j \qquad \text{(Figure 5.13)}.$$

Summing these approximations over all patches gives the approximation

$$\Delta V \approx \sum_{j=1}^{n} (v_j \cdot N_j)\, \Delta S_j. \tag{11}$$

Finally, we argue that as the number of patches becomes large (and their individual sizes become small) the approximation in (11) should converge to the desired rate. Since the right-hand side of approximation (11) has the form of the approximating sum in equation (1), we conclude from equations (8) and (11) that the rate at which the mass of the fluid is flowing across S (that is, the flux) is given by the surface integral

$$\frac{dM}{dt} = \iint_S \delta v(x, y, z) \cdot N(x, y, z)\, dS. \tag{12}$$

The integral in (12) is sometimes called the **rate of mass transport** across S in the direction of N.

Example 7

A fluid with mass density δ is emanating from the origin and flowing according to the central velocity field

$$v(x, y, z) = \lambda(x\boldsymbol{i} + y\boldsymbol{j} + z\boldsymbol{k})$$

where λ is constant. Find the rate of mass transport, $\dfrac{dM}{dt}$, across the square region $S = \{(x, y, z) \mid z = 2, -1 \le x \le 1, -1 \le y \le 1\}$.

Solution: The square S is horizontal, so we use the unit normal $N = \boldsymbol{k}$ at each point. Then $v(x, y, z) \cdot N = \lambda z$, so, by (12),

$$\frac{dM}{dt} = \iint\limits_{S} \delta\lambda z \; dS = \int_{-1}^{1} \int_{-1}^{1} 2\delta\lambda \; dx \; dy = 8\delta\lambda.$$

Note that the choice of $N = \boldsymbol{k}$ was arbitrary. If we had instead used $N = -\boldsymbol{k}$, the resulting integral would have yielded $\dfrac{dM}{dt} = -8\delta\lambda$. This simply says that the flow is in the direction *opposite* the vector $N = -\boldsymbol{k}$ (see Figures 5.14 and 5.15). $\diamond$

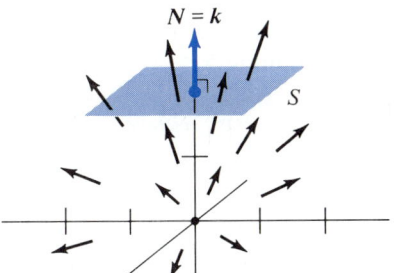

Figure 5.14 Flow across S in direction of $N = \boldsymbol{k}$.

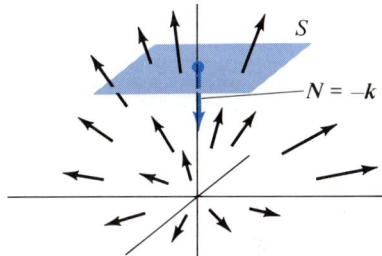

Figure 5.15 Flow in direction opposite to $N = -\boldsymbol{k}$.

A more general concept than that of equation (12) is obtained by letting F denote any vector field and $N = N(x, y, z)$ be a unit normal defined at each point on the surface S as described above. The flux of the vector field F across S in the direction of N is defined to be the surface integral

$$\text{Flux of } F \text{ across } S = \iint\limits_{S} F \cdot N \; dS. \tag{13}$$

Thus (12) is a special case of (13) with $F = \delta v$. However, the vector field in (13) need not be a velocity field for a fluid. It may, for example, be an electric field induced by one or more point charges, or a gradient field for a temperature function.

In working with either equation (12) or equation (13) you will need to recall that a normal $N(x, y, z)$ to the surface S that is the graph of $z = g(x, y)$ can be found by writing S as a level surface

$$f(x, y, z) = z - g(x, y) = 0$$

and forming

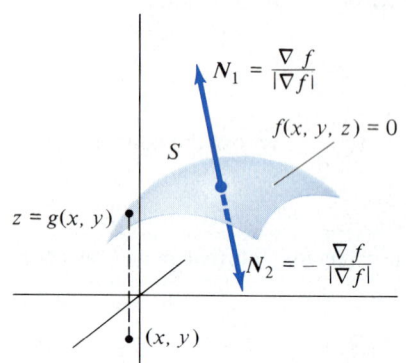

Figure 5.16 Choosing a unit normal to graph of $f(x, y, z) = z - g(x, y) = 0$.

$$N_1 = \frac{\nabla f}{|\nabla f|} = \frac{-\dfrac{\partial g}{\partial x}i - \dfrac{\partial g}{\partial y}j + k}{\sqrt{\left(\dfrac{\partial g}{\partial x}\right)^2 + \left(\dfrac{\partial g}{\partial y}\right)^2 + 1}} \tag{14}$$

or

$$N_2 = -\frac{\nabla f}{|\nabla f|} = \frac{\dfrac{\partial g}{\partial x}i + \dfrac{\partial g}{\partial y}j - k}{\sqrt{\left(\dfrac{\partial g}{\partial x}\right)^2 + \left(\dfrac{\partial g}{\partial y}\right)^2 + 1}}. \tag{15}$$

The choice between N_1 and N_2 is determined by whether you wish N to point upward (N_1) or downward (N_2) from the graph of $z = g(x, y)$ (Figure 5.16).

Before proceeding to our final example, we use equations (14) and (15) to express the flux integral (13) in a more manageable form. If the surface S is the graph of $z = g(x, y)$ and the vector field F has the component form

$$F(x, y, z) = M(x, y, z)i + N(x, y, z)j + P(x, y, z)k,$$

then the flux integral (13) corresponding to the upward unit normal N_1 in (14) is, according to Definition 5,

$$\iint_S F \cdot N \, dS = \iint_Q (Mi + Nj + Pk)$$

$$\left(\frac{-\dfrac{\partial g}{\partial x}i - \dfrac{\partial g}{\partial y}j + k}{\sqrt{\left(\dfrac{\partial g}{\partial x}\right)^2 + \left(\dfrac{\partial g}{\partial y}\right)^2 + 1}} \right) \left(\sqrt{\left(\dfrac{\partial g}{\partial x}\right)^2 + \left(\dfrac{\partial g}{\partial y}\right)^2 + 1} \right) dx \, dy$$

where Q is the region in the xy-plane over which S is defined. This simplifies to

$$\iint_S F \cdot N \, dS = \iint_Q \left[-M(x, y, g(x, y))\frac{\partial g}{\partial x} - N(x, y, g(x, y))\frac{\partial g}{\partial y} \right.$$
$$\left. + P(x, y, g(x, y)) \right] dx \, dy. \tag{16}$$

Similarly, the flux integral (13) corresponding to the downward unit normal N_2 in (15) is

$$\iint_S F \cdot N \, dS = \iint_Q \left[M(x, y, g(x, y))\frac{\partial g}{\partial x} + N(x, y, g(x, y))\frac{\partial g}{\partial y} \right.$$
$$\left. - P(x, y, g(x, y)) \right] dx \, dy. \tag{17}$$

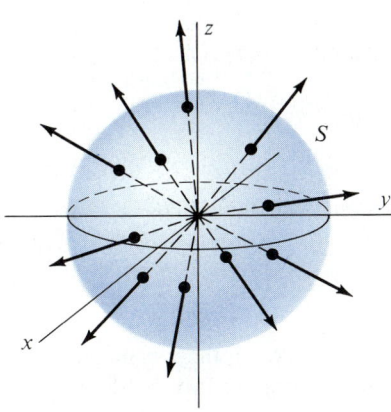

Figure 5.17 Flux of a central force field over a sphere is twice the flux over the upper hemisphere $z = \sqrt{R^2 - x^2}$.

Example 8

Calculate the flux over the sphere S with center $(0, 0, 0)$ and radius R associated with the (central) inverse square field

$$F(r) = \frac{\lambda r}{|r|^3}, \qquad r \neq 0, \qquad \lambda \text{ constant.}$$

That is,

$$F(x, y, z) = \lambda \cdot \frac{x i + y j + z k}{(x^2 + y^2 + z^2)^{3/2}}, \qquad x^2 + y^2 + z^2 \neq 0.$$

Solution: Since the force field F has the same symmetry with respect to the origin as does the sphere S, we will calculate the flux over the upper hemisphere only and then double the result (see Figure 5.17). Note that we are dealing with an improper integral as in Example 6. Using equation (16) with

$$M(x, y, z) = \frac{\lambda x}{(x^2 + y^2 + z^2)^{3/2}} = \frac{\lambda x}{R^3},$$

$$N(x, y, z) = \frac{\lambda y}{(x^2 + y^2 + z^2)^{3/2}} = \frac{\lambda y}{R^3},$$

$$P(x, y, z) = \frac{\lambda z}{(x^2 + y^2 + z^2)^{3/2}} = \frac{\lambda z}{R^3},$$

and

$$g(x, y) = \sqrt{R^2 - x^2 - y^2} = z$$

gives

$$\iint_S F \cdot N \, dS = 2 \iint_Q \left[-\frac{\lambda x}{R^3} \left(\frac{-x}{\sqrt{R^2 - x^2 - y^2}} \right) \right.$$

$$\left. -\frac{\lambda y}{R^3} \left(\frac{-y}{\sqrt{R^2 - x^2 - y^2}} \right) + \frac{\lambda \sqrt{R^2 - x^2 - y^2}}{R^3} \right] dx \, dy$$

$$= 2 \iint_Q \frac{\lambda}{R \sqrt{R^2 - x^2 - y^2}} \, dx \, dy$$

$$= 2 \int_0^{2\pi} \int_0^R \frac{\lambda}{R \sqrt{R^2 - r^2}} r \, dr \, d\theta \qquad \text{(switch to polar coordinates)}$$

$$= 2 \int_0^{2\pi} -\left[\frac{\lambda}{R} \sqrt{R^2 - r^2} \right]_{r=0}^{r=R} d\theta$$

$$= 4\pi\lambda. \qquad \Diamond$$

The surprising result of Example 8 is that *the flux of the inverse square field F over the sphere S is independent of the radius of S!* This result is true for any inverse square field, such as a gravitational field or an electric field. In the latter case the result of Example 8, together with another property of electric fields (Coulomb's

Law), gives **Gauss's Law:**

The flux of the electric field E through any closed surface S is

$$\iint\limits_{S} E \cdot N \, dS = 4\pi \sum_{j=1}^{n} q_j$$

where $q_1, q_2, \ldots, q_n$ are the point charges contained within S.

The Orientation of a Surface

When the surface S is the graph of a single function $z = g(x, y)$, there is no ambiguity about what is meant by the *upward* unit normal (as opposed to the downward unit normal), so a flux integral is easily determined to be of the form (16) or of the form (17). Also, in the cases where S is the graph of the function $y = g(x, z)$ or $x = g(y, z)$, the corresponding meaning of upward normal and the modifications of equations (16) and (17) should be clear. However, the situation becomes less clear as we think about surfaces of the form $S = S_1 \cup S_2 \cup \cdots \cup S_n$ where each S_j is as above. We need to devote just a few more lines to the meaning of N in the flux integral (13) for these more general cases.

According to our original motivation, flux is a measure of the *net* (positive minus negative) rate of flow across S. If more fluid is flowing upward across $S: z = g(x, y)$ than is flowing downward, flux is positive, although fluid may actually be flowing upward across some portions of S and downward across others. It is therefore crucial that N always point on the same "side" of S, since N provides the gauge with which the rate of flow is measured at each point. Thus, in order for the flux integral in (13) to make sense, the surface S must be **orientable,** that is, we must be able to identify two distinct sides of S and to assign a unique unit normal $N(x, y, z)$ that remains on one side or the other as (x, y, z) varies across S.

We have already seen that this is simple to do for surfaces of the form

$$S = \{(x, y, z) \mid z = g(x, y), (x, y) \in Q\}.$$

We just choose the unit normal with positive z-component (upward) or negative z-component (downward). For closed surfaces (such as a sphere or an ellipsoid) the two sides are the inside and the outside, and we must choose between the *inner* unit normal or the *outer* unit normal. (Figure 5.17 shows outer unit normals for a sphere.)

These two general situations (upper versus lower, or inner versus outer) encompass most of the surfaces with which you will need to deal, and we refer to any such surface as *orientable*. We shall not attempt to give a precise definition for this term (as is done in more advanced courses). However, you should know that not all surfaces are orientable. A classic example is the Möbius band, which is obtained by twisting one end of a long rectangular strip 180° and "gluing" it to the other end. As Figure 5.18 shows, a unit normal N beginning at point P and moving continuously once "around" the Möbius band arrives back at P pointing in the opposite direction.

In summary, we caution you to note that the surface integral in Definition 5 does not involve the notion of orientation for S. However, the flux integral in (13) does depend on an orientation for S. Reversing the orientation for S (and, hence, replacing N by $-N$) changes the sign of the integral in (13).

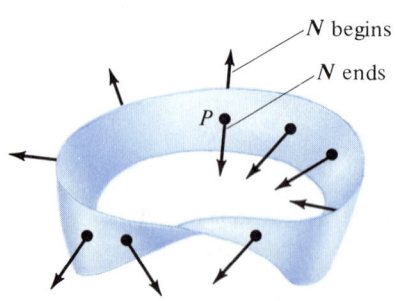

Figure 5.18 Möbius band is not orientable.

Exercise Set 19.5

1. Evaluate $\iint_S (x + 3y + z) \, dS$ over the portion of the plane $x - y + 2z = 4$ lying above the rectangle $Q = \{(x, y) \mid 0 \le x \le 1, 0 \le y \le 2\}$.

2. Evaluate the surface integral in Exercise 1 where S is the portion of the plane $x - y + 2z = 4$ lying above the unit circle.

3. Evaluate $\iint_S x^2z \, dS$ where S is the portion of the plane $2x + 3y + z = 6$ lying in the first octant ($x \ge 0$, $y \ge 0$, $z \ge 0$).

4. Evaluate $\iint_S y^2z \, dS$ where S is the portion of the cone $z^2 = x^2 + y^2$ lying between the planes $z = 1$ and $z = 4$.

5. Evaluate the surface integral $\iint_S (x^2 + 5y - z) \, dS$ where S is the graph of the cylinder $z = x^2$ over the square $-1 \le x \le 1$, $-1 \le y \le 1$.

6. Evaluate $\iint_S (x^2 - y^2 + 1) \, dS$ where S is the part of the plane $y = x + 2$ inside the cylinder $x^2 + y^2 = 4$.

7. Find the mass of the part of a sphere $x^2 + y^2 + z^2 = 9$ lying inside the cylinder $x^2 + y^2 = 4$ if the density per unit area δ is constant.

8. Find $\iint_S F \cdot N \, dS$ where $F = 2xi - yj + 3zk$, S is the part of the plane $x - y + 2z = 4$ bounded by the planes $y = 0$, $y = -4$, $x = 0$, and $x = 4$, and N is the upward unit normal.

9. Find $\iint_S (x + y + z^2) \, dS$ where S is the hemisphere $z = \sqrt{4 - x^2 - y^2}$.

10. Evaluate $\iint_S (y - x) \, dS$ where S is the part of the plane $6x + 4y + 2z = 8$ lying inside the cylinder $x^2 + y^2 = 4$.

11. Find the flux $\iint_S F \cdot N \, dS$ where $F = xi + yj + zk$, S is the upper half sphere $z = \sqrt{1 - x^2 - y^2}$, and N is the upward unit normal.

12. Use the result of Exercise 11 to find $\iint_S F \cdot N \, dS$ where $F = xi + yj + zk$, S is the entire unit sphere, and N is the outward unit normal.

13. Calculate the flux integral $\iint_S F \cdot N \, dS$ where $F = 2xi + yj + zk$, S is the surface of the paraboloid $z = x^2 + y^2$ bounded by the planes $z = 0$ and $z = 4$, and N is the outward unit normal.

14. Let $F = 2i - xj + yk$. Find the flux integral $\iint_S F \cdot N \, dS$ where S is the portion of the cylinder $x^2 + y^2 = 1$ lying between the planes $z = 0$ and $z = 1$, and N is the outward unit normal.

15. Find the flux outward through the sphere $x^2 + y^2 + z^2 = a^2$ for the vector field $F(x, y, z) = zk$. What is the flux if $F(x, y, z) = yj$? Can you give a geometric argument for these results?

16. Find $\iint_S F \cdot N \, dS$ where S is the ellipsoid $4x^2 + y^2 + z^2 = 4$, $F = x^2i + y^2j + z^2k$, and N is the outward unit normal. (This will require a geometric argument.)

17. Find the flux $\iint_S F \cdot N \, dS$ where $F = xi + yj + zk$, N is the outward unit normal, and S is the surface of the unit cube $\{(x, y, z) \mid 0 \le x \le 1, 0 \le y \le 1, 0 \le z \le 1\}$.

18. Suppose that $F(x, y, z) = axi + byj + czk$ and that S is part of the plane $ax + by + cz = d$. Show that the flux integral $\iint_S F \cdot N \, dS$ is $(a^2 + b^2 + c^2)A$ where A is the area of S, if N is the unit normal with $N \cdot ck \ge 0$.

19. Is the result of Example 8 the same if F is an inverse *cube* field $F(r) = \dfrac{r}{|r|^4}$? What is the result in this case?

20. Show that the *flux* of the vector field $F = Mi + Nj + Pk$ across a surface S can be written

$$\iint_S F \cdot N \, dS = \iint_S (P \cos \alpha + Q \cos \beta + R \cos \gamma) \, dS$$

where α, β, and γ are the direction numbers for the unit normal N.

For a surface S, the coordinates $(\bar{x}, \bar{y}, \bar{z})$ of the *centroid* are defined to be

$$\bar{x} = \frac{1}{A}\iint_S x \, dS, \quad \bar{y} = \frac{1}{A}\iint_S y \, dS, \quad \bar{z} = \frac{1}{A}\iint_S z \, dS$$

where A is the surface area of S. Use these formulas in Exercises 21–23.

21. Find the centroid of the hemisphere $z = \sqrt{a^2 - x^2 - y^2}$, $z \geq 0$. (*Hint:* Use symmetry as much as possible.)

22. Find the centroid of the cylinder $z = 1 - x^2$, $-2 \leq y \leq 2$, $-1 \leq x \leq 1$.

23. Find the centroid of the part of the spherical surface $x^2 + y^2 + z^2 = 1$ lying in the first octant.

24. Show that the surface integral, as defined by equation (1), can be written in the form

$$\iint_S f(x, y, z)\, dS = \iint_Q f(x, y, g(x, y))\, \sec \gamma \, dx\, dy$$

where $\gamma = \gamma(x, y)$ is the angle between the upward normal to the graph of $z = g(x, y)$ at (x, y, z) and the vector $\boldsymbol{k}$. (*Hint:* Refer to Section 19.5, Exercise 20.)

25. Show that if the function $f(r, \theta, z)$ is written using cylindrical coordinates for $\mathbb{R}^3$ and if the surface S is the graph of the function $z = g(r, \theta)$ in polar coordinates, then the formula in Definition 5 becomes

$$\iint_S f(r, \theta, z)\, dS = \iint_Q f(r, \theta, g(r, \theta))$$

$$\cdot \sqrt{r^2 + r^2 \left(\frac{\partial g}{\partial r}\right)^2 + \left(\frac{\partial g}{\partial \theta}\right)^2}\, dr\, d\theta.$$

19.6 STOKES' THEOREM

In this section we encounter a generalization of Green's Theorem, which we have earlier stated as

$$\int_C \boldsymbol{F} \cdot d\boldsymbol{r} = \iint_Q \left(\frac{\partial N}{\partial x} - \frac{\partial M}{\partial y}\right) dA = \iint_Q (\text{curl } \boldsymbol{F})\, dA. \tag{1}$$

Recall the setting of equation (1): $\boldsymbol{F}(x, y) = M(x, y)\boldsymbol{i} + N(x, y)\boldsymbol{j}$ is a vector field in the plane (with continuously differentiable components), C is a piecewise smooth simple closed curve enclosing the region Q, and curl $\boldsymbol{F}$ is the scalar function

$$\text{curl } \boldsymbol{F} = \frac{\partial N}{\partial x} - \frac{\partial M}{\partial y}. \tag{2}$$

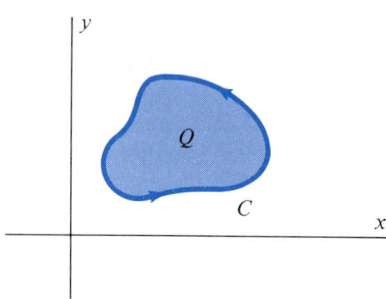

Figure 6.1 Curve C and region Q in Green's Theorem.

All this takes place in the plane, $\mathbb{R}^2$.

The generalization referred to above is Stokes' Theorem, which differs from Green's Theorem in three ways. First, the setting for Stokes' Theorem is $\mathbb{R}^3$. C will now be a closed space curve, and the vector field $\boldsymbol{F}$ will have the form

$$\boldsymbol{F}(x, y, z) = M(x, y, z)\boldsymbol{i} + N(x, y, z)\boldsymbol{j} + P(x, y, z)\boldsymbol{k}. \tag{3}$$

Second, whereas the integral on the right-hand side of equation (1) is a double integral over the plane region Q, the right-hand side of Stokes' Theorem will involve a surface integral evaluated over a smooth surface S bounded by the closed space curve C. Figures 6.1 and 6.2 illustrate the geometry associated with these two theorems.

Stokes' Theorem (which we shall state in more precise terms later) says this:

$$\int_C \boldsymbol{F} \cdot d\boldsymbol{r} = \iint_S (\text{curl } \boldsymbol{F}) \cdot \boldsymbol{N}\, dS \tag{4}$$

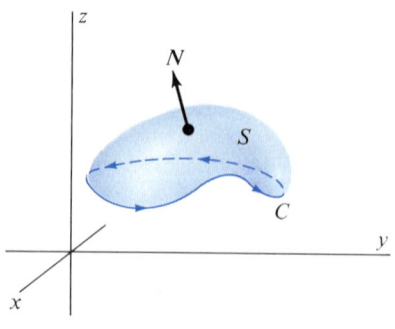

Figure 6.2 The surface S bounded by the space curve C in Stokes' Theorem.

where $\boldsymbol{N}$ is a unit normal for the surface S.

The strong similarity between equations (1) and (4) is obvious. However, the integrand in the surface integral in (4) has not yet been defined. This brings us to the third major difference between these two theorems.

Recall that the line integral in (1) was interpreted in Section 19.4 as the *circulation* around C of a fluid flowing with velocity field $\boldsymbol{F}$. This interpretation issued from the equation

$$\int_C \boldsymbol{F} \cdot d\boldsymbol{r} = \int_C \boldsymbol{F} \cdot \boldsymbol{T} \, ds \tag{5}$$

discussed earlier. Now equation (5) holds for vector fields of the form (3) defined on space curves as well as in the planar case. Retaining the notion of fluid flow as our primary motivation, we would like to define curl $\boldsymbol{F}$ for $\boldsymbol{F}$ in (3) in such a way that equation (4) can be interpreted as a statement about fluid flow (and, of course, so that equation (4) is true!). The following definition of curl $\boldsymbol{F}$ does this. We shall defer further physical interpretation to the end of this section.

DEFINITION 6

Let $\boldsymbol{F}$ be a vector field of the form

$$\boldsymbol{F}(x, y, z) = M(x, y, z)\boldsymbol{i} + N(x, y, z)\boldsymbol{j} + P(x, y, z)\boldsymbol{k}.$$

If all first order partial derivatives for each component function M, N, and P exist at (x, y, z), **curl $\boldsymbol{F}(x, y, z)$** is defined to be the vector

$$\text{curl } \boldsymbol{F}(x, y, z) = \left(\frac{\partial P}{\partial y} - \frac{\partial N}{\partial z}\right)\boldsymbol{i} + \left(\frac{\partial M}{\partial z} - \frac{\partial P}{\partial x}\right)\boldsymbol{j} + \left(\frac{\partial N}{\partial x} - \frac{\partial M}{\partial y}\right)\boldsymbol{k} \tag{6}$$

where each partial derivative is evaluated at (x, y, z).

It is important to note that curl $\boldsymbol{F}(x, y, z)$ is a *vector* in $\mathbb{R}^3$ while curl $\boldsymbol{F}(x, y)$ in (2) is a *number*. Now curl $\boldsymbol{F}(x, y, z) \cdot \boldsymbol{N}$ (the integrand in the surface integral in Stokes' Theorem) is a generalization of curl $\boldsymbol{F}(x, y)$ (the integrand in the double integral in Green's Theorem) in the following sense. If we restrict $\boldsymbol{F}$ in (3) to the xy-plane by setting $z \equiv 0$ and $P(x, y, z) \equiv 0$, equation (6) gives

$$\text{curl } \boldsymbol{F}(x, y, 0) = \left(\frac{\partial N}{\partial x} - \frac{\partial M}{\partial y}\right)\boldsymbol{k}. \tag{7}$$

Since a unit normal to the xy-plane is $\boldsymbol{N} = \boldsymbol{k}$, we obtain

$$[\text{curl } \boldsymbol{F}(x, y, 0)] \cdot \boldsymbol{N} = \left[\left(\frac{\partial N}{\partial x} - \frac{\partial M}{\partial y}\right)\boldsymbol{k}\right] \cdot \boldsymbol{k} = \frac{\partial N}{\partial x} - \frac{\partial M}{\partial y}$$

$$= \text{curl } \boldsymbol{F}(x, y), \text{ as claimed.}$$

You will often see the notation

$$\text{curl } \boldsymbol{F} = \nabla \times \boldsymbol{F}. \tag{8}$$

This results from thinking of the gradient ∇ as an "operator"

$$\nabla = \frac{\partial}{\partial x}\boldsymbol{i} + \frac{\partial}{\partial y}\boldsymbol{j} + \frac{\partial}{\partial z}\boldsymbol{k}$$

and forming the cross product*

---

*As in the mnemonic for remembering how to compute cross products, the determinant here is only formal notation, since only the third row of the matrix actually contains numbers.

$$\nabla \times F = \det \begin{bmatrix} \boldsymbol{i} & \boldsymbol{j} & \boldsymbol{k} \\ \dfrac{\partial}{\partial x} & \dfrac{\partial}{\partial y} & \dfrac{\partial}{\partial z} \\ M & N & P \end{bmatrix} \tag{9}$$

$$= \left(\frac{\partial P}{\partial y} - \frac{\partial N}{\partial z} \right) \boldsymbol{i} + \left(\frac{\partial M}{\partial z} - \frac{\partial P}{\partial x} \right) \boldsymbol{j} + \left(\frac{\partial N}{\partial x} - \frac{\partial M}{\partial y} \right) \boldsymbol{k}.$$

Example 1

For the vector field $F(x, y, z) = y\boldsymbol{i} - z^2\boldsymbol{j} + 2x\boldsymbol{k}$, $M = y$, $N = -z^2$, and $P = 2x$. From equations (8) and (9) we find that

$$\text{curl } F = \det \begin{bmatrix} \boldsymbol{i} & \boldsymbol{j} & \boldsymbol{k} \\ \dfrac{\partial}{\partial x} & \dfrac{\partial}{\partial y} & \dfrac{\partial}{\partial z} \\ y & -z^2 & 2x \end{bmatrix} = 2z\boldsymbol{i} - 2\boldsymbol{j} - \boldsymbol{k}. \qquad \diamondsuit$$

One final detail needs to be cleaned up before we state Stokes' Theorem. We shall need to require that the surface S be **simply connected.** Basically, this means that S contains no holes. Another way to say this is that, given any point P_0 on S, the boundary curve C can be "continuously contracted" across S to an arbitrarily small closed curve containing P_0. (If S had a hole you could not do this since the contraction of C would get "hung up" at the hole (see Figure 6.3).

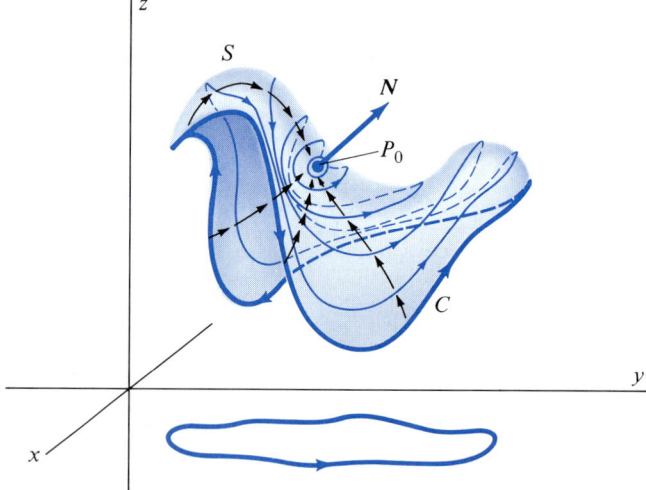

Figure 6.3 An oriented simply connected surface S. Contracting C continuously across S reveals the orientation of C induced by the orientation of S.

This notion of contracting C across S toward a point also allows us to clarify the issue of orientation. We shall require that the surface S be oriented. As defined in Section 19.5, this means that S comes equipped with a unit normal N that varies continuously across S, always remaining on the same side of S. In the statement of

Stokes' Theorem we must require that the orientation of the boundary curve C be the orientation *induced* on C by the orientation of S. Here's what we mean. Let N be the unit normal to S at the point P_0. If S is smooth, then very near P_0 the surface of S may be approximated by a plane tangent to S at P_0. Think of a very small circle C_S on this tangent plane with center P_0. This small circle approximates a small closed curve on S, and we can continuously contract C across S so that it takes the shape of this small closed curve. The point is that the orientation of all three curves (C, the small curve on S, and the small circle C_S on the tangent plane) must be the same. N determines an orientation for C_S by a variation of the right-hand rule: Think of wrapping the index finger of your right hand around C_S, keeping your thumb parallel to N. Your index finger points out the orientation of C_S (see Figure 6.4). This is the orientation induced by N on C.

We are now ready to state Stokes' Theorem.

THEOREM 4
Stokes' Theorem

Let S be a smooth, simply connected, oriented surface bounded by a piecewise smooth simple closed curve C with orientation induced by the orientation of S. Let

$$F(x, y, z) = M(x, y, z)i + N(x, y, z)j + P(x, y, z)k$$

be a vector field defined on an open box D containing S and C so that each of the component functions M, N, and P is continuous and has continuous partial derivatives on D. Then

$$\int_C F \cdot dr = \iint_S (\text{curl } F) \cdot N \, dS \tag{10}$$

where N is the unit normal to the oriented surface S.

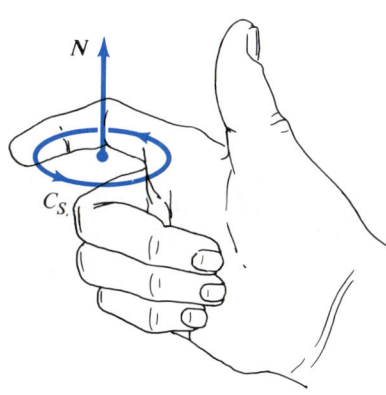

Figure 6.4 Right-hand rule determines orientation of C_S from direction of N.

Using equation (3), we may write equation (10) in the alternate form

$$\int_C M \, dx + N \, dy + P \, dz = \iint_S (\text{curl } F) \cdot N \, dS.$$

We may paraphrase Stokes' Theorem by saying that "the line integral of the vector field around the boundary of S equals the integral of the normal component of curl F over the surface S."

Note that the particular shape of S is unimportant—the surface integral over S is entirely determined by the line integral along its boundary. For example, the surface integrals of (curl F) $\cdot N$ over all of the surfaces in Figure 6.5 are the same, since each has the same boundary C.

A proof of Stokes' Theorem is beyond the scope of this text. (If you've noticed that the concepts of this chapter sometimes appear to be less precise than in earlier chapters, you are correct. We are now dabbling in ideas that require considerably more advanced mathematics to treat rigorously than we can hope to achieve in a first calculus course.) However, we can discuss why Stokes' Theorem should hold.

First consider the case of the tetrahedron sketched in Figure 6.6. We shall take the surface S to be the union of the three faces ABE, BDE, and DAE. Then S is bounded by the triangle ABD, which is the curve C. We orient S by taking the outward unit normal N at each point, which induces the orientation ABD on C.

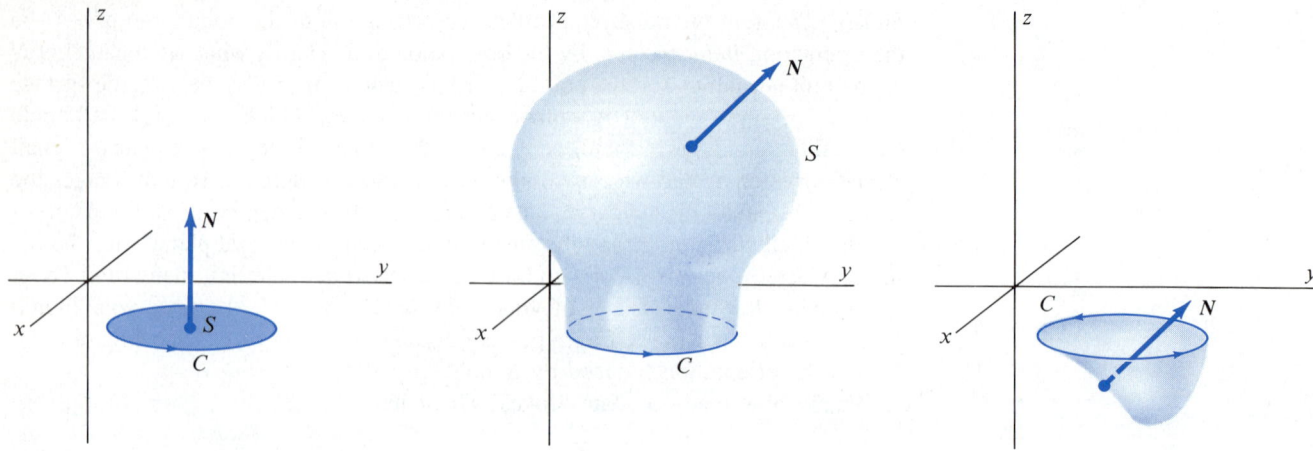

Figure 6.5 The integrals over all of these surfaces of (curl F) · N are the same: $\iint_S$ (curl F) · $N\,dS = \int_C F = dr$.

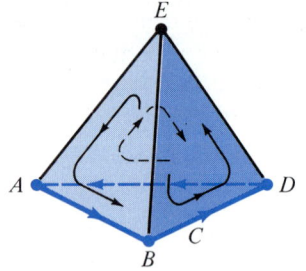

Figure 6.6 Stokes' Theorem holds for a tetrahedral surface.

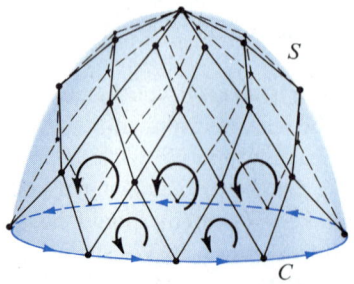

Figure 6.7 A polyhedral approximation to the surface S.

Since each face of the surface is a plane region, we apply Green's Theorem to each face and obtain the three equations

$$\int_{ABE} F \cdot dr = \iint_{ABE} (\text{curl } F) \cdot N\,dS, \tag{11}$$

$$\int_{BDE} F \cdot dr = \iint_{BDE} (\text{curl } F) \cdot N\,dS, \tag{12}$$

$$\int_{DAE} F \cdot dr = \iint_{DAE} (\text{curl } F) \cdot N\,dS. \tag{13}$$

(In each of equations (11) through (13), we have used the remark concerning equation (7) to write the scalar curl in vector form.)

Next, we wish to add the corresponding sides in equations (11) through (13). When we do so, the contributions to the resulting line integral along the edges AE, BE, and DE will be zero, since each of these edges is traversed twice, once in each direction. Also, the three surface integrals should sum to the surface integral over S. The result, then, is

$$\int_{ABD} F \cdot dr = \iint_S (\text{curl } F) \cdot N\,dS$$

which is equation (10).

Now consider a more complex polyhedron S, such as that in Figure 6.7. The idea is the same. When we apply Green's Theorem to each face and sum the corresponding sides of the resulting equations, the only nonzero contributions to the line integral occur along the bottom edges. Along all other edges, the line integral will have been evaluated twice, once in each direction. Referring to the bottom edge as C, we again obtain equation (10).

The argument for a smooth surface S is to first approximate S by a sequence of polyhedra, each of whose vertices lies on the surface S. As the number of vertices in the polyhedra becomes large, the polyhedra become increasingly accurate approximations to the surface. Since equation (10) holds for each polyhedron, it holds for the smooth surface "in the limit."

Example 2

Find $\int_C \mathbf{F} \cdot d\mathbf{r}$ where $\mathbf{F}$ is the vector field $\mathbf{F}(x, y, z) = y\mathbf{i} - z^2\mathbf{j} + 2x\mathbf{k}$ and C is the unit circle $C = \{(x, y, 0) \mid x^2 + y^2 = 1\}$ oriented in the counterclockwise direction.

Solution: We could evaluate the line integral directly (and, painfully) by parameterizing C by $x(t) = \cos t$, $y(t) = \sin t$, $0 \le t \le 2\pi$. Instead, we will apply Stokes' Theorem using any smooth surface S bounded by C. The simplest choice, of course, is the unit disc

$$S = \{(x, y, 0) \mid 0 \le x^2 + y^2 \le 1\}.$$

From Example 1 we know that

$$\text{curl } \mathbf{F} = 2z\mathbf{i} - 2\mathbf{j} - \mathbf{k}$$

and a unit normal giving S the required orientation is $\mathbf{N} = \mathbf{k}$. Thus

$$(\text{curl } \mathbf{F}) \cdot \mathbf{N} = (2z\mathbf{i} - 2\mathbf{j} - \mathbf{k}) \cdot \mathbf{k} = -1.$$

Thus, by Stokes' Theorem,

$$\int_C \mathbf{F} \cdot d\mathbf{r} = \iint_S (-1) \, dS = -\pi,$$

since π is the area of the unit disc. $\diamond$

Example 3

Verify Stokes' Theorem if $\mathbf{F}$ is the vector field

$$\mathbf{F}(x, y, z) = y\mathbf{i} - x^2\mathbf{j} + 2z^2\mathbf{k}$$

and S is the portion of the paraboloid $z = 4 - x^2 - y^2$ lying above the xy-plane. Take $\mathbf{N}$ to be the upward unit normal to S.

Solution: The curve C bounding S is the circle

$$C = \{(x, y, 0) \mid x^2 + y^2 = 4\}$$

which has parameterization $x(t) = 2 \cos t$, $y(t) = 2 \sin t$, $z(t) = 0$, $0 \le t \le 2\pi$. The line integral on the left side of equation (10) is therefore

$$\int_C \mathbf{F} \cdot d\mathbf{r} = \int_C y \, dx - x^2 \, dy + 2z^2 \, dz$$

$$= \int_0^{2\pi} (-4 \sin^2 t - 8 \cos^3 t) \, dt$$

$$= -4\pi.$$

To calculate the surface integral in (10), we first use (9) to find

$$\text{curl } F = \nabla \times F = \det \begin{bmatrix} i & j & k \\ \dfrac{\partial}{\partial x} & \dfrac{\partial}{\partial y} & \dfrac{\partial}{\partial z} \\ y & -x^2 & 2z^2 \end{bmatrix} = -(2x + 1)k.$$

Then, using equation (16), Section 20.5, we evaluate the surface integral as

$$\iint_S (\text{curl } F) \cdot N \, dS = \iint_S [-(2x + 1)k] \cdot N \, dS$$

$$= \iint_Q -(2x + 1) \, dx \, dy$$

$$= \int_0^{2\pi} \int_0^2 -(2r \cos \theta + 1) \, r \, dr \, d\theta \qquad \text{(switch to polar coordinates)}$$

$$= -4\pi,$$

which agrees with the value of the line integral. ◇

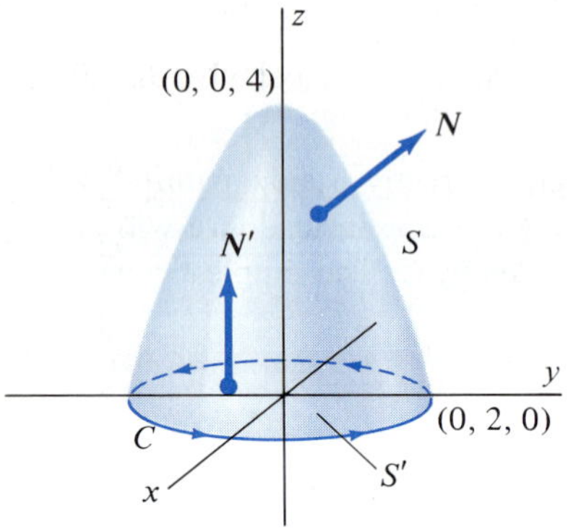

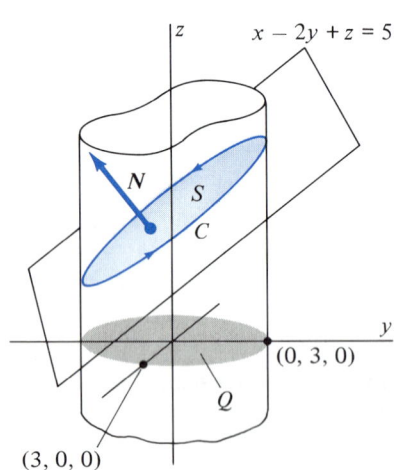

Figure 6.8 $\displaystyle\iint_S (\text{curl } F) \cdot N \, dS$ is the same as $\displaystyle\iint_{S'} (\text{curl } F) \cdot N' \, dS$, according to Stokes' Theorem.

REMARK: Notice that the calculation of the surface integral involved the same double integral as we would have obtained had the problem simply been to calculate $\displaystyle\iint_{S'} \text{curl } F \cdot N \, dS$ over the planar disc $S' = \{(x, y, 0) \mid 0 \le x^2 + y^2 \le 4\}$ (Figure 6.8). This is a direct illustration of the statement of Stokes Theorem—the value of the surface integral depends only on the line integral around the boundary.

Example 4

Find $\displaystyle\int_C F \cdot dr$ where the space curve C is the intersection of the plane $x - 2y + z = 5$ with the cylinder $x^2 + y^2 = 9$, oriented as in Figure 6.9, and F is the vector field $F(x, y, z) = (x^2 - 3y^2)i + (z^2 + y)j + (x + 2z^2)k$.

Figure 6.9 Line integral $\displaystyle\int_C F \cdot dr$ is evaluated as a surface integral $\displaystyle\iint_S (\text{curl } F) \cdot N \, dS$ over S.

Solution: Attempting to evaluate the line integral directly would be difficult at best, so we resort to Stokes' Theorem, taking S to be the portion of the plane enclosed by C. From the equation of the plane and the orientation of C, we see that a unit normal to S is

$$N = \frac{1}{\sqrt{6}}(i - 2j + k)$$

which is upward. Also,

$$\text{curl } F = \det \begin{bmatrix} i & j & k \\ \dfrac{\partial}{\partial x} & \dfrac{\partial}{\partial y} & \dfrac{\partial}{\partial z} \\ (x^2 - 3y^2) & (z^2 + y) & (x + 2z^2) \end{bmatrix} = -2zi - j + 6yk.$$

The equation for S becomes $z = g(x, y) = 5 - x + 2y$. By Stokes' Theorem and equation (16), Section 19.5, we have

$$\int_C \mathbf{F} \cdot d\mathbf{r} = \iint_S (\operatorname{curl} \mathbf{F}) \cdot \mathbf{N} \, dS$$

$$= \iint_Q \{-[-2(5 - x + 2y)](-1) - (-1)(2) + 6y\} \, dx \, dy$$

$$= \iint_Q (2x + 2y - 8) \, dx \, dy$$

$$= \int_0^{2\pi} \int_0^3 (2r \cos\theta + 2r \sin\theta - 8) r \, dr \, d\theta$$

$$= -72\pi.$$

(Note that we did not make use of the components of the unit normal $\mathbf{N}$—only the fact that it is oriented upward.) ◇

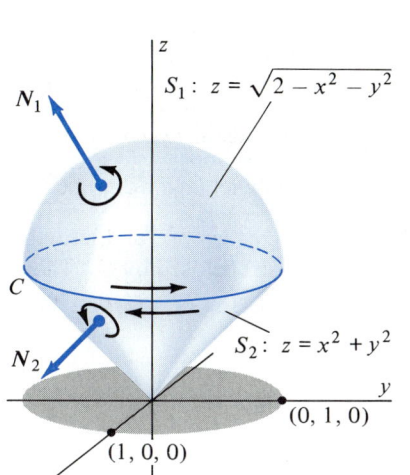

Figure 6.10 $\iint_S (\operatorname{curl} \mathbf{F}) \cdot \mathbf{N} \, dS = 0$ for $S = S_1 \cup S_2$.

Example 5

Calculate $\iint_S (\operatorname{curl} \mathbf{F}) \cdot \mathbf{N} \, dS$ where S is the closed surface shown in Figure 6.10 and $\mathbf{F}$ is a vector field satisfying the hypotheses of Stokes' Theorem.

Solution: The closed surface S may be written $S = S_1 \cup S_2$ where S_1 is the hemisphere $z = \sqrt{2 - x^2 - y^2}$ and S_2 is the graph of the cone $z = x^2 + y^2$ over the unit disc. These two surfaces share the common boundary curve

$$C = \{(x, y, 1) \mid x^2 + y^2 = 1\}.$$

However, the two surfaces induce opposite orientations on C. This is because the unit normal to S_1 is upward, inducing a counterclockwise orientation on C, while the unit normal to S_2 is downward, inducing a clockwise orientation on C. Thus, by Stokes' Theorem,

$$\iint_S (\operatorname{curl} \mathbf{F}) \cdot \mathbf{N} \, dS = \iint_{S_1} (\operatorname{curl} \mathbf{F}) \cdot \mathbf{N}_1 \, dS + \iint_{S_2} (\operatorname{curl} \mathbf{F}) \cdot \mathbf{N}_2 \, dS$$

$$= \int_C \mathbf{F} \cdot d\mathbf{r} + \int_{-C} \mathbf{F} \cdot d\mathbf{r}$$

$$= \int_C \mathbf{F} \cdot d\mathbf{r} - \int_C \mathbf{F} \cdot d\mathbf{r}$$

$$= 0. \qquad ◇$$

Curl F and Conservative Force Fields

For conservative planar vector fields $\mathbf{F}(x, y) = M\mathbf{i} + N\mathbf{j}$, we have previously seen that the condition that $\mathbf{F}$ be *conservative* in an open rectangle D is equivalent to the condition that curl $\mathbf{F}(x, y) = 0$ for all $(x, y) \in D$ (see Exercise 24, Section 19.4). We now sketch an argument for the same result in three dimensions.

First, suppose that curl $\mathbf{F}(x, y, z) = \mathbf{0}$ for all (x, y, z) in some simply connected region D in $\mathbb{R}^3$. Then if C is any closed path in D,

$$\int_C \mathbf{F} \cdot d\mathbf{r} = \iint_S \operatorname{curl} \mathbf{F} \cdot \mathbf{N} \, dS = \iint_S \mathbf{O} \cdot \mathbf{N} \, dS = 0$$

where S is any smooth surface in D bounded by C. Thus, $\int_C \mathbf{F} \cdot d\mathbf{r} = 0$ for any closed path in D, so $\mathbf{F}$ is conservative (Exercise 13, Section 19.3).

Conversely, suppose that $\mathbf{F}$ is conservative in D. Then (again by Exercise 13, Section 20.3) $\int_C \mathbf{F} \cdot d\mathbf{r} = 0$ for any closed curve C in D. Now let (x_0, y_0, z_0) be any point in D, let C_r be a small circle of radius r and center (x_0, y_0, z_0), and let $\mathbf{N}$ be any unit normal to the plane containing C_r. Let S_r be the disc enclosed by C_r. Then, by Stokes' Theorem

$$\int_{C_r} \mathbf{F} \cdot d\mathbf{r} = \iint_{S_r} (\operatorname{curl} \mathbf{F}) \cdot \mathbf{N} \, dS. \tag{14}$$

Now the line integral in (14) is zero, since $\mathbf{F}$ is conservative. Thus, $\iint_{S_r} (\operatorname{curl} \mathbf{F}) \cdot \mathbf{N} \, dS = 0$, no matter how small r is chosen. From this statement we conclude that

$$(\operatorname{curl} \mathbf{F}(x_0, y_0, z_0)) \cdot \mathbf{N} = 0. \tag{15}$$

(If the expression in (15) were either positive or negative, we could invoke the continuity of $\mathbf{F}$ to contract C_r to a small circle on which $(\operatorname{curl} \mathbf{F}) \cdot \mathbf{N}$ is either always positive or always negative. The resulting surface integral in (14) would then be nonzero, a contradiction of equation (14).)

Since $\mathbf{N}$ was arbitrary, it follows from (15) that $\operatorname{curl} \mathbf{F}(x_0, y_0, z_0) = \mathbf{0}$. Finally, since (x_0, y_0, z_0) was chosen arbitrarily in D, we conclude that $\operatorname{curl} \mathbf{F}(x, y, z) = \mathbf{0}$ for all $(x, y, z) \in D$ when $\mathbf{F}$ is conservative.

The preceding argument is not a rigorous proof. However, it does convey the general flavor of the arguments linking the ideas of conservative vector fields, independence of path for line integrals, and the condition $\operatorname{curl} \mathbf{F} = \mathbf{0}$. These relationships are summarized by the following theorem.

THEOREM 5

Let $\mathbf{F}(x, y, z) = M(x, y, z)\mathbf{i} + N(x, y, z)\mathbf{j} + P(x, y, z)\mathbf{k}$ be a vector field for which all of the component functions, together with all their first partial derivatives, are continuous in some open box D in $\mathbb{R}^3$. The following conditions are all equivalent:

(i) $\mathbf{F}$ is conservative in D.

(ii) $\int_C \mathbf{F} \cdot d\mathbf{r}$ is independent of path C in D.

(iii) $\int_C \mathbf{F} \cdot d\mathbf{r} = 0$ for every closed path C in D.

(iv) $\operatorname{curl} \mathbf{F}(x, y, z) = \mathbf{0}$ for all $(x, y, z) \in D$.

Theorem 5 gives the following condition for determining when a vector field $\mathbf{F} = M\mathbf{i} + N\mathbf{j} + P\mathbf{k}$ is conservative.

COROLLARY 2

Let F and D be as in Theorem 5. Then F is conservative in D if and only if each of the following equations holds for all (x, y, z) in D:

(i) $\dfrac{\partial P}{\partial y}(x, y, z) = \dfrac{\partial N}{\partial z}(x, y, z),$

(ii) $\dfrac{\partial M}{\partial z}(x, y, z) = \dfrac{\partial P}{\partial x}(x, y, z),$

(iii) $\dfrac{\partial N}{\partial x}(x, y, z) = \dfrac{\partial M}{\partial y}(x, y, z).$

The proof of Corollary 2 consists of applying Theorem 5, parts (i) and (iv) together with the definition of curl F (Definition 6).

A Physical Interpretation of Curl

As a last look at Stokes' Theorem, we use equation (5) to write it in the form

$$\int_C F \cdot T \, ds = \iint_S (\text{curl } F) \cdot N \, dS. \tag{16}$$

Returning to our primary motivation of fluid flow, we recall from Section 19.4 that if F is the velocity field of a moving fluid in which the curve C is submerged, the line integral in (16), called the **circulation** of F around C, is a measure of the tendency of the fluid to rotate, or circulate, around the curve C. Now let $P_0(x_0, y_0, z_0)$ be a point in the fluid, and let C_r be a circle with radius r and center P_0 (see Figure 6.11). Applying the Mean Value Theorem for double integrals (an obvious generalization of the Mean Value Theorem for functions of a single variable which we have not previously stated) to equation (16), we find that for some point $P_r = (x_r, y_r, z_r)$ in S_r (the disc enclosed by C_r)

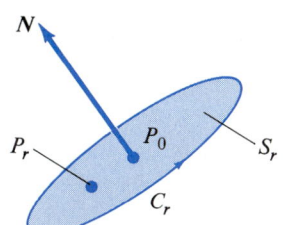

Figure 6.11

$$\int_{C_r} F \cdot T \, ds = \iint_{S_r} (\text{curl } F) \cdot N \, dS$$

$$= \pi r^2 [\text{curl } F(x_r, y_r, z_r)] \cdot N.$$

Letting $r \to 0$, we conclude that

$$[\text{curl } F(x_0, y_0, z_0)] \cdot N = \lim_{r \to 0} \frac{1}{\pi r^2} \int_{C_r} F \cdot T \, ds. \tag{17}$$

The interpretation of equation (17) is this. The right-hand side, as described earlier, is a measure of the tendency of the fluid to circulate, or rotate, about the small circle C_r. Since the normal N to the disc enclosed by this circle appears on the left-hand side of (17), this tendency to rotate will be the largest when C_r is positioned so that N is parallel to curl $F(x_0, y_0, z_0)$. Thus, curl $F(x_0, y_0, z_0)$ is a vector whose *magnitude* is a measure of the tendency of the fluid at (x_0, y_0, z_0) to rotate (as about the drain hole in a bathtub), and whose *direction* is along the axis about which the fluid has the maximal tendency to rotate.

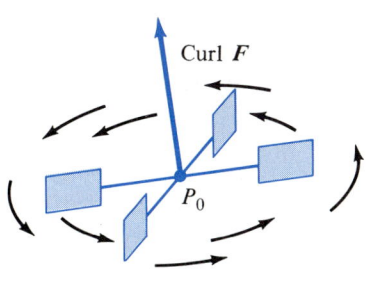

Figure 6.12 Paddle wheel interpretation of curl F. The wheel rotates most rapidly when the axis is parallel to curl F.

Figure 6.12 shows an interpretation of equation (17) in terms of a paddle wheel. The motion of the fluid will cause the paddle wheel, based at P_0, to rotate most quickly when the axis points parallel to curl $F(x_0, y_0, z_0)$. For obvious reasons, fluid fields for which curl $F = 0$ (such as conservative fields) are called *irrotational*.

It is important to note that what we have achieved in these last few observations is an interpretation of curl $F(x, y, z)$ as a *local* property of the vector field—the tendency of the fluid *at location* (x, y, z) to rotate. In general, this tendency will vary from point to point within the fluid. What Stokes' Theorem says in this context is that the collective measure of this tendency taken over the entire surface S (that is, the value of the surface integral) is completely determined by, and equals, the tendency of the fluid to circulate around the boundary C (the value of the line integral).

Exercise Set 19.6

In Exercises 1–4, calculate curl F.

1. $F(x, y, z) = xi - yj + z^2k$

2. $F(x, y, z) = xy^2i + xz^2j + yz^2k$

3. $F(x, y, z) = xyzi - \cos(xy)j + \sin(yz)k$

4. $F(x, y, z) = e^{xy}i - x^2z^2j + \sqrt{xy}\,k$

In Exercises 5–8, determine whether the given vector field is conservative.

5. $F(x, y, z) = yzi + xzj + xyk$

6. $F(x, y, z) = 2xyz^2i + x^2z^2j + 2x^2yzk$

7. $F(x, y, z) = \sin yi + \cos xj + \sqrt{xy}\,k$

8. $F(x, y, z) = yze^{xyz}i + xze^{xyz}j + xye^{xyz}k$

9. Verify Stokes' Theorem for the vector field $F(x, y, z) = zi + xj + yk$ and the surface of the hemisphere $z = \sqrt{1 - x^2 - y^2}$.

10. Use Stokes' Theorem to calculate the line integral $\int_C F \cdot dr$ where $F(x, y, z) = x^2y^2i + x^2z^2j + y^2z^2k$ and C is the perimeter of the rectangle with vertices $(1, 1, 0)$, $(1, 5, 0)$, $(3, 1, 0)$, and $(3, 5, 0)$ traversed in this order.

11. Verify Stokes' Theorem for $F(x, y, z) = x^2yi + y^2zj + xzk$ where C is the boundary of the rectangle $Q = \{(x, y, 0) \mid 0 \le x \le 1, 0 \le y \le 2\}$ oriented counterclockwise.

12. Use Stokes' Theorem to calculate the line integral $\int_C F \cdot dr$ where $F(x, y, z) = x^2yi + y^2zj + xzk$ and C is the intersection of the plane $x + 3y + z = 4$ with the cylinder $x^2 + y^2 = 1$ with orientation induced by the upward unit normal to the plane.

13. Verify Stokes' Theorem for the portion of the plane $x + 2y + z = 2$ in the first octant, the vector field $F(x, y, z) = zi + xj + yk$, and the upward unit normal.

14. Calculate $\int_C F \cdot dr$ where $F(x, y, z) = xzi + 2zj - xyk$ and C is the intersection of the plane $y = z + 2$ and the cylinder $x^2 + y^2 = 4$. The orientation on C is that induced by the upward unit normal to the plane.

15. Evaluate $\int_C F \cdot dr$ around the unit circle in the xy-plane, counterclockwise, where
$$F(x, y, z) = (\sqrt{x} + y)i + (e^y - x)j + (\sin z + y)k.$$

16. Calculate $\int_C F \cdot dr$ where $F(x, y, z) = xyzi + xzk$ and where C is the intersection of the paraboloid $z = x^2 + y^2$ and the plane $z = 4$. The orientation on C is that induced by the upward unit normal on the paraboloid.

17. Verify Stokes' Theorem for the surface $z = 9 - x^2 - y^2$, $z \ge 0$ and the vector field $F(x, y, z) = x^2i + y^2j + z^2k$.

18. Show that curl $F = 0$ for $F(x, y, z) = x^2i + y^2j + z^2k$. Can you give a geometric interpretation for this result?

19. Let $F(x, y) = M(x, y)i + N(x, y)j$ be a conservative vector field. Let $G(x, y, z) = F(x, y) + \lambda zk$ where λ is constant. Show that $G(x, y, z)$ is conservative. Is the result true if zk is replaced by $g(z)k$?

20. Let $F(x, y, z)$ be a vector field satisfying the hypotheses of Stokes' Theorem. Show that $\iint_S (\text{curl } F) \cdot N\, dS = 0$ where S is the unit sphere.

21. Use Stokes' Theorem to show that a line integral is independent of parameterization if orientation is preserved.

19.7 THE DIVERGENCE THEOREM

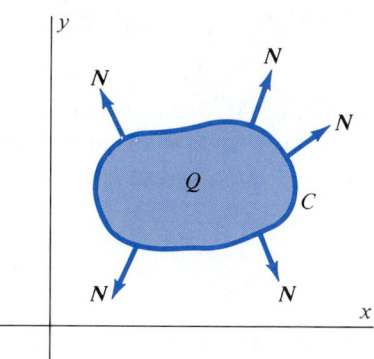

Figure 7.1 $\int_C \boldsymbol{F} \cdot \boldsymbol{N} \, dS$ is the flux of $\boldsymbol{F}(x, y)$ across C in $\mathbb{R}^2$.

In this last section, we generalize equation (28) of Section 19.4,

$$\int_C \boldsymbol{F} \cdot \boldsymbol{N} \, ds = \int\int_Q \text{div } \boldsymbol{F}(x, y) \, dA \qquad \text{(flux of } \boldsymbol{F} \text{ across } C), \tag{1}$$

to an appropriate equation for vector fields in $\mathbb{R}^3$. However, this time we begin with the physical interpretation of what we are after and see where the mathematics leads us.

Recall the interpretation for equation (1) in terms of fluid flow: $\boldsymbol{F}$ is a vector field in the plane giving the velocity vector $\boldsymbol{v}(x, y) = \boldsymbol{F}(x, y)$ for the steady state motion of a thin layer of fluid; C is a closed curve in the fluid layer; and $\boldsymbol{N}$ is the outward unit normal to C. In this setting the line integral $\int_C \boldsymbol{F} \cdot \boldsymbol{N} \, ds$ in (1) is interpreted as the **flux** of the fluid outward across C, that is, the net rate at which fluid is crossing the boundary C of Q (Figure 7.1). In Section 19.5 we have already encountered the notion of the flux of a vector field $\boldsymbol{F}$ across a closed surface S in $\mathbb{R}^3$.

$$\text{Flux} = \int\int_S \boldsymbol{F} \cdot \boldsymbol{N} \, dS, \tag{2}$$

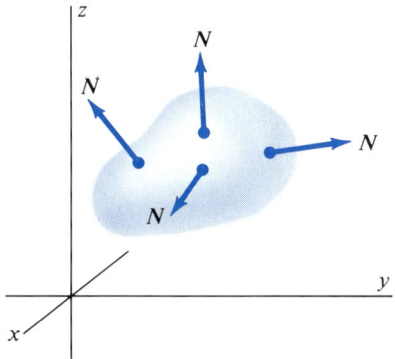

Figure 7.2 $\int\int_S \boldsymbol{F} \cdot \boldsymbol{N} \, dS$ is the flux of $\boldsymbol{F}(x, y, z)$ across S in $\mathbb{R}^3$.

where $\boldsymbol{N}$ is the outward unit normal to S at (x, y, z). That is, the natural generalization of the line integral in (1), at least in terms of fluid flow, is the surface integral in (2) (see Figure 7.2).

Now the right-hand side of equation (1) is a double integral evaluated over Q, the region bounded by C, of the scalar function div $\boldsymbol{F}$, and we can paraphrase equation (1) by saying that "div $\boldsymbol{F}$ is a measure of the local behavior of $\boldsymbol{F}$ that, when integrated over the enclosed region Q, determines the flux of $\boldsymbol{F}$ across the boundary C."

Thus, to properly generalize equation (1) to a closed surface S in $\mathbb{R}^3$, we need to find a scalar function div $\boldsymbol{F}$ that is a measure of the local behavior of $\boldsymbol{F}$ and that, when integrated over the region R enclosed by S, gives the flux of $\boldsymbol{F}$ across the surface S. That is, div $\boldsymbol{F}$ must satisfy the equation

$$\int\int_S \boldsymbol{F} \cdot \boldsymbol{N} \, dS = \int\int\int_R \text{div } \boldsymbol{F} \, dV. \tag{3}$$

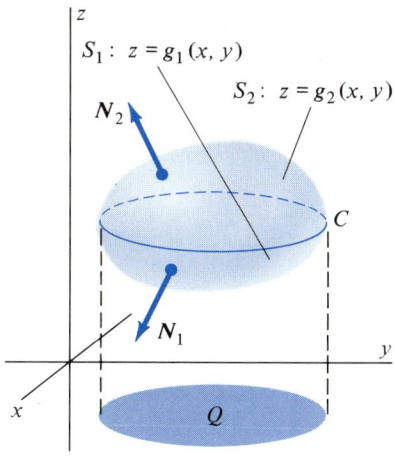

Figure 7.3 $S = S_1 \cup S_2$.

Although you can probably guess how we're going to define div $\boldsymbol{F}$, let's proceed "experimentally" by investigating the surface integral in (3). For starters, we express $\boldsymbol{F}$ in component form as

$$\boldsymbol{F}(x, y, z) = M(x, y, z)\boldsymbol{i} + N(x, y, z)\boldsymbol{j} + P(x, y, z)\boldsymbol{k}. \tag{4}$$

Also, we assume that S is a closed surface that is the union $S = S_1 \cup S_2$ of two smooth surfaces, which are graphs of the functions

$$S_1: \quad z = g_1(x, y), \qquad (x, y) \in Q$$

$$S_2: \quad z = g_2(x, y), \qquad (x, y) \in Q$$

with $g_1(x, y) \leq g_2(x, y)$ for all $(x, y) \in Q$. Let C be the boundary common to S_1 and S_2 (see Figure 7.3).

Using (4), we write the left side of equation (3) as

$$\iint\limits_{S} \mathbf{F} \cdot \mathbf{N} \, dS = \iint\limits_{S} (M\mathbf{i}) \cdot \mathbf{N} \, dS + \iint\limits_{S} (N\mathbf{j}) \cdot \mathbf{N} \, dS + \iint\limits_{S} (P\mathbf{k}) \cdot \mathbf{N} \, dS. \quad (5)$$

Let's now focus on the last integral on the right side of (5). Since the unit normal to S_2 is upward and the unit normal to S_1 is downward, we use both equation (16) and equation (17) of Section 19.5 to find that

$$\iint\limits_{S} [P(x, y, z)\mathbf{k}] \cdot \mathbf{N} \, dS \qquad\qquad\qquad (6)$$

$$= \iint\limits_{S_2} [P(x, y, z)\mathbf{k}] \cdot \mathbf{N}_2 \, dS + \iint\limits_{S_1} [P(x, y, z)\mathbf{k}] \cdot \mathbf{N}_1 \, dS$$

$$= \iint\limits_{Q} P(x, y, g_2(x, y)) \, dx \, dy + \iint\limits_{Q} -P(x, y, g_1(x, y)) \, dx \, dy$$

$$= \iint\limits_{Q} \{P(x, y, z)]_{z=g_1(x, y)}^{z=g_2(x, y)}\} \, dx \, dy$$

$$= \iint\limits_{Q} \left\{ \int_{g_1(x, y)}^{g_2(x, y)} \frac{\partial}{\partial z} P(x, y, z) \, dz \right\} dx \, dy$$

$$= \iiint\limits_{R} \frac{\partial}{\partial z} P(x, y, z) \, dx \, dy \, dz.$$

(Note that we have changed the order of integration from $dz \, dx \, dy$ to $dx \, dy \, dz$.) That is,

$$\iint\limits_{S} (P\mathbf{k}) \cdot \mathbf{N} \, dS = \iiint\limits_{R} \frac{\partial P}{\partial z} \, dx \, dy \, dz. \qquad (7)$$

Similarly, if S can be written as in Figure 7.4, we can show that

$$\iint\limits_{S} (N\mathbf{j}) \cdot \mathbf{N} \, dS = \iiint\limits_{R} \frac{\partial N}{\partial y} \, dx \, dy \, dz, \qquad (8)$$

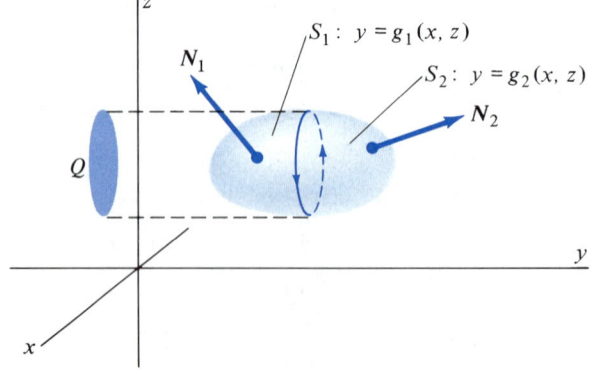

Figure 7.4 $S = S_1 \cup S_2$.

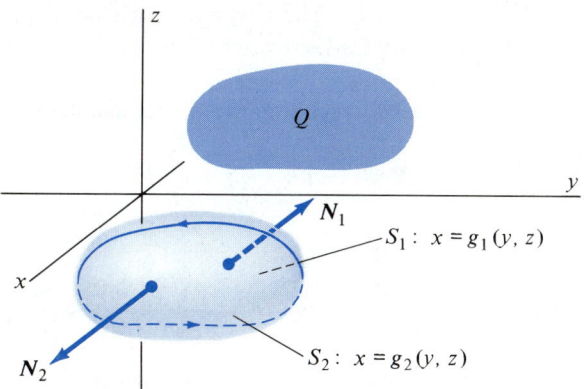

Figure 7.5 $S = S_1 \cup S_2$.

and if S can be written as in Figure 7.5, we obtain

$$\iint\limits_{S} (M\boldsymbol{i}) \cdot \boldsymbol{N} \, dS = \iiint\limits_{R} \frac{\partial M}{\partial x} \, dx \, dy \, dz. \tag{9}$$

Adding the corresponding sides of equations (7), (8), and (9), using equation (5), we obtain the equation

$$\iint\limits_{S} \boldsymbol{F} \cdot \boldsymbol{N} \, dS = \iiint\limits_{R} \left(\frac{\partial M}{\partial x} + \frac{\partial N}{\partial y} + \frac{\partial P}{\partial z} \right) dx \, dy \, dz. \tag{10}$$

Comparing equations (3) and (10) leads directly to the following definition.

DEFINITION 7

The **divergence of the vector field**

$$\boldsymbol{F}(x, y, z) = M(x, y, z)\boldsymbol{i} + N(x, y, z)\boldsymbol{j} + P(x, y, z)\boldsymbol{k}, \tag{11}$$

written div $\boldsymbol{F}$, is the scalar function

$$\operatorname{div} \boldsymbol{F}(x, y, z) = \frac{\partial}{\partial x} M(x, y, z) + \frac{\partial}{\partial y} N(x, y, z) + \frac{\partial}{\partial z} P(x, y, z). \tag{12}$$

With this definition of div $\boldsymbol{F}$, the preceding discussion shows that equation (3) holds for smooth closed surfaces that are simultaneously x-simple, y-simple, and z-simple. This is our desired generalization of the planar equation (1). A more general and precisely stated version of equation (3) is our final result.

THEOREM 6
Divergence Theorem

Let S be a closed piecewise smooth surface enclosing the region R. Let $\boldsymbol{F}(x, y, z) = M(x, y, z)\boldsymbol{i} + N(x, y, z)\boldsymbol{j} + P(x, y, z)\boldsymbol{k}$ be a vector field for which all of the components, together with all their first partial derivatives, are continuous throughout R. Then

$$\iint\limits_{S} \boldsymbol{F} \cdot \boldsymbol{N} \, dS = \iiint\limits_{R} \operatorname{div} \boldsymbol{F}(x, y, z) \, dx \, dy \, dz \tag{13}$$

where $\boldsymbol{N}$ is the outward unit normal to S.

REMARK 1: The Divergence Theorem was first discovered by the German mathematician Carl Friedrich Gauss and is often referred to as Gauss' Theorem.

REMARK 2: Using the operator notation

$$\nabla = \frac{\partial}{\partial x}i + \frac{\partial}{\partial y}j + \frac{\partial}{\partial z}k$$

we can write

$$\text{div } F = \nabla \cdot F = \frac{\partial M}{\partial x} + \frac{\partial N}{\partial y} + \frac{\partial P}{\partial z}.$$

With this notation, the Divergence Theorem may be written either as

$$\iint_S F \cdot N \, dS = \iiint_R \nabla \cdot F \, dV, \qquad dV = dx \, dy \, dz \tag{14}$$

or as

$$\iint_S F \cdot N \, dS = \iiint_R \left(\frac{\partial M}{\partial x} + \frac{\partial N}{\partial y} + \frac{\partial P}{\partial z} \right) dx \, dy \, dz. \tag{15}$$

The appeal of equation (14) is that it expresses Theorem 6 in what is referred to as "coordinate free" form. On the other hand, equation (15) is the most explicit form for use when working in Cartesian (rectangular) coordinates.

REMARK 3: Our proof of the Divergence Theorem addressed only the case of a surface enclosing a region which was simultaneously x-simple, y-simple, and z-simple. Figures 7.6 and 7.7 illustrate how this proof may be extended to include

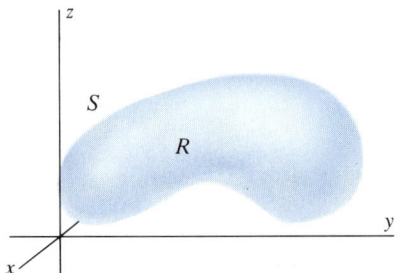

Figure 7.6 A region R that is not y-simple.

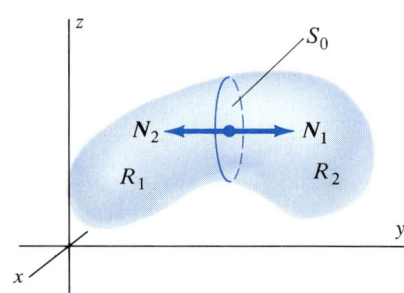

Figure 7.7 R_1 and R_2 are both y-simple. N_j is the outward unit normal for R_j, $j = 1, 2$.

more general regions. By constructing the surface S_0, we partition the region R into two regions, R_1 and R_2, both of which are simultaneously x-simple, y-simple, and z-simple. Letting S_1 denote the boundary of R_1 and S_2 the boundary of R_2, the previous proof shows that

$$\iint_{S_1} F \cdot N \, dS = \iiint_{R_1} \text{div } F \, dx \, dy \, dz \tag{16}$$

and

$$\iint\limits_{S_2} \mathbf{F} \cdot \mathbf{N} \, dS = \iiint\limits_{R_2} \operatorname{div} \mathbf{F} \, dx \, dy \, dz. \tag{17}$$

On the face S_0, the outward unit normal to S_1 is the *opposite* of the outward unit normal to S_2. Thus, the contributions over the face S_0 to the surface integrals in (16) and (17) cancel. Adding the corresponding sides of (16) and (17) then gives

$$\iint\limits_{S} \mathbf{F} \cdot \mathbf{N} \, dS = \iint\limits_{S_1} \mathbf{F} \cdot \mathbf{N} \, dS + \iint\limits_{S_2} \mathbf{F} \cdot \mathbf{N} \, dS$$

$$= \iiint\limits_{R_1} \operatorname{div} \mathbf{F} \, dx \, dy \, dz + \iiint\limits_{R_2} \operatorname{div} \mathbf{F} \, dx \, dy \, dz$$

$$= \iiint\limits_{R} \operatorname{div} \mathbf{F} \, dx \, dy \, dz.$$

The same argument can be applied to more complicated surfaces.

Before proceeding to examples, let's nail down a bit more precisely what the Divergence Theorem says about fluid flow. Of course, the left-hand side of equation (13) is the **net flux** (rate out minus rate in) of the fluid outward across the surface S. However, the volume integral on the right-hand side of (13) can be thought of as a sum taken throughout R of the product div $\mathbf{F} \, dV$, where dV is an infinitesimal element of volume containing the point (x, y, z). It is this local property of div $\mathbf{F}$ that we wish to understand better.

To do so, let $\Delta V_\epsilon = \dfrac{4}{3}\pi\epsilon^3$ be the volume of a small sphere S_ϵ of radius ϵ and center (x_0, y_0, z_0) contained within S. According to Theorem 6

$$\text{Flux of } \mathbf{F} \text{ out of } S_\epsilon = \iiint\limits_{R_\epsilon} \operatorname{div} \mathbf{F} \, dx \, dy \, dz \tag{18}$$

where R_ϵ is the region enclosed by the sphere S_ϵ. We now proceed with an argument analogous to that of Section 19.6 for the interpretation of curl $\mathbf{F}$. The mean value property for triple integrals guarantees the existence of a point $(x_\epsilon, y_\epsilon, z_\epsilon)$ in R_ϵ for which

$$\iiint\limits_{R_\epsilon} \operatorname{div} \mathbf{F} \, dx \, dy \, dz = \operatorname{div} \mathbf{F}(x_\epsilon, y_\epsilon, z_\epsilon) \, \Delta V_\epsilon. \tag{19}$$

Combining (18) and (19) we conclude that

$$\operatorname{div} \mathbf{F}(x_\epsilon, y_\epsilon, z_\epsilon) = \frac{\text{Flux of } \mathbf{F} \text{ out of } S_\epsilon}{\Delta V_\epsilon}. \tag{20}$$

Since $(x_\epsilon, y_\epsilon, z_\epsilon) \to (x_0, y_0, z_0)$ as $\epsilon \to 0$, we conclude that

$$\operatorname{div} \mathbf{F}(x_0, y_0, z_0) = \lim_{\epsilon \to 0} \frac{\text{Flux of } \mathbf{F} \text{ out of } S_\epsilon}{\Delta V_\epsilon}. \tag{21}$$

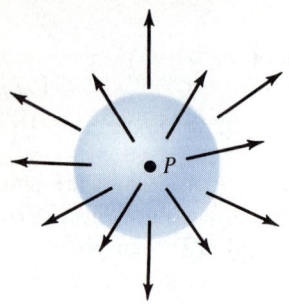

Figure 7.8 div $F > 0$; P is a source.

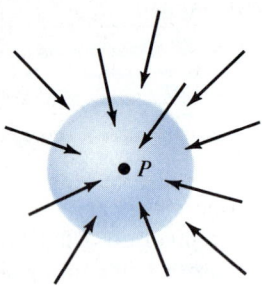

Figure 7.9 div $F < 0$; P is a sink.

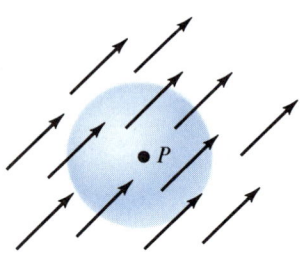

Figure 7.10 div $F = 0$; F is *incompressible* at P.

In other words, div $F(x_0, y_0, z_0)$ is the **flux per unit volume** at the point (x_0, y_0, z_0). Thus,

(i) If div $F(x_0, y_0, z_0) > 0$, more fluid is flowing out across a small sphere centered at (x_0, y_0, z_0) than is flowing in. In this case we say that the fluid is **expanding** at (x_0, y_0, z_0), or that (x_0, y_0, z_0) is a **source** for the vector field F (see Figure 7.8).

(ii) If div $F(x_0, y_0, z_0) < 0$, more fluid is flowing into a small sphere centered at (x_0, y_0, z_0) than is flowing out. The fluid is said, therefore, to be **contracting** at (x_0, y_0, z_0), and (x_0, y_0, z_0) is called a **sink** for F (see Figure 7.9).

(iii) If div $F(x_0, y_0, z_0) = 0$, the amount of fluid flowing out of the small sphere centered at (x_0, y_0, z_0) equals the amount flowing in. In this case the fluid is said to be **incompressible** at (x_0, y_0, z_0) (see Figure 7.10).

The powerful statement of the Divergence Theorem is that knowledge of the local property div $F(x, y, z)$, summed throughout the interior of S, determines the flux of F across the surface S, and conversely.

Example 1

Calculate the flux of the vector field

$$F(x, y, z) = (x - y^2)i + 3yj + (x^3 - z)k$$

outward over the sphere $S = \{(x, y, z) \mid x^2 + y^2 + z^2 = 9\}$.

Solution: Here

$$\text{div } F = \frac{\partial}{\partial x}(x - y^2) + \frac{\partial}{\partial y}(3y) + \frac{\partial}{\partial z}(x^3 - z)$$
$$= 1 + 3 - 1$$
$$= 3.$$

By the Divergence Theorem,

$$\text{Flux across } S = \iint_S F \cdot N \, dS = \iiint_R 3 \cdot dx \, dy \, dz = 3 \cdot \frac{4}{3}\pi \cdot 3^3 = 108\pi,$$

since the volume of the sphere of radius r is $\frac{4}{3}\pi r^3$. ◇

Example 2

Evaluate the surface integral

$$\iint_S F \cdot N \, dS$$

where $F(x, y, z) = xyi + xzj + yzk$ and N is the outward unit normal to the ellipsoid $S = \{(x, y, z) \mid x^2 + 4y^2 + z^2 = 1\}$.

Solution: We use the Divergence Theorem rather than trying to evaluate the surface integral directly. We have

$$\text{div } F = \frac{\partial}{\partial x}(xy) + \frac{\partial}{\partial y}(xz) + \frac{\partial}{\partial z}(yz) = 2y.$$

Letting R denote the region enclosed by the ellipsoid, we obtain

$$\iint\limits_{S} \mathbf{F} \cdot \mathbf{N} \, dS = \iiint\limits_{R} 2y \, dx \, dy \, dz$$

$$= \int_{-1}^{1} \int_{-(1/2)\sqrt{1-x^2}}^{(1/2)\sqrt{1-x^2}} \int_{-\sqrt{1-x^2-4y^2}}^{\sqrt{1-x^2-4y^2}} 2y \, dz \, dy \, dx.$$

Now this iterated integral can be evaluated directly. However, since the integrand satisfies the equation

$$\text{div } \mathbf{F}(x, -y, z) = -2y = -\text{div } \mathbf{F}(x, y, z),$$

and since the ellipsoid S is symmetric about the plane $y = 0$, the value of this integral is zero. $\diamond$

Example 3

Calculate the value of the surface integral $\displaystyle\iint\limits_{S} \mathbf{F} \cdot \mathbf{N} \, dS$ where $\mathbf{F}(x, y, z) = x^2 y \mathbf{i} +$

$xy\mathbf{j} + y^2 z^3 \mathbf{k}$, S is the cube formed by the planes $x = \pm 1$, $y = \pm 1$, and $z = \pm 1$, and $\mathbf{N}$ is the outward unit normal on S.

Solution: Rather than calculate the surface integral directly for each of the six faces, we use the Divergence Theorem. Since

$$\text{div } \mathbf{F} = \frac{\partial}{\partial x}(x^2 y) + \frac{\partial}{\partial y}(xy) + \frac{\partial}{\partial z}(y^2 z^3) = 2xy + x + 3y^2 z^2$$

we have, from Theorem 6,

$$\iint\limits_{S} \mathbf{F} \cdot \mathbf{N} \, dS = \int_{-1}^{1} \int_{-1}^{1} \int_{-1}^{1} (2xy + x + 3y^2 z^2) \, dx \, dy \, dz$$

$$= \int_{-1}^{1} \int_{-1}^{1} \left\{ \left[x^2 y + \frac{1}{2}x^2 + 3y^2 z^2 x \right]_{x=-1}^{x=1} \right\} dy \, dz$$

$$= \int_{-1}^{1} \int_{-1}^{1} 6y^2 z^2 \, dy \, dz$$

$$= \int_{-1}^{1} 2y^3 z^2 \Big]_{y=-1}^{y=1} \, dz$$

$$= \int_{-1}^{1} 4z^2 \, dz$$

$$= \frac{8}{3}. \qquad \diamond$$

Example 4

Find the flux of the vector field

$$\mathbf{F}(x, y, z) = y^2 z^3 \mathbf{i} + 4x^2 yz^2 \mathbf{j} + x^2 z^3 \mathbf{k}$$

outward across the unit sphere $S = \{(x, y, z) \mid x^2 + y^2 + z^2 = 1\}$.

Solution: Let R be the unit ball $R = \{(x, y, z) \mid x^2 + y^2 + z^2 \leq 1\}$. We have

$$\text{div } \mathbf{F} = \frac{\partial}{\partial x}(y^2 z^3) + \frac{\partial}{\partial y}(4x^2 y z^2) + \frac{\partial}{\partial z}(x^2 z^3) = 7x^2 z^2.$$

According to the Divergence Theorem,

$$\text{Flux across } S = \iint_S \mathbf{F} \cdot \mathbf{N} \, dS = \iiint_R 7x^2 z^2 \, dx \, dy \, dz.$$

To evaluate the triple integral, we use spherical coordinates:

$$x = \rho \cos\theta \sin\phi \qquad 0 \leq \rho \leq 1,$$
$$y = \rho \sin\theta \sin\phi \qquad 0 \leq \theta \leq 2\pi,$$
$$z = \rho \cos\phi \qquad 0 \leq \phi \leq \pi.$$

We obtain

$$\iiint_R 7x^2 z^2 \, dx \, dy \, dz = 7\int_0^1 \int_0^{2\pi} \int_0^{\pi} [\rho \cos\theta \sin\phi]^2 [\rho \cos\phi]^2 \rho^2 \sin\phi \, d\phi \, d\theta \, d\rho$$

$$= 7\int_0^1 \int_0^{2\pi} \int_0^{\pi} \rho^6 \cos^2\theta \sin^3\phi \cos^2\phi \, d\phi \, d\theta \, d\rho$$

$$= 7\int_0^1 \int_0^{2\pi} \int_0^{\pi} \rho^6 \cos^2\theta [1 - \cos^2\phi]\cos^2\phi \sin\phi \, d\phi \, d\theta \, d\rho$$

$$= 7\int_0^1 \int_0^{2\pi} \left\{ \rho^6 \cos^2\theta \left[-\frac{1}{3}\cos^3\phi + \frac{1}{5}\cos^5\phi \right]_0^{\pi} \right\} d\theta \, d\rho$$

$$= 7 \cdot \frac{4}{15}\int_0^1 \int_0^{2\pi} \rho^6 \cos^2\theta \, d\theta \, d\rho$$

$$= 7 \cdot \frac{4}{15} \cdot \pi \int_0^1 \rho^6 \, d\rho$$

$$= \frac{4\pi}{15}. \qquad \diamond$$

The Divergence Theorem finds wide application in physics, engineering, and applied mathematics. We have restricted our discussion to the example of fluid flow to keep the discussion straightforward and to allow you to focus primarily on the mathematics. The subject area of electricity and magnetism makes considerable use of each of the major theorems presented in this chapter (and involves another theorem due to Gauss).

Exercise Set 19.7

In Exercises 1–6, find div $\mathbf{F}$.

1. $\mathbf{F}(x, y, z) = x^2 \mathbf{i} + y^2 \mathbf{j} + z^2 \mathbf{k}$

2. $\mathbf{F}(x, y, z) = x^2 z \mathbf{i} + y^2 x \mathbf{j} + x z^2 \mathbf{k}$

3. $\mathbf{F}(x, y, z) = x \mathbf{i} + x y \mathbf{j} + x y z \mathbf{k}$

4. $\mathbf{F}(x, y, z) = (y - x)\mathbf{i} + (y - z)\mathbf{j} + (x - y)\mathbf{k}$

5. $\mathbf{F}(x, y, z) = \cos xy \mathbf{i} + e^{xyz}\mathbf{j} + y \sin(xz)\mathbf{k}$

6. $\mathbf{F}(x, y, z) = x \operatorname{Tan}^{-1} y \mathbf{i} - \sqrt{yz}\, \mathbf{j} - z \sec y \mathbf{k}$

7. Find the flux of the vector field $\mathbf{F}(x, y, z) = z \mathbf{i} + x \mathbf{j} + y \mathbf{k}$ out of the ellipsoid $x^2 + 9y^2 + 4z^2 = 1$.

8. Find the value of the surface integral $\displaystyle\iint_S \mathbf{F} \cdot \mathbf{N} \, dS$ where $\mathbf{F}$

is the vector field $F(x, y, z) = -y^2i + xj + zk$, S is the sphere $x^2 + y^2 + z^2 = 5$ and N is the outward unit normal.

9. Find the flux of the vector field $F(x, y, z) = 3xi - z^2yj + 2zk$ outward across the rectangular box with vertices $(1, 0, 0)$, $(1, 3, 0)$, $(-2, 0, 0)$, $(-2, 3, 0)$, $(1, 0, 5)$, $(1, 3, 5)$, $(-2, 0, 5)$, and $(-2, 3, 5)$.

10. Find the value of the surface integral $\iint\limits_{S} F \cdot N \, dS$ where F is the vector field $F(x, y, z) = 3xi - 4yj + 5zk$, and V is the volume of the solid enclosed by the smooth surface S.

11. Find the flux of the vector field $F(x, y, z) = 2xyi + z^2yj + xzk$ over the cube formed by the coordinate planes and the planes $x = 1$, $y = 1$, and $z = 1$.

12. Find the value of the surface integral $\iint\limits_{S} F \cdot N \, dS$ where $F(x, y, z) = (x + e^y)i + (e^{xz} - y)j + (xy + z)k$, S is the cylinder $\{(x, y, z) \mid x^2 + y^2 = 4, 0 \le z \le 2\}$ and N is the outward unit normal.

13. Find the flux of the vector field $F(x, y, z) = x^2y^2i + xy^3j + xyk$ outward across the cylinder $\{(x, y, z) \mid x^2 + y^2 = 4, 0 \le z \le 2\}$. (*Hint:* Use cylindrical coordinates.)

14. Find the value of the surface integral $\iint\limits_{S} F \cdot N \, dS$ where F is the vector field $F(x, y, z) = x^2yi + xy^2j + xyzk$, S is the surface of the quarter cylinder $C = \{(r, \theta, z) \mid 0 \le r \le 1, 0 \le \theta \le \pi/2, 0 \le z \le 1\}$.

15. Find the flux of the vector field $F(x, y, z) = x^3i + xz^2j + x^2zk$ outward across the sphere $S = \{(x, y, z) \mid x^2 + y^2 + z^2 = 4\}$. (Use spherical coordinates.)

16. Find the value of the surface integral $\iint\limits_{S} F \cdot N \, dS$ where F is the vector field $F(x, y, z) = x^2yi + xy^2j + xyzk$, S is the unit sphere $x^2 + y^2 + z^2 = 1$ and N is the outward unit normal. (Use spherical coordinates.)

17. Find the flux of $F(x, y, z) = 6xi - yj + 4k$ across the ellipsoid $x^2 + 4y^2 + z^2 = 4$. (Calculate the volume of the ellipsoid as a volume of revolution.)

18. Verify the Divergence Theorem for the vector field $F(x, y, z) = xi + yj + zk$ and the closed surface S of the cylinder $\{(x, y, z) \mid x^2 + y^2 \le 1, 0 \le z \le 2\}$.

In Exercises 19–21, verify the stated identities for differentiable vector fields $F(x, y, z)$ and $G(x, y, z)$.

19. $\nabla \times (F + G) = \nabla \times F + \nabla \times G$

20. $\nabla \cdot (F + G) = \nabla \cdot F + \nabla \cdot G$

21. $\nabla \cdot (F \times G) = (\nabla \times F) \cdot G - (\nabla \times G) \cdot F$

The **Laplacian of the scalar field** $\phi = \phi(x, y, z)$ is defined by the equation

$$\nabla^2\phi = \nabla \cdot (\nabla\phi) = \frac{\partial^2\phi}{\partial x^2} + \frac{\partial^2\phi}{\partial y^2} + \frac{\partial^2\phi}{\partial z^2}.$$

The equation $\dfrac{\partial^2\phi}{\partial x^2} + \dfrac{\partial^2\phi}{\partial y^2} + \dfrac{\partial^2\phi}{\partial z^2} = 0$ is called **Laplace's equation.** Functions which satisfy Laplace's equation are called **harmonic functions.**

22. Show that if $F = \nabla\phi$, then ϕ is harmonic if and only if div $F = 0$.

23. Determine which of the following vector fields are gradients of harmonic scalar functions.
    a. $F(x, y, z) = yzi + xzj + xyk$
    b. $F(x, y, z) = 2xye^zi + x^2e^zj + x^2ye^zk$
    c. $F(x, y, z) = x\sin(yz)i + z\cos yzj + y\cos yzk$

24. For the inverse square field $F(r) = \dfrac{r}{|r|^3}$, $|r| \ne 0$, show that div $F = 0$.

25. For the inverse square field of Exercise 24 show that $\iint\limits_{S} F \cdot N \, dS = 0$ if S is a surface containing a region R, as in the statement of the Divergence Theorem, so that $(0, 0, 0) \notin S \cup R$.

26. Let S be a sphere with center $(0, 0, 0)$. Show that $\iint\limits_{S} F \cdot N \, dS = 4\pi$ where F is the inverse square field of Exercise 24, and S is the outward unit normal. Why does this not contradict Theorem 6?

27. Show that $\iint\limits_{S} F \cdot N \, dS = 0$ if the vector field F is incompressible for all $(x, y, z) \in S$. (S, F, and N are as in Theorem 6.)

28. Show that if $\iint\limits_{S} F \cdot N \, dS > 0$, then S must contain at least one source in its interior. (S, F, and N are as in Theorem 6.) What if $\iint\limits_{S} F \cdot N \, dS < 0$?

29. True or false? If $\iint\limits_{S} F \cdot N \, dS = 0$, S may contain neither sources nor sinks. Explain.

30. Let F, S, N, and R satisfy the hypotheses of Theorem 6 in a region Ω. If $\iint\limits_{S} F \cdot N \, dS = 0$ for all closed surfaces S within Ω must $F \equiv 0$ for all $(x, y, z) \in \Omega$? Why or why not?

SUMMARY OUTLINE OF CHAPTER 19

◆ A **vector field** on $\mathbb{R}^3$ is a function (page 890)

$$F(x, y, z) = M(x, y, z)\mathbf{i} + N(x, y, z)\mathbf{j} + P(x, y, z)\mathbf{k}.$$

◆ The **work** done by the force field F in moving an object along a curve $C = \{r(t) \mid a \le t \le b\}$ is (page 896)

$$W = \int_C F \cdot dr = \int_a^b F(r(t)) \cdot r'(t)\, dt.$$

◆ The **line integral** of F over $C = \{r(t) = x(t)\mathbf{i} + y(t)\mathbf{j} + z(t)\mathbf{k} \mid a \le t \le b\}$ is (page 899)

$$\int_C F \cdot dr = \int_a^b F(r(t)) \cdot r'(t)\, dt$$

$$= \int_a^b [M(x(t), y(t), z(t))x'(t) + N(x(t), y(t), z(t))y'(t) + P(x(t), y(t), z(t))z'(t)]\, dt$$

$$= \int_C M\, dx + N\, dy + P\, dz.$$

◆ The **line integral** of the scalar function f **with respect to arc length** over the path C (above) is (page 904)

$$\int_C f(x, y, z)\, ds = \int_a^b f(x(t), y(t), z(t))\sqrt{[x'(t)]^2 + [y'(t)]^2 + [z'(t)]^2}\, dt.$$

◆ *Theorem 1:*   Under appropriate hypotheses, if $F = \nabla\phi$ and $C = \{r(t) \mid a \le t \le b\}$, then (page 907)

$$\int_C F \cdot dr = \phi(r(b)) - \phi(r(a)).$$

◆ *Theorem 2:* $\int_C F \cdot dr$ is independent of path if and only if F is **conservative** (i.e., $F = \nabla\phi$ for some scalar (page 910)
potential ϕ).

◆ *Green's Theorem:*   Let C be a simple closed path, oriented counterclockwise, that encloses Q. Then (page 915)

$$\int_C M\, dx + N\, dy = \iint_Q \left(\frac{\partial N}{\partial x} - \frac{\partial M}{\partial y} \right) dA.$$

◆ The **surface integral** of the scalar function f over the surface $S = \{(x, y, g(x, y)) \mid (x, y) \in Q\}$ is (page 925)

$$\iint_S f(x, y, z)\, dS = \iint_Q f(x, y, g(x, y))\sqrt{\left(\frac{\partial g}{\partial x}\right)^2 + \left(\frac{\partial g}{\partial y}\right)^2 + 1}\, dx\, dy.$$

◆ The **flux** of the vector field F across the surface S in the direction indicated by the unit normal N is (page 933)

$$\text{Flux} = \iint_S F \cdot N\, dS.$$

◆ The **curl** of the vector field $F = M\mathbf{i} + N\mathbf{j} + P\mathbf{k}$ is the **vector** curl (page 939)

$$F = \nabla \times F = \left(\frac{\partial P}{\partial y} - \frac{\partial N}{\partial z}\right)\mathbf{i} + \left(\frac{\partial M}{\partial z} - \frac{\partial P}{\partial x}\right)\mathbf{j} + \left(\frac{\partial N}{\partial x} - \frac{\partial M}{\partial y}\right)\mathbf{k}.$$

◆ *Stokes' Theorem:*   If the surface S is bounded by the curve C, then (page 941)

$$\int_C F \cdot dr = \iint_S (\text{curl } F) \cdot N\, dS.$$

◆ The **divergence** of the vector field $F = M\mathbf{i} + N\mathbf{j} + P\mathbf{k}$ is the **scalar** (page 951)

$$\text{div } F = \frac{\partial M}{\partial x} + \frac{\partial N}{\partial y} + \frac{\partial P}{\partial z}.$$

◆ **Divergence Theorem (Gauss):**   If the surface S encloses the region R,   (page 951)

$$\iint\limits_{S} \mathbf{F} \cdot \mathbf{N} \, dS = \iiint\limits_{R} \text{div } \mathbf{F} \, dV.$$

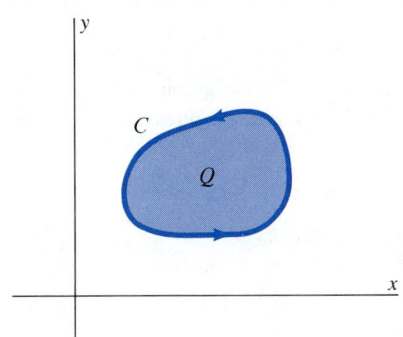

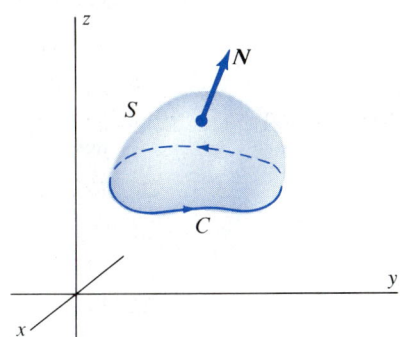

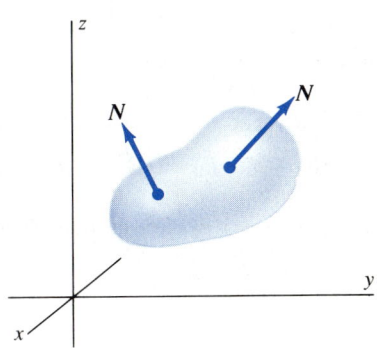

Green's Theorem

$$\int_{C} M \, dx + N \, dy = \iint\limits_{Q} \left(\frac{\partial N}{\partial x} - \frac{\partial M}{\partial y} \right) dA$$

Stokes' Theorem

$$\int_{C} \mathbf{F} \cdot d\mathbf{r} = \iint\limits_{S} (\text{curl } \mathbf{F}) \cdot \mathbf{N} \, dS$$

Divergence Theorem

$$\iint\limits_{S} \mathbf{F} \cdot \mathbf{N} \, dS = \iiint\limits_{R} \text{div } \mathbf{F} \, dV$$

REVIEW EXERCISES—CHAPTER 19

1. Compute the gradient vector field of the given function.
   a. $f(x, y, z) = y(\cos x)\ln z$
   b. $f(x, y, z) = x^3yz^2 + \sin xy$

2. Calculate the work done by the force field $\mathbf{F}(x, y) = x^2\mathbf{i} + 2xy\mathbf{j}$ when a particle is moved once around the triangle with vertices $(-1, 2)$, $(1, 0)$, and $(3, 2)$.

3. Show that the force field

   $$\mathbf{F}(x, y) = \frac{y}{x^2 + y^2}\mathbf{i} - \frac{x}{x^2 + y^2}\mathbf{j} \text{ is conservative.}$$

4. Show that curl $\mathbf{F} = \mathbf{0}$ where $\mathbf{F}$ is the force field in Exercise 3.

5. Let C be the unit circle. Show that $\iint\limits_{C} \mathbf{F} \cdot d\mathbf{r} \neq 0$ where $\mathbf{F}$ is the vector field in Exercise 3. Does this result together with the result of Exercise 4 contradict Stokes' Theorem? Why or why not?

6. For $\mathbf{F}(x, y, z) = xy\mathbf{i} + (y^2 - x^2)\mathbf{j} + x^2z^2\mathbf{k}$ find
   a. curl $\mathbf{F}$ **b.** div $\mathbf{F}$

7. Is the vector field $\mathbf{F}(x, y, z) = y^2z\mathbf{i} + 2xyz\mathbf{j} + xy^2\mathbf{k}$ conservative? Why or why not?

8. Show that $\iint\limits_{S} \mathbf{F} \cdot \mathbf{N} \, dS = 0$ if S is the unit sphere and $\mathbf{F}$ is the vector field $\mathbf{F}(\mathbf{r}) = \mathbf{r}_0$ where $\mathbf{r}_0$ is constant.

9. Find the flux of the vector field $\mathbf{F}(x, y, z) = zx\mathbf{i} + xy\mathbf{j} - x^2z\mathbf{k}$ outward over the tetrahedron with vertices $(0, 0, 0)$, $(0, 1, 0)$, $(1, 0, 0)$, and $(0, 0, 1)$.

10. Show that the vector field $\mathbf{F}(x, y, z) = x\mathbf{i} + y^2\mathbf{j} + z^3\mathbf{k}$ is irrotational.

11. Let $f(x, y, z)$ be a differentiable scalar function. Show that curl $(\nabla f) = \mathbf{0}$.

12. Let $\mathbf{F}(x, y, z)$ be a differentiable vector field. Show that div(curl $\mathbf{F}$) = 0.

13. True or false? If curl $\mathbf{F} = \mathbf{0}$ then $\mathbf{F} = \nabla\phi$ for some differentiable scalar function $\phi(x, y, z)$. Explain.

14. Find the value of the surface integral $\iint\limits_{S} \mathbf{F} \cdot \mathbf{N} \, dS$ where
    $\mathbf{F}(x, y, z) = xy\mathbf{i} + x^2z\mathbf{j} + xz\mathbf{k}$, S is the unit sphere and $\mathbf{N}$ is the outward unit normal.

15. Let $\mathbf{F}(x, y, z) = ax\mathbf{i} + by\mathbf{j} + cz\mathbf{k}$. Find the flux of $\mathbf{F}$ over the closed surface S in terms of the volume V of the region enclosed by S.

16. Find $\int_{C} \mathbf{F} \cdot d\mathbf{r}$ where $\mathbf{F}(x, y) = x\mathbf{i} - 3xy\mathbf{j}$ and C is the square with vertices $(0, 0)$, $(0, 2)$, $(2, 2)$, $(0, 2)$ oriented counterclockwise.

17. Find $\int_C F \cdot dr$ where $F(x, y) = x^2 i - xy j$ and C is the upper unit semicircle from $(1, 0)$ to $(-1, 0)$.

18. Find $\int_{(1, 2)}^{(3, 2)} F \cdot dr$ where $F(x, y) = 4x^3 y^2 i + 2x^4 y j$.

In Exercises 19–23, evaluate the given line integral.

19. $\int_C xz \, dx + yz \, dy + z \, dz$ where C is parameterized by $r(t) = t i + j + \sin t k$, $0 \le t \le \pi$.

20. $\int_C y \, dx + x \, dy$, C is the triangle with vertices $(0, 0)$, $(4, 0)$, and $(2, 4)$, taken in that order.

21. $\int_C F \cdot dr$, $F(x, y, z) = 2y i - x j + 3z^2 k$, $r(t) = t i - 3t j + k$, $0 \le t \le 1$.

22. $\int_C F \cdot dr$, $F(x, y) = e^{2y} i + \cos \pi x j$, C is the triangle with vertices $(0, 0)$, $(1, 0)$, and $(0, 1)$ taken in that order.

23. $\int_C F \cdot dr$, $F(x, y) = xy^2 i + x^3 y j$, C is the upper unit semicircle taken counterclockwise.

In Exercises 24–28, evaluate the given surface integral.

24. $\iint_S (x^2 + 2) \, dS$, S is the portion of the plane $x + y + z = 6$ lying inside the cylinder $x^2 + y^2 = 9$.

25. $\iint_S (x + y) \, dS$, S is the portion of the plane $x - 2y + z = 8$ lying above the square with vertices $(0, 0, 0)$, $(1, 0, 0)$, $(1, 1, 0)$, $(0, 1, 0)$.

26. $\iint_S (x^2 + y^2 + 1) \, dS$, S is the part of the plane $z = x + 4$ inside the cylinder $x^2 + y^2 = 4$.

27. $\iint_S x^2 z \, dS$, S is the portion of the cone $z^2 = x^2 + y^2$ lying between the planes $z = 1$ and $z = 4$.

28. $\iint_S \sqrt{x^2 + y^2} \, dS$, S is the part of the graph of $z = 2xy$ lying inside the cylinder $x^2 + y^2 = 1$.

In Exercises 29–33, find the flux $\iint_S F \cdot N \, dS$ of the given vector field outward over the given surface.

29. $F(x, y, z) = xy i + y j + 2z k$, S is the graph of the paraboloid $z = x^2 + y^2$ for $1 \le z \le 4$, N is the downward unit normal.

30. $F(x, y, z) = y i + x j$, S is the unit sphere $x^2 + y^2 + z^2 = 1$.

31. $F(x, y, z) = y i + x j$, S is the cylinder $x^2 + y^2 = 1$, $0 \le z \le 1$, including the top and bottom discs.

32. $F(x, y, z) = (x + y^3) i + (z^2 - y) j + (x^3 - y^2) k$, S is the unit sphere $x^2 + y^2 + z^2 = 1$.

33. $F(x, y, z) = ax i + by j + cz k$, S is the graph of $z = 4 - x^2 - y^2$, $z \ge 0$, N is the upward unit normal.

In Exercises 34–37, evaluate the given line integral.

34. $\int_C F \cdot dr$, $F(x, y, z) = (x + y) i + (z - y) j + z k$, C is the circle $x^2 + y^2 = 4$ oriented counterclockwise.

35. $\int_C F \cdot dr$, $F(x, y, z) = 2y^2 i - 3z j + 2x k$, C is the triangle with vertices $(0, 0, 0)$, $(2, 4, 2)$, and $(0, 4, 0)$, taken in that order.

36. $\int_C F \cdot dr$, $F(x, y, z) = 3y i - z j + 4x k$, C is the boundary of the part of the unit sphere $x^2 + y^2 + z^2 = 1$ lying in the first quadrant with orientation induced by the outward unit normal to the sphere.

37. $\int_C F \cdot dr$, $F(x, y, z) = (x^3 - y) i + (\cos y - x) j + \sin z k$, C is the triangle with vertices $(0, 0, 0)$, $(4, 2, 0)$, and $(2, 4, 2)$, taken in that order.

38. Let $F(x, y, z) = x^3 i + z j + y k$. Find the work done by F on an object that moves from $(1, 0, 0)$ to $(-1, 0, \pi)$ along a straight line.

39. Find the work done by F in Exercise 38 if the motion is along the helix $r(t) = \cos t i + \sin t j + t k$.

Chapter 20
Differential Equations

The purpose of this chapter is to indicate briefly how the theory and techniques of the calculus can be used to solve certain types of ordinary differential equations.

For a comprehensive treatment of the subject of differential equations, you are referred to courses and texts with this title. The treatment here, however, should enable you to appreciate the use made of differential equations in elementary courses in science and engineering.

20.1 FIRST ORDER LINEAR DIFFERENTIAL EQUATIONS

Recall that a differential equation is an equation involving one or more derivatives of an unknown function. Two examples of differential equations are

$$\frac{dy}{dt} = ky \tag{1}$$

and

$$f''(t) + 4f'(t) + f(t) = \cos t. \tag{2}$$

By a *solution* of a differential equation on an interval I we mean a function that is differentiable on I as many times as the equation requires, and that satisfies the equation. For example, the function $y = e^{kt}$ is a solution of equation (1) on $I = (-\infty, \infty)$ because $y = e^{kt}$ is differentiable for all $t \in (-\infty, \infty)$ and

$$\frac{dy}{dt} = \frac{d}{dt}(e^{kt}) = k(e^{kt}) = ky, \qquad -\infty < t < \infty.$$

The *general solution* of a differential equation is a function involving one or more constants that is a solution of the differential equation and from which all particular solutions can be obtained by properly specifying the choices for these constants. For example, the general solution of equation (1) is $y = Ce^{kt}$, since it can be shown that all solutions of equation (1) are of this form.

The *order* of a differential equation is the order of the highest derivative in the equation. Thus, equation (1) is a first order equation and equation (2) is a second order equation.

In Section 5.3 we showed that a first order differential equation of the form

$$\frac{dy}{dt} = \frac{f(t)}{g(y)}$$

may be solved by the method of *separation of variables*. This involves rewriting the equation as

$$g(y) \cdot \frac{dy}{dt} = f(t)$$

and "integrating both sides:"

$$\int g(y)\left(\frac{dy}{dt}\right) dt = \int f(t) \, dt$$

or

$$\int g(y) \, dy = \int f(t) \, dt.$$

In the remaining part of this section we show how this technique, and the method of integrating factors, may be used to solve first order linear equations of the general form

$$\frac{dy}{dt} + p(t)y = g(t)$$

where the functions p and g (which may be constants, including zero) are continuous on the interval of interest.

Separation of Variables

First, suppose we arbitrarily add a constant b to the right side of equation (1). We obtain the differential equation

$$\frac{dy}{dt} = ky + b.$$

By letting $a = -\dfrac{b}{k}$ we can write this equation as

$$\frac{dy}{dt} = k(y - a). \tag{3}$$

The interpretation of equation (3) is that the rate of increase or decrease of y is proportional to the *difference* between y and the constant a. Before turning to practical examples of this type, we observe that equation (3) may be solved by the method of *separation of variables*. The following example illustrates the use of this method.

Example 1

Find a solution of the differential equation

$$\frac{dy}{dt} = 2 - ay, \qquad a \neq 0. \tag{4}$$

Strategy

Separate variables.

Solution

Recall that **separating variables** means isolating y terms on one side of the equation and t terms on the other. To do so we must assume $2 - ay \neq 0$. We then divide both sides by $(2 - ay)$ and multiply both sides by dt to obtain

$$\frac{dy}{2 - ay} = dt.$$

Integrate both sides.

Integrating both sides then gives

$$\int \frac{dy}{2 - ay} = \int dt,$$

so

(Only one constant of integration need be displayed.)

$$-\frac{1}{a} \ln |2 - ay| = t + C \qquad (C \text{ arbitrary})$$

or

$$\ln |2 - ay| = -at - aC.$$

Thus

Solve for y by applying the exponential function to both sides.

$$|2 - ay| = e^{(-at - aC)}$$
$$2 - ay = \pm e^{-at - aC}$$
$$-ay = -2 \pm e^{-at - aC}$$
$$y = \frac{2}{a} + Ae^{-at}, \qquad A = \pm \frac{1}{a} e^{-aC} \tag{5}$$

A is determined by initial conditions.

where the constant A is to be determined from an initial condition. For example, if $y(0) = 1$ we obtain

$$1 = \frac{2}{a} + Ae^0, \qquad \text{so} \qquad A = \left(1 - \frac{2}{a}\right).$$

Verify that assumption on $2 - ay$ holds for solutions obtained.

To check that our assumption $2 - ay \neq 0$ holds, notice that from (5)

$$2 - ay = 2 - a\left[\frac{2}{a} + Ae^{-at}\right]$$
$$= -aAe^{-at},$$

which is not zero unless $A = 0$. However, if $A = 0$ equation (5) still gives the valid solution $y = \frac{2}{a}$. Thus, in all cases equation (5) represents a solution of (4). ◇

Mixing Problems

Examples 2 and 3 illustrate how differential equations of the above type arise as models in certain ''mixing'' problems.

Example 2

A tank in which chocolate milk is being mixed contains a mixture of 460 liters of milk and 40 liters of chocolate syrup initially. Syrup and milk are then added to the tank at the rates of 2 liters per minute of syrup and 8 liters per minute of milk. Simultaneously, the mixture is withdrawn at the rate of 10 liters per minute. Assuming perfect mixing of the milk and syrup, find the function giving the amount of syrup in the tank at time t.

Strategy

Label variables.

Use fact that $\dfrac{dy}{dt}$ is a *rate* to find a differential equation satisfied by y.

Find y by solving this differential equation.

Determine the constant A by applying *initial conditions*.

State the solution.

Verify that restrictions on y are met.

Solution

If y represents the amount of syrup in the tank, $\dfrac{dy}{dt}$ is the rate at which the amount of syrup is changing. This rate is the rate at which syrup enters the tank (2 liters per minute) minus the rate at which syrup leaves the tank $\left(\dfrac{y}{500} \times 10 \text{ liters per minute} \right)$. The function y therefore satisfies the differential equation

$$\frac{dy}{dt} = 2 - .02y.$$

This is a particular form of equation (4) treated in Example 1, with $a = .02$. From (5) the solution is

$$y = Ae^{-.02t} + 100.$$

To determine A we apply the initial condition that $y = 40$ when $t = 0$. We obtain the equation

$$40 = A + 100,$$

so

$$A = -60.$$

The desired solution is therefore

$$y = 100 - 60e^{-.02t} \text{ liters.} \tag{6}$$

Note that the requirement of Example 1 that

$$y \neq \frac{2}{a} = \frac{2}{.02} = 100$$

is met since $60e^{-.02t} > 0$ for all t. The graph of the solution appears in Figure 1.1 and values for $y(t)$ are listed in Table 1.1. ◇

Table 1.1 Calculated values of $y(t)$ in Figure 1.1

t	0	1	2	3	5	10	20	100	200
$y = 100 - 60e^{-.02t}$	40	41.2	42.4	43.5	45.7	50.9	59.8	91.9	98.9

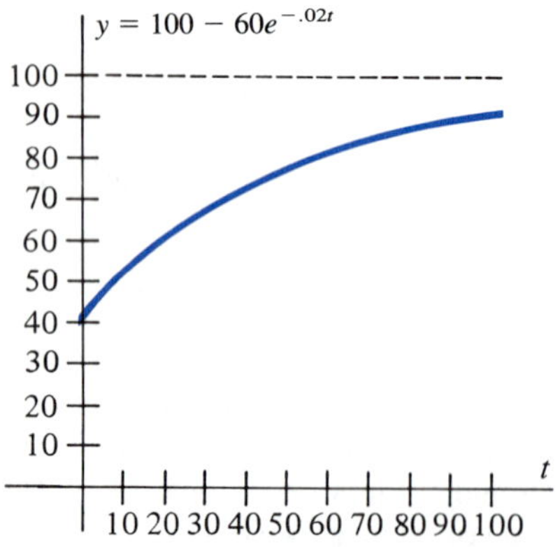

Figure 1.1 Solution of mixing problem in Example 2.

REMARK: Figure 1.1 illustrates an important property of solutions of differential equations of the form of equation (3), namely, that the solution approaches a horizontal asymptote (constant limit) as $t \to \infty$. For the solution y of Example 2 it is easy to see that

$$\lim_{t \to \infty} y = \lim_{t \to \infty} (100 - 60e^{-.02t}) = 100,$$

as Figure 1.1 suggests. This asymptote $y = 100$ is called the **steady state** part of the solution, while the term $60e^{-.02t}$ is referred to as the **transient** part of the solution. The result of Example 2 is precisely what intuition suggests—regardless of the initial concentration of syrup, the steady state concentration will be $100/500 = .2$, since 2 liters of syrup enter for every 8 liters of milk.

Example 3

Newton's law of cooling states that the rate at which an object gains or loses heat is proportional to the difference in temperature between the object and its surroundings. A bottle of soda at temperature 6°C is removed from a refrigerator and placed in a room at temperature 22°C. If, after 10 minutes, the temperature of the soda is 14°C, find its temperature after 20 minutes according to Newton's law.

Strategy

Label variables.

Interpret Newton's law as a differential equation for y.

Separate variables.

Integrate.

Apply exponential function to both sides.

Solve for y.

Use initial data to find A.

Use additional data to find k.

$$\ln(e^{-10k}) = -10k = \ln\left(\frac{1}{2}\right)$$
$$= -\ln 2,$$

Solution

Let $y(t)$ denote the temperature of the soda t minutes after it is removed from the refrigerator. Newton's law of cooling states that

$$\frac{dy}{dt} = k \cdot (22 - y).$$

Thus

$$\frac{dy}{22 - y} = k \, dt,$$

so

$$\int \frac{dy}{22 - y} = \int k \, dt.$$

Integrating on both sides we obtain

$$-\ln|22 - y| = kt + C$$
$$|22 - y| = e^{-kt-C}$$
$$22 - y = \pm e^{-kt-C}.$$

Thus

$$y = 22 - Ae^{-kt}, \qquad A = \pm e^{-C}. \tag{7}$$

To find A we apply the initial data* that $y_0 = 6°$ when $t = 0$. Inserting this information in (7) gives

$$6 = 22 - Ae^0,$$

so

$$A = 16. \tag{8}$$

To find k we use (7), (8), and the data $y(10) = 14$. We obtain the equation

$$14 = 22 - 16e^{-10k}.$$

Thus

$$e^{-10k} = \frac{-8}{-16} = \frac{1}{2},$$

so

$$k = \frac{\ln 2}{10}.$$

---

*Initial data refer to information about a function at *any* value of the independent variable, not necessarily $t = 0$.

so

$$k = \frac{-\ln 2}{-10} = \frac{\ln 2}{10}.$$

Find $y(20)$

$$e^{\ln 2^{-2}} = 2^{-2} = \frac{1}{4}.$$

Table 1.2

t	y in (9)
0	6.00
4	9.87
8	12.81
12	15.04
16	16.72
20	18.00
24	18.97
28	19.70
32	20.26
36	20.68

The function y is now completely determined as

$$y = 22 - 16e^{-\left(\frac{\ln 2}{10}\right)t}. \tag{9}$$

The temperature after 20 minutes is therefore

$$\begin{aligned}
y(20) &= 22 - 16e^{-\left(\frac{\ln 2}{10}\right)20} \\
&= 22 - 16e^{-2\ln 2} \\
&= 22 - 16e^{\ln 2^{-2}} \\
&= 22 - 16 \cdot \frac{1}{4} \\
&= 18°.
\end{aligned}$$

(See Table 1.2 and Figure 1.2.)   ◇

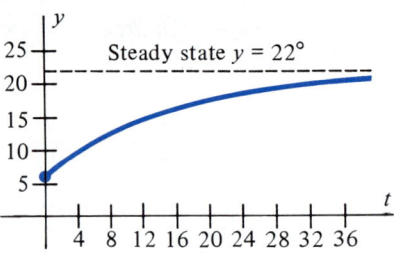

Figure 1.2 Graph of $y = 22 - 16e^{-\left(\frac{\ln 2}{10}\right)t}$.

Integrating Factors

A second generalization of equation (1) is to linear first order differential equations of the form

$$\frac{dy}{dt} + p(t)y = g(t) \tag{10}$$

where p and g are continuous functions of t. The differences between equation (10) and equation (1) involve these two functions. In equation (1) we have $p(t) \equiv -k$, and $g(t) \equiv 0$.

We may find solutions to equations of the form (10) if we let $u(t) = \int p(t)\, dt$ and note that *if we multiply both sides of* (10) *by the integrating factor*

$$e^{u(t)} = e^{\int p(t)\, dt} \tag{11}$$

it is easy to find the integral of the left side of equation (10). (In (11) the expression $\int p(t)\, dt$ denotes any *particular* antiderivative for p.) To demonstrate our remark we note that

$$\left[\frac{dy}{dt} + p(t)y\right] \cdot e^{u(t)} = \frac{dy}{dt}e^{\int p(t)\, dt} + yp(t)e^{\int p(t)\, dt} \tag{12}$$

$$= \frac{d}{dt}[ye^{\int p(t)\,dt}]$$

$$= \frac{d}{dt}[y \cdot e^{u(t)}].$$

To find a solution of the differential equation (10), we begin by multiplying both sides by an integrating factor e^u in (11) to obtain

$$\left[\frac{dy}{dt} + p(t)y\right]e^{u(t)} = g(t)e^{u(t)}.$$

Using (12) we can write this equation as

$$\frac{d}{dt}[ye^{u(t)}] = g(t)e^{u(t)},$$

so

$$ye^{u(t)} = \int g(t)e^{u(t)}\,dt + C$$

or

$$y = e^{-u(t)}\left\{\int g(t)e^{u(t)}\,dt + C\right\}. \tag{13}$$

The function y in (13) is the **general solution** of equation (10) because, as is shown in more advanced courses, every solution of (10) has the form (13) for some choice of the constant C.

Rather than remembering formula (13), you should simply remember that solving a differential equation of the form (10) requires the integrating factor $e^{u(t)} = e^{\int p(t)\,dt}$.

Example 4

Find the general solution of the differential equation

$$\frac{dy}{dt} + 2ty = t. \tag{14}$$

Solution: This equation has the form (10) with $p(t) = 2t$, $g(t) = t$. We therefore use the integrating factor

$$e^{u(t)} = e^{\int 2t\,dt} = e^{t^2}.$$

Multiplying both sides of (14) by e^{t^2} gives

$$e^{t^2} \cdot \frac{dy}{dt} + e^{t^2} \cdot 2t \cdot y = te^{t^2}.$$

We next rewrite the left side as a derivative

$$\frac{d}{dt}[e^{t^2} \cdot y] = e^{t^2} \cdot \frac{dy}{dt} + e^{t^2} \cdot 2t \cdot y$$

and integrate both sides:

$$e^{t^2}y = \int te^{t^2}\, dt$$

$$= \frac{1}{2}e^{t^2} + C.$$

Finally, we multiply both sides by e^{-t^2} to obtain

$$y = e^{-t^2}\left[\frac{1}{2}e^{t^2} + C\right]$$

$$= Ce^{-t^2} + \frac{1}{2}. \qquad \diamond$$

Example 5

Find the solution of $\dfrac{dy}{dt} - \dfrac{2}{t}y = t^3$, $t > 0$ satisfying the initial condition $y(1) = 2$.

Strategy

Find an integrating factor.

Solution

We use the integrating factor

$$e^{u(t)} = e^{\int -2/t\, dt} = e^{-2\ln t} = e^{\ln t^{-2}} = \frac{1}{t^2}.$$

Multiply both sides of the equation by the integrating factor.

We then obtain

$$\frac{1}{t^2} \cdot \frac{dy}{dt} - \frac{2}{t^3} \cdot y = t^3\left(\frac{1}{t^2}\right).$$

We can write this equation as

Write the left side as a derivative.

$$\frac{d}{dt}\left[\frac{1}{t^2}y\right] = t,$$

so

Integrate both sides.

$$\frac{1}{t^2}y = \int t\, dt = \frac{1}{2}t^2 + C.$$

Solve for y.

Thus

$$y = t^2\left[\frac{1}{2}t^2 + C\right]$$

$$= \frac{1}{2}t^4 + Ct^2.$$

Apply initial conditions to solve for C.

The initial condition $y(1) = 2$ gives

$$2 = \frac{1}{2} + C,$$

so

$$C = 2 - \frac{1}{2} = \frac{3}{2}.$$

The desired solution is

$$y(t) = \frac{1}{2}t^4 + \frac{3}{2}t^2.$$

◇

Exercise Set 20.2

In Exercises 1–4, find a particular solution of the initial value problem. Identify the steady-state and transient parts of the solution.

1. $\dfrac{dy}{dt} = y + 1$ $y(0) = 1$

2. $\dfrac{dy}{dt} = 2y + 2$ $y(0) = -1$

3. $\dfrac{dy}{dt} = -y + \pi$ $y(0) = \pi/2$

4. $\dfrac{dy}{dx} + y = 2$ $y(0) = 2$

In Exercises 5–14 find the general solution of the differential equation.

5. $y' + 2y = 4$

6. $y' + ay = b$

7. $y' - \dfrac{2}{t}y = t^4$

8. $y' = \dfrac{1}{t^2}y$

9. $y' - 2y = e^{2t} \sin t$

10. $y' = 2ty + 3t$

11. $y' + y = e^t$

12. $ty' - 2y = -2t$

13. $ty' + y = t$

14. $y' = y - 2e^{-t}$

15. Use the method of separation of variables to find a solution of the differential equation $\dfrac{dy}{dx} = axy$ where a is constant and $y > 0$.

16. A curve in the xy-plane has the property that at each point (x, y) the slope of the tangent equals xy, the product of the coordinates. Find an equation for such a curve containing the point $(0, 1)$ and sketch its graph. (*Hint:* Use the result of Exercise 15.)

17. A tank contains 200 liters of brine. The initial concentration of salt is 0.5 kilogram per liter of brine. Fresh water is then added at a rate of 3 liters per minute and the brine is drawn off from the bottom of the tank at the same rate. Assume perfect mixing.
   a. Find a differential equation for the number of kilograms of salt, $p(t)$, in the solution at time t.
   b. How much salt is present after 20 minutes?
   c. What is the steady state concentration of salt?

18. A bottle of water whose temperature is 26°C is placed in a room at temperature 12°C. If the temperature of the water after 30 minutes is 20°C, find its temperature after 2 hours.

19. A person initially places \$100 in a savings account that pays interest at a rate of 10% per year compounded continuously. If the person makes additional deposits of \$100 per year continuously throughout each year, find:
   a. a differential equation for $P(t)$, the amount on deposit after t years, and
   b. the amount on deposit after 7 years.

20. An electrical circuit contains a battery supplying E volts, a resistance R, and an inductance L. When the battery is connected, the resulting current I satisfies the differential equation

$$L\frac{dI}{dt} + RI = E, \qquad I(0) = 0.$$

   a. Find the solution I as a function of t.
   b. Obtain Ohm's Law, $I = E/R$, as the steady state part of the solution.

21. If, in the circuit described in Exercise 20, the inductance is replaced by a capacitance C, the equation becomes

$$R\frac{dI}{dt} + \frac{I}{C} = \frac{dE}{dt}.$$

   Find the solution $I(t)$, if the battery is switched off (i.e., $dE/dt = 0$ at the time $t = 0$.)

22. In the chemical reaction called inversion of raw sugar, the inversion rate is proportional to the amount of raw sugar remaining. If 100 kilograms of raw sugar is reduced to 75 kilograms in 6 hours, how long will it be until
   a. half the raw sugar has been inverted?
   b. 90% of the raw sugar has been inverted?

23. A pan of boiling water is removed from a burner and cools to 80°C in 3 minutes. Find its temperature after 10 minutes, according to Newton's law of cooling, if the surrounding temperature is 30°C.

24. A room containing 1000 cu ft of air is initially free of pollutants. Air containing 100 parts of pollutants per cu ft is pumped into the room at a rate of 50 cu ft per minute. The air in the room is kept perfectly mixed, and air is removed from the room at the same rate of 50 cu ft per minute. Find the function $P(t)$ giving the number of parts of pollutants per cubic foot in the room air after t minutes.

25. A tank contains a mixture of 500 gallons of salt brine (salt and water) initially containing 2 pounds of salt per gallon. A second mixture, containing 6 pounds of salt per gallon, is then pumped in at a rate of 10 gallons per minute. The resulting mixture is drawn off at the same rate. Assuming perfect mixing, find the function that describes the concentration of salt (pounds per gallon) in the mixture in the tank t minutes after mixing begins.

26. Influenza spreads through a university community at a rate proportional to the product of the proportion of those already infected and the proportion of those not yet infected. Assuming that those infected remain infected, find the proportion $P(t)$ infected after t days if initially 10% were infected and after 3 days 30% were infected.

27. An annuity is initially set up containing $10,000 to which interest is compounded continuously at a rate of 10% per year. The owner of the annuity withdraws $2000 per year in a continuous manner. When does the value of the annuity reach zero?

28. Certain biological populations grow in size according to the Gompertz growth equation

$$\frac{dy}{dt} = -ky \ln y$$

where k is a positive constant. Find the general solution of this differential equation.

29. The Weber-Fechner law in psychology is a model for the rate of change of the reaction y to a stimulus of strength s that is given by the differential equation

$$\frac{dy}{ds} = k\frac{y}{s}$$

where k is a positive constant. Find the general solution of this differential equation.

20.2 EXACT EQUATIONS

Certain types of first order differential equations may be solved by a technique discussed in Chapter 19 for vector functions. These are *exact* differential equations.

DEFINITION 1

Let the functions M and N be defined in an open rectangle R in $\mathbb{R}^2$. The differential equation

$$M(x, y) + N(x, y)\frac{dy}{dx} = 0, \qquad (x, y) \in R \tag{1}$$

is called *exact* if there exists a function f of two variables for which

$$M(x, y) = \frac{\partial f}{\partial x}(x, y) \text{ and } N(x, y) = \frac{\partial f}{\partial y}(x, y)$$

for all $(x, y) \in R$.

Example 1

The differential equation

$$e^{2y} + 2xe^{2y}\frac{dy}{dx} = 0$$

is exact since, for the function $f(x, y) = xe^{2y}$,

$$M(x, y) = e^{2y} = \frac{\partial}{\partial x}(xe^{2y}) = \frac{\partial f}{\partial x}$$

and

$$N(x, y) = 2xe^{2y} = \frac{\partial}{\partial y}(xe^{2y}) = \frac{\partial f}{\partial y}.$$

The general solution is therefore determined by the level curves for the function $f(x, y) = xe^{2y}$:

$$xe^{2y} = C.$$ ◇

Example 2

The differential equation

$$(3x^2 + 2y^2)\, dx + 4xy\, dy = 0$$

is exact, since the function $f(x, y) = x^3 + 2xy^2$ has the property that

$$M(x, y) = 3x^2 + 2y^2 = \frac{\partial}{\partial x}(x^3 + 2xy^2) = \frac{\partial f}{\partial x}$$

and

$$N(x, y) = 4xy = \frac{\partial}{\partial y}(x^3 + 2xy^2) = \frac{\partial f}{\partial y}.$$

The general solution is therefore

$$x^3 + 2xy^2 = C,$$

or

$$y = \pm\left(\frac{C - x^3}{2x}\right)^{1/2}.$$ ◇

Since Definition 1 may be paraphrased by saying that equation (1) is exact if the function f is a potential for the vector function $\boldsymbol{F}(x, y) = M(x, y)\boldsymbol{i} + N(x, y)\boldsymbol{j},$ Theorem 10 of Chapter 18 provides the following criterion for equation (1) to be exact.

THEOREM 1

Let the functions M and N have continuous partial derivatives in an open rectangle R in $\mathbb{R}^2$. The differential equation

$$M(x, y) + N(x, y)\,\frac{dy}{dx} = 0$$

is exact in R if and only if

$$\frac{\partial M}{\partial y}(x, y) = \frac{\partial N}{\partial x}(x, y)$$

for all (x, y) in R.

Example 3

The differential equation

$$xy^3\, dx + x^3y\, dy = 0$$

is not exact, since

$$\frac{\partial M}{\partial y}(x, y) = \frac{\partial}{\partial y}(xy^3) = 3xy^2 \neq 3x^2y = \frac{\partial}{\partial x}(x^3y) = \frac{\partial N}{\partial x}(x, y).$$ ◇

Example 4

Test the differential equation

$$(3x^2y + \cos x)\,dx + x^3\,dy = 0$$

for exactness. If it is exact, find a solution passing through the point $(1, \pi)$.

Solution: Here

$$M(x, y) = 3x^2y + \cos x; \qquad N(x, y) = x^3.$$

Since both M and N have continuous partial derivatives throughout $\mathbb{R}^2$, we may apply Theorem 1 to test for exactness. Since

$$\frac{\partial M}{\partial y}(x, y) = 3x^2 = \frac{\partial N}{\partial x}(x, y),$$

the equation is exact.

To solve the differential equation, we must find a potential f. To do so we integrate $M(x, y)$ partially with respect to x to find

$$f(x, y) = \int M(x, y)\,dx = \int (3x^2y + \cos x)\,dx$$

$$= x^3y + \sin x + h_1(y) + K_1.$$

Also,

$$f(x, y) = \int N(x, y)\,dy = \int x^3\,dy = x^3y + h_2(x) + K_2.$$

Equating these two expressions for f then gives

$$h_1(y) \equiv 0, \qquad h_2(x) = \sin x, \qquad \text{and} \qquad K_1 = K_2 = K.$$

Thus

$$f(x, y) = x^3y + \sin x + K$$

is the desired potential for any constant K. We therefore choose $K = 0$ and obtain the family of level curves

$$x^3y + \sin x = C \tag{2}$$

as the general solution for the differential equation. To find the particular curve passing through the point $(x_0, y_0) = (1, \pi)$, we set $x = 1$ and $y = \pi$ in (2) to find

$$1^3\,(\pi) + \sin(\pi) = \pi = C,$$

so $C = \pi$. A solution passing through $(1, \pi)$ is therefore determined by the equation

$$x^3y + \sin x = \pi$$

as

$$y = \frac{\pi - \sin x}{x^3}, \qquad x^3 \neq 0. \qquad\qquad \diamond$$

Example 4 demonstrates the general technique for solving exact equations. Once the equation is shown to be exact by use of Theorem 1, the solution $f(x, y) = C$ is found by integrating $M = \dfrac{\partial f}{\partial x}$ partially with respect to x, integrating $N = \dfrac{\partial f}{\partial y}$ partially with respect to y, and equating the resulting expressions for f. This is the same technique that was used in Chapter 17 to reconstruct a potential from its gradient.

Example 5

Solve the differential equation

$$\cos y \, dx + (3y^2 - x \sin y) \, dy = 0.$$

Solution: Writing the equation in the form

$$\cos y + (3y^2 - x \sin y) \, \frac{dy}{dx} = 0$$

and letting

$$M(x, y) = \cos y; \qquad N(x, y) = 3y^2 - x \sin y,$$

we verify that the equation is exact by noting that

$$\frac{\partial M}{\partial y} = -\sin y = \frac{\partial N}{\partial x}.$$

To find the solution $f(x, y) = C$, we first integrate M partially with respect to x to find that

$$f(x, y) = \int M(x, y) \, dx = \int \cos y \, dx = x \cos y + g(y).$$

Next, integrating N partially with respect to y gives

$$f(x, y) = \int N(x, y) \, dy = \int (3y^2 - x \sin y) \, dy$$

$$= y^3 + x \cos y + h(x).$$

Equating these two expressions for $f(x, y)$ gives

$$x \cos y + g(y) = y^3 + x \cos y + h(x),$$

so $g(y) = y^3$ and $h(x) \equiv 0$. Thus, $f(x, y) = y^3 + x \cos y$ and the general solution is

$$y^3 + x \cos y = C.$$

$\Diamond$

Exercise Set 20.2

In Exercises 1–10, determine whether the given equation is exact. If it is, find a family of solution (level) curves of the form $f(x, y) = C$.

1. $2xy \, dy + x^2 \, dy = 0$

2. $y^2 + xy \, \dfrac{dy}{dx} = 0$

3. $\dfrac{x \, dx}{\sqrt{x^2 + y^2}} + \dfrac{y \, dy}{\sqrt{x^2 + y^2}} = 0$

4. $\cos y^2 - 2xy \sin y^2 \, \dfrac{dy}{dx} = 0$

5. $\dfrac{x}{y^2} \, dx + \dfrac{y}{x^2} \, dy = 0$

6. $\dfrac{y}{1 + x^2 y^2} + \dfrac{x}{1 + x^2 y^2} \, \dfrac{dy}{dx} = 0$

7. $(x - 2xy^2) \, dx + (y + x^2 y) \, dy = 0$

8. $(y^2 - 2xy)\,dx + (2xy - x^2)\,dy = 0$

9. $\dfrac{x\,dx}{\sqrt{x + y^2}} + \dfrac{y\,dy}{\sqrt{x + y^2}} = 0$

10. $\dfrac{1}{\sqrt{x - y^2}} - \dfrac{2y}{\sqrt{x - y^2}}\dfrac{dy}{dx} = 0$

11. Find a solution of the differential equation in Exercise 1 passing through the point (1, 3).

12. Find a solution of the differential equation in Exercise 4 passing through the point (2, 0).

13. Find a solution of the differential equation in Exercise 8 passing through the point (1, 2).

14. Show that a separable differential equation $f(x)\,dx + g(y)\,dy = 0$ is always exact. What is the "solution" function $f(x, y)$?

15. Show that the differential equation $(1 + x^2y^2 + y)\,dx + x\,dy = 0$ is not exact. Then show that if this equation is multiplied through by the integrating factor $p(x, y) = \dfrac{1}{1 + x^2y^2}$ an exact differential equation results. Is a solution of this exact equation also a solution of the original equation?

20.3 SECOND ORDER LINEAR EQUATIONS

By adding a term involving the second derivative to the first order linear differential equation $y' = ky$ for natural growth, we obtain one type of *second order linear differential equation*,*

$$\frac{d^2y}{dt^2} + a\,\frac{dy}{dt} + by = 0 \tag{1}$$

where a and b are constants. From our experience with first order equations we might suspect that equation (1) has a solution of the form $y = e^{rt}$ where r is a constant. If this were true, the derivative of this proposed solution would be

$$\frac{dy}{dt} = re^{rt}, \qquad \text{and} \qquad \frac{d^2y}{dt^2} = r^2e^{rt}.$$

Substituting into equation (1) would therefore give

$$r^2e^{rt} + a(re^{rt}) + b(e^{rt}) = 0,$$

or

$$(r^2 + ar + b)e^{rt} = 0. \tag{2}$$

Since values of the exponential function $y = e^{rt}$ are always nonzero, the only way in which equation (2) can be true is if the *characteristic polynomial* $r^2 + ar + b$ equals zero. That is, r must be a (real) *root* of the quadratic polynomial $r^2 + ar + b$.

Let r be a real number. The function $y = e^{rt}$ is a solution of the differential equation

$$\frac{d^2y}{dt^2} + a\,\frac{dy}{dt} + by = 0$$

if and only if the constant r is a real root of the characteristic polynomial

$$r^2 + ar + b = 0.$$

*More precisely, equation (1) is referred to as a second order, constant-coefficient, linear, homogeneous equation. The term homogeneous refers to the fact that the right-hand side is zero. This is but one type of second order equation. There are many others, but we shall not pursue them here.

Thus, we expect to find two, one, or zero solutions of the differential equation (1) of the form $y = e^{rt}$, depending on the number of real roots r of the associated characteristic polynomial.

Example 1

For the differential equation

$$\frac{d^2y}{dt^2} - \frac{dy}{dt} - 2y = 0$$

the constant coefficients are $a = -1$ and $b = -2$, so the characteristic polynomial is

$$r^2 - r - 2 = (r - 2)(r + 1).$$

This characteristic polynomial has real roots $r_1 = 2$ and $r_2 = -1$. Two solutions of the differential equation are therefore

$$y_1 = e^{r_1 t} = e^{2t} \qquad \text{and} \qquad y_2 = e^{r_2 t} = e^{-t},$$

as you can verify. (We shall say more about the *general* solution of this equation later.) ◇

Example 2

The differential equation

$$\frac{d^2y}{dt^2} + 4\frac{dy}{dt} + 4y = 0$$

has characteristic polynomial $r^2 + 4r + 4 = (r + 2)^2$, which has only the single real root $r = -2$. Thus, the function $y_1 = e^{-2t}$ is a solution. ◇

REMARK 1: When the characteristic polynomial for equation (1) has only the single real root r, equation (1) will have both the functions $y_1 = e^{rt}$ and $y_2 = te^{rt}$ as solutions. Thus, the function $y_2 = te^{-2t}$ is also a solution of the equation in Example 2, as you can verify.

REMARK 2: When the characteristic polynomial $r^2 + ar + b$ has the *complex** root $r_1 = \alpha + i\beta$, $\beta \neq 0$, then the complex number $r_2 = \alpha - i\beta$ is also a root, and the differential equation (1) has the two solutions

$$y_1 = e^{\alpha t} \sin \beta t \qquad \text{and} \qquad y_2 = e^{\alpha t} \cos \beta t.$$

Example 3

For the differential equation

$$\frac{d^2y}{dt^2} - 2\frac{dy}{dt} + 5y = 0$$

---

*An introduction to complex numbers is presented in Appendix II. The ideas contained in that appendix are assumed in this discussion.

the characteristic polynomial is

$$r^2 - 2r + 5.$$

The roots of this polynomial, according to the quadratic formula, are

$$r = \frac{2 \pm \sqrt{(-2)^2 - 4(1)(5)}}{2}$$

$$= 1 \pm \frac{1}{2}\sqrt{-16}$$

$$= 1 \pm 2i, \qquad i^2 = -1.$$

Thus, the roots are $r_1 = \alpha + i\beta = 1 + 2i$ and $r_2 = \alpha - i\beta = 1 - 2i$ with $\alpha = 1$ and $\beta = 2$, so two solutions are

$$y_1 = e^t \sin 2t \qquad \text{and} \qquad y_2 = e^t \cos 2t,$$

which we leave for you to verify.   ◇

The General Solution of Equation (1)

If y_1 and y_2 are any two solutions of equation (1), then so is any *linear combination* of these solutions of the form

$$y(t) = Ay_1(t) + By_2(t) \tag{3}$$

where A and B are constants. (See Exercise 36.) If the functions y_1 and y_2 are not multiples of each other, the function y is called the *general solution* of equation (1) since it can be shown that any solution can be written in this form for an appropriate choice of the constants A and B.

Example 4

(i) The general solution of the differential equation $\dfrac{d^2y}{dt^2} - \dfrac{dy}{dt} - 2y = 0$ in Example 1 is

$$y = Ae^{2t} + Be^{-t}.$$

(ii) The general solution of the differential equation $\dfrac{d^2y}{dt^2} + 4\dfrac{dy}{dt} + 4y = 0$ in Example 2 is

$$y = Ae^{-2t} + Bte^{-2t}.$$

(iii) The general solution of the differential equation $\dfrac{d^2y}{dt^2} - 2\dfrac{dy}{dt} + 5y = 0$ in Example 3 is

$$y = Ae^t \sin 2t + Be^t \cos 2t$$
$$= e^t(A \sin 2t + B \cos 2t).$$ ◇

Initial Conditions

In general, two initial conditions are required to determine a unique solution of a second order differential equation, one for the solution y and one for its derivative $\dfrac{dy}{dt}$, *specified at the same number* t_0. Since the general solution of equation (1) has the form given by equation (3), two initial conditions will produce two linear equations for the constants A and B, from which these numbers may be determined.

Example 5

Find a solution of the initial value problem consisting of the differential equation

$$y'' - 2y' - 3y = 0 \qquad (4)$$

and the initial conditions

$$y(0) = 0$$
$$y'(0) = 4.$$

Strategy

First, find the general solution by

a. finding the characteristic polynomial

b. identifying its roots r_1 and r_2

c. combining the solutions $e^{r_1 t}$ and $e^{r_2 t}$ as in equation (3).

Next, substitute initial conditions into general solution y, and its derivative y', to obtain two equations in A and B.

Solve the two equations for A and B by substitution or by elimination.

Find particular solution by substituting for A and B in general solution.

Solution

The characteristic polynomial for equation (4) is

$$r^2 - 2r - 3 = (r - 3)(r + 1),$$

which has roots $r_1 = 3$ and $r_2 = -1$. The general solution of equation (4) is therefore

$$y(t) = Ae^{3t} + Be^{-t}, \qquad (5)$$

so the derivative y' has the form

$$y'(t) = 3Ae^{3t} - Be^{-t}. \qquad (6)$$

Substituting the initial condition $y(0) = 0$ into equation (5) gives the equation

$$A + B = 0. \qquad (7)$$

Substituting the initial condition $y'(0) = 4$ into equation (6) gives the equation

$$3A - B = 4. \qquad (8)$$

Substituting $B = -A$ from equation (7) into equation (8) gives

$$3A - (-A) = 4$$

or

$$4A = 4.$$

Thus, $A = 1$ and $B = -A = -1$.

Equation (5) now becomes

$$y(t) = e^{3t} - e^{-t},$$

the desired solution.   ◇

Modelling Oscillatory Motion

A version of equation (1) that occurs frequently in applications is

$$\frac{d^2y}{dt^2} + by = 0, \qquad b > 0. \qquad (9)$$

The characteristic polynomial for equation (9) is $r^2 + b$, which has the complex roots $r_1 = \sqrt{b}i$ and $r_2 = -\sqrt{b}i$. Since two solutions of this equation are $e^{\alpha t} \sin \beta t$

and $e^{\alpha t} \cos \beta t$ with $\alpha = 0$ and $\beta = \sqrt{b}$, we have:

> The differential equation
>
> $$\frac{d^2y}{dt^2} + by = 0, \qquad b > 0 \tag{10}$$
>
> has general solution
>
> $$y = A \sin \sqrt{b}\, t + B \cos \sqrt{b}\, t. \tag{11}$$

Example 6

Find the general solution of the differential equation

$$y'' + 9y = 0.$$

Solution: Since the differential equation has the form of equation (10), with $b = 9$ the general solution is

$$y = A \sin 3t + B \cos 3t. \qquad \diamondsuit$$

Because solutions of equation (9) are made up of sine and cosine functions, versions of this equation occur in models for the motion of vibrating physical systems involving springs, pendulums, etc. Here is a typical situation.

The Harmonic Oscillator

Imagine a mass m lying on a frictionless surface and attached to a spring (see Figure 3.1). When the mass is moved x units from its natural rest position, the spring will exert a restoring force, F_s, proportional to the displacement x (Hooke's Law). Since the restoring force acts in the direction opposite to the motion of the mass, we write

$$F_s = -kx \tag{12}$$

where the constant k is called the **spring constant** and depends on the particular spring.

Newton's second law of motion states that the sum of all forces acting on the mass must equal the product of the mass m and acceleration a. Since the net force* acting on the mass after it is displaced and released is F_s, we can write Newton's law as

$$F_s = ma, \qquad a = \text{acceleration}.$$

Since the variable x represents the displacement of the mass from its rest position, the time derivative $\dfrac{dx}{dt}$ is the velocity of the mass, and the second derivative $\dfrac{d^2x}{dt^2}$ is its acceleration. Assuming the displacement x to be a twice differentiable function of t, we can rewrite Newton's law as

$$F_s = m \cdot \frac{d^2x}{dt^2}. \tag{13}$$

---

*The force of gravity is exactly counteracted by the supporting force exerted by the surface, so the net *vertical* force is zero.

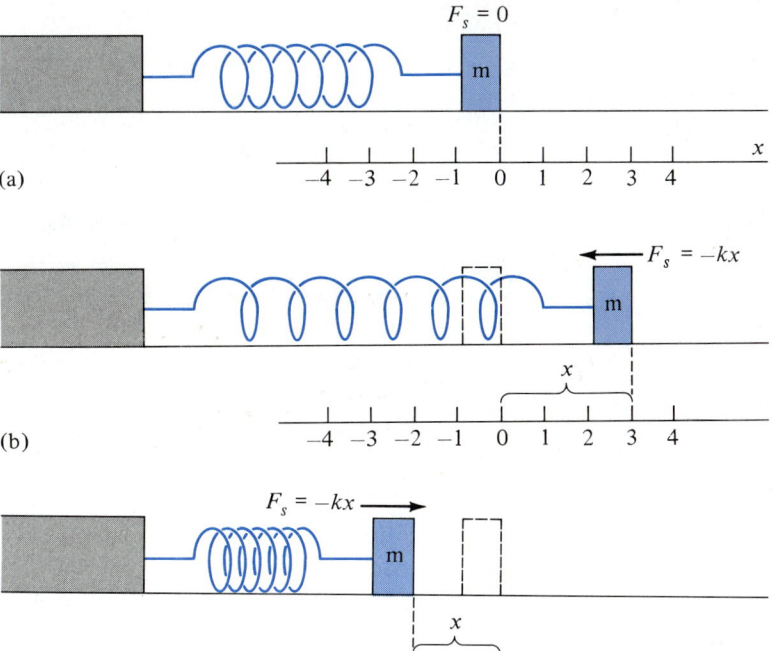

Figure 3.1 Mass-spring system (a) at equilibrium, (b) stretched, and (c) compressed. Displacement, x, is measured with respect to a particular point on the block.

Combining equations (12) and (13) now gives the differential equation

$$m \cdot \frac{d^2x}{dt^2} = -kx,$$

or

$$\frac{d^2x}{dt^2} + \frac{k}{m} \cdot x = 0. \tag{14}$$

Equation (14) is called the **harmonic oscillator equation,** and it provides the desired mathematical model for the motion of the mass for the situation described above.

The solution of equation (14) is given by equation (11):

$$x(t) = A \cos \sqrt{\frac{k}{m}} \, t + B \sin \sqrt{\frac{k}{m}} \, t. \tag{15}$$

Example 7

An 8-kg mass is attached to a spring as in Figure 3.1. The spring constant is $k = 2$ N/m. Find the resulting motion if the mass is displaced 10 cm from its equilibrium position and released from rest (zero initial velocity).

Strategy

Write differential equation (14), using given values of k and m.

Solution

Here $\dfrac{k}{m} = \dfrac{2}{8} = \dfrac{1}{4}$, so the differential equation (14) is

$$\frac{d^2x}{dt^2} + \frac{1}{4}x = 0.$$

Write general solution from (15).

From (15), with $\sqrt{k/m} = 1/2$, the general solution is

$$x(t) = A \cos \frac{t}{2} + B \sin \frac{t}{2}. \tag{16}$$

Set $x(0) = 10$ cm $= \dfrac{1}{10}$ m to find one equation in A and B.

To find A and B, we apply the given *initial conditions*

(i) $x(0) = \dfrac{1}{10}$ $\left(\text{initial displacement 10 cm} = \dfrac{1}{10} \text{ m}\right)$, and

(ii) $x'(0) = 0$ (initial velocity zero).

From equation (16) and condition (i) we have

$$\frac{1}{10} = A \cos 0 + B \sin 0 = A,$$

so $A = \dfrac{1}{10}$. Since, from (16),

Differentiate $x(t)$ to obtain equation for $x'(t)$.

$$\frac{dx}{dt} = -\frac{A}{2} \sin \frac{t}{2} + \frac{B}{2} \cos \frac{t}{2}, \tag{17}$$

condition (ii) gives

Set $x'(0) = 0$ to obtain second equation in A and B.

$$0 = -\frac{A}{2} \sin 0 + \frac{B}{2} \cos 0 = \frac{B}{2}.$$

Thus, $B = 0$. The solution is therefore

$$x(t) = \frac{1}{10} \cos \frac{t}{2} \text{ (meters)}. \qquad \diamond$$

Example 8

Repeat Example 7, except that the mass now is released with an initial velocity of 3 m/s.

Strategy

Same solution as for Example 7, except that the condition $x'(0) = 0$ is replaced by the condition $x'(0) = 3$.

Solution

As in Example 7 we have

$$x(t) = A \cos \frac{t}{2} + B \sin \frac{t}{2}.$$

Also, we still have $x(0) = \dfrac{1}{10}$, so $A = \dfrac{1}{10}$.

However, the condition that the initial velocity is 3 m/s means that $x'(0) = 3$. Equation (17) therefore gives

Set $x'(0) = 3$ and solve for B.

$$3 = -\frac{A}{2} \sin 0 + \frac{B}{2} \cos 0 = \frac{B}{2},$$

so $B = 6$. In this case the solution is therefore

$$x(t) = \frac{1}{10} \cos \frac{t}{2} + 6 \sin \frac{t}{2}.$$

$\diamond$

Exercise Set 20.3

In Exercises 1–6 find the general solution of the differential equation in the form $y = Ae^{r_1t} + Be^{r_2t}$ by finding two distinct real roots of the characteristic polynomial.

1. $\dfrac{d^2y}{dt^2} + 5\dfrac{dy}{dt} + 6y = 0$ **2.** $y'' + 3y' - 4y = 0$

3. $y'' - y = 0$ **4.** $\dfrac{d^2y}{dt^2} - \dfrac{dy}{dt} - 6y = 0$

5. $\dfrac{d^2y}{dt^2} + 3\dfrac{dy}{dt} - 10y = 0$ **6.** $y'' - 4y = 0$

In Exercises 7–10 find the general solution of the differential equation in the form $y = Ae^{rt} + Bte^{rt}$ by finding the single root r of the characteristic polynomial.

7. $y'' + 2y' + 1y = 0$ **8.** $y'' - 4y' + 4y = 0$

9. $\dfrac{d^2y}{dt^2} + 6\dfrac{dy}{dt} + 9y = 0$ **10.** $y'' - 2y' + 1y = 0$

In Exercises 11–14 find the general solution of the differential equation in the form $y = A \sin \sqrt{b}\, t + B \cos \sqrt{b}\, t$.

11. $\dfrac{d^2y}{dt^2} + 4y = 0$ **12.** $y'' + 16y = 0$

13. $y'' = -5y$ **14.** $\dfrac{d^2y}{dt^2} = -25y$

In Exercises 15–24 find the general solution of the differential equation.

15. $y'' - 3y' - 10y = 0$ **16.** $y'' + 8y' + 16y = 0$

17. $y'' - 4y' + y = 0$ **18.** $y'' + 6y' + 5y = 0$

19. $y'' - 2y' + 6y = 0$ **20.** $y'' - 6y' + 9y = 0$

21. $y'' - y' - 12y = 0$ **22.** $y'' - 2y' + 3y = 0$

23. $y'' + 2y' + 4y = 0$ **24.** $y'' + 4y' + 4y = 0$

In Exercises 25–30 find the solution of the initial value problem.

25. $y'' - y' - 6y = 0$
    $y(0) = 0$
    $y'(0) = 5$

26. $y'' - 5y' + 6y = 0$
    $y(0) = 2$
    $y'(0) = 0$

27. $y'' + 2y' + y = 0$
    $y(0) = 3$
    $y'(0) = 1$

28. $y'' - 4y' + 4y = 0$
    $y(0) = 2$
    $y'(0) = 5$

29. $y'' + 9y = 0$
    $y(0) = 3$
    $y'(0) = -3$

30. $y'' + y = 0$
    $y(0) = 0$
    $y'(0) = 2$

In Exercises 31–32 find the general solution by first converting the system of two first order equations into a second order equation.

31. $\dfrac{dy}{dt} = 4x$

    $\dfrac{dx}{dt} = -y$

32. $\dfrac{dy}{dt} = 2(x - 3)$

    $\dfrac{dx}{dt} = 8(y + 5)$

33. A 0.5 kg mass is attached to a spring with spring constant $k = 2$ N/m, as in Figure 3.1. The mass is pulled 10 cm from the equilibrium position and released with zero velocity. Find an equation describing the resulting motion.

34. A 0.5 kg mass is attached to a spring with spring constant $k = 100$ N/m, as in Figure 3.1. The mass is tapped while in its equilibrium position so as to give it an initial velocity of 10 cm/s.
  a. Find a function describing the resulting motion.
  b. When will the mass return to its equilibrium point for the first time?

35. Sketch the graph of the solution of the initial value problem

$$\frac{d^2x}{dt^2} + kx = 0$$

$$x(0) = 1$$
$$x'(0) = 0$$

for $k = 0, \pm 1, \pm 4, \pm 9$. What can you conclude about this solution as
  a. $k \to \infty$?
  b. $k \to -\infty$?
  c. $k \to 0^+$?
  d. $k \to 0^-$?

36. Verify that if y_1 and y_2 are solutions of the differential equation

$$y'' + ay' + by = 0,$$

then so is any function of the form

$$y(t) = Ay_1(t) + By_2(t).$$

20.4 NONHOMOGENEOUS LINEAR EQUATIONS

In Section 20.3 we noted that if y_1 and y_2 are solutions of the linear differential equation

$$\frac{d^2y}{dt^2} + a\,\frac{dy}{dt} + by = 0, \tag{1}$$

then so is any function y of the form

$$y(t) = C_1y_1(t) + C_2y_2(t) \tag{2}$$

where C_1 and C_2 are constants.

We say that the solutions y_1 and y_2 are *linearly independent* if they are not multiples of each other. It can be shown that when y_1 and y_2 are linearly independent solutions of equation (1), the general solution of equation (1) is given by the function y in line (2).

Equation (1) is called *homogeneous* because the term on the right-hand side is simply zero.

We next wish to determine the general solution of the *nonhomogeneous* differential equation of the form

$$\frac{d^2y}{dt^2} + a\,\frac{dy}{dt} + by = f(t) \tag{3}$$

for certain types of nonzero functions f.

We first note that the nonhomogeneous equation (3) does *not* have the property that the sum of two solutions of (3) is again a solution. (See Exercise 21.) The following theorem shows how the general solution of equation (3) is related to the general solution of the associated homogeneous equation (1). Its proof is sketched in Exercise 19.

THEOREM 2

Let y_p be any particular solution of the nonhomogeneous linear differential equation (3) and let y_g be the general solution of the associated homogeneous linear equation (1). Then the general solution of the nonhomogeneous equation (3) is

$$y(t) = y_g(t) + y_p(t). \tag{4}$$

Theorem 2 says that to find the general solution of the nonhomogeneous equation (3) we must do two things. First, find *any particular* solution y_p of equation (3). Then, find the *general* solution y_g of the "homogeneous part" given by equation (1). The general solution of (3) is the sum of these two solutions.

Example 1

The nonhomogeneous equation

$$y'' + y' - 2y = -4t \tag{5}$$

has the particular solution $y_p = 2t + 1$. To verify this note that

$$y_p' = 2 \qquad \text{and} \qquad y_p'' = 0,$$

so

$$y_p'' + y_p' - 2y_p = 0 + 2 - 2(2t + 1)$$
$$= -4t$$

as required.

To find the *general* solution we next note that the associated homogeneous equation

$$y'' + y' - 2y = 0 \qquad (6)$$

has characteristic polynomial $r^2 + r - 2 = (r + 2)(r - 1)$, with roots $r_1 = -2$ and $r_2 = 1$, so the general solution of the homogeneous equation (6) is

$$y_g = C_1 e^{-2t} + C_2 e^t.$$

The general solution of equation (5) is therefore

$$y = C_1 e^{-2t} + C_2 e^t + 2t + 1$$

according to Theorem 2. $\qquad\qquad\qquad\qquad\qquad\qquad\qquad\qquad\qquad\qquad \diamond$

The Method of Undetermined Coefficients

Although Theorem 2 specifies the form of the general solution of equation (3), it does not tell us how to find the particular solution y_p required to form the general solution.

For certain types of functions f, a simple observation enables us to find y_p. The idea is simply to think about which types of functions, when differentiated and combined according to the left side of equation (3), produce the function f on the right side of equation (3). We then substitute the ''general forms'' of such functions into the equation and see what happens.

For example, in the nonhomogeneous equation

$$y'' + 3y' - 4y = e^{2t} \qquad (7)$$

we suggest that only functions of the form $y = Ae^{2t}$ are candidates for a solution since derivatives of Ae^{2t} are again of the form Ae^{2t}. With $y = Ae^{2t}$ we have

$$y' = 2Ae^{2t} \qquad \text{and} \qquad y''(t) = 4Ae^{2t},$$

so substituting into the left side of equation (7) gives

$$y'' + 3y' - 4y = 4Ae^{2t} + 3(2Ae^{2t}) - 4Ae^t \qquad (8)$$
$$= 6Ae^{2t}.$$

Comparing the right sides of equations (7) and (8), we conclude that $y = Ae^{2t}$ is a solution of equation (7) if $e^{2t} = 6Ae^{2t}$. Thus, $6A = 1$, so $A = \frac{1}{6}$. A particular solution of equation (7) is therefore $y_p(t) = \frac{1}{6}e^{2t}$, as you can verify.

By the ''general form'' for a function f, we mean the most general form of a function y which, when combined with its derivatives $\dfrac{dy}{dt}$ and $\dfrac{d^2y}{dt^2}$ according to the left side of a linear second order differential equation

$$\frac{d^2y}{dt^2} + a\frac{dy}{dt} + by = f(t), \qquad (9)$$

can produce the function f on the right side. Here is a list of the general forms associated with the functions f for which the method we are describing works best.

Function f	*General Form*
$f(t) = t$	$y = At + B$
$f(t) = t^2$	$y = At^2 + Bt + C$
$f(t) = e^{kt}$	$y = Ae^{kt}$
$f(t) = \sin kt$	$y = A \sin kt + B \cos kt$
$f(t) = \cos kt$	$y = A \sin kt + B \cos kt$

The *method of undetermined coefficients* for finding a particular solution of equation (9) is to simply insert the general form for the function f into equation (9) and attempt to solve for the "undetermined coefficients" A, B, etc.

Example 2

In the differential equation

$$y'' + 3y' + 2y = t^2 \tag{10}$$

the general form for the function $f(t) = t^2$ is

$$y = At^2 + Bt + C. \tag{11}$$

Then

$$y' = 2At + B$$

and

$$y'' = 2A.$$

Inserting these expressions in equation (10) gives

$$2A + 3(2At + B) + 2(At^2 + Bt + C) = t^2$$

or

$$2At^2 + (6A + 2B)t + (2A + 3B + 2C) = t^2.$$

We therefore obtain the system of equations

$$\begin{aligned} 2A &= 1 \quad &\text{(coefficients of } t^2) \\ 6A + 2B &= 0 \quad &\text{(coefficients of } t) \\ 2A + 3B + 2C &= 0 \quad &\text{(constants)} \end{aligned}$$

which has solution $A = \frac{1}{2}$, $B = -\frac{3}{2}$, $C = \frac{7}{4}$. A particular solution of equation (10), given by equation (11), is therefore

$$y_p(t) = \frac{1}{2}t^2 - \frac{3}{2}t + \frac{7}{4}.$$

Since the homogeneous equation

$$y'' + 3y' + 2y = 0 \tag{12}$$

has characteristic polynomial $r^2 + 3r + 2 = (r + 2)(r + 1)$ with roots $r_1 = -2$ and $r_2 = -1$, the general solution of equation (12) is

$$y_g(t) = C_1 e^{-2t} + C_2 e^{-t},$$

and the general solution of equation (10) is

$$y(t) = C_1 e^{-2t} + C_2 e^{-t} + \frac{1}{2}t^2 - \frac{3}{2}t + \frac{7}{4}. \qquad \diamond$$

Example 3

In the differential equation

$$y'' - 2y' + y = \cos t \qquad (13)$$

the general form for the function $f(t) = \cos t$ is

$$y = A \sin t + B \cos t.$$

Then

$$y' = A \cos t - B \sin t$$

and

$$y'' = -A \sin t - B \cos t.$$

Inserting these expressions into equation (13) gives

$$(-A \sin t - B \cos t) - 2(A \cos t - B \sin t) + (A \sin t + B \cos t) = \cos t$$

or

$$2B \sin t - 2A \cos t = \cos t.$$

Thus, we obtain the equations

$$\begin{aligned} 2B &= 0 && \text{(coefficients of } \sin t) \\ -2A &= 1 && \text{(coefficients of } \cos t) \end{aligned}$$

with solution $A = -\frac{1}{2}$ and $B = 0$. A particular solution of equation (13) is therefore

$$y_p(t) = -\frac{1}{2} \sin t.$$

Since the corresponding homogeneous equation

$$y'' - 2y' + y = 0 \qquad (14)$$

has characteristic polynomial $r^2 - 2r + 1 = (r - 1)^2$ with the repeated root $r = 1$, the general solution of equation (14) is

$$y_g(t) = C_1 e^t + C_2 t e^t,$$

and the general solution of equation (13) is

$$y = C_1 e^t + C_2 t e^t - \frac{1}{2} \sin t. \qquad \diamond$$

Superposition of Solutions

When the function f on the right side of the nonhomogeneous equation

$$\frac{d^2 y}{dt^2} + a \frac{dy}{dt} + by = f(t)$$

is a *linear combination* of two or more functions of the form t^n, e^{kt}, $\sin kt$ or $\cos kt$

the following theorem enables us to obtain a particular solution as a linear combination of solutions of equations involving only one of these functions on their right-hand sides.

THEOREM 3

Let y_1 be a solution of the equation

$$\frac{d^2y}{dt^2} + a\frac{dy}{dt} + by = f_1(t), \tag{15}$$

and let y_2 be a solution of the equation

$$\frac{d^2y}{dt^2} + a\frac{dy}{dt} + by = f_2(t). \tag{16}$$

Then the function

$$y(t) = \alpha y_1(t) + \beta y_2(t) \tag{17}$$

is a solution of the equation

$$\frac{d^2y}{dt^2} + a\frac{dy}{dt} + by = \alpha f_1(t) + \beta f_2(t). \tag{18}$$

Theorem 3 says that we may find a solution of equation (18) by first finding solutions y_1 and y_2 of equations (15) and (16), and then forming the linear combination $y(t) = \alpha y_1(t) + \beta y_2(t)$ of these solutions. The proof of this theorem is given in Exercise 20.

Example 4

We leave it as an exercise for you to verify that the equation

$$\frac{d^2y}{dt^2} + 4y = e^t$$

has the particular solution $y_1(t) = \frac{1}{5}e^t$ and that the equation

$$\frac{d^2y}{dt^2} + 4y = t^2$$

has the particular solution $y_2(t) = \frac{1}{4}t^2 - \frac{1}{8}$.

Thus, according to Theorem 3,

(i) the equation

$$\frac{d^2y}{dt^2} + 4y = 6e^t$$

has solution

$$y(t) = 6y_1(t) = \frac{6}{5}e^t;$$

(ii) the equation

$$\frac{d^2y}{dt^2} + 4y = -4t^2$$

has solution

$$y(t) = -4y_2(t) = -4\left(\frac{1}{4}t^2 - \frac{1}{8}\right) = -t^2 + \frac{1}{2};$$

(iii) the equation

$$\frac{d^2y}{dt^2} + 4y = 10e^t + 8t^2 \tag{19}$$

has solution

$$y = 10y_1(t) + 8y_2(t)$$
$$= 10\left(\frac{1}{5}e^t\right) + 8\left(\frac{1}{4}t^2 - \frac{1}{8}\right)$$
$$= 2e^t + 2t^2 - 1.$$

We leave it as a further exercise for you to verify that we could have proceeded to find a particular solution of equation (19) directly by beginning with the general form

$$y = Ae^t + Bt^2 + Ct + D$$

and applying the method of undetermined coefficients. ◇

A Complicating Issue

There is a situation in which the method of undetermined coefficients, as it has been described thus far, fails to produce a particular solution of the nonhomogeneous equation

$$\frac{d^2y}{dt^2} + a\frac{dy}{dt} + by = f(t). \tag{20}$$

This occurs when the function $y = f(t)$ is a solution of the *homogeneous* equation

$$\frac{d^2y}{dt^2} + a\frac{dy}{dt} + by = 0. \tag{21}$$

In this case, inserting the general form for the function $y = f(t)$ in the left side of equation (20) will simply yield the zero function, because of equation (21). In this case we modify the method of undetermined coefficients as follows:

If the function $y = f(t)$ in equation (20) is a solution of the homogeneous equation (21), we seek a particular solution of equation (20) in the form $y = tg(t)$ where $g(t)$ is the general form for the function f.

We shall not "prove" this statement, because we have described the method of undetermined coefficients only as a way to *seek* particular solutions of equation (20). The following examples give you some indication, however, as to why this technique "works."

Example 5

For the nonhomogeneous equation

$$y'' + 2y' - 3y = e^t \tag{22}$$

the corresponding homogeneous equation is

$$y'' + 2y' - 3y = 0. \tag{23}$$

The characteristic polynomial for equation (23) is

$$r^2 + 2r - 3 = (r + 3)(r - 1),$$

which has roots $r_1 = -3$ and $r_2 = 1$, so the general solution of equation (23) is

$$y_g(t) = C_1 e^{-3t} + C_2 e^t. \tag{24}$$

Thus, the function $f(t) = e^t$ on the right side of equation (22) is a solution of equation (23) (with $C_1 = 0$ and $C_2 = 1$ in equation (24)).

If we were to seek a particular solution of equation (22) using the general form

$$y = Ae^t,$$

we would have $y' = Ae^t$ and $y'' = Ae^t$, and inserting these functions in equation (22) would give the "equation"

$$Ae^t + 2Ae^t - 3Ae^t = e^t. \tag{25}$$

Since "equation" (25) simplifies to the false statement $0 = e^t$, there is no number A for which $y = Ae^t$ is a solution of equation (22).

If, however, we begin with the general form

$$y = Ate^t \tag{26}$$

we obtain

$$y' = A(1 + t)e^t \quad \text{and} \quad y'' = A(2 + t)e^t.$$

Substituting these functions in equation (22) then gives

$$A(2 + t)e^t + 2A(1 + t)e^t - 3Ate^t = e^t,$$

so

$$A(2 + 2)e^t + A(1 + 2 - 3)te^t = e^t$$

or

$$4Ae^t = e^t.$$

Thus $A = \frac{1}{4}$, and a particular solution of equation (22) is

$$y_p(t) = \frac{1}{4} te^t$$

in equation (26). From equations (24) and (26) we may conclude that the *general* solution of equation (22) is

$$y(t) = C_1 e^{-3t} + C_2 e^t + \frac{1}{4} te^t. \qquad \diamond$$

Example 6

For the nonhomogeneous equation

$$y'' + y = \sin t \tag{27}$$

the associated homogeneous equation $y'' + y = 0$ has general solution

$$y_g(t) = C_1 \sin t + C_2 \cos t.$$

To find a particular solution of the nonhomogeneous equation (27) we must therefore use the general form

$$y = t(A \sin t + B \cos t)$$
$$= At \sin t + Bt \cos t.$$

We leave it as an exercise for you to verify that substituting this general form into equation (27) leads to the conclusion $A = 0$ and $B = -\frac{1}{2}$, so a particular solution of (27) is

$$y_p(t) = -\frac{1}{2}t \cos t,$$

and the general solution of (27) is

$$y(t) = C_1 \sin t + C_2 \cos t - \frac{1}{2}t \cos t. \qquad \diamondsuit$$

Example 7

In the nonhomogeneous equation

$$y'' - 4y' + 4y = e^{2t} \tag{28}$$

the associated homogeneous equation

$$y'' - 4y' + 4y = 0 \tag{29}$$

has characteristic polynomial $r^2 - 4r + 4 = (r - 2)^2$. Since $r = 2$ is a *repeated* root, the general solution of (29) is

$$y_g(t) = C_1 e^{2t} + C_2 t e^{2t}.$$

This means that not only is the function $f(t) = e^{2t}$ a solution of (29), *but so is the function $tf(t) = te^t$.* To find a particular solution of (28) we must therefore begin with the general form obtained by multiplying by yet another factor of t:

$$y = t(Ate^{2t}) = At^2 e^{2t} \tag{30}$$

gives

$$y' = 2Ate^{2t} + 2At^2 e^{2t}$$

and

$$y'' = 2Ae^{2t} + 8Ate^{2t} + 4At^2 e^{2t},$$

so

$$y'' - 4y' + 4y = (4A - 8A + 4A)t^2 e^{2t} + (8A - 8A)te^{2t} + 2Ae^{2t}$$

and we obtain the equation $2Ae^{2t} = e^{2t}$. Thus, $A = \frac{1}{2}$ and the particular solution of (28) of the form in line (30) is

$$y_p(t) = \frac{1}{2}t^2 e^{2t},$$

and the general solution of (28) is

$$y(t) = C_1 e^{2t} + C_2 te^{2t} + \frac{1}{2}t^2 e^{2t}. \qquad \diamondsuit$$

Exercise Set 20.4

In Exercises 1–12 find the general solution y of the given second order differential equation.

1. $y'' + 2y' - y = \sin 2t$

2. $y'' - 3y' + 4y = t$

3. $y'' + y = e^{-t}$

4. $y'' - 9y = t^2$

5. $y'' - 4y = 4t$

6. $y'' + 7y' + 12y = e^{2t}$

7. $y'' - 3y' - 10y = 5e^{2t}$

8. $y'' + y = t^2$

9. $y'' + 9y = \cos 3t$

10. $y'' + 2y' + y = 4e^{-t}$

11. $y'' - 3y' - 4y = 2e^{-t}$

12. $y'' + 5y' + 6y = 3e^{-3t}$

In Exercises 13–17 use Theorem 3 and the method of undetermined coefficients to find the general solution of the differential equation.

13. $y'' + 2y' + y = 3e^t + t^2$

14. $y'' - 2y' - 3y = 2e^{2t} + 3 \sin t$

15. $y'' - 2y' + y = 2 \cos 2t + \sin 2t + 2e^{-t}$

16. $y'' + y' = t^3 + 2t^2$

17. $y'' + y' + y = t^3 + 2t^2$

18. Solve the initial value problem
$$y'' - 2y' - 3y = 2e^t - 10 \sin t$$
$$y(0) = 2$$
$$y'(0) = 4.$$

19. Prove Theorem 2 by proving the following statements.

    a. If y_p and z_p are any two particular solutions of the nonhomogeneous equation
$$y'' + ay' + by = f(t),$$
then the function $y = y_p - z_p$ is a solution of the *homogeneous* equation
$$y'' + ay' + by = 0.$$
(*Hint:* Simply substitute the function $y = y_p - z_p$ into the left side of the homogeneous equation.)

    b. Let $y_g(t) = C_1y_1(t) + C_2y_2(t)$ be the general solution for the homogeneous equation. Conclude from part **a** that for some choice of the constants C_1 and C_2,
$$z_p(t) = C_1y(t) + C_2y_2(t) + y_p(t).$$

    c. Conclude from parts **a** and **b** that *any* particular solution z_p of the nonhomogeneous equation must have the form
$$z_p = y_g + y_p$$
where y_g is the general solution of the homogeneous equation.

20. Prove Theorem 3 by simply substituting $y(t) = \alpha y_1(t) + \beta y_2(t)$ into equation (18) and verifying the equality.

21. Demonstrate that the sum $y = y_1 + y_2$ of two distinct solutions of the nonhomogeneous equation
$$y'' + ay' + by = f(t), \qquad f(t) \neq 0$$
is not again a solution of this equation. For what differential equation *is* y a solution?

22. The method of undetermined coefficients may also be applied to certain *first* order nonhomogeneous equations. Verify the following facts required in its application.

    a. If a is a constant, the general solution of the first order homogeneous equation
$$y' + ay = 0$$
is
$$y_g(t) = Ce^{-at}.$$

    b. If y_p is any particular solution of the nonhomogeneous equation
$$y' + ay = f(t),$$
the general solution of the nonhomogeneous equation is
$$y = Ce^{-at} + y_p.$$
(*Hint:* As in Exercise 19, verify that the difference of any two particular solutions is a solution of the homogeneous equation in part **a.**)

    c. Conclude that if $f(t)$ is a function of the form t^n, e^{kt}, $\sin kt$, or $\cos kt$, and the method of undetermined coefficients can be used to find a particular solution of the equation
$$y' + ay = f(t),$$
then the general solution of this equation is as in part **b.**

Use the method of undetermined coefficients and the results of Exercise 22 to find the general solution of the differential equations in Exercises 23–30.

23. $y' - 3y = 6t - 14$

24. $y' + 2y = 3t$

25. $y' - 4y = 1 - 4t^2$

26. $y' + 5y = 10$

27. $y' - 4y = 2e^{3t}$

28. $y' + 2y = 2 \sin t + 11 \cos t$

29. $y' - 3y = 4e^t + 6t - 14$

30. $y' - 2y = -4t + 8 + 5 \sin t$

In Exercises 31–34 find the solution of the initial value problem.

31. $y' + 4y = 0$
$y(0) = 3$

32. $y' - 2y = 12 - 4t$
$y(0) = 8$

33. $y' + 3y = 5e^{2t}$
$y(0) = 7$

34. $y' - 2y = 2 \sin 4t + 16 \cos 4t$
$y(0) = 8$

20.5 POWER SERIES SOLUTIONS OF DIFFERENTIAL EQUATIONS

The theory of power series provides a method for solving certain types of differential equations. For differential equations of the form

$$f(x, y, y') = 0 \tag{1}$$

the idea is to express the (unknown) solution y as a power series in the independent variable x:

$$y = \sum_{k=0}^{\infty} a_k x^k. \tag{2}$$

Then, assuming (2) has a nonzero radius of convergence, we apply the theorem on differentiating power series to conclude that

$$y' = \sum_{k=0}^{\infty} k a_k x^{k-1}. \tag{3}$$

We next insert representations (2) and (3) into the differential equation (1). For each integer k, coefficients of x^k will appear in various places in equation (1). By equating coefficients of x^k on either side of equation (1) we can often find a set of equations which allow the coefficients $a_0, a_1, a_2, a_3, \ldots$ to be determined completely. In such cases the power series representation for the solution is obtained.

Example 1

Use the power series method to solve the differential equation

$$y' = y. \tag{4}$$

Solution: We assume that $y = \sum_{k=0}^{\infty} a_k x^k$, so that $y' = \sum_{k=0}^{\infty} k a_k x^{k-1}$, and both of these series have the same radius of convergence. Inserting these expansions in the differential equation (4) gives

$$\sum_{k=0}^{\infty} k a_k x^{k-1} = \sum_{k=0}^{\infty} a_k x^k$$

or

$$a_1 + 2a_2 x + 3a_3 x^2 + \cdots + k a_k x^{k-1} + \cdots$$
$$= a_0 + a_1 x + a_2 x^2 + \cdots + a_{k-1} x^{k-1} + \cdots. \tag{5}$$

Since the two series in equation (5) are equal, the coefficients of like powers of x must be the same. Thus,

$$
\begin{array}{ll}
a_1 = a_0 & \text{(constant terms)} \\
2a_2 = a_1 & \text{(x terms)} \\
3a_3 = a_2 & \text{(x^2 terms)} \\
\quad \vdots & \\
\end{array}
$$

$$ka_k = a_{k-1} \qquad (x^{k-1} \text{ terms})$$

.

.

.

Using these equations, we may solve for each coefficient a_2, a_3, a_4, . . . in terms of the one preceding and, therefore, in terms of a_0:

$$a_1 = a_0$$

$$a_2 = \frac{1}{2}a_1 = \frac{1}{2}a_0 = \frac{1}{2!}a_0$$

$$a_3 = \frac{1}{3}a_2 = \frac{1}{3 \cdot 2}a_0 = \frac{1}{3!}a_0$$

.

.

.

$$a_k = \frac{1}{k}a_{k-1} = \frac{1}{k(k-1) \cdot \ldots \cdot 3 \cdot 2}a_0 = \frac{1}{k!}a_0. \tag{6}$$

Returning to the expansion for y, we can now write

$$y = a_0 + a_0 x + \frac{1}{2!}a_0 x^2 + \frac{1}{3!}a_0 x^3 + \cdots + \frac{1}{k!}a_0 x^k + \cdots \tag{7}$$

$$= a_0\left\{1 + x + \frac{x^2}{2!} + \frac{x^3}{3!} + \cdots + \frac{x^k}{k!} + \cdots\right\},$$

which we recognize as the power series representation for the function $y = a_0 e^x$. Since we have previously verified that this power series converges for all values of x, the solution $y = a_0 e^x$ is valid for all x. The constant a_0 is determined by an initial condition for the equation (4). ◇

REMARK: The result of Example 1 should not have been unexpected, since we determined in Chapter 8 that the solution of the differential equation $y' = y$ is $y = Ce^x$. However, we see here that the use of power series provides an entirely different approach to solving differential equations.

The equation

$$a_k = \frac{1}{k}a_{k-1}$$

on the left side of equation (6) warrants special attention. It is referred to as the **recurrence relation** for the differential equation because it gives the general formula by which each coefficient in the series expansion for the solution (except the first) may be determined from its predecessor. If you can succeed in finding a recurrence relation, you will be able to generate all coefficients in the expansion for the solution of a differential equation. However, two issues then remain to be addressed:

(1) Does the power series obtained from the recurrence relation actually converge? If so, for which values of x? You will have succeeded in finding a legitimate solution of the differential equation only if the power series has a nonzero radius of convergence.

(2) Can a closed form expression* for the power series solution be recognized, as in Example 1 where the solution was recognized as the power series for a_0e^x? The answer to this question is not critical if you are willing to accept solutions in the form of power series. In fact, many important functions in mathematical physics arise as series solutions of differential equations and do not have closed form expressions (see Example 3).

Example 2

Use the power series method to solve the initial value problem

$$\begin{cases} y' = xy \\ y(0) = 1. \end{cases}$$

Strategy

Assume a power series form for the solution y. Differentiate to find series form for y'.

Solution

We assume that

$$y = \sum_{k=0}^{\infty} a_k x^k, \qquad \text{so} \qquad y' = \sum_{k=0}^{\infty} k a_k x^{k-1}.$$

Insert the expansions for y and y' into the differential equation.

This gives

$$\sum_{k=0}^{\infty} k a_k x^{k-1} = x \sum_{k=0}^{\infty} a_k x^k$$

$$= \sum_{k=0}^{\infty} a_k x^{k+1}$$

Write out the first few terms, including the general terms on both sides *for the same power of x*. (Here we arbitrarily picked x^{k-1}.)

Equate coefficients of like powers of x.

or

$$a_1 + 2a_2 x + 3a_3 x^2 + \cdots + k a_k x^{k-1} + \cdots$$
$$= a_0 x + a_1 x^2 + a_2 x^3 + \cdots + a_{k-2} x^{k-1} + \cdots,$$

so

$$a_1 = 0 \qquad \text{(constant terms)}$$

$$a_2 = \frac{1}{2} a_0 \qquad \text{(x terms)}$$

$$a_3 = \frac{1}{3} a_1 = 0 \qquad \text{(x^2 terms)}$$

$$\vdots$$

The recurrence relation is obtained by equating coefficients of x^{k-1}.

$$a_k = \frac{1}{k} a_{k-2} \qquad \text{(x^{k-1} terms)}$$

$$\vdots$$

---

*A closed form expression is one which does not involve an infinite process, such as summation. $f(x) = \dfrac{1}{1-x}$ is expressed in closed form, while $f(x) = \displaystyle\sum_{k=0}^{\infty} x^k$ is not.

Determine the form of all coefficients using the recurrence relation.

From the recurrence relation $a_k = \dfrac{1}{k}a_{k-2}$ and the fact that $a_1 = 0$, we see that all odd coefficients are zero: $0 = a_1 = a_3 = a_5 = \cdots$.

Also we can see that the even coefficients are

$$a_2 = \frac{1}{2}a_0$$

$$a_4 = \frac{1}{4}a_2 = \frac{1}{4 \cdot 2}a_0 = \frac{1}{2^2}\left(\frac{1}{2 \cdot 1}\right)a_0 = \frac{1}{2^2} \cdot \frac{1}{2!}a_0$$

$$a_6 = \frac{1}{6}a_4 = \frac{1}{6 \cdot 4 \cdot 2}a_0 = \frac{1}{2^3}\left(\frac{1}{3 \cdot 2 \cdot 1}\right)a_0 = \frac{1}{2^3} \cdot \frac{1}{3!}a_0$$

$$\vdots$$

$$a_{2k} = \frac{1}{2k}a_{2k-2} = \frac{1}{2k(2k-2) \cdot \ldots \cdot 2}a_0 = \frac{1}{2^k} \cdot \frac{1}{k!}a_0.$$

Write the series for y using coefficients found above.

Try to bring the series for y into the form of a known power series. (Here we use the fact that

$$e^u = \sum_{k=0}^{\infty} \frac{u^k}{k!}$$

for all u.)

The power series for the solution y is therefore

$$y = a_0 + a_1 x + a_2 x^2 + a_3 x^3 + a_4 x^4 + \cdots$$

$$= a_0 + \frac{1}{2}a_0 x^2 + \frac{1}{2^2} \cdot \frac{1}{2!}a_0 x^4 + \frac{1}{2^3} \cdot \frac{1}{3!}a_0 x^6 + \cdots + \frac{1}{2^k} \cdot \frac{1}{k!}a_0 x^{2k} + \cdots$$

$$= a_0\left\{1 + \frac{x^2}{2} + \frac{\left(\frac{x^2}{2}\right)^2}{2!} + \frac{\left(\frac{x^2}{2}\right)^3}{3!} + \cdots + \frac{\left(\frac{x^2}{2}\right)^k}{k!} + \cdots\right\}$$

$$= a_0 \sum_{k=0}^{\infty} \frac{(x^2/2)^k}{k!}.$$

This is the power series expansion for $y = a_0 e^{x^2/2}$, which converges for all values of x. The general solution for the differential equation $y' - xy = 0$ is therefore $y = a_0 e^{x^2/2}$. The constant a_0 is determined from the initial condition:

Apply initial condition to determine a_0.

$$1 = y(0) = a_0 e^0 = a_0.$$

The solution of the initial value problem is $y = e^{x^2/2}$.   ◇

Power series methods may be used in higher order differential equations as well. Note in the following example that the power series for y must be differentiated twice since the differential equation is of order 2.

Example 3

Find a power series solution for the differential equation

$$xy'' - y = 0.$$

Solution: We shall work with the equation in the form

$$xy'' = y. \tag{8}$$

We assume the solution y to have the form

$$y = \sum_{k=0}^{\infty} a_k x^k.$$

Then

$$y' = \sum_{k=0}^{\infty} k a_k x^{k-1}, \qquad \text{and} \qquad y'' = \sum_{k=0}^{\infty} k(k-1) a_k x^{k-2}.$$

Equation (8) becomes

$$x \sum_{k=0}^{\infty} k(k-1) a_k x^{k-2} = \sum_{k=0}^{\infty} a_k x^k,$$

or

$$2a_2 x + 3 \cdot 2a_3 x^2 + 4 \cdot 3 \cdot a_4 x^3 + \cdots + k(k-1) a_k x^{k-1} + \cdots$$
$$= a_0 + a_1 x + a_2 x^2 + a_3 x^3 + \cdots + a_{k-1} x^{k-1} + \cdots.$$

Thus

$$a_0 = 0$$

$$2a_2 = a_1 \text{ gives } a_2 = \frac{1}{2} a_1$$

$$3 \cdot 2 \cdot a_3 = a_2 \text{ gives } a_3 = \frac{1}{3 \cdot 2} a_2 = \frac{1}{3 \cdot 2^2} a_1 = \frac{1}{3(2!)^2} a_1$$

$$4 \cdot 3 \cdot a_4 = a_3 \text{ gives } a_4 = \frac{1}{4 \cdot 3} a_3 = \frac{1}{4 \cdot 3^2 \cdot 2^2} a_1 = \frac{1}{4(3!)^2} \cdot a_1$$

$$\vdots$$

$$k(k-1) a_k = a_{k-1} \text{ gives } a_k = \frac{1}{k(k-1)} a_{k-1} = \cdots = \frac{1}{k[(k-1)!]^2} \cdot a_1$$

The solution y is therefore

$$y = a_1 \left[x + \frac{1}{2} x^2 + \frac{1}{3 \cdot 2^2} x^3 + \frac{1}{4 \cdot (3!)^2} x^4 + \cdots + \frac{1}{k[(k-1)!]^2} x^k + \cdots \right]. \qquad (9)$$

To determine the radius of convergence for (9) we apply the Ratio Test.

$$\rho = \lim_{k \to \infty} \left| \frac{\dfrac{a_1}{(k+1)(k!)^2} x^{k+1}}{\dfrac{a_1}{k[(k-1)!]^2} x^k} \right| = \lim_{k \to \infty} \left(\frac{k}{k+1} \right) \left(\frac{1}{k^2} \right) |x| = 0$$

for all values of x, so the series in (9) converges for all x. We have therefore obtained a legitimate solution for the differential equation (8), although the solution is expressed in power series form and is not immediately recognizable as the power series for a known function in closed form. ◇

As you might suspect at this point, the theory associated with the use of power series in solving differential equations is not simple. In fact, the examples we have presented here were carefully chosen to convey the basic idea while avoiding the difficulties surrounding this method. These difficulties include the following:

(i) It is often possible only to obtain the first few terms of the series solution rather than the general term as we have succeeded in doing in these examples. This may leave the question of convergence unanswerable, although in many cases one can obtain at least an approximation to the solution.

(ii) The assumption that the solution y has a power series representation contains the implicit assumption that y has derivatives of all orders. Since the solution to a differential equation of degree n need have only n derivatives, the power series approach will necessarily fail to identify solutions to certain equations since it assumes too much.

Further work on power series methods in differential equations is left to more specialized courses. Should you take such a course you will find that the material in Chapter 13 constitutes an important foundation on which much of the work of that course depends.

Exercise Set 20.5

In Exercises 1–8, use the method of this section to find a power series form of the solution of the differential equation or initial value problem. Check your work by solving the equation by the method of separation of variables.

1. $y' + y = 0$

2. $y' + 2y = 0$

3. $y' - 6y = 0$

4. $y' + xy = 0$

5. $y' - 2xy = 0$

6. $y' + ax = 0$

7. $\begin{cases} y' + 4y = 0 \\ y(0) = 1 \end{cases}$

8. $\begin{cases} y' + xy = 0 \\ y(0) = 2 \end{cases}$

9. Find a power series solution for the second order differential equation $xy'' + y = 0$.

10. Find a power series solution for the initial value problem
$$\begin{cases} y'' + y = 0 \\ y(0) = 0 \\ y'(0) = 1. \end{cases}$$

(*Hint:* Let $y = \sum_{k=0}^{\infty} a_k x^k$. Since $y(0) = 0$, $a_0 = 0$. This observation simplifies the recurrence relation.)

11. Find a power series solution for the initial value problem
$$\begin{cases} y'' + y = 0 \\ y(0) = 1 \\ y'(0) = 0. \end{cases}$$

12. Conclude from Exercises 10 and 11 that the function $y = A \sin x + B \cos x$ is a solution of the differential equation $y'' + y = 0$.

20.6 APPROXIMATING SOLUTIONS OF DIFFERENTIAL EQUATIONS

There are many differential equations for which simple solutions cannot be found. In such cases, however, we can resort to certain approximation procedures to obtain information about the solution, just as we can use the Trapezoidal Rule or Simpson's Rule to approximate a definite integral when the corresponding antiderivative cannot be found.

To give you an idea of how such approximation procedures work, we discuss one such procedure, called Euler's method, in this section. It is named for the Swiss mathematician Leonhard Euler (1707–1783).

Euler's Method

Euler's method may be used to approximate the solution to an initial value problem of the form

$$\frac{dy}{dt} = f(t, y) \tag{1}$$

$$y(a) = y_0$$

on some interval $[a, b]$. (We shall assume that f and its partial derivatives are continuous for all t and y, although less restrictive conditions can be given.)

Figure 6.1 illustrates the basic idea to be pursued. Assume that $y = f(t)$ is a particular solution of the given differential equation. We divide the interval $[a, b]$ into n subintervals of equal length $h = \dfrac{b - a}{n}$, obtaining the $(n + 1)$ endpoints

$$a = t_0 < t_1 < t_2 < \cdots < t_n = b.$$

Then, beginning at the left endpoint $t_0 = a$ we use the given information about the solution y (namely, the initial value $y(a) = y_0$) and its derivative to approximate the value $y(t_1)$, and we call the approximation y_1. (Note, as suggested by Figure 6.1,

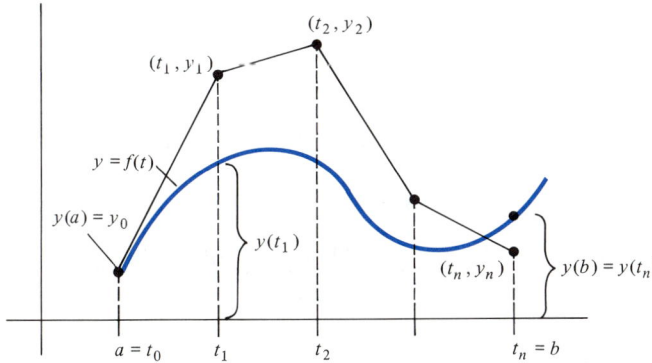

Figure 6.1 Approximation by Euler's method to the solution y of the initial value problem

$$y' = f(t, y), \quad y(a) = y_0.$$

that, in general, the approximation y_1 will not equal the function value $y(t_1)$, which we do not know.) Next, we use the assumed value y_1 and the number $t_1 = a + h$ to approximate $y(t_2)$, and we call this approximation y_2. Continuing in this way we finally approximate $y(t_n) = y(b)$ from the approximate value y_{n-1} and the number $t_{n-1} = a + (n - 1)h$. We therefore obtain *approximations* $y_1, y_2, \ldots, y_n$ to the (unknown) values of the solution $y(t_1), y(t_2), \ldots, y(t_n)$ at finitely many numbers t_j in $[a, b]$. The choice of the endpoint b and the number of intervals, n, depend upon which values of the solution you wish to approximate and with what accuracy.

Figure 6.2 illustrates how we get started in applying Euler's method. To find the approximation y_1 to the value $y(t_1)$, we follow the line tangent to the graph of y at the point $P_0 = (t_0, y_0)$ until we reach the point $P_1 = (t_1, y_1)$ with t-coordinate t_1. Since the slope of this line is $y'(t_0) = f(t_0, y_0)$, it follows that

$$\frac{y_1 - y_0}{t_1 - t_0} = f(t_0, y_0),$$

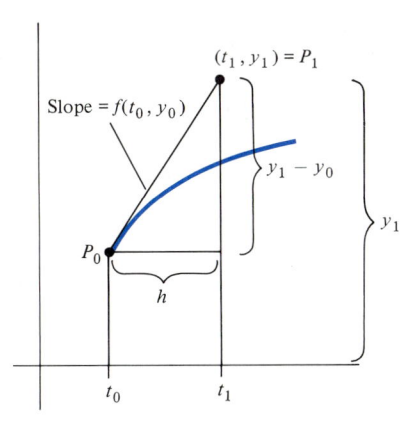

Figure 6.2 $y_1 = y_0 + f(t_0, y_0)h.$

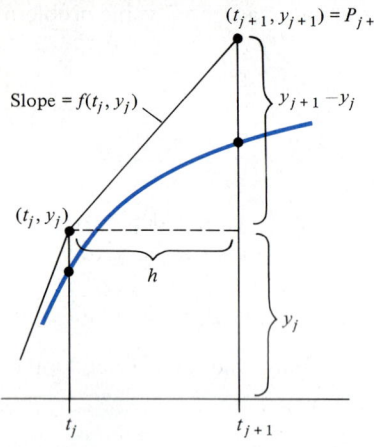

Figure 6.3 $y_{j+1} = y_j + f(t_j, y_j)h.$

and since $h = t_1 - t_0,$

$$y_1 = y_0 + f(t_0, y_0)h. \tag{2}$$

Once we have obtained the approximation y_1 from equation (2), we *pretend* that y_1 is the actual value of the solution y at $t = t_1$. Since we don't know the actual value $y(t_1)$, the approximation y_1 is the "next best thing." This "pretending" is necessary in order that we may carry out the same procedure to obtain the approximation y_2 for $y(t_2)$. Starting with the point (t_1, y_1) and the slope $f(t_1, y_1),$* we obtain the y-coordinate of the point $P_2 = (t_2, y_2)$ from the equation

$$\frac{y_2 - y_1}{t_2 - t_1} = f(t_1, y_1).$$

That is, since $h = t_2 - t_1,$

$$y_2 = y_1 + f(t_1, y_1)h.$$

Continuing in this way (see Figure 6.3), we obtain each succeeding approximation y_{j+1} from the preceding y_j and t_j by the equation

$$y_{j+1} = y_j + f(t_j, y_j)h \qquad j = 0, 1, \ldots, n - 1 \tag{3}$$

which is the summary statement of Euler's method.

The following examples show how a first order differential equation may be put in the form of equation (1), and how the solution of an associated initial value problem may be approximated using Euler's method.

Example 1

Use Euler's method to approximate the solution of the initial value problem

$$y' = y(4 - y) \tag{4}$$
$$y(0) = 1$$

on the interval $[0, 1]$ using 4 subintervals of equal length.

Solution: With $[0, 1]$ divided into 4 subintervals of length $h = \dfrac{1 - 0}{4} = \dfrac{1}{4}$, we have endpoints

$$t_0 = 0, \qquad t_1 = \frac{1}{4}, \qquad t_2 = \frac{1}{2}, \qquad t_3 = \frac{3}{4}, \qquad \text{and} \qquad t_4 = 1.$$

With $y_0 = y(0) = 1$ and $f(t, y) = y(4 - y)$, equation (3) gives

$$y_1 = y_0 + y_0(4 - y_0)h$$

$$= 1 + 1(4 - 1)\left(\frac{1}{4}\right)$$

$$= 1 + \frac{3}{4}$$

$$= 1.75$$

---

*The pretending occurs here in two places. Since we do not actually know $y(t_j)$, we do not know the actual slope $f(t_j, y(t_j))$, which we approximate by $f(t_j, y_j)$. Neither do we know the point of tangency, $(t_j, y(t_j))$, so we use (t_j, y_j) instead.

$$y_2 = y_1 + y_1(4 - y_1)h$$
$$= 1.75 + 1.75(4 - 1.75)(0.25)$$
$$\cong 2.7344,$$

etc. Table 6.1 shows the approximations y_1, y_2, y_3, y_4 together with the actual values $y(t_1), y(t_2), y(t_3), y(t_4)$ of the solution $y = \dfrac{4}{1 + 3e^{-4t}}$ of the initial value problem to four decimal place accuracy. (This solution is obtained by the method of Section 20.1.) ◇

Table 6.1 Approximate and actual values of the solution to the initial value problem in Example 1. ($h = 0.25$)

t_j	0	.25	.50	.75	1.0
y_j	1	1.75	2.7344	3.5995	3.9599
$y(t_j)$	1	1.9015	2.8449	3.4802	3.7917

Example 2

Approximate the solution to the initial value problem in Example 1 on the interval $[0, 1]$ using $n = 10$ subintervals of equal size.

Solution: This time we have $h = \dfrac{1 - 0}{10} = \dfrac{1}{10}$, so Euler's equation becomes

$$y_{j+1} = y_j + y_j(4 - y_j)(0.10), \qquad j = 0, 1, 2, \ldots, 9.$$

Table 6.2 shows the resulting approximations, together with the actual values of the solution, to four decimal place accuracy. (Note the improvement in accuracy compared with Example 1.) ◇

Table 6.2 Values of the approximations and actual solution to the initial value problem in Example 2. ($h = 0.10$)

t_j	0	.1	.2	.3	.4	.5	.6	.7	.8	.9	1.0
y_j	1	1.3	1.651	2.0388	2.4387	2.8194	3.1523	3.4195	3.6180	3.7562	3.8478
$y(t_j)$	1	1.3285	1.7036	2.1013	2.4911	2.8449	3.1443	3.3829	3.5642	3.6970	3.7917

Example 3

Use Euler's method to approximate the solution of the initial value problem

$$y' + 2y = e^t \tag{5}$$
$$y(0) = 1$$

on the interval $[0, 1]$.

Solution: We begin by moving all terms in equation (5) except the derivative y' to the right-hand side, obtaining

$$\frac{dy}{dt} = e^t - 2y.$$

Thus, $f(t, y) = e^t - 2y$ in equation (3) for Euler's method. We shall obtain two approximations.

(a) Using $n = 4$ subdivisions, we have $h = \dfrac{1 - 0}{4} = 0.25$, and equation (3) becomes

$$y_{j+1} = y_j + (e^{t_j} - 2y_j)(0.25), \qquad j = 0, 1, 2, 3.$$

Thus since $y_0 = y(0) = 1$ is given,

$$
\begin{aligned}
y_1 &= y_0 + (e^{t_0} - 2 \cdot y_0)(0.25) \\
&= 1 + (e^0 - 2 \cdot 1)(0.25) \\
&= 1 - 0.25 \\
&= 0.75, \\
y_2 &= y_1 + (e^{t_1} - 2y_1)(0.25) \\
&= 0.75 + [e^{.25} - 2(0.75)](0.25) \\
&= 0.6960,
\end{aligned}
$$

etc. Table 6.3 shows the same approximations y_1, y_2, y_3, and y_4 to four decimal place accuracy, as well as the actual values $y(t_1)$, $y(t_2)$, $y(t_3)$, and $y(t_4)$ for the actual solution $y = \dfrac{1}{3}(2e^{-2t} + e^t)$.

Table 6.3 Approximate and actual values for the initial value problem in Example 3. ($h = 0.25$)

t_j	0	.25	.5	.75	1.0
y_j	1	.75	.6960	.7602	.9093
$y(t_j)$	1	.8324	.7948	.8544	.9963

(b) The same solution, approximated using 10 subintervals of $[0, 1]$ of length $h = \dfrac{1 - 0}{10} = 0.1$, is obtained using the Euler equation

$$y_{j+1} = y_j + (e^{t_j} - 2y_j)(0.1), \qquad j = 0, 1, \ldots, 9.$$

Table 6.4 shows the results of this approximation. ◇

Table 6.4 Approximate and actual values of the solution of the initial value problem in Example 3. ($h = 0.1$)

t_j	0	.1	.2	.3	.4	.5	.6	.7	.8	.9	1.0
y_j	1	.9	.8305	.7866	.7642	.7606	.7733	.8009	.8421	.8962	.9629
$y(t_j)$	1	.9142	.8540	.8158	.7968	.7948	.8082	.8356	.8764	.9301	.9963

Euler's method is one procedure for approximating the solution of an initial value problem involving a first order differential equation. While we have used examples in which an actual solution may be obtained for purposes of comparison, Euler's method may be applied to differential equations whose solutions are not known.

Obviously there is a fair amount of tedium in applying Euler's method. Calculators and especially computers can reduce this tedium to a minimum. Program 9 in

Appendix I is a BASIC Computer program which was used to perform the calculations in Tables 6.1–6.4. You may find it useful in working the exercises.

Several sources of error enter into the calculations that are involved in Euler's method. Since we obtain only approximations y_j instead of the actual value $y(t_j)$ at each step, the slope calculation $\dfrac{dy}{dt} = f(t_j, y_j)$ contains an error due to the difference $|y_j - y(t_j)|$, and the point of assumed tangency, (t_j, y_j) contains an error due to the same factor. Finally, roundoff error in such calculations is inescapable. In general, however, we can say that accuracy decreases as we attempt to approximate values of the solution further away from the number t_0 where the initial condition is prescribed, and that accuracy increases as the stepsize h decreases. These important issues of accuracy are pursued in more advanced courses.

Exercise Set 20.6

In each of Exercises 1–10 use Euler's method to approximate the value of the solution to the initial value problem at the endpoints of 4 subintervals of $[0, 1]$ of size $h = 0.25$.

1. $y' = 2y$

    $y(0) = 1$

2. $y' = 9y(1 - y)$

    $y(0) = \dfrac{1}{2}$

3. $y' + 2y = 4$

    $y(0) = 1$

4. $\dfrac{dy}{dx} = \dfrac{y + 1}{x + 1}$

    $y(0) = 0$

5. $\dfrac{dy}{dx} = x(y + 1)$

    $y(0) = 1$

6. $\dfrac{dy}{dx} = 2 - 4y$

    $y(0) = 1$

7. $\dfrac{dy}{dt} + 2ty = t$

    $y(0) = \dfrac{3}{2}$

8. $e^y \cdot \dfrac{dy}{dt} - t^2 = 0$

    $y(0) = 0$

9. $\dfrac{dy}{dt} = 2y - 2t$

    $y(0) = \dfrac{3}{2}$

10. $\dfrac{dy}{dt} - 2y = e^{2t} \sin t$

    $y(0) = 0$

11. Find the actual solution of the initial value problem in Exercise 3. Compute the actual value of the solution at each endpoint and compare it with the value from Euler's method.

12. An investor opens a savings account paying 10% interest compounded continuously with an initial deposit of $1000. The investor makes deposits of $1000 per year in a continuous manner.
   a. Show that the function P, giving the amount on deposit after t years, satisfies the differential equation
   $P'(t) = 0.10P(t) + 1000$.
   b. Use Euler's method to approximate the values $P(1)$, $P(2)$, $P(3)$, and $P(4)$.

SUMMARY OUTLINE OF CHAPTER 20

◆ First order differential equations of the form   (page 962)

$$\frac{dy}{dt} = ky + b$$

may be solved by the method of *separation of variables*.

◆ First order differential equations of the form   (page 966)

$$\frac{dy}{dt} + p(t)y = g(t)$$

may be solved using the *integrating factor* $e^{\int p(t)\,dt}$.

◆ The differential equation   (page 970)

$$M(x, y) + N(x, y)\,\frac{dy}{dx} = 0, \qquad (x, y) \in R \tag{1}$$

is *exact* if there exists a function f for which

$$M(x, y) = \frac{\partial f}{\partial x}(x, y), \qquad N(x, y) = \frac{\partial f}{\partial y}(x, y).$$

◆ **Theorem:** Equation (1) is exact if and only if (page 971)

$$\frac{\partial M}{\partial y}(x, y) = \frac{\partial N}{\partial x}(x, y), \qquad (x, y) \in R.$$

◆ The second order homogeneous linear differential equation $y'' + ay' + by = 0$ has characteristic polynomial (page 974) $r^2 + ar + b$ and general solution

    (i) $y = C_1 e^{r_1 t} + C_2 e^{r_2 t}$ if $r^2 + ar + b$ has two distinct real roots r_1 and r_2;
    (ii) $y = C_1 e^{rt} + C_2 t e^{rt}$ if $r^2 + ar + b$ has a repeated real root r;
    (iii) $y = e^{\alpha t}[C_1 \sin \beta t + C_2 \cos \beta t]$ if $r^2 + ar + b$ has the complex roots $r = \alpha \pm \beta i;\ \beta \neq 0$.

◆ The second order linear homogeneous equation $y'' + by = 0, \qquad b > 0$ has general solution $y =$ (page 978) $C_1 \sin \sqrt{b} t + C_2 \cos \sqrt{b} t$.

◆ The nonhomogeneous second order linear equation $y'' + ay' + by = f(t)$ may be solved by the *method of undeter-* (page 983) *mined coefficients* when $f(t)$ is of the form t^n, e^{kt}, $\sin kt$ or $\cos kt$.

◆ A differential equation may sometimes be solved by assuming a *power series solution* of the form $y = \Sigma a_k x^k$, (page 991) substituting into the differential equation, and solving the resulting equations for the coefficients a_k.

◆ *Euler's Method* for approximating solutions of the differential equation $y' = f(t, y)$ with $y(a) = y_0$ is based on the (page 997) formula $y_{j+1} = y_j + f(t_j, y_j)h$, which gives the approximation y_{j+1} to the value $y(t_{j+1})$.

REVIEW EXERCISES—CHAPTER 20

In Exercises 1–30 find the general solution of the given differential equation.

1. $\dfrac{dy}{dx} = x - 3$

2. $\dfrac{dy}{dt} = 4 - \sqrt{t}$

3. $\dfrac{dy}{dt} = \dfrac{\sec \sqrt{t} \tan \sqrt{t}}{\sqrt{t}}$

4. $\dfrac{d^2 y}{dx^2} = \sqrt{1 + x}$

5. $\dfrac{dy}{dt} = 3ty^2$

6. $\dfrac{dy}{dt} = \dfrac{\sqrt{t + 1}}{y}$

7. $\dfrac{dy}{dx} = (1 + x)(2 + y)$

8. $ty' = \dfrac{t^2 + 3}{y}$

9. $y' + 2y = 4$

10. $y' = y(3 - y)$

11. $y' = y(4 + y)$

12. $ty' = y \ln t$

13. $y' + y \cos t = 0$

14. $y' = 4y \ln y$

15. $y' = 4y + 8$

16. $y' = 4y + 4t + 8$

17. $y'' + 4y' + 4y = 0$

18. $y'' - 3y' - 10y = 0$

19. $y'' - 2y' - 15y = \cos 2t$

20. $y'' + 9y = 5 + e^{2t}$

21. $y'' - 9y = 5 + e^{2t}$

22. $y' - 3y = -9 \sin 2t$

23. $t^2 y' + y = 0$

24. $\dfrac{dy}{dx} = y(2 - y)$

25. $\dfrac{d^2 y}{dt^2} - 5\dfrac{dy}{dt} - 14y = 2 + t$

26. $y'' + y' - 6y = 6 - e^{2t}$

27. $2xy^3 + 3x^2 y^2 \dfrac{dy}{dx} = 0$

28. $\cos y - x \sin y \dfrac{dy}{dx} = 0$

29. $2y - y^2 - 2x\dfrac{dy}{dx} = 0$

30. $(e^{xy} + xye^{xy})\dfrac{dy}{dx} + y^2 e^{xy} = 0$

In Exercises 31–38 find the solution of the initial value problem.

31. $\dfrac{dy}{dx} = \dfrac{x}{\sqrt{1 + x^2}}$

    $y(0) = 3$

32. $y' = 6y$

    $y(0) = 2$

33. $\dfrac{dy}{dt} = ty$

    $y(0) = 3$

34. $y' = 4(1 - y)$

    $y(0) = 1$

35. $\dfrac{d^2 y}{dt^2} = -9y$

    $y(0) = 3$
    $y'(0) = 9$

36. $y'' - 16y = 0$

    $y(0) = 2$
    $y'(0) = 8$

37. $y'' - 6y' + 9y = 0$
$y(0) = 0$
$y'(0) = 6$

38. $y' - 4y = -2e^{3x}$
$y(0) = 3$
$y'(0) = 10$

39. Find a differential equation satisfied by the function
a. $y = Ce^{2t}$
b. $y = C_1e^t + C_2e^{-t}$.

40. A tank contains 500 liters of brine with an initial concentration of salt of 1 kilogram of salt per liter. A second mixture of brine containing 2 kilograms of salt per liter is then added at a rate of 50 liters per minute. The tank is kept well mixed, and brine is drawn off the bottom at the same rate, 50 liters per minute. Find the concentration of salt t minutes after the second mixture begins entering the first.

41. An automobile radiator contains five gallons of pure anti-freeze. The owner begins adding fresh water at the rate of 1 gallon per minute, with the engine running to insure complete mixing, and draining the radiator at the same rate of 1 gallon per minute. Find the concentration $y(t)$ of antifreeze per gallon t minutes after the owner begins this process.

42. An investor places $5000 in a savings account paying 10% interest compounded continuously and pledges to make additional deposits of $2000 per year in a continual manner.
a. Find a differential equation for $P(t)$, the amount on deposit in this account t years after it is opened.
b. Use Euler's method to approximate $P(1)$, $P(2)$, $P(3)$, and $P(4)$.

43. A cold drink is removed from a refrigerator at temperature 40°F and is placed on a sunporch where the surrounding temperature is 90°F. After 5 minutes the temperature of the drink is 50°F. Find its temperature after 10 minutes.

44. Use a power series to find a solution to the initial value problem
$$\begin{cases} y' - xy = 0 \\ y(0) = 2. \end{cases}$$

45. Find a power series solution for the second order differential equation $xy'' - y = 0$.

46. Find a power series solution for the initial value problem
$$\begin{cases} xy'' + y = 0 \\ y(0) = 0 \\ y'(0) = 1. \end{cases}$$

Why is the condition $y(0) = 0$ *necessary?*

47. Find the general form of solutions to the differential equation
$$y'' - 4y' + 6y = 0.$$

48. Find the solution for the initial value problem
$$\begin{cases} y'' - 4y' + 8y = 0 \\ y(0) = 0 \\ y'(0) = 2. \end{cases}$$

Appendix I
Basic Computer Programs

Included here are nine BASIC computer programs that are referred to in various examples and exercises throughout the text. They are presented as "bare-bones" prototypes, which those with access to computing facilities (personal computers, programmable calculators, or large computers) can use in designing programs that actually operate on particular machines. Notation appearing in these programs includes the following:

1. $a * b$ means the product ab.
2. $a \uparrow b$ means the exponentiation a^b.
3. a/b means the quotient $\dfrac{a}{b}$.

Proper development of computer software requires full documentation, as well as the inclusion of checks to insure that the user does not attempt to supply inappropriate values to the program. (For example, in asking the user to specify the endpoints of an interval $[a, b]$, one should check to insure that $a < b$.) We have made no attempt to do either, since we wish to highlight only the algorithm involved in the program.

Program 1: Newton's Method for $f(x) = x^3 - 7$

```
 10 DEF FNF (T) = T↑3 - 7
 20 DEF FND (T) = 3*(T↑2)
 30 PRINT "how many iterations?"
 40 INPUT N
 50 PRINT "what is your first guess?"
 60 INPUT X
 70 FOR I = 1 TO N
 80   LET Z1 = FNF(X)
 90   LET Z2 = FND(X)
100   LET W = X - (Z1/Z2)
110   PRINT I,W
120   LET X = W
130 NEXT I
140 END
```

Comment: This program implements Newton's Method to locate a zero of the function $f(x) = x^3 - 7$. The value of the function is computed in line 10, the derivative $f'(x) = 3x^2$ is computed in line 20, and the formula for Newton's Method is implemented in line 100. The user supplies the number of iterations and a first guess at the root.

Program 2: Lower Riemann Sums for $f(x) = 3x^2 + 7$

```
10 DEF FNF (T) = 3*(T↑2) + 7
20 PRINT "enter interval endpoints a,b"
30 INPUT A,B
40 PRINT "how many subintervals?"
50 INPUT N
60 LET D = (B - A)/N
70 LET S = 0
80 FOR I = 1 TO N
90   LET X = A + (I - 1)*D
100   LET S = S + FNF(X)*D
110 NEXT I
120 PRINT S
130 END
```

Comment: This program computes a lower Riemann sum for the function $f(x) = 3x^2 + 7$ on the interval $[a, b]$ using n equal subintervals. The numbers a, b, and n are supplied by the user. The program computes the function value at left endpoints of the resulting subintervals, since $f(x)$ is an increasing function.

Program 3: Upper Riemann Sums for $f(x) = 3x^2 + 7$

```
10 DEF FNF(T) = 3*(T↑2) + 7
20 PRINT "enter interval endpoints a,b"
30 INPUT A,B
40 PRINT "how many subintervals?"
50 INPUT N
60 LET D = (B - A)/N
70 LET S = 0
80 FOR I = 1 TO N
90   LET X = A + I*D
100   LET S = S + FNF(X)*D
110 NEXT I
120 PRINT S
130 END
```

Comment: This program computes an upper Riemann sum for the function $f(x) = 3x^2 + 7$ on the interval $[a, b]$, if $a > 0$, using n equal subintervals. The numbers a, b, and n are supplied by the user. The program computes the function values at right endpoints of the resulting subintervals, since f is an increasing function for $x \geq 0$.

Program 4: Trapezoidal Rule Applied to $f(x) = \dfrac{1}{x}$

```
10 DEF FNF(T) = 1/T
20 PRINT "enter interval endpoints a,b"
```

```
 30 INPUT A,B
 40 PRINT "how many subintervals?"
 50 INPUT N
 60 LET D = (B - A)/N
 70 LET S = FNF(A)
 80 FOR I = 1 TO (N - 1)
 90  LET X = A + I*D
100  LET S = S + 2*FNF(X)
110 NEXT I
120 LET S = S + FNF(B)
130 LET S = S*(D/2)
140 PRINT S
150 END
```

Comment: This program implements the Trapezoidal Rule for approximate integration of the function $f(x) = \dfrac{1}{x}$ on the interval $[a, b]$ with $a > 0$. The numbers a, b, and n (the number of subintervals) are supplied by the user.

Program 5: Simpson's Rule Applied to the function $f(x) = \dfrac{1}{x}$

```
 10 DEF FNF(T) = 1/T
 20 PRINT "enter interval endpoints a,b"
 30 INPUT A,B
 40 PRINT "how many subintervals (an even number)
 50 INPUT N
 60 LET C = (B - A)/N
 70 LET D = C/3
 80 LET S = FNF(A)
 90 FOR I = 1 TO (N - 1) STEP 2
100  LET S = S + 4*FNF(A + I*C)
110 NEXT I
120 FOR I = 2 TO (N - 2) STEP 2
130  LET S = S + 2*FNF(A + I*C)
140 NEXT I
150 LET S = S + FNF(B)
160 LET S = S*D
170 PRINT S
180 END
```

Comment: This program implements Simpson's Rule for approximate integration for the function $f(x) = 1/x$ on the interval $[a, b]$ with $a > 0$. The numbers a, b, and n (the number of subintervals, *an even number*) are supplied by the user.

Program 6: Partial Sums of the Geometric Series

```
 10 PRINT "enter p,a,x,n"
 20 INPUT P,A,X,N
 30 LET S = 0
```

```
40 FOR K = P TO N
50   LET T = A*X↑K
60   LET S = S + T
70 NEXT K
80 PRINT S
90 END
```

Comment: This program computes the partial sum

$$\sum_{k=p}^{n} ax^k$$

where the constants p, a, x, and n are supplied by the user.

Program 7:   A Riemann Sum in Polar Coordinates for $f(\theta) = 1 + \sqrt{\sin \theta}$

```
10 DEF FNF(T) = 1 + SQR(SIN(T))
20 DEF FNG(T) = 0.5 * (T↑2)
30 PRINT "enter interval endpoints a,b"
40 INPUT A,B
50 PRINT "how many subintervals?"
60 INPUT N
70 LET D = (B - A)/N
80 LET S = 0
90 FOR I = 1 TO N
100   LET Y = FNF(A + I*D)
110   LET S = S + FNG(Y)*D
120 NEXT I
130 PRINT S
140 END
```

Comment: This program computes a Riemann sum, using n subintervals of equal length, for the integral

$$\int_a^b \frac{1}{2}(1 + \sqrt{\sin \theta})^2 \, d\theta$$

which approximates the area bounded by $f(\theta) = 1 + \sqrt{\sin \theta}$. The parameters a, b, and n are supplied by the user.

Program 8:   Riemann Sum for a Double Integral

```
10 PRINT "enter a,b,c,d"
20 INPUT A,B,C,D
30 PRINT "how many subintervals?"
40 INPUT N
50 LET D1 = (B - A)/N
60 LET D2 = (D - C)/N
70 LET D3 = D1*D2
80 LET S = 0
90 FOR J = 1 TO N
100   FOR K = 1 TO N
```

```
110    LET S = S + (8 - 2*(A + J*D1) - 4*(C + K*D2))
120    NEXT K
130 NEXT J
140 LET S = S*D3
150 PRINT S
160 END
```

Comment: This program computes a Riemann sum for the double integral

$$\int_{c}^{d} \int_{a}^{b} (8 - 2x - 4y) \, dx \, dy$$

using a grid consisting of n^2 rectangles of dimension $(b - a)(d - c)$. The parameters a, b, c, d, and n are supplied by the user.

Program 9: Euler's Method Applied to the Initial Value Problem

$$\frac{dy}{dt} = e^t - 2y$$

$$y(0) = 1$$

```
10 LET Y = 1
20 LET T = 0
30 FOR I = 1 TO 4
40 T = T + 0.25
50 Y = Y + (EXP(T) - 2*Y)*(0.25)
60 PRINT T,Y
70 NEXT I
80 END
```

Comment: This program gives an approximation to the stated initial value problem on the interval [0, 1] using $n = 4$ subintervals. The equation for Euler's method is implemented in step 50. Note that the right side of this equation involves current values of both T and Y.

Appendix II
Complex Numbers

In the set of real numbers, the quadratic equation

$$x^2 = -1 \qquad \qquad (1)$$

has no solution, since the square of a real number is never negative. The *complex numbers* constitute a number system containing the real number system (just as the real numbers contain the integers), and in this system equations such as equation (1) have solutions.

To form the complex number system, we first define the complex number i by the equation

$$i^2 = -1. \qquad \qquad (2)$$

Another way to write (2) is to say that $i = \sqrt{-1}$. Thus, the complex number i is a solution to equation (1), as is the number $-i = -\sqrt{-1}$. However, i and $-i$ are not the only complex numbers. The following definition fully determines the complex number system.

DEFINITION 1

The **complex numbers** are all numbers of the form

$$z = a + bi,$$

where a and b are real numbers, together with the following operations:

  (i) Scalar multiplication: $\lambda(a + bi) = \lambda a + (\lambda b)i, \qquad \lambda \in \mathbb{R}$
 (ii) Addition: $(a_1 + b_1 i) + (a_2 + b_2 i) = (a_1 + a_2) + (b_1 + b_2)i$
(iii) Multiplication: $(a_1 + b_1 i) \cdot (a_2 + b_2 i) = (a_1 a_2 - b_1 b_2) + (a_1 b_2 + a_2 b_1)i$

Notice the following about Definition 1:

(a) When $b = 0$, the complex number $z = a + 0i = a$ is a real number. Moreover, the definitions of addition and multiplication agree with the definitions of addition and multiplication of real numbers when $b_1 = b_2 = 0$. Thus, the set of complex numbers of the form $\{a + bi \mid a \in \mathbb{R}, b = 0\}$ is just the set $\mathbb{R}$ of real numbers.

(b) The definition of multiplication is obtained by assuming that the usual distributive and commutative laws hold for all expressions involving a_1, a_2, b_1, b_2, and i:

$$(a_1 + b_1i) \cdot (a_2 + b_2i) = a_1(a_2 + b_2i) + (b_1i)(a_2 + b_2i)$$
$$= a_1a_2 + a_1b_2i + a_2b_1i + b_1b_2i^2$$
$$= a_1a_2 - b_1b_2 + (a_1b_2 + a_2b_1)i.$$

For $z = a + bi$, we say that z is **pure imaginary** if $a = 0$.

Example 1

If $z_1 = 3 + 2i$ and $z_2 = 4 - i$, then

(i) $4z_1 = 4(3 + 2i) = 4 \cdot 3 + (4 \cdot 2)i = 12 + 8i.$

(ii) $z_1 + z_2 = (3 + 2i) + (4 - i) = (3 + 4) + (2 + (-1))i = 7 + i.$

(iii) $z_1 - z_2 = z_1 + (-z_2) = (3 + 2i) + (-1)(4 - i)$
$$= (3 + 2i) + (-4 + i)$$
$$= (3 + (-4)) + (2 + 1)i$$
$$= -1 + 3i.$$

(iv) $z_1z_2 = (3 + 2i)(4 - i) = 3 \cdot 4 - (2)(-1) + [3(-1) + 2 \cdot 4]i$
$$= 14 + 5i.$$ ◇

Geometric and Vector Interpretations

Since the complex number $z = a + bi$ is determined by a pair of real numbers, we refer to a as the **real part** (or real component) of z, and we refer to b as the **imaginary part** (imaginary component) of z. (Remember that b is a real number.) Plotting the real part of z on the x-axis (called the real axis) and the imaginary part of z on the y-axis (called the imaginary axis), we can plot the complex number $z = a + bi$ as the ordered pair (a, b) in the xy-plane (see Figure II.1). Note that the real numbers (considered as a subset of the complex numbers) correspond to points on the real axis, while pure imaginary numbers correspond to points on the imaginary axis.

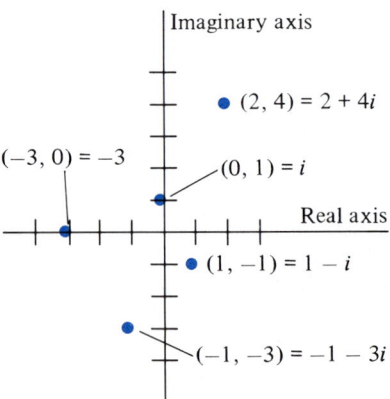

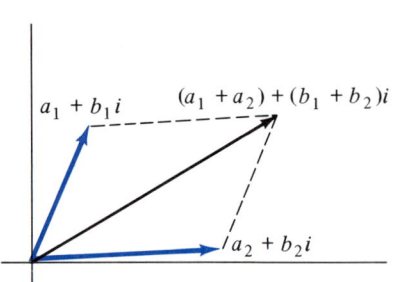

Figure II.1 Complex numbers $a + bi$ may be represented as points (a, b) in the complex plane.

Figure II.2 Addition of complex numbers corresponds to vector addition.

Since points (a, b) in the plane may be regarded as position vectors $\langle a, b \rangle$ originating at the origin $(0, 0)$, we may interpret the complex number $a + bi$ as the position vector $\langle a, b \rangle$. As Figure II.2 illustrates, the definition of addition for complex numbers agrees with the definition of vector addition. You can verify that the corresponding definitions of scalar multiplication agree as well. The vector interpretation leads directly to the concept of the *modulus* (also called the length, or absolute value) of a complex number.

DEFINITION 2

The **modulus of the complex number** $z = a + bi$ is the real number $|z| = \sqrt{a^2 + b^2}$.

Note that the modulus of a complex number is just its absolute value when the number is real. A concept related to the modulus is the *conjugate* of a complex number.

DEFINITION 3

The **conjugate of the complex number** $z = a + bi$ is the complex number $\bar{z} = a - bi$.

The relationship between the modulus and the conjugate is that

$$z\bar{z} = (a + bi)(a - bi) = a^2 + b^2 = |z|^2.$$

Figure III.3 shows that the conjugate $\bar{z}$ is just the reflection in the real axis of the complex number z.

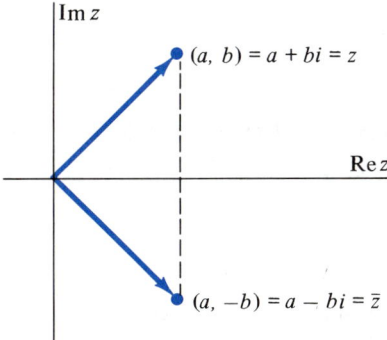

Figure II.3 The conjugate $\bar{z}$ of z is the reflection of z in the real axis.

Division for Complex Numbers

The concepts of modulus and conjugate allow us to define the operation of division. Since

$$z\bar{z} = |z|^2$$

it follows that

$$\left(\frac{1}{|z|^2}\right)(z\bar{z}) = 1,$$

so the reciprocal $\dfrac{1}{z}$, $z \neq 0$, is defined as

$$\frac{1}{z} = \frac{\bar{z}}{|z|^2}, \qquad z \neq 0.$$

Division is then defined as multiplication by a reciprocal.

DEFINITION 4

Let z_1 and z_2 be complex numbers with $z_2 \neq 0$. The quotient $\dfrac{z_1}{z_2}$ is defined to be

$$\frac{z_1}{z_2} = z_1\left(\frac{1}{z_2}\right) = \frac{z_1\bar{z}_2}{|z_2|^2}.$$

Again, this definition agrees with the definition of division in $\mathbb{R}$ when z_1 and z_2 are real.

Example 2

Let $z_1 = 1 + 3i$ and $z_2 = 2 - 4i$. Then

(i) $\bar{z}_1 = 1 - 3i$.

(ii) $z_1\bar{z}_1 = (1 + 3i)(1 - 3i) = (1 + 9) + (3 - 3)i = 10 = |z_1|^2$.

(iii) $\dfrac{1}{z_1} = \dfrac{\bar{z}_1}{|z_1|^2} = \dfrac{1}{10}(1 - 3i) = \dfrac{1}{10} - \dfrac{3}{10}i.$

(iv) $\dfrac{z_1}{z_2} = z_1\left(\dfrac{1}{z_2}\right) = \dfrac{z_1\bar{z}_2}{|z_2|^2} = \dfrac{1}{20}(1 + 3i)(2 + 4i)$

$$= \frac{1}{20}[(2 - 12) + (6 + 4)i]$$

$$= -\frac{1}{2} + \frac{1}{2}i. \qquad \diamond$$

Solutions of Quadratic Equations

We conclude by recalling that the solutions of the quadratic equation

$$ax^2 + bx + c = 0$$

are given by the quadratic formula as

$$x = -\frac{b}{2a} \pm \frac{\sqrt{b^2 - 4ac}}{2a}$$

when $b^2 - 4ac \geq 0$. Using complex numbers we may now write the roots as

$$x = -\frac{b}{2a} \pm \left(\frac{\sqrt{4ac - b^2}}{2a}\right)i$$

when $b^2 - 4ac < 0$.

Example 3

For the quadratic equation

$$x^2 - 2x + 3 = 0$$

the solutions are

$$x = -\frac{(-2)}{2} \pm \frac{\sqrt{(-2)^2 - 4(1)(3)}}{2}$$

$$= 1 \pm \frac{\sqrt{-8}}{2}$$

$$= 1 \pm \sqrt{2}\,i. \qquad \diamond$$

ANSWERS TO ODD-NUMBERED EXERCISES

CHAPTER 12

Exercise Set 12.1

1. $\left\{ \dfrac{1}{3}, \dfrac{2}{5}, \dfrac{3}{7}, \dfrac{4}{9}, \ldots \right\}$. Converges.

$$\lim_{n \to \infty} \frac{n}{2n+1} = \lim \frac{1}{2 + \dfrac{1}{n}} = \frac{1}{2}.$$

3. $\left\{ -1, -\dfrac{1}{3}, -\dfrac{1}{11}, 0, \ldots \right\}$. Converges to 0.

5. $\left\{ \dfrac{1}{2}, \dfrac{1}{5}, \dfrac{1}{10}, \dfrac{1}{17}, \ldots \right\}$. Converges to 0.

7. $\{\sqrt{5}, \sqrt{5}, \sqrt{5}, \sqrt{5}, \ldots\}$. Converges to $\sqrt{5}$.

9. $\left\{ 10, \dfrac{40}{1+\sqrt{2}}, \dfrac{60}{1+\sqrt{3}}, \dfrac{80}{3}, \ldots \right\}$. Diverges.

11. $\left\{ \dfrac{2}{3}, \dfrac{3+\sqrt{2}}{4}, \dfrac{3-\sqrt{3}}{5}, \dfrac{5}{6}, \ldots \right\}$. Converges to 0.

13. $\left\{ \sqrt{2}, \dfrac{1}{2}\sqrt{6}, \dfrac{2}{3}\sqrt{3}, \dfrac{1}{2}\sqrt{5}, \ldots \right\}$. Converges to 1.

15. $\left\{ 1, \cos \dfrac{1}{4}, \cos \dfrac{2}{9}, \cos \dfrac{3}{16}, \ldots \right\}$. Converges to $\cos 0 = 1$.

17. $\left\{ \dfrac{3}{2}, \dfrac{1+\sqrt{2}}{2\sqrt{2}}, \dfrac{2+3\sqrt{3}}{6\sqrt{3}}, \dfrac{5}{8}, \ldots \right\}$. Converges to $\dfrac{1}{2}$.

19. $\left\{ \dfrac{1}{2}, \dfrac{1}{6}, \dfrac{1}{12}, \dfrac{1}{20}, \ldots \right\}$. Converges.

$$\lim\left(\frac{1}{n} - \frac{1}{n+1} \right) = \lim\left(\frac{1}{n(n+1)} \right) = 0.$$

21. $\{\sqrt{2}-1, \sqrt{3}-2, 2-\sqrt{3}, \sqrt{5}-2, \ldots\}$.   Converges to 0.

23. $\left\{ \sqrt{3}, \dfrac{3}{2}, \dfrac{\sqrt{19}}{3}, \dfrac{\sqrt{33}}{4}, \ldots \right\}$. Converges to $\sqrt{2}$.

25. $\left\{ 1, \sqrt{2}, \dfrac{3}{2}, 4 \sin \dfrac{\pi}{8}, \ldots \right\}$. Converges to $\dfrac{\pi}{2}$.

27. $\left\{ 2, \dfrac{9}{4}, \dfrac{64}{27}, \dfrac{625}{256}, \ldots \right\}$. Converges to e.

39. Product is bounded, quotient need not be bounded.

Exercise Set 12.2

1. 2 **3.** 1 **5.** 0

7. Diverges **9.** 1 **11.** 0

13. 0 **15.** e^3 **17.** 1

23. $a_n = 2n - 5$.

25. $\{1, -1, 1, -1, 1, -1, \ldots\}$

27. No. An example is $\left\{ 1, -\dfrac{1}{2}, \dfrac{1}{3}, -\dfrac{1}{4}, \dfrac{1}{5}, -\dfrac{1}{6}, \ldots \right\}$.

Exercise Set 12.3

1. $-\dfrac{1}{2} + \dfrac{1}{4} - \dfrac{1}{8} + \dfrac{1}{16} - \cdots$

3. $\dfrac{3}{5} + \dfrac{5}{11} + \dfrac{9}{29} + \dfrac{17}{83} + \cdots$

5. $\ln\dfrac{1}{2} + \ln\dfrac{2}{3} + \ln\dfrac{3}{4} + \ln\dfrac{4}{5} + \cdots$

7. Converges to $\dfrac{7}{6}$.   **9.** Diverges.

11. Converges; $\dfrac{27}{23}$.

13. Converges to $\dfrac{1}{1 - \dfrac{1}{2+x}} = \dfrac{x+2}{x+1}$.

15. Converges to $-\dfrac{1}{2}$.   **17.** Diverges.

19. Converges to $\dfrac{1}{3}$.   **21.** Diverges.

23. Converges to $\dfrac{3}{3 - \sqrt{2}}$.

25. a. $\displaystyle\sum_{k=1}^{\infty} \dfrac{3}{10^k}$ **b.** $\dfrac{1}{3}$

27. a. $\displaystyle\sum_{k=1}^{\infty} \dfrac{92}{100^k}$ **b.** $\dfrac{92}{99}$

29. a. $\displaystyle\sum_{k=1}^{\infty} \dfrac{412}{1000^k}$ **b.** $\dfrac{412}{999}$

35. $5h$ **37.** No, if $c = 0$.

41. $S_n = S_{n-1} + a_n$

43.

S_5	S_{10}	S_{20}	S_{50}	S_{100}
2.736625514	2.96531694	2.999398543	2.999999997	3.

The actual sum is $\dfrac{1}{1 - \dfrac{2}{3}} = 3$.

45.

S_5	S_{10}	S_{20}	S_{50}	S_{100}
$-.3718688642$	-1.128682959	-1.295565042	-1.301540778	-1.301541025

The actual sum ≈ -1.301541025.

Exercise Set 12.4

1. Diverges   **3.** Diverges   **5.** Converges

7. Diverges   **9.** Converges   **11.** Diverges

13. Diverges   **15.** Converges   **17.** Diverges

19. Diverges   **23.** $n > \sqrt{\ln 25} \approx 1.794$

Exercise Set 12.5

1. Converges   **3.** Converges   **5.** Diverges

7. Diverges   **9.** Converges   **11.** Converges

13. Diverges   **15.** Converges   **17.** Diverges

19. Diverges   **21.** Converges   **23.** Diverges

25. Converges   **27.** Diverges   **29.** Converges

Exercise Set 12.6

1. Diverges; ratio test.

3. Converges; ratio test.

5. Converges; ratio test.

7. Converges; limit comparison test using $\displaystyle\sum \dfrac{1}{k^2}$.

9. Converges; root test.

11. Diverges, by Theorem 7.

13. Diverges; ratio test.

15. Converges; ratio test.

17. Converges; ratio test.

19. Diverges; ratio test.

21. All $x > 0$.

23. Converges for $x < 1$ and diverges for $x \geq 1$.

Exercise Set 12.7

1. Absolutely   **3.** Absolutely

5. Absolutely   **7.** Conditionally

9. Conditionally   **11.** Absolutely

13. Diverges   **15.** Conditionally

17. Absolutely   **19.** Absolutely

21. Diverges   **23.** Absolutely

25. Conditionally   **27.** Absolutely

29. Absolutely   **31.** All x

33. All x **35.** $n \geq 98$

Review Exercises—Chapter 12

1. Diverges   **3.** Diverges

5. Converges to 0   **7.** Diverges

9. Diverges

11. $\dfrac{6}{5}$

13. 2

15. $37 \displaystyle\sum_{k=1}^{\infty} \dfrac{1}{100^k} = \dfrac{37}{99}$

17. $\dfrac{1}{48}$

19. $\dfrac{10}{9}$

21. Converges

23. Diverges

25. Converges

27. Converges

29. Converges

31. Converges

33. Diverges

35. Converges

37. Diverges

39. Converges

41. Converges

43. Converges

45. Converges

47. Diverges

49. Converges

51. Converges

53. Converges

55. Converges

57. Converges

59. Converges conditionally

61. Diverges

63. Converges absolutely

65. Diverges

67. Converges absolutely

69. $\dfrac{1}{13}$

71. $\dfrac{1}{11}$

73. $-1 < c \le 1$

75. Σb_k must diverge

77. a. 90.6192 meters
b. 110 meters

79. a. Converges
b. Converges
c. Converges

CHAPTER 13

Exercise Set 13.1

1. $1 - x + \dfrac{1}{2}x^2 - \dfrac{1}{6}x^3 + \dfrac{1}{24}x^4$

3. $\dfrac{\sqrt{2}}{2} - \dfrac{\sqrt{2}}{2}\left(x - \dfrac{\pi}{4}\right) - \dfrac{\sqrt{2}}{4}\left(x - \dfrac{\pi}{4}\right)^2$
$\qquad + \dfrac{\sqrt{2}}{12}\left(x - \dfrac{\pi}{4}\right)^3 + \dfrac{\sqrt{2}}{48}\left(x - \dfrac{\pi}{4}\right)^4$
$\qquad\qquad - \dfrac{\sqrt{2}}{240}\left(x - \dfrac{\pi}{4}\right)^5 - \dfrac{\sqrt{2}}{1440}\left(x - \dfrac{\pi}{4}\right)^6$

5. $x - \dfrac{1}{2}x^2 + \dfrac{1}{3}x^3 - \dfrac{1}{4}x^4$

7. $x + \dfrac{1}{3}x^3$

9. $1 + x^2$

11. $1 - \dfrac{1}{2}x - \dfrac{1}{8}x^2 - \dfrac{1}{16}x^3$

13. $\sqrt{2} + \sqrt{2}\left(x - \dfrac{\pi}{4}\right) + \dfrac{3\sqrt{2}}{2}\left(x - \dfrac{\pi}{4}\right)^2$
$\qquad\qquad\qquad + \dfrac{11\sqrt{2}}{6}\left(x - \dfrac{\pi}{4}\right)^3$

15. $\dfrac{1}{2} - \dfrac{1}{4}x + \dfrac{1}{48}x^3$

19. True

Exercise Set 13.2

1. $e^{-x} = 1 - x + \dfrac{1}{2}x^2 - \dfrac{1}{6}x^3 + \dfrac{e^{-c}}{24}x^4$;
$\qquad\qquad\qquad\qquad c$ is between 0 and x

3. $\cos x = 1 - \dfrac{1}{2}x^2 + \dfrac{1}{24}x^4 - \dfrac{\sin c}{120}x^5$;
$\qquad\qquad\qquad\qquad c$ is between 0 and x

5. $\mathrm{Tan}^{-1}\, x = x - \dfrac{1}{3}x^3 + \dfrac{c - c^3}{(1 + c^2)^4}x^4$;
$\qquad\qquad\qquad\qquad c$ is between 0 and x

7. $\dfrac{1}{1 + x^2} = \dfrac{1}{2} - \dfrac{1}{2}(x - 1) + \dfrac{1}{4}(x - 1)^2$
$\qquad + 4\dfrac{c - c^3}{(1 + c^2)^4}(x - 1)^3$; c is between 1 and x

9. $\sec x = \sqrt{2} + \sqrt{2}\left(x - \dfrac{\pi}{4}\right) + \dfrac{3\sqrt{2}}{2}\left(x - \dfrac{\pi}{4}\right)^2$
$\qquad + \dfrac{(\sin c)(5 + \sin^2 c)}{6\cos^4 c}\left(x - \dfrac{\pi}{4}\right)^3$; c is between $\dfrac{\pi}{4}$ and x

11. $\sinh x = x + \dfrac{1}{6}x^3 + \dfrac{\cosh c}{5!}x^5$; c is between 0 and x

13. $\cosh x = \dfrac{5}{4} + \dfrac{3}{4}(x - \ln 2) + \dfrac{5}{8}(x - \ln 2)^2 + \dfrac{1}{8}(x - \ln 2)^3$
$\qquad + \dfrac{1}{24}(\cosh c)(x - \ln 2)^4$; c is between $\ln 2$ and x

21. $P_n(x) = 1 - x^2 + x^4 - x^6 + \cdots$, using all terms of
degree $\le n$.

Exercise Set 13.3

Note: In Problems 1–20, there are many ways to handle the R_n and make accuracy estimates.

1. $\ln(1.5) \approx 0.416\overline{6}$

3. $\sqrt{3.91} \approx 1.99737$

5. $\cos 1 \approx 0.540317$

7. $\sqrt[3]{10} \approx 2.152\overline{7}$

9. $|R_2(x)| \le 0.0000208$

11. $|R_1(x)| \le 0.00125$

13. $|R_1(x)| < 0.000118$

15. $|R_1(x)| < 0.00005$

17. $n = 6$

19. n is 2.

21. 2.7183

23. $|R_1(x)| < 0.005483$

27. 0.31

29. 0.9

Exercise Set 13.4

1. $(-1, 1)$

3. $(-\infty, \infty)$

5. $(-\infty, \infty)$

7. $[-1, 1)$

9. $(-1, 1]$

11. $[-1, 1]$

13. $(-\infty, \infty)$

15. Converges only for $x = 1$.

17. $\left(\dfrac{1}{3}, 1\right]$

19. $(-1, 1)$

21. $[2, 4]$

23. $(2 - e, 2 + e)$

25. $\left(-\dfrac{3}{7}, \dfrac{1}{7}\right)$

27. $(-1, 1)$

29. $[-2, 4)$

31. $[-e, e)$

33. $(-\infty, \infty)$

35. $[-1, 1]$

Exercise Set 13.5

1. $\sum_{k=0}^{\infty} 2^k x^k; \quad r = \dfrac{1}{2}$

3. $\sum_{k=0}^{\infty}(-1)^k 4^k x^{2k}; \quad r = \dfrac{1}{2}$

5. $\sum_{k=0}^{\infty}(-1)^k x^{2k+1}; \quad r = 1$

7. $-1 + \sum_{k=1}^{\infty}(-1)^{k+1} 2 x^k; \quad r = 1$

9. $\sum_{k=0}^{\infty} x^{4k}; \quad r = 1$

11. $-2 \sum_{k=1}^{\infty}(-1)^k k x^{k-1}; \quad r = 1$

13. $\sum_{k=1}^{\infty} k(-1)^{k+1} x^{2k-1}; \quad r = 1$

15. $\sum_{k=0}^{\infty}(2k + 1)x^{2k}; \quad r = 1$

17. $\sum_{k=1}^{\infty} 2k(-1)^{k+1} 4^k x^{2k-1}; \quad r = \dfrac{1}{2}$

19. $\sum_{k=1}^{\infty} 2k(-1)^{k-1} x^{k-1}; \quad r = 1$

21. $-\sum_{k=0}^{\infty} \dfrac{x^{k+1}}{k + 1}; \quad r = 1$

23. $2 \sum_{k=0}^{\infty} \dfrac{(-1)^k 4^k x^{2k+1}}{2k + 1}; \quad r = \dfrac{1}{2}$

25. $\sum_{k=0}^{\infty} \dfrac{(-1)^k x^{k+1}}{4^{k+1} \cdot k + 1} + \ln(4); \quad r = 4$

27. $\sum_{k=0}^{\infty}(-1)^k \dfrac{x^{2k+2}}{k + 1}; \quad r = 1$

Exercise Set 13.6

1. $\sum_{k=0}^{\infty} \dfrac{2^k x^k}{k!}; \quad$ all x

3. $\sum_{k=0}^{\infty} \dfrac{\sqrt{2}(-1)^{k(k+1)/2}\left(x - \dfrac{\pi}{4}\right)^k}{2(k!)}; \quad$ all x

5. $5 + 4(x - 2) + (x - 2)^2; \quad$ all x

7. $\sum_{k=1}^{\infty}(-1)^{k-1} \dfrac{x^k}{k 3^k} + \ln(3); \quad -3 < x \leq 3$

9. $\sum_{k=0}^{\infty} \dfrac{x^k (\ln 2)^k}{k!}; \quad$ all x

11. $1 + \sum_{k=1}^{\infty} \dfrac{\dfrac{3}{2}\left(\dfrac{3}{2} - 1\right) \cdots \left(\dfrac{3}{2} - k + 1\right) x^k}{k!}; \quad |x| < 1$

13. See Example 8.

15. $\sum_{k=0}^{\infty} \dfrac{(-1)^k x^{2k}}{(2k + 1)!}; \quad$ all x

17. $\sum_{k=1}^{\infty} \dfrac{\sqrt{2}}{2}\left(x - \dfrac{\pi}{4}\right)^k \left[\dfrac{(-1)^{(k-1)(k-2)/2}}{(k - 1)!} + \dfrac{\pi}{4} \dfrac{(-1)^{(k-1)k/2}}{k!}\right]$

$$+ \dfrac{\pi\sqrt{2}}{8}$$

19. $1 + \sum_{k=1}^{\infty} \dfrac{2^{2k-1} x^{2k}(-1)^k}{(2k)!}; \quad$ all x

21. $1 + x^2$. Use $\dfrac{d}{dx}(\tan x) = \sec^2 x$.

23. $\sum_{k=0}^{\infty} \dfrac{(-1)^k x^{4k}}{(2k)!}$

25. $\sum_{k=0}^{\infty} \dfrac{x^{2k+1}}{(2k)!}$

27. 0.035

29. 0.095

31. 0.485

33. $1 + x - \sum_{k=2}^{\infty}(-1)^k \dfrac{1 \cdot 3 \cdots (2k - 3)}{k!} x^k;$

$|x| < \dfrac{1}{2}$

35. $27 + \sum_{k=1}^{\infty} \dfrac{\dfrac{3}{2}\left(\dfrac{3}{2} - 1\right) \cdots \left(\dfrac{3}{2} - k + 1\right) x^k}{3^{k-3} k!}; \quad |x| < 3$

Review Exercises—Chapter 13

1. $2x - \dfrac{4}{3}x^3 + \dfrac{4}{15}x^5$

3. $\dfrac{\pi\sqrt{2}}{8} + \dfrac{\sqrt{2}}{2}\left(1 - \dfrac{\pi}{4}\right)\left(x - \dfrac{\pi}{4}\right)$

$\qquad - \dfrac{\sqrt{2}}{4}\left(2 + \dfrac{\pi}{4}\right)\left(x - \dfrac{\pi}{4}\right)^2 + \dfrac{\sqrt{2}}{12}\left(\dfrac{\pi}{4} - 3\right)\left(x - \dfrac{\pi}{4}\right)^3$

5. $1 + x^2$

7. $\dfrac{\pi}{4} + \dfrac{1}{2}(x - 1) - \dfrac{1}{4}(x - 1)^2 + \dfrac{1}{12}(x - 1)^3$

9. $1 + (\log 2)x + \dfrac{(\log 2)^2}{2}x^2 + \dfrac{(\log 2)^3}{6}x^3$

11. $\left| R_3\left(\dfrac{17\pi}{90}\right)\right| \le 9.89775 \cdot 10^{-7}$

13. $|R_2(35)| \le 0.000079$

15. $\left| R_3\left(\dfrac{1}{4}\right)\right| < 0.000488$

21. $(2, 4)$ **23.** $[1, 3)$ **25.** $(-1, 1]$

27. Replace $k - 1$ by k on the right.

29. $a_n = \dfrac{c^n a_0}{n!}(-1)^n$; $a_0 e^{-cx}$ **31.** $\sum_{k=0}^{\infty}(-1)^k x^{4k}$; $|x| < 1$

33. $1 + \displaystyle\sum_{k=1}^{\infty} \dfrac{\dfrac{1}{3}\left(\dfrac{1}{3} - 1\right) \cdots \left(\dfrac{1}{3} - k + 1\right)x^{2k}}{k!}$

35. $1 - \dfrac{1}{2\pi^2}\left(x - \dfrac{\pi^2}{4}\right)^2 + \dfrac{1}{\pi^2}\left(x - \dfrac{\pi^2}{4}\right)^3$

37. $\displaystyle\sum_{k=0}^{\infty} \dfrac{x^{2k+2}}{k!}$ **39.** $e^{1/2} \approx 1.649$

CHAPTER 14

Exercise Set 14.1

1. $P = (0, 1)$

3. $P = (0, 0)$

5. $P = (1, -1)$

7. $P = (3, 0)$

9. $P = \left(\sqrt{2}, \dfrac{\pi}{4}\right)$

11. $P = (-3, 0)$

13. $P = \left(-2, \dfrac{2\pi}{3}\right)$

15. $P = \left(-2\sqrt{2}, \dfrac{15\pi}{4}\right)$

17. Symmetric about the x-axis

19. Symmetric about the y-axis

21. x-axis, origin, y-axis

23. x-axis, y-axis, origin

25. $r = 2$

27. $r^2 = \dfrac{4}{1 + 3\cos^2 \theta}$

29. $r = \dfrac{6}{\cos \theta} = 6 \sec \theta$

31. $r = -2 \sin \theta$

33. $x^2 + y^2 - 4y = 0$

35. $x^4 + x^2 y^2 - y^2 = 0$

37. $x = 4$

39. $y^2 = 1 + 2x$

41.

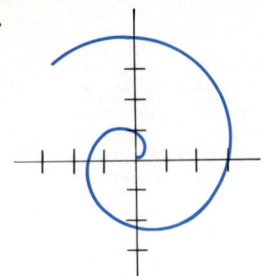

43.

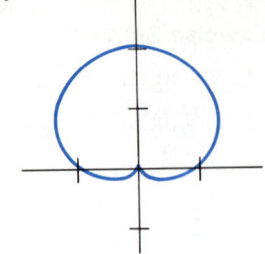

Exercise Set 14.2

1.

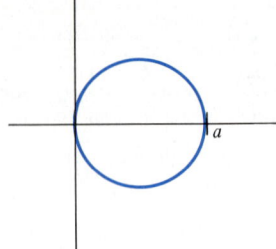

45.

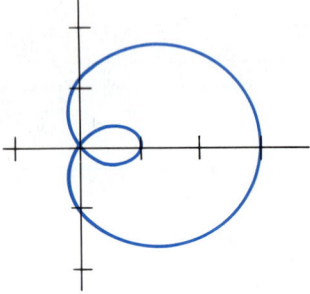

3.

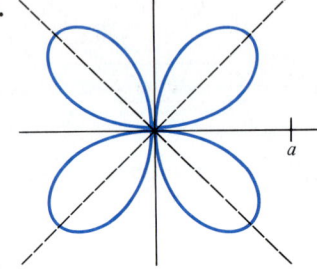

5.

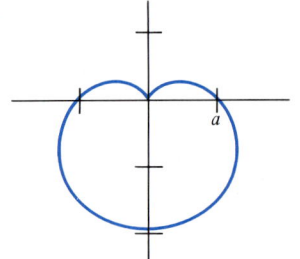

47.

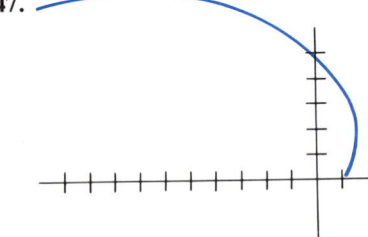

7.

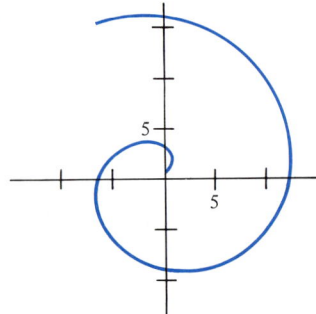

49.

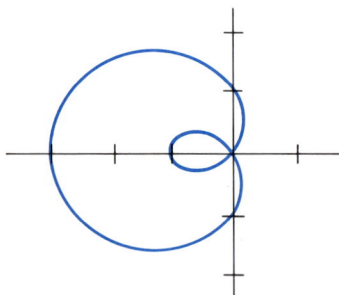

9.

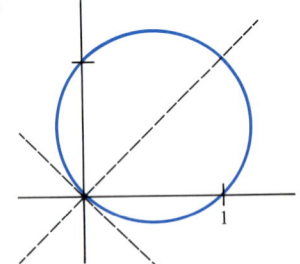

51. Center $\left(\sqrt{2}, \dfrac{3\pi}{4}\right)$ and radius $\sqrt{2}$

53. Center $\left(\dfrac{a}{2}, 0\right)$ and radius $\dfrac{a}{2}$

57. $\dfrac{9\pi}{4}$

59. $(0, 0)$, $(4, 0)$, $\left(4\sqrt{2}, \dfrac{\pi}{2}\right)$

11. $\left(\sqrt{2}, \dfrac{\pi}{4}\right)$, $\left(-\sqrt{2}, \dfrac{5\pi}{4}\right)$, $(0, \theta)$

13. $\left(\dfrac{3}{2}, \dfrac{\pi}{6}\right), \left(\dfrac{3}{2}, \dfrac{5\pi}{6}\right), (0, \theta)$

15. $\left(\dfrac{3a}{2}, \dfrac{\pi}{3}\right), \left(\dfrac{3a}{2}, \dfrac{5\pi}{3}\right), (0, \theta)$

17. $\left(a, \dfrac{\pi}{2}\right), \left(a, \dfrac{3\pi}{2}\right), (0, \theta)$

19. $\left(\dfrac{1}{\pi/4 + k\pi}\right), k = 0, \pm 1, \pm 2, \ldots$ **21.** $x^2 = a^2 \mp 2ay$

23. $y^2 = 1 - 2x$

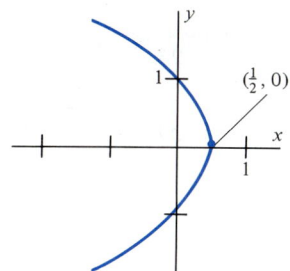

25. $\dfrac{\left(x + \dfrac{4}{3}\right)^2}{\left(\dfrac{8}{3}\right)^2} + \dfrac{y^2}{\left(\dfrac{4}{\sqrt{3}}\right)^2} = 1$ (ellipse)

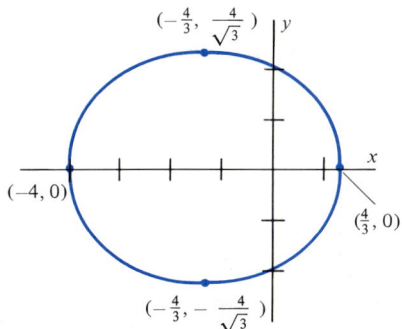

27. $y^2 = 4x + 4$ (parabola)

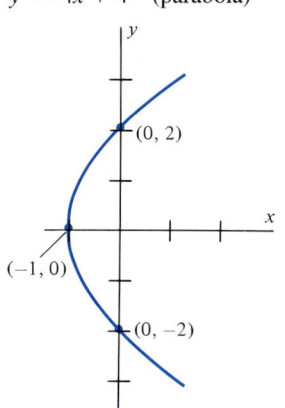

Exercise Set 14.3

1. $\dfrac{3\pi}{4}$ 　　　　　　　　　**3.** π

5. $\dfrac{9}{4}\pi + 4$ 　　　　　　　**7.** $\dfrac{33\pi}{8} + 4\sqrt{2} + \dfrac{1}{4}$

9. π 　　　　　　　　　　　**11.** $\dfrac{\pi a^2}{2}$

13. $\dfrac{\pi a^2}{4}$ 　　　　　　　　**15.** $\dfrac{\pi}{2}$

17. $\dfrac{3a^2\pi}{2} - 4a^2$ 　　　　　**19.** $\dfrac{\pi - 2}{2}$

21. $\dfrac{1}{4}e^{2\pi} - \dfrac{\pi^3}{6} - \dfrac{1}{4}$ 　　　　**23.** $2\pi + 3\sqrt{3}$

25. π

27.

	$n = 10$	$n = 100$	$n = 200$
a.	2.587376	2.494899	2.489281
b.	4.885851	4.964704	4.966249
c.	0.138355	0.091575	0.089359

Exercise Set 14.4

1. $y = x + 6$ 　　　　　　**3.** $y = \dfrac{2}{3}x + \dfrac{4}{3}$

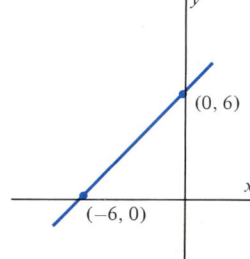

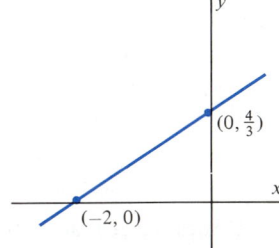

5. $y = x - 2;\ \ x \geq 1$ 　　**7.** $y = x^3$

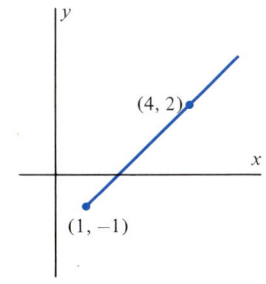

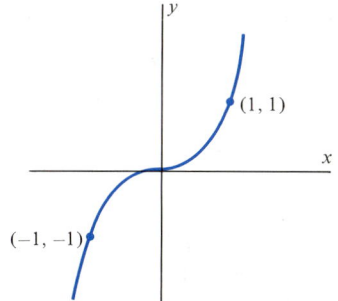

9. $x^2 - y^2 = 1$

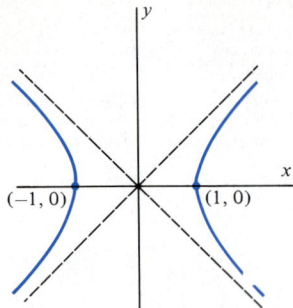

11. $y^2 = 4x^2(1 - x^2)$

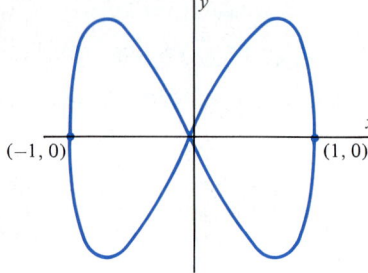

13. Slope $= 4$, $y = 4x - 3$

15. Slope $= -\sqrt{3}$, $y = -\sqrt{3}x + 2$

17. Slope $= -48$, $y = -48x + 29$

19. Slope $= -1$, $y = -x + 2$

21. 0   **23.** $\dfrac{5}{\sqrt{3}}$ **25.** $\dfrac{-2}{\pi}$

27. Vertical tangent: $(1, 0)$, $(-1, 0)$
Horizontal tangent: $(0, 1)$, $(0, -1)$

29. Vertical tangent at $(4, 2)$
Horizontal tangent at $(5, -1)$

31. Vertical tangent at $(6, 0)$
Horizontal tangent at $\left(\dfrac{27}{4}, \dfrac{1}{4}\right)$

33. $(1, 5)$

35. $x(\theta) = f(\theta) \cos \theta = a \sin 3\theta \cdot \cos \theta$
$y(\theta) = f(\theta) \sin \theta = a \sin 3\theta \cdot \sin \theta$

37. $(0, 0)$, $\left(\dfrac{3}{2}, \dfrac{\pi}{6}\right)$, $\left(\dfrac{3}{2}, \dfrac{5\pi}{6}\right)$

39. Vertical tangent at $(2, 0)$, $\left(\dfrac{1}{2}, \dfrac{2\pi}{3}\right)$, $\left(\dfrac{1}{2}, \dfrac{4\pi}{3}\right)$

41. a. $y = -(x^2 - 8x + 8)$
$y = x + 2$
  b. Yes. At $(2, 4)$ and $(5, 7)$.
  c. The particles collide at $t = -2$ and at $t = 1$ at the points $(2, 4)$ and $(5, 7)$.

Exercise Set 14.5

1. $\dfrac{8}{27}(2^{3/2} - 1)$ **3.** $2(3^{3/2} - 1)$ **5.** 36

7. $(5^{3/2} - 8^{3/2}) + (12\sqrt{2} - 3\sqrt{5} + 12 \ln(\sqrt{5} - 1)$
$- 12 \ln(\sqrt{8} - 2))$

9. 3   **11.** $\sqrt{2}(e^\pi - 1)$

13. $\dfrac{16\pi}{3}(8^{3/2} - 8)$ **15.** $\dfrac{\pi}{6}[27 - 5^{3/2}]$

17. $\pi\left[133\sqrt{5} - \dfrac{7}{2}\sqrt{2} + \dfrac{3}{2}\ln(\sqrt{5} + 2) - \dfrac{3}{2}\ln(\sqrt{2} + 1)\right]$

19. 4   **21.** $\dfrac{\sqrt{5}}{2}[e^{2\pi} - 1]$ **23.** 4π

25. 3π **27.** $\sqrt{2}(e^{2\pi} - 1)$ **29.** $\dfrac{2}{5}\sqrt{2}\pi(e^{2\pi} - 1)$

31. $S = \displaystyle\int_a^b 2\pi x(t)\sqrt{(x'(t))^2 + (y'(t))^2}\, dt$, $a \le t \le b$

33. $\dfrac{\pi}{64}[204\sqrt{2} - 36\sqrt{5} - \ln(17 + 12\sqrt{2}) + \ln(9 + 4\sqrt{5})]$

Review Exercises—Chapter 14

1. $x^2 + y^2 = 25$

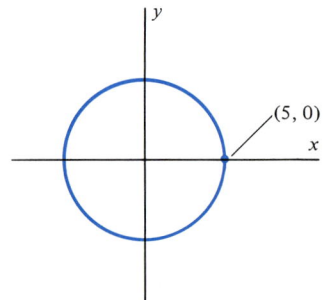

3. $r = \theta/\pi$

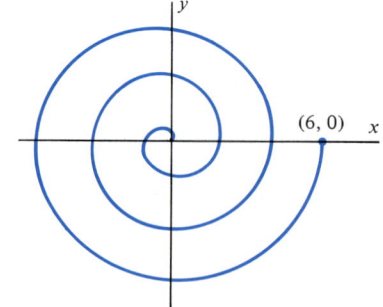

5. $y = 2x\sqrt{1 - x^2}$

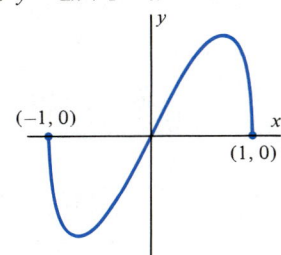

7. $y = x \tan \pi\sqrt{x^2 + y^2}$

9. $y^2 = 4x^2(1 - x^2)$

11. $x(t) = 3 \cos t$
$y(t) = 2 \sin t$

13. $x(t) = 2t^2 + 4t + 5$
$y(t) = t$

15.

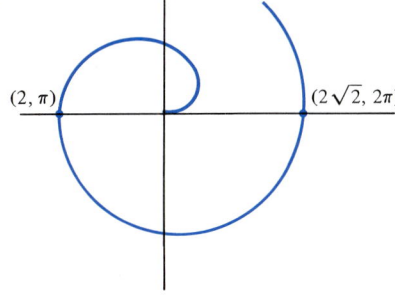

17. $x = \dfrac{y^2}{9} - 1$ (parabola)

19.

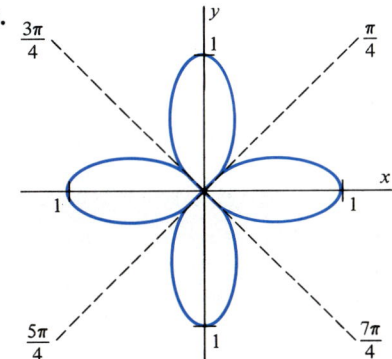

21.

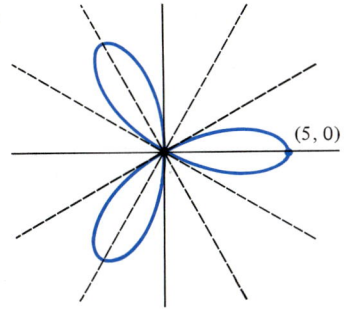

23.

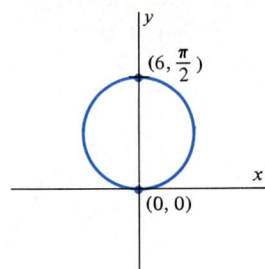

25.

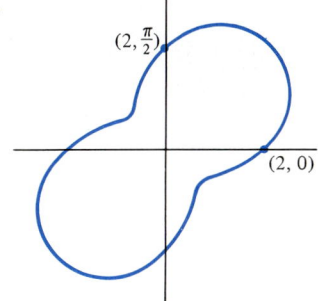

27. The y-axis.

29. 6π

31. $\dfrac{3a^2\pi}{2}$

33. 9π

35. $\dfrac{1}{27}(22^{3/2} - 13^{3/2})$

37. $\displaystyle\int_0^1 3\sqrt{1 + t^4}\, dt \approx 3.2683$

39. π

41. $\dfrac{\pi}{2}$

43. $2\sqrt{2}$

45. $\dfrac{\pi}{8} - \dfrac{1}{4}$

47. 4π

49. $\dfrac{8\pi}{3} - 2\sqrt{3}$

CHAPTER 15

Exercise Set 15.1

1.

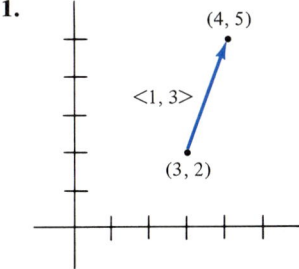

3.

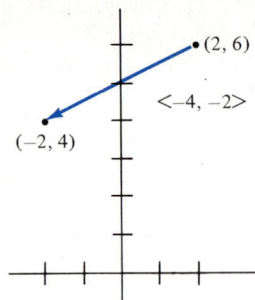

5.

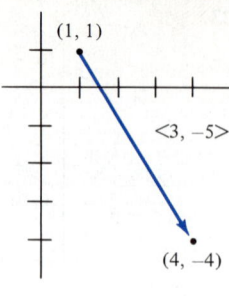

7. $\langle -1, 2 \rangle$

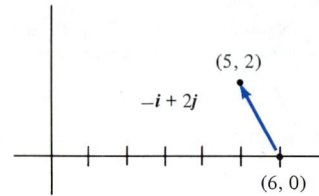

9. $\langle -3, 2 \rangle$

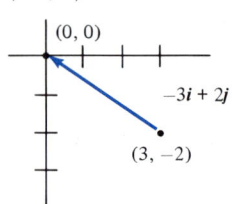

11. $\langle -3, -3 \rangle$

15. $\langle -6, -6 \rangle$

19. $\langle 10, 14 \rangle$

23. $\langle 7, 11 \rangle$

27. $\langle 19, 19 \rangle$

31. $\dfrac{\pi}{2}$

35. $\dfrac{5\pi}{3}$

39. $|\alpha|\sqrt{1 + \alpha^2}$

43. $3\sqrt{5}$

47. $\left\langle -\dfrac{1}{2}, \dfrac{\sqrt{3}}{2} \right\rangle$

49. $t\left\langle \dfrac{1}{\sqrt{10}}, \dfrac{3}{10} \right\rangle$

51. a. $D = (2, 3)$
 b. $D = (10, 3)$ or $(0, -1)$

61. When v and w have the same direction.

13. $\langle -3, -14 \rangle$

17. $\langle -1, 9 \rangle$

21. $\langle -1, 5 \rangle$

25. $\langle 4, 6 \rangle$

29. $\langle 12, 18 \rangle$

33. $\dfrac{5\pi}{6}$

37. $\sqrt{37}$

41. $\sqrt{2}$

45. $-i$

Exercise Set 15.2

1. 1

5. -15

9. a. 14
 c. -14
 e. 78
 g. $\dfrac{14}{\sqrt{34}}$
 i. $\left\langle -\dfrac{7}{5}, \dfrac{14}{5} \right\rangle$

11. a. $\dfrac{7}{\sqrt{26}}$
 c. $\dfrac{7}{\sqrt{13}}$
 e. $\left\langle \dfrac{17}{26}, -\dfrac{85}{26} \right\rangle$

13. b, and **c.**

15. $\left\langle \dfrac{1}{\sqrt{5}}, -\dfrac{2}{\sqrt{5}} \right\rangle$ and $\left\langle -\dfrac{1}{\sqrt{5}}, \dfrac{2}{\sqrt{5}} \right\rangle$

19. $\left\langle \dfrac{4}{5}, -\dfrac{2}{5} \right\rangle + \left\langle \dfrac{1}{5}, \dfrac{2}{5} \right\rangle$

21. $4\sqrt{2}$

25. v and w being parallel and of same direction.

3. -28

7. $-\dfrac{7}{\sqrt{170}}$

b. -28
 d. 42
 f. -56
 h. $-\dfrac{1}{\sqrt{13}}$
 j. $\left\langle \dfrac{16}{13}, \dfrac{24}{13} \right\rangle$

b. 2
 d. $\left\langle \dfrac{35}{26}, \dfrac{7}{26} \right\rangle$
 f. $\left\langle \dfrac{51}{13}, \dfrac{35}{13} \right\rangle$

23. $1000\sqrt{2}$ ft-lb

Exercise Set 15.3

1.

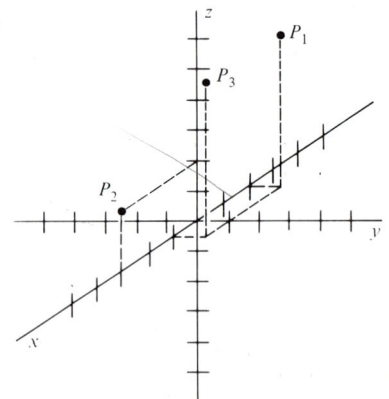

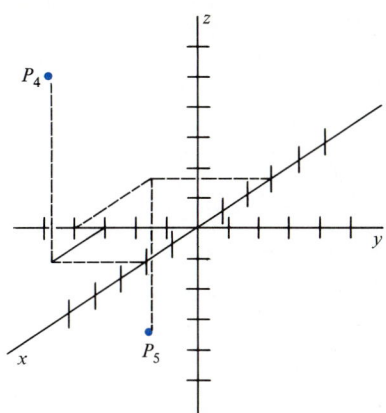

3. a. $2\sqrt{2}$ **b.** $6\sqrt{2}$ **c.** $\sqrt{22}$ **d.** $\sqrt{a^2 + b^2 + c^2}$

5. $(-3, 3, -3)$

7. Center $(0, 0, 2)$ and radius 3

9. Center $(2, -1, 3)$ and radius 4

11. Center $(3, -1, -2)$ and radius 5

13. $\langle -16, 29, 5 \rangle$ **19.** 1

15. $\langle 12, -13, 15 \rangle$ **21.** $\langle -12, 1, -7 \rangle$

17. $9\sqrt{6}$ **23.** -6

25. $\left\langle \dfrac{3}{\sqrt{17}} + \dfrac{5}{\sqrt{35}}, \dfrac{-2}{\sqrt{17}} - \dfrac{1}{\sqrt{35}}, \dfrac{2}{\sqrt{17}} + \dfrac{3}{\sqrt{35}} \right\rangle$

27. $\left\langle \dfrac{23}{7}, \dfrac{-23}{35}, \dfrac{69}{35} \right\rangle$ **29.** $\dfrac{-3}{\sqrt{6}}$

31. a. $\langle 2, 1, 2 \rangle$ **b.** $\langle 4, -1, -5 \rangle$ **c.** $\langle 6, 3, -1 \rangle$

33. a. $\cos \alpha = \dfrac{3}{\sqrt{14}}$, $\cos \beta = \dfrac{-1}{\sqrt{14}}$, $\cos \gamma = \dfrac{2}{\sqrt{14}}$

     b. $\cos \alpha = \dfrac{6}{\sqrt{41}}$, $\cos \beta = -\dfrac{2}{\sqrt{41}}$, $\cos \gamma = \dfrac{1}{\sqrt{41}}$

     c. $\langle -1, 2, -4 \rangle$

35. a. $\left\langle \dfrac{2}{3}, -\dfrac{4}{3}, \dfrac{4}{3} \right\rangle$ and $\left\langle -\dfrac{2}{3}, \dfrac{4}{3}, -\dfrac{4}{3} \right\rangle$

     b. $a = -3$

     c. $-\dfrac{5}{\sqrt{14}} i + \dfrac{10}{\sqrt{14}} j - \dfrac{15}{\sqrt{14}} k$

     d. $a = -\dfrac{3}{2}$, $b = \dfrac{1}{2}$

37. $t = 2$ **39. c.**

Exercise Set 15.4

1. $-7i + 8j + 11k$ **3.** $5i - 4j - k$

5. 12   **7.** $3i + 10j + k$

9. $-16i + 8k$ **11.** $37i + 30j - 11k$

13. 40   **15.** $5\sqrt{5}$

17. 8   **19.** $\dfrac{5}{2}\sqrt{2}$

21. $\dfrac{5\sqrt{10}}{2}$ **23.** $\dfrac{23}{2}$

25. k or $-k$

27. $\dfrac{i}{\sqrt{3}} - \dfrac{j}{\sqrt{3}} - \dfrac{k}{\sqrt{3}}$ and $-\dfrac{i}{\sqrt{3}} + \dfrac{j}{\sqrt{3}} + \dfrac{k}{\sqrt{3}}$

31. 43

Exercise Set 15.5

1. $x(t) = 1 + t$, $y(t) = 2 + t$, $z(t) = 3 - t$

3. $x(t) = 3t$, $y(t) = -t$, $z(t) = 5t$

5. $x(t) = -4 + t$, $y(t) = 2 + 3t$, $z(t) = 1 + 2t$

7. $x(t) = 1 - 10t$, $y(t) = 2 + 3t$, $z(t) = t$

9. $\dfrac{x - 7}{1} = \dfrac{y + 6}{4} = \dfrac{z - 3}{-2}$

11. $\dfrac{x - 0}{1} = \dfrac{y + 6}{8} = \dfrac{z - 4}{4}$

13. $\dfrac{x - 3}{3} = \dfrac{y + 1}{-1} = \dfrac{z - 5}{5}$

15. $x(t) = 3t$, $y(t) = 2t$, $z(t) = 5t$

17. $x(t) = 4t - 4$, $y(t) = -2t + 2$, $z(t) = 3t - 3$

19. $i + 4j - 2k$ **21.** $4i - 2j + 3k$

23. $-\dfrac{a \cdot b}{|b|^2}$ **25.** $\sqrt{3}$

27. $\dfrac{\sqrt{210}}{3}$ **29.** $\dfrac{1}{15}\sqrt{1270}$

31. d. $\dfrac{16}{\sqrt{53}}$ **33.** $(2, 1, 2)$

35. $x + 2y - z = 8$ **37.** $2x - 2y + 3z = 19$

39. $x + 2y + z = 1$ **41.** $29x - 6y - 15z = -16$

43. $x - y = -2$ **45.** $2i - 3j + k$

47. $3x - 6y + 2z = 31$

49. $x(t) = \dfrac{27}{6} + \dfrac{7}{6}t$, $y(t) = \dfrac{9}{6} + \dfrac{5}{6}t$, $z(t) = t$

51. $\theta = \text{Cos}^{-1} \dfrac{1}{\sqrt{57}}$ **53.** $\dfrac{27}{\sqrt{38}}$

55. $x(t) = 2 + 2t,\; y(t) = 4 + 3t,\; z(t) = -3 - 7t$

57. $3x - y - z = 7$

Exercise Set 15.6

1. a. $\left(\sqrt{2}, \dfrac{\pi}{4}, 0\right)$ **b.** $\left(2, \dfrac{\pi}{6}, 3\right)$

     c. $\left(\sqrt{2}, \dfrac{3\pi}{4}, -2\right)$ **d.** $\left(2, \dfrac{2\pi}{3}, 4\right)$

     e. $\left(3, \dfrac{\pi}{2}, -5\right)$ **f.** $\left(2, \dfrac{3\pi}{4}, \sqrt{2}\right)$

3. a. $\left(1, 0, \dfrac{\pi}{2}\right)$ **b.** $\left(2, \dfrac{\pi}{4}, \dfrac{\pi}{4}\right)$

     c. $\left(2, \dfrac{7\pi}{4}, \dfrac{\pi}{4}\right)$ **d.** $\left(2\sqrt{2}, \dfrac{\pi}{3}, \dfrac{\pi}{4}\right)$

     e. $\left(2\sqrt{2}, \dfrac{5\pi}{6}, \dfrac{3\pi}{4}\right)$ **f.** $\left(4, \dfrac{3\pi}{4}, \dfrac{\pi}{4}\right)$

5. a. $\left(1, 0, \dfrac{\pi}{2}\right)$ **b.** $\left(2, -\dfrac{\pi}{4}, \dfrac{\pi}{4}\right)$

     c. $\left(2\sqrt{2}, \dfrac{\pi}{3}, \dfrac{\pi}{4}\right)$ **d.** $\left(2\sqrt{2}, \dfrac{\pi}{4}, \dfrac{\pi}{4}\right)$

     e. $\left(2, \dfrac{5\pi}{3}, \dfrac{\pi}{2}\right)$ **f.** $\left(2\sqrt{2}, \dfrac{\pi}{6}, \dfrac{\pi}{4}\right)$

7. $z = 3$

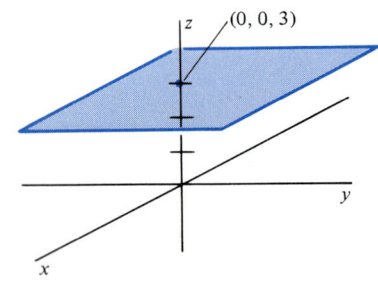

9. $z^2 = 4(x^2 + y^2)$

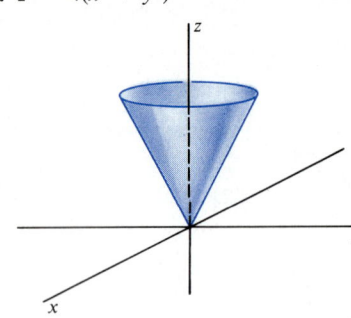

11. $x^2 + y^2 + z^2 = 4$

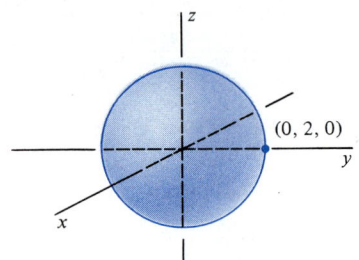

13. $x^2 - y^2 = \alpha^2$

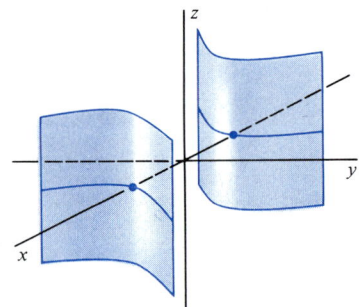

15. $r^2 = 9$ **17.** $r^2 = 9z$

19. $r^2 \cos^2 \theta + z^2 = 4$

21. $r^2 \cos \theta \sin \theta = 4 + ar \cos \theta$

23. $\rho^2 \sin^2 \phi = 9$

25. $\rho \sin^2 \phi = 9 \cos \phi$

27. $\rho^2 = \dfrac{4}{1 - \sin^2 \phi \sin^2 \theta}$

29. $\rho^2 \sin^2 \phi \cos \theta \sin \theta - a\rho \sin \theta \cos \theta = 4$

Exercise Set 15.7

	Equation	Name	Center	yz-sections	xz-sections	xy-sections
1.	$x^2 + y^2 + z^2 = 10$	sphere	$(0, 0, 0)$	circle*	circle*	circle*
3.	$\dfrac{x^2}{4} + \dfrac{y^2}{9} = z^2$	elliptic cone	$(0, 0, 0)$	hyperbola**	hyperbola**	ellipse*

5. $-\dfrac{x^2}{9} - \dfrac{y^2}{4} + z^2 = 1$	hyperboloid of two sheets	$(0, 0, 0)$	hyperbola	hyperbola	ellipse*
7. $\dfrac{(x+1)^2}{2} - \dfrac{(y-1)^2}{3} = z$	hyperbolic parabolid	none	parabola	parabola	hyperbola**
9. $\dfrac{(x+3)^2}{4} + \dfrac{(y-1)^2}{2} - \dfrac{z^2}{2} = 1$	hyperboloid of one sheet	$(-3, 1, 0)$	hyperbola	hyperbola	ellipse
11. $\dfrac{x^2}{9} + \dfrac{(y+2)^2}{6} = z$	elliptic paraboloid	none	parabola	parabola	ellipse*
13. $\dfrac{z^2}{4} + \dfrac{y^2}{9} = x$	elliptic paraboloid	none	ellipse*	parabola	parabola
15. $-x^2 + y^2 + z^2 = 1$	hyperboloid of one sheet	$(0, 0, 0)$	circle	hyperbola	hyperbola
17. $(x-3)^2 + (y+1)^2 + (z+3)^2 = 1$	sphere	$(3, -1, 3)$	circle*	circle*	circle*
19. $(x+1)^2 + (y-1)^2 = 36\left(z - \dfrac{85}{18}\right)$	paraboloid of revolution	none	parabola	parabola	circle*

21. Plane parallel to yz-plane

Direction of Ruling	Cross-section
23. z-axis	line
25. z-axis	hyperbola
27. z-axis	hyperbola
29. x-axis	parabola
31. y-axis	hyperbola
33. x-axis	parabola
35. y-axis	parabola

37. $y^2 = 4(x^2 + z^2)$

39. $y = x^2 + z^2$

Review Exercises—Chapter 15

1. $\sqrt{86}$

3. Center $(3, -2, 1)$, radius 2

5. $\dfrac{-1}{\sqrt{77}}$

7. $x(t) = 23t + 2$, $y(t) = -t + 1$, $z(t) = -9t - 3$

9. $7x - 9y + 3z = -22$

11. $8i + j - 2k$

13. $r(t) = (2 - 12t)i + 11tj + 9tk$

17.

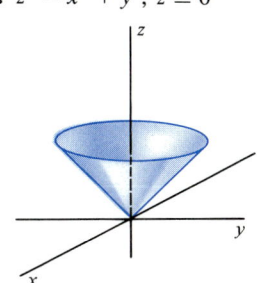

19. $\left(\dfrac{\sqrt{6}}{2}, \dfrac{\sqrt{6}}{2}, 1\right)$

21. $2x - y + 6z = 27$

23. a. $\dfrac{22}{5}$ **b.** $\dfrac{66}{25}j + \dfrac{88}{25}k$

25. $\dfrac{1}{2}\sqrt{118}$

27. $4x + z = 18$

29. $\dfrac{x-1}{3} = \dfrac{y-2}{7} = \dfrac{z-3}{-1}$

31. $z^2 = x^2 + y^2$, $z \geq 0$

*Section may degenerate to point or empty set.
**Section may degenerate to two lines.

33. $(x - 1)^2 + y^2 = 1$

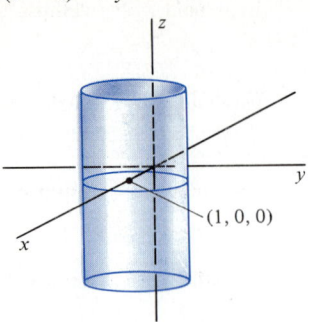

$(1, 0, 0)$

35. a. $\dfrac{10}{\sqrt{11}}$ **b.** $\dfrac{10}{7}i + \dfrac{15}{7}j - \dfrac{5}{7}k$

37. $\dfrac{x}{3} + \dfrac{y}{4} + \dfrac{z}{5} = 1$

39. $x(t) = 5,\ y(t) = -t + 2,\ z(y) = \dfrac{1}{2}t - 3$

41. $\sqrt{83}$

Equation	Name	Center	yz-section	xz-section	xy-section
45. $x^2 + \dfrac{y^2}{6} - \dfrac{z^2}{3} = 1$	hyperboloid of one sheet	$(0, 0, 0)$	hyperbola	hyperbola	ellipse
47. $\dfrac{y^2}{9} - \dfrac{x^2}{9} = z$	hyperbolic paraboloid	none	parabola	parabola	hyperbola*
49. $\dfrac{(x + 1)^2}{4} + \dfrac{y^2}{4} - \dfrac{(z - 2)^2}{9} = 1$	hyperboloid of one sheet	$(-1, 0, 0)$	hyperbola	hyperbola	circle

CHAPTER 16

Exercise Set 16.1

1.

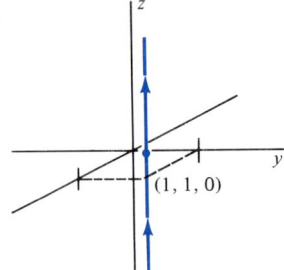

$(1, 1, 0)$

3.

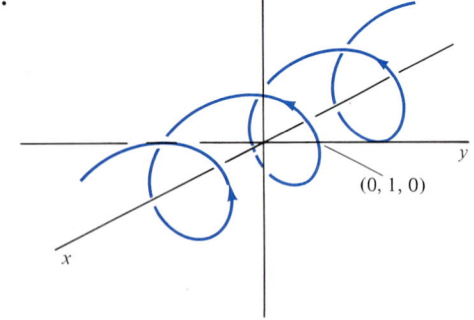

$(0, 1, 0)$

---

*Section may degenerate to two lines.

5.

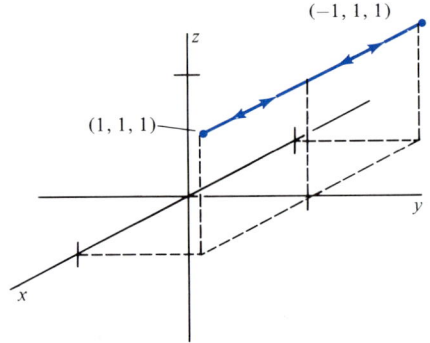

$(-1, 1, 1)$

$(1, 1, 1)$

7. $t \neq (2k + 1)\dfrac{\pi}{2},\quad k = 0, \pm 1, \pm 2, \ldots$

9. $t \geq 0$ and $t \neq 3$ **11.** $t \geq 0$

13. $(t + \sin t)i + (3t + \cos t)j + (t^2 + 1)k$

15. $(3t + 2 \sin t)i + (9t + 2 \cos t)j + (3t^2 + 2)k$

17. $t \sin t + 3t \cos t + t^2$

19. $(1 + t + 3\sqrt{t})i + (4 - 3t - t^2)j + (t + 3e^t)k$

21. $(1 + t - \sqrt{t}) \sin ti + (t - t^2) \sin tj + (t - e^t) \sin tk$

23. $\sqrt{t}(1 + t) + (1 - t)(1 - t^2) + te^t$

25. $\dfrac{\sqrt{2}}{2}i + j + \dfrac{\sqrt{2}}{2}k$ **27.** $i + k$

29. $6i + 7j + 14k$

31. $\left(-\dfrac{\pi}{2} + n\pi, \dfrac{\pi}{2} + n\pi\right)$, $n = 0, \pm 1, \ldots$

33. $[0, 1)$ and $(1, \infty)$ **35.** $[0, 1)$

37. $(-\infty, 1)$ and $(1, \infty)$

39. $600{,}000\pi$ One complete revolution takes $t = \dfrac{\pi}{500}$ seconds.

45. No conclusions in either case.

Exercise Set 16.2

1. $(0, \infty)$

3. $(-3, -2)$ and $(-2, \infty)$

5. $i + \dfrac{1}{2\sqrt{t}}j$

7. $\dfrac{1}{2\sqrt{t}}i - \dfrac{3}{2}t^{-5/2}j + \dfrac{2}{2t-1}k$

9. $\dfrac{1}{\sqrt{1-t^2}}i + \dfrac{t}{\sqrt{1+t^2}}j - 3t^2e^{-t^3}k$

11. $-\dfrac{1}{4}t^{-3/2}i + \dfrac{15}{4}t^{-7/2}j - \dfrac{4}{(2t-1)^2}k$

15. $(3\cos t - 2)i + (6t + 2\sin t)k$

17. $-\sin ti + (3t^2 + \sin^2 t - \cos^2 t)j - k$

19. $3e^{3t}\cos e^{3t}i + 6e^{6t}k$

23. $\dfrac{2}{3}t^{3/2}i + \dfrac{1}{2}e^{2t}j + \ln|t|k + C$

25. $(t\ln t - t)i + \ln|\ln t|j + C$

27. $(3 + t)i + \left(5 + \dfrac{t^3}{3}\right)j$

29. $\left(3 + \dfrac{t}{2} + \dfrac{1}{4}\sin 2t\right)i + \left(\dfrac{t}{2} - \dfrac{1}{4}\sin 2t - 6\right)j$
$\qquad\qquad\qquad\qquad\qquad + (3 - e^{-t})k$

31. $e^{-4t}i + 4e^{-4t}k$

33. $\left(\dfrac{b^2 - a^2}{2}\right)i + \left(\dfrac{b^3 - a^3}{3}\right)j - \left(\dfrac{b^4 - a^4}{4}\right)k$

35. $\dfrac{\sqrt{2}}{2}i + \dfrac{1}{2}j + \left(\dfrac{\pi}{8} + \dfrac{1}{4}\right)k$

39. False

Exercise Set 16.3

1. $48i + \dfrac{1}{4}j$ **3.** $\sqrt{2}i + 2j$

5. $-2i + j + 6k$ **7.** $\dfrac{1}{\sqrt{a^2 + b^2}}(-ai + bj)$

9. $t = \sqrt{2}$

11. $(64i + 2j + 2k) + t\left(48i + \dfrac{1}{4}j\right)$

13. $\dfrac{2 - x}{2} = y + 5 = \dfrac{z + 6}{6}$

15. a. $\dfrac{\pi}{4}$ **17.** $\dfrac{\pi}{2}$

b. $\dfrac{\pi}{4}$

19. $\dfrac{1}{2}\sqrt{6} + \dfrac{1}{2}\ln(2 + \sqrt{6}) - \dfrac{1}{4}\ln 2$

21. $\dfrac{1}{\sqrt{17}}(i + 4j)$ **23.** $\alpha = \dfrac{\sqrt{2}}{2}$

27. $\cos ti - \sin tj$

29. $(4i + j + 8k) + t(2i + 12k)$

Exercise Set 16.4

1. $v = 2j$, $a = 0$

3. $v = -3\sin 3ti + 3\cos 3tj$, $a = -9\cos 3ti - 9\sin 3tj$

5. $v = \dfrac{2}{t}i - 2\sin 2tj - \dfrac{1}{t^2}k$, $a = \dfrac{-2}{t^2}i - 4\cos 2tj + \dfrac{2}{t^3}k$

7. $v = e^t(\cos t - \sin t)i + e^t(\sin t + \cos t)j - \sin tk$,
$\quad a = -2e^t\sin ti + 2e^t\cos tj - \cos tk$

9. $|v| = 2\sqrt{41}$

11. $(3 + \sin t)i + (3 - \cos t)j$

13. $(\text{Tan}^{-1}\, t - 2)i + \left(\dfrac{1}{2}e^{t^2} + \dfrac{1}{2}\right)j + 4k$

15. $v = 2ti + tk$, $r = (t^2 + 3)i - j + \left(\dfrac{t^2}{2} + 4\right)k$

17. $v = \sin ti + (1 - \cos t)j + k$,
$\quad r = (4 - \cos t)i + (-1 + t - \sin t)j + (t + 2)k$

21. Speed $= |\alpha| \cdot$ radius

23. a. $50\sqrt{3}\, ti + \left(50t - \dfrac{1}{2}gt^2\right)j$ **d.** $\dfrac{5000\sqrt{3}}{9.8}$ m

b. $\dfrac{1250}{9.8}$ m **e.** 100 m/s

c. $\dfrac{100}{9.8}$ s

27. a. $\displaystyle\int_{0}^{50\sqrt{3}/g} \sqrt{2500 - 50\sqrt{3}gt + g^2t^2}\, dt$

b. 304.6

Exercise Set 16.5

1. $T(s) = \dfrac{1}{2}i + \dfrac{\sqrt{3}}{2}j,\quad r''(s) = 0,\quad \kappa(s) = 0$

3. $T(s) = \dfrac{-\sqrt{2}}{2}\sin(s)i + \dfrac{\sqrt{2}}{2}\cos(s)j + \dfrac{\sqrt{2}}{2}k,$

$r''(s) = \dfrac{-\sqrt{2}}{2}\cos(s)i - \dfrac{\sqrt{2}}{2}\sin(s)j,\quad \kappa(s) = \dfrac{\sqrt{2}}{2}$

5. 0

7. $\dfrac{e^t}{(1 + e^{2t})^{3/2}}$

9. $\dfrac{1}{2}$

11. $\dfrac{2}{(1 + 4x^2)^{3/2}}$

13. $\dfrac{|6x|}{(1 + 9x^4)^{3/2}}$

15. $\dfrac{|\cos x|}{(1 + \sin^2 x)^{3/2}}$

17. $\dfrac{2}{(1 + 4t^2)^{3/2}}$

19. $\dfrac{2e^t(1 + e^{2t})^{1/2}}{(2e^{2t} + 1)^{3/2}}$

21. $\left(\dfrac{\pi}{2}, 0\right)$

23. a. $\dfrac{2}{5^{3/2}}$ **b.** $\left(\dfrac{7}{2}, -4\right)$ **c.** 2

27. $\dfrac{3}{2^{3/2}}\dfrac{1}{(1 + \sin\theta)^{1/2}}$

29. $\dfrac{3}{2^{3/2}}\dfrac{1}{(1 - \cos\theta)^{1/2}}$

35. $a_T = \dfrac{2t}{\sqrt{t + 1}},\quad a_N\sqrt{\dfrac{16t^6 + 16t^4 + 4}{t^2 + 1}}$

Review Exercises—Chapter 16

1. $\dfrac{x^2}{a^2} + \dfrac{x^2}{b^2} = 1$

3. $[-1, 0) \cup (0, 1]$

5. $t = 2n\pi,\quad n = 0, 1, 2, \ldots$

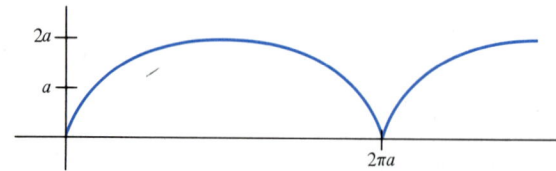

7. $3i + \sqrt{2}j + 2k$

9. $t = \dfrac{\pi}{2}, \dfrac{3\pi}{2}$

11. $t = \pm 3, \pm 2$

13. $r'(t) = e^t(t + 1)i + e^{-t}(1 - t)j$
$r''(t) = e^t(t + 2)i + e^{-t}(t - 2)j$

15. $r'(t) = t^{-1}i + \dfrac{1}{2}t^{-1/2}\, e^{\sqrt{t}}j$

$r''(t) = -t^{-2}i + \dfrac{1}{4}(t^{-1} - t^{3/2})e^{\sqrt{t}}j$

17. $i + (1 - 2e^{-1})j$

19. $(4i + 3j)\cosh 3t + C \sinh 3t$ for any constant vector C

21. $r(t) = 5 \cos\left(\dfrac{1}{5}t\right)i + 5 \sin\left(\dfrac{1}{5}t\right)j$

23. a. $\kappa(t) = \dfrac{1}{2}$

b. $T(t) = \dfrac{\sqrt{2}}{2}i - \dfrac{\sqrt{2}}{2}\sin tj + \dfrac{\sqrt{2}}{2}\cos tk$

25. $3\sqrt{13}$

27. $\dfrac{12}{(145)^{3/2}}$

33. $x = \dfrac{-\ln 2}{2}$

35. Approximately 4.14 m

CHAPTER 17

Exercise Set 17.1

1. $\{(x, y) \mid (x, y) \neq (0, 0)\}$

3. $\{(x, y) \mid y \geq x\}$

5. All $(x, y) \in \mathbb{R}^2$

7. All $(x, y, z) \in \mathbb{R}^3$

9. $\{(x, y, z) \mid xyz \neq 0\}$

11. a. $h(x, y) = \sqrt{x + y^2}$
b. $\{(x, y) \mid x + y^2 \geq 0\}$

13. $S(r, h) = 2\pi r^2 + 2\pi rh$

15. -5

17. $\dfrac{1}{\sqrt{2}}$

19. 0

21. 0

23.

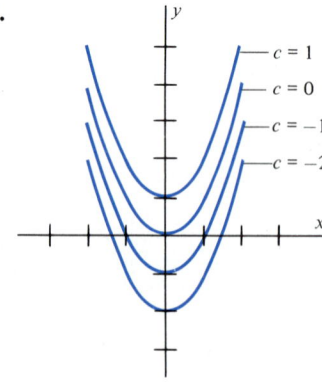

25.

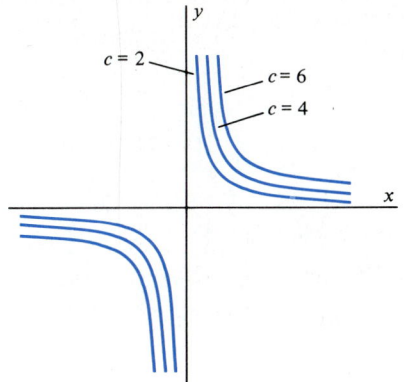

27.

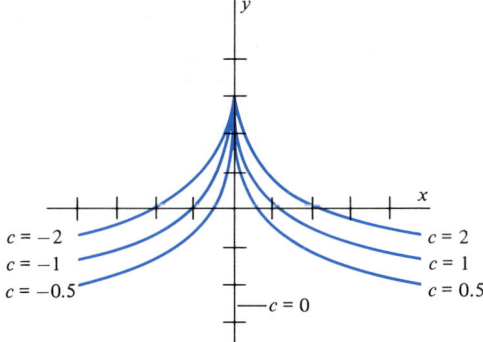

29.

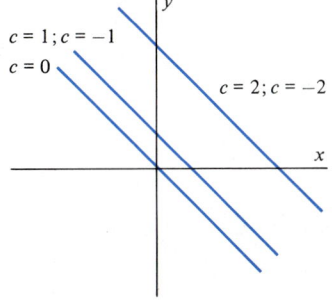

31.

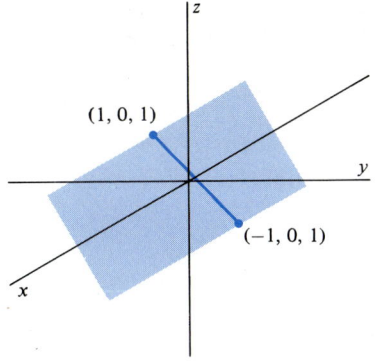

33.

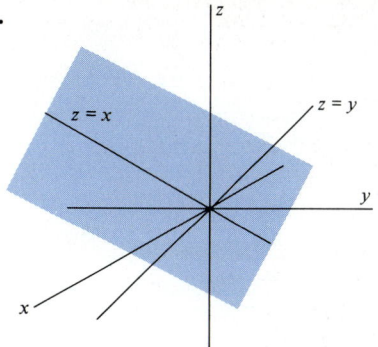

35.

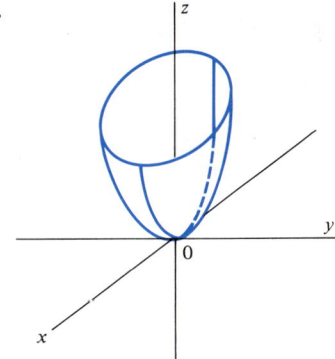

37.

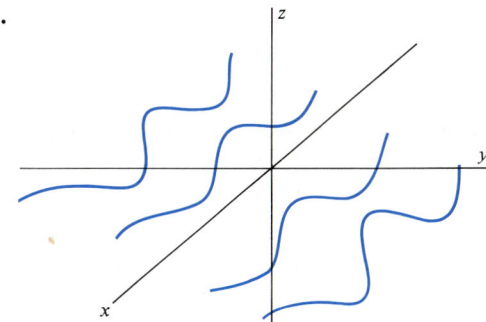

39. Does not exist

41. No

43. Yes

45. 0

47. False

Exercise Set 17.2

1. $\dfrac{\partial f}{\partial x} = y,\ \dfrac{\partial f}{\partial y} = x$

3. $\dfrac{\partial f}{\partial x} = \tan y^2,\ \dfrac{\partial f}{\partial y} = 2xy\sec^2 y^2$

5. $\dfrac{\partial f}{\partial x} = 2xe^{x^2+y^2},\ \dfrac{\partial f}{\partial y} = 2ye^{x^2+y^2}$

7. $\dfrac{\partial f}{\partial r} = \sin\!\left(\dfrac{\pi}{2} - \theta\right),\ \dfrac{\partial f}{\partial \theta} = -r\cos\!\left(\dfrac{\pi}{2} - \theta\right)$

9. $\dfrac{\partial f}{\partial x} = \dfrac{y^2 + 1}{xy^2 + x - y}, \dfrac{\partial f}{\partial y} = \dfrac{2xy - 1}{xy^2 + x - y}$

11. $\dfrac{\partial f}{\partial x} = yx^{y-1}, \dfrac{\partial f}{\partial y} = x^y \ln x$

13. $\dfrac{\partial f}{\partial r} = 2r \cos \theta, \dfrac{\partial f}{\partial \theta} = -r^2 \sin \theta$

15. $\dfrac{\partial f}{\partial x} = y^3, \dfrac{\partial f}{\partial y} = 3xy^2 - z^2, \dfrac{\partial f}{\partial z} = -2yz$

17. $\dfrac{\partial f}{\partial x} = \dfrac{2yz}{(x + y)^2}\left(\dfrac{x - y}{x + y}\right)^{z-1}$,

$\dfrac{\partial f}{\partial y} = \dfrac{-2xz}{(x + y)^2}\left(\dfrac{x - y}{x + y}\right)^{z-1}$

$\dfrac{\partial f}{\partial z} = \left(\dfrac{x - y}{x + y}\right)^z \ln\left(\dfrac{x - y}{x + y}\right)$

19. $\dfrac{\partial f}{\partial r} = \dfrac{s \ln t \, (s^2 + t)}{2r^{1/2} \, (s - 2r + t)^{3/2}}$

$\dfrac{\partial f}{\partial s} = \dfrac{r^{1/2} \ln t \, (-2r + t)}{(s^2 - 2r + t)^{3/2}}$

$\dfrac{\partial f}{\partial t} = \dfrac{-r^{1/2} s \, (s^2 - 2r + t - \frac{1}{2}t \ln t)}{t \, (s^2 - 2r + t)^{3/2}}$

21. 125

23. $\dfrac{1}{2\sqrt{3}}$

25. $\dfrac{\partial^2 f}{\partial r^2} = 2 \cos \theta, \quad \dfrac{\partial^2 f}{\partial r \, \partial \theta} = -2r \sin \theta, \quad \dfrac{\partial^2 f}{\partial \theta^2} = -r^2 \cos \theta$

27. $\dfrac{2}{x^2 + y^2 + z^2}$

29. a. $v_2 = \dfrac{A_1 v_1}{A_2}$

b. $\dfrac{\partial v_2}{\partial A_2} = \dfrac{-A_1 v_1}{(A_2)^2}$

c. $\dfrac{-100 \text{ cm/s}}{9 \text{ cm}^2}$

d. $\dfrac{3 \text{ cm}^2}{20 \text{ cm/s}}$

31. $\rho \sin \phi \cos \theta$ **33.** $\sin \phi \cos \theta$ **35.** $-\rho \sin \phi$

37. a. $\cos(x + y)^2 - \cos x^2$
b. $\cos(x + y)^2$

41. a. $\dfrac{\partial V}{\partial a} = \dfrac{a\sigma}{2\epsilon_0 \sqrt{a^2 + r^2}}$

b. $\dfrac{\partial V}{\partial r} = \dfrac{\sigma}{2\epsilon_0}\left(\dfrac{r}{\sqrt{a^2 + r^2}} - 1\right)$

47. $f(x, y) = \sin(x, y)$

Exercise Set 17.3

1. $2x + 6y - z = 10$ **3.** $x - 5y - z = 5$

5. $x - 2y - z = 4$ **7.** $x + y - z = 2$

9. $y + 2z = 2$ **11.** $y - z = 1$

13. $2x + ez = 0$

15. $r(t) = 3i - j + 4k + t(4i + 4j - k)$

17. $r(t) = i - 2j + 2k + t(-i + 2j - 2k)$

19. $r(t) = i + j + t(j - k)$

21. $(2, 3, 20)$

25. $18x + 16y - z = 25$

27. $x_0 x + y_0 y + z_0 z = r^2$

29. $-x - 2y + z + \dfrac{\pi}{2} = 0$

Exercise Set 17.4

1. Relative minimum: $(0, -2, 0)$

3. Saddle point $(-3, 2, 0)$

5. Saddle point $(0, 0, 9)$

7. Relative minimum: $(1, 3, 1)$

9. Saddle point $(0, 0, 0)$

11. Saddle point $(0, 0, 0)$

13. Saddle point $(0, 0, 0)$
Relative maximum: $\left(-\dfrac{4}{3}, -\dfrac{4}{3}, \dfrac{64}{27}\right)$

15. Saddle points $\left(0, \dfrac{\pi}{2} + n\pi, 0\right)$

19. The graph is unbounded.

23. $\ell = 2$ m, $w = 2$ m, $h = 2$ m

25. $\left(\dfrac{ad}{a^2 + b^2 + c^2}, \dfrac{bd}{a^2 + b^2 + c^2}, \dfrac{cd}{a^2 + b^2 + c^2}\right)$

27. $\ell = 25''$, $w = h = 14''$

29. Absolute minimum: $-2\sqrt{2} + 4$
Absolute maximum: $2\sqrt{2} + 4$

31. Absolute minimum: 3
Absolute maximum: 9

35. 86

37. a. $y = \dfrac{23}{20}x - \dfrac{19}{4}$

b.

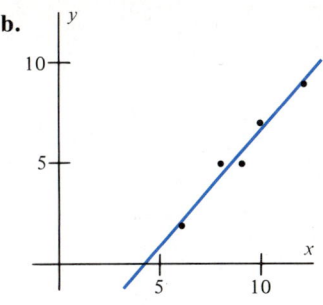

c. $y(4) = \dfrac{-3}{20}$

Exercise Set 17.5

1. $2xy^4\,dx + 4x^2y^3\,dy$

3. $\dfrac{x}{\sqrt{x^2+y^4}}\,dx + \dfrac{2y^3}{\sqrt{x^2+y^4}}\,dy$

5. $\dfrac{e^{\sqrt{x}}}{2\sqrt{x}}\cos y\,dx - e^{\sqrt{x}}\sin y\,dy$

7. $2xyz^3\,dx + x^2z^3\,dy + 3x^2yz^2\,dz$

9. $\dfrac{1}{y+3^z}\,dx - \dfrac{x}{(y+3^z)^2}\,dy - \dfrac{x3^z\ln 3}{(y+3^z)^2}\,dz$

11. $\dfrac{y^2+z^2-x^2+2xy}{(x^2+y^2+z^2)^2}\,dx + \dfrac{y^2-x^2-z^2-2xy}{(x^2+y^2+z^2)^2}\,dy$

$\qquad\qquad + \dfrac{2zy-2zx}{(x^2+y^2+z^2)^2}\,dz$

13. a. $f(2,3) = 48$
$\qquad f(2.1, 3.2) = 57.12$
b. $\Delta f = 9.12$

15. 5.076 **17.** 0.7441

19. 26.8 **21.** 55.44

23. $77.41 \le v \le 81.43$ **25.** -3%

27. 0.048 seconds   **29.** 10%

Exercise Set 17.6

1. $-16t + 4t^3$

3. $-\sin^3 t + 2\sin t\cos^2 t$

5. $\dfrac{1}{4\sqrt{t}\sqrt{\sqrt{t}+e^{t^2}}} + \dfrac{te^{t^2}}{\sqrt{\sqrt{t}+e^{t^2}}}$

7. $-6t^2 - 2t + 2\sin t\cos t$

9. $2e^{-3t}[a^2\cosh t + b^2\sinh t]$

11. $-2t^3\sin^4 t\cos t + 3t^2\sin^3 t\cos^2 t + 3t^3\sin^2 t\cos^3 t$

13. a. $2te^{2st} + 2st^2 - 2s + 2t$
    b. $2se^{2st} + 2s^2t + 2s - 2t$

15. a. $[y^2\cos xy^2 - 2xy](2s-t) + [2xy\cos xy^2 - x^2](2t^2s)$
    b. $[y^2\cos xy^2 - 2xy](-s) + [2xy\cos xy^2 - x^2](2s^2t)$

17. a. $y^3z^2\sin t - 3xy^2z^2t\sin s - 4sxy^3z$
    b. $sy^3z^2\cos t + 3xy^2z^2\cos s + 4txy^3z$

19. a. $2ue^{u^2}y^2 - 4xye^{2v}$
    b. $4xyue^{u^2} - 2x^2e^{2v}$

21. $6rt^2(1-\cos\theta) + \dfrac{r^2t\sin\theta}{\sqrt{1+t^2}}$

23. 116π cm^3/s

25. a. $D(t) = \sqrt{1+t^2}$

    b. $D'(t) = \dfrac{t}{\sqrt{1+t^2}}$

29. a. $12s^2t^2 - 8y^3 + 48s^2y^2$
    b. $2s^4 + 8y^3 + 48t^2y^2$

31. $2s^2t^2e^{rst} - 4s^2t^2e^{2rst} + rs^3t^3e^{rst}$

33. $e^{y^2}[-2yr\sin^2\theta - \sin\theta + 2yr\cos^2\theta$
$\qquad\qquad + x(2+4y^2)r\sin\theta\cos\theta + 2xy\cos\theta]$

Exercise Set 17.7

1. $6i + 9j$

3. $\dfrac{\sqrt{2}}{2}\left(1 - \dfrac{\pi}{2}\right)i + \dfrac{\sqrt{2}}{4}\pi j$

5. $6i + j + 4k$ **7.** $5i - 3j - k$

9. $\left(e + \dfrac{1}{e}\right)i + \dfrac{1}{4}\left(e + \dfrac{1}{e}\right)j + 2\left(e - \dfrac{1}{e}\right)k$

11. $-30\sqrt{2}$ **13.** $\dfrac{2\sqrt{3}-1}{18}$

15. $\dfrac{2}{\sqrt{3}}$ **17.** $\dfrac{3\sqrt{3}}{\sqrt{11}}$

19. $\dfrac{13\sqrt{2}}{6} + \dfrac{1}{3}\left(\dfrac{1}{e} - e\right) + \dfrac{\pi}{12}\left(\dfrac{1}{e} + e\right)$

21. $N: 2i + 3j + t(16i - 6j)$
$\qquad T: 2i + 3j + t(6i + 16j)$

23. $N: 4i + 4j + t(i + j)$
$\qquad T: 4i + 4j + t(i - j)$

25. $\sqrt{5}$ **27.** -1 **29.** $\dfrac{\sqrt{2}}{4}$

31. a. $(3, 1+\sqrt{2}), (3, 1-\sqrt{2})$
    b. $(1, 1), (5, 1)$

37. $2i + j$ **39.** $-i + j$

41. One example is $f(x, y) = \sqrt{x^2 + y^2}$ at $(0, 0)$.

Exercise Set 17.8

Maximum	Minimum

1. $f(0, \pm 1) = 4$ $f(\pm 1, 0) = 2$

3. None $f(1, -1) = 2$

5. $f\left(\dfrac{1}{\sqrt{2}}, \dfrac{1}{\sqrt{2}}\right)$ $f\left(\dfrac{1}{\sqrt{2}}, \dfrac{-1}{\sqrt{2}}\right)$

$= f\left(\dfrac{-1}{\sqrt{2}}, \dfrac{-1}{\sqrt{2}}\right) = \dfrac{1}{2}$ $= f\left(\dfrac{-1}{\sqrt{2}}, \dfrac{1}{\sqrt{2}}\right) = -\dfrac{1}{2}$

7. $f(2, 0) = 12$ $f(-2, 0) = -4$

9. $f\left(-\dfrac{2}{3}, \pm\dfrac{\sqrt{5}}{3}\right) = \dfrac{48}{9}$ $f(1, 0) = -3$

11. $f\left(\dfrac{1}{\sqrt{6}}, \dfrac{2}{\sqrt{6}}, -\dfrac{1}{\sqrt{6}}\right)$ $f\left(-\dfrac{1}{\sqrt{6}}, -\dfrac{2}{\sqrt{6}}, \dfrac{1}{\sqrt{6}}\right)$

$= \sqrt{6}$ $= -\sqrt{6}$

13. $f(2, 2, 2) = 6$ $f(-2, -2, -2) = -6$

15. $f\left(2, -1 + \dfrac{\sqrt{2}}{2}, \dfrac{3}{2}\right) = f\left(2, -1 - \dfrac{\sqrt{2}}{2}, \dfrac{3}{2}\right) = \dfrac{3}{4}$

17. $\left(3 - 9\dfrac{\sqrt{13}}{13}, -2 + 6\dfrac{\sqrt{13}}{13}\right)$

19. $(1, 0, 1)$

21. $(2, -1, 1)$

23. $\dfrac{2r}{\sqrt{3}} \times \dfrac{2r}{\sqrt{3}} \times \dfrac{2r}{\sqrt{3}}$

25. $r = \sqrt[3]{\dfrac{1}{\pi}}, \ h = 2\sqrt[3]{\dfrac{1}{\pi}}$

27. $x = 2, y = 3, z = 1$

29. $x = y$ and $y = z$ $x = 20$

31. $x = 2500, y = 0$ yields max. revenue

Exercise Set 17.9

1. $x - y + C$ **3.** None

5. None   **7.** $x^3 \cos y - x + y^2 + C$

9. $x^2 e^{xy} - \sqrt{y} + C$ **11.** $x^2 y^2 + y \cos x + C$

Review Exercises—Chapter 17

1. No

3.

5. $5i - 7j$ **7.** $-i - k$

9. $\dfrac{\partial f}{\partial x}(x, y) = \sin \sqrt{x^2 + y^2} + \dfrac{x^2 \cos \sqrt{x^2 + y^2}}{\sqrt{x^2 + y^2}},$

$\dfrac{\partial f}{\partial y}(x, y) = \dfrac{xy \cos \sqrt{x^2 + y^2}}{\sqrt{x^2 + y^2}}$

11. $\dfrac{784}{15}$

13. $\dfrac{\partial f}{\partial r} = 18x^2 + 4y^2 + 4xy - 2y,$

$\dfrac{\partial f}{\partial s} = 45x^2 + 10y^2 - 16sxy + 8sy$

15. $5x + 4y = 10$

19. $-\dfrac{36}{5}i - \dfrac{12}{5}j - (2 - \mathrm{Tan}^{-1} 2)k$

23. $\dfrac{2\sqrt{3}}{3} \times 2\sqrt{3} \times \dfrac{4\sqrt{3}}{3}$ **25.** $(0, 0, 0)$: saddle point

27. $(0, 0, 1)$: saddle point   **29.** $(0, 0, e)$: saddle point

31. a. 1
   b. 3
   c. $i + 3j$

33. $f(x, y) = x^2 e^y + \cos y + C$

35. R increases by 10%

39. Maximum: $5 + 6\sqrt{2}$ **41.** No
   Minimum: $5 - 6\sqrt{2}$

43. $\left(\dfrac{4}{\sqrt{14}}, \dfrac{8}{\sqrt{14}}, \dfrac{-12}{\sqrt{14}}\right)$ **49.** $\left(\dfrac{3}{7}, \dfrac{-9}{7}, \dfrac{6}{7}\right)$

51. $x + 2y + z = 6$ **53.** $3 - \dfrac{1}{2}\sqrt{3}$

CHAPTER 18

Exercise Set 18.1

1. 1   **3.** 7

5. $\dfrac{1}{2}$

7. $\dfrac{8}{3}$

9. $\dfrac{1}{3}$

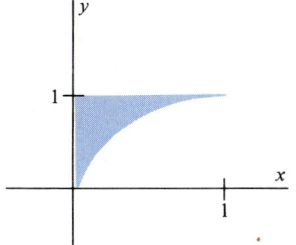

9. $\dfrac{1}{4}\left(e - \dfrac{1}{e}\right)$

11. $\dfrac{1}{2}$

13. $\dfrac{4\pi}{3}$

15. $\dfrac{5}{6}$

17. 8

19. $\dfrac{1}{2}\ln\sqrt{2}$

21. $\dfrac{\pi\sqrt{2}}{8} + 1 - \sqrt{2}$

23. 6

11. $\dfrac{1}{2}$

25. $\dfrac{1}{2}$

27. $8e(e - 1)$

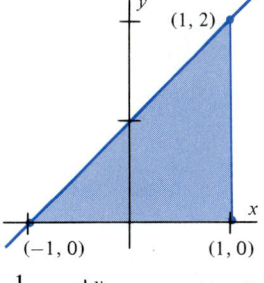

33. The integral is 24.064 to 3 places.

35. The integral is 24.636 to 3 places.

Exercise Set 18.2

1. $-\dfrac{1}{6}$

3. $\dfrac{13}{24}$

13. $\dfrac{2}{15}$

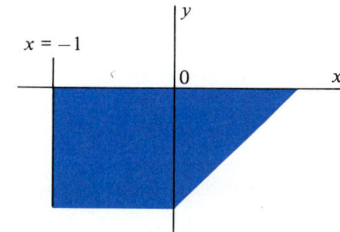

15. $\dfrac{1}{3}(1 + e)^{3/2} - \dfrac{4}{3}2^{3/2} + 2\sqrt{e}\sqrt{1 + e}$

$$+ 2\ln\dfrac{\sqrt{e} + \sqrt{1 + e}}{1 + \sqrt{2}}$$

17. $\dfrac{1}{24}$

5. 0

19. $\dfrac{3e^2 + e^{-2}}{4}$

21. $\dfrac{1}{3}$

23. 2

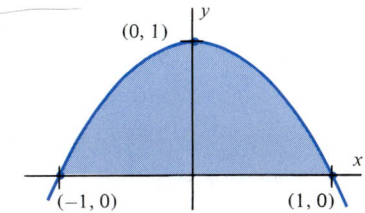

7. $\dfrac{1}{20}$

25. $\dfrac{1}{10}$

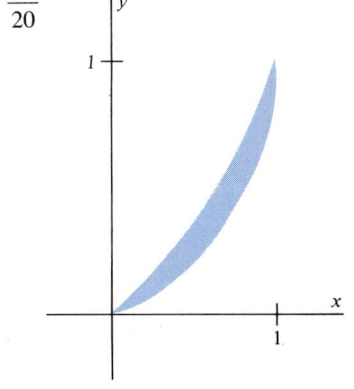

27. $\int_0^1 \int_{e^y}^e f(x, y)\, dx\, dy$

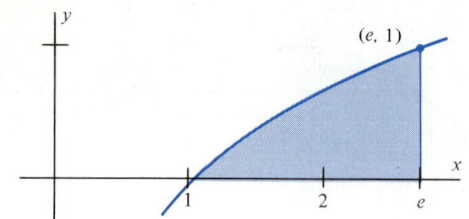

7. $\dfrac{4}{27}$

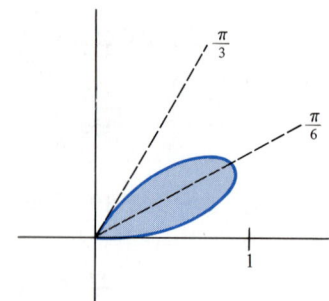

29. $\dfrac{9}{2}$

31. $\sqrt{2} - 1$

33. $\dfrac{1}{3}$

35. 6π

9. $\dfrac{\pi}{4}$

11. $2 + \dfrac{\pi}{4}$

37. 10π

39. $\dfrac{16}{3}$

13. $\dfrac{9\pi}{2}$

15. $\dfrac{\pi}{2}$

41. $\dfrac{2}{3}$

17. $\dfrac{\pi}{4}(e - 1)$

19. 2

21. 0

23. 16π

25. $2\sqrt{3}\,\pi$

27. 3π

Exercise Set 18.3

29. 16π

31. $\dfrac{\pi}{4}$

1. $\dfrac{2\pi}{3}$

33. $\dfrac{16\pi}{3}$

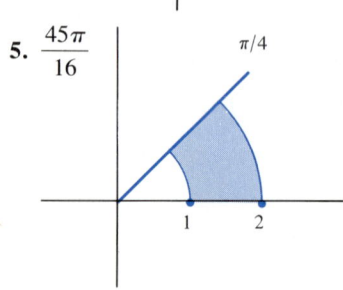

Exercise Set 18.4

1. 3

3. 8

3. $\pi(e^{a^2} - 1)$

5. $\sqrt{2} - 1$

7. $\dfrac{\pi}{4}$

9. 2π

11. $(\bar{x}, \bar{y}) = \left(\dfrac{11}{9}, \dfrac{5}{9}\right)$

13. $(\bar{x}, \bar{y}) = \left(\dfrac{3}{\pi}, \dfrac{3}{\pi}\right)$

15. $(\bar{x}, \bar{y}) = \left(\dfrac{17}{54}, \dfrac{13}{9}\right)$

17. $(\bar{x}, \bar{y}) = \left(0, \dfrac{8}{5}\right)$

19. $(\bar{x}, \bar{y}) = (2.6, 1.7)$

21. $(\bar{x}, \bar{y}) = \left(\dfrac{4a}{3\pi}, \dfrac{4a}{3\pi}\right)$

23. $(\bar{x}, \bar{y}) = \left(\dfrac{8}{5}, 2\right)$

25. 4λ

5. $\dfrac{45\pi}{16}$

Exercise Set 18.5

1. $\sqrt{3}$

3. $\dfrac{\pi}{6}(13^{3/2} - 1)$

5. $2\sqrt{3}$

7. $3\sqrt{2}\pi$

9. $\sqrt{2} + \ln(1 + \sqrt{2})$ **11.** $\dfrac{5\sqrt{2}}{12} + \dfrac{1}{4}\ln(3 + 2\sqrt{2})$

13. $\dfrac{\pi}{6}(7^3 - 29^{3/2})$ **15.** $4\pi(2 - \sqrt{3})$

17. $A \approx 7.904$

Exercise Set 18.6

1. $\dfrac{1}{8}$ **3.** $\dfrac{1}{2}$

5. $\dfrac{14}{3}$ **7.** 0

9. $\displaystyle\int_0^2 \int_{y/2}^1 \int_0^{x+y} f(x,\,y,\,z)\,dz\,dx\,dy$

11.

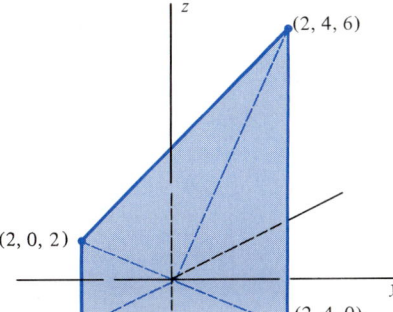

13. $\dfrac{2\pi}{3}$ **15.** 27π

17. 0 **19.** π

21. $\dfrac{405\pi}{8}$ **23.** 12

25. $(\bar{x},\,\bar{y},\,\bar{z}) = \left(0,\,\dfrac{3}{4},\,\dfrac{15}{8}\right)$

27. $(\bar{x},\,\bar{y},\,\bar{z}) = \left(\dfrac{3}{4},\,\dfrac{3}{4},\,\dfrac{3}{4}\right)$

29. $\displaystyle\int_0^2 \int_0^{1-y/2} \int_0^{4-4x-2y} (x-y+z)\,dz\,dx\,dy,$

$\displaystyle\int_0^1 \int_0^{4-4x} \int_0^{2-2x-z/2} (x-y+z)\,dy\,dz\,dx,$

$\displaystyle\int_0^4 \int_0^{2-z/2} \int_0^{1-y/2-z/4} (x-y+z)\,dx\,dy\,dz,$

$\displaystyle\int_0^2 \int_0^{4-2y} \int_0^{1-y/2-z/4} (x-y+z)\,dx\,dz\,dy,$

$\displaystyle\int_0^4 \int_0^{1-z/4} \int_0^{2-2x-z/2} (x-y+z)\,dy\,dx\,dz$

Exercise Set 18.7

1. $\dfrac{81\pi}{2}$ **3.** $(\bar{x},\,\bar{y},\,\bar{z}) = (0,\,0,\,6)$

5. $18\pi + 24$ **7.** $\dfrac{2\sqrt{2}\,4^4}{3^2} \approx 80$

9. $\dfrac{32}{9}$ **11.** 0

13. π **15.** $(\bar{x},\,\bar{y},\,\bar{z}) = \left(0,\,\dfrac{3}{8},\,\dfrac{3}{8}\right)$

17. $\dfrac{16\sqrt{2}\,\pi}{3}$ **19.** 2π

21. $V = \dfrac{16}{3}R^3$ **23.** $9\pi\left(1 - \dfrac{\sqrt{3}}{3}\right)$

25. $\dfrac{32\sqrt{2}\,\pi}{5}$ **27.** $\dfrac{40\pi}{3}$

29. $\dfrac{104\pi}{3}\left(1 - \dfrac{\sqrt{2}}{2}\right)$

Review Exercises—Chapter 18

1. $\dfrac{1}{15}$ **3.** $2\left(1 - \dfrac{1}{\sqrt{e}}\right)$

5. 64 **7.** $16\left(\dfrac{\pi}{2} - 1\right)$

9. 3π **11.** $\dfrac{abc}{6}$

13. 36π

15. $(\bar{x},\,\bar{y},\,\bar{z}) = \left(0,\,0,\,\dfrac{3}{4}\right)$

17. $\dfrac{2a^3}{9}(3\pi - 4)$ **19.** $4\sqrt{17}\,\pi$

21. $\dfrac{32\pi}{3}(8 - 3\sqrt{3})$ **23.** $\dfrac{977}{1017}$

25. $(\bar{x},\,\bar{y}) = \left(0,\,\dfrac{16}{3\pi}\right)$ **27.** $\dfrac{2\pi}{3}(e - 1)$

29. $(\bar{x},\,\bar{y}) = \left(\dfrac{5}{6},\,0\right)$ **31.** $\dfrac{81\pi}{4}$

33. $32\pi\rho$ **35.** $\dfrac{\pi}{2}(\sqrt{2} - 1)$

37. $\dfrac{256}{9}$

39. The integral is 2.2228 to 4 places.

41. 36π **43.** $8\sqrt{3}\,\pi$

CHAPTER 19

Exercise Set 19.1

1.

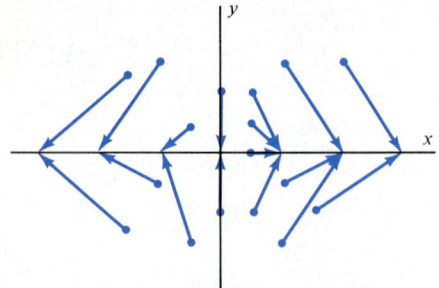

3.

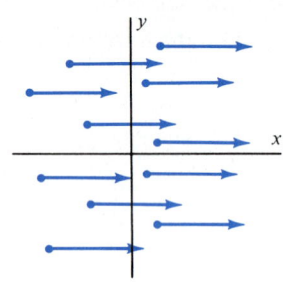

5.

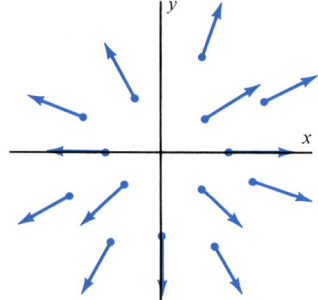

7.

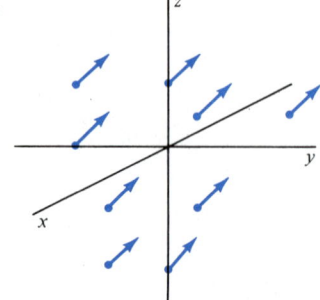

9.

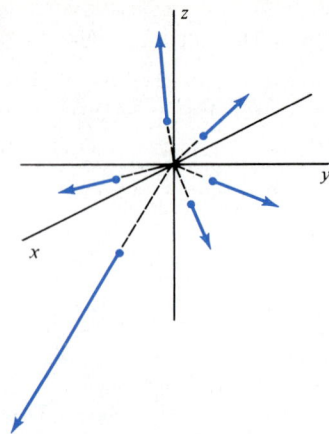

11. 4, 5, 9

13. $F = 2xi - 2yj$

15. $\dfrac{1}{\sqrt{x^2 + y^2 + z^2}}(xi + yj + zk)$

17. $e^{x-y}(zi - zj + k)$

19. Not a gradient

21. Not a gradient

23. $F = \nabla(e^{xyz})$

27. **a.** $F(0, 0, 0) = 4i$

b. $F(-2, 0, 0) = -\dfrac{26}{9}i$

c., d.

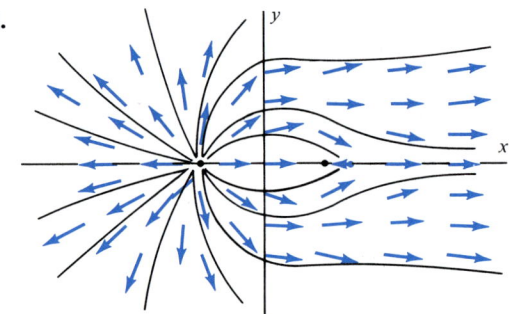

Exercise Set 19.2

1. 16

3. $\dfrac{1708}{3}$

5. $\dfrac{328}{15}$

7. $\dfrac{\pi^2}{4} + \pi$

9. 12

11. $-\pi$

13. -4

15. $\dfrac{\pi}{2}\left(\dfrac{\pi}{2} - 1\right)$

17. -1

19. 3

21. $\dfrac{1}{6} + \dfrac{\pi}{16}$

23. $\dfrac{-442}{3}$

25. $\dfrac{4}{3}$

29. $\dfrac{70\sqrt{5}}{3}$

21. $(\bar{x}, \bar{y}, \bar{z}) = \left(0, 0, \dfrac{a}{2}\right)$

31. $\dfrac{2}{3}(5^{3/2} - 1)$

33. 5π

23. $(\bar{x}, \bar{y}, \bar{z}) = \left(\dfrac{1}{2}, \dfrac{1}{2}, \dfrac{1}{2}\right)$

Exercise Set 19.3

For problems 1 through 10 $F = \nabla f$ with

f	$\int_c F \cdot dr$
1. xy	3
3. e^{xy}	$e^2 - 1$
5. $x^2 + y^2$,	0
7. xe^{y^2},	$2e$
9. $e^x \sin y$,	$e\dfrac{\sqrt{2}}{2}$

11. $-a$

19. No, the field is not conservative.

Exercise Set 19.4

1. $\dfrac{1}{2}$

3. $\dfrac{2}{3}$

5. $\dfrac{\pi}{8}$

7. $-\dfrac{11}{60}$

9. 0

11. -5π

13. 0

15. $-\dfrac{9\pi}{2}$

17. $\dfrac{1}{6}$

19. $\dfrac{3a^2\pi}{8}$

Exercise Set 19.5

1. $\dfrac{23\sqrt{6}}{4}$

3. $\dfrac{27\sqrt{14}}{5}$

5. 0

7. $48\pi(9 - 3\sqrt{5})$

9. $\dfrac{32\pi}{3}$

11. 2π

13. 16π

15. $\dfrac{4\pi a^3}{3}$

17. 3

19. No, not the same: $\dfrac{4\pi}{r}$.

Exercise Set 19.6

1. 0

3. $z\cos(yz)i + xyj + (y\sin xy - xz)k$

5. 0

7. Not conservative

9. π

11. $-\dfrac{2}{3}$

13. 4

15. -2π

17. 0

19. 0

Exercise Set 19.7

1. $2x + 2y + 2z$

3. $1 + x + yx$

5. $-y\sin(xy) + xze^{xyz} + xy\cos(xz)$

7. 0

9. -150

11. $\dfrac{11}{6}$

13. 0

15. $\dfrac{512\pi}{15}$

17. $\dfrac{80\pi}{3}$

23. a.

Review Exercises—Chapter 19

1. a. $-y\sin x \ln zi + \cos x \ln zj + \dfrac{y\cos x}{z}k$

b. $[3x^2yz^2 + y\cos(xy)]i + [x^3z^2 + x\cos(xy)]j + 2x^3yzk$

3. $F = \nabla\left(-\text{Tan}^{-1}\left(\dfrac{y}{x}\right)\right)$

5. -2π

7. $F = \nabla(xy^2z)$

9. $\dfrac{1}{15}$

13. True

15. $(a + b + c)V$

17. $-\dfrac{4}{3}$

19. π

21. $-\dfrac{3}{2}$

23. $\dfrac{2}{5}$

25. $\sqrt{6}$

27. $\dfrac{1023\sqrt{2}\,\pi}{5}$

29. $-\dfrac{15\pi}{2}$

31. 0

33. $8\pi(a + b + c)$

35. $-\dfrac{164}{3}$

37. 0

39. 0

CHAPTER 20

Exercise Set 20.1

1. $y = 2e^t - 1$

3. $y = \dfrac{-\pi}{2}e^{-t} + \pi$

5. $y = Ce^{-2t} + 2$

7. $y = \dfrac{1}{3}t^5 + Ct^2$

9. $y = e^{2t}(-\cos t + C)$

11. $y = \dfrac{1}{2}e^t + Ce^{-t}$

13. $y = \dfrac{1}{2}t + Ct^{-1}$

15. $y = ke^{ax^2/2}$

17. a. $\dfrac{dp}{dt} = -\dfrac{3}{200}p$

   b. 74.08 kg
   c. $p = 0$

19. a. $\dfrac{dp}{dt} = 0.10p + 100$
   b. $1215.13

21. $I = I_0e^{-t/RC}$

23. 52.8° C

25. $p = 6 - 4e^{-.02t}$

27. 6.93 years

29. $y = Cs^k$

Exercise Set 20.2

1. $x^2y = C$

3. $\sqrt{x^2 + y^2} = C$

5. Not exact

7. Not exact

9. $x^2 + y^2 = C$

11. $x^2y = 3$

13. $xy^2 - x^2y = 2$

Exercise Set 20.3

1. $y = Ae^{-3t} + Be^{-2t}$

3. $y = Ae^t + Be^{-t}$

5. $y = Ae^{2t} + Be^{-5t}$

7. $y = Ae^{-t} + Bte^{-t}$

9. $y = Ae^{-3t} + Bte^{-3t}$

11. $y = A \sin 2t + B \cos 2t$

13. $y = A \sin \sqrt{5}t + B \cos \sqrt{5}t$

15. $y = Ae^{5x} + Be^{-2x}$

17. $y = Ae^{(2+\sqrt{3})x} + Be^{(2-\sqrt{3})x}$

19. $y = e^x(A \sin \sqrt{5}x + B \cos \sqrt{5}x)$

21. $y = Ae^{4x} + Be^{-3x}$

23. $y = e^{-x}(A \sin \sqrt{3}x + B \cos \sqrt{3}x)$

25. $y = e^{3t} - e^{-2t}$

27. $y = 3e^{-t} + 4te^{-t}$

29. $y = -\sin 3t + 3 \cos 3t$

31. $x = A \sin 2x + B \cos 2x$

   $y = -2A \cos 2x + 2B \sin 2x$

33. $x(t) = \dfrac{1}{10} \cos 2t$

35. a. Oscillating with increasing frequency
   b. Exponential function with increasingly large exponent
   c and **d** approach constant solution $x(t) = 1$

Exercise Set 20.4

1. $y = C_1e^{(-1+\sqrt{2})t} + C_2e^{(-1-\sqrt{2})t} - \dfrac{5}{41} \sin 2t - \dfrac{4}{41} \cos 2t$

3. $y = C_1 \sin t + C_2 \cos t + \dfrac{1}{2}e^{-t}$

5. $y = C_1e^{2t} + C_2e^{-2t} - t$

7. $y = C_1e^{5t} + C_2e^{-2t} - \dfrac{5}{12}e^{2t}$

9. $y = \left(C_1 + \dfrac{1}{6}t\right) \sin 3t + C_2 \cos 3t$

11. $y = C_1e^{4t} + \left(C_2 - \dfrac{2}{5}t\right)e^{-t}$

13. $y = (C_1 + C_2t)e^{-t} + \dfrac{3}{4}e^t + t^2 - 4t + 6$

15. $y = (C_1 + C_2t)e^t - \dfrac{11}{25} \sin 2t - \dfrac{2}{25} \cos 2t + \dfrac{1}{2}e^{-t}$

17. $y = e^{-t/2}\left(C_1 \sin \frac{1}{2}\sqrt{3}t + C_2 \cos \frac{1}{2}\sqrt{3}t\right)$
$$+ t^3 - t^2 - 4t + 6$$

23. $y = Ce^{3t} - 2t + 4$

25. $y = Ce^{4t} + t^2 + \frac{1}{2}t - \frac{1}{8}$

27. $y = Ce^{4t} - 2e^{3t}$

29. $y = Ce^{3t} - 2e^t - 2t + 4$

31. $y = 3e^{-4t}$

33. $y = 5e^{-3t} + e^{2t}$

Exercise Set 20.5

1. $y = a_0 \sum_{k=0}^{\infty} \frac{(-1)^k x^k}{k!} = a_0 e^{-x}$

3. $y = a_0 \sum_{k=0}^{\infty} \frac{6^k x^k}{k!} = a_0 e^{6x}$

5. $y = a_0 \sum_{k=0}^{\infty} \frac{x^{2k}}{k!} = a_0 e^{x^2}$

7. $y = \sum_{k=0}^{\infty} \frac{(-1)^k 4^k x^k}{k!} = e^{-4x}$

9. $y = a_1 \sum_{k=1}^{\infty} \frac{(-1)^k x^k}{k[(k-1)!]^2}$

11. $y = \sum_{k=0}^{\infty} \frac{x^{2k}(-1)^k}{(2k)!} = \cos x$

Exercise Set 20.6

1.

t_j	0	.25	.5	.75	1.0
y_j	0	1.5	2.25	3.375	5.0625
$y(t_j)$	1	1.6487	2.7183	4.4817	7.3891

$$y = e^{2t}$$

3.

t_j	0	.25	.50	.75	1.0
y_j	1	1.5	1.75	1.875	1.9875
$y(t_j)$	1	1.3934	1.6321	1.7769	1.8647

$$y = 2 - e^{-2t}$$

5.

t_j	0	.25	.5	.75	1.0
y_j	1	1	1.125	1.3906	1.8389
$y(t_j)$	1	1.0635	1.2663	1.6496	2.2974

$$y = 2e^{\frac{1}{2}x^2} - 1$$

7.

t_j	0	.25	.50	.75	1.0
y_j	1.5	1.5	1.375	1.1563	0.9102
$y(t_j)$	1.5	1.4394	1.2788	1.0698	0.8679

$$y = e^{-t^2} + \frac{1}{2}$$

9.

t_j	0	.25	.5	.75	1.0
y_j	1.5	2.25	3.25	4.625	6.5625
$y(t_j)$	1.5	2.3987	3.7183	5.7317	8.8890

$$y = t + \frac{1}{2} + e^{2t}$$

11. See answer to Exercise 3.

Review Exercises—Chapter 20

1. $y = \frac{1}{2}x^2 - 3x + C$

3. $y = 2 \sec \sqrt{t} + C$

5. $y = \frac{-2}{3t^2 + C}$

7. $y = Ce^{(1+x)^2/2} - 2$

9. $y = Ce^{-2t} + 2$

11. $y = \frac{4}{Ce^{-4t} - 1}$

13. $y = Ce^{-\sin t}$

15. $y = Ce^{4t} - 2$

17. $y = Ce^{-4t} - 1$

19. $y = C_1 e^{5t} + C_2 e^{-3t} - \frac{4}{377}\sin 2t - \frac{19}{377}\cos 2t$

21. $y = C_1 e^{3t} + C_2 e^{-3t} - \frac{5}{9} - \frac{1}{5}e^{2t}$

23. $y = Ce^{1/t}$

25. $y = C_1 e^{7t} + C_2 e^{-2t} - \frac{1}{14}t - \frac{23}{196}$

$Cx^{-2/3}$

29. $y = \dfrac{2x}{C + x}$

31. $y = \sqrt{x^2 + 1} + 2$

33. $y = 3e^{t^2/2}$

35. $y = 3 \sin 3t + 3 \cos 3t$

37. $y = 6te^{3t}$

39. a. $y' = 2y$
   b. $y'' - y = 0$

41. $y = e^{-t/5}$

43. $58°$ F

45. $y = C_1\left(x + \dfrac{x^2}{2} + \dfrac{x^3}{3!2!} + \dfrac{x^4}{4!3!} + \dfrac{x^5}{5!4!} + \cdots\right)$

47. $y = e^{2t}\left(A \sin \sqrt{2}t + B \cos \sqrt{2}t\right)$

INDEX

TABLE OF INTEGRALS (continued)

Forms Involving Trigonometric Functions

69. $\displaystyle\int \sin^2 u \, du = \frac{u}{2} - \frac{1}{4}\sin 2u + C$

70. $\displaystyle\int \sin^3 u \, du = -\frac{1}{3}\cos u(\sin^2 u + 2) + C$

71. $\displaystyle\int \cos^2 u \, du = \frac{1}{2}u + \frac{1}{4}\sin 2u + C$

72. $\displaystyle\int \cos^3 u \, du = \frac{1}{3}\sin u(\cos^2 u + 2) + C$

73. $\displaystyle\int \frac{du}{1 \pm \sin u} = \pm\tan\left(\frac{\pi}{4} \pm \frac{u}{2}\right) + C$

74. $\displaystyle\int \frac{du}{1 + \cos u} = \tan\left(\frac{u}{2}\right) + C$

75. $\displaystyle\int \frac{du}{a + b\sin u}$
$\quad \begin{cases} \dfrac{2}{\sqrt{a^2 - b^2}}\,\mathrm{Tan}^{-1}\dfrac{a\tan\left(\frac{u}{2}\right) + b}{\sqrt{a^2 - b^2}} + C, & a^2 > b^2 \\[3ex] \dfrac{1}{\sqrt{b^2 - a^2}}\ln\dfrac{a\tan\left(\frac{u}{2}\right) + b - \sqrt{b^2 - a^2}}{a\tan\left(\frac{u}{2}\right) + b + \sqrt{b^2 - a^2}} + C, & a^2 < b^2 \end{cases}$

76. $\displaystyle\int \frac{du}{a + b\cos u}$
$\quad \begin{cases} \dfrac{2}{\sqrt{a^2 - b^2}}\,\mathrm{Tan}^{-1}\dfrac{\sqrt{a^2 - b^2}\tan\left(\frac{u}{2}\right)}{a + b} + C, & a^2 > b^2 \\[3ex] \dfrac{1}{\sqrt{b^2 - a}}\ln\left(\dfrac{\sqrt{b^2 - a^2}\tan\left(\frac{u}{2}\right) + a + b}{\sqrt{b^2 - a^2}\tan\left(\frac{u}{2}\right) - a - b}\right) + C, & a^2 < b^2 \end{cases}$

77. $\displaystyle\int \sin(mu)\sin(nu) \, du = \frac{\sin(m - n)u}{2(m - n)} - \frac{\sin(m + n)u}{2(m + n)} + C, \quad m^2 \neq n^2$

78. $\displaystyle\int \cos(mu)\cos(nu) \, du = \frac{\sin(m - n)u}{2(m - n)} + \frac{\sin(m + n)}{2(m + n)} + C, \quad m^2 \neq n^2$

79. $\displaystyle\int \tan^3 u \, du = \frac{1}{2}\tan^2 u + \ln|\cos u| + C$

80. $\displaystyle\int \tan^4 u \, du = \frac{1}{3}\tan^3 u - \tan u + u + C$

81. $\displaystyle\int \sin(mu)\cos(nu) \, du = -\frac{\cos(m - n)u}{2(m - n)} - \frac{\cos(m + n)u}{2(m + n)} + C, \quad m^2 \neq n^2$

82. $\displaystyle\int \sin^2 u \cos^2 u \, du = -\frac{1}{32}\sin 4u + \frac{u}{8} + C$

Forms Involving Inverse Trigonometric Functions

83. $\displaystyle\int \mathrm{Sin}^{-1} u \, du = u\,\mathrm{Sin}^{-1} u + \sqrt{1 - u^2} + C$

84. $\displaystyle\int \mathrm{Cos}^{-1} u \, du = u\,\mathrm{Cos}^{-1} u - \sqrt{1 - u^2} + C$

85. $\displaystyle\int \mathrm{Tan}^{-1} u \, du = u\,\mathrm{Tan}^{-1} u - \frac{1}{2}\ln(1 + u^2) + C$

86. $\displaystyle\int \mathrm{Cot}^{-1} u \, du = u\,\mathrm{Cot}^{-1} u + \frac{1}{2}\ln(1 + u^2) + C$

87. $\displaystyle\int \mathrm{Sec}^{-1} u \, du = u\,\mathrm{Sec}^{-1} u - \ln|u + \sqrt{u^2 - 1}| + C$

88. $\displaystyle\int \mathrm{Csc}^{-1} u \, du = u\,\mathrm{Csc}^{-1} u + \ln|u + \sqrt{u^2 - 1}| + C$

89. $\displaystyle\int (\mathrm{Sin}^{-1} u)^2 \, du = u(\mathrm{Sin}^{-1} u)^2 - 2u + 2\sqrt{1 - u^2}(\mathrm{Sin}^{-1} u) + C$

90. $\displaystyle\int (\mathrm{Cos}^{-1} u)^2 \, du = u(\mathrm{Cos}^{-1} u)^2 - 2u - 2\sqrt{1 - u^2}(\mathrm{Cos}^{-1} u) + C$

91. $\displaystyle\int u\,\mathrm{Sin}^{-1}(au) \, du = \frac{1}{4a^2}[(2a^2u^2 - 1)\mathrm{Sin}^{-1}(au) + au\sqrt{1 - a^2u^2}] + C$

92. $\displaystyle\int u\,\mathrm{Cos}^{-1}(au) \, du = \frac{1}{4a^2}[(2a^2u^2 - 1)\mathrm{Cos}^{-1}(au) - au\sqrt{1 - a^2u^2}] + C$